Handbook of Research on Recent Developments in Electrical and Mechanical Engineering

Jamal Zbitou
University of Hassan 1st, Morocco

Catalin Iulian Pruncu
Imperial College London, UK

Ahmed Errkik
University of Hassan 1st, Morocco

A volume in the Advances in Computer and Electrical Engineering (ACEE) Book Series

IGI Global
DISSEMINATOR OF KNOWLEDGE

Published in the United States of America by
IGI Global
Engineering Science Reference (an imprint of IGI Global)
701 E. Chocolate Avenue
Hershey PA, USA 17033
Tel: 717-533-8845
Fax: 717-533-8661
E-mail: cust@igi-global.com
Web site: http://www.igi-global.com

Library of Congress Cataloging-in-Publication Data

Names: Zbitou, Jamal, 1976- editor. | Pruncu, Catalin Iulian, 1984- editor. | Errkik, Ahmed, 1960- editor.
Title: Handbook of research on recent developments in electrical and mechanical engineering / Jamal Zbitou, Catalin Iulian Pruncu, and Ahmed Errkik, editors.
Description: Hershey, PA : Engineering Science Reference, [2020] | Includes bibliographical references.
Identifiers: LCCN 2019016786| ISBN 9781799801177 (hardcover) | ISBN 9781799801184 (ebook)
Subjects: LCSH: Electrical engineering. | Mechanical engineering.
Classification: LCC TK145 .H29 2020 | DDC 621.3--dc23 LC record available at https://lccn.loc.gov/2019016786

This book is published in the IGI Global book series Advances in Computer and Electrical Engineering (ACEE) (ISSN: 2327-039X; eISSN: 2327-0403)

British Cataloguing in Publication Data
A Cataloguing in Publication record for this book is available from the British Library.

For electronic access to this publication, please contact: eresources@igi-global.com.

Advances in Computer and Electrical Engineering (ACEE) Book Series

Srikanta Patnaik
SOA University, India

ISSN:2327-039X
EISSN:2327-0403

Mission

The fields of computer engineering and electrical engineering encompass a broad range of interdisciplinary topics allowing for expansive research developments across multiple fields. Research in these areas continues to develop and become increasingly important as computer and electrical systems have become an integral part of everyday life.

The **Advances in Computer and Electrical Engineering (ACEE) Book Series** aims to publish research on diverse topics pertaining to computer engineering and electrical engineering. **ACEE** encourages scholarly discourse on the latest applications, tools, and methodologies being implemented in the field for the design and development of computer and electrical systems.

Coverage

- Algorithms
- Analog Electronics
- Applied Electromagnetics
- VLSI Fabrication
- Computer Architecture
- Microprocessor Design
- Optical Electronics
- Electrical Power Conversion
- VLSI Design
- Computer Hardware

IGI Global is currently accepting manuscripts for publication within this series. To submit a proposal for a volume in this series, please contact our Acquisition Editors at Acquisitions@igi-global.com or visit: http://www.igi-global.com/publish/.

The Advances in Computer and Electrical Engineering (ACEE) Book Series (ISSN 2327-039X) is published by IGI Global, 701 E. Chocolate Avenue, Hershey, PA 17033-1240, USA, www.igi-global.com. This series is composed of titles available for purchase individually; each title is edited to be contextually exclusive from any other title within the series. For pricing and ordering information please visit http://www.igi-global.com/book-series/advances-computer-electrical-engineering/73675. Postmaster: Send all address changes to above address.

Titles in this Series

For a list of additional titles in this series, please visit:
https://www.igi-global.com/book-series/advances-computer-electrical-engineering/73675

Novel Practices and Trends in Grid and Cloud Computing
Pethuru Raj (Reliance Jio Infocomm Ltd. (RJIL), India) and S. Koteeswaran (Vel Tech, India)
Engineering Science Reference • copyright 2019 • 374pp • H/C (ISBN: 9781522590231) • US $255.00 (our price)

Blockchain Technology for Global Social Change
Jane Thomason (University College London, UK) Sonja Bernhardt (ThoughtWare, Australia) Tia Kansara (Replenish Earth Ltd, UK) and Nichola Cooper (Blockchain Quantum Impact, Australia)
Engineering Science Reference • copyright 2019 • 243pp • H/C (ISBN: 9781522595786) • US $195.00 (our price)

Contemporary Developments in High-Frequency Photonic Devices
Siddhartha Bhattacharyya (RCC Institute of Information Technology, India) Pampa Debnath (RCC Institute of Information Technology, India) Arpan Deyasi (RCC Institute of Information Technology, India) and Nilanjan Dey (Techno India College of Technology, India)
Engineering Science Reference • copyright 2019 • 369pp • H/C (ISBN: 9781522585312) • US $225.00 (our price)

Applying Integration Techniques and Methods in Distributed Systems and Technologies
Gabor Kecskemeti (Liverpool John Moores University, UK)
Engineering Science Reference • copyright 2019 • 351pp • H/C (ISBN: 9781522582953) • US $245.00 (our price)

Handbook of Research on Cloud Computing and Big Data Applications in IoT
B. B. Gupta (National Institute of Technology Kurukshetra, India) and Dharma P. Agrawal (University of Cincinnati, USA)
Engineering Science Reference • copyright 2019 • 609pp • H/C (ISBN: 9781522584070) • US $295.00 (our price)

Multi-Objective Stochastic Programming in Fuzzy Environments
Animesh Biswas (University of Kalyani, India) and Arnab Kumar De (Government College of Engineering and Textile Technology Serampore, India)
Engineering Science Reference • copyright 2019 • 420pp • H/C (ISBN: 9781522583011) • US $215.00 (our price)

Renewable Energy and Power Supply Challenges for Rural Regions
Valeriy Kharchenko (Federal Scientific Agroengineering Center VIM, Russia) and Pandian Vasant (Universiti Teknologi PETRONAS, Malaysia)
Engineering Science Reference • copyright 2019 • 432pp • H/C (ISBN: 9781522591795) • US $205.00 (our price)

IGI Global
DISSEMINATOR OF KNOWLEDGE

701 East Chocolate Avenue, Hershey, PA 17033, USA
Tel: 717-533-8845 x100 • Fax: 717-533-8661
E-Mail: cust@igi-global.com • www.igi-global.com

Table of Contents

Detailed Table of Contents

This book chapter mainly focuses on analytical analysis for the branch-line coupler in which this method provides an explicit solution in the coupler design. Generally, the directional coupler is one of the fundamental components for Microwave Integrated Circuit (MIC), especially the equal power-split coupler that is used for signal monitoring, power measurement, power division, and balanced-type components such as balanced mixers. In this chapter, several applications of the branch-line coupler are also described. The analytical and design formulations of the coupler are derived based on ABCD matrix, transmission line principle, and even-odd mode decomposition. Although the simple analytical analysis is not sufficiently implemented in complex coupler structure, it is capable of providing an initial design guideline for the coupler dimensions. The initial design of the coupler dimensions based on analytical analysis can be gradually modified and optimized to achieve the desired size or performance of the coupler using advanced numerical simulation.

Metamaterials are artificially-engineered materials which possess unique properties not found in natural materials. The properties are derived from the structural designs of metamaterials and they allow the structure to manipulate electromagnetic waves and achieve desired responses in a certain frequency range. This chapter reviews past achievements, recent developments, and future trends on electromagnetic metamaterials in microwave regime. The chapter first briefly introduces electromagnetic metamaterials from a general prospect including the definition, historical overview, and classification of metamaterials. Furthermore, three selected applications of metamaterials which are microwave absorbers, sensors, and energy harvesters are discussed based on their operation principles, designs, and characteristics.

from 1.17 GHz to 3 GHz is presented in this chapter. The wider bandwidth circuits are designed by using transmission lines which are intrinsically wideband circuits. With carefully designed biasing and broadband matching networks, unconditionally stability and excellent matching performance are fulfilled over the overall operating bandwidth with a maximum power gain of 17.34 dB and a saturated output power of 17 dBm. Considering the wider bandwidth of the proposed BPA, the latter compares favorably with the contemporary state-of-the-art.

This chapter introduces an overview of coplanar technology and the general techniques and process to improve the response and characteristics of microwave components. A new circular defected ground structure (DGS) with shaped coplanar line is investigated for compact stopband filter (SBF) for microwave and millimeter wave applications. With this structure, the response of resonant element in 20 GHz exhibits the bandstop function. The proposed DGS is also modified by introducing four symmetrical slots with L-configuration in conductor line of a coplanar circuit to improve separately the stopband and passband performances. An equivalent circuit model derived for the proposed structures will be provided.

The most heard sound exists as a mixture of several audio sources. All human beings have the ability to concentrate on a single source of their interest and ignore the other sources as disturbing background noise. To apply this powerful gift to a machine, it must obligatory pass through the source separation process. If there is not enough information about the process of mixture of those sources and their nature as well, the problem is known by Blind Source Separation BSS. This thesis is dedicated to study the BSS as a solution for human machine interaction. The objective consists in recovering one or several source signals from a given mixture signal. Recently, the science research is towards artificial intelligence and machine learning applications. The proposed approach for the separation will be to apply a Deep Neural Network method based on Keras. Extracting features from the audio with signal processing techniques and machine learning to learn a representation from the audio for the compression tasks and the suppression of the noise will improve the state-of-the-art.

Chapter 8

Yassine Yazid, National School of Applied Sciences of Tangier, Morocco & Abdelmalek Essaadi University, Morocco
Imad Ezzazi, National School of Applied Sciences of Fes, Morocco & Université Sidi Mohamed Ben Abdellah, Morocco
Mounir Arioua, National School of Applied Sciences of Tetouan, Morocco & Abdelmalek Essaadi University, Morocco
Ahmed El Oualkadi, National School of Applied Sciences of Tangier, Morocco & Abdelmalek Essaadi University, Morocco

Since the appearance of WSN, the energy efficiency has been widely considered as a critical issue due to the limited battery-powered nodes. In this regard, communication process is the most energy demanding in sensor nodes. Subsequently, using energy-aware routing protocols in order to decrease the communications costs as much as possible and increase the network lifetime is of paramount importance. In this chapter, we have mainly focused on the most recent-based clustered routing algorithms for heterogeneous WSNs, namely Selected Election Protocol (SEP), and Distributed Energy Efficient Clustering Protocol (DEEC). In addition, we have proposed an efficient clustered routing protocol based on Zonal SEP algorithm. Indeed, we have evaluated the performance of the proposed protocol according to different scenarios in order to guarantee the best distribution of heterogeneous nodes in the network. The results have shown that the proposed clustered routing approach outperforms the existed Z-SEP protocol in terms of energy efficiency and stability.

Chapter 9

Nabila Aoulass, University Abdelmalek Essaadi, Morocco
Otman Chakkour, University Abdelmalek Essaadi, Morocco

NMF method aim to factorize a non-negative observation matrix X as the product X =G.F between two non-negative matrices G and F, respectively the matrix of contributions and profiles. Although these approaches are studied with great interest by the scientific community, they often suffer from a lack of robustness with regard to data and initial conditions and can present multiple solutions. The work of this chapter aims to examine the different approaches of NMF, thus introducing the constraint of sparsity in order to avoid local minima. The NMF can be informed by introducing desired constraints on the matrix F (resp G) such as the sum of 1 of each of its lines. Applications on images made it possible to test the interest of many algorithms in terms of precision and speed.

Chapter 10

Sivaraman P., TECH Engineering, India
Sharmeela C., Anna University, India

A solar micro inverter is a small-size inverter designed for single solar PV module instead of group of solar PV modules. Each module is equipped with a micro inverter to convert the DC electricity into AC electricity and the micro inverter is placed/installed below the module. The advantages of micro inverters are: reduced effect of shading losses, module degradation and soiling losses, enabled module independence, different rating of micro inverter can be connected in parallel to achieve the desired capacity, additional modules can be included at time which allows the good scalability, string design and sizing are avoided, failure of any micro inverter does not affect the overall power generation, individual MPPT controller for each module increases the power generation, any orientation and tilt angle allows higher design flexibility, lower DC voltage increasing the safety, easy to design, handle and install, requires less maintenance, draws attention of design engineers, contractors, etc.

Solid-state lighting technology is rapidly gaining acceptance in lighting industry street lighting, traffic lighting, decorative lighting, projection displays, display backlighting, automotive lighting, and so on. Differing from conventional light sources that use tungsten filament, plasma, or gases to generate light, solid-state lighting is based on organic or inorganic light emitting diodes (LEDs), and has the potential to generate light with almost 100 % efficiency. LED luminaires have a long lifetime and are environmentally friendly with no toxic mercury contained. However, the success of these luminaires depends on system design, which comprises an understanding of several factors such as performance and control. In this chapter, we shall touch upon some technological advancements in the field of solid-state lighting technologies and challenges that limit their market penetration for consumer lighting.

In this chapter, we model by Finite Element Method (FEM) the Lamb waves' propagation and their interactions with symmetric and asymmetric delamination in sandwich skin. Firstly, a theoretical model is established to obtain the equation of lamb modes propagation. Secondly, dispersion curves are plotted using Matlab program for the laminate [0]4. The simulations were then carried out using ABAQUS CAE by exciting the fundamental A0 Lamb mode in the frequency 300 kHz. The delamination was then estimated by analyzing the signal picked up at two sensors using two techniques: Two Dimensional Fast Fourier Transform (2D-FFT) to identify the propagating and converted modes, and Wavelet Transform (WT) to measure the arrival times. The results showed that the mode A0 is sensible to symmetric and asymmetric delamination. Besides, based on signal changes with the delamination edges, a localization method is proposed to estimate the position and the length of the delamination. In the last section, an experimental FEM verification is provided to validate the proposed method.

presentation of the main functions and types of coatings, the chapter progresses by offering information regarding the methods, the area of application, and the performance properties for each of the presented coating technologies. Furthermore, the quality and precision aspects, as well as the applicable standards, are discussed. Some practical examples regarding the performance improvement of different mechanical components by applying precision coatings are additionally presented. The final goal of this chapter is to enable the readers to choose the proper system for their application, by providing all the relevant information regarding various coating types.

Additive Manufacturing or three-dimensional printing refers to a process of building lighter, stronger three-dimensional parts, manufactured layer by layer. Additive manufacturing uses a computer and CAD software which passes the program to the printer to build the desired shape. Metals, thermoplastic polymers, and ceramics are the preferred materials used for additive manufacturing. Fused deposition modeling is one additive manufacturing technique involving the use of thermoplastic polymer for creating desired shape. Carbon fibers can be added into polymer to strengthen the composite without adding additional weight. Present work deals with the manufacturing of Carbon fiber-reinforced Polylactic Acid composites prepared using fused deposition modeling. Mechanical and thermo-mechanical properties of composites are studied as per ASTM standards and using sophisticated instruments. It is observed that there is enhancement in thermo-mechanical properties of composites due to addition reinforcement which is discussed in detail.

Resin Transfer Molding has become one of the most efficient processes to manufacture composite parts. Among the steps in composite part processing is the curing reaction. In the majority of cases, this reaction is of exothermic nature accompanied by a rise in temperature in the laminate. This leads to the appearance of a thermal gradient. This research aims to study the thermal gradient generated. The objective is to minimize the temperature excess in the composite. By means of a one-dimensional numerical study using the finite differential method, we have showed that the energy balance depends not only on the temperature and on the degree of curing but also on several other factors, namely: the volume fraction of the fibres, the temperature cycle, and the reinforcement thickness. Authors have shown in this study the effect of increasing temperature on the optimization of the curing cycle. The chapter also investigated the effect of thickness variation on temperature distribution in the composite. A comparison of the authors' results with literature achievements showed agreement.

Preface

Today, mechanical and electrical engineering are two research domains which attract many researches due to highly potential to improve the human life. The Mechanical Engineering field is covered by many areas as heat transfer, manufacturing, mechanical design, system dynamics and control, thermodynamics, energy Systems, and the study on the fundamentals of fluid mechanics. Hence, to understand the engineering phenomenon is required a certain knowledge in mathematics, physics, chemistry. Further, the Electrical Engineering field numerous applications as well. For examples, electrical engineers work in several fields including aeronautics, biomedical, energy, computer, electrical and electronic systems, as well as telecommunications. They participate to the industrial design and implementation of projects. The contents of Handbook of Research on Recent Developments in Electrical and Mechanical Engineering represents the efforts of many researchers working in electrical and mechanical field. The objective of this handbook is to share experiences and research studies in different attractive areas of electrical and mechanical engineering.

The state of art of the investigated research field was presented briefly in each chapter. It makes this manuscript an ideal handbook for university that focus in mechanical courses, microwave active and passive circuit courses, training courses, engineers, PhD students, and researchers. The themes covered in this book are divided into fourth sections:

The first section is about "Microwave Passive & Active circuits" that presents some studies on computer aided design of planar circuits; the introduction to electromagnetic metamaterials and their applications; some research work on measurement technique, and microwave passive and active circuits design as filters, antennas, power amplifier and RF-DC rectifiers.

The second one is "Information Technology and Signal Processing" that contains chapters around Autoencoders in Deep Neural Network. There we reported a chapter which presents wireless sensor networks (WSNs) with a comparative study of clustering-based routing protocols in terms of their energy efficiency, network lifetime, throughput and stability performances. The aim of the third chapter were to describe the dictionary learning by factorization techniques in Non negative matrices (NMF) for the separation of signals.

The third section describes the research studies focused on "Solid-state lighting technology & Renewable Energy". This section deals with solid-state lighting technology particularly LEDs, and other works on renewable energy.

The fourth section is about mechanical engineering, particularly a chapter on Lamb waves which are an attractive tool used to control long distances such as pipes. Another study presents propagation characteristics of anti-plane shear horizontal surface wave in a homogeneous micro-polar layered structure by introducing interfacial complexity. We reported as well as a research works on the antilock braking system, performance improvement of mechanical components by precision coating, 3D printers and Thermal improvement of curing in resin transfer molding process.

The book is organized into 18 chapters. A brief description of each chapter were made in order to easy accommodate the reader with the content of this manuscript:

In chapter 1, the planar branch-line coupler circuits were reviewed and analyzed comprehensively in terms of analytical modeling development, numerical simulation using commercial software, the recent trend of modified structures of the branch-line coupler, and its applications in microwave devices. The analytical analysis is the easiest method in simple branch-line coupler design. The analytical equations are explicit and capable of determining the characteristic impedance of each branch line for the coupler at the desired coupling level as well as the suitability of broadband S-parameters analysis. However, nowadays most planar branch-line couplers have been modified to miniature size, better bandwidth, thus a lot of modified structures of circuit's branch line have been proposed, such as slow wave structure, meandering line, cascading circuit, and proximity feed gap-coupled. To modify the branch-line coupler circuit, the numerical method usually takes over from the analytical method in design work due to analytical analysis is unable to solve the complicated modified branch-line circuit. Typically, microwave office software (AWR), Advanced Design System (ADS), and Computer Simulation Technology (CST) are used in modified branch-line coupler design. In fact, the branch line coupler circuit has many applications in microwave devices, such as complex ratio measuring unit, 2-way 90-degree power splitter, 3-dB coupler, and double balanced mixer. In this chapter, the branch-line applications will also be briefly reviewed.

Chapter 2 deals with Metamaterials which are artificial engineered materials that possess unique properties not found in natural materials. The properties are derived from the structural designs of metamaterials and they allow the structure to manipulate electromagnetic waves and achieve desired responses in certain frequency range. This chapter reviews past achievements, recent development and future trend on electromagnetic metamaterials in microwave regime. The chapter first briefly introduces electromagnetic metamaterials from a general prospect including the definition, historical overview and classification of metamaterials. Furthermore, five selected applications of metamaterials which are microwave absorbers, filters, sensors, antennas and energy harvesters will be discussed in details based on their designs, characteristics and operation theories.

Chapter 3 discusses various types of Microwave complex ratio measurement (MCRM) circuits. They have been described in detail that includes previous studies, basic theory, and general calibration methods for MCRM circuits. This chapter attempts to provide concise and comprehensive information to researchers who are involved in the construction of MCRM circuit. MCRM is usually used to construct RF reflectometer. The advantages of the MCRM circuit are simple, inexpensive, not easy to damage, do not need to use many electronic components in the construction of the reflectometer. In addition, MCRM allows the signal through it steadily and insensitive to operating temperature. Normally, the temperature of MCRM circuit does not increase significantly when the circuit is operated for a long period.

Chapter 4 introduces Wireless power transmission (WPT) which has become a novel alternative technology to solve all these power supply problems. A wireless power transmission system consists of a block that converts continuous energy into microwave energy capable of transmitting in free space through a transmitting antenna. Reception is provided by a receiving antenna followed by an RF-DC rectification system. Each element of the WPT system can be characterized by its efficiency. The key element of a WPT system is called Rectenna (for rectifying antenna), a conventional rectenna circuit consists of a receiving antenna followed by an RF-DC conversion circuit with a non-linear characteristic. This circuit usually contains one or more Schottky diodes, an HF input filter, a DC output filter and a resistive load that models the consumption of the powered system. The challenge is to optimize the entire rectenna, with the objective of maximizing the DC output and efficiency of RF-DC conversion. Optimization must be carried out over the entire circuit, hence the need to use global analysis methods combining electromagnetic simulation and circuit. Numerical modeling presented is a fundamental and decisive tool throughout this work. The main aim being to take into account all possible couplings between the different parts of the circuit. Further, the objective of this chapter was to present experimentally an innovative, compact and high-efficiency rectifier circuit at 2.45 GHz for the power supply of low consumption devices. This rectifier has been designed by using Advanced Design System. The bridge topology was employed on an FR4 substrate with dielectric constant er=4.4, substrate thickness h=1.6 mm and the loss tangent is 0.025. A good matching input impedance was observed and high conversion efficiency was obtained. Simulation results have been validated through realization and measurements.

Chapter 5 introduces the design of new structures of broadband power amplifiers which span a wide range of areas, among which microwave crop drying and quarantine in agriculture, medical diathermy, medical imaging, heating, electronic warfare, telecommunications, tracking and navigation systems. The design of BPAs, as that for any other power amplifier circuit, is basically subdivided in a chain of methodical steps, from the identification of BPA specifications up to the concluding characterization and measurements of the fabricated circuit, to prove fulfilment of the design requirements. Throughout the design of BPAs, essential and commonly contrasting specifications have to be at the same time fulfilled. On the one hand, wide bandwidth, high output power as well as high power gain are typically needed to minimize the number of PA stages, and consequently reduce the size of the whole unit. Furthermore, in order to ensure adequate signal amplification unaccompanied by affecting the data content, as well as preserve the suitable quality for the transmitted signal, high linearity must be assured. There are several techniques and circuit topologies deployed to realize broadband power amplifiers. In this chapter, the broadband power amplifier considerations will be briefly presented and described. Then, the broadband impedance matching techniques will be described. Finally, the proposed broadband power amplifier circuit design as well as the simulated results will be presented.

In chapter 6, an overview of coplanar waveguide technology is presented. Varieties of defected ground structure with the recent DGS units which developed to replace the Electromagnetic Band Gap (EBG) circuits in the goal to improve response and characteristics of microwave components such as filters are introduced. In this chapter, a circular defected ground structure (DGS) with shaped coplanar line is investigated for compact stopband filter (SBF) for microwave and millimeter wave applications. With this structure, the first proposed response of resonant element in 20 GHz exhibits the bandstop function. The proposed DGS is also modified by introducing four symmetrical slots with L-configuration in conductor line of a coplanar circuit to improve separately the stopband and passband performances. The insertion loss can be reduced by introducing four symmetrical slots with L-configuration in conductor line of a coplanar circuit to improve separately the stopband and passband performances. Additionally, the operating frequency ranges are extended to millimeter-waves with little increase in radius of circular defected ground. It combined between those requirements: simple structure and design, easy fabrication and good performance.

In chapter 7 we have reported some studies on "Blind Audio source Separation (BASS)". In this way, this chapter is dedicated to study the BSS as a solution for human machine interaction. The objective consists in recovering one or several source signals from a given mixture signal. Mixtures can be classified according to the nature of the environment (instantaneous or convolutive), the number of sources compared with the number of acoustical sensors which determines the nature of the mixture (over-determined, determined and underdetermined) and time-varying or time-invariant conditions. The purpose of BSS still remains the same, which is to recover the signals sources, without knowledge of the mixture, using only the observations. There are a lot of methods that can solve this kind of problem, like the Independent Component Analysis (ICA) and the FAST-ICA algorithm.

Chapter 8 deals with wireless sensor networks (WSN) which became an emerging technology since its increasing usage in various domains and different applications such medical systems, environment monitoring, military applications, surveillance and recently they are induced to respond to several Internet of Things applications requirements. That is all because of their advantages which are principally reflected in the absence of infrastructure, the ease deployment of the network, low maintenance and faster communication. This chapter presents a list of comparative studies of certain clustering-based routing protocols in terms of their energy efficiency, network lifetime, throughput and stability performances. In one hand, it will focus mainly in the recent based routing algorithms belonging to two different clustered routing protocols families dedicated for heterogeneous WSNs namely respectively Selected Election Protocol (SEP) and Distributed Energy Efficient Clustering Protocol (DEEC). It is about eight protocols named SEP, E-SEP, T-SEP, Z-SEP, DEEC, E-DEEC, D-DEEC and T-DEEC. In other hand, a new approach inspired by the SEP protocol family is presented.

Chapter 9 is focus on the NMF methods. The aim is to factorize a non-negative observation matrix X as the product X =G.F between two non-negative matrices G and F, respectively the matrix of contributions and profiles. Although, these approaches are studied with great interest by the scientific community, they often suffer from a lack of robustness with regard to data and initial conditions and can present multiple solutions. The work of this chapter aims to examine the different approaches of NMF, thus introducing the constraint of sparsity in order to avoid local minima. The NMF can be informed by introducing desired constraints on the matrix F (resp G) such as the sum of 1 of each of its lines. Applications on images made it possible to test the interest of many algorithms in terms of precision and speed.

Chapter 10 focuses on the Solar energy, which is the one of the major reliable renewable sources. Solar photovoltaic system is used to convert the light energy (photons) into electrical energy from the sunlight. A simple solar PV system consists of solar PV panel and inverter to convert the sunlight into electricity. Solar panel is a device that used to convert the photons in the sun light into DC electricity and inverter is used to convert the DC electricity into AC electricity. This chapter will introduce some standards in this field and an introduction of solar micro inverter which is designed for single solar PV module instead of group of solar PV modules. Each module shall be equipped with a micro inverter to convert the DC electricity into AC electricity and the micro inverter is placed/installed below the module.

Chapter 11 presents some of the recent technological developments in the field of LED lighting technology and challenges that limit their performance. Solid-state lighting technology is rapidly gaining acceptance in lighting industry with a growing list of applications, such as, street lighting, traffic lighting, decorative lighting, projection displays, display backlighting, automotive lighting and so on. It differs from conventional light sources that use tungsten filament, plasma or gases to generate light, solid-state lighting is based on organic or inorganic light emitting diodes (LEDs), and has the potential to generate light with almost 100% efficiency. LED lighting fixtures have long lifetime and are environmental friendly with no toxic mercury contained. However, the success of these fixtures depends on the system design, which comprises an understanding of electrical and photometrical characteristics under temperature effect on device performance.

Chapter 12 is focused on the study of Lamb waves' propagation. In this chapter Finite Element Method (FEM) is used to study the Lamb waves' propagation and their interactions with symmetric and asymmetric delamination in sandwich skin. Firstly, a theoretical model is established to obtain the equation of lamb modes propagation. Secondly dispersion curves are plotted using Matlab program for the laminate. The simulations were then carried out using ABAQUS CAE by exciting the fundamental A0 Lamb mode in the frequency 300 kHz. The delamination was after that estimated by analyzing the signal picked up at two sensors using two technics: Two Dimensional Fast Fourier Transform (2D-FFT) to identify the propagating and converted modes and Wavelet Transform (WT) to measure the arrival times. The results confirmed that the mode A0 is sensible to symmetric and asymmetric delamination. Besides, based on the signal that changes with the delamination edges, a localization method is proposed to estimate the position and the length of the delamination. In the last section of this chapter, an experimental FEM verification is provided to validate the proposed method.

In chapter 13 is presented the concept of Microgrid. From which the main resources of these systems are transformed in renewable energy resources. On the other hand, the use of DC-based systems potentially has significant efficiency and cost advantages to a range of power system applications. One of the systems which are used from DC power is DC Microgrids. In the Microgrids, the loads are supplied from the local Distributed Generations (DGs) sources; also, the majority of loads and DGs in Microgrids are DC, hence, using DC power in Microgrids is more suitable. Besides, the main problems of these systems are modeling and protection of DC Microgrids. This chapter study different modeling techniques of DC Microgrid components such as Photovoltaic (PV), Wind Turbine (WT), lines and converters. Therefore, the challenges and different methods for identifying the location of fault and methods for reliable and secure operation of the system is introduced. Moreover, one of the challenging problems of DC Microgrids is locating the fault location. Therefore, different fault location method is proposed in this chapter.

Chapter 14 includes the theoretical investigation of the propagation characteristics of SH-type wave and a new type of dispersive surface wave in an irregular layer overlying a half-space, both constituted by distinct homogeneous micropolar isotropic elastic materials. At the common interface of the layered structure, two types of irregularities viz. rectangular and parabolic shaped, are studied in two distinct cases. Existence of new type of dispersive surface wave along with its dispersion relation has been deduced in the closed-form by adopting a distinct mathematical treatment for various cases concerned with the presence and absence of microrotational components in the composite structure. It is also examined that the dispersion equation of new type of dispersive surface wave vanishes identically in classical elastic case.

Chapter 15 is focused on the Antilock Braking System (ABS). It is an active safety system that aims to reduce the number of road accidents. The ABS investigated is an electronically controlled system that allows the driver to keep up control of the vehicle amid crisis braking while at the same time keeping the wheels under control. Besides, by keeping brake weight just underneath the point of making a wheel lock up. The ABS guarantee that the most extreme braking power is utilized to stop the vehicle while ensuring the minimum possible braking distance. The formal methods are mathematically based techniques that allows to specify and verify the systems. The use of those methods can greatly increase our understanding of a system by revealing inconsistencies, ambiguities, and incompleteness that might otherwise go undetected. The formal method used in this paper is Event-B, which is based on set theory as a modeling notation.

Chapter 16 presents the summary of latest performance for Improvement of Mechanical Components by Precision Coating. Here, we present the main coating technologies and their practical applications. Furthermore, the quality and precision aspects, as well as the applicable standards are discussed. Some practical examples regarding the performance improvement of different mechanical components by applying precision coatings are additionally presented. The final goal of this chapter is to enable the readers to choose the proper system for their application, by providing all the relevant information regarding various coating types.

Chapter 17 presents the Additive Manufacturing process. It is a three dimensional printing methods that permits to manufacture lighter, stronger three-dimensional parts by using so calling manufactured layer by layers of material deposition. The Additive Manufacturing uses a computer and CAD software's which passes the program to the printer to build the desired shape. Metals, thermoplastic polymers, ceramics are the preferred material used for additive manufacturing. Fused deposition modeling is one the additive manufacturing technique that involves the use of thermoplastic polymer for creating desired shape. Carbon fibers can be added into polymer to strengthen the composite without adding additional weight. The present chapter deals with the manufacturing of Carbon fiber reinforced Polylactic Acid composites prepared using fused deposition modeling. Mechanical and thermo-mechanical properties of composites are studied as per ASTM standards and using sophisticated instruments. It is observed that there is enhancement in thermo-mechanical properties of composites due to addition reinforcement which is discussed in detail.

Chapter 18 is focused on the Resin Transfer Molding (RTM) method which has become one of the most efficient processes to manufacture medium size reinforced composite parts. Among the main steps in processing the composite parts is the curing reaction. In the majority of cases this reaction is of exothermic nature and accompanied by a rise in temperature in the center of the laminate, this leads to the appearance of a thermal gradient. This chapter presents a study on the effect of increasing temperature on the optimization of the curing cycle, furthermore, and another study on the effect of thickness variation on temperature distribution in the composite. Further, this research work aims also to elucidate the thermal gradient phenomenon generated as a result of a cross linking reaction. The objective is to minimize the temperature excess in the composite laminate.

Jamal Zbitou
University of Hassan 1st, Morocco

Catalin Iulian Pruncu
University of Birmingham, UK & Imperial College London, UK

Ahmed Errkik
University of Hassan 1st, Morocco

Chapter 1
Computer-Aided Design and Applications of Planar Branch-Line Coupler Circuits

Kok Yeow You
https://orcid.org/0000-0001-5214-7571
Universiti Teknologi Malaysia, Malaysia

Nadera Najib Al-Areqi
Universiti Teknologi Malaysia, Malaysia

Chia Yew Lee
Universiti Teknologi Malaysia, Malaysia

Yeng Seng Lee
https://orcid.org/0000-0003-3395-7338
Universiti Malaysia Perlis, Malaysia

ABSTRACT

This book chapter mainly focuses on analytical analysis for the branch-line coupler in which this method provides an explicit solution in the coupler design. Generally, the directional coupler is one of the fundamental components for Microwave Integrated Circuit (MIC), especially the equal power-split coupler that is used for signal monitoring, power measurement, power division, and balanced-type components such as balanced mixers. In this chapter, several applications of the branch-line coupler are also described. The analytical and design formulations of the coupler are derived based on ABCD matrix, transmission line principle, and even-odd mode decomposition. Although the simple analytical analysis is not sufficiently implemented in complex coupler structure, it is capable of providing an initial design guideline for the coupler dimensions. The initial design of the coupler dimensions based on analytical analysis can be gradually modified and optimized to achieve the desired size or performance of the coupler using advanced numerical simulation.

DOI: 10.4018/978-1-7998-0117-7.ch001

INTRODUCTION

Computer-aided design (CAD) is an important software tool in today's engineering system design. Recently, many new high technology systems, such as new mobile phones, can be designed and modified in a short time by using CAD tool without actual building the phone prototype. Normally, the CAD is particularly used in designing electronic system, which is also called as electronic design automation (EDA). Recently, the EDA tool has an important role in the communication industry sector as well as research-based universities. Hence, in this new era, graduated engineers are required to have skills in engineering design. Thus, some universities are incorporating commercial electronic design automation (EDA) into their electrical and computer engineering curriculum (You et al., 2017).

During the 70s and early 80s, most engineering systems were developed based on analytical analysis or experimental testing due to less sophisticated computer technology during that time. Analytical analysis is the simplest and fastest method, yet less accurate in engineering design. On the other hand, the engineering design based on experimental work method is most reliable, nonetheless requires high cost and time consuming. Furthermore, during the 80s and early 90s, several numerical methods, such as Finite-Difference Time-Domain Method (FDTD), Method of Moments (MoM), Finite Element Method (FEM), and Boundary Element Method (BEM) was studied and started applied in engineering system design. After the end of the 90's until the date, many numerical-based commercial EDA simulators have already exist, such as Advanced Design System (ADS), Microwave office (AWR), IE3D, High Frequency Electromagnetic Field Simulation (HFSS), Computer Simulation Technology (CST), COMSOL Multiphysics, FEKO, and Sonnet. Hence, the risks of trial-and-error in the experiment design can be diminished as well as saving the cost and time.

This book chapter mainly focuses on analytical analysis for the branch-line coupler in which this method provides an explicit solution in the coupler design. Generally, the directional coupler is one of the fundamental components for Microwave Integrated Circuit (MIC); especially the equal power-split coupler is used for signal monitoring, power measurement, power division, and balanced-type components such as balanced mixers. In this chapter, several applications of the branch-line coupler are also described. The analytical and design formulations of the coupler are derived based on *ABCD* matrix, transmission line principle, and even-odd mode decomposition (Reed and Wheeler, 1956). Although the simple analytical analysis is not sufficiently implemented in complex coupler structure, it is capable of providing an initial design guideline for the coupler dimensions. The initial design of the coupler dimensions based on analytical analysis can be gradually modified and optimized to achieve the desired size or performance of the coupler using advance numerical simulation. Hence, the time used in coupler design can be reduced significantly.

Regular Branch-Line Couplers

Principle of Branch-Line Couplers

The basic branch-line coupler is four-port passive microwave devices consists of four 90° ($\lambda/4$) transmission lines, namely two horizontal and two vertical transmission lines with characteristic impedance Z_1 and Z_2, respectively as shown in Figure 1 (a). Typically, a single branch-line coupler is 3-dB directional coupler with a 90° phase difference between the two output ports (port-2 and port-3) and its bandwidth is limited to the range of 10%. The bandwidth can be enhanced by increasing the branch sections. For

Figure 1. (a) The basic structure of single branch line coupler. (b) Signal flow along the branch line of the coupler. (c) Planar form branch-line coupler fabrication.

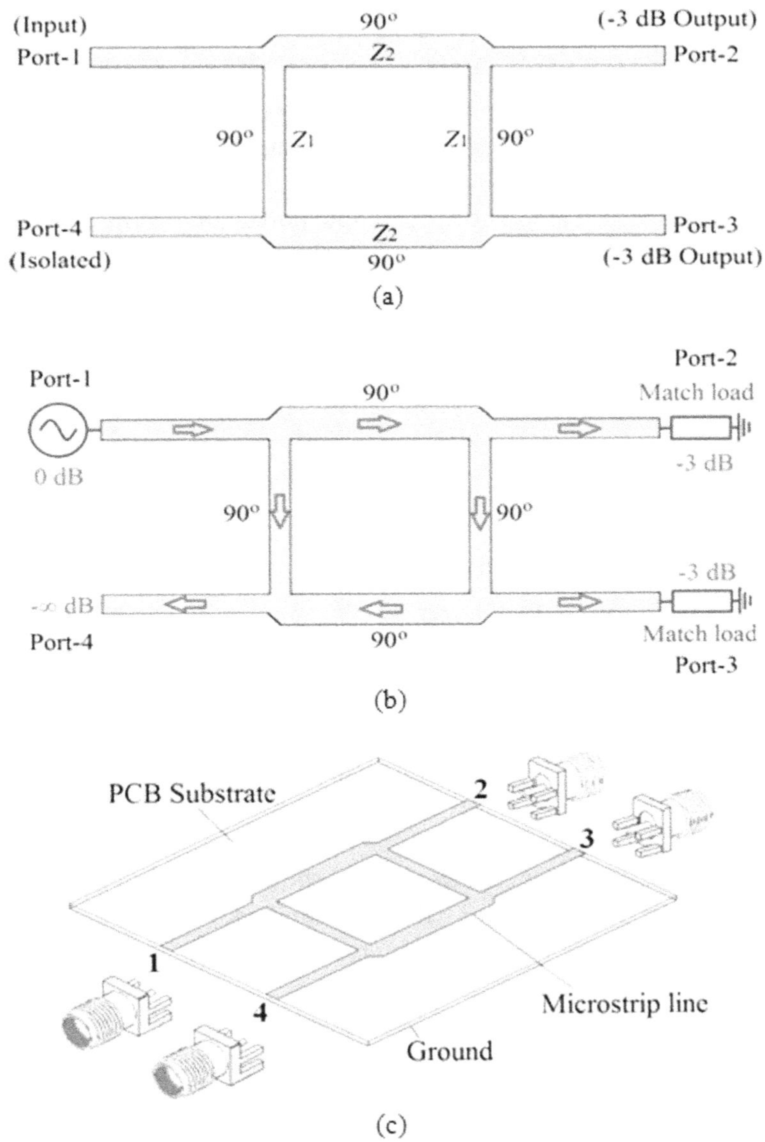

planar form branch-line coupler, the circuit structure in Figure 1 (a) is normally printed on printed circuit board (PCB) as illustrated in Figure 1 (c).

By appropriately selecting the characteristic impedance, (Z_o, Z_1 and Z_2) of the branch lines, uneven power split between the two output ports (Port-2 and Port-3) at the centre frequency, f_o can be achieved. Normally, there are five parameters which describe the characteristics and performances of the regular branch-line directional coupler as:

1. Reflection coefficient (return loss), $|S_{11}|$ at port-1
2. Transmission coefficient (insertion loss), $|S_{21}|$ at port-2
3. Coupling, $|S_{31}|$ at port-3
4. Isolation, $|S_{41}|$ at port-4
5. Phase difference, $|\phi_{21}-\phi_{31}|$ between port-2 and port-3

Ideally, at center frequency, f_o, the power applied to the input port (Port-1) is transferred equally to the two output ports (Port-2 and Port-3), with the port-4 in isolation ($|S_{41}| = 0$ or or $-\infty$ dB). Input port-1 match ($|S_{11}| = 0$ or $-\infty$ dB) is perfect at f_o. The upper output port-2 ($|S_{21}| = 1/\sqrt{2}$ or -3 dB) leads the lower output port-3 ($|S_{31}| = 1/\sqrt{2}$ or -3 dB) by phase shift of 90°. Thus, the S-parameter matrix of the coupler in Figure 1, is given in Equation (1). Ideally, the reflection coefficient, $|S_{11}| = 0$ (Match) at port-1 and no signal is coupled to port-4, $|S_{41}| = 0$ (Isolated).

$$\begin{bmatrix} S_{11} & S_{12} & S_{13} & S_{14} \\ S_{21} & S_{22} & S_{23} & S_{24} \\ S_{31} & S_{32} & S_{33} & S_{34} \\ S_{41} & S_{42} & S_{43} & S_{44} \end{bmatrix} = \begin{bmatrix} 0 & S_{12} & S_{13} & 0 \\ S_{21} & 0 & 0 & S_{24} \\ S_{31} & 0 & 0 & S_{34} \\ 0 & S_{42} & S_{43} & 0 \end{bmatrix} = \frac{-1}{\sqrt{2}} \begin{bmatrix} 0 & j & 1 & 0 \\ j & 0 & 0 & 1 \\ 1 & 0 & 0 & j \\ 0 & 1 & j & 0 \end{bmatrix} \tag{1}$$

For instance, when a microwave signal (0 dB input power) applied to port-1, the entire signal power is transmitted and spitted equally between output port-2 (-3 dB = 50% of input power) and port-3 (-3 dB = 50% of input power) in which the phase difference between both the output signals is 90°. In this circumstance, port-4 plays a role as 'isolated' port and does not receive any signal power ($-\infty$ dB). This is due to the incident signal from port-1 has two transmission paths to port-4, which are 90° and 270° paths and this will cause 180° delay difference between the two transmit signals to cancel one another as shown in Figure 1 (b).

Vice versa, if there is an impedance mismatch at port-2, the transmitted signal power is partly reflected back from port-2. The reflected signal power will be transmitted backward and divided proportionally between port-1 and port-4. However, the reflected signal from port-2 is not fed to port-3 (Crane Aerospace & Electronics Microwave Solutions, 2012).

The 90° phase difference between the output port-2 and port-3 makes the coupler circuit useful in many applications, such as duplexer, dual-fed circular polarization circuit, double-balanced mixer, power divider/combiner, six-port complex ratio measuring unit (CRMU), antenna beam forming network circuit, IQ-modulator/demodulation, mono-pulse comparators, single-sideband up-conversion, image reject down-conversion, and phase shifters.

Cascaded Branch-Line Couplers and Its Line Impedances

As mentioned in Subsection 1.1.1, the single section branch-line coupler in Figure 1 has narrow operating bandwidth (Typically 10% bandwidth). Normally, multi-section branch-line coupler is implemented to increase the bandwidth of the couplers, since as well known that the cascaded branch line can enhance the bandwidth (Reed, 1958). The multi-section branch-line coupler model proposed by Reed (1958) is shown in Figure 2 and its coresponding line impedances, Z are listed in Table 1.

Figure 2. Multi-section branch-line couplers proposed by Reed (1958).

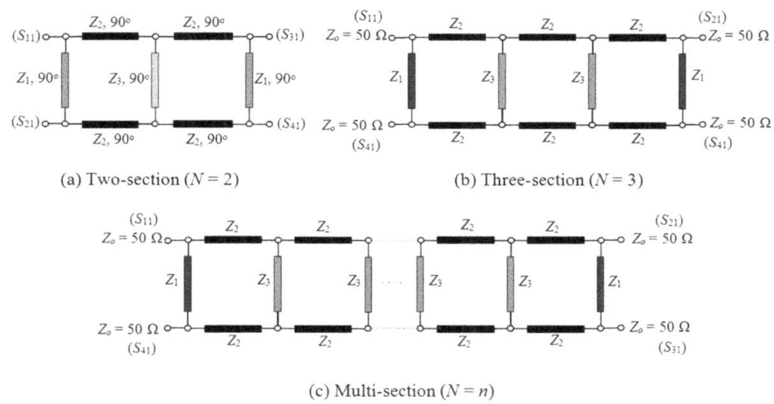

Muraguchi et al. (1983) was used as a numerical method to optimize the characteristic impedances, *Z* of the couplers. The distinction between Muraguchi et al. (1983) and Reed (1958) coupler model is the impedance, *Z* of the horizontal branch line for each section which is not always constant. Figure 3 shows the coupler model proposed by Muraguchi et al. (1983) and its optimized impedances, Z values are listed in Table 2.

Levy et al. (1968) had obtained the characteristic impedance, *Z* of branch-line coupler using Butterworth and Chebyshev methods, respectively. The values of *Z* are listed in Tables 3, 4, 5, and 6, respec-

Table 1. Characteristic impedances, Z of each branch line in Figure 2 (Reed, 1958).

Number of sections *N*	Characteristic Impedance, *Z* of Each Branch Line, (Ω)								
	0 dB coupling			3 dB coupling			10 dB coupling		
	Z_1	Z_2	Z_3	Z_1	Z_2	Z_3	Z_1	Z_2	Z_3
2	50	50	50	120.7	50	70.7	308.6	50	158.1
				120.7	35.4	35.4			
3	100	50	50	213.1	50	92.4	529.1	50	220.8
4	80.9	50	80.9	239.5	50	131.2	616.5	50	312.1
5	161.8	50	80.9	341.5	50	157.3	844.6	50	381.1
6	112.4	50	112.4	363.9	50	193.6			
7	224.7	50	112.4	469.9	50	221.5			
8	144.0	50	144.0	489.2	50	256.7			
9	288.0	50	144.0	597.4	50	285.4			
10	175.7	50	175.7						
11	351.4	50	175.7						
12	207.5	50	207.5						
13	415.0	50	207.5	851.8	50	413.1			
22	366.3	50	366.3						
23	733.1	50	366.3						

Figure 3. Single-, two-, and three-section branch-line couplers (Muraguchi et al., 1983).

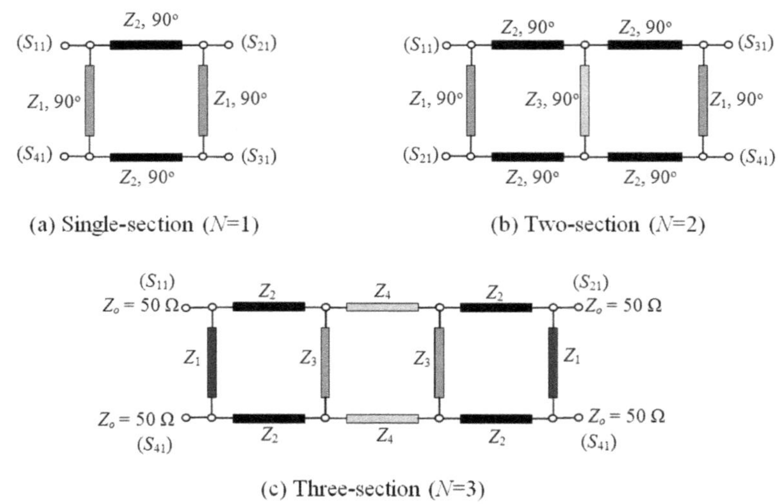

tively. Besides the number of branch section, N, Levy et al. (1968) also indicated that line impedance, Z values of the coupler can be selected based on the desired operating bandwidth, VSWR, directivity, and coupling level. The electrical length of each branch line for the coupler is 90° (π/2). Examples $N = 5$ and $N = 6$ branch-line couplers proposed by Levy et al. (1968) are shown in Figure 4. However, normally, microstrip branch-line couplers (PCB-based couplers) are capable of fabricating only up to three-section ($N = 3$) coupler. Owing to higher section ($N > 3$) coupler will cause the higher impedance, Z (narrow width of microstrip line) existed for outer vertical branch-line (as described in Subsection 1.1.3).

Dimensions of Branch-Line Couplers

Once the characteristic impedance, Z of each microstrip branch line of the coupler had been determined, the width, W of each microstrip branch line at desired 2 center frequency, f_o can be calculated (Pozar, 2012):

$$W = \begin{cases} \dfrac{8h\exp(A)}{\exp(2A)-2} & \text{for } \dfrac{W}{h} < 2 \\ \dfrac{2h}{\pi}\left\{B-1-\ln(2B-1)+\dfrac{\varepsilon_r-1}{2\varepsilon_r}\left[\ln(B-1)+0.39-\dfrac{0.61}{\varepsilon_r}\right]\right\} & \text{for } \dfrac{W}{h} > 2 \end{cases} \quad (2)$$

where

$$A = \frac{Z_n}{60}\sqrt{\frac{\varepsilon_r+1}{2}}+\frac{\varepsilon_r-1}{\varepsilon_r+1}\left(0.23+\frac{0.11}{\varepsilon_r}\right) \text{ and } B = \frac{377\pi}{2Z_n\sqrt{\varepsilon_r}}$$

Table 2. Characteristic impedances, Z of each branch line in Figure 3 (Muraguchi et al., 1983)

Number of sections N	Characteristic Impedance, Z of Each Branch Line, (Ω)				Bandwidth (%) (For $\|S_{11}\| \leq -20$ dB; Imbalance coupling $\leq \pm 0.43$ dB)	Max $\|S_{11}\|$ (dB)	Coupling imbalance at f_o (dB)
	Z_1	Z_2	Z_3	Z_4			
1	50.0	35.4			10	-	0
	59.5	41.1			9	-23.2	0.4
	57.3	39.9			10	-25.4	0.3
	55.1	38.6			10	-28.3	0.2
	52.0	36.6			10	-36.2	0.1
	50.5	35.7			10	-48.5	0
2	120.5	36.3	37.2		29	-	0
	99.9	34.7	41.1		27	-20.5	0.2
	102.9	27.0	25.3		33	-20.9	0.2
	105.2	26.9	24.3		32	-22.3	0.1
	105.0	29.4	29.4		36	-21.9	0.2
	105.0	31.2	33.3		37	-21.8	0.2
	105.6	34.9	41.7		36	-21.9	0.3
	106.7	38.3	50.0		35	-22.3	0.3
	106.7	42.4	62.5		35	-21.8	0.5
	108.7	43.1	62.5		34	-23.3	0.4
	110.4	43.7	62.5		33	-24.7	0.3
	105.5	44.3	71.4		37	-20.7	0.6
	108.0	45.5	71.4		35	-22.4	0.5
	110.0	46.3	71.4		33	-24.0	0.4
	106.6	48.0	83.3		34	-21.0	0.7
	109.0	49.2	83.3		31	-22.8	0.6
	110.8	50.2	83.3		29	-24.4	0.4
3	263.9	40.0	52.8	30.3	32	-	0
	170.5	36.8	57.4	28.8	34	-22.1	0
	153.0	44.3	91.0	39.4	44	-20.5	0.3
	162.4	44.1	85.3	39.2	40	-22.7	0.2
	157.5	45.0	91.1	42	38	-23.0	0.1
	166.7	39.4	67.0	30.7	44	-20.5	0.3
	156.3	43.0	83.9	36.8	44	-20.5	0.3
	147.1	47.2	106.3	44.8	43	-20.5	0.3
	138.9	52.0	136.1	54.9	43	-20.4	0.3
	131.6	57.4	175.0	67.4	43	-20.2	0.3
	142.9	53.7	142.9	56.5	45	-20.2	0.4
	142.9	53.9	142.9	58.3	43	-21.4	0.3
	138.9	52.5	138.9	55.9	43	-20.5	0.3
	138.9	52.6	138.9	57.1	41	-21.0	0.2
	138.9	52.7	138.9	58.1	40	-21.2	0.2
	135.1	51.3	135.1	55.3	40	-20.0	0.2
	135.1	51.3	135.1	56.4	39	-20.1	0.1

Figure 4. Examples N = 5 and N = 6 branch-line couplers (Levy et al., 1968).

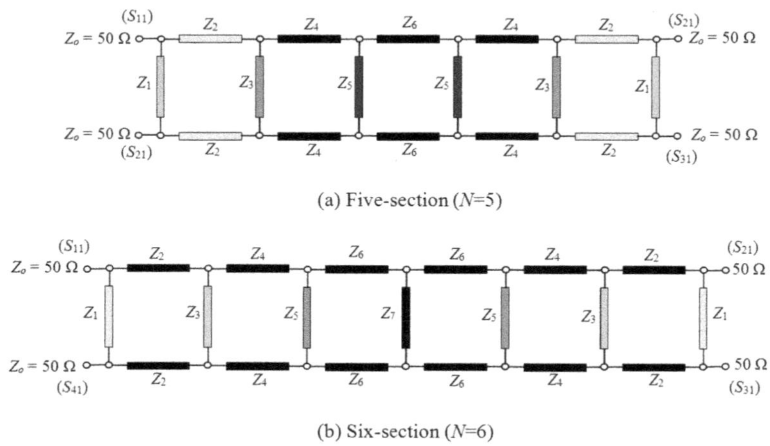

(a) Five-section (N=5)

(b) Six-section (N=6)

where Z_n is the characteristic impedance of each microstrip branch line and the subscript n = 0, 1, 2, 3,..., N represents the number of branch line. Symbols ε_r and h are the dielectric constant and thickness of used substrate. Once the values of W are obtained from Equation (2), the length, l (corresponding to $\lambda/4$ length) of each branch line can be calculated as (Pozar, 2012):

$$l = \frac{c}{4f_o\sqrt{\varepsilon_{eff}}} - 0.412h\left[\frac{\left(\varepsilon_{eff}+0.3\right)\left(W/h+0.264\right)}{\left(\varepsilon_{eff}-0.258\right)\left(W/h+0.8\right)}\right] \quad (3)$$

The velocity of light in vacuum, c = 299792458 ms^{-1} and the effective relative permittivity, ε_{eff} of the branch microstrip line is given as (Pozar, 2012):

$$\varepsilon_{eff} = \begin{cases} \frac{\varepsilon_r+1}{2} + \frac{\varepsilon_r-1}{2}\left[\frac{1}{\sqrt{1+12h/W}} + 0.04\left(1-\frac{W}{h}\right)^2\right] & \text{for } \frac{W}{h} \le 1 \\ \frac{\varepsilon_r+1}{2} + \frac{\varepsilon_r-1}{2}\left(\frac{1}{\sqrt{1+12h/W}}\right) & \text{for } \frac{W}{h} > 1 \end{cases} \quad (4)$$

As known that, a very high value of characteristic impedance, Z can be achieved by very thin microstrip line width, W. From Table 2, the required characteristic impedance, Z of the vertical branch line for the conventional three section coupler (N = 3) is between 135 Ω to 264 Ω in Butterworth and Chebyshev designs (Muraguchi et al., 1983), which normally corresponding to less than 0.5 mm of the microstrip line width based on the thickness and material of the substrate as shown in Table 7. In fact, higher value of impedance, Z for microstrip line can be achieved by fabricating on thick and low dielectric PCB substrate. In general, the smallest microstrip line width, W that can be fabricated in usual laboratory without significant error is 0.4 mm.

Table 3. Values of characteristic impedance, Z for branch-line coupler obtained using Butterworth method (Levy et al., 1968)

$N = 2$	**Coupling level**									
	2 dB	**2.5 dB**	**3 dB**	**4 dB**	**5 dB**	**6 dB**	**8 dB**	**10 dB**	**15 dB**	**20 dB**
Z_1 (Ω)	101.2	110.8	120.5	140.7	162.4	186.1	240.9	308.1	558.0	998.0
Z_2 (Ω)	31.3	34.0	36.3	39.8	42.3	44.1	46.5	47.9	49.4	49.8
Z_3 (Ω)	24.6	30.9	37.2	50.1	63.5	77.5	108.5	144.9	274.1	496.0
VSWR	1.29	1.22	1.19	1.12	1.09	1.07	1.04	1.03	1.01	1.01
% BW	26	26	28	28	28	30	32	34	36	38
$N = 3$										
Z_1 (Ω)	237.9	250	263.9	296.2	333.8	376.5	478.9	609.0	1101	1976
Z_2 (Ω)	35.7	38.1	40.0	42.7	44.6	46.0	47.7	48.6	49.6	49.9
Z_3 (Ω)	35.0	43.9	52.8	70.6	88.7	107.6	148.9	197.3	369.0	664.9
Z_4 (Ω)	23.4	27.1	30.3	35.3	38.9	41.6	45.0	47.0	49.1	49.7
VSWR	1.28	1.23	1.20	1.15	1.12	1.08	1.05	1.04	1.02	1.01
% BW	40	42	42	44	46	46	50	52	56	58
$N = 4$										
Z_1 (Ω)	500.5	512.3	531.3	583.4	649.4	727.8	922.5	1174	2146	3876
Z_2 (Ω)	40.1	41.9	43.3	45.3	46.6	47.5	48.6	49.2	49.8	49.9
Z_3 (Ω)	68.3	81.3	94.0	119.0	144.5	171.1	230.2	300.3	554.3	996
Z_4 (Ω)	23.3	27.3	30.6	35.7	39.4	42.0	45.4	47.3	49.2	49.8
Z_5 (Ω)	25.5	34.6	43.9	63.0	82.7	102.9	146.6	196.7	370.9	667.6
VSWR	1.33	1.24	1.22	1.16	1.12	1.10	1.07	1.05	1.02	1.02
% BW	52	52	54	56	58	60	62	64	68	72
$N = 5$										
Z_1 (Ω)	980.4	988.1	1016	1104	1225	1374	1748	2232	4167	7576
Z_2 (Ω)	43.5	44.8	45.8	47.1	48.0	48.5	49.2	49.5	49.9	50.0
Z_3 (Ω)	140.4	157.3	174.0	207.8	243.5	281.8	369.8	477.1	875.7	1577
Z_4 (Ω)	26.7	30.5	33.6	38.2	41.4	43.6	46.4	47.9	49.4	49.8
Z_5 (Ω)	31.5	42.9	54.6	78.1	101.9	126.4	178.7	238.5	446.8	802.6
Z_6 (Ω)	18.2	22.7	26.6	32.7	37.1	40.3	44.5	46.8	49.1	49.7
VSWR	1.34	1.28	1.22	1.19	1.14	1.11	1.08	1.06	1.03	1.02
% BW	60	62	62	66	68	68	72	74	80	82
$N = 6$										
Z_1 (Ω)	1838	1845	1887	2049	2273	2551	3289	4237	8065	15151
Z_2 (Ω)	45.9	46.8	47.5	48.3	48.8	49.2	49.6	49.8	49.9	50.0
Z_3 (Ω)	276.2	294.1	313.9	358.4	409.2	466.4	602.4	774.0	1429	2591
Z_4 (Ω)	31.2	34.7	37.3	41.2	43.7	45.4	47.5	48.6	49.6	49.9
Z_5 (Ω)	53.0	68.7	84.1	114.5	144.8	175.7	242.4	319.7	593.8	1066
Z_6 (Ω)	17.9	22.5	26.5	32.8	37.3	40.6	44.7	46.9	49.2	49.8
Z_7 (Ω)	23.6	34.8	46.7	71.4	96.8	122.7	177.5	239.1	450.0	806.5
VSWR	1.39	1.30	1.25	1.19	1.15	1.12	1.08	1.06	1.03	1.02
% BW	66	68	70	72	74	76	80	82	88	92

Table 4. Values of characteristic impedance for N = 2 branch-line coupler obtained using Chebyshev method (Levy et al., 1968).

	Coupling (dB)	Directivity (dB)	VSWR	Characteristic Impedance (Ω) (For $N = 2$)		
				Z_1	Z_2	Z_3
BW= 20%	3	30.4	1.045	115.5	36.0	38.2
	4	31.2	1.03	135.8	39.6	51.4
	5	31.9	1.02	157.5	42.1	65.1
	6	32.5	1.015	181.1	44.0	79.4
	8	33.5	1.007	235.4	46.4	110.8
	10	34.4	1.004	301.9	47.8	147.7
	15	35.9	1.001	548.8	49.4	278.6
	20	36.8	1.000	982.3	49.8	503.5
BW= 40%	3	17.7	1.236	99.9	34.7	41.1
	4	18.5	1.150	121.1	38.8	55.8
	5	19.3	1.102	142.9	41.6	70.6
	6	19.9	1.073	166.2	43.7	85.8
	8	21.0	1.040	219.3	46.3	118.8
	10	21.9	1.024	283.9	47.8	157.4
	15	23.5	1.009	522.5	49.3	293.6
	20	24.5	1.004	872.6	49.8	578.7
BW= 60%	10	14.2	1.075	254.8	47.6	179.5
	15	15.9	1.029	479.4	49.3	326.4
	20	16.9	1.013	872.6	49.8	578.7

However, for usual industry precision standard, the achievable microstrip line width, W can be smaller up to 0.2 mm. From the Table 7, for $W = 0.2$ mm, the corresponding maximum impedance, Z value capable of being fabricated is limited to around 150 Ω to 220 Ω with 1.5 mm to 3 mm thick of substrate. In Subsection 1.1.2, Tables 2, 3, and 6, it is found that the values of vertical branch line impedance, Z for the conventional higher order multi-section coupler ($N > 4$) are almost higher than 220 Ω, which is impractical to realize a microstrip line using regular PCB process. The comparison circuit area size and width, W of the microstrip branch-line coupler for $N = 1$, $N = 2$, and $N = 3$ branch-line coupler is shown in Figure 5.

Kim *et al.* (2000) had suggested that a high impedance microstrip line of branch line coupler could be realized with a much greater microstrip line width, W using defected ground structure (DGS). The 150 Ω of characteristic impedance, Z with $W = 1.0$ mm of microstrip line had been achieved by adding the DGS, while $W = 1.0$ mm corresponds to $Z = 82$ Ω of conventional microstrip line on the RT/Duroid 5880 substrate with 0.787 mm of thickness (Kim *et al.*, 2000). On the other hand, Tang et al. (2006) has successfully fabricated impedance microstrip line of 231 Ω with $W = 0.4$ mm on 1.5 mm thick of FR-4 substrate using defected ground structure (DGS) for four-branch line coupler ($N = 4$). Normally, the microstrip line width, W for $Z = 231$ Ω should be 0.006 mm using the same thickness of FR-4 substrate with regular circuit ground. However, the DGS structure provides addition effective inductance which is

Table 5. Values of characteristic impedance for N = 3 branch-line coupler obtained using Chebyshev method (Levy et al., 1968).

	Coupling (dB)	Directivity (dB)	VSWR	Characteristic Impedance (Ω) (For $N = 3$)			
				Z_1	Z_2	Z_3	Z_4
BW= 20%	3	50.3	1.004	252.8	39.7	53.1	30.2
	4	51.4	1.003	285.2	42.5	71.1	35.2
	5	52.3	1.002	322.6	44.5	89.5	38.8
	6	53.1	1.001	364.7	45.9	108.5	41.5
	8	54.4	1.001	466.0	47.6	150.2	45.0
	10	55.5	1.000	593.1	48.6	199.0	47.0
	15	57.4	1.000	1075	49.6	372.0	49.1
	20	58.6	1.000	1938	49.9	669.3	49.7
BW= 40%	3	31.6	1.043	220.5	38.7	54.1	29.7
	4	32.8	1.027	253.2	41.9	73.0	34.9
	5	33.7	1.019	289.7	44.0	92.0	38.6
	6	34.6	1.013	330.5	45.5	111.8	41.4
	8	35.9	1.007	427.0	47.5	154.7	45.0
	10	37.1	1.004	548.2	48.5	204.6	47.0
	15	39.1	1.001	1006	49.6	381.1	49.1
	20	40.3	1.000	1818	49.9	684.9	49.7
BW= 60%	3	20.0	1.197	170.5	36.8	57.4	28.8
	4	21.2	1.124	203.3	40.6	77.9	34.3
	5	22.3	1.084	238.4	43.2	98.3	38.2
	6	23.2	1.059	276.9	44.9	119.2	41.1
	8	24.7	1.033	366.3	47.1	164.3	44.8
	10	25.9	1.020	477.6	48.3	216.5	46.9
	15	28.0	1.007	894.5	49.5	400.0	49.1
	20	29.2	1.003	1634	49.9	715.3	49.7
BW= 80%	5	13.5	1.290	175.9	41.8	115.3	37.7
	6	14.4	1.202	210.3	44.0	138.0	40.8
	8	16.1	1.110	289.7	46.6	186.4	44.7
	10	17.4	1.067	387.3	48.1	241.9	46.9
	15	19.6	1.025	750.8	49.5	437.4	49.1
	20	21.0	1.011	1393	49.8	774.0	49.7

possible to increase the characteristic impedance, Z of the microstrip line and remain a wide microstrip line width, W. Besides, microstrip lines with DGS exhibit an increased group velocity delay along the transmission line in which is capable of reducing the size of the coupler circuit.

Table 6. Values of characteristic impedance for N = 4 branch-line coupler obtained using Chebyshev method (Levy et al., 1968)

	Coupling (dB)	Directivity (dB)	VSWR	Characteristic Impedance (Ω) (For N = 4)				
				Z_1	Z_2	Z_3	Z_4	Z_5
BW= 20%	3	70.5	1.000	507.6	43.1	93.1	30.4	44.3
	4	71.8	1.000	559.9	45.1	118.4	35.6	63.8
	5	72.9	1.000	625.0	46.5	144.0	39.3	83.7
	6	73.9	1.000	702.2	47.4	170.7	42.0	104.2
	8	75.5	1.000	892.9	48.6	229.9	45.4	148.3
	10	76.8	1.000	1139	49.2	300.1	47.3	198.9
	15	79.0	1.000	2083	49.8	554.3	49.2	374.5
	20	80.4	1.000	3788	49.9	996.0	49.8	673.9
BW= 40%	3	45.6	1.008	440.1	42.2	90.7	29.7	45.6
	4	47.0	1.005	491.6	44.6	116.6	35.1	66.1
	5	48.2	1.003	553.7	46.1	142.7	39.0	86.9
	6	49.1	1.002	626.6	47.2	169.8	41.8	108.2
	8	50.8	1.001	805.2	48.4	229.6	45.3	153.9
	10	52.1	1.001	1033	49.1	300.3	47.2	206.0
	15	54.5	1.000	1916	49.8	554.9	49.2	386.4
	20	55.9	1.000	3497	49.9	996.0	49.8	693.5
BW= 60%	3	30.1	1.055	337.6	40.5	87.6	28.5	47.9
	4	31.6	1.034	388.5	43.5	114.9	34.3	70.3
	5	32.9	1.023	446.8	45.4	142.0	38.4	92.9
	6	33.9	1.016	512.8	46.6	170.0	41.4	115.9
	8	35.7	1.009	672.0	48.2	230.9	45.1	164.4
	10	37.1	1.005	874.1	49.0	302.5	47.1	219.3
	15	39.6	1.002	1650	49.7	559.3	49.2	408.5
	20	41.1	1.001	3030	49.9	1004	49.8	729.9
BW= 80%	4	19.6	1.163	264.7	41.4	117.0	32.9	77.3
	5	21.1	1.106	317.5	44.0	146.4	37.5	103.4
	6	22.2	1.073	375.1	45.7	175.9	40.8	129.2
	8	24.2	1.039	510.2	47.7	239.1	44.8	182.6
	10	25.8	1.023	680.3	48.7	312.5	47.0	242.0
	15	28.4	1.008	1326	49.7	574.1	49.2	444.8
	20	30.0	1.004	2475	49.9	1027	49.8	788.6
BW= 100%	6	12.1	1.293	232.6	44.0	206.3	39.8	153.1
	8	14.3	1.148	341.3	46.9	272.5	44.4	215.0
	10	16.0	1.086	474.8	48.3	348.7	46.8	281.5
	15	19.0	1.030	978.5	49.6	621.1	49.1	505.6
	20	20.7	1.013	1880	49.9	1096	49.7	885.0

Table 7. Maximum value of characteristic impedance, Z for width, W = 0.2 mm microstrip line.

PCB substrate		PCB thickness (mm)	Max value of Z (Ω) (for W = 0.2 mm)
FR-4 $\varepsilon_r = 4.4 \pm 0.3$ $\tan\delta = 0.02$	Design specification: $\varepsilon_r = 4.2$ $\tan\delta = 0.022$	0.8	123
		1.6	149
RT Duroid/5880 $\varepsilon_r = 2.20 \pm 0.02$ $\tan\delta = 0.0009$	Design specification: $\varepsilon_r = 2.2$ $\tan\delta = 0.001$	0.381	125
		0.508	138
		0.787	159
		1.575	192
		3.17	225
RT Duroid/5870 $\varepsilon_r = 2.33 \pm 0.02$ $\tan\delta = 0.0012$	Design specification: $\varepsilon_r = 2.33$ $\tan\delta = 0.002$	0.381	122
		0.508	135
		0.787	155
		1.575	188
RO 3003 $\varepsilon_r = 3.00 \pm 0.04$ $\tan\delta = 0.001$	Design specification: $\varepsilon_r = 3.0$ $\tan\delta = 0.002$	0.25	93
		0.50	122
		0.75	139
		1.52	169
RO 3035 $\varepsilon_r = 3.50 \pm 0.05$ $\tan\delta = 0.0015$	Design specification: $\varepsilon_r = 3.6$ $\tan\delta = 0.002$	0.25	86
		0.50	113
		0.75	129
		1.52	157
RO 3006 $\varepsilon_r = 6.15 \pm 0.15$ $\tan\delta = 0.002$	Design specification: $\varepsilon_r = 6.5$ $\tan\delta = 0.003$	0.25	66
		0.64	95
		1.28	116
RO 3010 $\varepsilon_r = 10.2 \pm 0.3$ $\tan\delta = 0.0022$	Design specification: $\varepsilon_r = 11.2$ $\tan\delta = 0.003$	0.25	51
		0.64	74
		1.28	90

Modified Branch-Line Couplers

Port Extension Branch-Line Couplers

Kim *et al.* (2010) had designed a dual band single-section branch-line coupler using port extensions as shown in Figure 6. The electrical length, θ_o for all branch lines of the coupler is identical. In fact, the value of θ_o is determined by choosing the desired dual band frequencies at f_1 (low center frequency) and f_2 (high center frequency) as:

Figure 5. Cascading branch-line coupler.

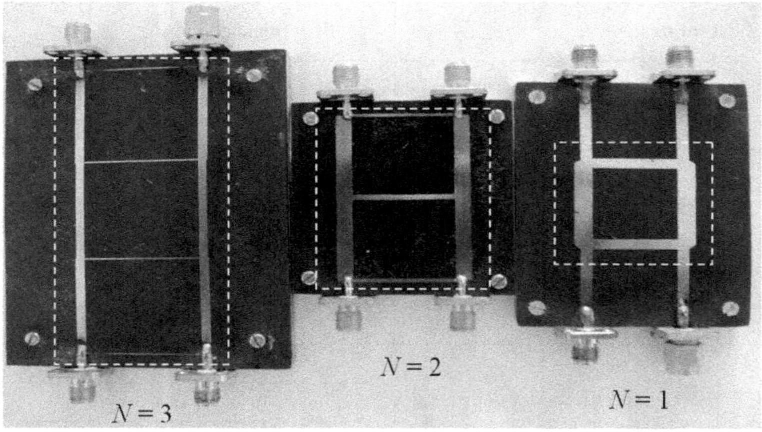

$$\theta_o = \frac{m\pi}{1+\left(f_2/f_1\right)},\ m = 1, 2, 3, \ldots \quad (5)$$

The analytical solutions of the line impedances (Z_o, Z_1, and Z_2) for the coupler at center frequencies f_1 and f_2 are given as (Kim et al., 2010):

$$Z_o = 50\sqrt{\alpha - \sqrt{\alpha^2 - 1}} \quad (6a)$$

$$Z_1 = \frac{Z_o\left(Z_o^2 + \cot^2\theta_o\right)\left(1+\sqrt{2}\right)}{1 - Z_o^2} \quad (6b)$$

$$Z_2 = \frac{Z_1}{\sqrt{2}} \quad (6c)$$

where α in (6a) is given as:

$$\alpha = \left(2.5+\sqrt{2}\right) + \left(1.5+\sqrt{2}\right)\cot^2\theta_o \quad (7)$$

Lee and Lee (2012) had implemented similar port extensions structure with fixed $\theta_o = 90^\circ$ in their coupler designs as shown in Figure 7. The proposed analytical design formulations of the line impedances (Z_o, Z_1, and Z_2) are simple and capable of enhancing the bandwidth (up to 50%) of the coupler with excellent flat coupling (±0.5 dB) which covered the wide operating frequency. In addition, the coupling

Figure 6. Schematic circuit proposed by Kim et al. (2010).

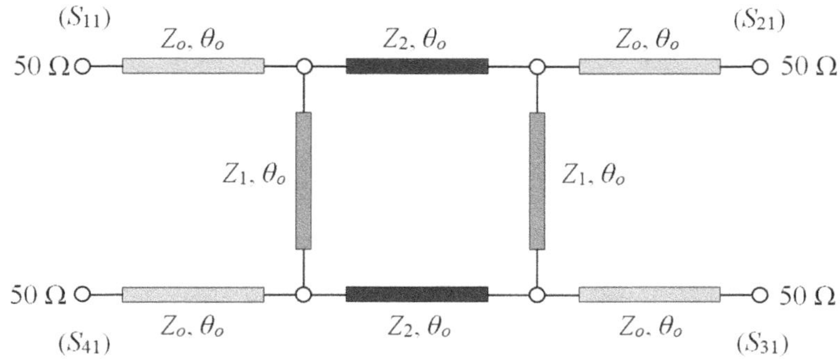

factor, C term is included in the formulations in order to provide arbitrary coupling levels design as (Lee and Lee, 2012):

$$Z_1 = \frac{Z_o^2}{50}\left(\frac{|S_{21}|}{|S_{31}|}\right) = \frac{Z_o^2}{50}\sqrt{10^{C/10} - 1} \tag{8a}$$

and

$$Z_2 = \frac{Z_o^2}{50}\sqrt{\frac{\left(|S_{21}|/|S_{31}|\right)^2}{1+\left(|S_{21}|/|S_{31}|\right)^2}} = \frac{Z_o^2}{50}\sqrt{1 - 10^{-C/10}} \tag{8b}$$

where $C = -|S_{31}|$ (in unit dB) is the coupling level. Since θ_o is fixed to be 90°, thus, the selected value of Z_o plays an important role to achieve a wide bandwidth. Lee and Lee (2012) had found that the bandwidth of coupler can be achieved up to 42% ($|S_{11}|$ and $|S_{41}|$ better than -20 dB) with $Z_o = 36\ \Omega$.

In fact, in 1978, Riblet had designed the single-section branch-line coupler using port extension structure with added open-circuit stub in order to achieve very flat coupling level (coupling imbalance $< \pm 0.05$ dB) and bandwidth of 30%. Figure 8 shows the schematic circuit of designed coupler (Riblet, 1978). The analytical design equations of the proposed coupler are given as:

$$\theta_o = \cos^{-1}\left\{\frac{1}{\sqrt{2}}\cos\left[\frac{\pi}{2}\left(1+\frac{\Delta f}{2f_o}\right)\right]\right\} \tag{9a}$$

Figure 7. Schematic circuit proposed by Lee and Lee (2012).

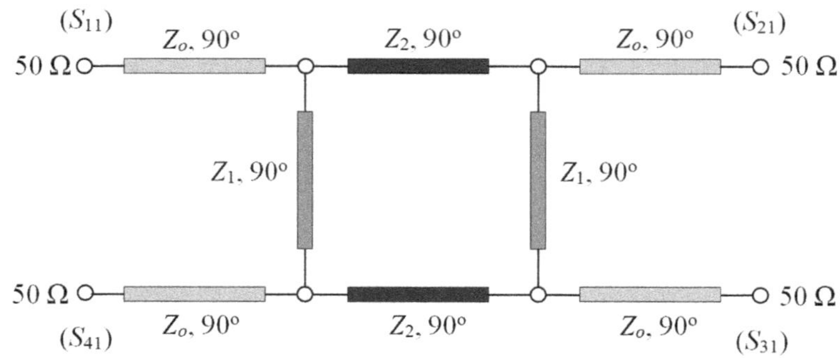

$$Z_2 = \frac{Z_1}{\sqrt{\left(\left|S_{21}\right|/\left|S_{31}\right|\right)^2 + 1}} \tag{9b}$$

$$Z_S = \frac{Z_P}{\sqrt{\frac{Z_1 \sin\theta_o}{50\cos^2\theta_o\left(\left|S_{21}\right|/\left|S_{31}\right|\right)} - \frac{Z_P}{50}\tan^2\theta_o - 1}} \tag{9c}$$

$$\frac{Z_P \sin\theta_o}{50} - \left(1 + \frac{Z_P}{Z_S}\right)\left(\frac{50\sin\theta_o}{Z_P} - \frac{50}{Z_S}\frac{\cos^2\theta_o}{\sin\theta_o}\right) = -\sqrt{\frac{\sqrt{\left(\left|S_{21}\right|/\left|S_{31}\right|\right)^2 + 1} + 1}{\sqrt{\left(\left|S_{21}\right|/\left|S_{31}\right|\right)^2 + 1} - 1}} \tag{9d}$$

The line impedance, Z_1, coupling ratio, $|S_{21}|/|S_{31}|$, and bandwidth, Δf are the free parameters and will be chosen in order to achieve the desired performance of the coupler. Normally, $Z_1 \approx 50\ \Omega$ and $\theta_o \approx 90°$ ($\pi/2$).

Besides, Mayer, (1990; 1992), Paul *et al.*, (1991), Knochel, (1999), had modified the coupler in Figure 8 by adding 180° length of series branches and 180° length of open-circuit stub at center frequency to further improve bandwidth and stabilize coupling of the coupler as shown in Figure 9. The proposed three sets of line impedances for the coupler in Figure 9 (b) by Mayer, (1992) are listed in Table 8.

Added Open-Circuit Stub Branch-line Coupler

The regular branch-line coupler can be miniaturized using added open-circuit stub on the horizontal branch line of the coupler (Eccleston and Ong, 2003; Sun et al., 2005). The design analytical formulations of the coupler with added open-circuit stub were described by Chun and Hong (2005, 2006). Figure 10 shows the regular coupler is miniaturized by adding open-stubs.

Figure 8. Schematic circuit proposed by Riblet (1978)

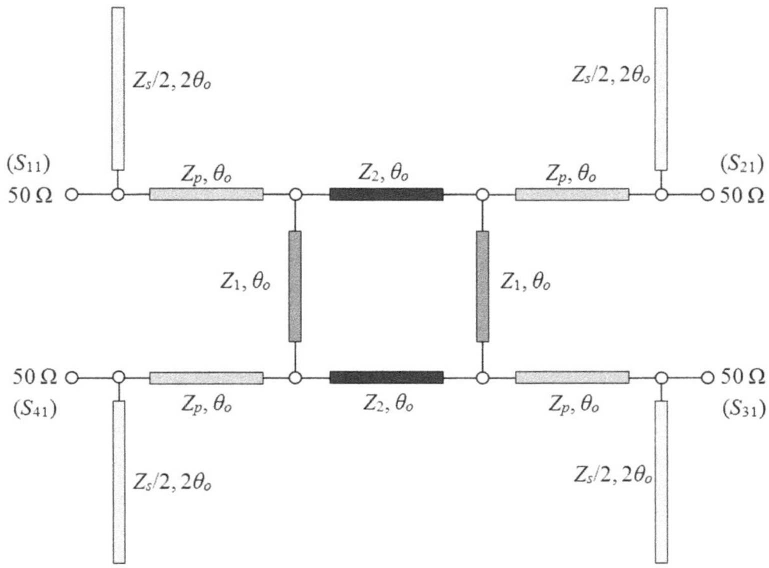

In fact, the characteristic impedances, Z of the branch line with added stub are derived from regular transmission line as (refer to Figure 10):

$$Z_2' = \frac{Z_2 \sin\theta_2}{\sin\theta_2'} \tag{10}$$

where Z_2 and Z_2' are the characteristic impedances of the regular branch line and corresponding branch line with added open-stub, respectively. On the other hand, θ_2 and θ_2' are the electrical lengths of the regular branch line and equivalent branch line with open-stub, respectively. Normally, the values of Z_2 and θ_2 of the regular branch line are known. While, θ_2' is a freedom parameter and normally it is given

Table 8. Three sets of line impedances for Figure 9 (b) (Mayer, 1992).

	Line impedance, Z set (Ω)		
	Set 1	**Set 2**	**Set 3**
Z_1	114.0	101.0	87.8
Z_2	37.3	76.8	39.1
Z_3	84.5	104.6	70.1
Z_4	43.0	88.5	59.4
Z_5	80.9	69.0	86.7
Z_6	86.0	88.2	94.9

Figure 9. Wide-band branch-line coupler (Mayer, 1990, 1992; Paul et al., 1991; Knochel, 1999).

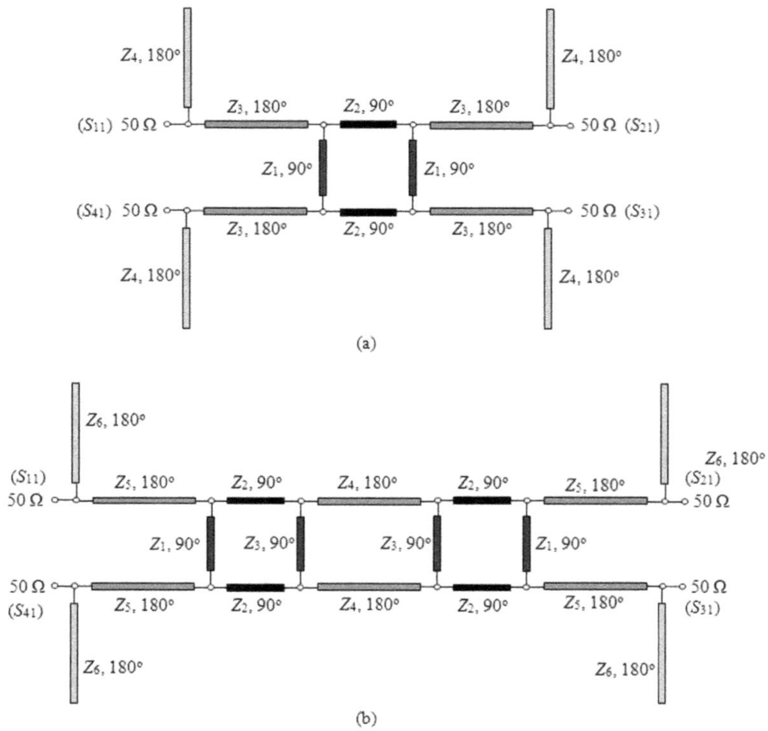

as $\theta_2' \approx 0.42\theta_2$ (Chun and Hong, 2006). The higher value of Z_2', the smaller size of coupler can be designed.

After that, the determined Z_2, θ_2, and θ_2' are used to calculate the susceptance, B_s of the open-stub as:

Figure 10. (a) Regular branch-line coupler and (b) miniaturized branch-line coupler with open-circuit stubs.

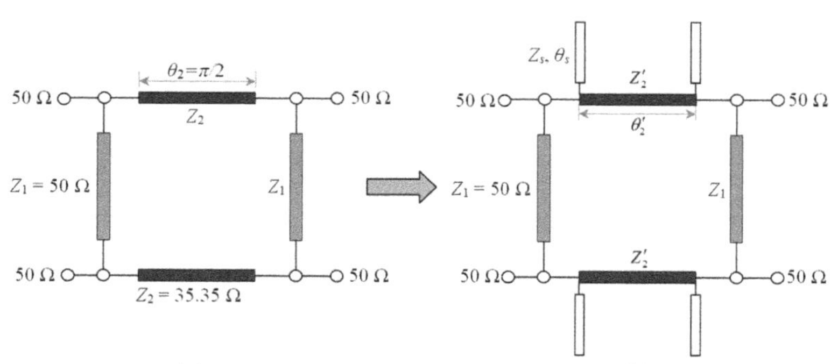

$$B_s = \frac{\cos\theta_2' - \cos\theta_2}{Z_2 \sin\theta_2} \tag{11}$$

Once the B_s is calculated, the impedance, Z_s and electrical length, θ_s of the open-stub can be explicitly found from B_s as:

$$B_s = \frac{\tan\theta_s}{Z_s} \tag{12}$$

The Z_s and θ_s are the freedom parameters. If the higher value of Z_s is chosen, the corresponding θ also has higher value. In general, the chosen value of Z_s is within 30 Ω to 70 Ω based on desired size reduction of the designed coupler.

The coupler can be further miniaturized by adding more open-stub on the horizontal branch-line, as shown in Figure 11, in which the regular branch line is requested to separate more section.

1.2.3 Coupled Feed Branch-line Coupler

Arriola et al. (2011) had designed a single-section branch-line with bandwidth up to 49% using coupled-feed line as shown in Figure 12. The designed coupler is capable of remaining return loss, $|S_{11}|$ and isolation, $|S_{41}|$ better than 20 dB, as well as imbalance coupling of 0.5 dB over the operating frequency. The electrical length, θ of the branch line can be determined as:

$$\theta = \frac{\pi}{2}\frac{f^+}{f_o} \tag{13}$$

where f_o is the center frequency. On the other hand, f^+ is the operating frequency at upper cutoff (above the center frequency).

Figure 11. (a) Regular branch-line coupler and (b) miniaturized branch-line coupler with more open-circuit stubs.

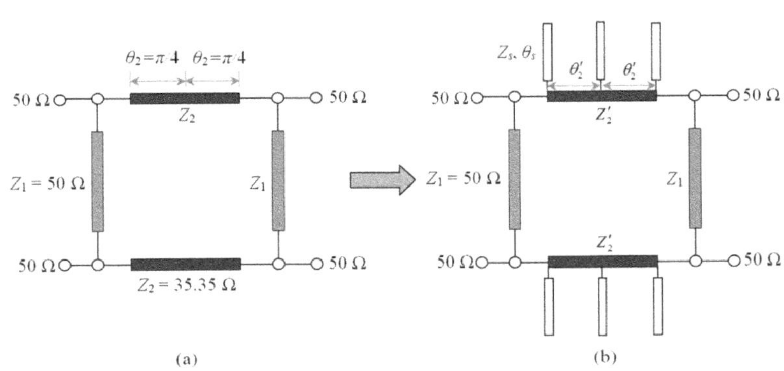

Figure 12. Coupled-feed line single-section branch-line coupler (Arriola et al., 2011).

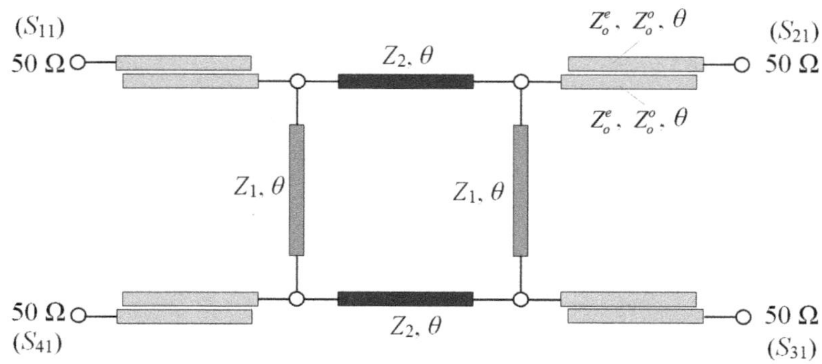

The even- and odd-mode impedances (Z_o^o and Z_o^e) of the feed line in unit Ω can be found as (Arriola et al., 2011):

$$z = \sqrt{\frac{\left(8\sqrt{2}+12\right)\cos^2\theta+4}{\sin\theta\cos^2\theta} - 4\tan^2\theta} \tag{14a}$$

$$Z_o^o = 50\left(\frac{z}{2}-1\right) \tag{14b}$$

$$Z_o^e = 100 + Z_o^o \tag{14c}$$

The line impedances (Z_1 and Z_2) of the center branch lines for the coupler are same with regular single-section branch-line coupler as:

$$Z_1 = 50\Omega \text{ and } Z_2 = 35.35\,\Omega \tag{15}$$

Dual Transmission Lines Branch-Line Coupler

Tang *et al.* (2008) had used dual transmission lines for each branch line in the coupler as shown in Figure 13 (b), in order to miniaturize the size of regular single-section branch-line coupler up to 63.9%.

The new line impedances, (Z_1' and Z_2') of the dual transmission lines are determined by referring to the characteristic impedances (Z_1, Z_2) of the regular coupler in Figure 13 (a) as (Tang et al., 2008):

Figure 13. Miniaturized coupler based on characteristic impedance of the regular coupler: (a) Regular branch-line coupler and (b) dual transmission lines branch-line coupler.

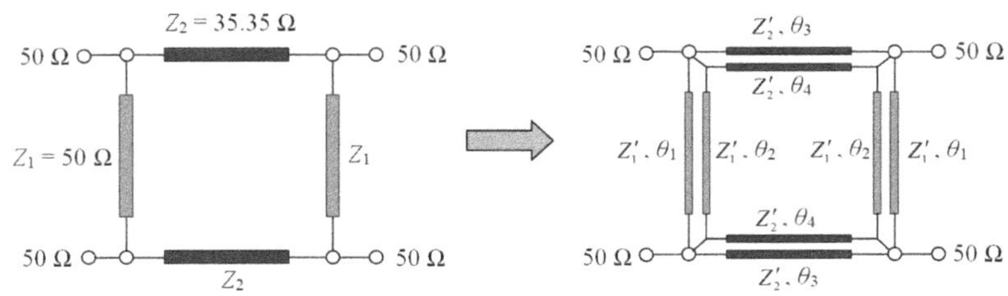

(a) Regular branch-line coupler (b) Dual transmission lines branch-line coupler

$$\begin{aligned} Z_1' &= Z_1 \frac{\cos\theta_2 - \cos\theta_1}{\sin\theta_1 \cos\theta_2} \\ &= 50 \frac{\cos(\pi - \theta_1) - \cos\theta_1}{\sin\theta_1 \cos(\pi - \theta_1)} \end{aligned} \tag{16a}$$

and

$$\begin{aligned} Z_2' &= Z_2 \frac{\cos\theta_4 - \cos\theta_3}{\sin\theta_3 \cos\theta_4} \\ &= 35.35 \frac{\cos(\pi - \theta_3) - \cos\theta_3}{\sin\theta_3 \cos(\pi - \theta_3)} \end{aligned} \tag{16b}$$

where $\theta_2 = \pi - \theta_1$ and $\theta_4 = \pi - \theta_3$ with conditions: $0 < \theta_1 < \pi/2$ and $0 < \theta_3 < \pi/2$.

Crossover Branch-Line Couplers

A crossover branch-line coupler is also a four-port matched passive device, which similar to the regular branch-line coupler. Recently, the crossover coupler is widely used in applications such as Butler matrix for antenna arrays (as described in Section 4). Initially, the crossover coupler was designed through cascading two branch-line couplers by Wight et al. (1976). The crossover coupler has two signal paths which allowed to cross each other, but, at the same time, both of the signals are isolated ($S_{11} = S_{21} = S_{41} = 0$ and $S_{31} \neq 0$, $|S_{31}| = 1$). Thus, the *S*-parameter matrix for crossover application can be written as:

$$\begin{bmatrix} S_{11} & S_{12} & S_{13} & S_{14} \\ S_{21} & S_{22} & S_{23} & S_{24} \\ S_{31} & S_{32} & S_{33} & S_{34} \\ S_{41} & S_{42} & S_{43} & S_{44} \end{bmatrix} = \begin{bmatrix} 0 & 0 & S_{13} & 0 \\ 0 & 0 & 0 & S_{24} \\ S_{31} & 0 & 0 & 0 \\ 0 & S_{42} & 0 & 0 \end{bmatrix} = e^{j\phi} \begin{bmatrix} 0 & 0 & 1 & 0 \\ 0 & 0 & 0 & 1 \\ 1 & 0 & 0 & 0 \\ 0 & 1 & 0 & 0 \end{bmatrix} \tag{17}$$

Figure 14. Single-, two-, and three-section crossover branch-line couplers proposed by Yao et al., 2011.

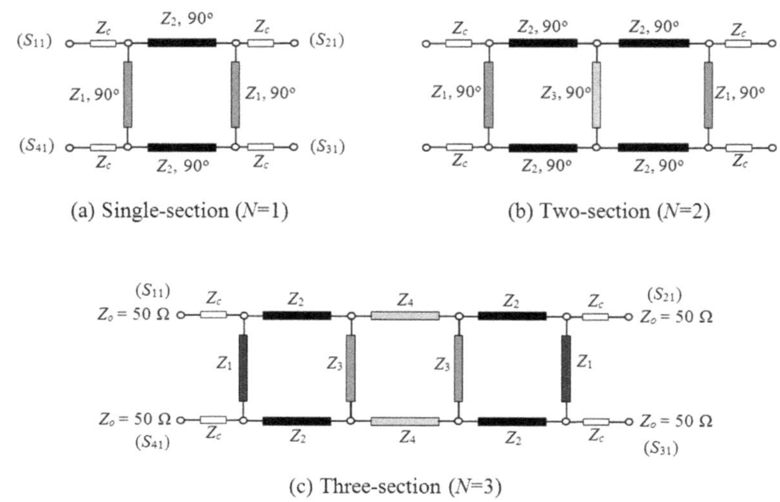

Yao et al. (2011) was comprehensively studied the single-, two-, three-, and four-section crossover coupler as shown in Figure 14. From (17), the analytical crossover design formulations of the characteristic impedance, Z for the branch-line couplers are listed in Table 9, and its impedance, Z values are listed in Table 10 (Yao et al., 2011).

Lin et al. (2013) had designed a dual-band crossover using two-section branch-line coupler and its design formulations are also listed in Table 9. The two-section crossover branch-line coupler is shown in Figure 15.

Arbitrary Phase Branch-Line Couplers

Normally, the length of each microstrip branch line in the coupler is fixed to be 0°, 90°, or 180°. Hence, Toker et al. (2001) had proposed a branch-line couplers using unequal line lengths which are more suitable for microwave-integrated-circuit (MIC) and monolithic-microwave integrated-circuit (MMIC) applications as shown in Figure 16 (a). By varying the length of branch lines, the corresponding characteristic impedances of the lines (width of the lines) are more flexible to be chosen to suit the MIC and

Figure 15. Two-section crossover branch-line coupler proposed by Lin et al. (2013).

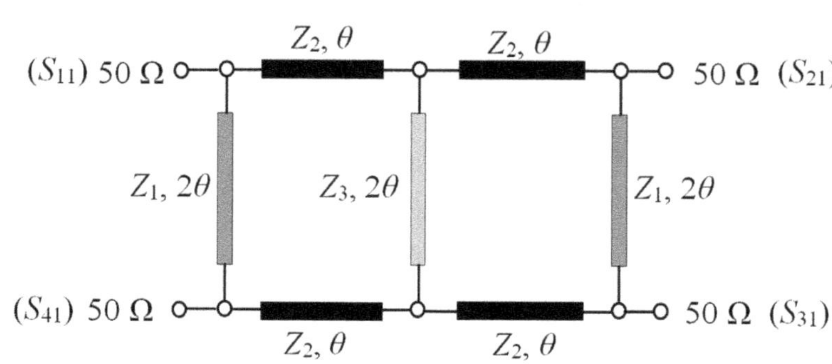

Table 9. Crossover design formulations of the characteristic impedance, Z (Yao et al., 2011).

Crossover Branch-Line Coupler	Characteristic Impedance of Each Branch Line	
	Yao et al., 2011	**Lin et al., 2013**
Single-section	$Z_1 = Z_2 = \left(0.064Z_o\right)^2$	-
Two-section	$Z_1 = \frac{Z_o^2}{50}$ and $Z_3 = \frac{50Z_2^2}{Z_o^2}$	$Z_1 = \frac{2Z_2Z_3 + Z_2^2}{2Z_3}$, $Z_3 = \frac{-Z_2\cos 2\theta}{4\cos^2\theta}$, $Z_2 = Z_o\sqrt{-\sin^2 2\theta - 2\cos 2\theta}$ and $\theta = \frac{\pi f_1}{f_1 + f_2}$
Three-section	$Z_3 = \frac{2Z_2^2}{Z_1}$ and $Z_4 = Z_3$	-
Four-section	$Z_1 = \frac{Z_o^2}{100}$, $Z_3 = \frac{50Z_2^2}{Z_o^2}$, and $Z_5 = \frac{\left(Z_oZ_4\right)^2}{150Z_2^2}$	-

MMIC specifications. The S-parameter matrix of the coupler in Figure 16 (a), is given in Equation (1), in which identical with regular single-section branch-line coupler (as described in Subsection 1.1.1).

Later, Wong et al. (2012) had proposed asymmetric unequal line lengths branch-line coupler as shown in Figure 16 (b) in order to produce nonstandard phase difference condition in the coupler design which is capable of applying in a Butler matrix and antenna feeding circuits. Hence, the S-parameters for Figure 16 (b) at center frequency, f_o can be written as (Wong et al., 2012):

$$\begin{bmatrix} S_{11} & S_{12} & S_{13} & S_{14} \\ S_{21} & S_{22} & S_{23} & S_{24} \\ S_{31} & S_{32} & S_{33} & S_{34} \\ S_{41} & S_{42} & S_{43} & S_{44} \end{bmatrix} = \begin{bmatrix} 0 & 0 & S_{13} & S_{14} \\ 0 & 0 & S_{23} & S_{24} \\ S_{31} & S_{32} & 0 & 0 \\ S_{41} & S_{42} & 0 & 0 \end{bmatrix} = \frac{-je^{j\theta}}{\sqrt{2}} \begin{bmatrix} 0 & 0 & 1 & e^{j\psi} \\ 0 & 0 & e^{j\psi} & 1 \\ 1 & e^{j\psi} & 0 & 0 \\ e^{j\psi} & 1 & 0 & 0 \end{bmatrix} \quad (18)$$

Branch-line coupler with $S_{11} = 0$, $S_{21} = 0$, $|S_{31}| \neq 0$, and $|S_{41}| \neq 0$, can be implemented as a duplex which is described in Section 4.

Table 10. Values of characteristic impedance, Z for crossover coupler (Yao et al., 2011).

Impedance (Ω)	Number of Sections N for Branch-line Structure							
	N =1	N =2	N =3	N =4		N =5	N =6	N =7
Z_c	79.0	71.3	50.0	43.0	50.0	40.1	36.9	35.3
Z_1	20.0	100	80.0	66.2	99.3	73.0	82.5	73.5
Z_2	20.0	48.0	36.5	25.2	38.5	21.2	23.3	21.1
Z_3		25.1	35.0	28.6	43.8	25.2	38.6	44.3
Z_4			32.1	25.5	44.6	25.4	25.3	25.0
Z_5				49.0	89.4	94.3	79.0	93.8
Z_6						42.9	37.3	34.8
Z_7							100	99.5
Z_8								36.4
Bandwidth (For isolation and return loss better than 20 dB)	1%	9%	23%	33%		42%	49%	55%

The proposed analytical design formulations at center frequency, f_o by Toker et al. (2001) and Wong et al. (2012) are listed in Table 11. Later, Park (2013) had re-expressed the formulations of Wong et al. (2012) in comparative explicit form. The design freedom of re-expressed formulation had been improved by Wu et al. (2013) by adding the coupling factor, C term in the formulation as listed in Table 11.

Basic Theory of Analytical Analysis

ABCD Matrix

The branch line coupler is a symmetrical network circuit in which can be simply analyzed as a two-port network using the *ABCD* matrix. Figure 17 shows the relationship between the total voltage, V and current, I at the network ports.

The *ABCD* matrix of the network circuit is given as:

Figure 16. (a) Symmetric and (b) asymmetric unequal transmission line lengths branch-line coupler.

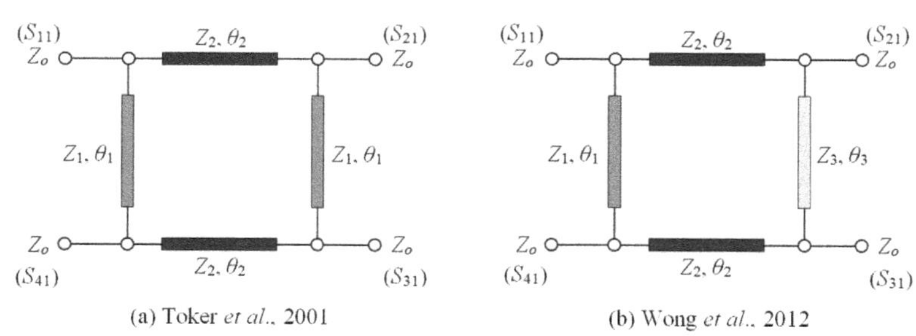

Table 11. Comparison between analytical design formulas for arbitrary phase branch-line couplers

Reference	Analytical Design Formulations
Toker et al., 2001 (refer to Figure 16 (a))	$Z_1 = \frac{\lvert S_{21}\rvert^2 Z_o}{\sqrt{1-\lvert S_{21}\rvert^4}\,\lvert \sin\theta_1 \rvert}$, $Z_2 = \frac{\lvert S_{21}\rvert^2 Z_o}{\sqrt{\lvert S_{21}\rvert^4 + \left(1-\lvert S_{21}\rvert^4\right)\sin^2\theta_1}}$, $\theta_1 = \frac{\pi}{1+\left(f_2/f_1\right)}$, and $\theta_2 = \pi - \theta_1$. where $\lvert S_{21}\rvert$ is the desired linear magnitude of transmission coefficient. Parameters f_1 and f_2 are the dual-band operating center frequency at f_1 and f_2.
Wong *et al.*, 2012 (refer to Figure 16 (b))	$Z_1 = Z_3,\ Z_2 = Z_o\left[\frac{Z_3^2 + Z_1^2}{2Z_3^2 Z_1^2} - 1\right]^{-\frac{1}{2}},\ \theta_2 = \frac{\pi}{2}$, and $\theta_3 = \pi - \theta_1$ The selected value of θ_1 must obey the condition of $\pi/2 < \theta_1 < \pi$, $\tan\frac{\theta_1}{2}\tan\frac{\theta_3}{2} = 1$, and $Z_1 < Z_o$.
Park, 2013 (refer to Figure 16 (b))	$Z_1 = \frac{Z_o \sin\psi}{\sqrt{1+\sin^2\psi}},\ Z_2 = Z_o\sin\psi,\ Z_3 = Z_1,\ \theta_2 = \frac{\pi}{2},\ \theta_3 = \pi - \theta_1$, and $\psi = a\cos\left(-\sqrt{2}\cos\theta_1\right)$ The selected value of Ψ must obey the condition of $0 < \Psi < \pi/2$.
Wu et al., 2013 (refer to Figure 16 (b))	$Z_1 = Z_o \sin\psi\sqrt{\frac{1-C^2}{C^2+\sin^2\psi\left(1-C^2\right)}},\ Z_2 = Z_o\sin\psi\frac{\sqrt{1-C^2}}{C}$, $Z_3 = Z_1,\ \theta_2 = \frac{\pi}{2},\ \theta_3 = \pi - \theta_1$, and $\psi = a\cos\left(\frac{-\cos\theta_1}{\sqrt{1-C^2}}\right)$ The selected value of Ψ must obey the condition of $0 < \Psi < \pi/2$. The coupling factor, $C = -\lvert S_{31}\rvert$ (in unit dB)

$$\begin{bmatrix} V_1 \\ I_1 \end{bmatrix} = \begin{bmatrix} A & B \\ C & D \end{bmatrix}\begin{bmatrix} V_2 \\ -I_2 \end{bmatrix} \tag{19}$$

where V_1 and V_2 are the total voltages at port-1 and port-2, respectively. On the other hand, I_1 and I_2 are the total currents flowing into the network at port-1 and port-2, respectively.

Transmission Lines and Circuit Theory

The *ABCD* matrix of each branch line circuit part is shown in Figure 18.

Figure 17. A two-port network.

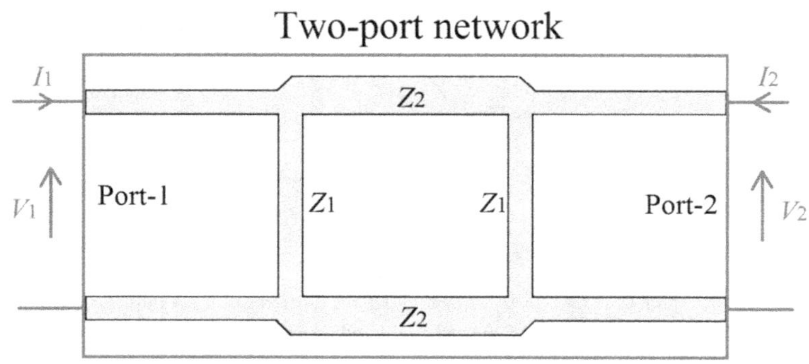

Hence, the *ABCD* matrix in (19) can be written as:

$$\begin{bmatrix} A & B \\ C & D \end{bmatrix} = \begin{bmatrix} 1 & 0 \\ \frac{1}{Z_1} & 1 \end{bmatrix} \begin{bmatrix} \cos\left(\beta_2 l_2\right) & jZ_2 \sin\left(\beta_2 l_2\right) \\ \frac{j}{Z_2} \sin\left(\beta_2 l_2\right) & \cos\left(\beta_2 l_2\right) \end{bmatrix} \begin{bmatrix} 1 & 0 \\ \frac{1}{Z_1} & 1 \end{bmatrix} \tag{20}$$

For the case of a vertical branch line with characteristic impedance of Z_{o1} and length of l_1 terminated by an open end (even mode), the input impedance, Z_1 at the other end of the branch line is purely imaginary as:

$$Z_1 = -jZ_{o1} \cot\left(\beta_1 l_1\right) \tag{21}$$

Similarly, for Z_{o1} of the impedance branch line terminated to the ground or short circuit (odd mode), the input impedance, Z_1 at the other end of the line is also purely imaginary as:

$$Z_1 = jZ_{o1} \tan\left(\beta_1 l_1\right) \tag{22}$$

Figure 18. The shunt and series impedance of the two-port network.

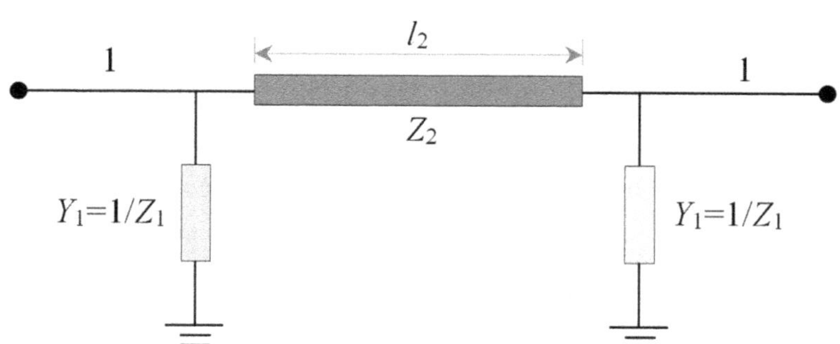

Even-Odd Mode Decomposition

The branch-line coupler circuit in Figure 1 (a) can be decomposed into the superposition of symmetric (even mode) and anti-symmetric (odd mode) excitations. Let's consider two signal sources have same magnitudes and phases in which are applied to port-1 and port-4, a maximum voltage will be existed at the center of the symmetry branch line. This condition is equivalent of the open circuit with infinite impedance, $Z_1 = \infty\ \Omega$ at the symmetric plane (so-called even mode) as shown in Figure 19 (a). Similarly, when two signals with same magnitudes and out of 180° phase shift are applied to port-1 and port-4, a minimum voltage will occur at the center of the symmetric branch line. This is the equivalence of the short circuit with infinite impedance, $Z_1 = 0\ \Omega$ (so-called odd mode) as shown in Figure 19 (b) (Reed and Wheeler, 1956).

All methods described in Subsection 1.2 will be used in formulation derivation of the branch-line coupler (in Section 2).

COMPUTER AIDED DESIGN (CAD) USING ANALYTICAL APPROACHES

Characteristic Impedance of Microstrip Branch-Line Coupler

Single Section Branch-Line Coupler ($N = 1$)

Reed *et al.* (1956) had firstly analyzed the four-port branch-line coupler based on a half portion of the coupler circuit using even-odd decomposition. In this analysis, the coupler is assumed to be separated into two symmetric even and odd mode equivalent circuits as shown in Figure 20.

The physical microstrip line length, l_2 for horizontal branch line is corresponding to $\lambda/4$ length. On the other hand, each vertical branch line, $l_1/2$ becomes $\lambda/8$ length, in which terminated with open or a short circuit for the even- or the odd-mode, respectively.

Based on Figure 20 (a), by multiplying the *ABCD* parameters of each horizontal $\lambda/4$ line and vertical $\lambda/8$ line, the *ABCD* parameter matrix of the coupler for even mode can be written as:

Figure 19. Decomposition of the single section branch-line coupler ($N = 1$) into (a) even- and (b) odd-mode excitation.

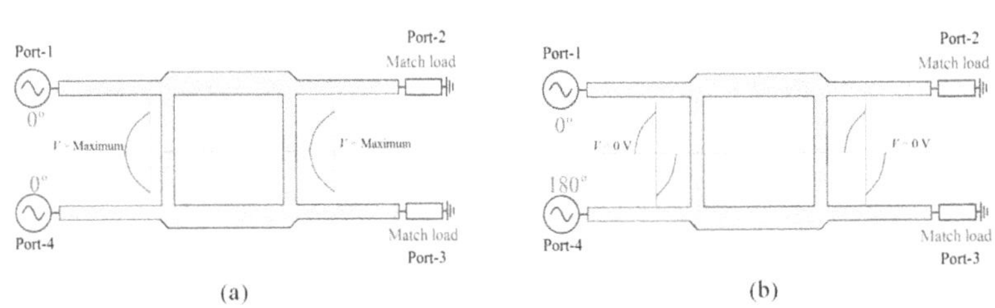

Figure 20. Single section branch-line coupler (N = 1) in (a) even-mode and (b) odd-mode.

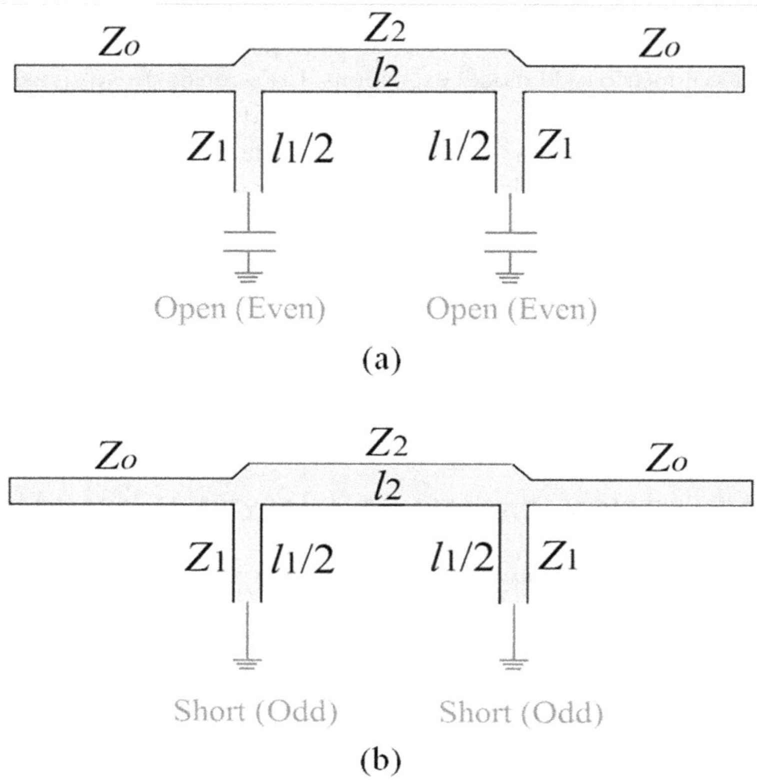

$$\begin{bmatrix} A_e & B_e \\ C_e & D_e \end{bmatrix} = \underbrace{\begin{bmatrix} 1 & 0 \\ \dfrac{j}{Z_1 \cot\left(\beta_1 l_1/2\right)} & 1 \end{bmatrix}}_{Vertical\ Branch} \underbrace{\begin{bmatrix} \cos\left(\beta_2 l_2\right) & jZ_2 \sin\left(\beta_2 l_2\right) \\ \dfrac{j \sin\left(\beta_2 l_2\right)}{Z_2} & \cos\left(\beta_2 l_2\right) \end{bmatrix}}_{Horizontal\ Branch} \underbrace{\begin{bmatrix} 1 & 0 \\ \dfrac{j}{Z_1 \cot\left(\beta_1 l_1/2\right)} & 1 \end{bmatrix}}_{Vertical\ Branch} \tag{23}$$

From Equation (23), the *ABCD* parameters for the even mode are given as:

$$A_e = \cos\left(\beta_2 l_2\right) - \frac{Z_2 \sin\left(\beta_2 l_2\right)}{Z_1 \cot\left(\beta_1 l_1/2\right)} \quad ; \qquad B_e = jZ_2 \sin\left(\beta_2 l_2\right) \quad ;$$

$$C_e = \frac{2j \cos\left(\beta_2 l_2\right)}{Z_1 \cot\left(\beta_1 l_1/2\right)} + \frac{j \sin\left(\beta_2 l_2\right)}{Z_2} - \frac{jZ_2 \sin\left(\beta_2 l_2\right)}{Z_1^2 \cot^2\left(\beta_1 l_1/2\right)}$$

$$D_e = \cos\left(\beta_2 l_2\right) - \frac{Z_2 \sin\left(\beta_2 l_2\right)}{Z_1 \cot\left(\beta_1 l_1/2\right)}$$

Similarly, based on Figure 20 (b), the *ABCD* parameter matrix of the coupler for odd mode can be written as:

$$\begin{bmatrix} A_o & B_o \\ C_o & D_o \end{bmatrix} = \underbrace{\begin{bmatrix} 1 & 0 \\ \dfrac{-j}{Z_1 \tan\left(\beta_1 l_1/2\right)} & 1 \end{bmatrix}}_{Vertical\ Branch} \underbrace{\begin{bmatrix} \cos\left(\beta_2 l_2\right) & jZ_2 \sin\left(\beta_2 l_2\right) \\ \dfrac{j \sin\left(\beta_2 l_2\right)}{Z_2} & \cos\left(\beta_2 l_2\right) \end{bmatrix}}_{Horizontal\ Branch} \underbrace{\begin{bmatrix} 1 & 0 \\ \dfrac{-j}{Z_1 \tan\left(\beta_1 l_1/2\right)} & 1 \end{bmatrix}}_{Vertical\ Branch} \tag{24}$$

By solving the matrix in (24), the *ABCD* parameters for the odd mode are given as:

$$A_o = \cos\left(\beta_2 l_2\right) + \frac{Z_2 \sin\left(\beta_2 l_2\right)}{Z_1 \cot\left(\beta_1 l_1/2\right)} \quad ; \qquad B_o = jZ_2 \sin\left(\beta_2 l_2\right) \quad ;$$

$$C_o = \frac{-2j \cos\left(\beta_2 l_2\right)}{Z_1 \tan\left(\beta_1 l_1/2\right)} + \frac{j \sin\left(\beta_2 l_2\right)}{Z_2} - \frac{jZ_2 \sin\left(\beta_2 l_2\right)}{Z_1^2 \tan^2\left(\beta_1 l_1/2\right)}$$

$$D_o = \cos\left(\beta_2 l_2\right) + \frac{Z_2 \sin\left(\beta_2 l_2\right)}{Z_1 \tan\left(\beta_1 l_1/2\right)}$$

Once the *ABCD* parameters for the even and odd modes are obtained, the reflection coefficient, Γ for both modes can be calculated as:

$$\Gamma_e = \frac{A_e + \left(B_e/Z_o\right) - C_e Z_o - D_e}{A_e + \left(B_e/Z_o\right) + C_e Z_o + D_e} \tag{25}$$

$$\Gamma_o = \frac{A_o + \left(B_o/Z_o\right) - C_o Z_o - D_o}{A_o + \left(B_o/Z_o\right) + C_o Z_o + D_o} \tag{26}$$

While, the transmission coefficient, T for both modes can be determined as:

$$T_e = \frac{2}{A_e + \left(B_e/Z_o\right) + C_e Z_o + D_e} \tag{27}$$

$$T_o = \frac{2}{A_o + \left(B_o / Z_o\right) + C_o Z_o + D_o} \tag{28}$$

where Z_o in (27) and (28) are equal to 50 Ω. Finally, the reflection coefficient, S_{11} at port-1 can be found by summing (superposition) both the cases of reflection coefficient (Γ_o and Γ_e). Similarly, insertion loss, S_{21} (at port-2), coupling, S_{31} (at port-3), and isolation, S_{41} (at port-4) are also the resultant superposition signals due to even and odd mode conditions.

$$S_{11} = S_{22} = S_{33} = S_{44} = \frac{\Gamma_e + \Gamma_o}{2} \tag{29a}$$

$$S_{21} = S_{12} = S_{43} = S_{34} = \frac{T_e + T_o}{2} \tag{29b}$$

$$S_{31} = S_{42} = S_{13} = S_{24} = \frac{T_e - T_o}{2} \tag{29c}$$

$$S_{41} = S_{32} = S_{23} = S_{14} = \frac{\Gamma_e - \Gamma_o}{2} \tag{29d}$$

It should be noted that $\beta_1 l_1 = \beta_2 l_2 = 90°$, however, $\beta_1 \neq \beta_2$ due to different effective dielectric constants, ε_{eff} being different for different microstrip line impedances, $Z_1 \neq Z_2$, thus each branch line will have different physical lengths, $l_1 \neq l_2$.

Recently, Kim et al. (2010) had derived the analytical model of the single section branch-line coupler, which only based on a quarter portion of coupler circuit using even-odd mode decomposition. The derived formulations are explicit; in other words, without involving complex optimization procedures or solving an *ABCD*-matrix. Figure 21 shows a single branch coupler with two fold symmetries along the horizontal and vertical directions.

The equivalent circuit of the single branch coupler can be analyzed using even-odd decomposition by assuming the open- or short-circuit terminations along the horizontal and vertical of the quarter branch lines as shown in Figure 22.

Based on the equivalent quarter circuits, the input impedance, Z_{in} at input port-1 for the four possibly cases (in Figure 22) can be expressed as:

$$Z_{in}^{ee} = Z_o \frac{Z_{ee} + jZ_o \tan\left(\beta_o l_o\right)}{Z_o + jZ_{ee} \tan\left(\beta_o l_o\right)} \tag{30a}$$

Figure 21. Schematic of the single-section branch-line coupler (N = 1)

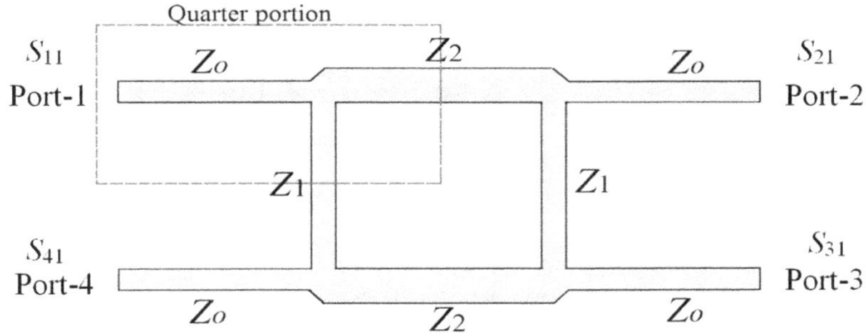

$$Z_{in}^{eo} = Z_o \frac{Z_{eo} + jZ_o \tan\left(\beta_o l_o\right)}{Z_o + jZ_{eo} \tan\left(\beta_o l_o\right)} \quad (30b)$$

$$Z_{in}^{oe} = Z_o \frac{Z_{oe} + jZ_o \tan\left(\beta_o l_o\right)}{Z_o + jZ_{oe} \tan\left(\beta_o l_o\right)} \quad (30c)$$

$$Z_{in}^{oo} = Z_o \frac{Z_{oo} + jZ_o \tan\left(\beta_o l_o\right)}{Z_o + jZ_{oo} \tan\left(\beta_o l_o\right)} \quad (30d)$$

where

$$Z_{ee} = \frac{-jZ_1 Z_2 \cot\left(\frac{\beta_1 l_1}{2}\right)\cot\left(\frac{\beta_2 l_2}{2}\right)}{Z_1 \cot\left(\frac{\beta_1 l_1}{2}\right) + Z_2 \cot\left(\frac{\beta_2 l_2}{2}\right)} \quad Z_{oe} = -j\frac{Z_1 Z_2 \tan\left(\frac{\beta_1 l_1}{2}\right)\cot\left(\frac{\beta_2 l_2}{2}\right)}{Z_1 \tan\left(\frac{\beta_1 l_1}{2}\right) - Z_2 \cot\left(\frac{\beta_2 l_2}{2}\right)}$$

Figure 22. Quarter circuit of the single-branch line coupler with (a) horizontal even -vertical even, (b) horizontal odd-vertical even, (c) horizontal even-vertical odd, and (d) horizontal odd-vertical odd.

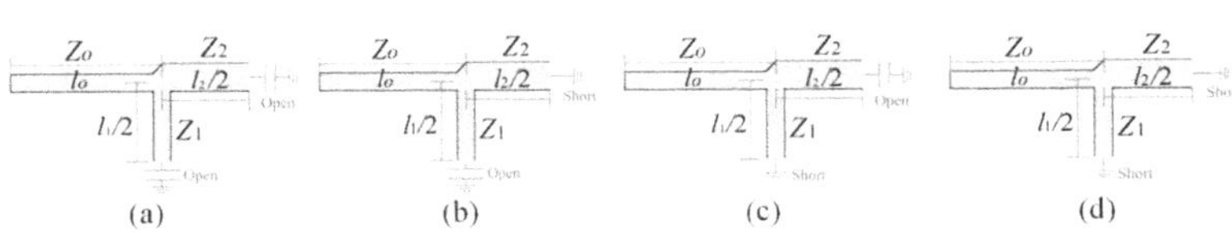

$$Z_{eo} = j\frac{Z_1 Z_2 \cot\left(\frac{\beta_1 l_1}{2}\right)\tan\left(\frac{\beta_2 l_2}{2}\right)}{Z_1 \cot\left(\frac{\beta_1 l_1}{2}\right) - Z_2 \tan\left(\frac{\beta_2 l_2}{2}\right)} \quad Z_{oo} = j\frac{Z_1 Z_2 \tan\left(\frac{\beta_1 l_1}{2}\right)\tan\left(\frac{\beta_2 l_2}{2}\right)}{Z_1 \tan\left(\frac{\beta_1 l_1}{2}\right) + Z_2 \tan\left(\frac{\beta_2 l_2}{2}\right)}$$

It should be noted that $\beta_o l_o = \beta_1 l_1 = \beta_2 l_2 = 90^o$. Symbol Z_o, Z_1, and Z_2 are the characteristic impedances of horizontal and vertical branch line, respectively as shown in Figure 21. The corresponding reflection coefficient, Γ at port-1 for the four conditions are given as:

$$\Gamma_{ee} = \frac{Z_{in}^{ee} - Z_o}{Z_{in}^{ee} + Z_o} \tag{31a}$$

$$\Gamma_{eo} = \frac{Z_{in}^{eo} - Z_o}{Z_{in}^{eo} + Z_o} \tag{31b}$$

$$\Gamma_{oe} = \frac{Z_{in}^{oe} - Z_o}{Z_{in}^{oe} + Z_o} \tag{31c}$$

$$\Gamma_{oo} = \frac{Z_{in}^{oo} - Z_o}{Z_{in}^{oo} + Z_o} \tag{31d}$$

Finally, the full scattering parameters (so-called *S*-parameters) at port-1, -2, -3 and -4 can be found by adding or subtracting the four reflection coefficients from (31a) to (31d) (Collin, 1992):

$$S_{11} = S_{22} = S_{33} = S_{44} = \frac{\Gamma_{ee} + \Gamma_{oe} + \Gamma_{oo} + \Gamma_{eo}}{4} \tag{32a}$$

$$S_{21} = S_{12} = S_{43} = S_{34} = \frac{\Gamma_{ee} + \Gamma_{oe} - \Gamma_{oo} - \Gamma_{eo}}{4} \tag{32b}$$

$$S_{31} = S_{42} = S_{13} = S_{24} = \frac{\Gamma_{ee} - \Gamma_{oe} + \Gamma_{oo} - \Gamma_{eo}}{4} \tag{32c}$$

$$S_{41} = S_{32} = S_{23} = S_{14} = \frac{\Gamma_{ee} - \Gamma_{oe} - \Gamma_{oo} + \Gamma_{eo}}{4} \tag{32d}$$

The derived formulas (30a)-(30d) and (32a)-(32b) are valid for broad operating frequency range. At center frequency, f_o (resonant condition), the coupler should be matched at the input port and $\beta_1 l_1 = \beta_2 l_2 = 90^o$. Thus, from (32a) and (32d), the ideal values of reflection, S_{11} and isolation, S_{41} are given by:

$$S_{11} = S_{41} = \left(\frac{Z_{in}^{oo} - Z_o}{Z_{in}^{oo} + Z_o} + \frac{Z_{in}^{oe} - Z_o}{Z_{in}^{oe} + Z_o}\right) = \left(\frac{Z_{in}^{ee} - Z_o}{Z_{in}^{ee} + Z_o} + \frac{Z_{in}^{eo} - Z_o}{Z_{in}^{eo} + Z_o}\right) = 0 \tag{33}$$

Hence, the characteristic impedances, Z_1 and Z_2 for arbitrary coupling levels at center frequency, f_o can be found as (Lee and Lee, 2012):

$$\begin{aligned} Z_1 &= Z_o\sqrt{K} \\ &= Z_o\sqrt{10^{C/10} - 1} \end{aligned} \tag{34a}$$

and

$$\begin{aligned} Z_2 &= Z_o\sqrt{\frac{K}{1+K}} \\ &= Z_o\sqrt{1 - 10^{-C/10}} \end{aligned} \tag{34b}$$

where $C = -|S_{31}|$ (in unit dB) is the coupling level. On the other hand, K is the power division ratio of signal power at the through port-2 to that at coupled port-3.

$$K = \left(\frac{|S_{21}|}{|S_{31}|}\right)^2 \tag{35}$$

The S-parameters of the regular single-section branch-line coupler at center frequency of 2.4 GHz are calculated using Equations (29) and (32) and compared with measurement results as shown in Figure 23. Clearly, (29) and (32) are given same calculated S-parameters results.

Double Sections Branch-Line Coupler (N = 2)

The operating bandwidth of single section branch-line coupler can be enhanced with double or multi section branch-line coupler as described in Section 1. By applying even and odd mode analysis, the two-section branch-line coupler as shown in Figure 24 can be separated into two half circuits in which

Figure 23. Calculated and measured S-parameter results at center frequency of 2.4 GHz

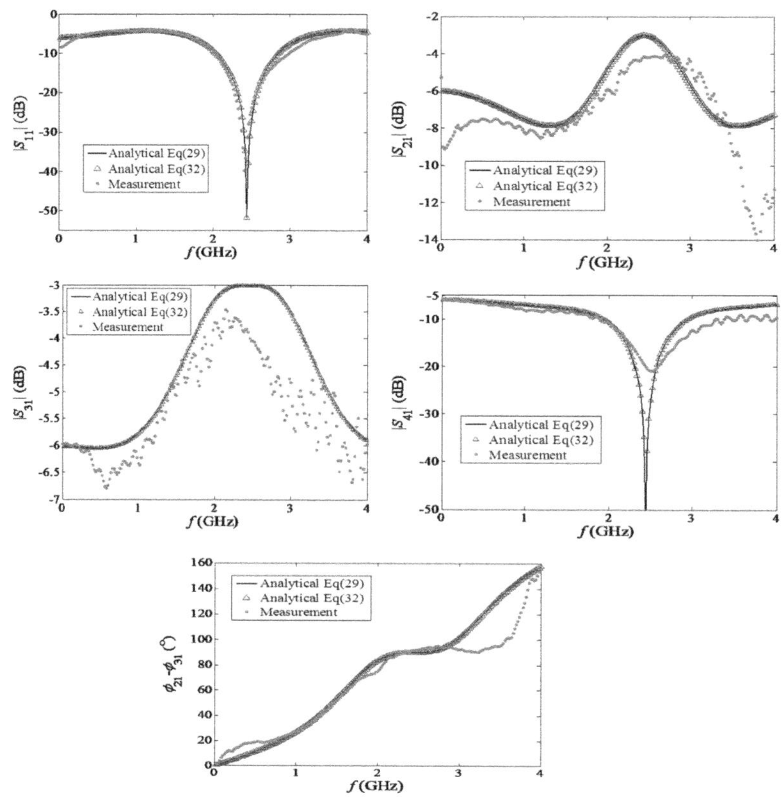

the vertical microstrip branch lines are terminated by open and short circuits, respectively. The four equivalent quarter circuits are as shown in Figure 25.

The analytical impedance formulations (Z_{in}^{oe} , Z_{in}^{ee} , Z_{in}^{eo} , Z_{in}^{oo}) of the four quarter circuits are given as:

Figure 24. Schematic of the two-section branch-line coupler (N = 2)

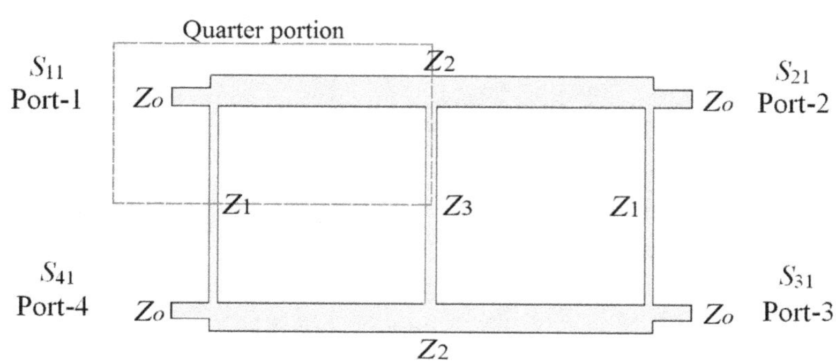

Figure 25. Quarter circuit of the double-branch line coupler with (a) even-even, (b) odd-even, (c) even-odd, and (d) odd-odd modes.

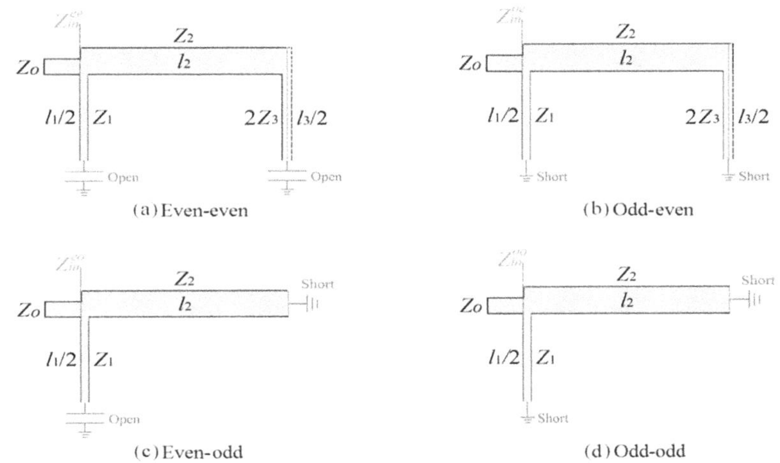

$$Z_{in}^{oe} = \frac{jZ_o' Z_1 \tan\left(\frac{k_1 l_1}{2}\right)}{jZ_1 \tan\left(\frac{k_1 l_1}{2}\right) + Z_o'} \tag{36a}$$

$$Z_{in}^{ee} = \frac{-jZ_e Z_1 \cot\left(\frac{k_1 l_1}{2}\right)}{-jZ_1 \cot\left(\frac{k_1 l_1}{2}\right) + Z_e} \tag{36b}$$

$$Z_{in}^{eo} = \frac{jZ_1 Z_2 \cot\left(\frac{\beta_1 l_1}{2}\right) \tan\left(\beta_2 l_2\right)}{Z_1 \cot\left(\frac{\beta_1 l_1}{2}\right) - Z_2 \tan\left(\beta_2 l_2\right)} \tag{36c}$$

$$Z_{in}^{oo} = \frac{jZ_1 Z_2 \tan\left(\frac{\beta_1 l_1}{2}\right)\tan\left(\beta_2 l_2\right)}{Z_1 \tan\left(\frac{\beta_1 l_1}{2}\right) + Z_2 \tan\left(\beta_2 l_2\right)} \tag{36d}$$

where Z_o' and Z_e in (36a) and (36b) are respectively written as:

$$Z_o' = \frac{jZ_2\left[2Z_3 \tan\left(\frac{k_3 l_3}{2}\right) + Z_2 \cot\left(k_2 l_2\right)\right]}{Z_2 - 2Z_3 \tan\left(\frac{k_3 l_3}{2}\right)\tan\left(k_2 l_2\right)} \tag{37a}$$

$$Z_e = \frac{jZ_2\left[-2Z_3 \cot\left(\frac{k_3 l_3}{2}\right) + Z_2 \tan\left(k_2 l_2\right)\right]}{Z_2 + 2Z_3 \cot\left(\frac{k_3 l_3}{2}\right)\tan\left(k_2 l_2\right)} \tag{37b}$$

From (36) to (37), the corresponding input reflection coefficient, Γ_{oe}, Γ_{ee}, Γ_{eo}, and Γ_{oo} at input port-1 can be written as:

$$\Gamma_{oe} = \frac{Z_{in}^{oe} - Z_o}{Z_{in}^{oe} + Z_o},\ \Gamma_{ee} = \frac{Z_{in}^{ee} - Z_o}{Z_{in}^{ee} + Z_o},\ \Gamma_{eo} = \frac{Z_{in}^{eo} - Z_o}{Z_{in}^{eo} + Z_o},\ \Gamma_{oo} = \frac{Z_{in}^{oo} - Z_o}{Z_{in}^{oo} + Z_o}$$

where the Z_o (= 50 Ω) is the characteristic impedance at the termination port. Finally, the scattering parameters of the coupler (reflection coefficient, S_{11}, transmission coefficient, S_{21} coupling, S_{31}, and isolation, S_{41}) at the four termination ports are expressed in term of the input reflection coefficients as:

$$S_{11} = \frac{\Gamma_{ee} + \Gamma_{eo} + \Gamma_{oe} + \Gamma_{oo}}{4} \tag{38a}$$

$$S_{21} = \frac{\Gamma_{ee} - \Gamma_{eo} + \Gamma_{oe} - \Gamma_{oo}}{4} \tag{38b}$$

Figure 26. Comparison S-parameters between analytical calculation, numerical simulation, and measurement for two-section branch-line coupler.

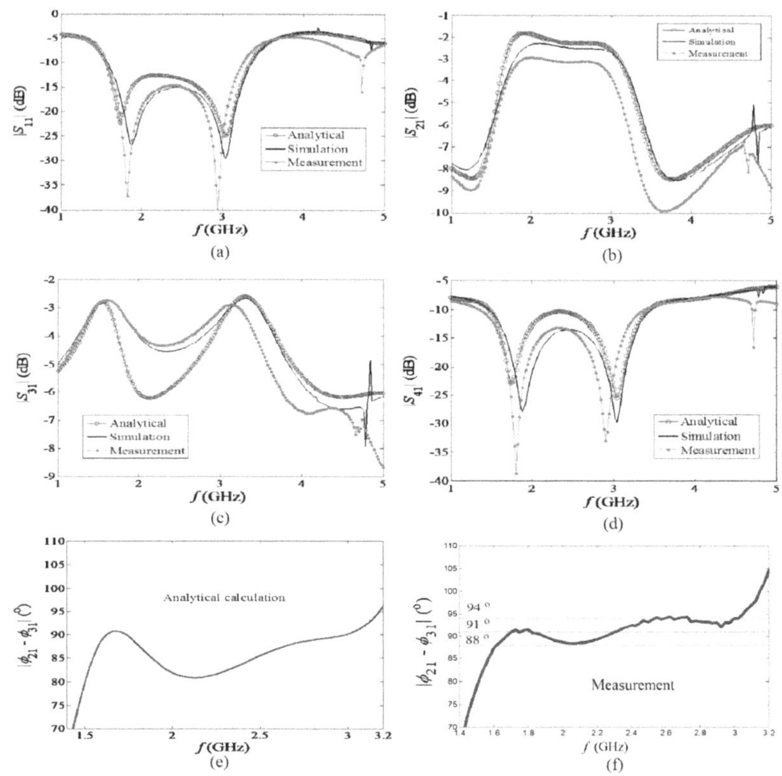

$$S_{31} = \frac{\Gamma_{ee} - \Gamma_{eo} - \Gamma_{oe} + \Gamma_{oo}}{4} \tag{38c}$$

$$S_{41} = \frac{\Gamma_{ee} + \Gamma_{eo} - \Gamma_{oe} - \Gamma_{oo}}{4} \tag{38d}$$

Accuracy comparison between analytical (using Equations (36) to (38)), numerical simulation (AWR simulator), and measurement results of the regular two-section branch-line coupler is shown in Figure 26.

Chiu *et al.* (2010) proposed a two-section branch-line coupler with a port assignment ($|S_{11}| = |S_{31}| = 0$, $|S_{21}| \neq 0$, $|S_{41}| \neq 0$) different from regular two-section coupler. The proposed two-section branch-line coupler circuit can be designed by using phase inverter as shown in Figure 27 (b) (Chiu *et al.*, 2010).

The two-section branch-line coupler with 180° phase inverter can be modeled by separating the coupler into two half circuits in which the vertical microstrip branch lines are terminated by open and short circuits, respectively, as shown in Figure 28. (Chiu *et al.*, 2010).

From Figure 28, the *ABCD* matrix of the even-odd-even mode half-circuit can be derived as:

Figure 27. (a) Regular two-section branch-line coupler (|S_{11}| = |S_{41}| = 0, |S_{21}| ≠ 0, |S_{31}| ≠ 0) and (b) branch-line coupler with 180° phase inverter at the vertical center branch line (|S_{11}| = |S_{31}| = 0, |S_{21}| ≠ 0, |S_{41}| ≠ 0).

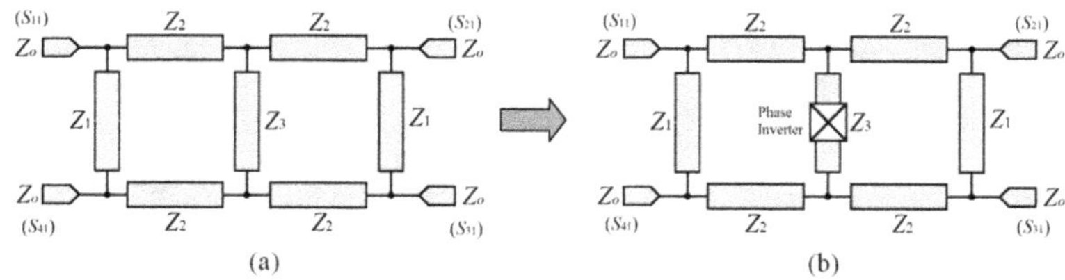

$$\begin{bmatrix} A_{eoe} & B_{eoe} \\ C_{eoe} & D_{eoe} \end{bmatrix} = \underbrace{\begin{bmatrix} 1 & 0 \\ \dfrac{j}{Z_1 \cot\left(\beta_1 l_1/2\right)} & 1 \end{bmatrix}}_{Vertical\ Branch} \underbrace{\begin{bmatrix} a_1 & a_2 \\ a_3 & a_4 \end{bmatrix}}_{Horizontal\ Branch} \underbrace{\begin{bmatrix} 1 & 0 \\ \dfrac{-j}{Z_1 \tan\left(\beta_1 l_1/2\right)} & 1 \end{bmatrix}}_{Vertical\ Branch} \underbrace{\begin{bmatrix} a_1 & a_2 \\ a_3 & a_4 \end{bmatrix}}_{Horizontal\ Branch} \underbrace{\begin{bmatrix} 1 & 0 \\ \dfrac{j}{Z_1 \cot\left(\beta_1 l_1/2\right)} & 1 \end{bmatrix}}_{Vertical\ Branch} \tag{39}$$

Figure 28. Half circuit of the branch line coupler with (a) even-odd-even mode and (b) odd-even-odd mode.

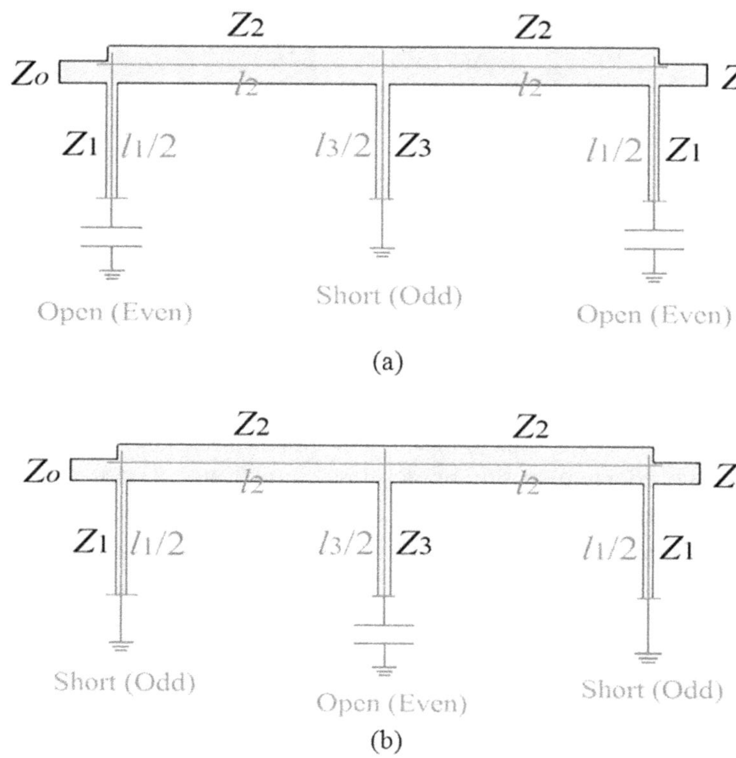

Similarly, the odd-even-odd mode *ABCD* matrix is given as:

$$\begin{bmatrix} A_{oeo} & B_{oeo} \\ C_{oeo} & D_{oeo} \end{bmatrix} = \underbrace{\begin{bmatrix} 1 & 0 \\ \dfrac{-j}{Z_1 \tan\left(\beta_1 l_1/2\right)} & 1 \end{bmatrix}}_{Vertical\ Branch} \underbrace{\begin{bmatrix} a_1 & a_2 \\ a_3 & a_4 \end{bmatrix}}_{Horizontal\ Branch} \underbrace{\begin{bmatrix} 1 & 0 \\ \dfrac{j}{Z_1 \cot\left(\beta_1 l_1/2\right)} & 1 \end{bmatrix}}_{Vertical\ Branch} \underbrace{\begin{bmatrix} a_1 & a_2 \\ a_3 & a_4 \end{bmatrix}}_{Horizontal\ Branch} \underbrace{\begin{bmatrix} 1 & 0 \\ \dfrac{-j}{Z_1 \tan\left(\beta_1 l_1/2\right)} & 1 \end{bmatrix}}_{Vertical\ Branch} \quad (40)$$

where parameters a_1, a_2, a_3, and a_4 in (39) and (40) are expressed as:

$$a_1 = \cos\left(\beta_2 l_2\right);\ a_2 = jZ_2 \sin\left(\beta_2 l_2\right);\ a_3 = \frac{j\sin\left(\beta_2 l_2\right)}{Z_2};\ a_4 = \cos\left(\beta_2 l_2\right)$$

From (39) and (40), the *ABCD* parameters can be expressed independently as:

$$A_{eoe} = \left\{ \begin{array}{l} \cos^2\left(\beta_2 l_2\right) - \sin^2\left(\beta_2 l_2\right) + \dfrac{Z_2 \sin\left(\beta_2 l_2\right)\cos\left(\beta_2 l_2\right)}{Z_3 \tan\left(\beta_3 l_3/2\right)} - \dfrac{2Z_2 \sin\left(\beta_2 l_2\right)\cos\left(\beta_2 l_2\right)}{Z_1 \cot\left(\beta_1 l_1/2\right)} \\ -\dfrac{Z_2^2 \sin^2\left(\beta_2 l_2\right)}{Z_1 Z_3 \cot\left(\beta_1 l_1/2\right)\tan\left(\beta_3 l_3/2\right)} \end{array} \right\} \quad (41a)$$

$$B_{eoe} = j2Z_2 \sin\left(\beta_2 l_2\right)\cos\left(\beta_2 l_2\right) + \frac{jZ_2^2 \sin^2\left(\beta_2 l_2\right)}{Z_3 \tan\left(\beta_3 l_3/2\right)} \quad (41b)$$

$$C_{eoe} = \left\{ \begin{array}{l} \dfrac{j2\cos^2\left(\beta_2 l_2\right)}{Z_1 \cot\left(\beta_1 l_1/2\right)} + \dfrac{j2\sin\left(\beta_2 l_2\right)\cos\left(\beta_2 l_2\right)}{Z_2} + \dfrac{j2Z_2 \sin\left(\beta_2 l_2\right)\cos\left(\beta_2 l_2\right)}{Z_1 Z_3 \cot\left(\beta_1 l_1/2\right)\tan\left(\beta_3 l_3/2\right)} - \dfrac{j\cos^2\left(\beta_2 l_2\right)}{Z_3 \tan\left(\beta_3 l_3/2\right)} \\ -\dfrac{j2\sin^2\left(\beta_2 l_2\right)}{Z_1 \cot\left(\beta_1 l_1/2\right)} - \dfrac{j2Z_2 \sin\left(\beta_2 l_2\right)\cos\left(\beta_2 l_2\right)}{Z_1^2 \cot^2\left(\beta_1 l_1/2\right)} - \dfrac{jZ_2^2 \sin^2\left(\beta_2 l_2\right)}{Z_1^2 Z_3 \cot^2\left(\beta_1 l_1/2\right)\tan\left(\beta_3 l_3/2\right)} \end{array} \right\} \quad (41c)$$

$$D_{eoe} = \left\{ \begin{array}{l} \cos^2\left(\beta_2 l_2\right) - \sin^2\left(\beta_2 l_2\right) - \dfrac{2Z_2 \sin\left(\beta_2 l_2\right)\cos\left(\beta_2 l_2\right)}{Z_1 \cot\left(\beta_1 l_1/2\right)} - \dfrac{Z_2^2 \sin^2\left(\beta_2 l_2\right)}{Z_1 Z_3 \cot\left(\beta_1 l_1/2\right)\tan\left(\beta_3 l_3/2\right)} \\ + \dfrac{Z_2 \sin\left(\beta_2 l_2\right)\cos\left(\beta_2 l_2\right)}{Z_3 \tan\left(\beta_3 l_3/2\right)} \end{array} \right\} \tag{41d}$$

and

$$A_{oeo} = \left\{ \begin{array}{l} \cos^2\left(\beta_2 l_2\right) - \sin^2\left(\beta_2 l_2\right) + \dfrac{2Z_2 \sin\left(\beta_2 l_2\right)\cos\left(\beta_2 l_2\right)}{Z_1 \tan\left(\beta_1 l_1/2\right)} - \dfrac{Z_2 \sin\left(\beta_2 l_2\right)\cos\left(\beta_2 l_2\right)}{Z_3 \cot\left(\beta_3 l_3/2\right)} \\ - \dfrac{Z_2^2 \sin^2\left(\beta_2 l_2\right)}{Z_1 Z_3 \tan\left(\beta_1 l_1/2\right)\cot\left(\beta_3 l_3/2\right)} \end{array} \right\} \tag{42a}$$

$$B_{oeo} = j2Z_2 \sin\left(\beta_2 l_2\right)\cos\left(\beta_2 l_2\right) - \frac{jZ_2^2 \sin^2\left(\beta_2 l_2\right)}{Z_3 \cot\left(\beta_3 l_3/2\right)} \tag{42b}$$

$$C_{oeo} = \left\{ \begin{array}{l} \dfrac{-j2\cos^2\left(\beta_2 l_2\right)}{Z_1 \tan\left(\beta_1 l_1/2\right)} + \dfrac{j2\sin\left(\beta_2 l_2\right)\cos\left(\beta_2 l_2\right)}{Z_2} + \dfrac{j2Z_2 \sin\left(\beta_2 l_2\right)\cos\left(\beta_2 l_2\right)}{Z_1 Z_3 \tan\left(\beta_1 l_1/2\right)\cot\left(\beta_3 l_3/2\right)} + \dfrac{j\cos^2\left(\beta_2 l_2\right)}{Z_3 \cot\left(\beta_3 l_3/2\right)} \\ + \dfrac{j2\sin^2\left(\beta_2 l_2\right)}{Z_1 \tan\left(\beta_1 l_1/2\right)} - \dfrac{j2Z_2 \sin\left(\beta_2 l_2\right)\cos\left(\beta_2 l_2\right)}{Z_1^2 \tan^2\left(\beta_1 l_1/2\right)} + \dfrac{jZ_2^2 \sin^2\left(\beta_2 l_2\right)}{Z_1^2 Z_3 \tan^2\left(\beta_1 l_1/2\right)\cot\left(\beta_3 l_3/2\right)} \end{array} \right\} \tag{42c}$$

$$D_{oeo} = \left\{ \begin{array}{l} \cos^2\left(\beta_2 l_2\right) - \sin^2\left(\beta_2 l_2\right) + \dfrac{2Z_2 \sin\left(\beta_2 l_2\right)\cos\left(\beta_2 l_2\right)}{Z_1 \tan\left(\beta_1 l_1/2\right)} - \dfrac{Z_2^2 \sin^2\left(\beta_2 l_2\right)}{Z_1 Z_3 \tan\left(\beta_1 l_1/2\right)\cot\left(\beta_3 l_3/2\right)} \\ - \dfrac{Z_2 \sin\left(\beta_2 l_2\right)\cos\left(\beta_2 l_2\right)}{Z_3 \cot\left(\beta_3 l_3/2\right)}. \end{array} \right\} \tag{42d}$$

Once the *ABCD* parameters for the even and odd modes are obtained, the reflection coefficient, Γ for both modes can be calculated as:

$$\Gamma_{eoe} = \frac{A_{eoe} + \left(B_{eoe}/Z_o\right) - C_{eoe}Z_o - D_{eoe}}{A_{eoe} + \left(B_{eoe}/Z_o\right) + C_{eoe}Z_o + D_{eoe}} \tag{43}$$

$$\Gamma_{oeo} = \frac{A_{oeo} + \left(B_{oeo}/Z_o\right) - C_{oeo}Z_o - D_{oeo}}{A_{oeo} + \left(B_{oeo}/Z_o\right) + C_{oeo}Z_o + D_{oeo}} \tag{44}$$

While, the transmission coefficient, T for both modes can be determined as:

$$T_{eoe} = \frac{2}{A_{eoe} + \left(B_{eoe}/Z_o\right) + C_{eoe}Z_o + D_{eoe}} \tag{45}$$

$$T_{oeo} = \frac{2}{A_{oeo} + \left(B_{oeo}/Z_o\right) + C_{oeo}Z_o + D_{oeo}} \tag{46}$$

Once even-odd-even and odd-even-odd modes *ABCD* parameters are determined, the *S*-parameters (S_{11}, S_{21}, S_{31}, and S_{41}) of the coupler can be calculated as:

$$S_{11} = \frac{\Gamma_{eoe} + \Gamma_{oeo}}{2} \tag{47a}$$

$$S_{21} = \frac{T_{eoe} + T_{oeo}}{2} \tag{47b}$$

$$S_{31} = \frac{T_{eoe} - T_{oeo}}{2} \tag{47c}$$

$$S_{41} = \frac{\Gamma_{eoe} - \Gamma_{oeo}}{2} \tag{47d}$$

When perfect impedance matching at each port and $S_{11} = 0$ at center frequency, the (47a) can be written as (Chiu and Xue, 2010):

$$\frac{Z_2^2}{Z_o Z_3} - \frac{Z_o}{Z_1}\left(2 + \frac{Z_2^2}{Z_1 Z_3}\right) = 0 \tag{48a}$$

and power division ratio, K is given as:

$$K = \left|\frac{S_{21}}{S_{41}}\right|^2 = \frac{Z_o^2 Z_3^2}{Z_2^4} \tag{48b}$$

By solving (48a) and (48b), the characteristic impedances, Z_1 and Z_2 for arbitrary coupling levels at center frequency, f_o can be found as (Chiu, 2014; Chiu and Xue, 2010):

$$Z_1 = \left(\sqrt{K} + \sqrt{K+1}\right) Z_o \tag{49a}$$

$$Z_2 = Z_o \tag{49b}$$

$$Z_3 = \frac{Z_2^2 \sqrt{K}}{Z_o} \tag{49c}$$

The calculated S-parameters results using Equations (41) to (49) is shown in Figure 29.

Three Branch-Line Coupler (*N* = 3)

Similarly with the analytical modeling of one-section branch-line coupler (Kim et al., 2010), the analytical theory has been extended in order to model the three-section branch-line coupler. Figure 30 shows a three-section coupler and its characteristic impedance (Z_1, Z_2 and Z_3) for each branch line. The scattering analysis of the coupler can be simplified by analyzing the quarter portion of the coupler circuit using even-odd decomposition.

In this work, only four load termination conditions are needed to be considered for the quarter circuit as shown in Figure 31. The end of the horizontal and vertical microstrip branch line are assumed to be horizontal even-vertical even-vertical even [Figure 31(a)], horizontal even-vertical odd-vertical odd [Figure 31(b)], horizontal odd-vertical even-vertical even [Figure 31(c)], and horizontal odd-vertical odd-vertical odd [Figure 31(d)] mode.

Based on the transmission line theory, the input impedance, Z_{in} at the input port for the four possible cases are given by:

(50a)

Figure 29. Calculated $|S_{11}|$, $|S_{21}|$, $|S_{31}|$, $|S_{41}|$, *and* $|\phi_{21} - \phi_{41}|$ *of the coupler (Chiu et al., 2010) for power division ratio, K = 1, K= 2, and K = 4, respectively.*

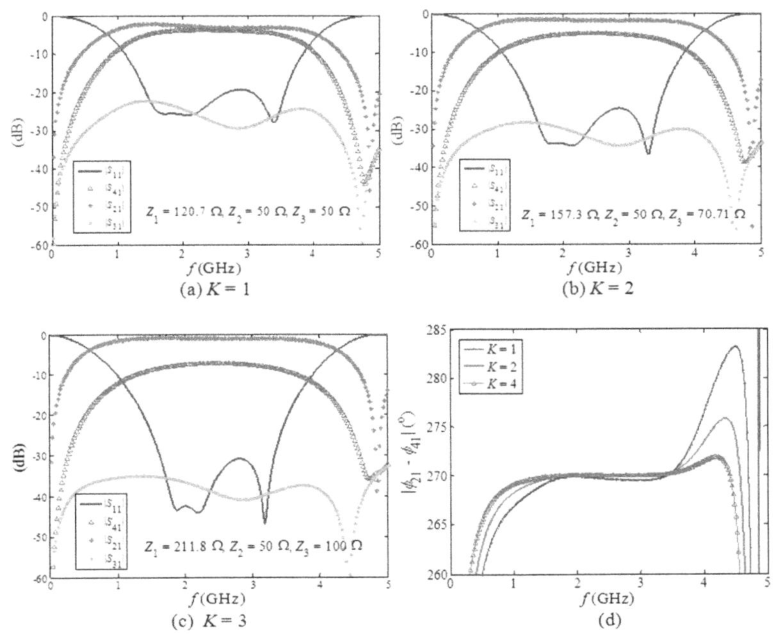

(50b)

$$Z_{in}^{ooo} = \frac{jZ_1 Z_2 \tan\left(\beta_1 l_1/2\right)\left[Z_C + jZ_2 \tan\left(\beta_2 l_2\right)\right]}{Z_2\left[Z_C + jZ_2 \tan\left(\beta_2 l_2\right)\right] + jZ_1 \tan\left(\beta_1 l_1/2\right)\left[Z_2 + Z_C \tan\left(\beta_2 l_2\right)\right]} \tag{50c}$$

Figure 30. Schematic of the three-section branch-line coupler (N = 3).

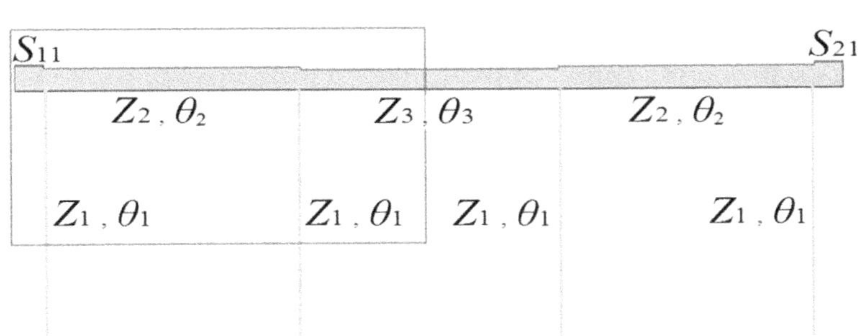

Figure 31. Quarter circuit of the coupler with (a) horizontal even-vertical even-vertical even, (b) horizontal odd-vertical even-vertical even, (c) horizontal odd-vertical odd-vertical odd, and (d) horizontal even-vertical odd-vertical odd mode.

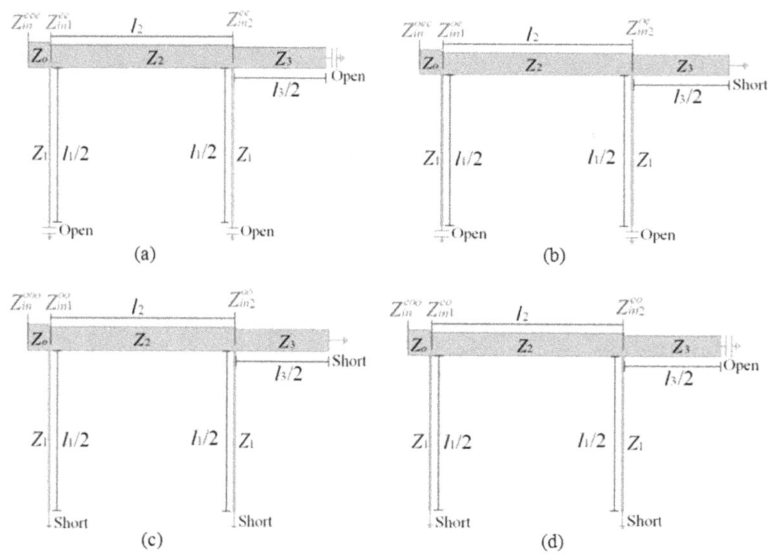

$$Z_{in}^{eoo} = \frac{jZ_1Z_2 \tan\left(\beta_1 l_1/2\right)\left[Z_D + jZ_2 \tan\left(\beta_2 l_2\right)\right]}{Z_2\left[Z_D + jZ_2 \tan\left(\beta_2 l_2\right)\right] + jZ_1 \tan\left(\beta_1 l_1/2\right)\left[Z_2 + Z_D \tan\left(\beta_2 l_2\right)\right]} \tag{50d}$$

where Z_A, Z_B, Z_C and Z_D in (50a)-(50d) are expressed as:

$$Z_A = \frac{-jZ_3Z_1 \cot\left(\beta_3 l_3/2\right)\cot\left(\beta_1 l_1/2\right)}{Z_3 \cot\left(\beta_3 l_3/2\right) + Z_1 \cot\left(\beta_1 l_1/2\right)} \tag{51a}$$

$$Z_B = \frac{-jZ_3Z_1 \tan\left(\beta_3 l_3/2\right)\cot\left(\beta_1 l_1/2\right)}{Z_3 \tan\left(\beta_3 l_3/2\right) - Z_1 \cot\left(\beta_1 l_1/2\right)} \tag{51b}$$

$$Z_C = \frac{jZ_3Z_1 \tan\left(\beta_3 l_3/2\right)\tan\left(\beta_1 l_1/2\right)}{Z_3 \tan\left(\beta_3 l_3/2\right) + Z_1 \tan\left(\beta_1 l_1/2\right)} \tag{51c}$$

$$Z_D = \frac{-jZ_3Z_1 \cot\left(\beta_3 l_3/2\right)\tan\left(\beta_1 l_1/2\right)}{-Z_3 \cot\left(\beta_3 l_3/2\right) + Z_1 \tan\left(\beta_1 l_1/2\right)} \tag{51d}$$

where the Z_1, Z_2 and Z_3 are the characteristic impedances of the vertical and horizontal branch lines, respectively. The l_1, l_2, and l_3 are the corresponding lengths of the branch lines. The superscripts *eee*, *eoo*, *oee*, and *ooo* of the Z_{in} in (50a)-(50d) are represented as the horizontal even-vertical even-vertical even, horizontal even-vertical odd-vertical odd, horizontal odd-vertical even-vertical even, and horizontal odd-vertical odd-vertical odd mode, respectively. Symbols β_1, β_2, and β_3 are the propagation constants of the vertical and horizontal branch line and both are microstrip line width-dependent parameters.

From (50a) to (50d), the corresponding input reflection coefficients, Γ_{eee}, Γ_{eoo}, Γ_{oee}, and Γ_{ooo} at input port-1 can be written as:

$$\Gamma_{eee} = \frac{Z_{in}^{eee} - Z_o}{Z_{in}^{eee} + Z_o},\ \Gamma_{eoo} = \frac{Z_{in}^{eoo} - Z_o}{Z_{in}^{eoo} + Z_o},\ \Gamma_{oee} = \frac{Z_{in}^{oee} - Z_o}{Z_{in}^{oee} + Z_o},\ \Gamma_{ooo} = \frac{Z_{in}^{ooo} - Z_o}{Z_{in}^{ooo} + Z_o}$$

where the Z_o (= 50 Ω) is the characteristic impedance at the termination port. Finally, the scattering parameters of the coupler (reflection coefficient, S_{11}, transmission coefficient, S_{21} coupling, S_{31}, and isolation, S_{41}) at the four termination ports are expressed in term of the input reflection coefficients as:

$$S_{11} = \frac{\Gamma_{eee} + \Gamma_{oee} + \Gamma_{ooo} + \Gamma_{eoo}}{4} \tag{52a}$$

$$S_{21} = \frac{\Gamma_{eee} - \Gamma_{oee} - \Gamma_{ooo} + \Gamma_{eoo}}{4} \tag{52b}$$

$$S_{31} = \frac{\Gamma_{eee} - \Gamma_{oee} + \Gamma_{ooo} - \Gamma_{eoo}}{4} \tag{52c}$$

$$S_{41} = \frac{-\Gamma_{eee} - \Gamma_{oee} + \Gamma_{ooo} + \Gamma_{eoo}}{4} \tag{52d}$$

At resonant condition, the coupler should be matched at the input port. Thus, from (52a) and (52d), the ideal values of reflection, S_{11} and isolation, S_{41} are given by:

$$S_{11} = S_{41} = \left(\frac{Z_{in}^{ooo} - Z_o}{Z_{in}^{ooo} + Z_o} + \frac{Z_{in}^{eoo} - Z_o}{Z_{in}^{eoo} + Z_o}\right) = \left(\frac{Z_{in}^{eee} - Z_o}{Z_{in}^{eee} + Z_o} + \frac{Z_{in}^{oee} - Z_o}{Z_{in}^{oee} + Z_o}\right) = 0 \tag{53}$$

In addition, the branch lines, $l_3/2$, l_2, and $l_1/2$ are expected to have physical lengths corresponding to $\lambda/8$, $\lambda/4$, and $\lambda/8$, respectively. Instead the (50) into (53), yields (You *et al.*, 2018)

$$Z_1^2 Z_2^4 Z_3^2 + Z_o^2 Z_2^4 Z_3^2 - 2\left(Z_o Z_1 Z_2 Z_3\right)^2 + Z_o^2 Z_1^4 Z_3^2 - \left(Z_1 Z_2\right)^4 - \left(Z_o Z_1\right)^2 Z_2^4 = 0 \tag{54}$$

Rearranged the (54), the function of Z_3 in terms of parameters Z_1 and Z_2 can be expressed as:

$$Z_3 = \sqrt{\frac{Z_2^4\left(Z_1^4 - Z_o^2 Z_1^2\right)}{Z_2^4\left(Z_1^2 + Z_o^2\right) - Z_1^2 Z_o^2\left(2Z_2^2 - Z_1^2\right)}} \tag{55}$$

As known that, the relationship of S_{21} and S_{31} with its coupling factor, C can be written as (Wu *et al.*, 2013):

$$S_{21} = \sqrt{1 - C^2} \text{ and } S_{31} = jC$$

Thus, the ratio of S_{31} to S_{21} with the aid of (52b), (52c), and (55) at resonant stage is expressed as:

$$\frac{S_{31}}{S_{21}} = \frac{C}{\sqrt{1 - C^2}} = \frac{Z_o Z_1\left(Z_1^2 - 2Z_2^2\right)}{Z_o^2 Z_1^2 - Z_o^2 Z_2^2 + Z_1^2 Z_2^2} \tag{56}$$

Rearranged the (55), yields

$$Z_2 = \sqrt{\frac{Z_o Z_1^3\sqrt{1 - C^2} - C\, Z_o^2 Z_1^2}{C\,\left(Z_1^2 - Z_o^2\right) + 2Z_o Z_1\sqrt{1 - C^2}}} \tag{57}$$

Insert (57), (55) and the desired value of C into (54), then, the Z_1 can be solved by routine for finding zero.

Performance comparisons between the analytical [using Equations (50) to (52)], simulation, and measurement results of three-section branch-line coupler is shown in Figure 32. The simulated results are derived using AWR Microwave Office simulator for coupler calculations and measurements validation. The result of simulations and measurements are fairly close to each other. The design specification for conventional coupler is according to Muraguchi et al. (1983).

Figure 32. Analytical calculation, simulation, and measurement results of three-section branch-line coupler.

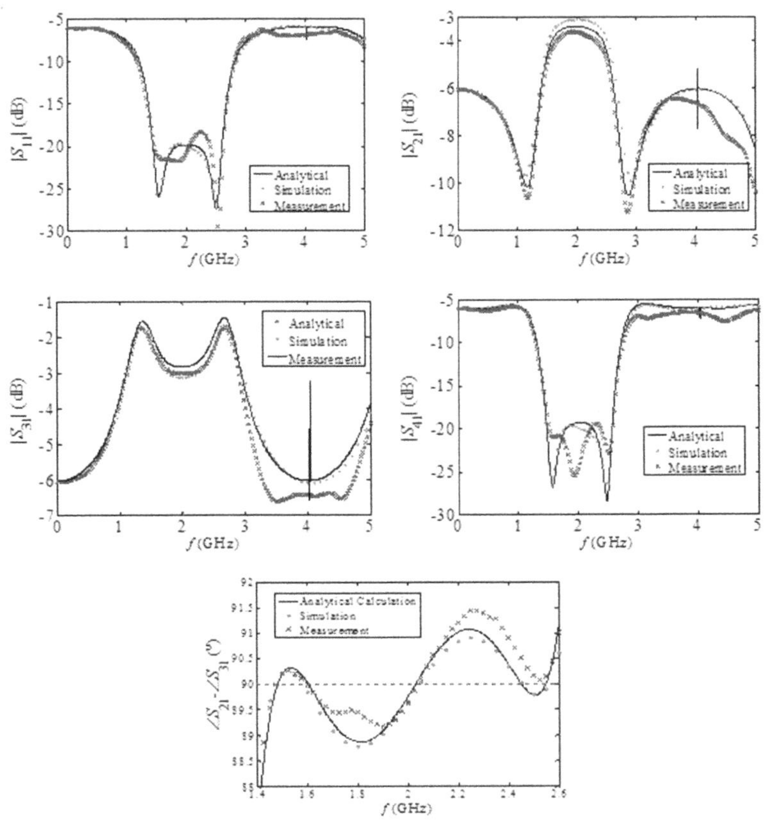

ANALYTICAL-BASED AND NUMERICAL-BASED COMPUTER AIDED DESIGN (CAD)

Collaboration Between Analytical Calculation and Numerical Simulation

As mentioned in Section 2, for the conventional branch-line coupler, its drawback is that at low frequencies, it consumes a significant amount of circuit area and provides a narrow fractional bandwidth of 10 to 20%. Even though cascading branch sections can increase the bandwidth of the coupler, it will also lead to a larger size (Muraguchi et al., 1983). In addition, the cascading branch-line coupler has higher value of characteristic impedance, Z for the vertical branch line and caused the small microstrip line width, W, which is difficult to fabricate on PCB substrate.

Despite this, the small size and single layer printed circuit board of the branch-line coupler is required for portable RF instruments since the space to install coupler is limited and the close proximity of RF components is unavoidable. There is a demand for the branch-line coupler, which is minimizing the size of the coupler, while the performance of the small coupler remains unchanged or improved. Thus, different modification concepts have been used to reduce the length of microstrip line of the conventional coupler, such as meandering microstrip line and slow-wave microstrip line structure (Liao et al., 2005; Chun and Hong, 2006; Chen et al., 2009; Kurgan and Kitliński, 2009; Krishna et al., 2010; Kurgan et al., 2012; You et al., 2018). Besides, in recent years, there has been an explosive growth of interest in the development of wideband microwave circuits, in which the characteristics of the circuits are relatively independent of operating frequency over a large bandwidth.

Nevertheless, the complex slow-wave patterns (SW), defected ground structure (DGS), and meandering line (ML) that are used for branch-line coupler design, appeared to be too complex, difficult to model by analytical equations. Hence, the numerical-based computer aided design (CAD) tools need to be used to overcome this issue. The design procedure of the branch coupler can be demonstrated in four steps:

Step 1

First, the characteristic impedance, Z at each branch microstrip line is determined based on manually analytical calculation. Next, the desired specifications of the designed branch-line coupler are considered, such as center frequency, f_o operating bandwidth, and used PCB substrate. Once the impedances, Z and specification of designed coupler are obtained, the corresponding length, l and width, W (initial dimensions) of the microstrip branch lines for the coupler are estimated using equations (2) to (4) (or TX-Line program calculator). The guide-user interface (GUI) of the TX-Line calculator is shown in Figure 33.

The performance of coupler is roughly observed based on S-parameters versus operating frequency using the analytical calculations as described in Section 2. The observed S-parameters are the return loss $|S_{11}|$, insertion loss $|S_{21}|$, coupling $|S_{31}|$, isolation $|S_{41}|$, and phase difference $|\phi_{21}-\phi_{31}|$.

Step 2

The initial designed coupler circuit structure is re-simulated using open source or commercial numerical-based CAD. The simulated S-parameters can be compared to the analytical calculation results in **Step 1**. Minor adjustments to coupler circuit structure may be done in the simulation in order to optimize the performance of coupler. For instance, if the center frequency requires to be slightly shifted to a lower value, then the length, l of the microstrip branch line should be slightly extended and vice versa. On the

Figure 33. Guide-user interface (GUI) of the TX-Line calculator.

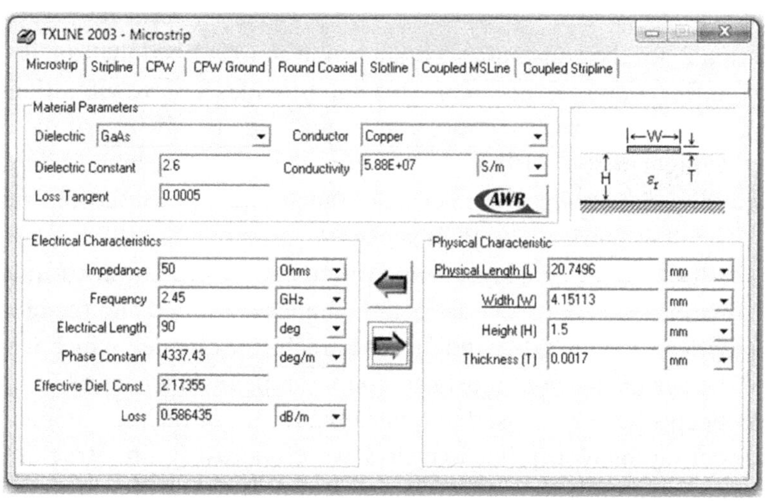

other hand, if the performance of the S-parameters would like to be further improved, then the branch line width should be slightly modified.

Step 3

The initial circuit structure in **Step 2** is miniaturized by bending the microstrip branch line into loop line form (so-called meandering line). The total electrical length of the meandering line in miniaturized coupler should be same with the electrical length of the regular straight branch line in initial circuit structure. The effective electrical length, l_{eff} of the meandering line may be predicted as:

$$l_{eff} = l + \Delta l \tag{58}$$

where l is the actual physical length of the meandering line, while Δl is the lying electrical length caused by the bend in the branch line as (Getsinger, 1993):

$$\Delta l = \alpha W\left(\frac{\pi}{4} - 1\right) - 120\alpha\left\{\frac{1 - \exp\left(-Z_n\sqrt{\varepsilon_{eff}}/60\right)}{Z_n\sqrt{\varepsilon_{eff}}}\right\} \tag{59}$$

Symbol α in (59) is the number of bend sections in the meandering line. The S-parameters of the miniaturized coupler is re-simulated. The length's tolerance of the meandering line can be minor adjusted to obtain the desired performance.

Step 4

The circuit structure in **Step 3** is further miniaturized by adding slow-wave structure to achieve the desired circuit size. The slow-wave structure can be in the form of a series of open-stub or short-stub attached on the microstrip branch line. Normally, the total size reduction is expected to be more than 50% compared to the conventional coupler size. The performance of coupler is optimized by minor adjustment of the length, width, and distance gap between the stub as well as the number of the stub on the microstrip branch line. The effective electrical length, l_{eff} of the open stub can be predicted using Equation (60), but the lying electrical length, Δl caused by the fringing field at the open end of the microstrip line is given as (Kirschning et al., 1981):

$$\Delta l = h\left(\frac{\eta_1\eta_2\eta_3}{\eta_4}\right) \tag{60}$$

where η_1, η_2, η_3, and η_4 in (60) are given as:

$$\eta_1 = 0.434907 \frac{\varepsilon_{eff}^{0.81} + 0.26\left(W/h\right)^{0.8544} + 0.236}{\varepsilon_{eff}^{0.81} - 0.189\left(W/h\right)^{0.8544} + 0.87} \tag{61a}$$

$$\eta_2 = 1 + \frac{0.5274 \arctan\left[0.084\left(W/h\right)^{\frac{1.9413}{\beta}}\right]}{\varepsilon_{eff}^{0.9236}} \tag{61b}$$

$$\eta_3 = 1 - 0.218 \exp\left(-7.5W/h\right) \tag{61c}$$

$$\eta_4 = 1 + 0.0377 \arctan\left[0.067\left(W/h\right)^{1.456}\right]\left\{6 - 5\exp\left[0.036\left(1 - \varepsilon_r\right)\right]\right\} \tag{61d}$$

The parameter β in (61b) is written as:

$$\beta = 1 + \frac{\left(W/h\right)^{0.371}}{2.358\varepsilon_r + 1} \tag{62}$$

The W, h, and ε_{eff} are the line width of the open-stub, thickness of PCB substrate, and microstrip effective dielectric constant. All parameters may be calculated using Equations (2) and (4).

Miniaturization of Conventional Branch-Line Couplers

The physical length, l of conventional microstrip line in Figure 24 can be reduced using capacitive slow-wave structures as shown in Figure 34 (You *et al.*, 2019). The repeated discontinuous line structure consists of capacitive and inductive lines which will increase the value of inductance, ΔL and capacitance, ΔC of the microstrip transmission line. This means that the value of the phase velocity, ν for the transmission line is reduced. The smaller value of the v, results in smaller electrical wavelength, λ (= ν/f). Thus, the new physical length of the modified coupler is required to achieve the desired quarter-wavelength at certain center frequency; which is smaller in discontinuous microstrip line structure compared with regular straight microstrip line.

You et al. (2018) was designed compact three-section coupler using slow-wave structures and meandering lines in which this coupler was achieved 53% reduced circuit size (compare to regular three-section branch-line coupler) and fractional bandwidth of 77% as shown in Figure 36. The values of the coupler's dimensions are listed in Table 13. The simulated and measured S-parameters of the miniaturized coupler are shown in Figure 37.

Figure 34. Schematic and dimensions (in unit mm) of the modified double-section branch-line coupler.

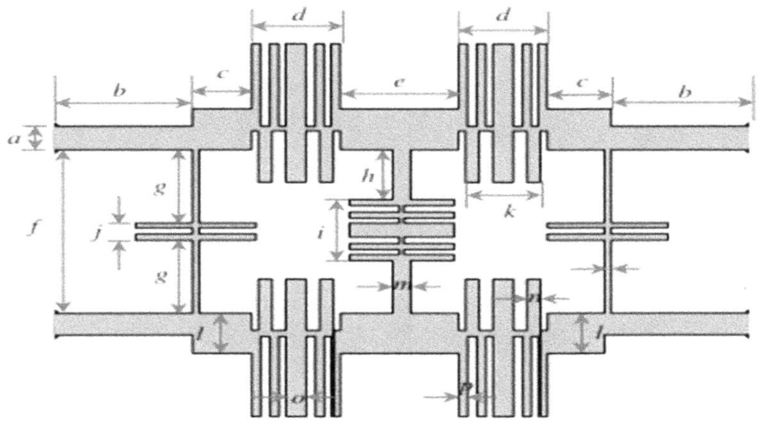

Al-Areqi et al. (2018) had further reduced the size of the three-section coupler using more meandering line, irregular open-stub, and thin RT/Duroid substrate. Figure 38 and Figure 39 show the modified three-section coupler ($N = 3$) using meandering lines, open-stub, and diamond stub slow-wave structures (Al-Areqi et al., 2018). The dimensions of both the couplers are given in millimeter unit and tabulated in Table 14 and Table 15. The dimensions of both three-section branch-line couplers are printed on RT/Duroid 5880 substrate with thickness $h = 0.38$ mm. The proposed coupler has fractional bandwidth of 83.33% from 1.5 GHz to 3.1 GHz and achieves up to 83.3% size reduction.

APPLICATIONS

As mentioned in Section 1, hybrid branch-line circuit has played an important role in various designs of RF and microwave devices. Let's use an amplifier driver as an example, when a microwave signal source connected to port-1 and both outputs (port-2 and port-3) are respectively terminated with mismatch loads. The signal power from port-1 is transmitted and divided between both outputs port-2 and port-3;

Table 12. Dimensions of the modified branch-line coupler in Figure 34.

Dimensions			
Symbols	**Values (mm)**	**Symbols**	**Values (mm)**
a	1.9	i	5.2
b	8.2	j	1.5
c	3.5	k	4.4
d	5.2	l	3.5
e	7.0	m	1.2
f	14.1	n	1.8
g	6.3	o	1.2
h	4.4	p	0.5

Figure 35. Measurements results of conventional and miniaturized couplers: (a) return loss, $|S_{11}|$, (b) insertion loss, $|S_{21}|$, (c) coupling $|S_{31}|$, and (d) isolation, $|S_{41}|$ (You et al., 2019).

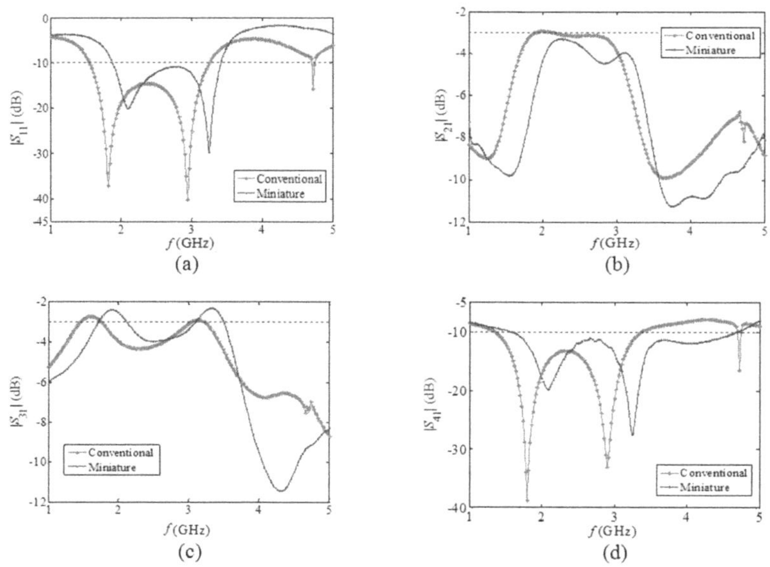

part of the incident signal power will be reflected back from both ports due to mismatch impedance. In this circumstance, both the reflected signals are combined and received at the isolated port-4, while both the reflected signals are eliminated at input port-1 due to phase shift cancellation (180° phase dif-

Figure 36. (a) Schematic and dimensions (in unit mm) of the modified three-section branch-line coupler. (b) Actual coupler circuit (You et al., 2018).

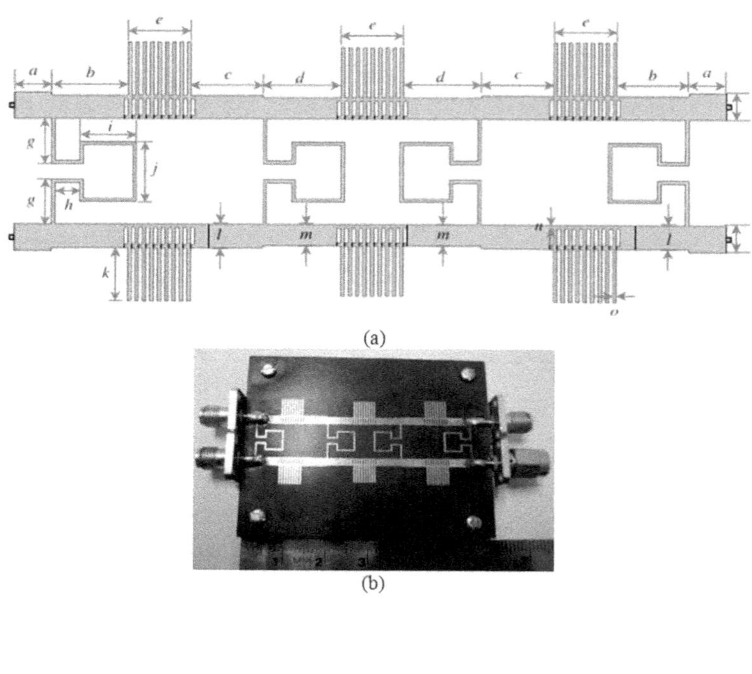

Table 13. Dimensions of the modified branch-line coupler in Figure 36 (You et al., 2018).

Dimensions			
Symbols	**Values (mm)**	**Symbols**	**Values (mm)**
a	3.0	*i*	4.6
b	5.9	*j*	5.4
c	5.6	*k*	4.8
d	6.0	*l*	2.1
e	5.1	*m*	1.9
f	2.4	*n*	0.3
g	4.0	*o*	0.3
h	2.0		

ference). Hence, the match to transmit the signal in amplifier is unaffected although both outputs have been connected to the arbitrary mismatch load.

Figure 37 Simulation and measurement S-parameters results of miniaturized couplers (You et al., 2018).

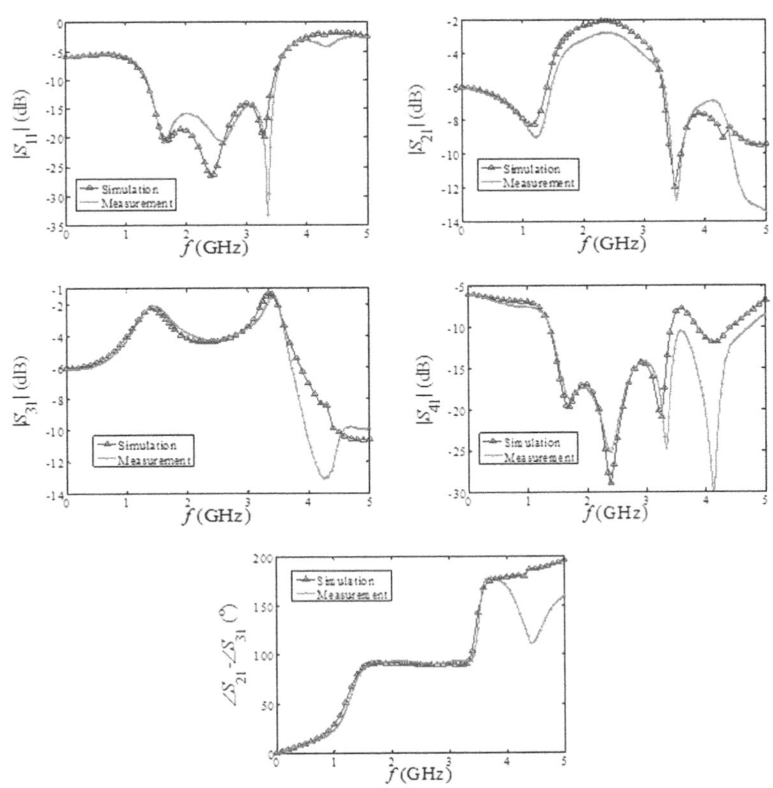

Figure 38. (a) The three-section coupler (N =3) with open-stub slow-wave structure. (b) Quarter part of the modified three-section coupler circuit. (c) Actual coupler circuit(Al-Areqi et al., 2018).

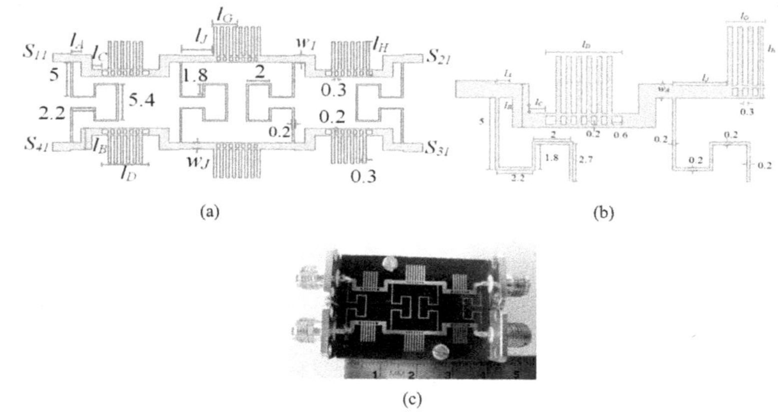

Figure 39. (a) The three-section coupler (N =3) with diamond-stub slow-wave structure. (b) Single diamond stub (Zhou et al., 2010). (c) Seven series diamond stubs part in the modified three-section coupler circuit. (d) Actual coupler circuit (Al-Areqi et al., 2018).

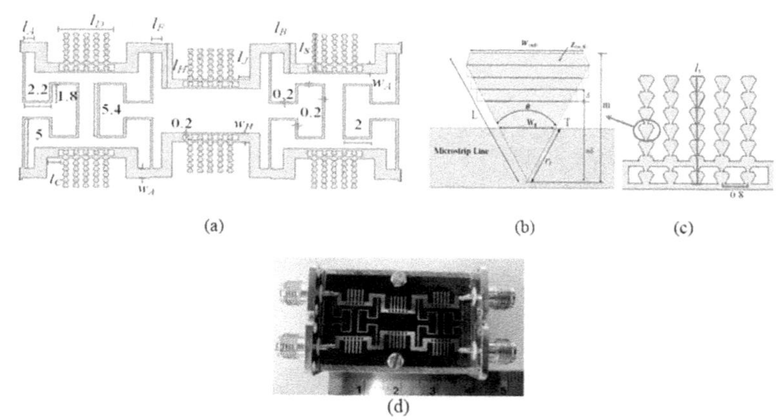

Table 14. Dimensions of the modified branch-line coupler in Figure 38 (Al-Areqi et al., 2018).

Dimensions			
Symbols	**Values (mm)**	**Symbols**	**Values (mm)**
l_A	1.0	W_A	1.0
l_B	3.0	W_J	0.9
l_C	1.0	l_G	2.25
l_D	4.5	l_H	5.0
l_J	3.25		

Table 15. Dimensions of the modified branch-line coupler in Figure 39 (Al-Areqi et al., 2018).

Dimensions			
Symbols	**Values (mm)**	**Symbols**	**Values (mm)**
l_A	1.0	l_G	2.25
l_B	3.0	l_H	4.5
l_C	1.0	l_J	0.9
l_D	4.5	w_A	1.0
L_F	0.9	w_h	0.2

Figure 40. Simulations and measurements results of modified couplers using open stubs (a) insertion loss, $|S_{21}|$& coupling $|S_{31}|$, (b) return loss, $|S_{11}|$, (c) isolation (Al-Areqi et al., 2018).

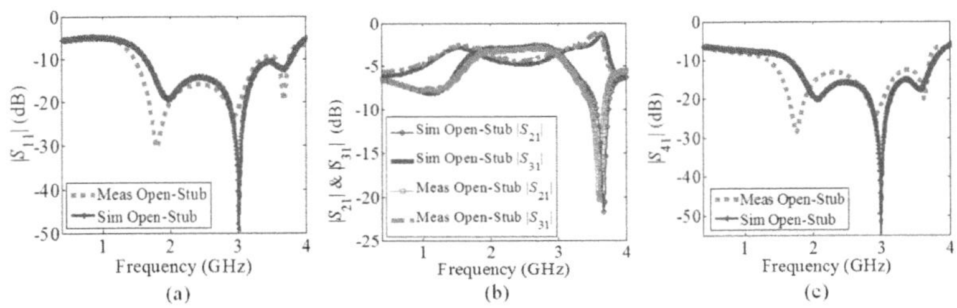

Duplexer

Normally, radar and RF communications systems use a common single antenna for both transmit and receive, in order to reduce size and complexity. Hence, the duplexer is needed in the systems to allow bi-directional (transmit and receive signal) communication over a single path. The duplexer isolates the receiver from the transmitter while permitting them to share a common antenna as shown in Figure 42.

At transmitter port, T_x, the non-reciprocal power amplifier (PA) modules deter the appearance of received power from antenna to the transmitter, T_x. Thus, the T_x is isolated during the receive function. At the receiver, R_x port, small amount of received power from antenna is amplified to the desired levels using a low-noise amplifier (LNA), in order to retain the desired signal-to-noise ratio of the received signal.

Figure 41. Simulations and measurements results of modified couplers using series diamond-stubs (a) insertion loss, $|S_{21}|$& coupling $|S_{31}|$, (b) return loss, $|S_{11}|$, (c) isolation (Al-Areqi et al., 2018).

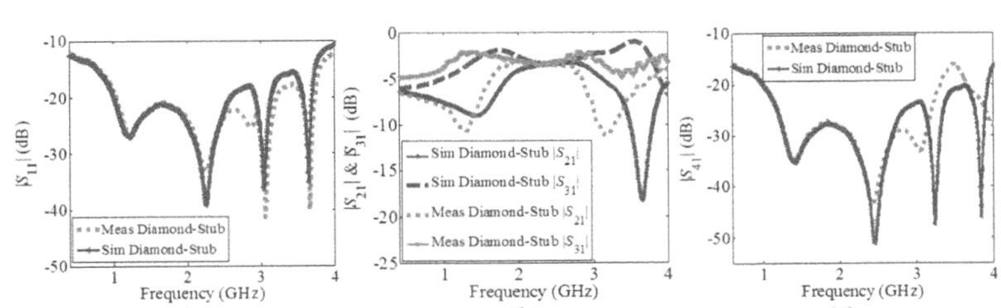

Figure 42. Simple duplexer

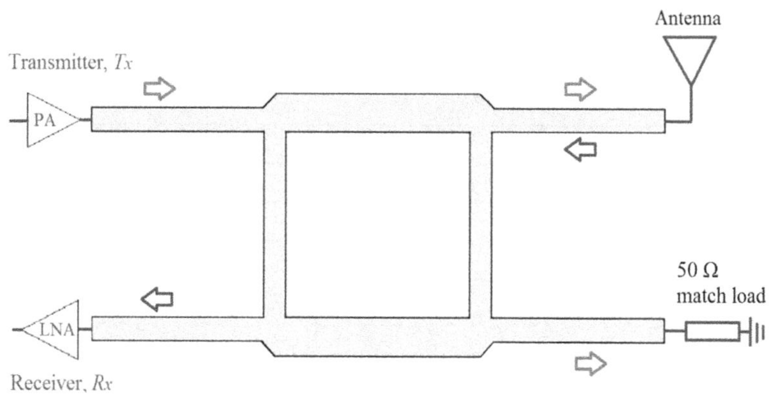

The branch-line circuit can also be used as a crossover network (as discussed in Section 1.3) as shown in Figure 43. The signal is excited to the antenna. Any leakage signal from the generator will cross the network to the 50 Ω match load port. On the other hand, the reflected signal from the antenna is diagonally across the network circuit to the receiver, thus the incident and reflected signals are isolated.

Dual-Fed Circular Polarization Circuit

Dual-fed circular polarization circuit is regularly applied on antenna design in which it has the ability to reduce polarization losses of the antennas, due to linearly polarized antennas that need strict alignment between transmitter and receiver. On the other hand, the circularly polarized antennas can communicate for random orientations between transmitter and receiver. In addition, this type of antenna will reduce any radio interference due to a multipath signal process. Hence, the antenna with changeable circularly polarization mode is requested. For simple solution, the antenna is excited by orthogonal dual-feed branch-line circuit with equal amplitude, but relative phase shift of $\pm 90^{\circ}$ at center frequency to produce circular polarization as shown in Figure 44.

The hybrid branch-line inputs will produce left hand circular polarization (LHCP) and right hand circular polarization (RHCP). If the antenna is fed at RHCP port, the LHCP port will seem like terminated with a matched load and vice versa (Bancroft, 2006). Thus, the circularly polarized antenna allows more radio channels in the same frequency without interference in which the polarization of the antennas is interleaved on channels that are next to each other.

Double-Balanced Mixer

Mixers are used to convert a signal from one frequency to another by combining the original radio signal (RF) with a local oscillator (LO) signal in a Schottky-barrier diode. The desired output frequency is called the intermediate frequency (IF). In a communication receiver application, the lower value of the output frequency (IF) is easier to amplify and process than original input RF frequency. There are two types of RF mixer that are balanced, namely single balanced mixer and double balanced mixer. The double balanced mixer is capable of providing good isolation between both RF and LO with output IF

Figure 43. Duplexer for antenna measurement application.

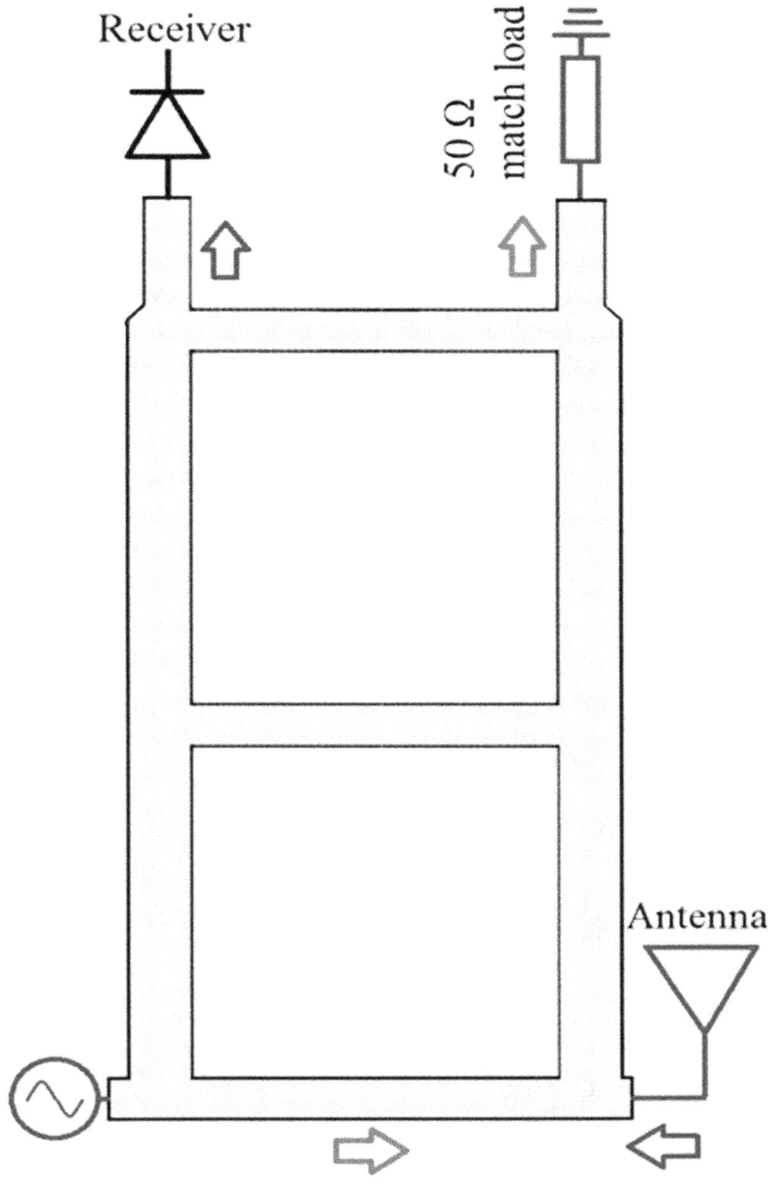

Figure 44. Circularly polarized (a) circular and (b) rectangular patch antennas.

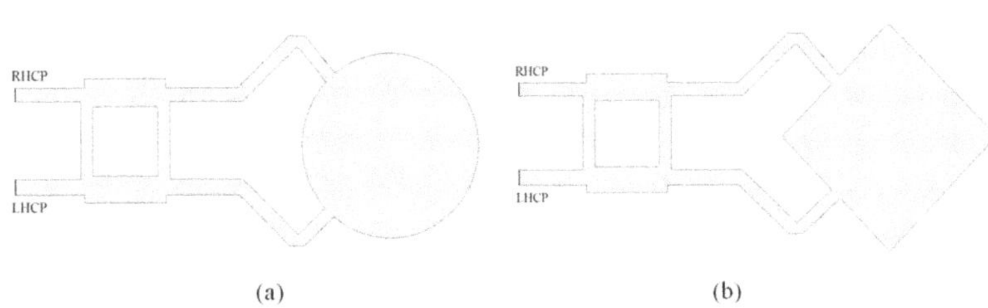

compared to single balanced mixer. The microwave double-balanced mixer can be created by hybrid branch-line circuit as shown in Figure 45. The mixer recombines the RF and LO inputs to two mixing diodes in order to create the multiplication of the two input frequencies and cancel many of the unwanted harmonics signals. In fact, mixers can act as modulators, phase detectors, and frequency discriminators.

Six-port Complex Ratio Measuring Unit (CRMU)

Microwave complex ratio measurement (MCRM) system is one of the important tools used in the microwave component design. In fact, MCRM system is a vector instrument that can be used to measure the complex reflection coefficient, Γ of the device under test (DUT) over a relatively broad frequency range. In general, the magnitude, |Γ| and phase shift, θ of reflection coefficient for DUT is determined based on the number of n power ratios between the measured output power amplitudes, P_n ($n = 1, 2, 3$, and 4) and input reference power amplitudes, P_{ref}. One input port of MCRM circuit is applied by power source and another port is terminated with DUT. The remaining output ports (normally three ports or four ports) will be connected to power detectors in which the power ratios of the DUT are measured. For instance, a six-port complex ratio measuring system consists of a dual directional coupler, three 3-dB 90° hybrid coupler, and two power dividers are shown in Figure 46. Hence, this kind of circuit is so-called five-port or six-port network circuits. The measured power, P_n at each port is the result of interference between the input and output amplitude signals. Finally, the measured power ratios, P_n/P_{ref} are used to predict the complex reflection coefficient, Γ of the DUT using some mathematical operations and circuit calibration process.

Antenna Beam Forming Network Circuit

Beam forming network is used to feed a phased array of antenna elements in order to control the direction of the antennas radiation beams. The antenna array is capable of producing multiple directional beams simultaneously with orthogonal phase-modes. Hence, multiple mobile stations using same operating frequency can be achieved when fed through beam forming network circuit. Generally, the beam forming network is also called Butler matrix ($n \times n$ matrix). For instance, 2 × 2 Butler matrix network as shown in Figure 47, only uses common branch-line coupler. The beam forming network circuit can be economically created using branch-line crossover coupler and phase shifter (45° delay microstrip line) as shown in Figure 48 (Innok *et al.*, 2012).

Figure 45. Microwave double-balanced mixer

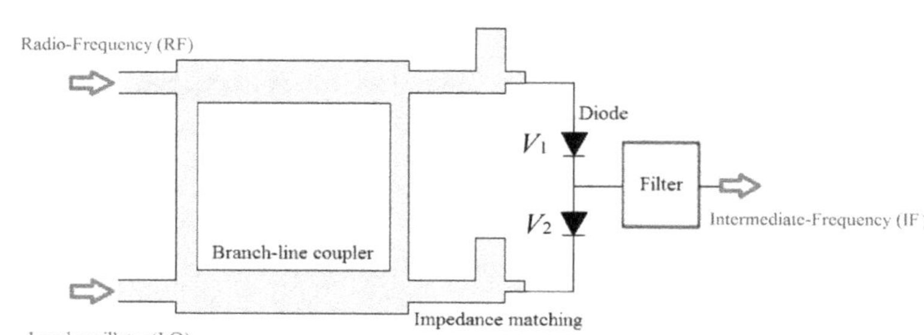

Figure 46. Six-port reflectometer

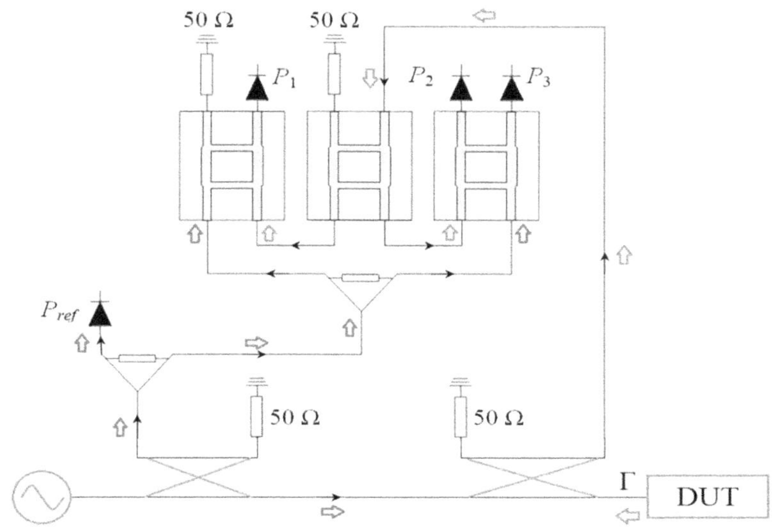

CONCLUSION

In this chapter, many analytical formulas for different types of branch-line couplers have been reviewed. Indeed, there is still much information related to coupler study that is not mentioned in this chapter. In fact, this chapter aims to facilitate RF designers in the branch-line coupler design, and assist the fresh designers to understand the current coupler development and its applications.

REFERENCE

Al-Areqi, N. N., You, K. Y., Dimon, M. N., Khamis, N. H., & Lee, C. Y. (2018). *Progress in Electromagnetics Research, C*, 82, 199–207. Miniaturization of three-section branch-line coupler using diamond-series stubs microstrip line.

Figure 47. (a) 2 × 2 Butler matrix using branch-line coupler and (b) produced beam pattern.

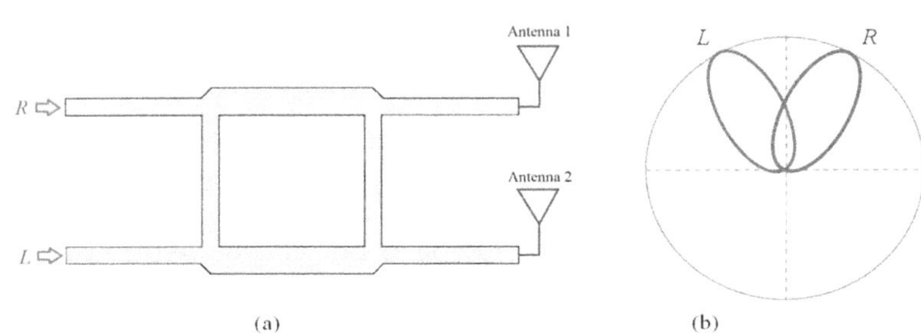

Figure 48. (a) 4 × 4 Butler matrix in microstrip circuit and (b) produced beam pattern.

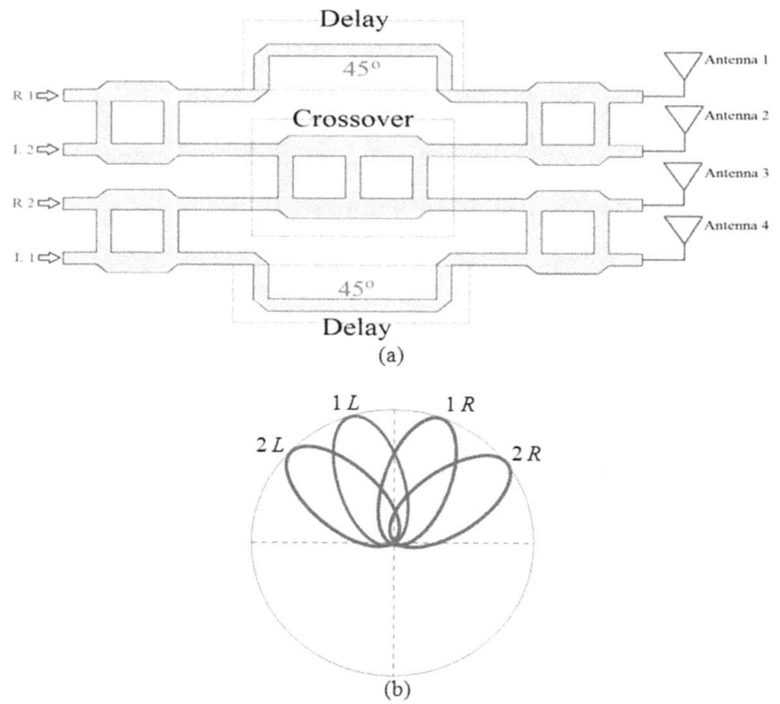

Arriola, W. A., Lee, J. Y., & Kim, I. S. (2011). Wideband 3 dB branch line coupler based on λ/4 open circuited coupled lines. *IEEE Microwave and Wireless Components Letters*, *21*(9), 486–488. doi:10.1109/LMWC.2011.2138687

Bancroft, R. (2006). *Microstrip and Printed Antenna Design*. New Delhi, India: Prentice-Hall of India.

Bekasiewicz, A., & Koziel, S. (2015). Miniaturised dual-band branch-line coupler. *Electronics Letters*, *51*(10), 769–771. doi:10.1049/el.2015.0751

Bhat, B., & Koul, S. K. (1989). *Stripline-Like Transmission Lines for Microwave Integrated Circuits*. New York, NY: Wiley.

Bree, G. (2009). Transmission line and lumped element quadrature couplers. *High Frequency Electronics*. 44-48.

Chen, W. L., Wang, G. M., & Zhang, C. X. (2009). Miniaturization of wideband branch-line couplers using fractal-shaped geometry. *Microwave and Optical Technology Letters*, *51*(1), 26–29. doi:10.1002/mop.24002

Chiu, L., & Xue, Q. (2010). Investigation of a wideband 90° hybrid coupler with an arbitrary coupling level. *IEEE Transactions on Microwave Theory and Techniques*. *58*(4), 1022-1029.

Chun, Y. H., & Hong, J. S. (2006). Compact wide-band branch-line hybrids. *IEEE Transactions on Microwave Theory and Techniques*. *54*(2), 704–709.

Collin, R. E. (1992). *Foundations for Microwave Engineering* (2nd ed.). New York, NY: IEEE Press.

Crane Aerospace & Electronics Microwave Solutions. (2012). *Quadrature hybrids*. West Caldwell, NJ: Merrimac Industries, 1–6.

Eccleston, K. W., & Ong, S. H. M. (2003). Compact planar microstripline branch-line and rat-race couplers. *IEEE Transactions on Microwave Theory and Techniques*, *51*(10), 2119–2125. doi:10.1109/TMTT.2003.817442

Getsinger, W. J. (1993). End-effects in quasi-TEM transmission lines. *IEEE Transactions on Microwave Theory and Techniques, 41*(4), 666–672.

Ghosh, D., & Kumar, G. (2015). A four branch microstrip coupler with improved bandwidth and isolation. In *Twenty First National Conference on Communications*, Mumbai, India.

Innok, A., Uthansakul, P., & Uthansakul, M. (2012). Angular beamforming technique for MIMO beamforming system. *International Journal of Antennas and Propagation*, *2012*, 1–10. doi:10.1155/2012/638150

Iran-Nejad, V., Lotfi-Neyestanak, A. A., & Shahzadi, A. (2012). Compact broadband quadrature hybrid coupler using planar artificial transmission line. *Electronics Letters*, *48*(25), 1602–1603. doi:10.1049/el.2012.3252

Kim, H. C., Lee, B. J., & Park, M. J. (2010). Dual-band branch-line coupler with port extensions. *IEEE Transactions on Microwave Theory and Techniques, 58*(3), 651–655.

Kirschning, M., Jansen, R. H., & Koster, N. H. L. (1981). Accurate model for open end effect of microstrip lines. *Electron*ics Letters, *17*(3), 123–125.

Knöchel, R. (1999). Broadband flat coupling two-branch and multibranch directional couplers. *IEEE MTT-S International Microwave Symposium Digest*, *3*, 1327–1330.

Krishna, V. V., Patel, B., & Sanyal, S. (2010). Harmonic suppressed compact wideband branch-line coupler using unequal length open-stub units. *International Journal of RF and Microwave Computer-Aided Engineering*, *21*(1), 115–119.

Kurgan, P., & Kitliński, M. (2009). Novel doubly perforated broadband microstrip branch-line couplers. *Microwave and Optical Technology Letters*, *51*(9), 2149–2152. doi:10.1002/mop.24566

Kurgan, P., Filipcewicz, J., & Kitlinski, M. (2012). Development of a compact microstrip resonant cell aimed at efficient microwave component size reduction. *IET Microwaves, Antennas & Propagation, 6*(12), 1291-1298.

Lee, S. K., & Lee, Y. S. (2012). Wideband branch-line couplers with single-section quarter-wave transformers for arbitrary coupling levels. *IEEE Microwave and Wireless Components Letters, 22*(1), 19-21.

Levy, R., & Lind, L. F. (1968). Synthesis of symmetrical branch-guide directional couplers. *IEEE Transactions on Microwave Theory and Techniques*, *16*(2), 80–89. doi:10.1109/TMTT.1968.1126612

Liao, S. S., Sun, P. T., Chin, N. C., & Peng, J. T. (2005). A novel compact-size branch-line coupler. *IEEE Transactions on Microwave and Wireless Components Letters*, *15*(9), 588–590. doi:10.1109/LMWC.2005.855378

Lim, J. S., Kim, C. S., Park, J. S., Ahn, D., & Nam, S. W. (2000). Design of 10 dB 90° branch line coupler using microstrip line with defected ground structure. *Electronics Letters*, *36*(21), 1784–1785. doi:10.1049/el:20001238

Lin, F., Chu, Q. X., & Wong, S. W. (2013). Dual-band planar crossover with two-section branch-line structure. *IEEE Transactions on Microwave Theory and Techniques*, *61*(6), 2309–2316. doi:10.1109/TMTT.2013.2261084

Mayer, B. (1990). New broadband branch-line coupler. *Electronics Letters*, *26*(18), 1477–1478. doi:10.1049/el:19900948

Mayer, B. (1992). *Wide-band branch line coupler*. USA: U.S. Patent, Number: 5,132,645.

Mayer, B., & Knöchel, R. (1990). Branchline-couplers with improved design flexibility and broad bandwidth. *IEEE In*ternational Digest on Microwave Symposium. 391-394. IEEE.

Muraguchi, M., Yukitake, T., & Naito, Y. (1983). Optimum design of 3-dB branch-line couplers using microstrip lines. *IEEE Transactions on Microwave Theory and Techniques*, *31*(8), 674–678. doi:10.1109/TMTT.1983.1131568

Park, M. J. (2013). Comments on 'quasi-arbitrary phase-difference hybrid coupler'. *IEEE Transactions on Microwave Theory and Techniques*, *61*(3), 1397–1398. doi:10.1109/TMTT.2013.2241779

Paul, D. K., Gardner, P., & Prasetyo, B. Y. (1991). Broadband branchline coupler for S band. *Electronics Letters*, *27*(15), 1318–1319. doi:10.1049/el:19910828

Pozar, D. M. (2012). *Microwave Engineering* (4th ed., pp. 147–149). USA: Wiley.

Reed, J. (1958). The multiple branch waveguide coupler. *I.R.E. Transactions on Microwave Theory and Techniques*, *6*(4), 398–403. doi:10.1109/TMTT.1958.1125213

Reed, J., & Wheeler, G. J. (1956). A method of analysis of symmetrical four-port networks. *I.R.E. Transactions on Microwave Theory and Techniques*, *4*(4), 246–252. doi:10.1109/TMTT.1956.1125071

Riblet, G. P. (1978). A directional coupler with very flat coupling. *IEEE Transactions on Microwave Theory and Techniques*, *26*(2), 70–74. doi:10.1109/TMTT.1978.1129315

Sun, K. O., Ho, S. J., Yen, C. C., & van der Weide, D. (2005). A compact branch-line coupler using discontinuous microstrip lines. *IEEE Microwave and Wireless Components Letters, 15*(8), 519-520.

Tang, C. W., Chen, M. G., Lin, Y. S., & Wu, J. W. (2006). Broadband microstrip branch-line coupler with defected ground structure. *Electronics Letters*, *42*(25), 1458–1460. doi:10.1049/el:20063025

Tang, C. W., Chen, M. G., & Tsai, C. H. (2008). Miniaturization of microstrip branch-line coupler with dual transmission lines. *IEEE Microwave and Wireless Components Letters*, *18*(3), 185–187. doi:10.1109/LMWC.2008.916798

Tang, C. W., Tseng, C. T., & Hsu, K. C. (2014). Design of wide passband microstrip branch-line couplers with multiple sections. *IEEE Transactions on Components, Packaging, and Manufacturing Technology*, *4*(7), 1222–1227. doi:10.1109/TCPMT.2014.2320499

Then, Y. L., You, K. Y., Dimon, M. N., Chong, J. C., & Tan, T. S. (2013). Compact microstrip S-band 90° hybrid coupler. *IEEE Symposium on Wireless Technology & Applications*, 202-206.

Toker, C., Saglam, M., Ozme, M., & Gunalp, N. (2001). Branch-line couplers using unequal line lengths. *IEEE Transactions on Microwave Theory and Techniques*, *49*(4), 718–721. doi:10.1109/22.915448

Wight, J. S., Chudobiak, W. J., & Makios, V. (1976). A microstrip and stripline crossover structure. *IEEE Transactions on Microwave Theory and Techniques*, *24*(5), 270. doi:10.1109/TMTT.1976.1128838

Wong, Y. S., Zheng, S. Y., & Chan, W. S. (2012). Quasi-arbitrary phase-difference hybrid coupler. *IEEE Transactions on Microwave Theory and Techniques*, *60*(6), 1530–1539. doi:10.1109/TMTT.2012.2187918

Wu, Y. L., Shen, J. Y., & Liu, Y. N. (2013). Comments on 'quasi-arbitrary phase-difference hybrid coupler. *IEEE Transactions on Microwave Theory and Techniques*. *61*(4), 1725–1727.

Yao, J. J. (2010). Nonstandard hybrid and crossover design with branch-line structures. *IEEE Transactions on Microwave Theory and Techniques*, *58*(12), 3801–3808.

Yao, J. J., Lee, C., & Yeo, S. P. (2011). Microstrip branch-line couplers for crossover application. *IEEE Transactions on Microwave Theory and Techniques*, *59*(1), 87–92. doi:10.1109/TMTT.2010.2090695

Yoon, H. J., & Min, B. W. (2017). Two section wideband 90° hybrid coupler using parallel-coupled three-line. *IEEE Microwave and Wireless Components Letters*, *27*(6), 548–550. doi:10.1109/LMWC.2017.2701304

You, K. Y., Chung, J. C., Malek, M. F. A., Lee, Y. S., & Mirza, S. (2019). Miniaturized two-section branch-line coupler using open-stub slow-wave structure. *8th International Conference on Innovation in Electronics and Communication Engineering* (Accepted).

You, K. Y., Lee, C. W., & Lee, C. Y. (2018). Commercial and open source electromagnetic simulators for education, research and industrial design. *International Journal of Advances in Microwave Technology*, *2*(3), 117–121.

You, K. Y., Nadera, A. A., Chong, J. C., Lee, K. Y., Cheng, E. M., & Lee, Y. S. (2018). Analytical modeling of conventional and miniaturization three-section branch-line couplers. *Journal of Electrical Engineering & Technology*, *13*(2), 858–867.

Young, L. (1962). Synchronous branch guide directional couplers for low and high power applications. *I.R.E. Transactions on Microwave Theory and Techniques*, *10*(6), 459–475. doi:10.1109/TMTT.1962.1125554

Zhou, B., Wang, H., & Sheng, W. X. (2010). A modified UWB wilkinson power divider using delta stub. *Progress In Electromagnetics Research Letters*, *19*, 49–55. doi:10.2528/PIERL10101805

Zong, B. F., Wang, G. M., Zhang, C. X., & Wang, Y. W. (2014). Miniaturised branch-line coupler with ultra-wide high suppression stopband. *Electronics Letters*, *50*(19), 1365–1367. doi:10.1049/el.2014.1150

Chapter 2
Electromagnetic Metamaterials in Microwave Regime

Man Seng Sim
https://orcid.org/0000-0001-7776-2239
Universiti Teknologi Malaysia, Malaysia

Kok Yeow You
https://orcid.org/0000-0001-5214-7571
Universiti Teknologi Malaysia, Malaysia

Fahmiruddin Esa
Universiti Tun Hussein Onn Malaysia, Malaysia

ABSTRACT

Metamaterials are artificially-engineered materials which possess unique properties not found in natural materials. The properties are derived from the structural designs of metamaterials and they allow the structure to manipulate electromagnetic waves and achieve desired responses in a certain frequency range. This chapter reviews past achievements, recent developments, and future trends on electromagnetic metamaterials in microwave regime. The chapter first briefly introduces electromagnetic metamaterials from a general prospect including the definition, historical overview, and classification of metamaterials. Furthermore, three selected applications of metamaterials which are microwave absorbers, sensors, and energy harvesters are discussed based on their operation principles, designs, and characteristics.

INTRODUCTION TO ELECTROMAGNETIC METAMATERIALS

Electromagnetic metamaterials are artificial materials engineered to have a geometrical structure which possesses unique electromagnetic properties at a certain range of frequency. The novel macroscopic properties are originated from both the properties of the constituent materials and their designed geometry. The effective properties of metamaterials do depend on its properties of their constituents. The

DOI: 10.4018/978-1-7998-0117-7.ch002

structures are usually in periodic arrangement and the unit cell size is small compared to the wavelength. Resonances with the incident electromagnetic waves are normally occurred at the operating frequencies.

The history of electromagnetic metamaterials begins with the development of artificial materials with desired electromagnetic responses in the 19th century. For example, artificial dielectrics were designed using metallic arrays of small spheres, disc or round wires (Kock, 1948). Artificial magnetics were introduced by using an electrically conducting loops loaded with capacitor (split ring resonator) (Schelkunoff & Friis, 1952). Artificial chiral composites were used as microwave absorber (Varadan, Varadan, & Lakhtakia, 1987).

In 2000, metamaterials with negative refractive index became the significant breakthrough of the research on artificial electromagnetic materials (Smith & Kroll, 2000). Negative refractive index was found to be obtained by combining two structures which individually exhibit negative permittivity and negative permeability. This phenomenon is a resonant effect and it causes the media to be dispersive and dissipative (Ramakrishna, 2005). Different basic elements of metamaterials (electric dipoles, magnetic dipoles or chiral particles) can be combined to realize desired response.

The current flourishing state of metamaterials studies have proven the applicability of metamaterials in various applications including absorbers, sensors, energy harvesters, antennas, lenses and filters. Besides microwave frequency range, research of metamaterials in different frequency ranges of the electromagnetic spectra have been carried out. Based on the properties of metamaterials, there are several types of metamaterials such as double negative (DNG) materials, single negative (SNG) materials, artificial magnetic conductors (AMC), and frequency selective surfaces (FSS). Their characteristics and applications are tabulated in Table 1.

Applications of Electromagnetic Metamaterials

Three selected applications of electromagnetic metamaterials—absorbers, sensors, and energy harvesters—are reviewed in this section.

Table 1. Classification of metamaterials

Metamaterials	Characteristics	Applications
Double negative materials (DNG) (also known as left-handed metamaterials (LHM), negative-index materials (NIM) or backward-wave media)	• Negative value of permittivity and permeability simultaneously • Negative refractive index • Reverse propagation	Absorbers, Antennas, Waveguides
Single negative (SNG) materials (or more specifically known as ε-negative (ENG) materials or μ-negative materials (MNG))	• Negative value of permittivity or permeability	Antennas
Artificial magnetic conductors (AMC) (also known as high impedance surface (HIS))	• Behave as perfect magnetic conductor although composed by non-magnetic materials • Provide zero-degree reflection phases at resonant frequency	Antennas, Waveguides
Frequency selective surfaces (FSS)	Tailor frequency selectiveness	Filters, Antennas, Microwave ovens

Metamaterial Absorbers

Introduction of Metamaterial Absorbers

Microwave absorbers effectively absorb incident microwave energy of certain frequency range with negligible reflection and transmission. There are several types of commercial microwave absorbers such as particle-filled composite absorbers, Salisbury screens and Jaumann absorbers. Particle-filled composite absorbers are commonly used and they are made by introducing particles such as carbon and ferrites into a dielectric polymeric matrix (Deng & Han, 2007). The absorption mechanisms of composite absorbers are dielectric and magnetic loss of the constituent materials and the scattering effect originated from their designed shapes such as pyramid, wedge and convoluted. Salisbury screens and Jaumann absorbers use the concept of destructive interference for wave cancellation (Knott & Lunden, 1995). These conventional absorbers usually require a minimum thickness of quarter wavelength to operate (Venneri, Costanzo, & Massa, 2018).

Metamaterial microwave absorbers (MMAs) usually consist of array of metallic resonators which exhibits unique properties which allow the structure to manipulate the incident electromagnetic waves (Koschny, Markos, Smith, & Soukoulis, 2003). The basic and common design of MMAs is a three-layered metal-dielectric-metal structure which can be fabricated by printed circuit board process. The designed metallic patterns are etched on both side of the dielectric substrate. The thickness of the MMA structure is very thin compared with the operating wavelength, λ. Compared with conventional microwave absorbers, MMAs are thinner, lighter in weight, robust and more flexible in design. Metamaterial absorbers can be achieved in Terahertz (Ju et al., 2018), mid-infrared (Zhang et al., 2013) and visible (Hedayati et al., 2011) regions of electromagnetic wave by scaling their geometrical sizes. There are various type of MMAs based on their absorption characteristics, such as single-band (J. Zhong et al., 2012), dual-band (Tak, Jin, & Choi, 2016), multi-band (Sim, You, Esa, Dimon, & Khamis, 2018), tunable (Zhao et al., 2015) and broadband MMA (Soheilifar & Sadeghzadeh, 2014). Criteria of a good MMA are high absorption, wide angle of incident, polarization insensitive, thin and small unit cell size (Agarwal, Behera, & Meshram, 2016).

Applications of MMAs include standard absorption rate (SAR) reduction, electromagnetic interference (EMI) shielding, radar cross section (RCS) reduction in stealth technology, sensing, energy harvesting, anechoic chamber for wireless communication testing and measurements. A triangular MMA is used to reduce the SAR value which is the amount of energy received by human tissue when exposed to a phone device (Faruque, Islam, & Misran, 2012). A cross-shaped MMA, which operates at ~5.75 GHz, reduces the monostatic and bistatic RCS of a slot antenna (Liu, Cao, Gao, & Zheng, 2013). Square and ring-based MMA is used as a pressure, temperature, density and humidity sensors (Bakır, Karaaslan, Unal, Akgol, & Sabah, 2017). An MMA can be applied as an energy harvester which exhibits high absorption level to the radiating electromagnetic energy in our surrounding environment (Karaaslan, Bağmancı, Ünal, Akgol, & Sabah, 2017).

Operation Principles of Metamaterial Absorbers

There are three possible interactions (which are reflection, transmission and absorption) between an incident electromagnetic wave and a material. Other phenomena such as diffraction and scattering are insignificant as the MMAs are subwavelength-scaled (Ghosh, Nguyen, & Lim, 2019). The interactions can be described using equation (1) as

$$\alpha + R + T = 1 \tag{1}$$

where α is absorptivity, R is reflectivity and T is transmission. Therefore, absorptivity, α can be expressed using equation (2) as

$$\alpha = 1 - \left|S_{11}\right|^2 - \left|S_{21}\right|^2 \tag{2}$$

where S_{11} and S_{21} represent the reflection coefficient and transmission coefficient, respectively. It can be noted that perfect absorption can be achieved when the reflection and transmission are both negligible.

Most of the recent published works propose MMA structure with a metal back plate. The transmission, T becomes zero and absorptivity, α depends only on the reflectivity, R. To minimize R, impedance matching condition between the metamaterial structure and the free space have to be fulfilled. The reflection coefficient, S_{11} can be expressed as

$$S_{11} = \frac{Z_{in} - Z_o}{Z_{in} + Z_o} \tag{3}$$

where Z_{in} is input impedance of the MMA and Z_o is free space impedance. S_{11} becomes zero when the input impedance is perfectly matched to the free space impedance ($Z_{in} = Z_o$). In the case of MMA structure without metal back plate, dielectric loss of the substrate is important for the decay of electromagnetic wave (Li et al., 2015).

Figure 1 (a-c) shows a split ring and multi-width line MMA structure (Sim et al., 2018). Based on Figure 1(d), it can be seen that both S_{11} and S_{21} have negligible magnitude at 5.8 GHz. In other words, both the reflectivity, R and transmission, T have near zero values at 5.8 GHz. The absorptivity, α which are determined using equation (2) achieves near unity value as shown in Figure 1(e). 5.8 GHz is one of the industrial, scientific and medical (ISM) radio bands for applications such as power transfer in microwave transmitting devices (Mohammed, Ramasamy, & Shanmuganantham, 2010), rectennas in solar power satellites (Matsumoto, 2002), object tracking in radiolocation devices (Strobel, Eickhoff, Ziroff, & Ellinger, 2010). Therefore, this MMA structure has possible applications in microwave heating, energy harvesting and tracking system.

Electric field distributions on the front layer of the MMA as shown in Figure 1(g) shows that the field concentrates on top and bottom parts of the inner ring. For the multi-width line structure at the back layer, the electric field mainly forms on the slits structure. Based on the surface current distribution plot in Figure 1(h), the surface current concentrates on the left and right side of the ring. The surface charges are induced by the strong electric fields around the ring resonators. The current flows in opposite direction for the inner and outer ring near the gap structure between both rings. The same phenomenon for the anti-parallel current flow is observed at the slit structure at the back layer.

Figure 1. Metamaterial microwave absorber's: (a) perspective view, (b) front view, (c) back view, (d) side view, (e) reflection coefficient, S_{11} and transmission coefficient, S_{21} as a function of frequency, (f) absorptivity, α, reflectivity, R and transmission, T as a function of frequency (g) electric field distribution (h) surface current distribution. (Dimensions in mm: a = 2.4, b = 0.2, d = 0.3, g = 0.8, l = 10.8, p = 16.8, r = 8.2, t = 1.6, w = 2.2)

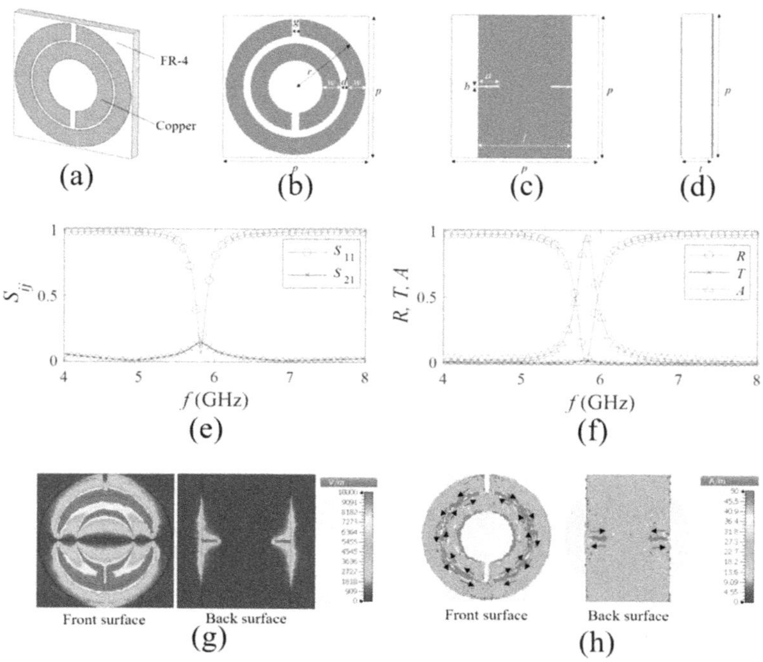

Recent Development of Metamaterial Absorbers

There are several focus on the recent research of MMAs, such as multi-band, tunable and bandwidth-enhanced MMAs. To achieve multi-band absorption characteristics, the first approach is placing different resonators together on a same plane to create a hybrid unit cell (Chaurasiya, Ghost, Bhattacharyya, & Srivastava, 2015). For example, a MMA structure combines two different closed-loop resonators (CLRs) in a unit cell and achieves triple band absorption at 3.1, 4.6 and 9.5 GHz (Sharma, Ghosh, & Srivastava, 2016). The absorption peak at 3.1 and 9.5 GHz are both contributed by the outer CLR structure (Minkowski fractal-based square loop) whereas the peak at 4.6 GHz is originated from the inner CLR structure (a structure combining a closed square loop and a square loop with splits).

The second approach to get multi-band characteristics is scaling the resonator to different sizes which contribute to different absorption frequencies (Karaaslan et al., 2018). For example, a fractal-based Jerusalem cross structures are scaled to four different sizes and four absorption peaks at 8.3, 9.8, 11.5 and 13.2 GHz are achieved (G. D. Wang et al., 2014). The third approach is by placing the resonators in rotational arrangement in which each position contributes to its respective absorption frequency (Yu, Liu, Dong, Zhou, & Zhou, 2017). Four identical circular-shaped spiral structures which are placed in different rotational angles possess absorption at 9.8, 12.2, 15.3 GHz (Huang et al., 2013). The final approach involves 3-dimensional placement of the resonators on different planes. For example, a three-layered structure, each etched with split square rings exhibits absorption at 2.8, 5.4 and 10.5 GHz (Karaaslan et al., 2017). The advantages and disadvantages of the four approaches for multi-band MMAs are summarized in Table 2.

Recent research has also been done on offering tunability or flexibility in the control of the resonant frequencies of a MMA. Three approaches including electrical, thermal and mechanical approaches are used to tune or adjust the resonant frequencies. For the electrical technique, electric lumped elements are incorporated to the MMA structure. For example, varactor diodes are used to tune the absorption frequency of a microstrip line resonator-based MMA by changing the supplied bias voltage (Yuan et al., 2015). Another ELC resonator-based MMA can be tuned by applying forward or reverse bias to the diode (Zhu, Huang, Feng, Zhao, & Jiang, 2010). For the thermal approach, the absorption performance of a MMA can be tuned under different environment temperature because the dielectric substrate is made of water in which its permittivity is of temperature dependence (Pang et al., 2017). Mechanical approach makes use of a dielectric slab to change the effective permittivity of the structure. The absorption peak frequency can be shifted by placing a dielectric slab in parallel to the MMA's surface (Zhu, Huang, Rukhlenko, Wen, & Premaratne, 2012). The advantages and disadvantages of these approaches are tabulated in Table 3.

One of the limitations of MMAs is having a narrow absorption bandwidth because they are resonance based. There are many studies on enhancing the bandwidth of MMAs. Structures with adjacent absorption frequencies can be combined on a same plane to improve the absorption bandwidth. By combining four ring-dish structures having adjacent absorption peaks in Ku band, the MMA structure achieves a bandwidth of 3.7 GHz for absorption higher than 80% (Vu et al., 2018). By combining two slightly geometrical variants of a swastika-like structure, the MMA structure achieves a bandwidth of 0.3 GHz above 85% absorption (Ghosh, Bhattacharyya, & Srivastava, 2014). An MMA combining different orientation of the E-shaped resonators possesses absorption higher than 70% between 13 GHz and 14 GHz (Li et al., 2015).

Table 2. Multi-band metamaterial absorbers

Approaches	Advantages	Disadvantages
Resonators of different shapes are combined	• Compact in geometry	• Difficulty during the design process to avoid the coupling effects between the resonators • Reduced absorption if the arrangement is not proper
Resonators of same shape but different sizes are combined	• The resonators possess similar characteristics but at different frequencies	• Difficulty in the arrangement of the resonators on a same plane since different sizes of resonators perform best on different periodicity (distance between two adjacent resonators)
Resonators of same shape are arranged in rotational position	• Ease in arrangement of the resonators on a same plane since the periodicity can be fixed	• Complex structure which leads to longer drawing and simulation time
Resonators are arranged on different layers vertically or on different planes	• Some achieves broader bandwidth and wider incident angles	• Thick • Fabrication complexity

Table 3. Tunable metamaterial absorbers

Approaches	Operation Principle	Advantages	Disadvantages
Electrical	Change the capacitance and inductance of the equivalent circuit model	Easy to control	External sources of electrical energy is required
Thermal	Change the dielectric properties of the substrate using temperature	Negligible change in the thickness of the MMA structure	• Not sensitive to the change of temperature • Time consuming and waste of energy
Mechanical	Change the effective permittivity of the structure	• Numerous design possibilities without having to change the MMA structure • External sources of energy is not required	Sensitive to air gap between the MMA and the dielectric substance

Future Research of Metamaterial Absorbers

Metamaterial absorbers operating in the microwave frequency range should overcome problems relevant to the thickness, unit cell size, angle of incidence, polarization sensitivity, flexibility, bandwidth and thermal stability. One of the future trend is to create a multifunctional MMAs structure and apply the MMAs in various applications.

Metamaterial Sensors

Introduction to Metamaterial Sensors

Microwave sensors measure properties of a sample based on microwave interaction with matter. They usually comprise of a signal generator, a receiver and an output device. The signal generator produces microwave signal at a certain frequency range to be transmitted through the sample. A physical change in the characteristics of the sample causes a change in response to some electromagnetic excitation. The receiver such as antenna receives the transmitted microwave signals. The signals are then converted to a signal or voltage output response. By investigating some standard samples with physical characteristics known, the relationship between the changes in sensing parameters of samples and the output response can be obtained (You, Abbas, & Khalid, 2010). The output devices are used to display the sensing parameters such as moisture content, permittivity, density, chemical concentration and pressure.

For metamaterial microwave sensors, a metamaterial-inspired resonator is added in close proximity to the samples. The physical size, shape, resonance and certain unique properties of the resonator make them useful in sensing. One of the reasons is that metamaterial resonators are able to localize and enhance electromagnetic fields which provides high sensitivity (Withayachumnankul, Jaruwongrungsee, Tuantranont, Fumeaux, & Abbott, 2013). The electromagnetic fields produced by the resonator enable the detection of small amount of changes. For example, the changes can be caused by liquid samples of different concentration, dielectric solid samples of different permittivity, or substances of different moisture content. The resonance frequency of metamaterials is greatly dependence of their own geometrical dimensions. This allows various modification in their structural design to detect external stimuli in which the external stimuli such as pressure, temperature or physical vibration changes one of the structural parameters and therefore changes the resonance frequency.

Metamaterial sensors have various potential applications including characterization of material, medical diagnosis or bio-sensing, detection of changes in the external stimulus such as air humidity, temperature and pressure. For industrial applications, the materials can be characterized by using metamaterial sensors according to their characteristics such as permittivity, density, concentration, moisture content or strain. Therefore, metamaterial sensors are useful in many manufacturing processes such as evaluating and controlling manufacturing quality, classification and labelling of chemical products, detection of liquid chemicals and their concentration, and detection of faulty (presence of cracks) in products. For medical and biological applications, metamaterial sensors are potentially used for blood analysis, measurement of a substance's concentration in a body tissue, detection of the binding of biomolecules and disease diagnostics. For detection of external stimulus, metamaterial sensors can be used to detect movement, vibration, humidity, temperature and pressure.

The criteria of good metamaterial sensors are they are able to

1. produce a measurable signal to accurately track the resonance frequencies
2. pertain the linearity of sensing
3. resolve small changes (good sensor sensitivity)
4. have a low operating frequency to reduce absorption by the substrate and background

One of the advantages of metamaterial microwave sensors is they are non-destructive to the samples (Zhang, Zhao, Cao, & Mao, 2018). For example, in moisture content measurement, the microwave signals do not affect the samples, and the changes in the moisture content are detected by the shift of resonance frequency produced by the resonator. On the other hand, the conventional method involves the drying of the samples to compare the mass of the samples before and after drying. Besides that, metamaterial sensors are non-contact to the samples, since a reasonable air gap between the sample and resonator can be made provided that the electric field is reachable to the samples. Furthermore, centimeter wavelength of microwave provides large penetration depth and a considerable volume of samples can be investigated. Metamaterial resonators are sensitive to variations in capacitive and inductive effects and this make the metamaterial sensors precise. On-time and rapid measurement can be achieved using metamaterial sensors.

Metamaterial microwave sensors are sensitive to small changes. This can be problematic when there are a few changing variables caused by the environmental variations. For example, temperature changes can cause the changes in the dielectric permittivity of a sample, which affect the accuracy of the sensors. Physical vibrations can change the relative distance or position of the samples from the resonator, and this also affects the resonance frequency. However, these variations can be reduced by calibration of the sensors before measurement and proper control of environmental variation. Another limitation of metamaterial microwave sensors is the need of high cost microwave components such as coaxial cables, adapters and electrical components. The equipment dimensions are also large, depending on the operating frequency band. Metamaterial microwave sensors have poor specificity to a particular material for certain sensors. For example, moisture sensor designed for almond kernel would require a calibration if it is to measure the moisture content for other samples.

Operation Principles of Metamaterial Sensors

The operation principles of metamaterial sensors are based on frequency shift, frequency splitting, amplitude modulation and coupling modulation (Su, Mata-Contreras, Vélez, & Martín, 2017). The first operation principle of metamaterial sensors is based on the shift of resonance frequency. It is one of the common and useful approach to characterize materials by comparing the resonance frequency of the metamaterial resonator with and without the sample materials. Based on Figure 2 (a), a split ring resonator is loaded to a microstrip transmission line and coupling capacitance exists between them. Resonance occurs at a frequency depends on the geometrical dimensions of the resonator and the dielectric substrate used. This resonance frequency, f_{air} can be seen as the notch when the transmission coefficient, S_{21} reaches a minimum value, as shown in Figure 2 (b). When a sample is placed on top of the resonator, this changes the effective permittivity of the resonator and the resonance frequency shifts to a new resonance frequency, f_{sample}. The magnitude of the shift of resonance frequency, $f_{\mathrm{sample}} - f_{\mathrm{air}}$ depends on the dielectric properties of the samples. Sensors based on resonance shift can be used to determine the dielectric constant, moisture content and temperature of a particular substance.

To obtain accurate measurements, the sensors are usually calibrated using standard samples with known properties. For example, for permittivity sensor, dielectric samples with known dielectric constant are measured and their respective resonance frequencies are then used to make a calibration curve. The same calibration process is carried out for other sensing parameters such as the moisture content of tea leaves, temperature of processed food and concentration of liquid chemical. Most of these sensors can be technically considered as permittivity sensors since the dielectric properties of the samples depends on the sensing parameters. Once the calibration curve is obtain, the sensors are able to measure certain parameters of a particular substance accurately.

Without calibration curve, the dielectric constant of the sample, ε_{sample} can be calculated depending on the capacitance, C_c and inductance, L_c of the resonator. Taking complementary split ring resonator (CSRR) as an example, the capacitance and inductance of CSRR without sample are C_c and L_c, respectively. The capacitance of the CSRR loaded with sample, C_c' can be expressed as equation (4) (Su, Mata-Contreras, Paris, Fernandez-Prieto, & Martin, 2018).

$$C_c' = C_c\left(\frac{\varepsilon_{substrate} + \varepsilon_{sample}}{\varepsilon_{substrate} + 1}\right) \tag{4}$$

where $\varepsilon_{substrate}$ is the dielectric constant of substrate and ε_{sample} is the dielectric constant of sample. The resonance frequency of the CSRR without sample, f_{air} and the resonance frequency of the CSRR with sample, f_{sample} are expressed as equation (5) and (6), respectively.

$$f_{air} = \frac{1}{\sqrt{L_c\left(C + C_c\right)}} \tag{5}$$

Figure 2. Operation principle of metamaterial sensing: (a) Frequency shift, (b) Frequency dependence of transmission coefficient with and without sample

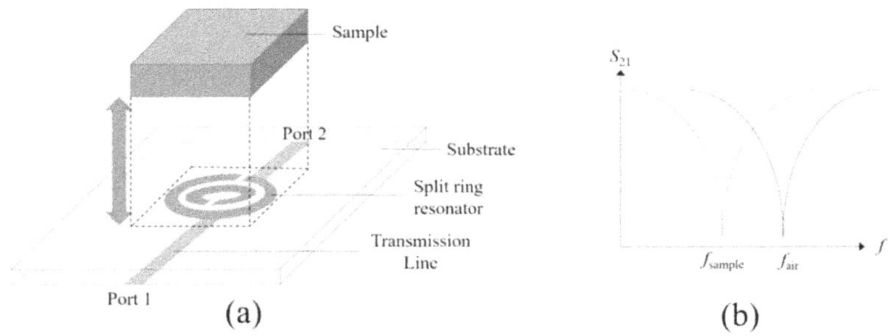

$$f_{sample} = \frac{1}{\sqrt{L_c \left(C + C_c{}'\right)}} \tag{6}$$

where L_c is the inductance of the CSRR, C is the capacitance between conductive transmission line and the inner region of the CSRR. By combining equation (4-6), the dielectric constant of the sample, ε_{sample} can be calculated using equation (7) as (Su et al., 2017):

$$\varepsilon_{sample} = 1 + \frac{1 + \varepsilon_{substrate}}{L_c C_c} \left(\frac{1}{f_{sample}^{\;\;2}} - \frac{1}{f_{air}^{\;\;2}} \right) \tag{7}$$

where $\varepsilon_{substrate}$ is the dielectric constant of substrate,, f_{sample} and f_{air} are the resonance frequency of the metamaterial structure with and without sample, respectively.

Various designs of metamaterial sensors can be made based on the sensing environment, type and dimension of samples. Without the implementation of transmission line, the metamaterial resonator with samples can be measured in free space using a pair of antennas. A metamaterial absorber reported in (Sim et al., 2018) can be used as a permittivity sensor for dielectric materials. Figure 3 (a) shows the absorption against frequency plot for planar samples having different dielectric constant, $\varepsilon_{\text{sample}}$ which are individually being placed at the front of the split ring resonator. An increase in the dielectric con-

Figure 3. The absorption peaks for (a) MMA with dielectric samples; (b) MMA with acrylic sample with different separation distance

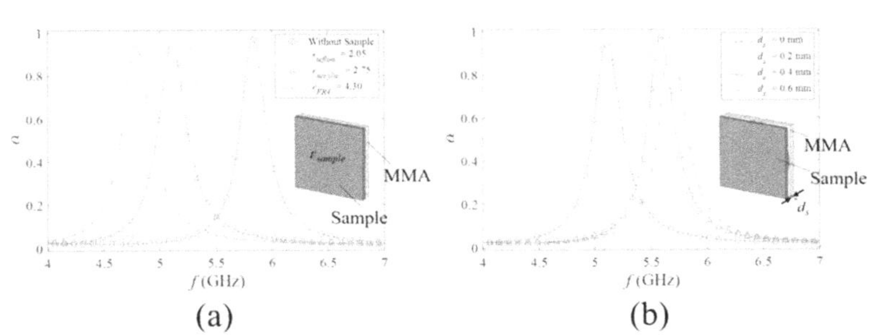

stant of 1 shifts approximately 0.2 GHz of the absorption frequency to lower frequency. In addition, the metamaterial absorber can be modified into a displacement or pressure sensor. Based on Figure 3 (b), the absorption frequencies vary based on the lateral separation distance, d_s between the sample and split ring resonator.

The advantages of sensors based on frequency shift are simple, ease in fabrication and low cost. Accurate measurements can be achieved provided that the calibration process has been carried out using standard samples with known properties. Sensing parameters such as dielectric constant, moisture content, temperature and relative position of a substance can be measured using this working principle. Metamaterial sensors are sensitive to the environmental changes. Therefore, other variables caused by the external factors are required to be fixed and controlled. For example, since the dielectric properties of a dielectric material are affected by temperature, calibration curves at different temperature are required for a permittivity sensor. This allows the reference to the accurate calibration curve based on the temperature at the time when the measurements are taking place.

The second operation principle of metamaterial sensors is based on frequency splitting. Two identical resonators are loaded on a transmission line symmetrically. An example of the ELC resonator is illustrated in Figure 4 (a), showing a symmetrical structure. This resonator structure exhibits a resonance at a single frequency, in other words, there is only a single notch (transmission zero, $S_{21} = 0$) as shown in Figure 4 (b). When one of the resonator structures is loaded with a sample on top of it, the resonance frequency split into two and two notches appear. In this case, the unloaded resonator is considered to have resonance at f_{air} and the loaded resonator contributes to resonance at f_{sample}. The difference between these two notches, $f_{sample} - f_{air}$ can be used to compare the magnitude of the sensing parameters. In another case, the resonator structures can both be loaded with samples. One of the sample will be the reference sample, and if only a single notch appear, the sample is identical to the reference sample. The resonance frequency of the resonators splits if loaded with different samples due to asymmetry. This allows the sensors to be used as a comparator which is useful in the detection of defects of samples.

Metamaterial sensors based on frequency splitting find applications in characterizing material. As reported in (Su & Vélez, 2016), a splitter/combiner microstrip line which is loaded with a pair of complementary split ring resonator (CSRR) is used to investigate the permittivity of dielectric samples. The frequency splitting phenomenon also indicates the symmetry disruption on the sample loading when a sample is compared with the reference sample. The defects or anomalies of the sample can be detected. The larger the variation of the difference between the two notches, $\Delta(f_{sample} - f_{air})$ for the dielectric samples, the better the sensitivity of the permittivity sensor. Besides that, the structure in (Su, Mata-Contreras, Vélez, & Martín, 2016) implements stepped impedance resonators (SIRs) at a distance

Figure 4. Operation principle of metamaterial sensing: (a) Frequency split, (b) Frequency dependence of transmission coefficient with and without sample

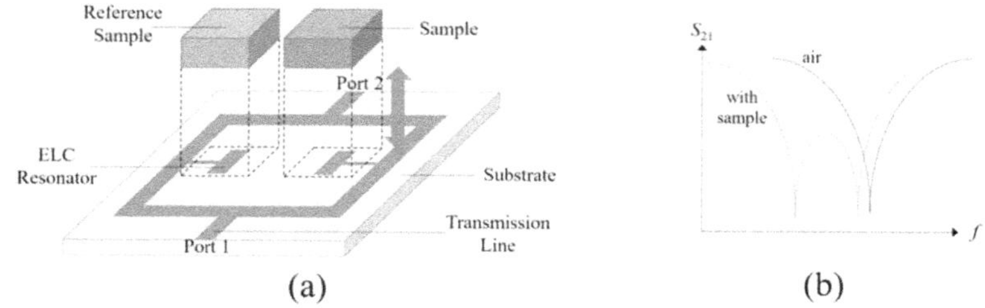

of half wavelength from the T-junctions of the splitter/combiner microstrip line. This structure also can be used as a sensor and comparator for dielectric samples. One of the limitations of these sensors is the coupling effect between the two resonators which reduces the sensitivity of the sensor. Another phenomenon which reduces the sensitivity is the result of interference when both the resonator structures are loaded with different samples.

The third operation principle of metamaterial sensors is based on coupling modulation. Coupling modulation is the control of the level of coupling between a transmission line and a resonator. The level of coupling can be changed by geometrical misalignment of the resonator or disruption of symmetry. Based on Figure 5 (a), a split ring resonator is located at a distance from the transmission line. When the split ring resonator is moved horizontally (increasing x), the level of coupling between the resonator and the line decreases, as illustrated in Figure 5 (b). The depth of the notch (normally expressed in dB) depends on the magnitude of coupling. By measuring the depth of notch, the sensors can be used to measure alignment, position and displacement.

A diamond-shaped split ring resonator reported in (Horestani et al., 2013) shows the change in both resonant frequency and magnitude of the reflection coefficient, S_{11} received by the input port of the transmission line, when the displacement changes. Higher dynamic range and linearity is obtained compared to previous studies on displacement sensing. Another displacement sensor based on broadside-coupled split ring resonator is reported by (Horestani et al., 2014). The proper design of placing two pairs of the resonators and the line along both x and y-axis allows the displacement sensing along both directions. Coupling modulation metamaterial sensors is robust against environmental variations since these external factors will not change the geometrical misalignment. These sensors are susceptible to electromagnetic interference (EMI) and this can be solved by shielding the sensors to avoid EMI. Since the geometrical size of a metamaterial can be scaled, the resonance can be tailored to a new frequency to reduce the effect of EMI.

The forth operation principle of metamaterial sensors is amplitude modulation. The amplitude of transmission coefficient, S_{21} depends on the level of coupling between the transmission line and the resonator (Naqui, Coromina, Karami-Horestani, & Fumeaux, 2015). If the level of coupling is high, the signals are reflected back to port 1 (input port) and S_{21} will be minimized. Amplitude modulation sensors are useful to measure angular position and angular velocity. Figure 6 (a) illustrates a metamaterial sensor, which consists of a transmission line and a resonator on both sides of a dielectric substrate, and two resonators on the rotating object. When the resonator on the substrate are aligned with the resonator on the rotating object, resonance occurs. The resonance occurs twice in a cycle (depending on the number

Figure 5. Operation principle of metamaterial sensing: (a) Amplitude modulation, (b) Frequency dependence of reflection coefficient with increasing value of horizontal displacement, x

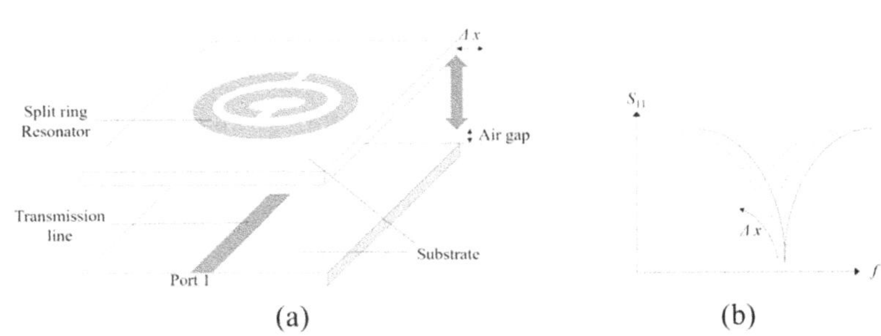

Figure 6. Operation principle of metamaterial sensing: (a) Coupling modulation, (b) Time dependence of transmission coefficient

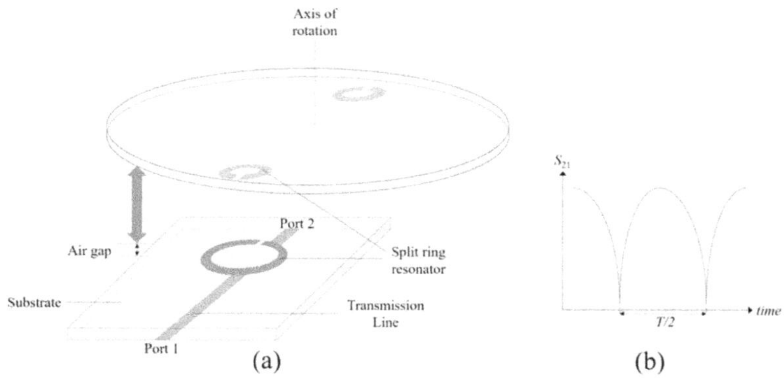

of resonators on the rotating object), and the time to complete a cycle, T can therefore be determined from the S_{21} against time plot as shown in Figure 6 (b). In this case, $T/2$ equals to the time between two adjacent transmission minimums or maximums. In general, the angular velocity can be obtained using equation (8) as (Naqui & Martin, 2016):

$$\omega = \frac{d\theta}{dt} = \frac{2\pi}{T_m n} \tag{8}$$

where θ is the angular position, t is time, T_m is the time between two adjacent transmission minimums, and n is the number of resonators on the rotating object. The angular displacement, θ of the rotating object after rotating for a duration of time, t is calculated using equation (9) as: $\Delta\theta = \omega\Delta t$ (9)

The angular velocity sensor reported in (Naqui & Martin, 2016) consists of a static substrate printed with split ring resonator and a rotor printed with a circular array of split ring resonators. The sensor prototype has been successfully fabricated by etching 300 resonators on a rotor and the concept is proven by measuring and counting the pulse using an oscilloscope. Another velocity sensor proposed by (Naqui, Member, & Mart, 2013) uses a transmission line etched on a substrate and a circular electric-LC (ELC) resonator etched on a rotor. Due to the non-symmetrical shape of the resonator, different angular orientation of the resonator provides different level of coupling with the transmission line, which affects the notch magnitude. The angular displacement can be plotted as a function of the notch amplitude. The angular velocity can be simply obtained from the period. The advantage of amplitude modulation velocity sensors is able to obtain quasi-instantaneous angular velocities. This can be achieved by distributing sufficient number of resonators along the circular rotating object. However, this might lead to cross-coupling effects between the adjacent resonators.

The four operation principles are summarized in Table 4 for comparison in sensing variables, advantages and disadvantages. Permittivity sensors are usually based on frequency shift or frequency splitting. These sensors can be calibrated to characterize samples based on their moisture content, temperature or concentration because these sensing parameters affect the dielectric properties of the samples. Coupling modulation and amplitude modulation sensors are useful in determining the displacement and velocity because the resonators are sensitive to their alignment or relative position with each other.

Table 4. Operation principles of metamaterial sensors

Operation Principles	Sensing Variables	Advantages	Disadvantages
Frequency shift	• Permittivity • Pressure • Moisture content • Temperature	• Simple in design • Low cost	• Typically required calibration for accurate measurements • Sensitive to environmental variations
Frequency splitting	• Permittivity	• Can be used as a comparator which is useful in the detection of defects of samples	• Coupling between the two resonators reduce the sensitivity of the sensor
Coupling modulation	• Alignment • Displacement	• Robust against environmental variations	• Susceptible to electromagnetic interference (EMI)
Amplitude modulation	• Angular position • Angular velocity	• Quasi-instantaneous angular velocities can be obtained • High reliability	• Many resonators are required for good accuracy of measurements and this might lead to cross-coupling effects

Recent Development of Metamaterial Sensors

Recently, there is much research on designing sensing devices for different materials using different metamaterial-based resonators. Figure 7 illustrates some of the basic designs of metamaterial sensors for different types of samples (solid and liquid) and different mechanisms of electromagnetic transmission (with and without microstrip transmission line). Based on Figure 7 (a-b), the samples are on top of the resonator. Microstrip transmission line is used for the transmission of the electromagnetic energy. Based on Figure 7 (c-d), the samples are placed at the middle (as part of the substrate layer) in between the resonator and back metal plate. Free space transmission of the electromagnetic wave is employed. There are many mechanical improvements being implemented for the sensing of different type of samples. Table 5 shows some of the metamaterial-based sensors for different sensing variables using different resonator structures.

Future Research of Metamaterial Sensors

Metamaterial sensors face some challenges such as interference from external noise, improvements in sensitivity and accuracy, selecting substrate for compact design, miniaturization and applicability in real life applications. Future research of metamaterial sensors would be working on designing sensing devices with better performance compared to current sensors on the market. Sensitivity and resolution of the sensors can be greatly enhanced using metamaterials.

In conclusion, metamaterial sensors help in developing a new generation of sensing technologies. Some of the important criteria for a marketable sensing product to be achieved are compact, robust, simple, sensitive, fast response time, repeatable, non-destructive to the samples and stable. Designing the devices can be done with ease using the simulation software and fabrication technique of microwave metamaterial is advanced. Despite the challenges, metamaterial-based sensing devices have optimistic future for various applications including medical, industrial and communication.

Figure 7. Metamaterial sensing: (a) Coupling modulation, (b) Time dependence of transmission coefficient

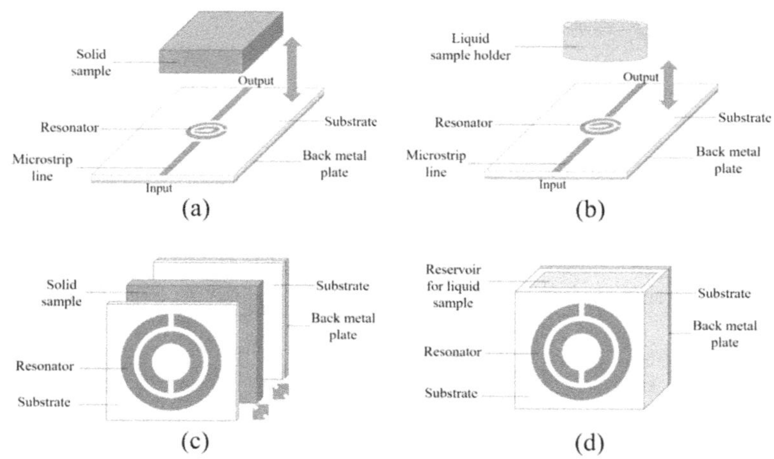

Metamaterial Energy Harvesters

Introduction of Metamaterial Harvesters

Table 5. Examples of metamaterial sensors

Structure	Sensing Variables	Advantages	Disadvantages	Ref.
Square and circular ring resonator	• Temperature of marrowbone • Dielectric constant of different substrate • Moisture content of almond kernel	• Multi-functional • Non-destructive • Real-time	Mechanical mechanism is required to make sure both substrates layers are in parallel position for practical applications	(Bakır, Karaaslan, Unal, Akgol, & Sabah, 2017)
Split ring cross resonator (SRCR)	Concentrations of chemicals (ethanol, methanol, hexane)	Small volume of chemical is required	Not suitable for acetone, propanol and benzene because the adhesive laminating film reacted with these chemicals, causing the microfluidic channel to be stuck	(Yoo, Kim, & Lim, 2016)
Split ring resonator and slot structure	Dielectric constant of thin planar dielectric samples	Near-perfect absorption peak at ~5.8 GHz gives measurable signal to accurately track the shift	Air gap between the sensor and solid samples affect the accuracy	(Sim et al., 2018)
Split ring resonator	Mechanical vibration	Vibration changes the spacing between two rings which are etched separately on two identical substrate, and this leads to shift in resonance frequency	Calibration required to set the initial resonance frequency	(Sikha Simon et al., 2017)
Broadside-coupled split ring resonator	Angular velocity	• Robust in space environments • Contactless	• Cross-coupling effects reduce the sensitivity • Difficulty in fixing the air gap between the rotor and sensing resonator	(Naqui & Martin, 2016)

Energy harvesting is a process to harvest and convert ambient energies such as kinetic, electromagnetic and thermal energy into electrical energy. Energy harvesting is useful to supply power for electronics devices or to recharge a secondary battery. Unlike chemical-based batteries, energy harvesting, which reduces the amount of disposal of used batteries, is more environmental-friendly. A conversion medium is required to convert the energy from the environment into electrical energy. AC voltage is the direct output and it is adjusted into DC voltage. The conversion efficiency of an energy harvester depends on the conversion medium. The conversion medium made by natural materials have limitations in term of efficiency. Therefore, electromagnetic metamaterials are introduced as the conversion media.

Electromagnetic metamaterial-based energy harvesters involve the use of metamaterial structures to carry out wireless power transmission (WPT). Rectenna (rectifier antenna) is a receiving antenna used to convert electromagnetic energy to electrical energy. Metamaterial or metasurface-based microwave rectenna is an interesting research topic recently. For example, a multiband receiving antenna based on corrugated ring resonator is proposed to operate at 0.51, 0.73 and 0.90 GHz with directivity 1.49. 2.65 and 4.04 dBi, respectively (Zhou et al., 2016). Furthermore, metamaterial absorbers can be designed to be a harvester by delivering its absorbed power to the resistive loads. The conversion efficiency, η is calculated as

$$\eta = \frac{P_{load}}{P_{in}} \tag{10}$$

where P_{load} and P_{in} are power dissipated on the resistive load and power incident on the energy harvester, respectively (H. T. Zhong, Yang, Tan, & Yu, 2016). However, not all metamaterial absorbers are suitable to be used as energy harvester because their main dissipation mechanism is due to the dielectric loss of the lossy substrate (Almoneef & Ramahi, 2015).

Operation Principles of Metamaterial Harvesters

Metamaterial harvesters require a metamaterial-based antenna, energy conversion mechanism, AC to DC converter and DC to DC converter. The metamaterial-based antenna structure captures the microwave energy. Energy conversion mechanism converts the wave energy to AC. The rectifier converts the AC to DC power. DC to DC converter changes the voltage level suitable to be used for the devices. The block diagrams of metamaterial harvester is shown in Figure 8.

Recent Development of Metamaterial Harvesters

There is much research being done on designing microwave harvesters using metamaterial structures. Metamaterial microwave absorbers have been used and modified to harvest the electromagnetic energy. The absorbed energy is dissipated in the resistive loads which are incorporated to the structure. For example, resistive loads are added to an octagonal shaped microwave absorber to prove its applicability in energy harvesting which operates at 5.5 GHz (Alkurt et al., 2018). An electric-LC (ELC) resonator and a complementary patch based metamaterial absorber is able to channel almost all absorbed power to the resistive loads at 2.65 GHz (Cheng et al., 2016). An absorber based on circular ring resonators converts at least 80% of the absorbed power in the frequency range from 7.8 to 14.0 GHz (Bakır et al., 2017).

Figure 8. Block diagrams of metamaterial energy harvester

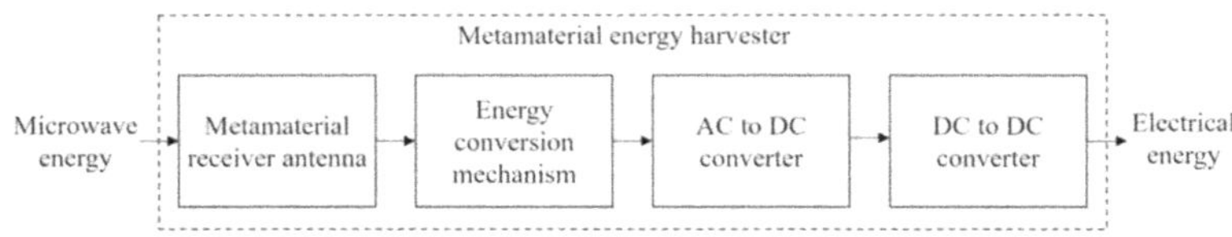

Future Research of Metamaterial Harvesters

Improvement works are required to increase the return loss and conversion efficiency on metamaterial harvesters. One of the future trend of metamaterial harvesting research is developing active tunable metamaterial harvester. This allows the metamaterial harvester to adapt to the changes of the operating environment which will affect the efficiency. There are many research works on SRR structure due to the ease in design and multi-band characteristics. More research on other structures can be done to achieve miniaturized and efficient harvester. Multi-functional metamaterial structure will be designed. For example, metamaterial structures can also be used as a filter to remove unwanted certain range of frequency and the filtered energy is converted to useful electrical energy. In conclusion, the use of metamaterial structure in energy harvesting improves the performance of the energy harvesters in term of return loss, efficiency and geometrical size.

CONCLUSION

In conclusion, a brief history, recent development and future trend in the area of electromagnetic metamaterials are reviewed. Three applications of the electromagnetic metamaterials, which are metamaterial microwave absorbers, sensors and energy harvesters, in microwave frequency range have been discussed. The operation principles of these metamaterial applications provide understanding and inspire more related research works to enhance and improve the current state of the arts. Electromagnetic metamaterials are useful in manipulating the incoming wave and therefore being applied in many applications. The findings in the electromagnetic metamaterials field have affected and inspired new modes of research in other areas including optics, acoustics, mechanics and material science.

Table 6. Metamaterial microwave energy harvesters

Structure	Dimension (mm^2)	Operating frequency (GHz)	Efficiency (%)	Ref.
Split ring resonator	5.9 × 5.9	5.8	~76	(Ramahi, Almoneef, Alshareef, & Boybay, 2012)
Two face-to-face split rings with via holes	7.5 × 7.5	3.0	97	(Almoneef & Ramahi, 2015)
Split-loop resonator	28.0 × 14.0	2.45	97	(Wang, Xu, Geyi, & Ma, 2016)
Wheel-shaped resonator with two orthogonal dipoles	15.7 × 15.7	2.5	98	(Almoneef, Erkmen, Al Dhaeebi, & Ramahi, 2018)

ACKNOWLEDGMENT

This research received no specific grant from any funding agency in the public, commercial, or not-for-profit sectors.

REFERENCES

Agarwal, M., Behera, A. K., & Meshram, M. K. (2016). Wide-angle quad-band polarisation-insensitive metamaterial absorber. *Electronics Letters*, *52*(5), 340–342. doi:10.1049/el.2015.4134

Alkurt, F. O., Altintas, O., Bakir, M., Tamer, A., Karadag, F., Bagmanci, M., ... Akgol, O. (2018). Octagonal shaped metamaterial absorber based energy harvester. *Materials Science*, *24*(3), 253–259. doi:10.5755/j01.ms.24.3.18625

Almoneef, T. S., Erkmen, F., Al Dhaeebi, M. A., & Ramahi, O. M. (2018). Dual polarized metasurface for microwave energy harvesting and wireless power transfer. *12th European Conference on Antennas and Propagation (EuCAP 2018)*, 337, 1-3. 10.1049/cp.2018.0696

Almoneef, T. S., & Ramahi, O. M. (2015). Metamaterial electromagnetic energy harvester with near unity efficiency. *Applied Physics Letters*, *106*(15). doi:10.1063/1.4916232

Bakır, M., Karaaslan, M., Unal, E., Akgol, O., & Sabah, C. (2017). Microwave metamaterial absorber for sensing applications. *Opto-Electronics Review*, *25*(4), 318–325. doi:10.1016/j.opelre.2017.10.002

Bakır, M., Karaaslan, M., Unal, E., Akgol, O., & Sabah, C. (2017). Microwave metamaterial absorber for sensing applications. *Opto-Electronics Review*, *25*(4), 318–325. doi:10.1016/j.opelre.2017.10.002

Bakır, M., Karadağ, F., Tetik, E., Bağmancı, M., Altıntaş, O., & Karaaslan, M. (2017). Wideband metamaterial absorber based on CRRs with lumped elements for microwave energy harvesting. *The Journal of Microwave Power and Electromagnetic Energy*, *52*(1), 45–59.

Chaurasiya, D., Ghost, S., Bhattacharyya, S., & Srivastava, K. V. (2015). An ultrathin quad-band polarization-insensitive wide-angle metamaterial absorber. *Microwave and Optical Technology Letters*, *57*(3), 697–702. doi:10.1002/mop.28928

Cheng, Y. Z., Fang, C., Zhang, Z., Wang, B., Chen, J., & Gong, R. Z. (2016). A compact and polarization-insensitive perfect metamaterial absorber for electromagnetic energy harvesting application. *2016 Progress In Electromagnetics Research Symposium, PIERS 2016 - Proceedings*, 1, 1910–1914. 10.1109/PIERS.2016.7734828

Deng, L., & Han, M. (2007). Microwave absorbing performances of multiwalled carbon nanotube composites with negative permeability. *Applied Physics Letters*, *91*(2). doi:10.1063/1.2755875

Faruque, M. R. I., Islam, M. T., & Misran, N. (2012). Design analysis of new metamaterial for EM absorption reduction. *Progress in Electromagnetics Research*, *124*, 119–135. doi:10.2528/PIER11112301

Ghosh, S., Bhattacharyya, S., & Srivastava, K. V. (2014). Bandwidth-enhancement of a polarization insensitive metamaterial absorber. *Microwave and Optical Technology Letters*, *56*(2), 350–355. doi:10.1002/mop.28122

Ghosh, S., Nguyen, T. T., & Lim, S. (2019). Recent progress in angle-insensitive narrowband and broadband metamaterial absorbers. *EPJ Applied Metamaterials*, *6*(12), 1–15.

Hedayati, M. K., Javaherirahim, M., Mozooni, B., Abdelaziz, R., Tavassolizadeh, A., Sai, V., ... Elbahri, M. (2011). Design of a perfect black absorber at visible frequencies Using Plasmonic Metamaterials. *Advanced Materials*, *23*(45), 5410–5414. doi:10.1002/adma.201102646 PMID:21997378

Horestani, A. K., Member, S., Fumeaux, C., Member, S., Al-sarawi, S. F., & Abbott, D. (2013). Displacement sensor based on diamond-shaped tapered split ring resonator. *IEEE Sensors Journal*, *13*(4), 1153–1160. doi:10.1109/JSEN.2012.2231065

Horestani, A. K., Naqui, J., Shaterian, Z., Abbott, D., Fumeaux, C., & Martín, F. (2014). Two-dimensional alignment and displacement sensor based on movable broadside-coupled split ring resonators. *Sensors and Actuators A: Physical*, *210*, 18–24. doi:10.1016/j.sna.2014.01.030

Huang, X., Yang, H., Yu, S., Wang, J., Li, M., & Ye, Q. (2013). Triple-band polarization-insensitive wide-angle ultra-thin planar spiral metamaterial absorber. *Journal of Applied Physics*, *113*(21). doi:10.1063/1.4809655

Ju, Z., Xu, G., Wei, Z., Li, J., Zhao, Q., & Huang, J. (2018). A single-patterned five-band terahertz metamaterial absorber based on multiple resonance mechanisms. *Modern Physics Letters B*, *32*(3).

Karaaslan, M., Bağmancı, M., Ünal, E., Akgol, O., Altıntaş, O., & Sabah, C. (2018). Broad band metamaterial absorber based on wheel resonators with lumped elements for microwave energy harvesting. *Optical and Quantum Electronics*, *50*(5), 1–18. doi:10.100711082-018-1484-2

Karaaslan, M., Bağmancı, M., Ünal, E., Akgol, O., & Sabah, C. (2017). Microwave energy harvesting based on metamaterial absorbers with multi-layered square split rings for wireless communications. *Optics Communications*, *392*, 31–38. doi:10.1016/j.optcom.2017.01.043

Knott, E. F., & Lunden, C. D. (1995). The two-sheet capacitive Jaumann absorber. *IEEE Transactions on Antennas and Propagation*, *43*(11), 1339–1343. doi:10.1109/8.475112

Kock, W. E. (1948). Metallic Delay Lenses. *The Bell System Technical Journal*, *27*(1), 58–82. doi:10.1002/j.1538-7305.1948.tb01331.x

Koschny, T., Markos, P., Smith, D. R., & Soukoulis, C. M. (2003). Resonant and anti-resonant frequency dependence of the effective parameters of metamaterials. *Physical Review E*, *68*(6), 1–4. doi:10.1103/PhysRevE.68.065602

Li, L., Wang, J., Du, H., Wang, J., Qu, S., & Xu, Z. (2015). A band enhanced metamaterial absorber based on E-shaped all-dielectric resonators. *AIP Advances*, *5*(1). doi:10.1063/1.4907050

Liu, T., Cao, X., Gao, J., Zheng, Q., Li, W., & Yang, H. (2013). RCS Reduction of Waveguide Slot Antenna With Metamaterial Absorber. *IEEE Transactions on Antennas and Propagation*, *61*(3), 1479–1484. doi:10.1109/TAP.2012.2231922

Matsumoto, H. (2002). Research on solar power satellites and microwave power transmission in Japan. *IEEE Microwave Magazine*, *3*(4), 36–45. doi:10.1109/MMW.2002.1145674

Mohammed, S. S., Ramasamy, K., & Shanmuganantham, T. (2010). Wireless Power Transmission – A Next Generation Power Transmission System. *International Journal of Computers and Applications*, *1*(13), 100–103.

Naqui, J., Coromina, J., Karami-Horestani, A., Fumeaux, C., & Martín, F. (2015). Angular displacement and velocity sensors based on coplanar waveguides (CPWs) loaded with s-shaped split ring resonators (S-SRR). *Sensors (Basel)*, *15*(5), 9628–9650. doi:10.3390150509628 PMID:25915590

Naqui, J., & Martı, F. (2013). Transmission lines loaded with bisymmetric resonators and their application to angular displacement and velocity sensors. *IEEE Transactions on Microwave Theory and Techniques*, *61*(12), 4700–4713. doi:10.1109/TMTT.2013.2285356

Naqui, J., & Martín, F. (2016, May). Application of broadside-coupled split ring resonator (BC-SRR) loaded transmission lines to the design of rotary encoders for space applications. In *2016 IEEE MTT-S International Microwave Symposium (IMS)* (pp. 1-4). IEEE. 10.1109/MWSYM.2016.7540017

Pang, Y., Wang, J., Cheng, Q., Xia, S., Zhou, X. Y., Xu, Z., ... Qu, S. (2017). Thermally tunable water-substrate broadband metamaterial absorbers. *Applied Physics Letters*, *110*(10), 104103. doi:10.1063/1.4978205

Ramahi, O. M., Almoneef, T. S., AlShareef, M., & Boybay, M. S. (2012). Metamaterial particles for electromagnetic energy harvesting. *Applied Physics Letters*, *101*(17). doi:10.1063/1.4764054

Ramakrishna, S. A. (2005). Physics of negative refractive index materials. *Reports on Progress in Physics*, *68*(2), 449–521. doi:10.1088/0034-4885/68/2/R06

Schelkunoff, S. A., & Friis, H. T. (1952). *Antennas: theory and practice* (Vol. 639). New York, NY: Wiley.

Sharma, S. K., Ghosh, S., & Srivastava, K. V. (2016). An ultra-thin triple-band polarization-insensitive metamaterial absorber for S, C and X band applications. *Applied Physics A*, *122*(12).

Sikha Simon, K., Chakyar, S. P., Andrews, J., & Joseph, V. P. (1849). Metamaterial split ring resonator as a sensitive mechanical vibration sensor. In AIP Conference Proceedings (Vol. 20021, No. 2017).

Sim, M. S., You, K. Y., Esa, F., Dimon, M. N., & Khamis, N. H. (2018). Multiband metamaterial microwave absorbers using split ring and multiwidth slot structure. *International Journal of RF and Microwave Computer-Aided Engineering*, *28*(7), 1–13. doi:10.1002/mmce.21473

Smith, D. R., & Kroll, N. (2000). Negative refractive index in left-handed materials. *Physical Review Letters*, *85*(14), 2933–2936. doi:10.1103/PhysRevLett.85.2933 PMID:11005971

Soheilifar, M. R., & Sadeghzadeh, R. A. (2014). Design, fabrication and characterization of stacked layers planar broadband metamaterial absorber at microwave frequency. *AEÜ International Journal of Electronics and Communications*, *69*(1), 126–132. doi:10.1016/j.aeue.2014.08.005

Strobel, A., Eickhoff, R., Ziroff, A., & Ellinger, F. (2010). Comparison of Pulse and FMCW based Radiolocation for Indoor Tracking Systems. In 2010 Future Network & Mobile Summit (pp. 1–8). IEEE.

Su, L., Mata-Contreras, J., Vélez, P., Fernández-Prieto, A., & Martín, F. (2018). Analytical method to estimate the complex permittivity of oil samples. *Sensors, 18*(4).

Su, L., Mata-Contreras, J., Vélez, P., & Martín, F. (2016). Configurations of splitter/combiner microstrip sections loaded with stepped impedance resonators (SIRs) for sensing applications. *Sensors, 16*(12).

Su, L., Mata-Contreras, J., Vélez, P., & Martín, F. (2017). A review of sensing strategies for microwave sensors based on metamaterial-inspired resonators: Dielectric characterization, displacement, and angular velocity measurements for health diagnosis, telecommunication, and space applications. *International Journal of Antennas and Propagation*, *2017*, 1–13. doi:10.1155/2017/5619728

Su, L., & Vélez, P. (2016). Splitter/combiner microstrip sections loaded with pairs of complementary split ring resonators differential sensing applications. *IEEE Transactions on Microwave Theory and Techniques*, *64*(12), 4362–4370. doi:10.1109/TMTT.2016.2623311

Tak, J., Jin, Y., & Choi, J. (2016). A dual-band metamaterial microwave absorber. *Microwave and Optical Technology Letters*, *58*(9), 2052–2057. doi:10.1002/mop.29977

Varadan, V. K., Varadan, V. V., & Lakhtakia, A. (1987). On the possibility of designing anti-reflection coatings using chiral composites. *Journal of Wave-Material Interaction*, *2*(3), 71–81.

Venneri, F., Costanzo, S., & Di Massa, G. (2018). Multi-band fractal microwave absorbers. In *World Conference on Information Systems and Technologies* (pp. 1488–1493). Springer International Publishing.

Vu, D. Q., Le, D. H., Dinh, H. T., Trinh, T. G., Yue, L., Le, D. T., & Vu, D. L. (2018). Broadening the absorption bandwidth of metamaterial absorber by coupling three dipole resonances. *Physica B, Condensed Matter*, *532*, 90–94. doi:10.1016/j.physb.2017.03.046

Wang, G. D., Liu, M. H., Hu, X. W., Kong, L. H., Cheng, L. L., & Chen, Z. Q. (2014). Multi-band microwave metamaterial absorber based on coplanar Jerusalem crosses. *Chinese Physics B*, *23*(1). doi:10.1088/1674-1056/23/1/017802

Wang, S. Y., Xu, P., Geyi, W., & Ma, Z. (2016). Split-loop resonator array for microwave energy harvesting. *Applied Physics Letters*, *109*(20). doi:10.1063/1.4967917

Withayachumnankul, W., Jaruwongrungsee, K., Tuantranont, A., Fumeaux, C., & Abbott, D. (2013). Metamaterial-based microfluidic sensor for dielectric characterization. *Sensors and Actuators. A, Physical*, *189*, 233–237. doi:10.1016/j.sna.2012.10.027

Yoo, M., Kim, H. K., & Lim, S. (2016). Chemical Electromagnetic-based ethanol chemical sensor using metamaterial absorber. *Sensors and Actuators. B, Chemical*, *222*, 173–180. doi:10.1016/j.snb.2015.08.074

You, K. Y., Abbas, Z., & Khalid, K. (2010). Application of Microwave Moisture Sensor for Determination of Oil Palm Fruit Ripeness. *Measurement Science Review*, *10*(1), 7–14.

Yu, D., Liu, P., Dong, Y., Zhou, D., & Zhou, Q. (2017). A sextuple-band ultra-thin metamaterial absorber with perfect absorption. *Optics Communications*, *396*, 28–35. doi:10.1016/j.optcom.2017.03.026

Yuan, H., Zhu, B. O., Feng, Y., Yuan, H., Zhu, B. O., & Feng, Y. (2015). A frequency and bandwidth tunable metamaterial absorber in x-band. *Journal of Applied Physics*, *117*(17). doi:10.1063/1.4919753

Zhang, N., Zhou, P., Cheng, D., Weng, X., Xie, J., & Deng, L. (2013). Dual-band absorption of mid-infrared metamaterial absorber based on distinct dielectric spacing layers. *Optics Letters*, *38*(7), 1125–1127. doi:10.1364/OL.38.001125 PMID:23546265

Zhang, Y., Zhao, J., Cao, J., & Mao, B. (2018). Microwave Metamaterial Absorber for Non-Destructive Sensing Applications of Grain. *Sensors, 18*(1912), 1–10.

Zhao, X., Fan, K., Zhang, J., Seren, H. R., Metcalfe, G. D., Wraback, M., ... Zhang, X. (2015). Optically tunable metamaterial perfect absorber on highly flexible substrate. *Sensors and Actuators. A, Physical*, *231*, 74–80. doi:10.1016/j.sna.2015.02.040

Zhong, H. T., Yang, X. X., Tan, C., & Yu, K. (2016). Triple-band polarization-insensitive and wide-angle metamaterial array for electromagnetic energy harvesting. *Applied Physics Letters*, *109*(25), 1–5. doi:10.1063/1.4973282

Zhong, J., Huang, Y., Wen, G., Sun, H., Wang, P., & Gordon, O. (2012). Single-/dual-band metamaterial absorber based on cross-circular-loop resonator with shorted stubs. *Applied Physics. A, Materials Science & Processing*, *108*(2), 329–335. doi:10.100700339-012-6989-0

Zhou, Y. J., Yang, L., Xiao, Q. X., Pan, T. Y., Ma, H. F., & Tan, C. (2016, July). Plasmonic metamaterials based subwavelength multiband antenna for wireless energy harvesting. In *2016 IEEE MTT-S International Microwave Workshop Series on Advanced Materials and Processes for RF and THz Applications (IMWS-AMP)* (pp. 1-4). IEEE. 10.1109/IMWS-AMP.2016.7588367

Zhu, B., Huang, C., Feng, Y., Zhao, J., & Jiang, T. (2010). Dual band switchable metamaterial electromagnetic absorber. *Progress In Electromagnetics Research B*, *24*, 121–129. doi:10.2528/PIERB10070802

Zhu, W., Huang, Y., Rukhlenko, I. D., Wen, G., & Premaratne, M. (2012). Configurable metamaterial absorber with pseudo wideband spectrum. *Optics Express*, *12*(19), 1563–1568. PMID:22418545

ADDITIONAL READING

Bogue, R. (2017). Sensing with metamaterials: A review of recent developments. *Sensor Review*, *37*(3), 305–311. doi:10.1108/SR-12-2016-0281

Chen, T., Li, S., & Sun, H. (2012). Metamaterials application in sensing. *Sensors (Basel)*, *12*(3), 2742–2765. doi:10.3390120302742 PMID:22736975

Chen, Z., Guo, B., Yang, Y., & Cheng, C. (2014). Metamaterials-based enhanced energy harvesting: A review. *Physica B, Condensed Matter*, *438*, 1–8. doi:10.1016/j.physb.2013.12.040

Ghosh, S., Nguyen, T. T., & Lim, S. (2019). Recent progress in angle-insensitive narrowband and broadband metamaterial absorbers. *EPJ Applied Metamaterials*, *6*, 12. doi:10.1051/epjam/2019010

Lee, Y. P., Rhee, J. Y., Yoo, Y. J., & Kim, K. W. (2016). *Metamaterials for perfect absorption* (Vol. 236). Singapore: Springer. doi:10.1007/978-981-10-0105-5

Marqués, R., Martin, F., & Sorolla, M. (2011). *Metamaterials with negative parameters: theory, design, and microwave applications* (Vol. 183). John Wiley & Sons.

Sim, M. S., You, K. Y., Esa, F., Dimon, M. N., & Khamis, N. H. (2018). Multiband metamaterial microwave absorbers using split ring and multiwidth slot structure. *International Journal of RF and Microwave Computer-Aided Engineering*, *28*(7), e21473. doi:10.1002/mmce.21473

Su, L., Mata-Contreras, J., Vélez, P., & Martín, F. (2017). A review of sensing strategies for microwave sensors based on metamaterial-inspired resonators: Dielectric characterization, displacement, and angular velocity measurements for health diagnosis, telecommunication, and space applications. *International Journal of Antennas and Propagation*, ▪▪▪, 2017.

Vivek, A., Shambavi, K., & Alex, Z. C. (2018). A review: Metamaterial sensors for material characterization. *Sensor Review*.

KEY TERMS AND DEFINITIONS

Absorber: A structure which absorbs the incident electromagnetic energy with negligible reflection and transmission.

Absorption Bandwidth: The range of frequencies in which the absorptivity reaches above a particular level.

Energy Harvester: A device which captures energy from external sources for other usages including power up electronics devices or recharge batteries.

Metamaterial: A material which exhibit unique characteristics which are originated from its engineered geometrical structure and depend on the properties of its constituent materials.

Resonator: A structure that exhibits resonance at a specific frequency.

Sensor: A device which is used to detect the changes such as temperature, moisture content, concentration, permittivity, position and velocity.

Tunable: Able to be tuned, adjusted, switched or shifted to a new desired operating frequency.

Chapter 3
Microwave Complex-Ratio-Measuring Circuits:
Alternative Solutions to Microwave Vector Instruments

Kok Yeow You
https://orcid.org/0000-0001-5214-7571
Unversiti Teknologi Malaysia, Malaysia

Chia Yew Lee
Universiti Teknologi Malaysia, Malaysia

Nadera Najib AL Areqi
Universiti Teknologi Malaysia, Malaysia

Kim Yee Lee
https://orcid.org/0000-0001-8195-2203
Universiti Tunku Abdul Rahman, Malaysia

Ee Meng Cheng
Universiti Malaysia Perlis, Malaysia

Yeng Seng Lee
https://orcid.org/0000-0003-3395-7338
Universiti Malaysia Perlis, Malaysia

ABSTRACT

This chapter reviews the microwave complex ratio measuring (MCRM) circuits which are used for complex reflection coefficient measurement. This MCRM circuit is relatively simple and cost-effective. There are various structures for the MCRM circuit, such as multi-probe transmission line circuits, five-port ring circuits, six-port hybrid coupler-based circuits, switched-reflector circuits, dual-generator circuits, and Wheatstone bridge-based circuits. Each structure of the circuits has its own advantages and disadvantages. In this chapter, the MCRM circuit calibration process has been described in detail.

DOI: 10.4018/978-1-7998-0117-7.ch003

INTRODUCTION

RF and microwave technology are rapidly expanding due to fourth industrial revolution (Industry 4.0), development of Internet of things (IoT), and 5th Generation wireless communication technology (5G) that will be implemented in the future. Recently, the commercial rollouts 5G technology is located at 3.4 to 4.2 GHz and 4.4 to 5 GHz (sub-6 GHz spectrum bands) as well as 28 GHz of high frequency bands. Therefore, microwave equipments, such as reflectometer systems are increasingly important today. For instance, two up to date commercial reflectometers which were used for the complex reflection coefficient measurements are shown in Figure 1.

Microwave complex ratio measurement (MCRM) system is one of the important alternative tools used in the microwave measurements and components design. In fact, MCRM system is a vector system that can be used to measure the complex reflection coefficient of the device under test (DUT) over a relatively broad frequency range. The MCRM circuit has been studied and developed rapidly by some researchers since the 70's (Hoer, 1972; Engen and Hoer, 1972; Caldecott, 1973; Engen, 1976, 1977), due to its simple circuit design and it has long-term stability of operation. Several types of MCRM circuits have been proposed and configured, such as multi-probe transmission line circuits, five-port ring circuits, six-port hybrid coupler-based circuits, switched-reflector circuits, dual-generator circuits, and Wheatstone bridge-based circuits. In fact, the combination of two MCRM systems will produce a vector network analyzer which is capable of measuring reflection and transmission coefficients simultaneously of the DUT. In this book chapter, those types of MCRM circuits and its calibration process will be described in detail.

MICROWAVE COMPLEX RATIO MEASURING (MCRM) CIRCUITS

Microwave complex ratio measuring (MCRM) circuit is a part of passive circuit in vector measurement instruments, such as vector reflectometer. In general, the MCRM circuit is a multi-port circuit in which it is capable of measuring the magnitude, $|\Gamma|$ and phase shift, θ of reflection coefficient for device under test (DUT), based on the number of n power ratios between the measured output power amplitudes, P_n (n = 1, 2, and 3) and input reference power amplitudes, P_{ref}. One input port of MCRM circuit is applied by power source and another port is terminated with DUT. The remaining output ports (normally three

Figure 1. (a) CABAN R54 Vector Reflectometer (85 MHz to 5.4 GHz). (b) Arinst VR 23-6200 (23 MHz to 6.2 GHz).

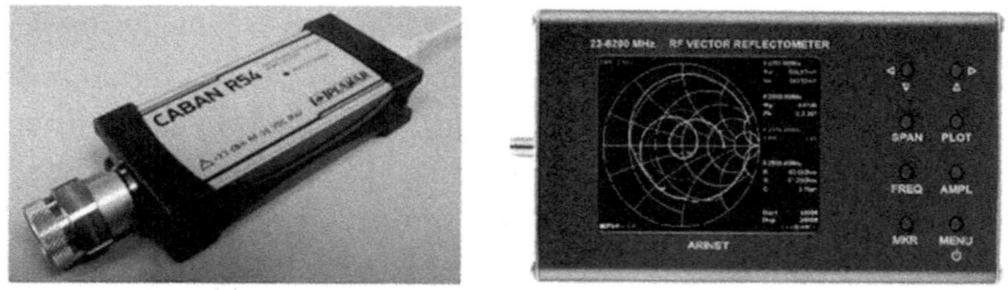

(a) (b)

ports or four ports) will be connected to power detectors in which the power ratios of the DUT are measured. For instance, a six-port complex ratio measuring system is shown in Figure 2.

This kind of circuit is denominated as five-port or six-port network circuit. The measured power, P_n at each port is the result of interference between the input and output amplitude signals. Finally, the measured power ratios, P_n/P_{ref} are used to predict the complex reflection coefficient, Γ of the DUT using some mathematical operations and circuit calibration process. For arbitrary linear multi-port MCRM circuit, the relationship between measured power ratios, P_n/P_{ref} and the complex reflection coefficient, Γ of the DUT, in which is desired to be determined, can be expressed in bilinear equation as:

$$\frac{P_n}{P_{ref}} = \eta_n \left|\frac{1+\alpha_n\Gamma}{1+\alpha_o\Gamma}\right|^2 \quad n = 1, 2, \text{ and } 3 \tag{1}$$

where η_n is the scalar constant, while α_o and α_n are the unknown complex parameters which can be determined by a proper calibration procedure. If all circuit ports are perfectly matched, the α_o in (1) will be zero, the (1) can be simplified as:

$$\frac{P_n}{P_{ref}} = \eta_n \left|1+\alpha_n\Gamma\right|^2 \quad n = 1, 2, \text{ and } 3 \tag{2}$$

In additional, the dependence of q_n on Γ can be minimized using linear equation (2) rather than bilinear equation (1) (Engen, 1977). Equation (2) can be rewritten in a more familiar complex-circle form:

$$r_n^2 = \left|\Gamma - q_n\right|^2 \quad n = 1, 2, \text{ and } 3 \tag{3}$$

where the complex constant, q_n and the scalar constant, r_n are the center and radius, respectively, of the P_n/P_{ref} circle in the Γ-plane as shown in Figure 3. Implicitly, q_n represent the complex calibration constants, which its values are found using calibration process. In fact, the radius circle, r_n^2 represents the power ratio, P_n/P_{ref}, thus the (3) can be expressed as:

$$\frac{P_n}{\eta_n P_{ref}} = \left|\Gamma - q_n\right|^2 \quad n = 1, 2, \text{ and } 3 \tag{4}$$

Figure 2. Arbitrary six-port microwave complex ratio measuring (MCRM) system.

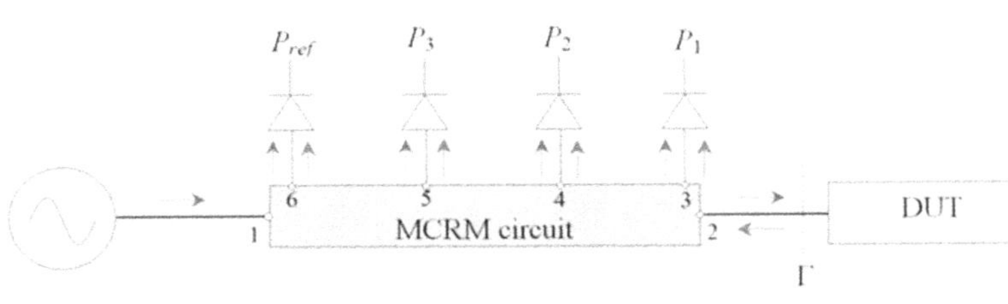

Hence, in order to solve the unknown complex reflection coefficient, Γ ($= \Gamma' + j\Gamma''$) of the DUT, at least three power ratios (P_1/P_{ref}, P_2/P_{ref}, and P_3/P_{ref}) from the four power measurements (P_1, P_2, P_3, and P_{ref}) are requested to be obtained. Three power ratios correspond to three q-circles (q_1-, q_2-, and q_3-circles) in Γ-plane (Γ'' versus Γ') as shown in Figure 3. Ideally, the three q-circles are intersected at a common Γ point in order to yield Γ unambiguously. In addition, the magnitudes of $|q_1|$, $|q_2|$, and $|q_3|$ should be equal and constant in order to cover the operating frequency range, as well as their angular separations should be 120° as shown in Figure 3(a) (Engen, 1977). However, in practice, the three q-circles may be distributed ($|q_1| \approx |q_2| \approx |q_3| \approx 1.5$) as shown in Figure 3(b) due to imperfect components and errors exist in the power measurements.

The use of the equation (4) is described and analyzed in detail in Sections 3 and 4. The main advantage of this circuit is that it only involves scalar amplitude measurements and no phase measurements, whereby it has less electronic circuit component requirements and reduces the cost of the electronic components. Normally, the MCRM are implemented in microwave measurement system, such as reflectometer, directional finder, radar systems, antenna measurement, multiport transceiver, and microwave material characterization. There are several kinds of MCRM circuits which have been previously investigated, such as multi-probe transmission line circuits, six-port hybrid coupler-based circuits, and five-port ring circuits.

VARIOUS KINDS OF MCRM CIRCUITS

Transmission Line Form MCRM Circuits

Multi-probe transmission line circuit is a transmission line with three attached fixed probes at separate points along the line (Caldecott, 1973) as shown in Figure 4 (a). The directional coupler is used to monitor the incident power, P_{ref} (as reference amplitudes). Each of the probes is used to detect the incident and reflected signals simultaneously. Normally, the three probes are separately spaced by $\lambda/6$ (60°) at the center frequency. Hence, the phase difference between the three probes is $\lambda/3$ (120°) at center frequency (optimum condition). However, the first and third probe shall be spaced $\leq \lambda/2$ (180°) in order to avoid phase ambiguities. The phases of the q-points are respectively changed with different operating frequencies. The operating bandwidth of the circuit is limited by the phase differences between the three probes.

Figure 3. (a) Ideal and (b) practice q-circle in the Γ-plane interpretation of the six-port circuit.

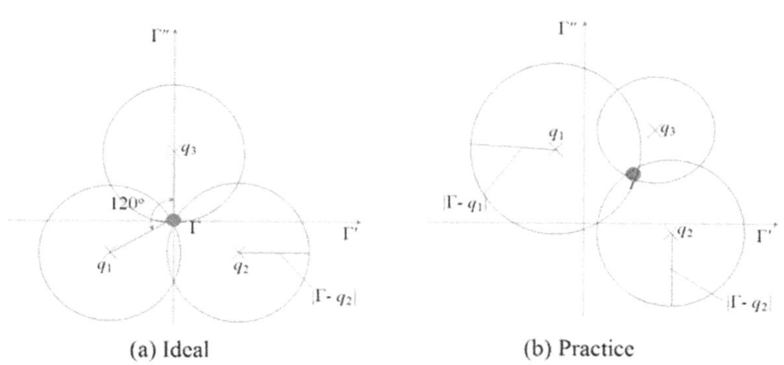

Figure 4. (a)Multi-probe transmission line circuit. (b) Fabricated simple multi-probe circuit.

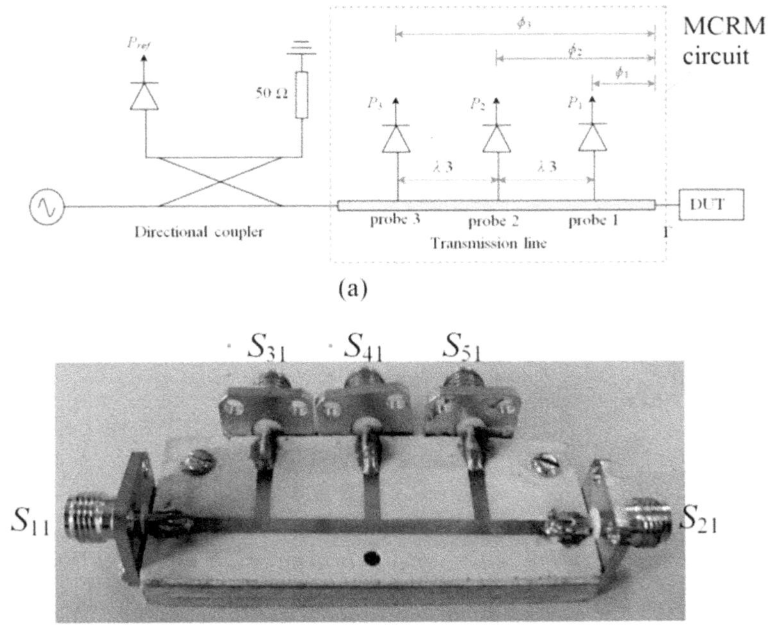

(a)

(b)

To be specific, the phase differences between any two probes should not be less than $\lambda/8$ (45°) (Caldecott, 1973; Bilik, 2002). Hence, this kind of MCRM circuit is only applicable for narrow operating bandwidth.

Figure 5 shows the phase difference between the three q-points in the Γ-plane at lower limit frequency, center frequency, and upper limit operating frequency, respectively. The interference magnitude of incident and reflected signals detected by the probe give rise to the complex power ratios in the Γ-plane, which is called q-point. Hence, three q-point (q_1, q_2, and q_3) can be obtained from the three probes, respectively. The phase of the q-point is corresponding to the total travel distance in which incident signal travel from the probe to DUT and back to the probe along the transmission line. Ideally, the q_1, q_2, and q_3 have equal magnitudes but different phase shift at center frequency, f_o.

A simple trial multi-probe transmission line circuit is fabricated as shown in Figure 4 (b) and tested using Vector Network Analyzer (VNA). Figure 6 shows the measured S-parameters results obtained from VNA. For wideband applications, phase differences between S_{31} and S_{41}, S_{41} and S_{51}, S_{31} and S_{51} must be constant throughout the frequency range. Unfortunately, in Figure 6 (f), those phase differences are increased (rather than constant) with operating frequency. Hence, this trial multi-probe circuit can only be used for any single operating frequency in the range of 1.37 GHz to 3.57 GHz. Furthermore, if possible, the multi-probe circuit should be designed and optimized so that the value of $|S_{11}|$ is better than -20 dB within the operating frequency.

From equation (2), the standing wave powers, P_1, P_2, and P_3 at probe 1, probe 2, and probe 3 on the transmission line can be written as:

Figure 5. q_1, q_2, and q_3 points in the Γ-plane at lower limit frequency, center frequency, and upper limit operating frequency (Bilik, 2002).

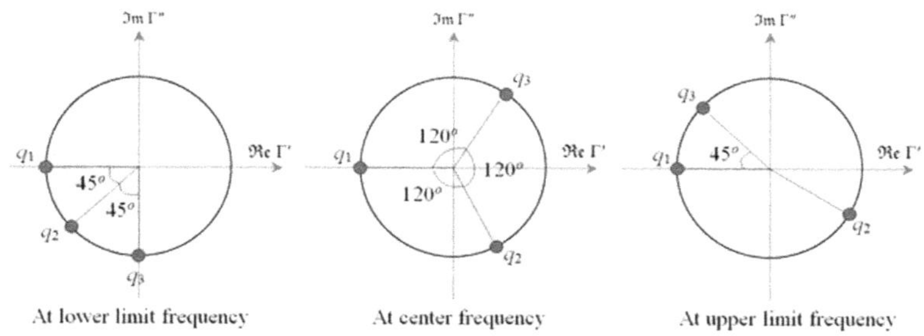

$$\frac{P_1}{\eta_1 P_{ref}} = \left|1 + \left|\Gamma\right| e^{j(\theta-\phi_1)}\right|^2 \tag{5a}$$

Figure 6. Measured S-parameters of trial multi-probe transmission line circuit in Figure 4 (b).

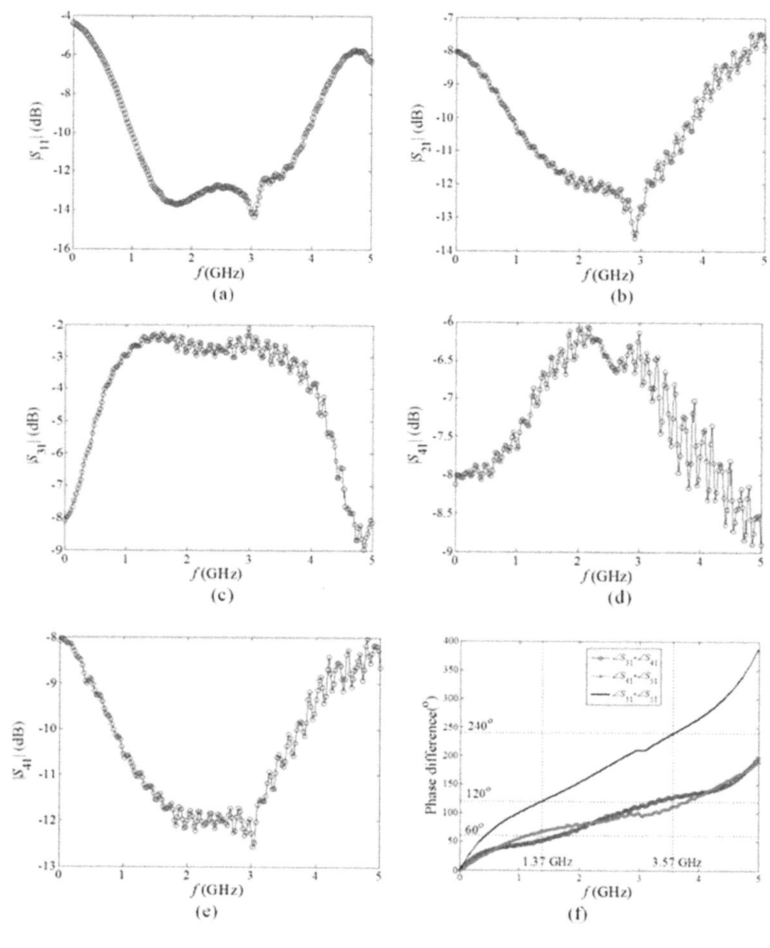

$$\frac{P_2}{\eta_2 P_{ref}} = \left|1 + \left|\Gamma\right| e^{j(\theta - \phi_2)}\right|^2 \tag{5b}$$

$$\frac{P_3}{\eta_3 P_{ref}} = \left|1 + \left|\Gamma\right| e^{j(\theta - \phi_3)}\right|^2 \tag{5c}$$

where α_n in (2) is assumed to be equal to $e^{-j\phi n}$. The values of η_1, η_2, and η_3 in (5) can be found by measuring the P_1, P_2, P_3, and P_{ref} by connecting 50 Ω match load at measurement port. For perfectly matched load, the $\Gamma = 0$. The modulus of (5) will yield

$$\frac{P_1}{\eta_1 P_{ref}} = 1 + \left|\Gamma\right|^2 + 2\left|\Gamma\right|\left(\sin\theta\sin\phi_1 + \cos\theta\cos\phi_1\right) \tag{6a}$$

$$\frac{P_2}{\eta_2 P_{ref}} = 1 + \left|\Gamma\right|^2 + 2\left|\Gamma\right|\left(\sin\theta\sin\phi_2 + \cos\theta\cos\phi_2\right) \tag{6b}$$

$$\frac{P_3}{\eta_3 P_{ref}} = 1 + \left|\Gamma\right|^2 + 2\left|\Gamma\right|\left(\sin\theta\sin\phi_3 + \cos\theta\cos\phi_3\right) \tag{6c}$$

where $|\Gamma|$ and θ are the unknown magnitude and phase of the reflection coefficient for the DUT at the measurement port. Symbols ϕ_1, ϕ_2, and ϕ_3 represent the standing wave phases at positions of the probe 1, probe 2, and probe 3 relative to the measurement port. If the phase difference, $(\phi_3 - \phi_2) = (\phi_2 - \phi_1) = \lambda/3$ at center frequency, the solution of the (6a) to (6c) at center frequency, f_o can be simplified as:

$$\left|\Gamma\right|\cos\theta = -\frac{1}{\sqrt{3}P_{ref}}\left[\frac{P_1}{\eta_1}\left(\sin\phi_3 - \sin\phi_2\right) + \frac{P_2}{\eta_2}\left(\sin\phi_1 - \sin\phi_3\right) + \frac{P_3}{\eta_3}\left(\sin\phi_2 - \sin\phi_1\right)\right] \tag{7a}$$

$$\left|\Gamma\right|\sin\theta = \frac{1}{\sqrt{3}P_{ref}}\left[\frac{P_1}{\eta_1}\left(\cos\phi_3 - \cos\phi_2\right) + \frac{P_2}{\eta_2}\left(\cos\phi_1 - \cos\phi_3\right) + \frac{P_3}{\eta_3}\left(\cos\phi_2 - \cos\phi_1\right)\right] \tag{7b}$$

$$\left(1 + \Gamma^2\right) = \frac{1}{P_{ref}}\left(\frac{P_1}{\eta_1} + \frac{P_3}{\eta_3} - \frac{P_2}{\eta_2}\right) \tag{7c}$$

By dividing (7b) with (7a), yields

$$\theta = \arctan\left\{-\left[\frac{\left(P_1/\eta_1\right)\left(\cos\phi_3-\cos\phi_2\right)+\left(P_2/\eta_2\right)\left(\cos\phi_1-\cos\phi_3\right)+\left(P_3/\eta_3\right)\left(\cos\phi_2-\cos\phi_1\right)}{\left(P_1/\eta_1\right)\left(\sin\phi_3-\sin\phi_2\right)+\left(P_2/\eta_2\right)\left(\sin\phi_1-\sin\phi_3\right)+\left(P_3/\eta_3\right)\left(\sin\phi_2-\sin\phi_1\right)}\right]\right\} \tag{8}$$

and from (7c), the $|\Gamma|$ is written as:

$$|\Gamma| = \sqrt{\frac{\left(P_1/\eta_1\right)+\left(P_3/\eta_3\right)-\left(P_2/\eta_2\right)-P_{ref}}{P_{ref}}} \tag{9}$$

Later, the performance improvement and different calibration techniques of the multi-probe circuit have been demonstrated by several researchers (Hu, 1980; Li & Bosisio, 1982; Martin et al., 1982; Chang et al., 1990 ; Cripps, 2007). Hu, (1980; 1983) and Martin et al. (1982) had applied the multi-probe transmission line circuit for measuring high power antenna impedance. Li and Bosisio (1982) had calibrated the multi-probe transmission line circuit using a match load and four offset short circuits from 2 GHz to 4 GHz. Later, Chang et al. (1990) had proposed a microstrip-based multi-probe circuit with operating frequency from 9 GHz to 11 GHz as shown in Figure 7. The microstrip-based multi-probe circuit was calibrated using three-standard calibration technique (open circuit, short termination, and match load). Madonna et al. (1999) had improved the operating bandwidth of the microstrip-based multi-probe circuit by adding two extra probes along the microstrip transmission line circuit and optimized the spacing between the probes as shown in Figure 8. In addition, the regular tips of the tapered probes were modified into T-shaped in order to improve the coupling factor between the probes and main transmission line. Cripps, (2007) reviewed and implemented the multi-probe transmission line circuit into antenna

Figure 7. Microstrip-based three-probe transmission line circuit (Chang et al., 1990).

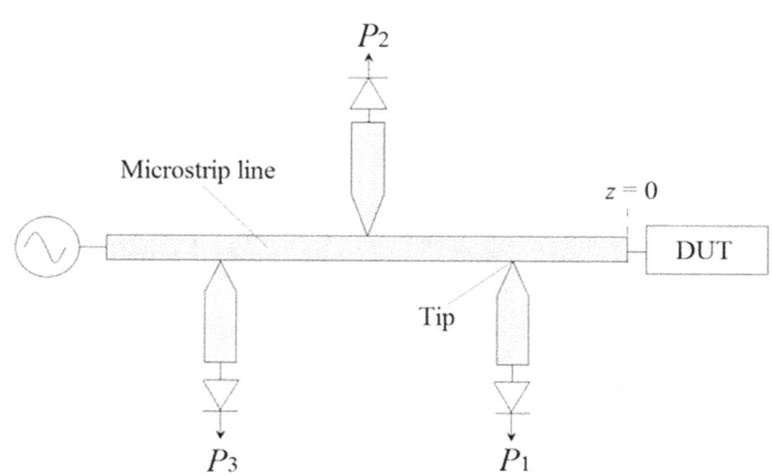

measurement. Moreover, Ülker and Weikle, (2001) had applied this kind of reflectometer for reflection measurement up to W-band (75 to 110 GHz).

Ring-Shaped MCRM Circuits

The first five-port ring-based circuit was implemented for complex reflection measurement by Riblet and Hansson, (1981) as shown in Figure 9. The measurement system consists of five-port ring circuit and 10-dB directional coupler to detect incident wave at reference port (Port-3). Figure 10 shows several fabricated five-port ring circuits, while, Figure 11 illustrates the structure of regular five-port ring circuit.

Based on Figure 9, the measured ratio power ratio, P_4/P_{ref}, P_5/P_{ref}, and P_6/P_{ref} at port-4, port-5, and port-6 can be written as (Riblet and Hansson, 1982):

$$\frac{P_4}{P_{ref}} = \eta_4 \frac{1 + X_4^2 \left|\Gamma\right|^2 + 2X_4 \left|\Gamma\right| \cos\left(\phi_4 + \theta\right)}{1 + Z^2 \left|\Gamma\right|^2 + 2Z \left|\Gamma\right| \cos\left(\phi_Z + \theta\right)} \tag{10a}$$

$$\frac{P_5}{P_{ref}} = \eta_5 \frac{1 + X_5^2 \left|\Gamma\right|^2 + 2X_5 \left|\Gamma\right| \cos\left(\phi_5 + \theta\right)}{1 + Z^2 \left|\Gamma\right|^2 + 2Z \left|\Gamma\right| \cos\left(\phi_Z + \theta\right)} \tag{10b}$$

Figure 8. Microstrip-based five-probe transmission line circuit with modified T-shaped tip (Madonna et al., 1999).

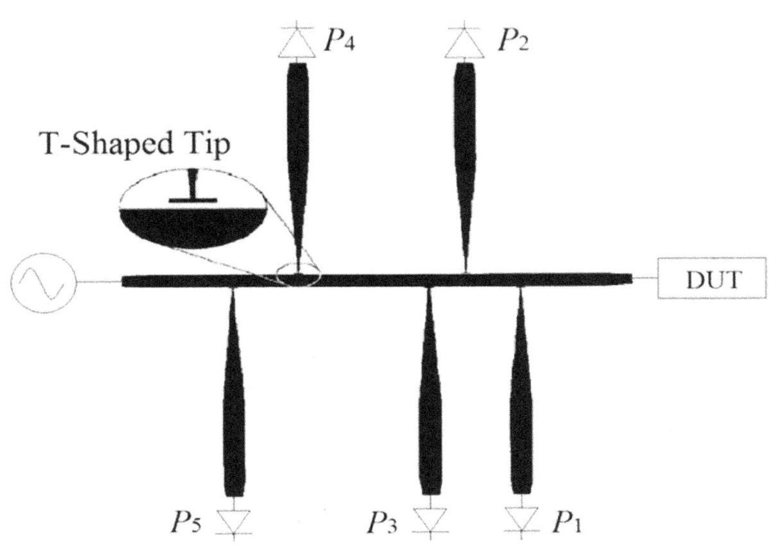

Figure 9. Five-port ring-based measurement system (Riblet and Hansson, 1981).

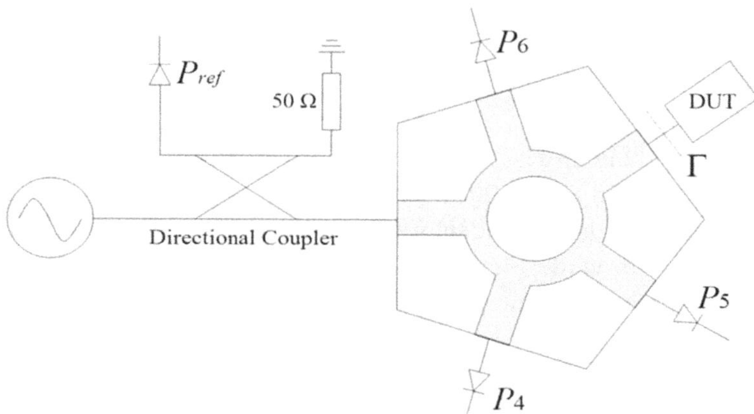

$$\frac{P_6}{P_{ref}} = \eta_6 \frac{1 + X_6^2 \left|\Gamma\right|^2 + 2X_6 \left|\Gamma\right| \cos\left(\phi_6 + \theta\right)}{1 + Z^2 \left|\Gamma\right|^2 + 2Z \left|\Gamma\right| \cos\left(\phi_Z + \theta\right)} \tag{10c}$$

where $|\Gamma|$ and θ are the magnitude and phase of the reflection coefficient to be determined at measurement port. The η_4, η_5, and η_6 can be found from the measurement of match load on the measurement port ($\Gamma = 0$). If all circuit ports are perfectly matched, the Z in (10) will be zero and Equations (10a) to (10c) can be simplified as:

Figure 10. Fabricated five-port ring circuits.

Figure 11. Regular five-port ring circuit (Dong et al., 1984).

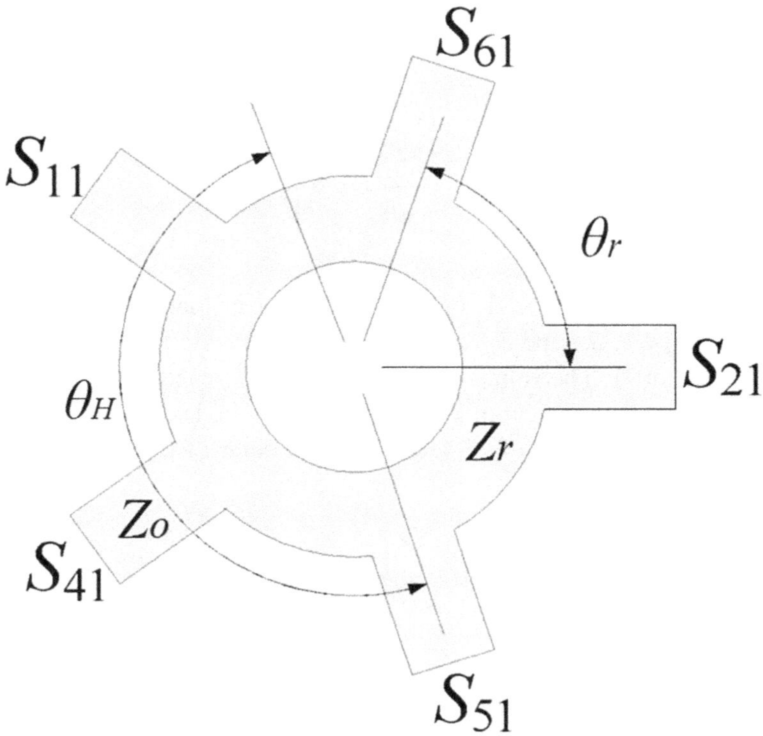

$$\frac{P_4}{\eta_4 P_{ref}} = 1 + X_4^2 \left|\Gamma\right|^2 + 2X_4 \left|\Gamma\right| \left\{\cos\phi_4 \cos\theta - \sin\phi_4 \sin\theta\right\} \tag{11a}$$

$$\frac{P_5}{\eta_5 P_{ref}} = 1 + X_5^2 \left|\Gamma\right|^2 + 2X_5 \left|\Gamma\right| \left\{\cos\phi_5 \cos\theta - \sin\phi_5 \sin\theta\right\} \tag{11b}$$

$$\frac{P_6}{\eta_6 P_{ref}} = 1 + X_6^2 \left|\Gamma\right|^2 + 2X_6 \left|\Gamma\right| \left\{\cos\phi_6 \cos\theta - \sin\phi_6 \sin\theta\right\} \tag{11c}$$

where X_4, X_5, X_6, $X_4\cos\phi_4$, $X_5\cos\phi_5$, $X_6\cos\phi_6$, $X_4\sin\phi_4$, $X_5\sin\phi_5$, and $X_6\sin\phi_6$ are the unknown constants which will be determined using four calibration standards, namely an open circuit, a short circuit, and two offset open circuits (Riblet and Hansson, 1982). Belfort et al. (1982) and Dong et al. (1984) had analyzed the five-port ring based on *S*-parameters as shown in Figure 11. Belfort and Cullen (1982) was expressed the S_{21}, S_{41}, S_{51}, and S_{61} as a function of S_{11} as:

$$S_{21} = S_{51} = \frac{1}{2}\left(1 - \sqrt{3}\left|S_{11}\right|\sin\theta_1\right)\exp\left\{-j\frac{\pi}{3} - j\frac{\left|S_{11}\right|}{\sqrt{3}}\cos\theta_1\right\} \tag{12a}$$

$$S_{61} = S_{41} = \frac{1}{2}\left(1 + \sqrt{3}\left|S_{11}\right|\sin\theta_1\right)\exp\left\{j\frac{\pi}{3} + j\frac{\left|S_{11}\right|}{\sqrt{3}}\cos\theta_1\right\} \tag{12b}$$

where S_{11} is the complex reflection coefficient at port-1 (Linear magnitude = $|S_{11}|$ and phase shift = θ_1). On the other hand, S_{21}, S_{41}, S_{51} and S_{61} are the transmission coefficients at port-2, port-4, port-5, and port-6 by referring to port-1. The values of S_{21}, S_{41}, S_{51} and S_{61} can be determined, once the value of S_{11} is known. Normally, linear magnitude of $|S_{11}|$ is requested to be less than 0.1. From Equations (12a) and (12b), the possible positions q_i (i = 4, 5, and 6) of the center of the q-circles can be calculated as (Belfort & Cullen, 1982):

$$q_6 = 2\left(1 + \left|S_{11}\right|\cos\theta_1\right)\exp\left\{j\frac{\pi}{3} + j\frac{4}{\sqrt{3}}\left|S_{11}\right|\cos\theta_1\right\} \tag{13a}$$

$$q_5 = 2\left(1 + \left|S_{11}\right|\cos\theta_1\right)\exp\left\{-j\frac{\pi}{3} - j\frac{4}{\sqrt{3}}\left|S_{11}\right|\cos\theta_1\right\} \tag{13b}$$

$$q_4 = 2\left\{1 + 5.57\left|S_{11}\right|\cos\left(\theta_1 + 68.95^o\right)\right\}\exp\left\{j\pi - j2\sqrt{3}\left|S_{11}\right|\cos\left(\theta_1 - 60^o\right)\right\} \tag{13c}$$

As mentioned, that parameter q_i is a complex number. Hence, the imaginary part of q_i versus real part of q_i can be plotted as Figure 12. The phase shift, θ_1 is varied from 0 to 360° in Figure 12, since the value of θ_1 is changed with operating frequency.

For ideal case $|S_{11}| = 0$, the q_4, q_5, and q_6 are constant and distributed separately in triangular form. In addition, the magnitudes of $|q_4| = |q_5| = |q_6| = 2$. However, in practical situations, there is no circuit with a perfect match as well as not always good matched over the operating bandwidth, thus, the value of $|S_{11}|$ is impossible to be zero. Here, several practice cases that $|S_{11}| = 0.05$ (-26 dB), $|S_{11}| = 0.1$ (-20 dB), and $|S_{11}| = 0.3$ (-10 dB) are considered. For small value of $|S_{11}| = 0.05$ with phase θ_1 varies from 0 to 360°, the excursions of the q-point from their ideal locations are not very significant. In this situation, high accuracy of reflection coefficient, Γ measurement for the ring circuit is still achievable using the appropriate calibration method.

Based on port connection arrangement in Figure 9, the q_4 point is most sensitive to the error that exists on $|S_{11}|$ compared to q_5 and q_6. For $|S_{11}| = 0.3$, the loop of q_4, q_5 and q_6 are progressively larger. From the loop, it is found that magnitudes of $|q_i|$ are not constant, especially for large values of $|S_{11}|$. This condi-

Figure 12. Imaginary part of q_i versus real part of q_i with constant (a) $|S_{11}| = 0$, (b) $|S_{11}| = 0.05$, (c) $|S_{11}| = 0.1$, (d) $|S_{11}| = 0.3$, and the θ_1 varies from 0 to 360°.

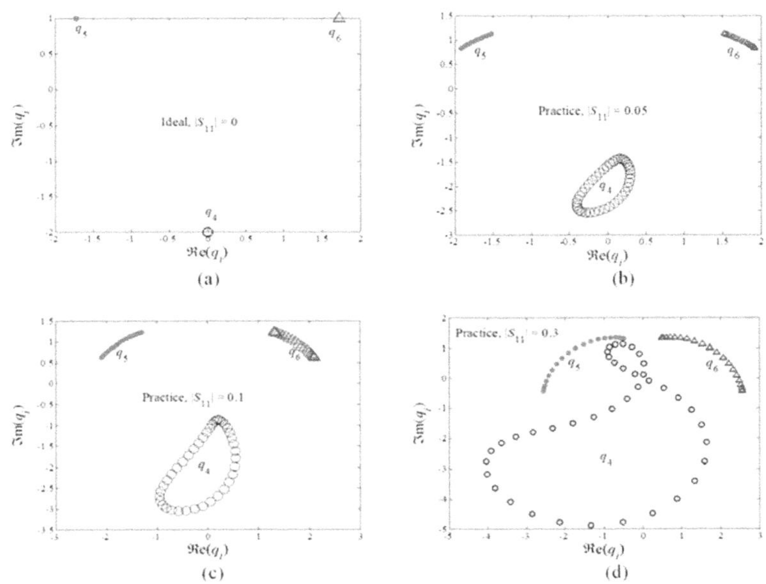

tion may result in unambiguous reflection coefficient, although the ring circuit is calibrated. Therefore, the ring circuit must be designed to achieve level of $|S_{11}| <0.1$ (<-20 dB) for reflectometer application.

The ideal reflection and transmission coefficients (S_{11}, S_{21}, S_{41}, S_{51} and S_{61}) for each port of primal five-port ring circuit were expressed by Dong et al. (1984) as:

$$S_{11} = \frac{e^{j\theta_1} + 2(e^{j(\theta_1+\theta_H+\theta_r)} + e^{j(\theta_1+\theta_H-\theta_r)})}{5} \tag{14a}$$

$$S_{21} = S_{51} = \frac{e^{j\theta_1} + 2\left\{e^{j(\theta_1+\theta_H+\Delta\theta_r)} \cos\left(2\pi/5\right) + e^{j(\theta_1+\theta_H-\theta_r)} \cos\left(4\pi/5\right)\right\}}{5} \tag{14b}$$

$$S_{41} = S_{61} = \frac{e^{j\theta_1} + 2\left\{e^{j(\theta_1+\theta_H+\theta_r)} \cos\left(4\pi/5\right) + e^{j(\theta_1+\theta_H-\theta_r)} \cos\left(2\pi/5\right)\right\}}{5} \tag{14c}$$

where θ_r is the phase difference between adjacent port and θ_1 is the initial phase shift in (14). The θ_H and θ_r (in radian) in equations (14a) to (14c) can be estimated as (Dong et al., 1984):

$$\cos^{-1}\left(\frac{1-5\left|S_{11}\right|}{4}\right) \leq \theta_r \leq \cos^{-1}\left(\frac{1+5\left|S_{11}\right|}{4}\right) \tag{15}$$

$$\theta_H = \pi - \cos^{-1}\left[\frac{1-2\left|S_{11}\right|^2}{8\cos\left(\theta_r\right)} + 2\cos\left(\theta_r\right)\right] \tag{16}$$

In fact, the value of θ_1 is insensitivity to the S_{11}, S_{21} and S_{41}. Thus, $e^{j\theta 1}$ can be assumed as unity (without any initial phase shift, $\theta_1 \approx 0$). Equations (14a) to (14c) can be used as computer-aided design (CAD) tool to design dimensions of the five-port ring circuit in conjunction with microstrip realization equation (Pozar, 2012). Firstly, the Equations (14a) to (14c) are requested to be rewritten in frequency domain as:

$$S_{11} = \frac{\left\{1 + 2\exp\left[\left(\alpha + j\frac{2\pi f}{c}\sqrt{\varepsilon_{eff}}\right)\left(l_r + \Delta l_r\right)\right] + 2\exp\left[j\frac{2\pi f}{c}\sqrt{\varepsilon_{eff}}\left(l_r - \Delta l_r\right)\right]\right\}}{5} \tag{17a}$$

$$S_{21} = \frac{\left\{1 + 2\exp\left[j\frac{2\pi f}{c}\sqrt{\varepsilon_{eff}}\left(l_r + \Delta l_r\right)\right]\cos\left(2\pi/5\right) + 2\exp\left[j\frac{2\pi f}{c}\sqrt{\varepsilon_{eff}}\left(l_r + \Delta l_r\right)\right]\cos\left(4\pi/5\right)\right\}}{5} \tag{17b}$$

$$S_{41} = \frac{\left\{1 + 2\exp\left[j\frac{2\pi f}{c}\sqrt{\varepsilon_{eff}}\left(l_r + \Delta l_r\right)\right]\cos\left(4\pi/5\right) + 2\exp\left[j\frac{2\pi f}{c}\sqrt{\varepsilon_{eff}}\left(l_r + \Delta l_r\right)\right]\cos\left(2\pi/5\right)\right\}}{5} \tag{17c}$$

The effective relative permittivity, ε_{eff} is given as (Pozar, 2012):

$$\varepsilon_{eff} = \frac{\varepsilon_{r1} + \varepsilon_{r2}}{2} + \frac{\varepsilon_{r1} - \varepsilon_{r2}}{2}\frac{1}{\sqrt{1 + 12\, d/w}} \tag{18}$$

where ε_r is the relative permittivity of substrate. While, h is the thickness of substrate and W is the width of the microstrip ring line, which can be calculated as:

$$\frac{W}{h} = \begin{cases} \dfrac{8\exp(A)}{\exp(2A)-2} & \text{for } \dfrac{W}{h} < 2 \\ \dfrac{2}{\pi}\left\{ \begin{aligned} & B - 1 - \ln(2B-1) \\ & + \frac{\varepsilon_{r1}-1}{2\varepsilon_{r1}}\left[\ln(B-1) + 0.39 - \frac{0.61}{\varepsilon_{r1}}\right] \end{aligned} \right\} & \text{for } \dfrac{W}{h} > 2 \end{cases} \tag{19}$$

where

$$A = \frac{Z}{60}\sqrt{\frac{\varepsilon_{r1}+1}{2}} + \frac{\varepsilon_{r1}-1}{\varepsilon_{r1}+1}\left(0.23 + \frac{0.11}{\varepsilon_{r1}}\right) \text{ and } B = \frac{377\pi}{2Z_o\sqrt{\varepsilon_r}}$$

Here, Z is the characteristic impedance of the microstrip ring line. At resonant frequency so-called center frequency, f_o, the reflection coefficient, $|S_{11}| \approx 0$, thus, $\Delta\theta \approx 1.309$ rad (75o) and $\theta \approx 3.142$ rad (180o). The physical length, l_r and Δl_r (in meters) in Equations (17a) to (17c) are estimated as:

$$l_r = \frac{3.142c}{2\pi f_r\sqrt{\varepsilon_{eff}}} \text{ and } \Delta l_r = \frac{1.309c}{2\pi f_r\sqrt{\varepsilon_{eff}}}$$

where c is the speed of light in free space. For instance, the calculated S-parameters (S_{11}, S_{21}, S_{41}, S_{51}, and S_{61}) of regular ring using Equations (17) and compared with measurements are shown in Figure 14. The measurement data are obtained using a fabricated regular five-port ring circuit in which it has an operating center frequency, $f_o = 3$ GHz and printed on Rogers RO4360 substrate ($\varepsilon_r = 6.15 - j\,0.025$ and h = 1.524 mm) as shown in Figure 13. The value of Z_r of microstrip line is fixed to be 45.5 Ω and section length, Δl_r is required to be equal to $0.211c/(f_o\sqrt{\varepsilon_{eff}})$ in order to achieve a matched circuit condition ($|S_{11}| \approx 0$ or $|S_{11}| \approx -\infty$ dB) at resonant frequency, f_o (Dong et al., 1984). Hence, using equation (19), the values of W and Δl_r are given as 2.63mm and 8.80mm, respectively.

From Figure 14, clearly, there are good agreement between analytical calculations and measurements for $|S_{11}|$. The measured $|S_{21}|$, $|S_{41}|$, $|S_{51}|$, $|S_{61}|$, and phase difference between adjacent ports deviate slightly from the calculation as shown in Figure 14 (b) and Figure 14 (c), which is deduced that the deviation may cause by the abrupt transmission line bend at junction between feed-line and ring.

Riblet and Hansson, 1981 proposed a multi-section feedline ring circuits as shown in Figure 15. In 1982, De Ronde combined the ring and the star structures for the five-port circuit design as shown in Figure 16, in order to obtain a broader bandwidth. However, the De Ronde's report did not provide information about the optimum value of the impedances and electrical lengths of the five-port microstrip ring line. Fortunately, in 1984, Dong et al. had optimized the five-port circuit's impedance and electric length of the five-port circuit as presented in Table 1. Dong et al. (1984) also introduced double ring and star-shaped five-port circuit as shown in Figure 17, which further broaden the bandwidth of the five-port circuit. Over a few years, Tsang et al. (1991) had designed the five-port circuit in taper star shape instead of the ring as shown in Figure 18 (a). In 2001, the bandwidth of the five-port circuit which proposed by Dong et al. (1984) was further improved by Yeo and Choong that changed the multi-section feedline

Figure 13. Fabricated regular five-port ring circuit.

Figure 14. Analytical calculations and measurements of ring circuit: (a) $|S_{11}|$ (b) $|S_{21}|$, $|S_{41}|$, $|S_{51}|$, $|S_{61}|$, and (c) phase difference.

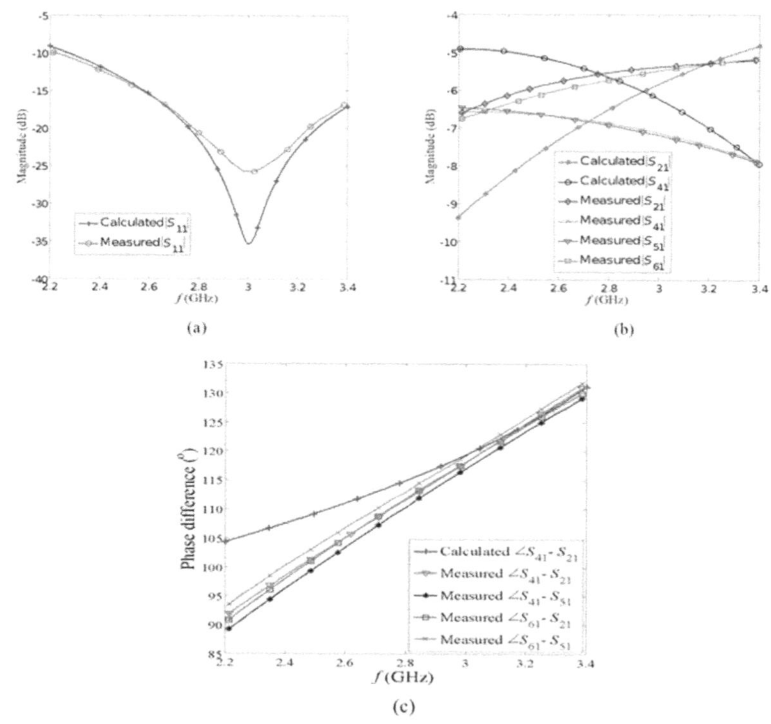

Figure 15.Multi-section feed-line ring circuit (Riblet & Hansson, 1981).

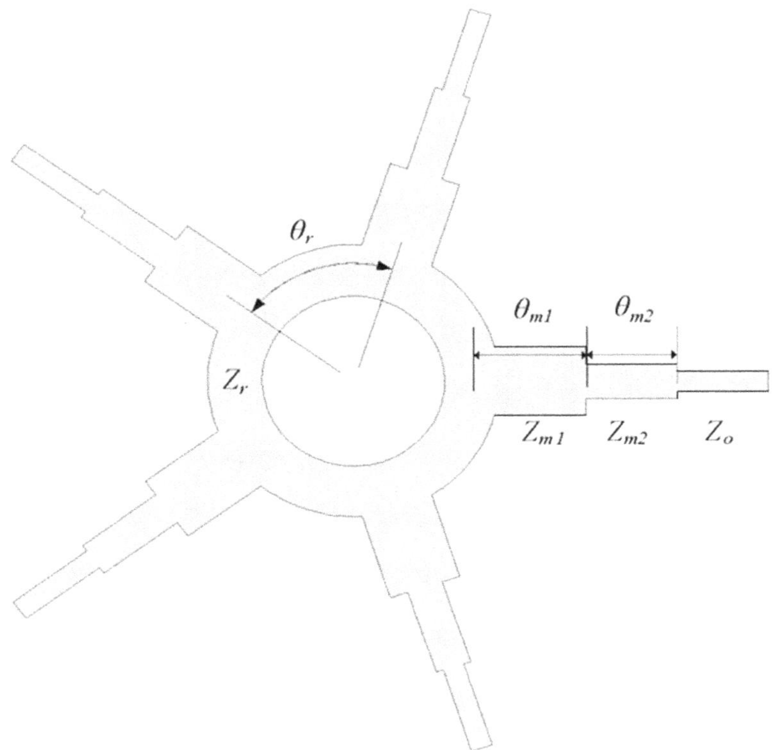

into taper shape as shown in Figure 18 (b). The summary in Table 1 showed that the bandwidth five port circuit's operating frequency increase as well as the structure size due to the increase in lumped elements or section lines. The section line could be represented in lumped element, hence more matching section lines in the ring circuit require more lumped elements to represent the ring circuit.

Recently, the five-port ring-based reflectometer is used together with sensory devices in agricultural and medical applications, such as moisture detection in oil palm fruits (Lee et al., 2011). Lee et al. (2016, 2018) was implemented the five-port ring reflectometer for *in vitro* human breast tumor detection and the measurement set-up is shown in Figure 19. On the other hand, Hossain et al. (2019) was designed a broadband and compact ring circuit for microwave brain imaging applications.

Hybrid Coupler-Based MCRM Circuits

Engen, (1977) had proposed that the complex reflection coefficient measurement can be achieved using six-port hybrid coupler-based network circuits. Until now, this kind of network circuit is still widely studied by some researchers (Ciminelli et al., 2000; Mohra, 2004; Ji Jun and Swee Ping, 2008; Li et al., 2010; Fang et al., 2011). This structure consists of a 6-dB directional coupler, three 3-dB 90° hybrid coupler (Q), and a 3-dB 180° (H) hybrid coupler as shown in Figure 20. However, this architecture required more microwave circuit elements to realize which is bigger in size and 6-dB directional coupler is not commonly available.

Figure 16. Multi-section feed-line star-shaped ring circuit (De Ronde, 1982).

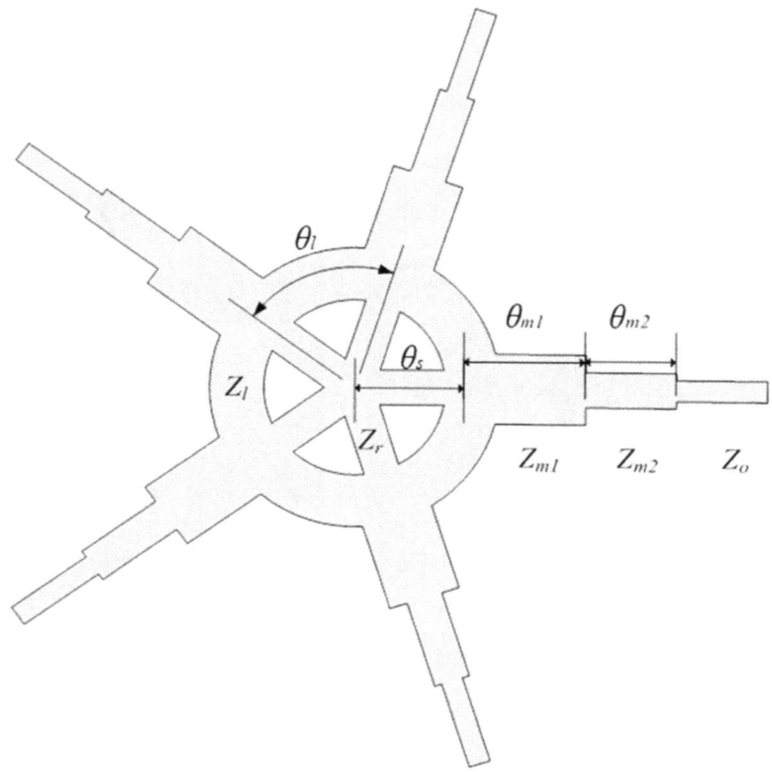

Al-Areqi and You (2018) was miniaturized MCRM part of Figure 20 as well as improved its bandwidth. The miniaturized MCRM circuit is designed using a three-section branch-line coupler and 3 dB-Wilkinson power divider with enhanced series-diamond stubs and meandering lines as shown in Figure 21.

For this type of MCRM circuit, q_i of port-i (i = 4, 5, and 6) can be determined by Equation (20) employing the obtained measured S_{21} and S_{i1} (Al-Areqi and You, 2018).

$$q_i = -\frac{S_{i1}}{S_{i2}S_{21}} \tag{20}$$

The S_{i1}, S_{i2}, and S_{21} can be tested using vector network analyzer. Ideally, based on (20), the magnitude values of q_4, q_5, and q_6 should be within 1 to 2 (in the neighbourhood 1.5). Figure 22 shows the magnitudes and phase shifts of q_i (i = 4, 5, and 6) over the frequency range 1.6 GHz to 3.3 GHz.

Switched-Reflector Circuits

The switch-reflector MCRM circuit was first proposed by Morawski and Sypniewski (1987) that reduce the use of four diode detectors in regular MCRM into two diode detectors. However, this type of implementation requires additional switch element into the circuits. In general, this circuit consists of two

Figure 17. Multi-section feed-line double star-shaped ring circuit (Ding et al., 1984).

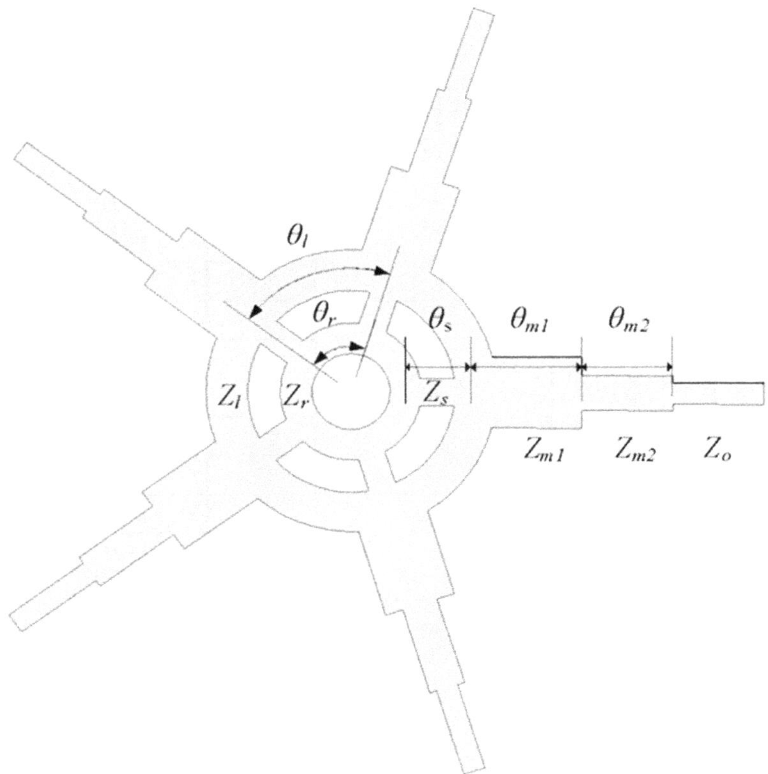

directional couplers, two diode detectors, a signal generator, and a reflection phase modulator (RPM) as shown in Figure 23. The reflection phase modulator (RPM), consisting of three switchable circuits with reflection coefficients Γ. For planar switch-reflector circuit, the three switchable circuits are an open-ended microstrip lines with lengths differing by 60°. The forward signal from signal generator is coupled

Figure 18. (a) Five-port star-taper ring circuit (Tsang et al., 1991). (b) Five-port star-taper double circuit with 76% of operating bandwidth (Yeo and Choong, 2001).

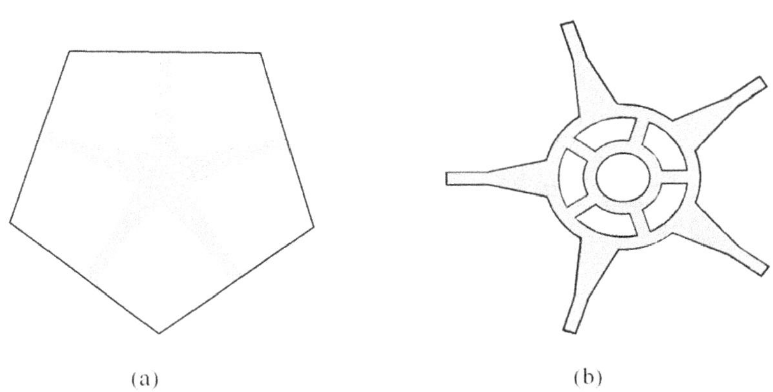

Table 1. Ring circuit's bandwidth, number of lumped-element, and optimum design parameters.

Type of ring circuit	Bandwidth (%)	Number of lumped-element	Impedance (Ω)		Electrical length (°)	
Regular	10	10	Z_r	45.45	θ_r	75.96
			Z_o	50.00		
Multi-section feedline. (Riblet and Hansson, 1981)	35	20	Z_r	16.67	θ_r	75.71
			Z_{m1}	32.03	θ_{m1}	90.00
			Z_{m2}	57.34	θ_{m2}	90.00
			Z_o	50.00		
Multi-section feedline star-shaped ring. (De Ronde, 1982)	57	25	Z_s	40.03	θ_s	71.14
			Z_l	12.50	θ_l	89.42
			Z_{m1}	16.43	θ_{m1}	83.48
			Z_{m2}	38.97	θ_{m2}	90.00
			Z_o	50.00		
Multi-section feedline star-shaped double ring (Ding et al., 1984)	64	30	Z_r	14.29	θ_r	68.98
			Z_s	13.27	θ_s	41.72
			Z_l	111.11	θ_l	121.43
			Z_{m1}	22.13	θ_{m1}	66.60
			Z_{m2}	40.19	θ_{m2}	66.60
			Z_o	50.00		

Figure 19. Five-port ring reflectrometer-sensor system (Lee et al., 2018).

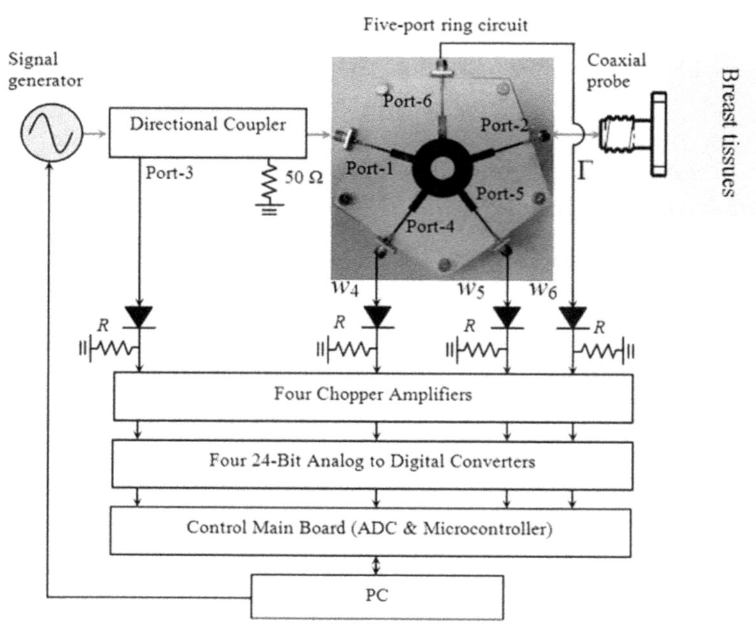

Figure 20. Hybrid coupler-based MCRM circuits.

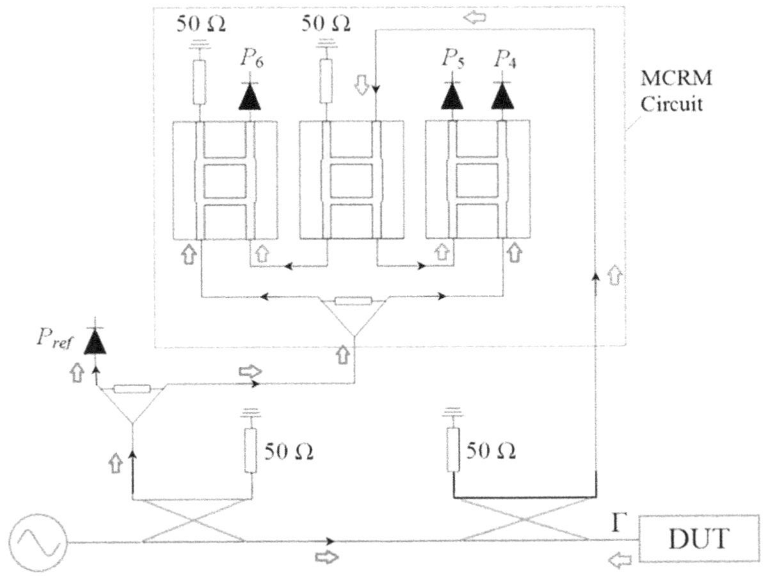

at the first directional coupler down to the reference detector, P_3. The second detector, P_4 detects the combination of a wave reflected from DUT and the wave reflected from RPM. The phase shift is depending on the state of the switch. The disadvantage of this circuit is that it has narrow operation bandwidth.

Dual-Generator Circuits

The dual generators structure circuit required phase shift difference between two signal generators excluding 0° and 180° phase shift difference. The implementation of this circuit requires two signal generators that are synchronized and phase shifted that fed into the resistive elements as shown in Figure 24. Raffaj and Bilik (1993) have practically implemented and experimentally verified using two Direct Digital Synthesizer (DDS) generators in the frequency range from DC to 10 MHz. The drawback of this circuit is limited for low frequency operation.

Wheatstone Bridge-Based Circuits

The first Wheatstone bridge-based circuit was implemented by Raffaj *et al.* (1988) that operated in the frequency range from 100 kHz to 100 MHz (Bilik et al., 1990; 1991). The basic schematic of the Wheatstone bridge-based circuit is shown in Figure 25. The circuit consists of power divider circuit and combination circuit. Later, this kind of circuit has been modified and implemented by several researchers (Raffaj et al., 1988; Bilik et al., 1990; Bilik et al., 1991; Sebechlebsky et al., 1991; Bezek et al., 1991; Bilik & Bezek, 1995; Wiedmann et al., 1997; Hesselbarth et al., 1997).

Figure 21. (a) Compact MRCM circuit schematic. (b) Fabricated circuit (Al-Areqi and You 2018).

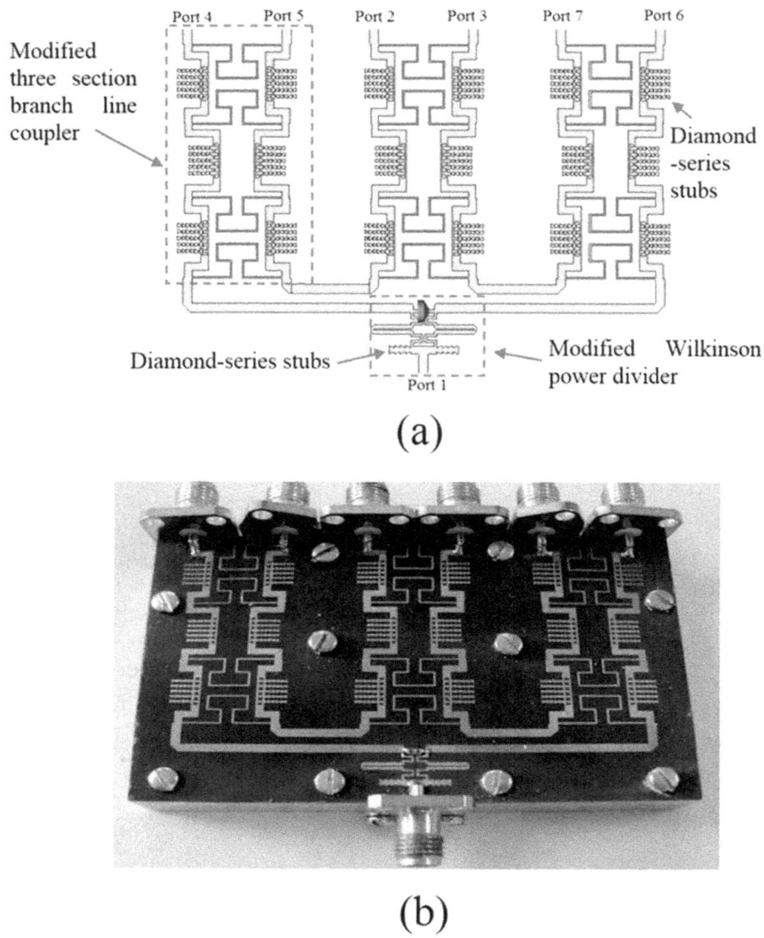

MICROWAVE COMPLEX RATIO MEASURING (MCRM) CIRCUIT CALIBRATIONS

Normally, the ring-shaped and hybrid coupler-based MCRM circuits have a same calibration procedure. There are two parts of calibration process, which are diode power detectors calibration and MCRM circuit

Figure 22. The magnitude and phase shift of q_i points (Al-Areqi and You 2018).

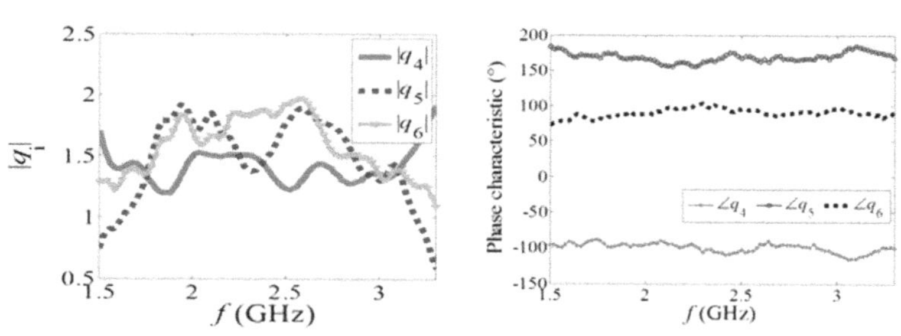

Figure 23. Switch-reflector MCRM circuit.

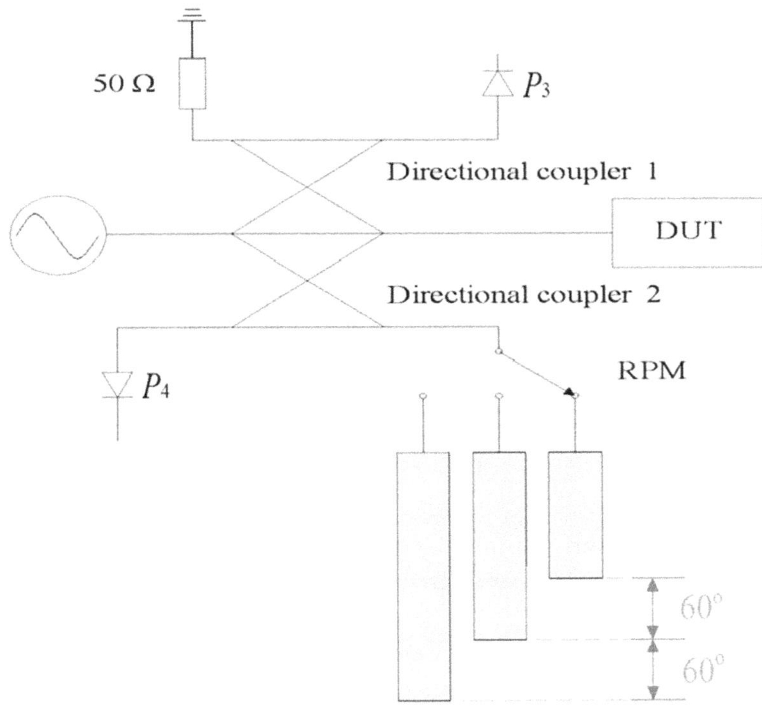

calibration. Both of the calibration procedures are separated to five sub-processes as:

Figure 24. Dual generators structure circuit.

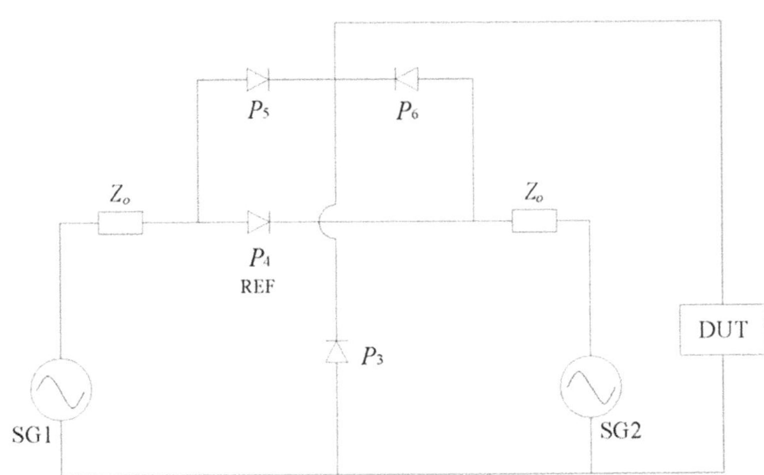

Figure 25. Wheatstone bridge-based circuit.

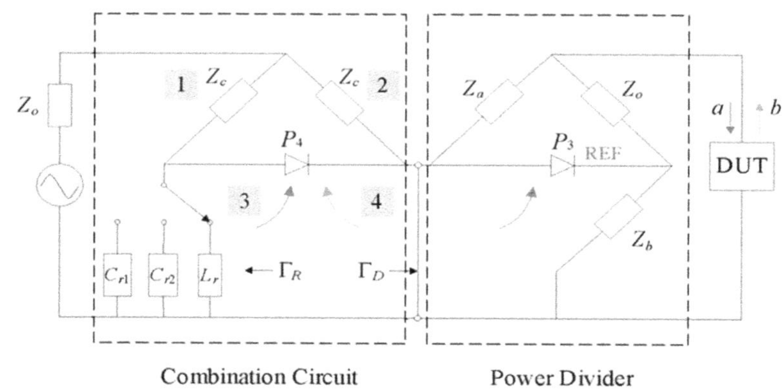

Step 1

The calibration of the diode detectors was initially done by setting the value of resistor, R in four chopper amplifiers part as shown in Figure 26, in order to minimize the deviation from square law characteristic of the diode detectors.

In fact, the output voltage, V_{out} of the diode detector deviates more from the square law function as higher power signal input, P_{in} to the diode detector as shown in Figure 27.

As the values of the load resistance reduced, the diode output voltage, V_{out} also becomes lower, and the curve becomes nearer to the square law region. However, as the resistance value was further reduced, subsequently the curve began to bow upward and the deviation started in the opposite direction. Trial and error method was applied to find the optimum load resistance value for diode detector. A mathematical model for an ideal square-law detector can be written as:

$$P_{in} = K \times V_{out} \tag{21}$$

where K is a constant value. For practical square-law detector, the (21) can be modified as (Hoer et al., 1976):

Figure 26. Diode detector with load resistor R.

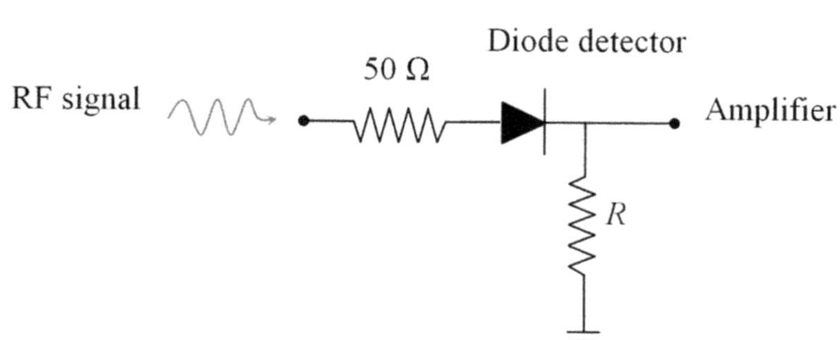

Figure 27. The deviation between ideal and actual diode output voltage (Hoer et al., 1976).

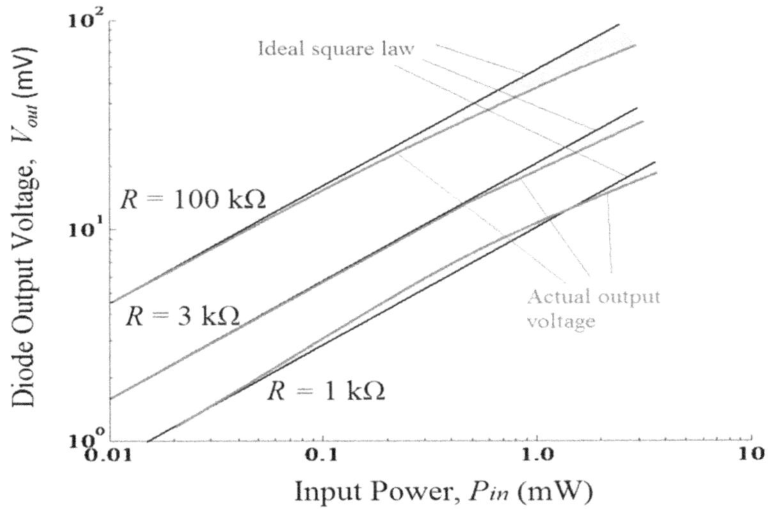

$$P_{in} = K \times \left\{ \gamma \left(V_{out} - V_o \right) \right\} e^{\beta} \tag{22}$$

where γ is a scale value chosen so that maximum value of $\gamma(V_{out} - V_o) \approx 0.5$ in order to avoid ill-condition when $\gamma(V_{out} - V_o) = 1$, as well as loss the precision of measured $\gamma(V_{out} - V_o)$ when $\gamma(V_{out} - V_o) \approx 0$ (refer to Figure 27). Symbol β is a nth order polynomial function which is given as:

$$\kappa = 1 + \xi_1 \left(V_{out} - V_o \right) + \xi_2 \left(V_{out} - V_o \right)^2 + \cdots + \xi_n \left(V_{out} - V_o \right)^n \tag{23}$$

On the other hand, when $P_{in} = 0$ mW, the value of $V_o = V_{out}$.

Step 2

An open-circuit kit is connected to the measurement port (port connected to DUT). The maximum supply power from the signal generator was initially set to +10 dBm and then reduced by 1 dBm of interval until to -20 dBm. The detected output voltage, V_{out} diode values from the four diode detectors were stored for each corresponding supply power level.

Step 3

The **Step 2** is repeated using short-circuit, offset short-circuit and 50 Ω match load kits, respectively. Once all detected output voltage, V_{out} values for those kits are obtained; the non-linear properties of the four diode detectors were calibrated. At the same time, the corresponding detected powers, P_{in} are calculated from the values of V_{out} using (22).

Step 4

Transform the calibrated detected powers (P_{ref}, P_4, P_5, and P_6) from equation (4) at port-3, port-4, port-5 and port-6, respectively into reflection coefficients, w using (24) (Engen,1978):

$$|w|^2 = P_4/P_{ref} \tag{24a}$$

$$|w - q_5|^2 = a_1\left(P_5/P_{ref}\right) \tag{24b}$$

$$|w - q_6|^2 = a_2\left(P_6/P_{ref}\right) \tag{24c}$$

where q_5, q_6, a_1, and a_2 are the unknown constants. The q-circle in Figure 3 can be transformed into w-plane interpretation as shown in Figure 28. The relationship between equations (24a), (24b), and (24c) is represented by w-plane configuration (see Figure 28). From Figure 18, the w can be explicitly expressed as:

$$w = \frac{P_4 - a_1 P_5 + q_5^2}{2q_5} + j\left\{\frac{P_4 - a_2 P_6 + u_6^2 + v_6^2}{2v_6} - \left(\frac{P_4 - a_1 P_5 + q_5^2}{2v_6 q_5}\right)u_6\right\} \tag{25}$$

where the q_5 and q_6 are given as:

$$q_5 = \sqrt{a_5} \tag{26a}$$

$$q_6 = u_6 + jv_6 \tag{26b}$$

The real and imaginary parts of q_6 are written as:

$$u_6 = \frac{a_4 + a_5 - a_3}{2\sqrt{a_5}} \tag{27a}$$

$$v_6 = \pm\sqrt{a_4 - u_6^2} \tag{27b}$$

The unknown values of a_1, a_2, a_3, a_4, and a_5 in (25) to (27) will be predicted in **Step 5** and **Step 6**.

Figure 28. w-plane.

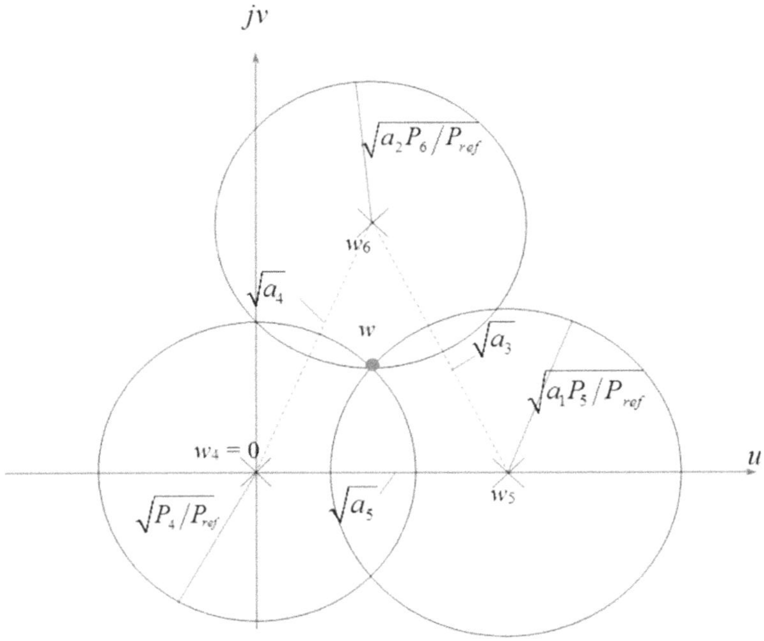

Step 5

The sliding short kit was connected at the measurement port (port-2) and the position of the sliding short was changed sequentially as shown in Figure 29.

Up to 16 different distance positions of the offset short are measured. The power values of the diode detectors (P_3, P_4, P_5, and P_6) were recorded at each position in order to determine the minimum power, P_{min} and the maximum power, P_{max} values at port-4, -5 and -6 using optimization technique (Wiedmann et al., 1999). Then, the initial constant values (a_1 and a_2) were estimated as (Neumeyer, 1990; Stumper, 1990; Wiedmann et al., 1999):

$$a_1 = \left(\frac{\sqrt{P_{4\ max}} \pm \sqrt{P_{4\ min}}}{\sqrt{P_{6\ max}} \pm \sqrt{P_{6\ min}}} \right)^2 \tag{28a}$$

Figure 29. Sliding short kit.

$$a_2 = \left(\frac{\sqrt{P_{4\,max}} \pm \sqrt{P_{4\,min}}}{\sqrt{P_{5\,max}} \pm \sqrt{P_{5\,min}}} \right)^2 \quad (28b)$$

In this step, the performance of circuit can be tested by pairing two power measurements P_i and P_j $(i \neq j)$ at the 16 different distance positions of the offset short. For instance, the P_4 is plotted versus corresponding P_6 for each offset short distance position as shown in Figure 30. Ideally, a broad ellipse relationship between P_4 and P_6 at certain operating frequency can be formed as shown in Figure 30 (a). Furthermore, maximum and minimum values for P_4 and P_6 can be clearly defined. If the flat ellipse form is obtained as shown in Figure 30 (b), this means that the MCRM circuit has higher sensitivity to power measurement error in which we undesirable. Similar graphs can also be plotted for pair P_4 and P_5.

Step 6

The obtained initial constant a_1 and a_2 values were optimized to increase the accuracy of the SPR calibration using the optimization mathematical model by Potter and Hjipieris, (1993). From the values of a_1, a_2, P_{4max}, P_{4min}, P_{5max}, P_{5min}, P_{6max}, and P_{6min}, other unknown values can be found (Wiedmann et al., 1999):

$$a_3 = \left[\frac{\left(a_2 P_{6\,max} - a_1 P_{5\,min}\right) - \left(a_2 P_{6\,min} - a_1 P_{5\,max}\right)}{2\left(\sqrt{P_{4\,max}} \pm \sqrt{P_{4\,min}}\right)} \right]^2 \quad (29a)$$

Figure 30. Pair of P_4 and P_6 at different distance positions of the offset short:(a) broad ellipse case and (b) flat ellipse case.

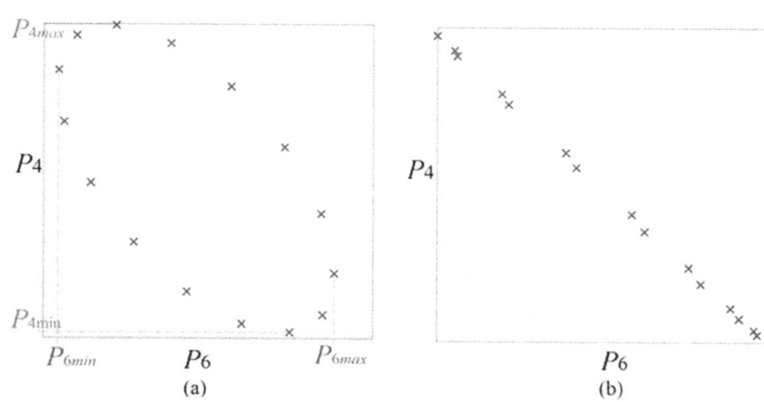

$$a_4 = \left[\frac{\left(P_{4\,max} - a_2 P_{6\,min}\right) - \left(P_{4\,min} - a_2 P_{6\,max}\right)}{2\left(\sqrt{P_{4\,max}} \pm \sqrt{P_{4\,min}}\right)} \right]^2 \tag{29b}$$

$$a_5 = \left[\frac{\left(a_1 P_{5\,max} - P_{4\,min}\right) - \left(a_1 P_{5\,min} - P_{4\,max}\right)}{2\left(\sqrt{P_{4\,max}} \pm \sqrt{P_{4\,min}}\right)} \right]^2 \tag{29c}$$

Step 7

Once the values of w were obtained from Equation (25), the further calibration process using three standard kits (open-circuit, short-circuit and match load) was done to translate reference plane from w-plane back to measurement Γ-plane at port-2 (Engen, 1978):

$$\Gamma = \frac{w - c_1}{c_2 w + c_3} \tag{30}$$

where c_1, c_2, and c_3 are the unknown complex constants. Let w_{open}, w_{short}, and w_{load} represent the measured reflection coefficients for the open, short, and match-load standards. The w_{open} is measured at certain frequency range by terminating the open end of the port-2 with an open-circuit terminator. Later, the measurement is repeated to obtain the w_{short} by terminating the port-2 (measurement-port) with a short terminator. Finally, the w_{load} is obtained by replacing short terminator with 50 Ω match-load terminator.

The three calibration terminators are shown in Figure 31. The c_1, c_2, and c_3 are expressed as:

$$c_1 = w_{load} \tag{31a}$$

$$c_2 = \frac{w_{short} + w_{open} - 2w_{load}}{w_{short} - w_{open}} \tag{31b}$$

$$c_3 = \frac{w_{open} w_{load} + w_{short} w_{load} - 2w_{open} w_{short}}{w_{short} - w_{open}} \tag{31c}$$

After all the values of c_1, c_2, and c_3 are determined, the reflection coefficient, Γ at measurement-port (port-2) can be found from the measured w using Equation (30).

Figure 31. Three calibration kits (Open, short and 50 Ω load).

Ideally, the sensitivity ratio in Equation (32) should be equal to unity. However, the higher the sensitivity ratio (>1), the higher the measurement resolution (good reflection measurement sensitivity) (Bannister et al., 1992).

$$Sensitivity\ Ratio = \frac{\left|c_3\right|}{\left|c_3\right| + \left|c_1\right|} \tag{32}$$

CONCLUSION AND SUMMARY

In this chapter, various types of MRCM circuits have been described in detail, including previous studies, basic theory, and general calibration methods for MRCM circuits. This chapter attempts to provide concise and comprehensive information to researchers who are involved in the construction of MRCM circuit.

MRCM is usually used to construct RF reflectometer. The advantages of the MRCM circuit are simple, inexpensive, not easy to damage, do not need to use many electronic components in the construction of the reflectometer. In addition, MRCM allows the signal through it steadily and insensitive to operating temperature. Normally, the temperature of MRCM circuit does not increase significantly when the circuit is operated for a long period.

In fact, two MRCM circuits can build a simple vector network analyzer. A brief summary about the various MCRM circuits are tabulated in Table 2.

REFERENCES

Al-Areqi, N. N., & You, K. Y. (2018). Broadband and compact complex ratio measuring unit. *Microwave and Optical Technology Letters*, *60*(12), 3039–3045. doi:10.1002/mop.31445

Table 2. The advantages and disadvantages of the different MCRM circuits.

MCRM circuits	Advantages	Disadvantages
Hybrid coupler-based	Provides stable and precise phase measurement.	Required more microwave circuit elements.
Switched-reflector	Need only one or two diode detectors.	Requires additional switch element.
Dual-generator	Simple structure.	Suitable for lower frequencies. Need two signal generators.
Multi-probe transmission line	Simple structure and easy to fabricate.	Narrow operating frequency.
Wheatstone bridge-based	Able to achieve ultra wideband performance.	Low reproducibility rate.
Five-port ring	High reproducibility rate. Relative wide bandwidth. Low fabrication cost.	Narrow operating frequency.

Bannister, D. J., Ide, J. P., & Perkins, M. (1992). Improved six-port reflectometer based on symmetrical five-port junction. *Electronics Letters, 28*(4), 406–408. doi:10.1049/el:19920255

Belfort, A. J., & Cullen, A. L. (1982). First-order theory of the five-port symmetrical star junction. *Electronics Letters, 18*(19), 841–842. doi:10.1049/el:19820572

Bergeault, E., Huyart, B., Geneves, G., & Jallet, L. (1991). Characterization of diode detectors used in six-port reflectometers. *IEEE Transactions on Instrumentation and Measurement, 40*(6), 1041–1043. doi:10.1109/19.119790

Bezek, J., Raffaj, V., & Bilik, V. (1991). Application of the lumped six-port reflectometer to water level measurement. *Proc. 9th Microwave Conf, MIKON-91*, Poland, 374-378.

Bilik, V. (1992a). Calibrating the six-port reflectometer using a matched load and four unity-reflection standards. Part 1: Perfectly matched load. *Wuxiandian Gongcheng, 1*, 10–13.

Bilik, V. (1992b). Calibrating the six-port reflectometer using a matched load and four unity-reflection standards. Part 2: Sliding termination. *Wuxiandian Gongcheng, 1*, 2–8.

Bilík, V. (2002). Six-port measurement technique: principles, impact, applications. Radioelektronika-2002 Conference, Slovakia. 123-155.

Bilik, V., & Bezek, J. (1995). Thin-film lumped six-port vector reflectometer for 10-1900MHz. *Proc. MITEKO 1995*, Pardubice, Czech Republic. 181-184.

Bilik, V., Raffaj, V., & Bezek, J. (1991). Miniature broadband lumped six-port reflectometer. *Journal of Communication, 42*, 7–14.

Bilik, V., Raffaj, V., & Bezek, J. A. (1990). New extremely wideband lumped six-port reflectometer. *20th European Microwave Conference*. 1473-1478.

Caldecott, R. (1973). The generalized multiprobe reflectometer and its application to automated transmission line measurements. *IEEE Transactions on Antennas and Propagation, 21*(4), 550–554. doi:10.1109/TAP.1973.1140527

Chang, K., Li, M. Y., & Sauter, T. H. (1990). Low-cost microwave/millimeter-wave impedance measuring scheme using a three-probe microstrip circuit. *IEEE Transactions on Microwave Theory and Techniques*, *38*(10), 1455–1460. doi:10.1109/22.58685

Chen, Z., & Xu, B. (1987). Linearization of Diode Detector Characteristics. *IEEE MTT-S International Microwave Symposium Digest*. 265-267. 10.1109/MWSYM.1987.1132380

Ciminelli, C., D'Orazio, A., De Sario, M., Petruzzelli, V., & Prudenzano, F. (2000). A new structure for a six port reflectometer by using a five port ring. *10th Mediterranean Electrotechnical Conference, MELECON 2000*. 181-184.

Collier, R. J., & El-Deeb, N. A. (1979). On the use of a microstrip three-line system as a six-port reflectometer. *IEEE Transactions on Microwave Theory and Techniques*, *27*(10), 847–853. doi:10.1109/TMTT.1979.1129747

Cripps, S. C. (2007). Microwave bytes - VNA tales. *IEEE Microwave Magazine*, *8*(5), 28–44. doi:10.1109/MMM.2007.904719

Cullen, A. L., & Yeo, S. P. (1986). Six-port reflectometer theory for symmetrical five-port junction with directional coupler. *IEE Proceedings H-Microwaves, Antennas and Propagation*. *133*(4), 277-287.

De Ronde, F. C. (1982). Octave-wide matched symmetrical, reciprocal, 4- and 5 ports. *IEEE MTT-S International Microwave Symposium Digest*. 521-523. 10.1109/MWSYM.1982.1130778

Dong, I. K., Araki, K., & Naito, Y. (1984). Properties of the symmetrical circuit and its broad-band five-port design. *IEEE Transactions on Microwave Theory and Techniques*, *32*(1), 51–57. doi:10.1109/TMTT.1984.1132611

El-Deeb, N. A. (1983). The calibration and performance of a microstrip six-port reflectometer. *IEEE Transactions on Microwave Theory and Techniques*, *31*(7), 509–514. doi:10.1109/TMTT.1983.1131538

Engen, G. F. (1977a). An improved circuit for implementing the six-port technique of microwave measurements. *IEEE MTT-S International Microwave Symposium Digest*. 53-55. 10.1109/MWSYM.1977.1124355

Engen, G. F. (1977b). The six-port reflectometer: Alternative network analyzer. *IEEE Transactions on Microwave Theory and Techniques*, *25*(12), 1075–1080. doi:10.1109/TMTT.1977.1129277

Engen, G. F. (1978). Calibrating the six-port reflectometer by means of sliding terminations. *IEEE Transactions on Microwave Theory and Techniques*, *26*(12), 951–957. doi:10.1109/TMTT.1978.1129527

Engen, G. F., & Hoer, C. A. (1972). Application of an arbitrary 6-port junction to power-measurement problems. *IEEE Transactions on Instrumentation and Measurement*, *21*(4), 470–474. doi:10.1109/TIM.1972.4314069

Fang, X. T., Zhang, X. C., & Tong, C. M. (2011). A novel miniaturized microstrip six-port junction. *Progress In Electromagnetics Research Letters*, *23*, 129–135. doi:10.2528/PIERL11032302

Fitzpatrick, J. (1978). Error models for systems measurement. *Microwave Journal*, *21*(5), 63–66.

Ghannouchi, F. M., & Mohammadi, A. (2009). *The six-port technique with microwave and wireless applications*. Norwood, MA: Artech House.

Hasan, A., Helaoui, M., & Ghannouchi, F. M. (2012). Dynamic linearization of diodes for high speed and peak power detection applications. *IEEE MTT-S International Microwave Symposium Digest.* 1-3. 10.1109/MWSYM.2012.6259710

Hesselbarth, J., Wiedmann, F., & Huyart, B. (1997). Two new six-port reflectometers covering very large bandwidths. *IEEE Transactions on Instrumentation and Measurement*, *46*(4), 966–969. doi:10.1109/19.650809

Hoer, C. A. (1972). The six-port coupler: A new approach to measuring voltage, current, power, impedance, and phase. *IEEE Transactions on Instrumentation and Measurement*, *21*(4), 466–470. doi:10.1109/TIM.1972.4314068

Hoer, C. A. (1979). Calibrating a six-port reflectometer with four impedance standards. United States: National Bureau of Standards (NBS) Technical Note 1012.

Hoer, C. A., Roe, K. C., & Allred, C. M. (1976). Measuring and minimizing diode detector nonlinearity. *IEEE Transactions on Instrumentation and Measurement*, *25*(4), 324–329. doi:10.1109/TIM.1976.6312236

Hossain, T. M., Jamlos, M. F., Jamlos, M. A., Soh, P. J., Ibrahim, S. Z., Isa, M. M., ... Ghazali, M. F. (2019). Compact broadband triple-ring five-port reflectometer for microwave brain imaging applications. *IEEE Access: Practical Innovations, Open Solutions*, *7*. doi:10.1109/ACCESS.2019.2891058

Hu, C. J. (1980). A novel approach to the design of multiple-probe high-power microwave automatic impedance measuring schemes. *IEEE Transactions on Microwave Theory and Techniques*, *28*(12), 1422–1428. doi:10.1109/TMTT.1980.1130260

Hu, C. J. (1983). Microwave automatic impedance measuring schemes using three fixed probes. *IEEE Transactions on Microwave Theory and Techniques*, *31*(9), 756–762. doi:10.1109/TMTT.1983.1131586

Hunter, J. D., & Somlo, P. I. (1985). An explicit six-port calibration method using five standards (short papers). *IEEE Transactions on Microwave Theory and Techniques*, *33*(1), 69–72. doi:10.1109/TMTT.1985.1132950

Judah, S. K. (1985). Calibration of multiport reflectometers. *IEE Proceedings H-Microwaves, Antennas and Propagation. 132*(7), 468-470.

Koelpin, A., Vinci, G., Laemmle, B., Kissinger, D., & Weigel, R. (2010). The six-port in modern society. *IEEE Microwave Magazine*, *11*(7), 35–43. doi:10.1109/MMM.2010.938584

Lee, C. Y., You, K. Y., Abbas, Z., Lee, K. Y., Lee, Y. S., & Cheng, E. M. (2018). S-band five-port ring reflectometer-probe system for *in vitro* breast tumor detection. *International Journal of RF and Microwave Computer-Aided Engineering*, *28*(3), 1–13. doi:10.1002/mmce.21198

Lee, C. Y., You, K. Y., Tan, T. S., Then, Y. L., Lee, Zahid, L., Lim, W. L., & Lee, C. H. (2016). Enhanced five-port rin circuit reflectometer for synthetic breast tissue dielectric determination. *Progress In Electromagnetic Research C. 69,* 83-95.

Lee, K. Y., Abbas, Z., Jusoh, M. A., You, K. Y., & Cheng, E. M. (2011). Determination of moisture content in oil palm fruits using a five-port reflectometer. *Sensors (Basel)*, (11), 4073–4085. PMID:22163837

Li, C. J., Zhang, H. Q., & Wang, P. S. (2010). A novel six-port circuit based on four quadrature hybrids. *International Journal of RF and Microwave Computer-Aided Engineering*, *20*(1), 128–132.

Li, S., & Bosisio, R. G. (1982). Calibration of multiport reflectometers by means of four open/short circuits. *IEEE Transactions on Microwave Theory and Techniques*, *30*(7), 1085–1090. doi:10.1109/TMTT.1982.1131200

Lu, H. C., & Chu, T. H. (1999). Microwave diversity imaging using six-port reflectometer. *IEEE Transactions on Microwave Theory and Techniques*, *47*(1), 84–87. doi:10.1109/22.740082

Madonna, G., Ferrero, A., & Pirola, M. (1999). Design of a broadband multiprobe reflectometer. *IEEE Transactions on Instrumentation and Measurement*, *48*(2), 622–625. doi:10.1109/19.769672

Martin, E., Margineda, J., & Zamarro, J. M. (1982). An automatic network analyzer using a slotted line reflectometer. *IEEE Transactions on Microwave Theory and Techniques*, *30*(5), 667–670. doi:10.1109/TMTT.1982.1131119

Mohra, A. S. (2004). Six-port reflectometer based on four 0°/180° microstrip ring couplers. *Microwave and Optical Technology Letters*, *40*(2), 167–170. doi:10.1002/mop.11318

Morawski, T., & Sypniewski, M. (1987). AM/PM switched three port reflectometer. *Proc. 2nd Intern. Symp. on Measurement of Electrical Quantities IMEKO 87*. 33-40.

Neumeyer, B. (1990). A new analytical method for complete six-port reflectometer calibration. *IEEE Transactions on Instrumentation and Measurement*, *39*(2), 376–379. doi:10.1109/19.52518

Potter, C., & Bullock, A. (1993). Nonlinearity correction of microwave diode detectors using a repeatable attenuation step. *Microwave Journal*, *36*, 272, 274, 277–279.

Potter, C. M., & Hjipieris, G. (1993). A robust six- to four-port reduction algorithm. *IEEE MTT-S International Microwave Symposium Digest*. 1263-1266. 10.1109/MWSYM.1993.277103

Qiao, L., & Yeo, S. P. (1995). Improved implementation of four-standard procedure for calibrating six-port reflectometers. *IEEE Transactions on Instrumentation and Measurement*, *44*(3), 632–636. doi:10.1109/19.387297

Raffaj, V., Bezek, J., Bilik, V., & Beles, M. (1988). Experiments with lumped 6-port reflectometer in the 100 kHz - 100 MHz range. *Proc. Elektrotechnika*, *88*, 101–103.

Raffaj, V., & Bilik, V. (1993). All-resistive six-port reflectometer with two-phase feeding. *Proc.MIOP-93*. 60-64.

Riblet, G. P., & Hansson, E. R. B. (1981). The use of a matched symmetrical five-port junction to make six-port measurements. *IEEE MTT-S International Microwave Symposium Digest*. 151-153. 10.1109/MWSYM.1981.1129852

Riblet, G. P., & Hansson, E. R. B. (1982). Aspects of the calibration of a single six-port using a load and offset reflection standards. *IEEE Transactions on Microwave Theory and Techniques*, *30*(12), 2120–2125. doi:10.1109/TMTT.1982.1131394

Schneider, M. V. (1969). Microstrip lines for microwave integrated circuits. *The Bell System Technical Journal*, *48*(5), 1421–1444. doi:10.1002/j.1538-7305.1969.tb04274.x

Sebechlebsky, S., Raffaj, V., Bilik, V. (1991). Synthesis of reflectors for wide-band lumped six-port reflectometers. *Elektrotechnicky casopis. 42,* 449-470.

Shihe, L., & Bosisio, R. G. (1982). Calibration of multiport reflectometers by means of four open/short circuits. *IEEE Transactions on Microwave Theory and Techniques*, *30*(7), 1085–1090. doi:10.1109/TMTT.1982.1131200

Somlo, P. I., & Hunter, J. D. (1982). A six-port reflectometer and its complete characterization by convenient calibration procedures. *IEEE Transactions on Microwave Theory and Techniques*, *30*(2), 186–192. doi:10.1109/TMTT.1982.1131045

Staszek, K., Gruszczynski, S., & Wincza, K. (2013). Theoretical limits and accuracy improvement of reflection-coefficient measurements in six-port reflectometers. *IEEE Transactions on Microwave Theory and Techniques*, *61*(8), 2966–2974. doi:10.1109/TMTT.2013.2269053

Staszek, K., Gruszczynski, S., & Wincza, K. (2015). Measurement accuracy enhancement in six-port reflectometers. *IEEE Microwave and Wireless Components Letters*, *25*(8), 553–555. doi:10.1109/LMWC.2015.2440784

Stumper, U. (1990). Finding initial estimates needed for the Engen method of calibrating single six-port reflectometers. *IEEE Transactions on Microwave Theory and Techniques*, *38*(7), 946–949. doi:10.1109/22.55790

Susman, L. (1984). Calibration of a six-port reflectometer using projective geometry concepts. *Electronics Letters*, *20*(1), 9. doi:10.1049/el:19840007

Szymanski, P., & Wiatr, W. (2004). A broadband multistate interferometer for impedance measurement. *15th International Conference on Microwaves, Radar and Wireless Communications, MIKON-2004*,755-758. 10.1109/MIKON.2004.1357150

Tsang, K. F., Lee, K. Y., Chan, S. P., & Chan, W. S. (1991). Design of broadband five-port circuits. *China., 1991 International Conference on Circuits and Systems*. 980-983. 10.1109/CICCAS.1991.184529

Ülker, S., & Weikle, R. M. (2001). A millimeter-wave six-port reflectometer based on the sampled-transmission line architecture. *IEEE Microwave and Wireless Components Letters*, *11*(8), 340–342. doi:10.1109/7260.941783

Wiedmann, F., Huyart, B., Bergeault, E., & Jallet, L. (1997). New structure for a six-port reflectometer in monolithic microwave integrated-circuit technology. *IEEE Transactions on Instrumentation and Measurement*, *46*(2), 527–530. doi:10.1109/19.571902

Wiedmann, F., Huyart, B., Bergeault, E., & Jallet, L. (1999). A new robust method for six-port reflectometer calibration. *IEEE Transactions on Instrumentation and Measurement*, *48*(5), 927–931. doi:10.1109/19.799649

Woods, D. (1979). Analysis and calibration theory of the general 6-port reflectometor employing four amplitude detectors. *Proceedings of the Institution of Electrical Engineers*, *126*(2), 221–228. doi:10.1049/piee.1979.0051

Yao, J. J., & Yeo, S. P. (2008). Six-port reflectometer based on modified hybrid couplers. *IEEE Transactions on Microwave Theory and Techniques*, *56*(2), 493–498. doi:10.1109/TMTT.2007.914626

Yeo, S. P., & Ang, A. L. (1988). Alternative approach to the calibration of six-port reflectometers using four standards. *IEE Proceedings H-Microwaves, Antennas and Propagation*. 135, 353-355.

Yeo, S. P., & Choong, F. C. (2001). Matched symmetrical five-port microstrip coupler. *IEEE Transactions on Microwave Theory and Techniques*, *49*(8), 1498–1500. doi:10.1109/22.939932

Yeo, S. P., & Lee, K. H. (1990). Improvements in design of six-port reflectometer comprising symmetrical five-port waveguide junction and directional coupler. *IEEE Transactions on Instrumentation and Measurement*, *39*(1), 184–188. doi:10.1109/19.50441

Yeo, S. P., Yeo, T. S., & Ng, K. M. (1989). First-order eigenmode analysis of symmetrical five-port microstrip ring coupler. *Microwave and Optical Technology Letters*, *2*(3), 91–94. doi:10.1002/mop.4650020305

Chapter 4
Study and Design of New Rectenna Structures for Wireless Power Transmission Applications

Abdellah Taybi
https://orcid.org/0000-0002-7845-7532
University of Hassan 1st., Morocco

Abdelali Tajmouati
University of Hassan 1st, Morocco

Jamal Zbitou
University of Hassan 1st, Morocco

Mohamed Latrach
Microwave Group ESEO Angers, France

ABSTRACT

This chapter presents many research works that have been carried out to deal with the problem of power supply to remote sensors. A 2.45 GHz voltage multiplier rectifier was validated to deliver 18V of output voltage with a conversion efficiency of 69%. Another rectenna was fabricated at 5.8 GHz of the Industrial Scientific Medical band and reach a measured voltage of 7.4V at 18 dBm. The third structure is about a series rectifier working at 2.45 GHz associated with a microstrip low pass filter which produces a supplying voltage of 11.23V. Added to the aforementioned results, the objective in this work is to design, optimize and realize two structures: A dual band patch antenna working at 2.45 GHz and 5.8 GHz, and a compact rectifier circuit at 2.45 GHz for the power supply of low-consumption devices. This rectifier has been designed using Advanced Design System. The bridge topology was employed on an FR4 substrate. A good matching input impedance was observed and high conversion efficiency was obtained. Simulation results have been validated through realization and measurements.

DOI: 10.4018/978-1-7998-0117-7.ch004

INTRODUCTION

Electronics has undeniably evolved in recent years. Progress, particularly in digital electronics and circuit integration has led to more efficient, miniature and energy-efficient systems. Technological developments, combining advances in computing and digital technologies and their increasing integration into multiple objects, have led to the development of a new paradigm of systems qualified as cyber-physical systems. These systems are now being deployed on a massive scale due to the expansion of Internet of Things (IoT) applications. Cyber-physical systems are based, among other things, on the deployment of massive array of autonomous wireless communicating sensors, these have several advantages:

- Flexibility in the choice of location. They allow access to dangerous or difficult to access areas.
- Postage for cables that are heavy, bulky and expensive.
- Elimination of problems related to cables (wear, sealing, etc.)
- Easy deployment of sensor networks.

However, these wireless sensors require power autonomy to operate. Conventional techniques such as batteries or cells, ensure the operation of the sensors only for a limited time and require periodic change. This presents an obstacle in the case where wireless sensors are placed in a location where access is impossible. It is therefore necessary to find another way to permanently supply energy to these wireless sensor networks.

Wireless energy transmission (Shams & Ali, 2007) by RF wave and ambient RF energy recovery are very promising solutions for the sustainable power supply of portable communicating systems. Although the energy densities used in this case are relatively low compared to other micro-sources, RF energy offers superior performance in terms of mobility, flexibility and ease of deployment, which opens the doors to a very wide range of applications. Research on wireless power transmission began at the end of the 19th century when Hertz and Marconi noticed that energy could be transported from one point to another without the existence of any conductive means. The theoretical foundations of this phenomenon were established by Maxwell through equations established in 1862. At the beginning of the 20th century, Nikola Tesla (Tanuj Kumar Mandal, 2006) was already working on the "Wardenclyffe Tower", the prototype of a transmitter for what he called "World Wireless System", that would be able to provide wireless power to a remote receiver. Advances in wireless communications have greatly contributed to the development of wireless energy transmission techniques. The modern development of wireless transmission systems was largely stimulated by the advances in telecommunications and radar remote sensing that took place during World War II. It is the use of microwaves that relaunches the perspective of wireless energy transmission. In the 1960s, William C. Brown began experimenting with wireless energy transmission using microwave tubes such as magnetrons and klystrons (Brown, W. C., 1984). W. C. Brown was also the first to develop a rectifying antenna or "rectenna" in 1963 for the reception of electromagnetic waves and their conversion into direct voltage (DC). In 1964, as part of the RAMP Project, W. C. Brown's team demonstrated an airborne platform flying at an altitude of 18 meters powered exclusively through a microwave beam from the ground (Brown, W. C., 1965).

Indeed, electromagnetic energy (Paris & Hurd, 1969) is nowadays omnipresent on our planet (Harrist, 2001), so using it as an energy source for electronic systems seems to be a plausible and feasible idea. This work is part of this concept, its objective is the design and manufacture of electromagnetic energy recovery systems for the supply of wireless sensor networks. The electromagnetic energy recovery circuit is called "Rectenna" (Zbitou & Latrach, 2006; Matsunaga, Nishiyama, & Toyoda, 2015). This word is the association of two entities that are "antenna" and "rectifier". The antenna recovers the ambient electromagnetic energy and the rectifier converts it into a continuous signal (DC) that will be used then to power the wireless sensors. In this manuscript, several rectennas will be presented, for frequencies ranging from 2.45 GHz to 5.8GHz.

The proposed chapter is divided on two sections. The first part is devoted to the study of the antenna of the energy recovery system. First, we recall the main characteristics of an antenna and its impact on the recovered radiofrequency power. This will allow us to choose some geometries that we will design and then optimize by simulation under ADS. The operation of these antenna will be evaluated experimentally and commented on.

The second part will deal with the recovery system. It describes the operating principle of RF/DC conversion. It presents a comparison of diode characteristics and possible rectifier topologies. This comparison will allow us to select the most suitable diode and topology for maximum detection sensitivity and RF/DC conversion efficiency. This part describes the adaptation of the rectifier to the antenna and its influence on the recovered power. The choice of the load and its influence on the efficiency of the RF/DC conversion is also discussed in this section.

SECTION I: MICROSTRIP DUAL BAND ANTENNA FOR RECTENNA APPLICATIONS

Wireless Power Transmission

A wireless power transmission system consists of a block that converts continuous energy into microwave energy capable of transmitting in free space through a transmitting antenna. Reception is ensured by a receiving antenna followed by an RF-DC rectification system. Each element of the WPT system can be characterized by its efficiency (Ohira, 2013).

To recover the radiated energy from the ambient environment, it is sufficient to proceed in the same way as for the recovery of a radio signal, by receiving it from one or more receiving antennas, followed by an RF/DC conversion system. A matching circuit between the rectifier and the antenna improves the efficiency of the RF/DC conversion and minimizes reflection losses on the antenna, in addition to the adaptation, this module must play the role of a band pass filter. Indeed, the rectifier, diode for example, is a non-linear element that generates harmonics when the RF signal is rectified, these harmonics move on either side of the diode. It is necessary to prevent these harmonics from reaching the antenna, especially the $2f_0$ and $3f_0$ harmonics, which are generally the most important, as they may be re-radiated by the antenna, resulting in radiation losses. On the other side of the rectifier component, there is a low-pass filter, its role is to filter all the components from the fundamental signal to nf_0, and to keep only the DC component, which represents the DC signal useful for powering the wireless sensor, this sensor is generally modelled by a load (resistance) or the energy is stored in a battery for later use.

In the case of energy transfer by radiation, two scenarios exist: The first scenario is Wireless Power Transfer (WPT). In this case, a dedicated source is used for transmission, the transmitting antenna is generally directional, the rectenna is placed in line of sight opposite the transmission antenna. The experiments given in the introduction, aimed at making airborne platforms fly, are an example of this situation. The advantage of this type of energy transfer is the prior knowledge of the frequency, arrival

Figure 1. Wireless power transmission system

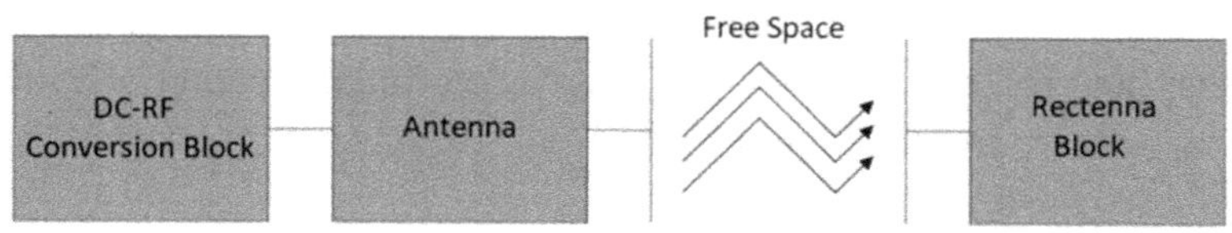

direction of the electromagnetic wave and its polarization. From this information, the location of the rectenna and the received power density can be estimated in advance. On the other hand, this forces the receiving antenna to be placed in specific locations in order to receive a maximum power density. The remote feeding of tags is a similar example to WPT. In this case, the approach will be to design a single-band rectenna, with a directive antenna having the same polarization as the transmitting antenna, in order to maximize efficiency.

The second scenario is the recovery of ambient electromagnetic energy (EH: Energy Harvesting) (Dolgov, Zane, & Popovic, 2010). This scenario is relevant, for example, in urban areas where electromagnetic energy sources are becoming more and more available. In this case, electromagnetic energy can come from any direction. The advantage of the EH is the freedom of positioning of the rectenna. On the other hand, due to the limitation to electromagnetic field exposures imposed by law and the non-directionality of the transmitting antennas, the power densities available for reception are low. In addition, the power density also depends on the environment. This is the biggest challenge for electromagnetic energy recovery systems. If we add the technological limitations of electronic components, it becomes very difficult to achieve very high efficiencies. However, the diversity of the frequencies used has a great advantage. Energy recovery can be performed at several frequencies simultaneously, which would increase efficiency.

So, as a summary, in the case of remote power supply defined above, the distance between the radiant source and the receiver is studied so that the latter recovers the transmitted signal with a sufficient power level to operate the self-powered system. In the case of electromagnetic energy recovery in the surrounding environment, the levels of power are very low, variable from one place to another and spread over a wide band frequency.

Micro Strip Patch Antenna Analysis

An antenna (Lee & Luk, 2010; Ojha & Peters, 2010) is a passive element that transforms electrical power into electromagnetic power and, conversely, transfers the energy of a wave propagating in the surrounding environment to a receiving system. Antennas are essential for wireless communication systems. They are the only element that can ensure the emission or reception of electromagnetic waves. They can take several shapes and sizes depending on the operating frequency and the type of application.

The main functions of the antenna are therefore:

- The transmission or reception of energy in privileged directions.
- The correct matching between the terminal electronic circuits (front-end) and the environment of propagation.
- The most accurate possible transmission of the desired information.

The antenna initially contains a thin metallic conductor (about 17.5 to 35 μm thick) of random shape (the radiating element), which is placed on a dielectric substrate whose underside contains a fully metalized ground plane. In the microwave and millimeter wave range, printed antennas usually use a single layer of substrate and are very popular. In any geometric shape (Figure 2), they are installed on a dielectric substrate and a ground plane. There are several shapes of these radiating elements: square, rectangular, circular, elliptical or more complex. We note that their dimension is in the order of $\lambda/2$ to λ. Because they reliably meet system requirements, printed antennas are in high demand today. Due to their shape and dimensions, they can easily be integrated into mobile devices (aircraft, car, missile, mobile phone, etc.) and they have advantages over conventional antennas (horn and wired dipole).

Among many analysis models for micro strip antennas the most common is the transmission line model (the easiest of all that gives good insight into physical phenomena, but is less accurate and it is more difficult to model the coupling), the cavity model, on the other hand there is electromagnetic modeling which mainly includes solving integral equations by the moment method or the finite element method. The antenna can be considered as a resonant cavity transmission line with two open ends at a length of $\lambda/2$ where the fields at the edges of the patch and the mass are exposed to the upper space and create the radiation. The finished dimensions of the radiating element (length and width) cause the field lines at the edges of the patch. The concentration of the field lines is a function of the patch dimensions and substrate thickness, for the main electric field plane E the field lines are a function of the ratio of the patch length L to the substrate thickness h (L/h), as well as the relative dielectric constant of the substrate. Although for micro strip antennas the field lines are reduced (due to $L/h \gg 1$), its influences on the antenna resonance frequency must be taken into account.

Bandwidth

The bandwidth can be defined by one or more physical criteria. Therefore, it does not have a single definition and it is important to specify the criterion used each time (impedance, polarization or radia-

Figure 2. Different micro strip antenna shapes.

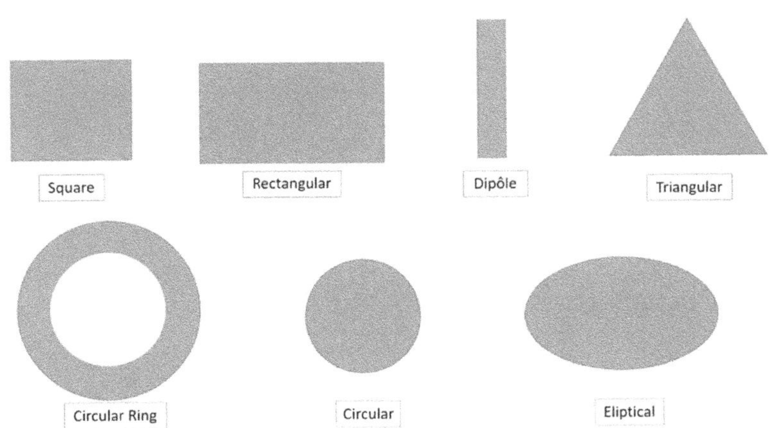

tion). The bandwidth is the frequency range (between f_{MIN} and f_{MAX}) in which the specified criterion is met. In general, the relative bandwidth is given by the following expression:

$$BP = \frac{2\left(F_{max} - F_{min}\right)}{F_{max} + F_{min}} \times 100 \tag{1}$$

The most common definitions are given in terms of radiation, polarization and most generally in terms of impedance. The input impedance of an antenna varies with frequency. The range of frequencies in which this impedance remains adapted to the generator is defined as the bandwidth in terms of impedance. This is usually expressed from the reflection coefficient (S11) or the standing wave ratio (SWR). The impedance matching can be translated by a reflection coefficient lower than a specific level (generally -10dB or -15dB) or a SWR typically lower than 2 or 1.5. In addition, the bandwidth is inversely proportional to the antenna quality factor, given by the relationship next:

$$BP = \frac{ROS - 1}{Q\sqrt{ROS}} \tag{2}$$

The quality factor is defined by the ratio between the stored energy and the energy lost by the circuit.

$$Q = \frac{E_{stored}}{2\pi E_{lost}} \tag{3}$$

In general, the quality factor represents the total losses of the antenna, including losses due to the radiation that is being maximized, ohmic conductance, dielectric losses and surface waves that should be minimized. Thus, the total quality factor is influenced by all these losses and is generally written as follows:

$$\frac{1}{Q_t} = \frac{1}{Q_{rad}} + \frac{1}{Q_c} + \frac{1}{Q_d} + \frac{1}{Q_{sw}} \tag{4}$$

Q_t = Total quality factor
Q_{rad} = Quality factor due to radiation
Q_c = Quality factor due to conductance (ohmic)
Q_d = Dielectric quality factor
Q_{sw} = Quality factor due to surface waves

Figure 3. Rectangular micro strip antenna configuration.

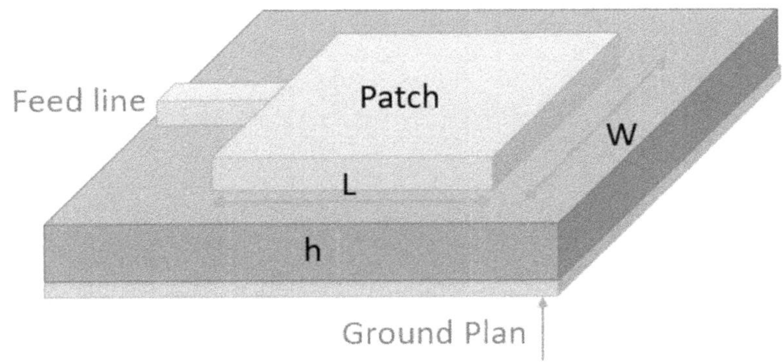

Calculation of Width (W)

The width of the patch has a minor effect on the resonance frequencies and radiation pattern of the antenna. On the other hand, it plays a role in the input impedance (on the edge) of the antenna and the bandwidth at its resonances. To allow good antenna performance, a practical width W is (Balanis, 2005):

$$W = \frac{C_0}{2F_0}\sqrt{\frac{2}{1+\varepsilon_r}} \tag{5}$$

Where C_0 is the free-space velocity of light i.e. 3×10^{-8} m/s and ε_r is the dielectric constant of material.

Calculation of Length (L)

The length of the patch determines the resonance frequencies of the antenna. Do not forget to deduct twice the length ΔL which corresponds to the fringing fields (Balanis, 2005):

$$L = \frac{\lambda_{eff}}{2} - 2\Delta L \tag{6}$$

The length L must be slightly less than the wavelength in the dielectric. At this frequency of operation, λ depends on the effective dielectric constant (ε_{reff}). The fringing fields effect is electrically made; the patch seems larger than its physical dimension. There is therefore an increase of ΔL on each side of the patch:

$$\Delta L = 0.412h\frac{\left(\varepsilon_{reff} + 0.3\right)\left(\frac{W}{h} + 0.264\right)}{\left(\varepsilon_{reff} - 0.258\right)\left(\frac{W}{h} + 0.8\right)} \tag{7}$$

The value of this effective dielectric constant is given by:

$$\varepsilon_{reff} = \frac{\varepsilon_r + 1}{2} + \frac{\varepsilon_r - 1}{2}\left[1 + 12\frac{h}{W}\right]^{-1/2} \quad (8)$$

Substrate Choice

The substrate plays a dual role in microstrip technology. It is both a dielectric material, where the circuits are etched, and a mechanical part, because it supports the structure. Their electromagnetic properties are essentially characterized by three physical quantities: conductivity, relative dielectric constant, the angle of loss. The substrates used in the design of printed antennas are numerous and must be compatible with the patch material. In the case where the antenna patch material is normal like copper, the relative permittivity of the substrate varies from 1 to 12. Sometimes, it is preferable to use high thickness and low permittivity dielectric substrates in order to have high efficiency and high bandwidth. But, the losses by surface wave increase and the antenna becomes more and more cumbersome. On the contrariwise, the use of thin substrates with high permittivities is recommended for microwave circuits allowing the minimization of surface waves and unwanted radiations. However, efficiency and bandwidth decrease due to the high loss in the substrate.

In microwave, the main characteristics of a substrate are:

- The thickness h.
- The dielectric constant.
- The coefficient of thermal expansion (in ppm/°C).
- The dissipation factor (delta tangent).
- Thermal conductivity (in W/m. °K).

In our case, we have used an FR4 substrate with dielectric permittivity constant 4.4, thickness of 1.6 mm and loss tangent of 0.025. FR4 epoxy glass substrates are the material of choice for most PCB applications. The material is very low cost and has excellent mechanical properties, making it ideal for a wide range of electronic component applications. As more and more microwave systems aimed at consumer markets are developed, there is a considerable interest in minimizing the cost of these systems. Substantial cost savings could be realized by using FR4 in place of costly substrates for microwave circuits and antennas.

Feeding Technics

Feeding mechanism is a key point when studying printed antennas. Indeed, energy is supplied to the radiating element in a way that can directly influence its radiation and modify its performance. The power supply to the antenna depends also on how the antenna is integrated into the device. The methods of feeding the patch antennas can be classified into two categories:

- Contact feeding method (by probe or micro strip line).

- Proximity feeding method (electromagnetic coupling by line or slot).

Coaxial or probe feeding is a technique widely used to feed micro strip antennas. In this case, the inner conductor of the coaxial connector passes through the dielectric and is soldered to the patch, while the outer conductor is connected to the ground plane.

The second method is aperture coupled feeding, in this type of power supply, the radiating patch and the micro strip power supply line are separated into ground plane. The conjunction between the patch and the power supply line is made by an opening or slot in the ground plane. Generally, high permittivity materials are used for the lower substrate while low dielectric constant materials are used for the upper substrate to optimize patch radiation. This type of feeding is difficult to design because of the multiple layers, which increase the antenna thickness.

The proximity coupled feeding method is based on two dielectric substrates used such that the power supply line is between the two substrates and the radiating patch is on the upper substrate. Among the disadvantages of this feeding method are the difficulty of manufacturing because of the two dielectric layers requiring proper alignment, as well as the difficulty in integrating active devices. The last and efficient way to feed a patch antenna is the direct feeding by a micro strip line. In this type of supply, a copper strip is connected directly to the edge of the micro strip patch. The copper line is smaller in width compared to the patch. The advantage of this type of feeding is that it can be etched on the same side of the antenna, it is easy to manufacture and simple to adapt to the resonance. It is one of the most widely used techniques in printed antennas, and it is also the one employed in our antenna structure.

Dual Band Patch Antenna Design for Rectenna Applications

The use of a dual band patch antenna (Lee, Yang & Kishk, 2008; **Error! Hyperlink reference not valid.** &**Error! Hyperlink reference not valid.**, 2016; Gupta, Dhillon, Khera & Marwaha 2013) in rectenna system is the appropriate solution to harvest and benefit from the maximum of useful energy to be rectified. In this part, a novel miniature dual band patch antenna is analyzed and presented. The structure is simulated at 2.45 GHz and 5.8 GHz of the Industrial Scientific Medical band used for rectenna applications. The dual band characteristic is obtained by introducing slots in the rectangular patch antenna. The antenna parameters have been investigated and optimized by using Advanced Design System (ADS) from Agilent Technologies. This antenna has an area of 13×11.2 mm^2. The structure has been mounted on an FR4 substrate with dielectric permittivity constant 4.4, thickness of 1.6 mm and loss tangent of 0.025. The application targeted by this antenna being the reception of radio frequencies in the ISM band, we have opted for planar technology antennas (patch antennas) because their manufacturing processes are well known and, in addition, they are low cost. A small size and low weight are the other two advantages of this technology. On the other hand, we preferred to use a patch antenna powered by a micro strip line because this type of power supply is simple to design.

Input Impedance

The design and optimization of the proposed antenna structure was done using Advanced Design System software (ADS). The latter is a tool that allows to perform different types of simulations dedicated to RF antennas and circuits such as: Momentum, Harmonic Balance, SParameter and LSSP. As shown in the

Figure 4. The proposed patch antenna.

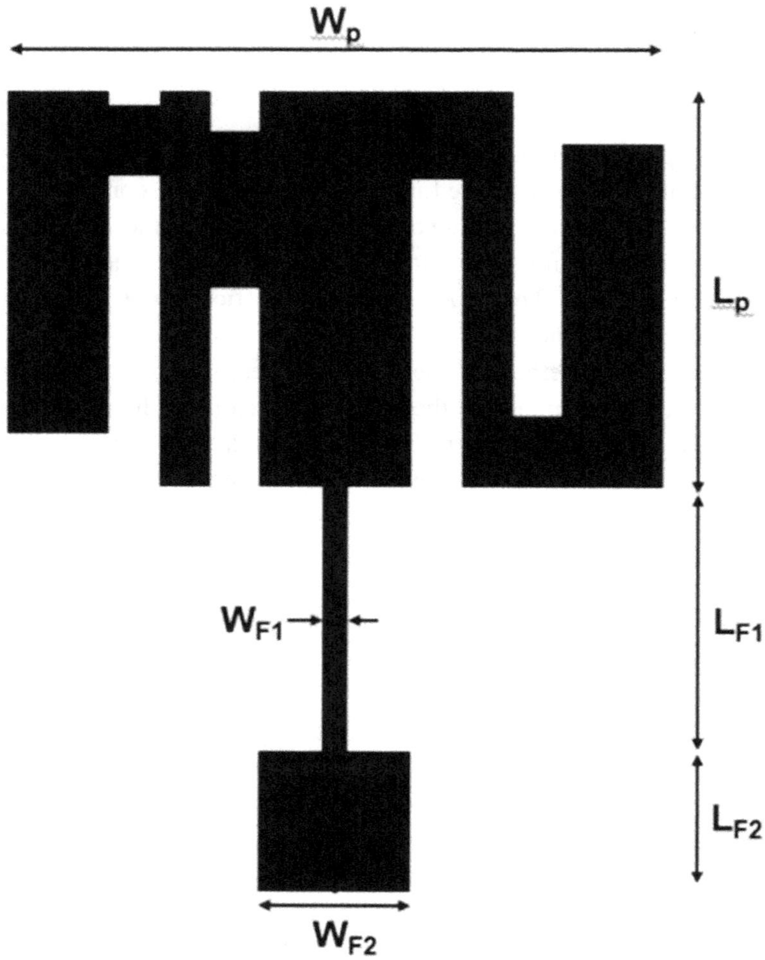

Figure 5, the antenna resonates on both 2.45 Ghz and 5.8 Ghz frequencies with good level of reflection coefficients.

The Figure 6 describes the simulated 3D radiation pattern of the proposed antenna which show a stable and directional radiation pattern for the two resonant frequency bands.

To validate the results obtained by simulations using Advanced design system, we have manufactured the proposed antenna, in our project we used photosensitive plates composed of three distinct layers. The first layer, usually green, is a resin sensitive to UV radiation (photosensitive resin). The second is a thin layer of copper (35 um) which is an excellent conductor. And the last layer is made of an insulating and heat-resistant substance (FR4 of relative permittivity $\varepsilon r = 4.4$, thickness $h = 1.6$ mm and tangential losses 0.025). The following is a summary of the process we used to manufacture the antenna, this process has six steps:

- Printing of the mask on tracing paper.
- Insolation of the photosensitive resin using ultraviolet (UV) radiation.

Figure 5. Simulated reflection coefficient versus frequency.

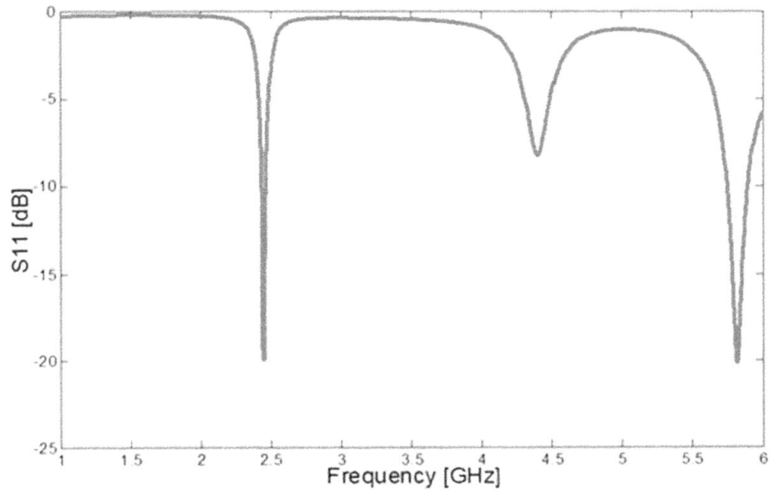

- Revelation: plate development (dissolution of the photosensitive resin exposed to UV radiation).
- Etching the antenna by removing copper not protected by photosensitive resin.
- The transition to the eliminator.
- Soldering of SMA ports.

The Figure 8 shows the measurement results obtained after calibration of the achieved antenna. We have tested both the reflection coefficient by using a Network Analyzer and radiation pattern in E-plane and H-plane in an anechoic chamber. We can conclude that the antenna is well matched at 2.45 GHz and 5.8 GHz of the Industrial Scientific Medical Band and confirm the simulated one using ADS. The measured radiation pattern for both frequencies in E-plane and H-plane validates the proper functioning and directivity of the proposed antenna.

Figure 6. 3D radiation pattern of the proposed antenna

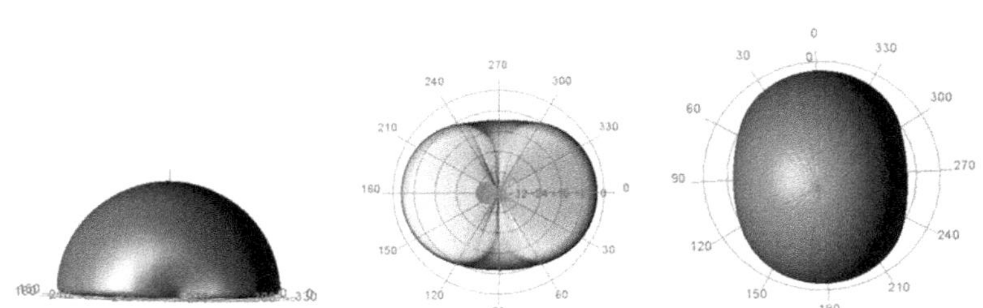

Figure 7. Fabricated Antenna

SECTION II: MICROSTRIP 2.45GHZ BRIDGE RECTIFIER FOR RECTENNA APPLICATIONS

The technological evolution and the mastering of wireless communication systems allow objects to have a knowledge of their environments through sensors. This concept is called in the literature "Internet

Figure 8. Simulated and measured reflection coefficient versus frequency of the proposed patch antenna

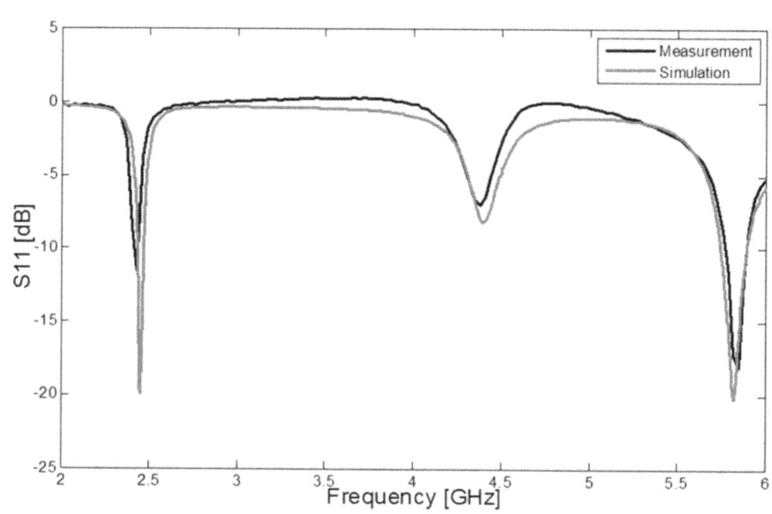

Figure 9. The measured E-plan antenna radiation pattern at 2.45 GHz for azimuth plan of 0, 45 and 90°

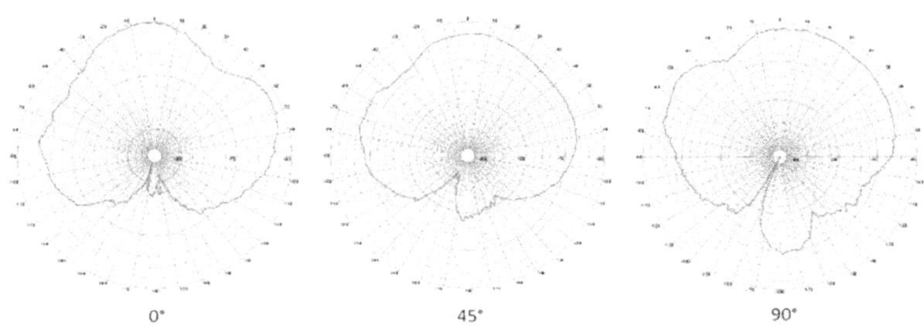

of Things". These sensors recover various data that are then used locally or transmitted wirelessly for analysis and processing. This evolution of sensors has reinforced the feasibility of "smart cities". For better integration of these sensors into the environment, it is necessary to reduce their size and ensure their energy autonomy. Conventional battery or battery power techniques remain constraining, or even difficult to consider for some applications. They are limited in autonomy, require periodic replacements and their recycling is expensive.

Wireless power transmission (WPT) has become a novel alternative technology to solve all these power supply problems.

The key element of a WPT system is called Rectenna (for rectifying antenna), a conventional rectenna circuit consists of a receiving antenna followed by an RF-DC conversion circuit with a non-linear characteristic. This circuit usually contains one or more Schottky diodes, an HF input filter, a DC output filter and a resistive load that models the consumption of the powered system. The HF input filter is located between the receiving antenna and the conversion circuit, the diode(s) of the RF-DC converter generate not only DC power but also higher order harmonics. These harmonics can degrade the operation of the circuit, they are undesirable and must be blocked. In order to miniaturize the rectenna circuits and improve their efficiency, the HF input filter can be directly integrated into the receiving antenna by using antennas with harmonic rejection. The DC output filter is a low-pass filter that is most often composed of a parallel capacity block. The role of this filter is to isolate the DC charge from the RF side. It lets the DC component pass through and filters all higher order harmonics, including the fundamental component. The output capacity can be replaced by one or more radial or rectangular stubs.

The first works on rectennas were focused in the ISM band at 2.45 GHz, most of these works were intended for space applications. The choice of this frequency was not random, it is mainly explained by the low attenuation that this frequency presents through the atmosphere. In 1960, Raytheon had

Figure 10. The measured E-plan antenna radiation pattern at 5.8 GHz for azimuth plan of 0, 45 and 90°

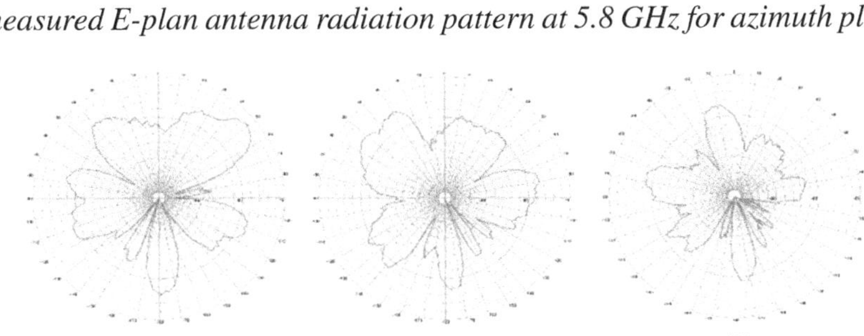

developed a rectenna at 2.45 GHz. The circuit developed contains a half-wave dipole antenna associated with an RF-DC conversion circuit containing a single diode connected to a resistive load. One of the key parameters of rectenna circuits is the conversion efficiency. It is most often defined as the ratio between the continuous power recovered or measured across the load and the RF power received by the radiating element. This conversion efficiency has continued to increase since the first work was carried out in this area. However, the most important performance was achieved by Brown at Raytheon in 1977 (Raytheon Company, 1977). W. C. Brown used a Schottky Gallium Platinum Arsenide (GaAs-Pt) diode, an aluminum dipole antenna and a transmission line to achieve a 90.6% conversion efficiency with an input power of 8 W. In 1982, W. C. Brown and J. F. Triner developed a printed version of the circuit at 2.45 GHz, engraved on a thin film. The yield obtained was 85% (Brown & Triner, 1982). In 1991, K. Chang and T. W. Yoo developed an integrated 35 GHz rectenna with a 33% efficiency for input powers of about 60 mW (Yoo & Chang, 1991). The circuit consists of a dipole antenna and a Schottky DMK6606 diode (Alpha Industries) connected in parallel. At the same frequency, other works have been reported subsequently in the literature, including a rectenna combining an antenna array and a Schottky MA4E1317 diode connected in series (Ren, Li, & Chang, 2007). The circuit achieves an efficiency of 35% when the power density is 30 mW/cm^2. The change from the traditionally used 2.45 GHz frequency to 35 GHz is due to the need to have structures that occupy less space and thus be able to put more rectennas on the same surface. However, the components used to generate and convert energy at this frequency are less efficient and more expensive. The disadvantages encountered at 35 GHz, which were mentioned above, led the researchers to use a new frequency at 5.8 GHz. This makes it possible to reduce the size of the circuits compared to the 2.45 GHz frequency, without losing efficiency. In 1992 was reported the first rectenna in the C-band at 5.87 GHz with 80% efficiency (Bharj, Camisa, Grober, Wozniak, & Pendleton, 1992). The rectenna was measured inside a waveguide with an input power of 700 mW. The circuit uses a printed dipole and a Schottky diode. Further work at 5.8 GHz was subsequently reported. In 1998, J.O. McSpadden and al. developed a rectenna with a dipole printed at 5.8 GHz, an efficiency of 82% was measured (McSpadden, Fan & Chang, 1998). The rectenna is developed in CPS (Coplanar Stripline) technology and uses a Schottky MA40150-119 diode.

The challenge is to optimize the entire rectenna, with the objective of maximizing the DC output and efficiency of RF-DC conversion. Optimization must be carried out over the entire circuit, hence the need to use global analysis methods combining electromagnetic simulation and circuit. Numerical modelling will be a fundamental and decisive tool throughout this work, the aim being to take into account all possible couplings between the different parts of the circuit

The incident electromagnetic energy is picked up by a receiving antenna and sent to the RF-DC converter as a high frequency sinusoidal wave. The rectification stage is the key element of a Rectenna system which transforms this wave into voltage and direct current. The voltage level obtained at the output of the rectenna is often too low to be able to directly power an electronic circuit or to charge a battery; especially when the transmitter-receiver distance is large. A voltage step-up stage is often used to obtain a sufficient DC voltage level (Huang, Tsai, & Chen, 2010; Richelli, Colalongo, Tonoli, & Kovacs, 2009), while DC-DC conversion strategies of type MPPT are often used to ensure power transfer between the rectenna and the load (Brunton, Rowley, Kulkarni, & Clarkson, 2010). In parallel, by following the general trend towards the miniaturization of electronic devices, new micro-scale batteries rechargeable are developed and can represent viable energy sources for isolated sensors that can be recharged wirelessly (Alahmad & Hess, 2008; Chen et al., 2010).

RF-DC conversion structures are generally built around diodes or transistors mounted as diodes. The diodes are characterized by a threshold voltage that must be exceeded in order to put them into conduction. When the incident power level is high, the threshold voltage of the diodes is not a limiting factor, as the amplitude of the incident signal is significantly higher than the threshold voltage of the diodes (Merabet et al., 2009a). In the case of very low levels of incident power, losses due to diode threshold voltages become predominant. On the other hand, the Schottky diode (Alexandre Douyére, 2008) is a better candidate in this context. Indeed, its threshold voltage varies between 0.1V and 0.3V, it is more sensitive to low voltage levels delivered by the antenna and can provide a DC voltage from low power densities. In addition, the Schottky diode has a very fast switching time due to its very low junction capacity, which justifies its highly responded use in microwave applications.

Other ways, the frequency limitation of a Schottky diode is much higher than that of a simple PN junction. This limitation can be extended beyond 100GHz with Schottky diodes based on gallium arsenide.

The equivalent small signal model of a Schottky diode without a package, connected in parallel with an R_L load, is shown in Figure 11:

With Rs: semiconductor resistance.

Rj: resistance of the junction.
Cj: junction capacity.
RL: load resistance.
Vj: voltage at the terminals of the metal semiconductor junction.

The junction resistance Rj can be expressed by the following relationship:

$$R_j = \frac{nkT}{q\left(I_s + I_b\right)} \tag{9}$$

Figure 11. Small signal model of a Schottky diode.

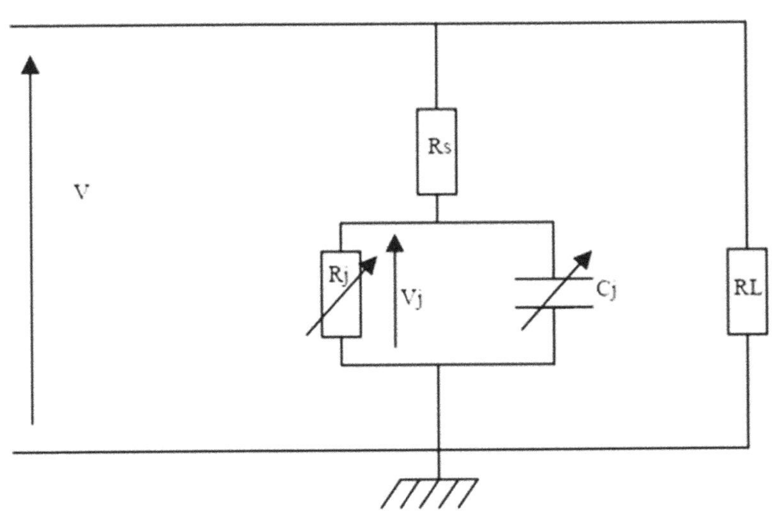

Figure 12. Rectifier series topology.

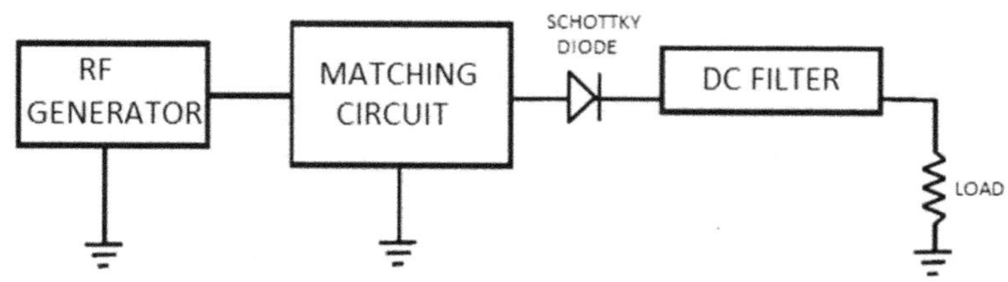

With n: ideality factor.

k =1.38 10^{-23}J/°K: Boltzmann constant
T: junction temperature in °K
q =1.6 10^{-19}: electrical charge of an electron
Is: saturation current
Ib: polarization current

Depending on the application, several diode rectifier topologies are possible (Marian, Vollaire, Verdier, & Allard, 2011):

Series Topology

This is the simplest topology; it uses a simple diode in series to rectify the RF signal (Figure 12). This is a single alternance rectification, the low-pass filter, usually a capacitor, allows the DC component to be recovered. This type of rectifier is most commonly used for low input power levels, in the order of -15 dBm. If the conversion efficiency is to be maximized at a particular frequency, an adaptation circuit is used upstream of the diode. It can be composed of localized components such as capacitors and inductances or distributed elements such as stubs. The simplicity of this topology and its low power results make it the most widespread in rectenna design.

Figure 13. Rectifier parallel topology.

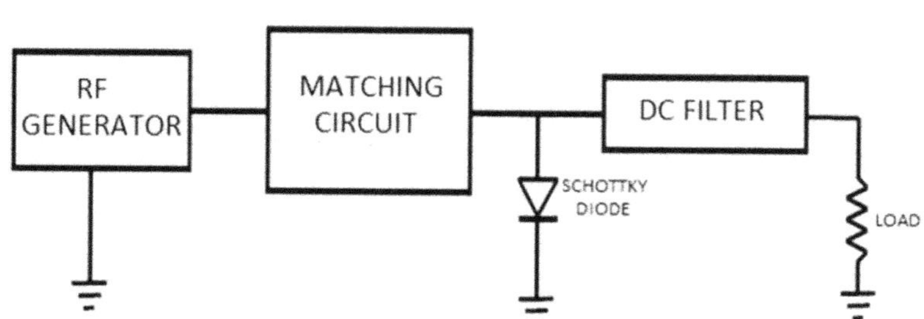

Figure 14. Rectifier voltage doubler topology.

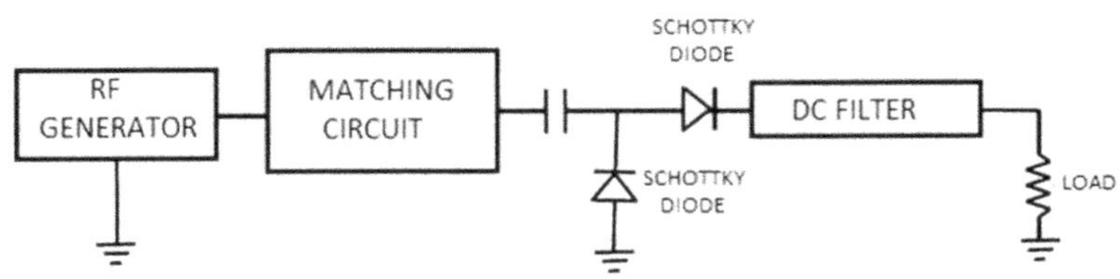

Shunt Topology (Parallel)

It is also a single alternance rectifier topology, the diode is connected in parallel, with the anode or cathode connected to the ground.

Voltage Doubler Topology

The voltage doubler can be seen as the combination of the two topologies described above, with additional input storage capacity for one of the two alternances. This is a double alternance rectification, for positive alternance for example, diode D1 is in pass mode and diode D2 is blocked, so the signal is transferred to the load. In the case of negative alternance, diode D2 is in pass mode and diode D1 is blocked, the rectified signal is stored in the capacitance at the input and transmitted at the same time than the next positive alternance.

Bridge Topology

The "diode bridge" topology can also be used for RF signal rectification. As for the doubler or multiplier, it is a double alternating rectification.

Figure 15. Rectifier bridge topology.

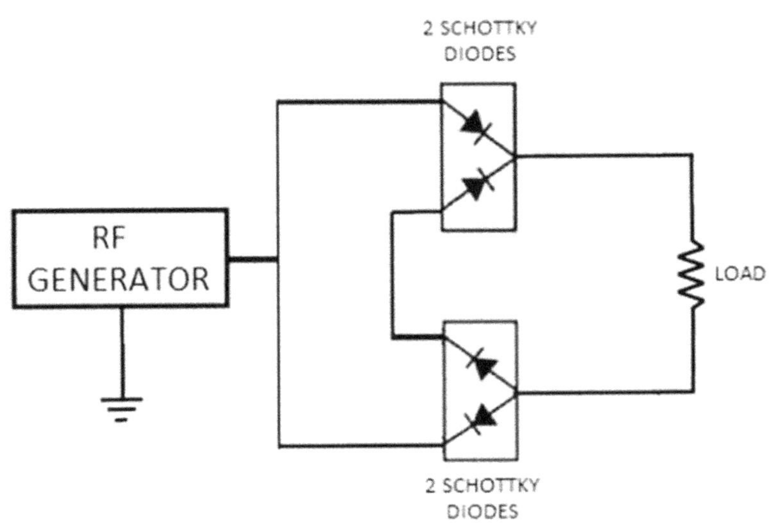

As the frequency increases, microwave devices are integrating more and more components and functions. An electronic device can contain both power components, radiating elements, transmission lines, etc. Moreover, the components can have very different dimensions. In many cases, the simulation and separate optimization of each function is insufficient. Large differences between experimental results and simulations can be observed. These differences are often explained by numerical modelling, which does not take into account all physical phenomena, including couplings and interactions between the various components of the device. To analyze a microwave device, there are several approaches depending on how the problem is addressed and the accuracy sought. Indeed, there is the electromagnetic approach which is based on Maxwell's equations and which takes into account all the electromagnetic phenomena present in the device. On the other hand, there is also the circuit approach which is based on equivalent circuit models. These models, where simplifications are often introduced, are only valid in a specific area and under certain conditions. This shows the importance of rigorous modelling that takes into account all physical phenomena in a microwave device. The microwave device that is the subject of this chapter is the rectenna. Rigorous design and optimization requires a combination of microwave electronics and electrical engineering skills. Since the objective is to maximize the RF-DC conversion efficiency, it is difficult to consider the separate optimization of the different components that make up a rectenna circuit. Indeed, to take into account the couplings and interactions between the different components, simulation and optimization must be carried out over the entire circuit and in the same simulation environment. Global electromagnetic-circuit analysis methods, which combine the laws of electromagnetism and circuit theory, are more than desirable in our case. Another difficulty in modeling is related to the dimensions of the different components. The components of a rectenna are of very varied dimensions. For example, we go from a half-wave radiating element to a localized component (diode, resistance...) which is sometimes less than one hundredth of the wavelength. In addition, a very fine mesh based on the smallest element is difficult to consider for obvious reasons of time and computing capacity. To solve these types of problems, more flexible and manageable models are often used. Many electromagnetic simulation tools exist. These tools are suitable for distributed structures whose dimensions are in the order of wavelength. The main difference between these simulation tools lies in the method of solving Maxwell's equations. Indeed, they can be resolved either in the time domain or in the frequency domain. Temporal analysis allows broadband characterization in a single simulation and allows non-linear elements to be taken into account. Its main disadvantage is the calculation time, which can sometimes be very important. The most commonly used time methods are: the finite time domain difference method (FDTD) and the transmission line method (TLM). On the other hand, there is frequency analysis, which is generally faster. However, for a broadband characterization, as many simulations as frequency points in the band are required.

In addition, non-linear elements are difficult to model in the frequency domain. Among the most commonly used frequency methods, we can mention for example the moment method (MoM) (GIBSON, 2008) and the finite element method (FEM). Circuit analysis methods, to study rectenna devices, have been proposed in the literature. However, since the RF-DC conversion circuit is non-linear and therefore generates higher order harmonics, it is not always easy to formulate a perfectly correct analytical model.

Rectennas circuits are most often characterized by two efficiencies, the RF-DC conversion efficiency and the overall efficiency. The first efficiency describes the rectifier's ability to provide continuous electrical power to the load from the RF energy provided to it by the receiving system or any other microwave energy source. This efficiency is the main objective in a conversion circuit optimization process. However, since the conversion process is non-linear in nature, optimization is performed at an input power point and efficiency is optimal over a narrow range around that point. Indeed, the impedance of the Schottky diode and its efficiency depend closely on the voltage applied to its terminals. Variations in this impedance lead to mismatches that directly affect the conversion efficiency. The latter is defined as the ratio between the DC output power, taken from the load terminals, and the RF input power. Depending on the input power considered, several definitions of performance exist in the literature. The RF power can be considered as the maximum power that the source or receiving antenna will be able to deliver on a load of 50 Ω, and in this case the reflection losses are taken into account. It can also be considered as the RF power actually transmitted to the diode, without taking into account reflection losses. The other efficiency, which is more global, describes the ability of the complete rectenna circuit to convert received RF energy into DC energy. The power that the antenna is likely to receive is often calculated by the Friis equation. This definition of efficiency is the most common in the literature, since it is not easy to access the value of the RF power actually transmitted to the converter. However, the efficiency can be calculated not on the basis of the power that the antenna receives in the best case but by taking into account the actual power received.

The efficiency (η) of the microwave rectifier is defined by:

$$\eta = P_{dc} / P_r = V_{dc}^{\ 2} / P_r R_L \tag{10}$$

Where P_{DC} is the dc power produced at the load resistance (R_L) of the rectifier and P_r is the power received at the antenna of rectenna or any other source of microwave energy. P_r is calculated from the Friis transmission equation wich gives the amount of power an antenna received under ideal conditions from another antenna. The Power from isotropic antenna falls off as R^2, so that the power density (p) would be:

$$p = p_t / 4\pi R^2 \tag{11}$$

Multiplying by the gain of the transmitting antenna gives a real antenna pattern

$$p = (p_t / 4\pi R^2)\, G_t \tag{12}$$

If receiving antenna has an effective aperture of A_{eff} the power received by this antenna (P_r) is

$$P_r = p.\, A_{eff} \tag{13}$$

Thus

$$P_r = (p_t / 4\pi R^2)\, G_t\, A_{eff} \tag{14}$$

The effective aperture of an antenna can be written as:

$$A_{eff} = (\lambda^2/4\pi).\, G \tag{15}$$

So, we conclude that the friis transmission equation could be expressed as follow:

$$P_r = P_t . G_t . G_r . (\lambda/4\pi R)^2 \quad (16)$$

Where P_t is the power of the transmitter, λ is the wavelength used, R is the distance separating the transmitter and the receiver, G_t and G_r are the transmitter and receiver antenna gain, respectively.

Simulation Tool

The ADS software is developed by Agilent Technologies. It is dedicated to the simulation and design of RF electronic circuits and systems. It offers a wide range of circuit and electromagnetic simulation environments, in both time and frequency domains. In this work, the ADS software was used to simulate and optimize the RF-DC conversion circuit and antennas by means of its two simulators Harmonic Balance and Momentum. It also contains tools such as:

- "Impedance matching" which allows to generate an optimized adaptation circuit from the input and output impedances;
- "Smith chart" which allows to design an adaptation circuit based on localized components, lines or both, from the source and load impedances;
- "LineCalc" which allows to calculate the dimensions of the lines from the substrate parameters used and the characteristic impedance of the Z_0 line (generally equal to 50Ω).

It also contains generic models of diodes and bipolar transistors or MOSFETs. For the rectifier, a Schottky diode from the general trade is used. It is sufficient to manually integrate the diode parameters found in the technical documentation into the ADS generic diode model or to directly integrate the model from the diode by downloading its library and importing it on ADS.

Harmonic balance is a frequency analysis method; it takes into account all non-linearities in a circuit. It offers a whole set of optimization methods of random and iterative types. The device to be studied is divided into linear and non-linear sub-circuits. The linear sub circuit is analyzed in the frequency domain, while the non-linear sub circuit is analyzed in the time domain. The results of the time analysis are then transposed into the frequency domain by the Fourier Transform (TF). A solution is found if the interconnection currents between the linear and non-linear subsystems, for each harmonic, are the same. The different currents must be balanced for each harmonic.

Momentum is an electromagnetic simulator based on the moment method, it is dedicated to the simulation of planar RF circuits (antennas, distributed filters ...). It has the following advantages:

- Simulation of topologies whose circuit model is outdated or non-existent.
- Consideration of couplings between components.
- Adaptive frequency sampling for fast and accurate simulations.
- Integrated optimization tools.
- Coupling with circuit simulators as harmonic balance.

Manufacturing Process

The manufacturing process of the antennas and conversion circuits was done using the micro strip technology well known by its low cost. It is based on printing a thin layer of copper on a substrate (In

Figure 16. (a) The proposed rectifier topology (b) Simulated S11 versus frequency.

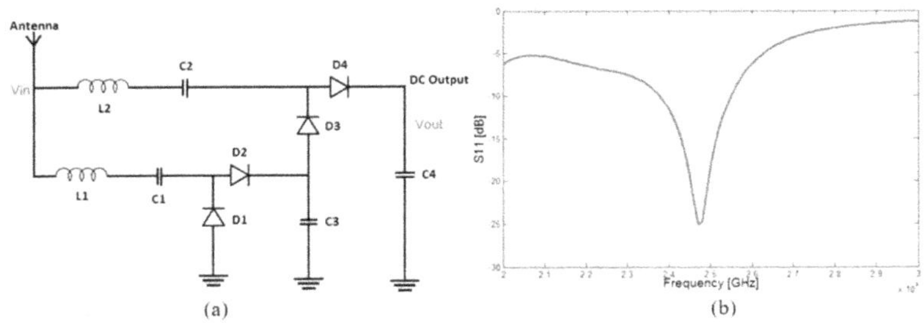

our case: FR4) thus forming the desired circuit or antenna with indeed on a ground plane on the other side of the substrate. In our project we used photosensitive plates composed of three distinct layers. The first layer, usually green, is a resin sensitive to UV radiation (photosensitive resin). The second is a thin layer of copper (35 um) which is an excellent conductor. And the last layer is made of an insulating and heat-resistant substance (FR4 of permittivity relative εr = 4.4, thickness h = 1.6 mm and tangential losses 0.025).

Measurement Setup

As a test bench, an AGILENT 8753E Network Analyzer from 30KHz to 6GHz was used to test the antennas for the extraction of S11 reflection coefficients and an anechoic chamber for the extraction of radiation patterns. An anechoic chamber (or deaf chamber) is an experimental room whose walls absorb electromagnetic waves (pyramids of polyurethane foam loaded with a carbon-based complex), reproducing free field conditions and therefore not causing an echo that could disturb measurements.

To be able to perform the test in this anechoic chamber, we need a reference antenna that we use to characterize the receiving antenna under study.

The measuring bench used to test the Rectifier circuits consists of:

- An AGILENT 8753E network analyzer from 30KHz to 6GHz for the recovery of S11 reflection coefficients
- An ANRITSU 68347C 10 MHZ-20 GHZ RF generator to supply the circuits with the RF signal at the desired frequency.
- A multimeter to measure the voltage and current at the output of the circuit.

Achieved and Published Rectenna Circuits

- A New Configuration of A High Output Voltage 2.45 GHz Rectifier For Wireless Power Transmission Applications (Taybi, Tajmouati, Zbitou, Errkik, Latrach & El Abdellaoui, 2018)

This structure comes to deal with the problematic of power supply in many zones where the use of batteries or wires is difficult due to many reasons. It's about a voltage multiplier rectifier using five Schottky diodes HSMS2820 (HSMS-282X series, 2001) mounted on an FR4 substrate with dielectric constant of 4.4 and having a thickness of 1.6mm. The micro strip technology has been applied and many

Figure 17. Fabricated voltage multiplier rectifier.

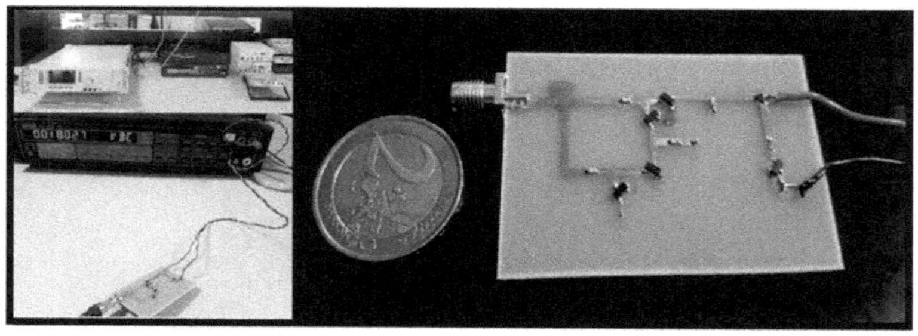

optimization steps have been performed in order to reach the desired results in terms of output voltage and conversion efficiency. As a final step, we have fabricated the proposed structure to validate the obtained simulation results.

The rectifier is loaded by a 2KOhm resistive element to replace the device to be powered. Measured results seems to be in total agreement with the simulated ones performed by the Advanced Design System solver. This structure has a good matching input impedance and important conversion efficiency with an input power varying between -20dBm to 24dBm and the structure start to detect at a low power level.

- <u>A New Design of a Micro strip Rectenna at 5.8 GHz for Wireless Power Transmission Applications</u> (Taybi, Tajmouati, Zbitou, Errkik, & Latrach, 2019).

Figure 18. Simulated and measured results of the rectifier (a) S11 versus Frequency (b) Conversion Efficiency versus Input power (c) Output voltage versus input power.

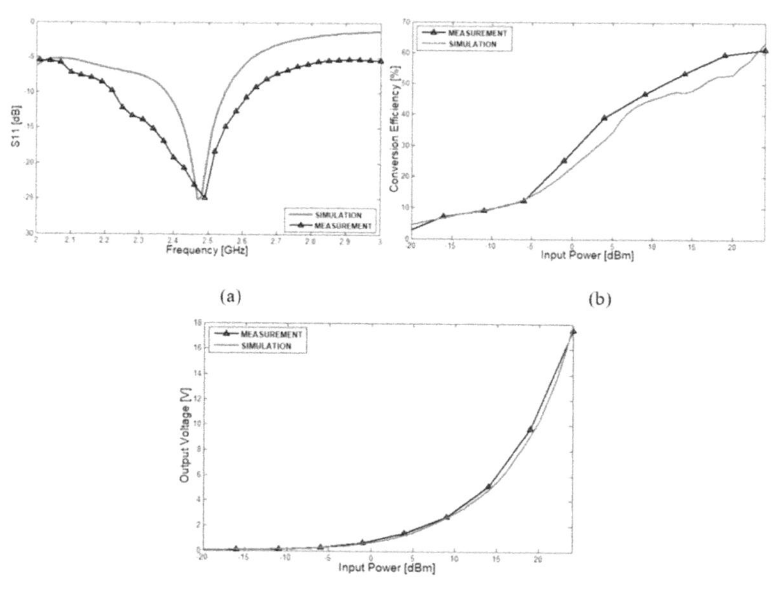

Figure 19. The proposed patch antenna

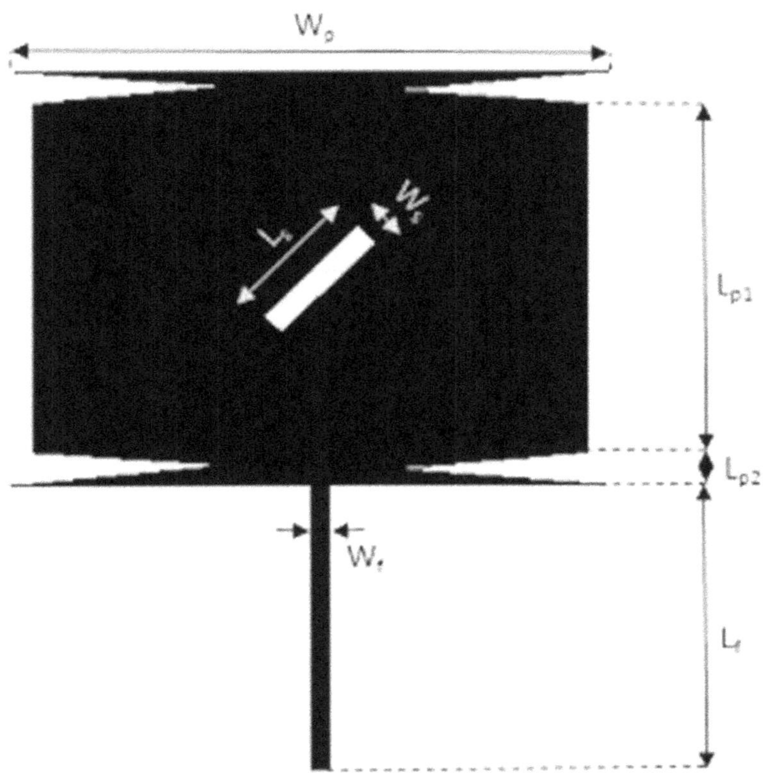

A novel 5.8GHz Rectenna system is presented in this part. In one hand, a miniature 5.8 GHz micro strip antenna have been designed, optimized and simulated using Momentum Solver of Advanced Design

Figure 20. Simulated S11 versus frequency.

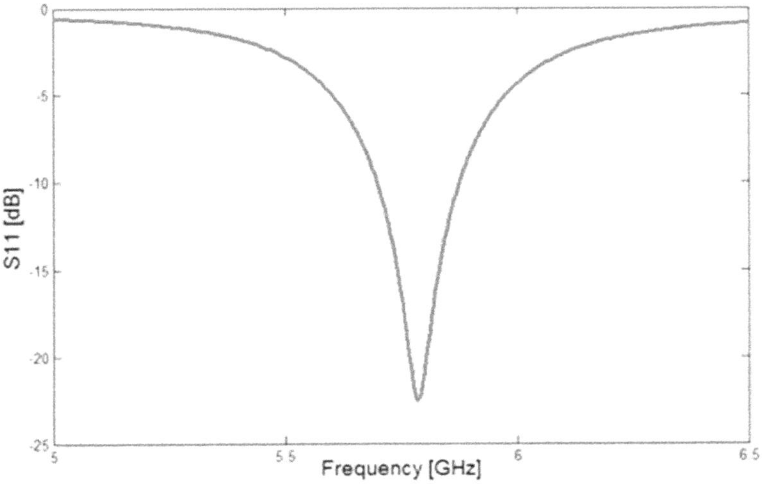

Figure 21. The proposed rectifier topology

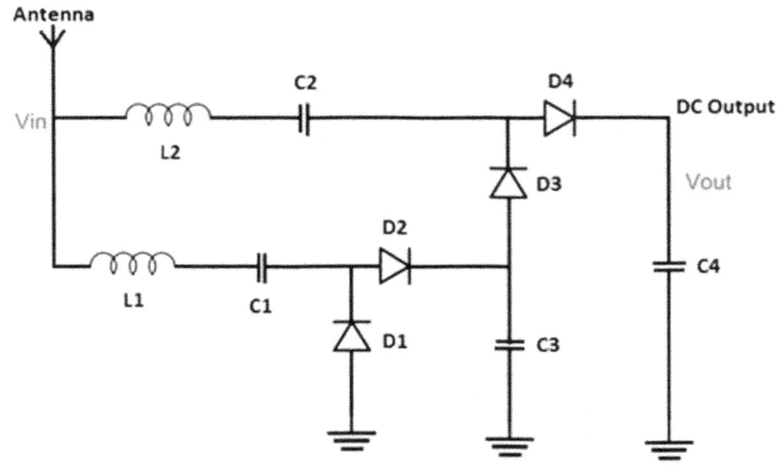

Figure 22. (a) Simulated rectifier efficiency versus input power, (b) Simulated output voltage versus input power of the rectifier.

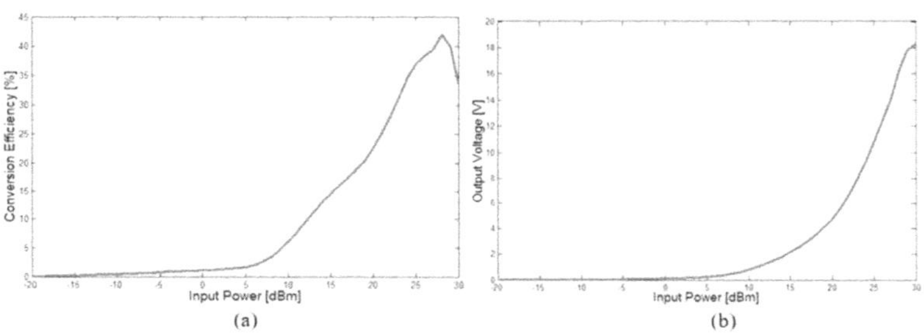

System. In the other hand, a micro strip rectifier working at 5.8 GHz of the Industrial Scientific Medical band (ISM) have been investigated. Five Schottky diodes implemented in a voltage multiplier topology

Figure 23. Fabricated Rectenna system

Figure 24. Measured results of the rectenna. (a) S11 versus frequency (for Rectifier) (b) S11 versus frequency (for Antenna) (c) Output voltage versus input power (d) Conversion efficiency versus input power.

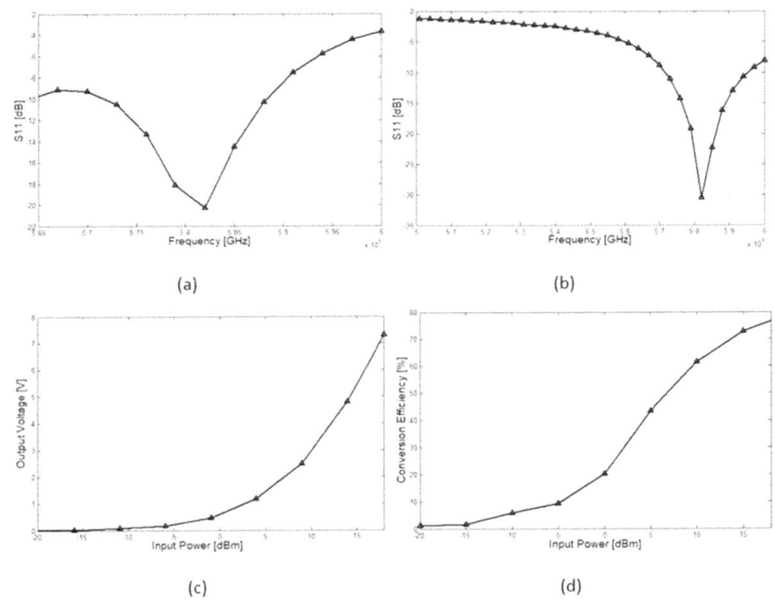

have been used to convert the collected RF power by the antenna. In addition, a 1 kOhm load resistance have been employed to replace the device to be powered according to the targeted application. Both structures were manufactured and tested to confirm simulated results.

- A New Design of High Output Voltage Rectifier For Rectenna System at 2.45 GHz (Taybi, Tajmouati, Zbitou, Errkik, Latrach, & El Abdellaoui, 2019)

The results presented in this section involves two important parts of a Rectenna system. First, a 2.45GHz rectifier based on Schottky diodes mounted in series topology have been designed and optimized. Then, a stepped impedance low pass filter has been integrated to stop the unwanted harmonics generated by the non-linear characteristic of a Schottky diode. The stepped impedance filter is a low-pass filter that consists of a cascade of alternating transmission lines of low and high impedance characteris-

Figure 25. The measured E-plan antenna radiation pattern at 5.8 GHz for azimuth plan of 0, 45 and 90°

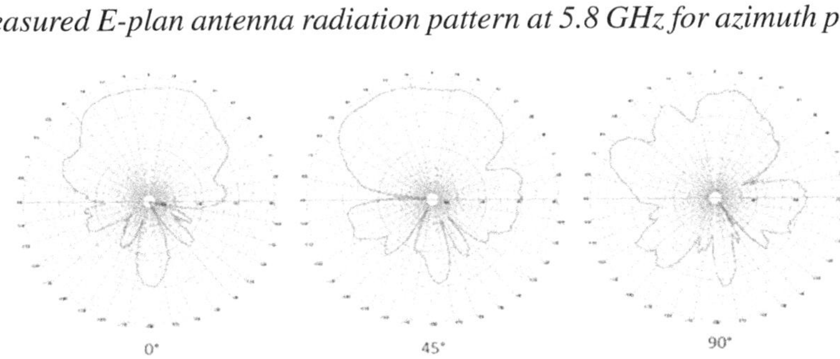

Figure 26. (a) S11 The proposed rectifier topology (b) The proposed filter configuration

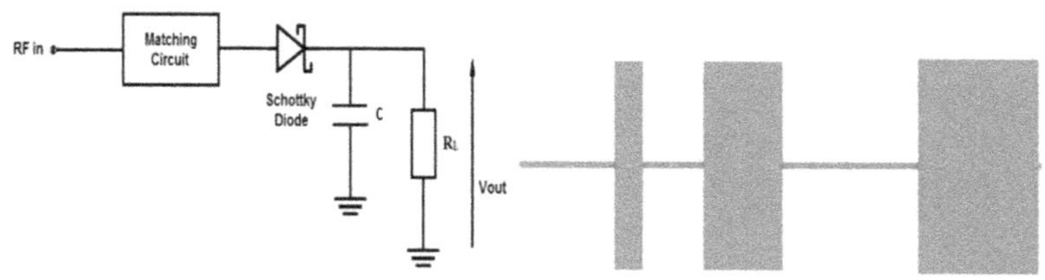

(a) (b)

tic. These lines, shorter than the wavelength, act as semi-localized elements. Indeed, a high impedance line is equivalent to a serial inductance while a low impedance line replaces a parallel capacitance. The line widths are determined by the choice of characteristic impedances. A low Zc value ensures a better approximation of the localized capacity. Similarly, a significant value of ZL is equivalent to a localized series inductance but the manufacturing constraints become stronger and losses increase. Some a priori design information must be provided about the micro strip lines, because expressions for inductance and capacitance depend upon both characteristic impedance and length. It would be practical to initially fix the characteristic impedances of high- and low-impedance lines by consideration of:

- ZL < Z0 < ZH, where ZL and ZH denote the characteristic impedances of the low and high impedance lines, respectively, and Z0 is the source impedance, which is usually 50 ohms for micro strip filters.
- A lower ZL results in a better approximation of a lumped-element capacitor, but the resulting line width WC must not allow any transverse resonance to occur at operation frequencies.
- A higher ZH leads to a better approximation of a lumped-element inductor, but ZH must not be so high that its fabrication becomes inordinately difficult as a narrow line, or its current-carrying capability becomes a limitation.

The two aforementioned structures have been printed using the micro strip technology on an FR4 substrate with dielectric constant of 4.4 and having a thickness of 1.6mm. The experimental results confirm the proper functioning of the final structure and a good agreement with the simulated ones by Advanced Design System was observed.

Figure 27. Final Rectifier Design

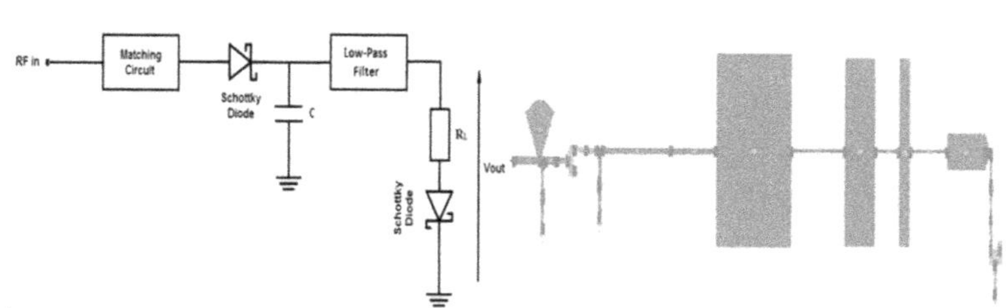

Figure 28. Simulation results (a) Conversion efficiency versus input power (b) Output voltage versus input power.

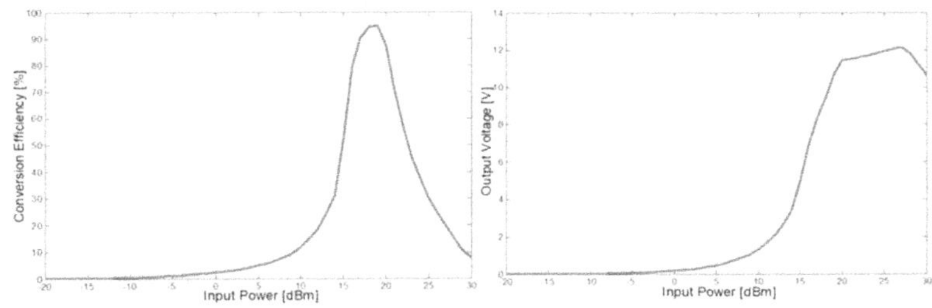

(a) (b)

MICRO STRIP 2.45GHZ BRIDGE RECTIFIER FOR RECTENNA APPLICATIONS

Design

The proposed rectifier structure is shown in figure 31, it consists of 5 HSMS2820 Schottky diodes built-in with bridge topology in order to convert the RF energy collected by the antenna. For this purpose, based on the theories on patch rectifiers, and with series of optimizations made following experiments with Advanced Design System software, we were able to obtain an efficient rectifier whose performances are

Figure 29. (a) measurement setup (b) measured output voltage (c) fabricated rectifier.

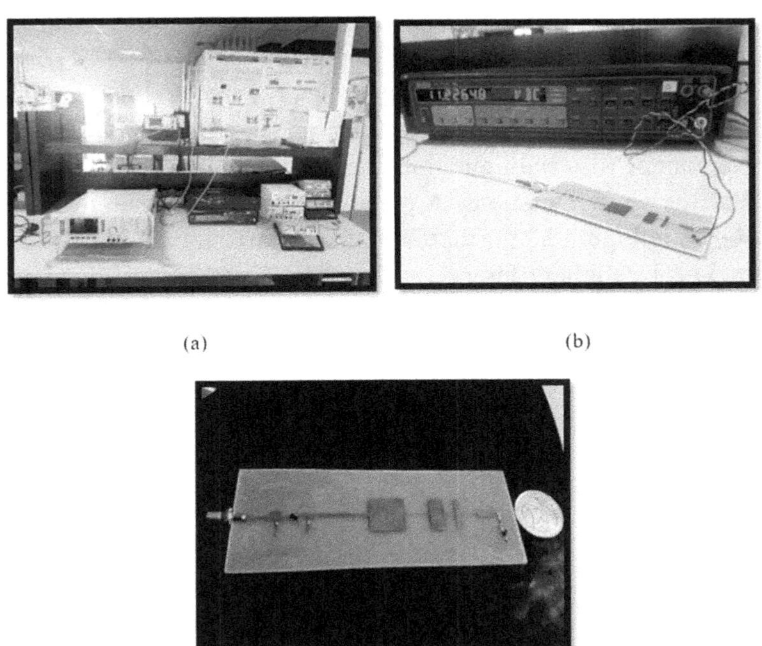

(a) (b)

(c)

Figure 30. Simulated and measured output voltage versus input power of the rectifier.

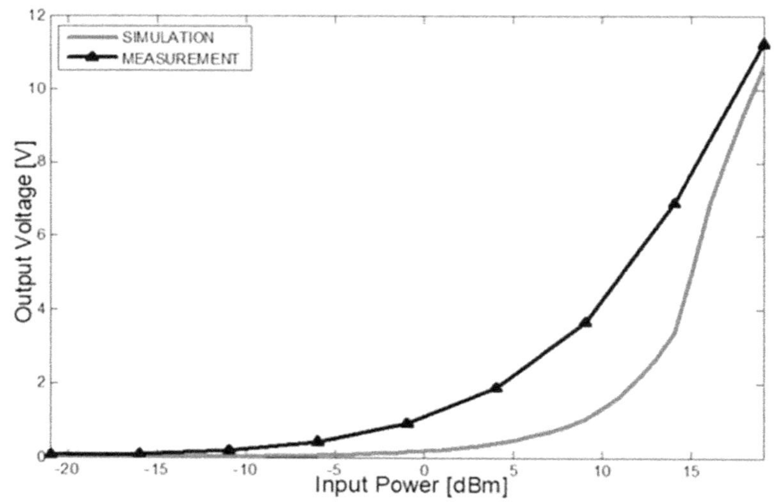

interesting for wireless power transmission applications. We have reached an important output voltage of 5.7V at 30dBm, in addition, the structure was well matched to operate at 2.45 GHz of the Industrial Scientific Medical Band (ISM) by optimizing the lengths and widths of the micro strip lines ensuring the interconnection between the different parts of the circuit. All simulation steps have been carried out by the Harmonic Balance Solver and Large Signal S-Parameters.

To investigate and validate the bridge rectifier structure, it was necessary to proceed the fabrication of the prototype. We have respected the same process presented at the dual-band antenna manufacturing part. After soldering the Schottky HSMS2820 diodes, the capacitors and the load resistance, the circuit have been tested by using an RF generator ANRITSU 68347C 10 MHZ-20 GHZ to supply the circuits with the RF signal at the desired frequency and a multimeter to measure the output voltage level of the circuit. The input impedance matching has been verified by means of a Network Analyzer from Agilent Technologies. However, due to material limitation, the maximum input power that we could reach was 24 dBm instead of 30dBm with simulations. A photograph of the measurement setup and the manufactured circuit are shown in Figure 33, furthermore, a comparison of simulated and measured results is presented in Figure 34. The slight difference is generally due to fabrication tolerance since all steps were done by human hands.

Table 1. Performance Comparison with Other Recent Researches

Source	Operating frequency	Technology	Output Voltage
(Chuc & Duong, 2016)	2.45 GHz	Voltage Doubler	10.75V (20 dBm)
(Zhang, Zhang, Wu, Liu & Zhang, 2017)	2.45 GHz	Voltage Doubler	2.3V (Max)
This paper	2.45 GHz	Single Diode	11.23V (20 dBm)

Figure 31. Proposed rectifier design.

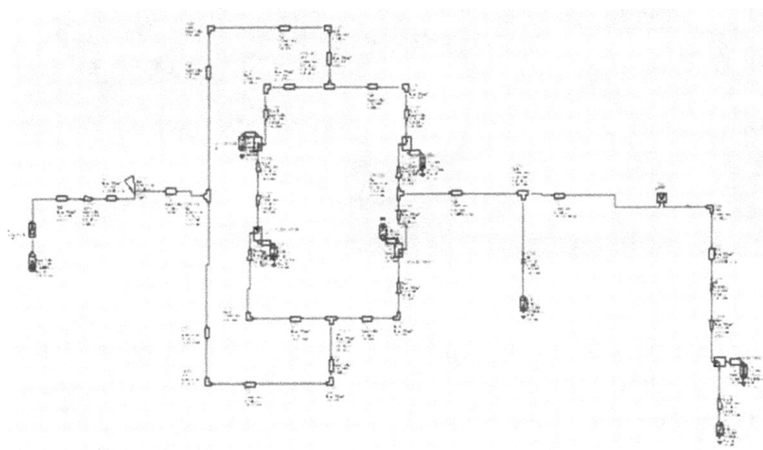

CONCLUSION

In this conclusion we will review the important results that have been presented in this manuscript. Firstly, we have presented and explained a general overview of Wireless Power Transmission (WPT), then, we had to devote several lines to explain the operating principle of a rectenna system. In the same section, a complete theoretical study was presented for the design of a patch antenna and this was followed by a presentation of the proposed dual band patch antenna that resonates at 2.45 GHz and 5.8 GHz. Design and simulation have been carried out using Advanced Design System from Agilent Technologies. Hence,

Figure 32. Simulation results (a) Conversion efficiency versus input power (b) Output voltage versus input power (c) reflection coefficient versus frequency

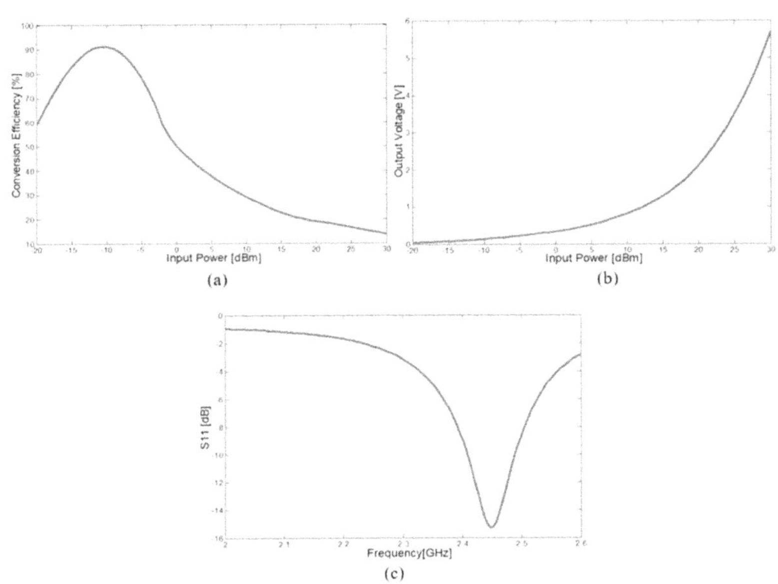

Figure 33. (a) fabricated rectifier (b) measured output voltage.

we were required to achieve the structure to validate the results obtained by simulations, the antenna have been printed on an FR4 substrate with dielectric permittivity constant 4.4, thickness of 1.6 mm and loss tangent of 0.025 and tested by means of a Network Analyzer to extract the reflection coefficient then in an anechoic chamber to obtain the radiation pattern of the antenna. The final fabricated structure shows a good accordance between simulation and measurement results. The second section have been dedicated to the rectifier part of a rectenna system. In one hand, a summary of the achieved and published rectifier structures using different topologies and configurations have been provided. In the other hand, a new 2.45 GHz micro strip bridge rectifier have been designed and optimized using ADS software. The structure consists of 5 HSMS 2820 Schottky diodes mounted on the basis of a bridge topology. The micro strip lines have been optimized with the optimization function available in ADS to obtain a well-matched rectifier circuit. Good performances have been observed in terms of matching input impedance, output voltage and conversion efficiency, we have reached an important output voltage of 5.7V. The structure has been validated throughout measurements and considered adequate to overcome the power supply problems observed in several cases.

REFERENCES

Advanced Design System (ADS). (n.d.). Retrieved from http://www.home.agilent.com/agilent/home.jsp

Alahmad, M. A., & Hess, H. L. (2008). Evaluation and Analysis of a New Solid State Rechargeable Microscale Lithium Battery. *IEEE Transactions on Industrial Electronics*, *55*(9), 3391–3401. doi:10.1109/TIE.2008.925318

Figure 34. Simulated and measured results of the rectifier (a) S11 versus Frequency (b) Output voltage versus input power.

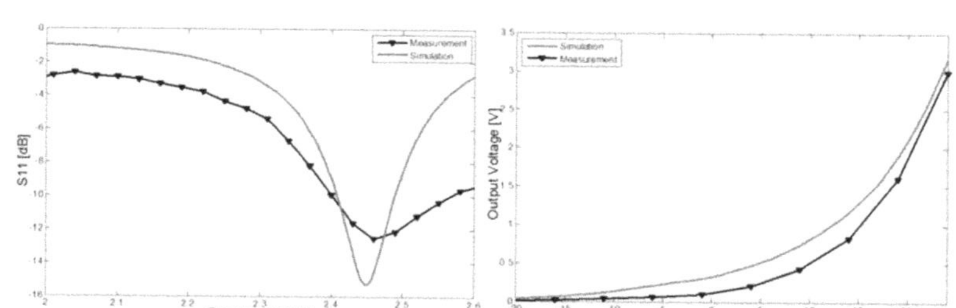

Balanis, C. A. (2005). *Antenna Theory, Analysis and Design* (3rd ed.). New York, NY: John Wiley & Sons.

Bharj, S. S., Camisa, R., Grober, S., Wozniak, F., & Pendleton, E. (1992). High efficiency Cband 1000 element rectenna array for microwave powered applications. *IEEE MTTS Int. Microwave Symp. Dig., Albuquerque, NM*, 301.

Brown, W. C. (1965). Experimental airborne microwave supported platform. *Raytheon, Burlington, MA, USA, Tech. Rep. RADC-TR-65-188. Contract (New York, N.Y.), AF30*(602), 3481.

Brown, W. C. (1984). The history of power transmission by radio waves. *IEEE Transactions on Microwave Theory and Techniques, MTT-32*(9), 1230–1242. doi:10.1109/TMTT.1984.1132833

Brown, W. C., & Triner, J. F. (1982). Experimental thin-film, etched-circuit rectenna. *IEEE MTT-S Int. Microwave Symp. Dig., Dallas, TX*, pp. 185-187. 10.1109/MWSYM.1982.1130655

Brunton, S. L., Rowley, C. W., Kulkarni, S. R., & Clarkson, C. (2010). Maximum Power Point Tracking for Photovoltaic Optimization Using Ripple-Based Extremum Seeking Control. *IEEE Transactions on Power Electronics, 25*(10), 2531–2540.

Chen, P. H., Ishida, K., Zhang, X., Okuma, Y., Ryu, Y., Takamiya, M., & Sakurai, T. (2010). 0. 18-V Input Charge Pump with Forward Body Biasing in Startup Circuit using 65nm CMOS. In *Custom Integrated Circuits Conference (CICC), IEEE*, pp. 2–5.

Chuc, D. H., & Duong, B. G. (2016). Design and Fabrication of Rectifying Antenna Circuit for Wireless Power Transmission System Operating at ISM Band. *Iranian Journal of Electrical and Computer Engineering, 6*(4), 1522–1528.

Dolgov, A., Zane, R., & Popovic, Z. (2010). Power Management System for Online.

Douyére, A. (2008). *Methodology for synthesis and optimization of rectenna bounded to an operating domain. Physics*. France: University of Reunion.

El Hamraoui, A., Abdelmounim, E. H., Zbitou, J., Bennis, H., & Latrach, M. (2016). A new compact CPW-fed dual-band monopole antenna for RFID applications. *In Proceedings of The 5th International Conference on Multimedia Computing and Systems (ICMCS)* (pp. 492-496). Marrakech, Morocco. IEEE.

Gibson, W. C. (2008). The Method of Moments in Electromagnetics. *Chapman* & Hall/CRC.

Gupta, S., Dhillon, S. S., Khera, P., & Marwaha, A. (2013). Dual Band U-Slotted Microstrip Patch Antenna for C Band and X Band Radar Applications. *5th International Conference on Computational Intelligence and Communication Networks (CICN)*, pp. 41-45. 10.1109/CICN.2013.18

Harrist, D. W. (2001). Wireless Battery Charging System Using Radio Frequency Energy Harvesting. *Thesis BS, University of Pittsburgh.*

HSMS-282X series. (2001). Surface mount RF Schottky barrier diode. *Agilent Technol.,* Palo Alto, CA, Tech. Data.

Huang, M., Tsai, Y., & Chen, K. (2010). Sub-1 V Input Single-Inductor Dual-Output (SIDO) DC-DC Converter with Adaptive Load-Tracking Control (ALTC) for Single-Cell-Powered Systems. *IEEE Transactions on Power Electronics, 25*(7), 1713–1724.

Lee, K. F., & Luk, K. M. (2010). *Microstrip Patch Antennas*. London, U.K.: Imperial College Express. doi:10.1142/p669

Lee, K. F., Yang, S. L. S., & Kishk, A. (2008). Dual- and multi-band U-slot patch antennas. *IEEE Antennas Wireless Commun. Lett.*, *7*, 645–648. doi:10.1109/LAWP.2008.2010342

Low Power RF Energy Harvesting Optimization. (n.d.). *IEEE Transactions on Circuits and Systems. I, Regular Papers*, *57*(7), 1802–1811. doi:10.1109/TCSI.2009.2034891

Marian, V., Vollaire, C., Verdier, J., & Allard, B. (2011). Rectenna circuit topologies for contactless energy transfer. Eindhoven, The Netherlands.

Matsunaga, T., Nishiyama, E., & Toyoda, I. (2015). 5.8-GHz Stacked Differential Rectenna Suitable for Large-Scale Rectenna Arrays With DC Connection. *IEEE Transactions on Antennas and Propagation*, *63*(12), 5944–5949. doi:10.1109/TAP.2015.2491319

McSpadden, J. O., Fan, L., & Chang, K. (1998). Design and experiments of a high conversion-efficiency 5.8-GHz rectenna. IEEE Trans. On Microwave Theory and Techniques, 46(12), 2053-2060.

Merabet, B., Cirio, L., Takhedmit, H., Costa, F., Vollaire, C., Allard, B., & Picon, O. (2009a). Low-cost converter for harvesting of microwave electromagnetic energy. In 2009 IEEE Energy Conversion Congress and Exposition, pp. 2592–2599.

Ohira, T. (2013). Power efficiency and optimum load formulas on RF rectifiers featuring flow-angle equations. *IEICE Electronics Express*, *10*(11), 1–9. doi:10.1587/elex.10.20130230

Ojha, J. R., & Peters, M. (2010). *Patch Antennas and Microstrip Lines*. InTech Germany.

Paris, D. T., & Hurd, F. K. (1969). *Basic Electromagnetic Theory*. New York, NY: McGraw-Hill Book Co.

Raytheon Company, Wayland. (1977). Electronic and mechanical improvement of the receiving terminal of a free-space microwave power transmission system. *MA, Tech. Rep. PT-4964, NASA Rep.CR135-194.*

Ren, Y. J., Li, M. Y., & Chang, K., (2007). 35 GHz rectifying antenna for wireless power transmission. *Electronics Letters*, *43*(11).

Richelli, A., Colalongo, L., Tonoli, S., & Kovacs-Vajna, Z. (2009). A 0.2-1.2V DC/DC Boost Converter for Power Harvesting Applications. *IEEE Transactions on Power Electronics*, *24*(5-6), 1541–1546. doi:10.1109/TPEL.2009.2013224

Shams, K. M. Z., & Ali, M. (2007). Wireless power transmission to a buried sensor in concrete. *IEEE Sensors Journal*, *7*(12), 1573–1577. doi:10.1109/JSEN.2007.908230

Tanuj Kumar Mandal. (2006). Wireless Transmission of Electricity – Development and Possibility. *Sixth International Symposium Nikola TESLA*, SASA, Belgrade, Serbia, Oct. 18–20.

Taybi, A., Tajmouati, A., Zbitou, J., Errkik, A., & Latrach, M. (2019). A new design of a microstrip rectenna at 5.8 GHz for wireless power transmission applications. *Iranian Journal of Electrical and Computer Engineering*, *9*(2), 1258–1266.

Taybi, A., Tajmouati, A., Zbitou, J., Errkik, A., Latrach, M., & El Abdellaoui, L. (2018). A New Configuration of a High Output Voltage 2.45 GHz Rectifier for Wireless Power Transmission Applications. *Telecommunication, Computing, Electronics and Control Journal*, *16*(5), 1939–1946.

Taybi, A., Tajmouati, A., Zbitou, J., Errkik, A., Latrach, M., & El Abdellaoui, L. (2019). A new design of high output voltage rectifier for rectenna system at 2.45 GHz. *Indonesian Journal of Electrical Engineering and Computer Science*, *13*(1), 226–234. doi:10.11591/ijeecs.v13.i1.pp226-234

Yoo, T. W., & Chang, K., (1991). 35 GHz Integrated Circuit Rectifying Antenna with 33% Efficiency. *Electronics Letters*, *27*(23), 2117.

Zbitou, J., Latrach, M., & Toutain, S. (2006). Hybrid Rectenna and Monolithic Integrated Zero-Bias Microwave Rectifier. *IEEE Transactions on Microwave Theory and Techniques*, *54*(1), 147–152. doi:10.1109/TMTT.2005.860509

Zhang, B. H., Zhang, J. W., Wu, Z. P., Liu, C. G., & Zhang, B. (2017). A 2.45 GHz Dielectric Resonator Rectenna for Wireless Power Transmission. *International conference of Electronics, Communication and Aerospace Technology (ICECA)*. Coimbatore, India. pp. 20-22. 10.1109/APCAP.2017.8420562

Chapter 5
GaAs Solid State Broadband Power Amplifier for L and S Bands Applications

Mohamed Ribate
University of Hassan 1st, Morocco

Rachid Mandry
University of Hassan 1st, Morocco

Larbi El Abdellaoui
University of Hassan 1st, Morocco

Mohamed Latrach
ESEO, France

ABSTRACT

This chapter provides an insight view of the Broadband Power Amplifier (BPA) design. Basically, the aim of the BPA is to increase the power level of the signal presents at its input terminal up to a prefixed power level at its output terminal in the operating frequency band. The research of a GaAs single stage solid state broadband power amplifier based of ATF13876 which operates in the frequency band ranging from 1.17 GHz to 3 GHz is presented in this chapter. The wider bandwidth circuits are designed by using transmission lines which are intrinsically wideband circuits. With carefully designed biasing and broadband matching networks, unconditionally stability and excellent matching performance are fulfilled over the overall operating bandwidth with a maximum power gain of 17.34 dB and a saturated output power of 17 dBm. Considering the wider bandwidth of the proposed BPA, the latter compares favorably with the contemporary state-of-the-art.

INTRODUCTION

Today, every person of us uses communication systems at home, on the moving or anywhere. We re-

DOI: 10.4018/978-1-7998-0117-7.ch005

ceive RF signals from satellites in our home on our satellite TV receivers, as well as at our smart phones from global positioning systems (GPS). We are driving cars that equipped with new tools that employ microwave technology such as radars for collision detection (Ribate, et al., 2018). Other types of radars are used in detection, identification and navigation applications. Basically, almost every device of these communication systems has some sort of transceiver, and inherently a power amplifier operates in one of the following bands determined by IEEE Standard 521-1984: L, S, C, X or Ku bands (IEEE, 1984).

However, with the rapid evolution of wireless communications, driven mostly by system-level requirements, and the subsequent growing demand to transmit mounting quantity of data, which resulted in the adoption of complex modulation schemes such as Orthogonal Frequency-Division Multiplexing (OFDM) and Wideband Code-Division Multiplexing Access (WCDMA), a growth request of broadband applications pop up (Coleman, 2004). As a consequence, the compelling necessity to evolve efficient and cheaper power amplifies in order to satisfy these demands has pushed researchers towards sophisticated architectures and solutions.

In general, whether characterized as narrowband, broadband, low noise, high efficiency, high power or otherwise, the design of all power amplifiers is ultimately equivalent to the design of an appropriate circuit able to convert the DC power supplied to an active device, into Microwave/ RF energy, thereafter transferred to an external load (Colantonio, Giannini, & Limiti, 2009). Various classes of operation and techniques have been developed and investigated towards this purpose, attempting to reduce overall DC power consumption and optimize power conversion as well as increasing system efficiency and spreading the operating bandwidth. Among such approaches, the Broadband Power Amplifiers (BPA), thanks of its numerous features including low cost, low complexity and broad bandwidth. In addition, BPAs can significantly reduce the hardware research & develop costs due to compatibility for the new and old wireless standards (Kazimierczuk, 2015).

Essentially, Broadband Power Amplifiers span a wide range of areas, among which microwave crop drying and quarantine in agriculture, medical diathermy, medical imaging, heating, electronic warfare, telecommunications, tracking and navigation systems represents just a few examples. Given such tremendously diversified fields, PA requirements may greatly vary in design, operating and technological specifications, which resulted in the broad variety of power amplifier realizations from microwave heating tubes to power amplifies composing hyperthermia devices, and from travelling-wave tube amplifiers used in satellite payloads to solid state power amplifiers deployed in personal wireless communication handsets.

The design of BPAs, as that for any other power amplifier circuit, is basically subdivided in a chain of methodical steps, from the identification of BPA specifications up to the concluding characterization and measurements of the fabricated circuit, to prove fulfilment of the design requirements. Throughout the design of BPAs, essential and commonly contrasting specifications have to be at the same time fulfilled. On the one hand, wide bandwidth, high output power as well as high power gain are typically needed to minimize the number of PA stages, and consequently reduce the size of the whole unit. Furthermore, in order to ensure adequate signal amplification unaccompanied by affecting the data content, as well as preserve the suitable quality for the transmitted signal, high linearity must be assured (Bahl, 2009).

There are several techniques and circuit topologies deployed to realize broadband power amplifiers. These techniques include the Real Frequency Technique (RFT), parallel or shunt resistive feedback, the traditional resistive / reactive matching, the traveling-wave approach and balanced configurations. The microwave power amplifier that reactively matched employs completely reactive matching circuits at the input and the output of the active device. Either transmission lines or lumped capacitors and inductors can be utilized. While the technique gives good noise and power fulfillment, it achieves a moderate bandwidth. However, owing to the instability of the active devices and gain roll-off, it is difficult to synthesis a BPA using this technique.

The resistive/reactive technique uses resistors in its matching circuits in order to reach a broad bandwidth with good gain flatness. The widespread topology of this technique is to employ inductors or transmission lines in shunt with series resistors. Whereas the resistive/reactive topologies are fitting for bandwidths less than 1 octave, this technique has poor VSWR performance and necessitates more matching elements to attain ultra-broad bandwidth.

The resistive feedback technique uses a resistance from the drain to the gate. This can stabilize the active device and match the input and the output impedances to the desired 50 ohms. This technique gives good VSWR performance, good gain flatness and multi octave bandwidth despite the lower power density and higher noise figure. The balanced topology has good VSWR performance and good gain flatness up around a two-octave bandwidth, but the noise figure is a bit higher and PAE is a little lower than a single stage power amplifier owing the use of the coupler that introduces additional loss. The distributed or traveling-wave power amplifier approach has wider bandwidth, good gain flatness and low VSWR, with restricted output power, poor PAE and large size.

The RFT was primarily introduced by Carlin and subsequently developed by Amstuz and Komiak, the technique is an optimum solution for the design of BPAs (Jarry & Beneat, 2016). Basically, the RFT is a CAD (Computer-Aided Design) procedure that utilizes the Levenberg-Marquardt optimization algorithm and the measured scattering parameters for the active device in order to solve the least-squares problems (Yarman, 2010). In general, during the design of BPA, we commonly deal with mathematical functions used in the design of classical matching circuits. On the contrary, when the RFT applied to the design of BPA, the technique admits measured scattering parameters for characterizing the active devices. Furthermore, the RFT does not need a priori determination of equalizer topology or an analytical formula for the transfer function of the overall system. The RFT has the feature of commonness, and it simple formalism permits optimizing numerous parameters of single or multistage power amplifiers. Thus, the RFT is perhaps the finest circuit synthesis procedure available for the design of BPA.

In this chapter a broadband power amplifier operating in L and S bands from 1.16 GHz to 3 GHz is introduced. The broadband impedance matching techniques and the broadband biasing schemes will be briefly presented and described. Then, the power amplifier design considerations including scattering parameters and stability analysis will be described. Finally, the proposed broadband power amplifier circuit design as well as the simulated results will be reported.

Broadband Impedance Matching Techniques

Impedance matching or transformation is an important part of the design process of microwave and RF power amplifiers. Impedance transformation is essential to achieve an optimum power gain, output power, linearity (dynamic rang) and efficiency as well as evolving the sensitivity of receivers. In addition, it permits a maximum transfer of the RF energy available from the generator to the active device, from the active device to the load or between devices. There are various matching schemes that can be used in order to match two mismatched impedances. The main factors to be considered in the chosen of a matching technique are ease of implementation, complexity, frequency response and bandwidth (Grebennikov, Kumar, & Yarman, 2016).

In this section, impedance matching techniques for broadband applications are presented. The techniques for designing distributed as well as lumped matching circuits are presented by means of examples. The impedance matching techniques illustrated her are in common applicable to solve impedance matching problems using planar transmission lines such as microstrip lines, strip lines, coplanar waveguides or coaxial lines and parallel stubs.

Theoretical Limits on the Bandwidth

Before starting the study of the broadband impedance matching circuits, it is worth considering the theoretical limits that restricting the performance of a given impedance transformation network. This can be described by the help of the Bode-Fano criterion, that was firstly introduced in the 1930s by Bode and Fano, who found that the integral of the reflection coefficient is restricted by a constant. (Bode, 1950) (Fano, 1960). The criterion set for different types of load impedances a theoretical limit on the bandwidth that can be reached when a lossless matching network terminated in an arbitrary load. In other words, the so-called Bode-Fano criterion gives the minimum magnitude of the reflection coefficient that can be achieved when an arbitrary source is matched with an arbitrary load. Such an optimum technique is essential because it provides the upper limit of performance and permits us a trade-off between bandwidth, return loss and circuit complexity as well as gives us a benchmark for comparisons of useful design. Figure 1.a shows a parallel RC load impedance matched with a lossless network. For this case, the Bode-Fano criterion is given by:

$$\int_0^\infty ln\left|\frac{1}{\Gamma(\omega)}\right| d\omega \leq \frac{\pi}{RC} \tag{1}$$

$$\int_0^\infty \ln\frac{1}{|\Gamma(\omega)|} d\omega \leq \frac{\pi}{RC} \text{ (2.a)}$$

$$\int_0^\infty \frac{1}{\omega^2}\ln\frac{1}{|\Gamma(\omega)|} d\omega \leq \pi RC \text{ (2.b)}$$

$$\int_0^\infty \frac{1}{\omega^2}\ln\frac{1}{|\Gamma(\omega)|} d\omega \leq \frac{\pi L}{R} \text{ (2.c)}$$

$$\int_0^\infty \ln\frac{1}{|\Gamma(\omega)|} d\omega \leq \frac{\pi R}{L} \text{ (2.d)}$$

Figure 1. Bode-Fano limits for 4 types of simple loads. (a) Matching parallel RC, (b) Matching series RC, (c) Matching parallel LR and (d) Matching series LR.

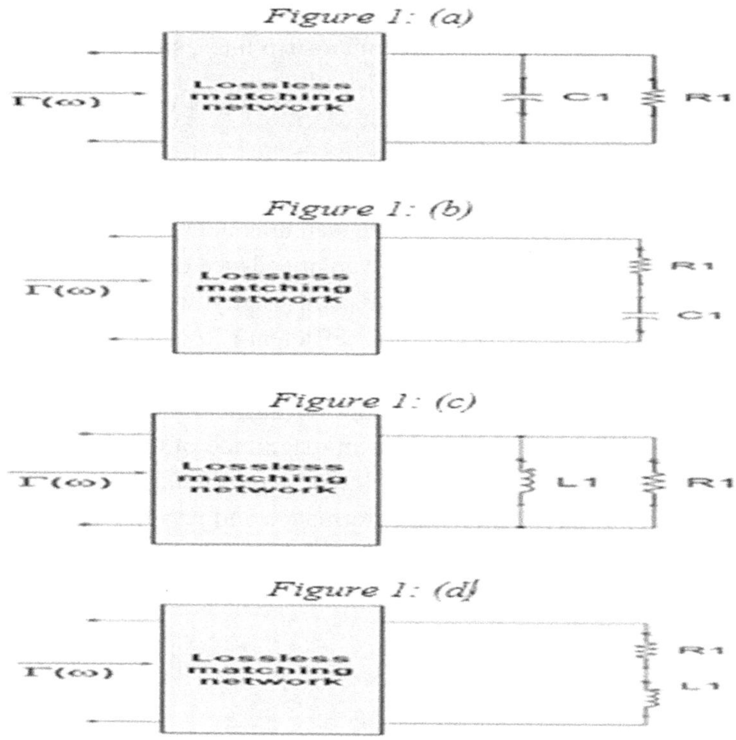

Where (ω) is the reflection coefficient of the arbitrary matching network.

Admit that a designer wish to design a matching circuit with a minimum return loss and is constant over the requested frequency band (ω_a to ω_b) and unity anywhere else as shown in figure 2. The best return loss that can be reached is given by:

$$\int_{\omega a}^{\omega b} \ln\left|\frac{1}{\Gamma(\omega)}\right| d\omega = \frac{\pi}{RC} = \Delta\omega \ln\left|\frac{1}{\Gamma_m}\right| \text{(3.a)}$$

And $\Gamma_m = e^{-\pi \frac{Q_C}{Q_L}}$ (3.b)

$$Q_C = \frac{\omega_0}{\Delta_\omega}; \Delta_\omega = \omega_b - \omega_a; BW = {}^{1}\!/_{Q_C} \text{ (4.a)}$$

$$Q_L = \omega_0 RC \text{ (4.b)}$$

Figure 2. Representation of Bode-Fano criterion; " $_m$ is the maximum tolerable return loss in passband; " $_m$ is the passband which is the optimum response.

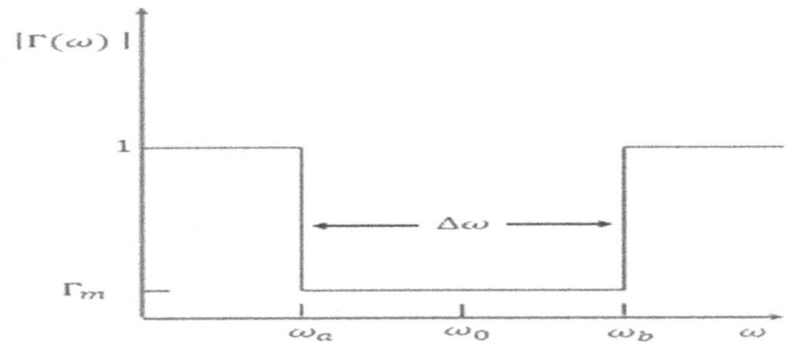

$$\omega_0 = \sqrt{\omega_a \omega_b} \quad (4.c)$$

$$Q_L BW = \frac{\pi}{\ln\left(1/\Gamma_m \right)} \quad (4.d)$$

Where Q_L and Q_C are the load quality factor and circuit quality factor.
The conclusions that can be drawn from the Bode-Fano criterion are summarized below:

- For a given load impedance (for example a fixed RC product), a broader bandwidth ($\Delta\omega$) can be reached only with the sacrifice of return loss in the passband. In other words, the wider bandwidth is the higher value of reflection coefficient.
- The return loss in the passband cannot be zero except if $\Delta\omega=0$. This perfect match can be obtained only at a single frequency.
- From (3) and (4), as R and/or C increase, the quality of the match ($1/\Gamma_m$ or/and $\Delta\omega$) decrease. As a consequence, higher quality circuits (Qc) are basically difficult to match than lower quality circuits.
- When R = ∞, the quality load (Q_L) and the minimum reflection coefficient Γ_m becomes infinite. As a result, a lossless circuit cannot match a reactive load.
- When C = 0, the quality load (Q_L) and the minimum return loss Γ_m are zero. Multi section impedance transformers with unlimited numbers have infinite theoretical bandwidth.

One Port and Two Port Circuits

The ordinary impedance matching problem is examined as a one port in essence (Poole, Darwazeh, 2016), wherever a real or complex load impedance Z_L is to be transformed to a real source Z_S as depicted in figure 3. As already outlined in detail in the previously subsection, Bode and Fano results exhibit that a perfectly impedance transformation network for a complex load can be a band pass filter that break

Figure 3. One-port circuit with power waves.

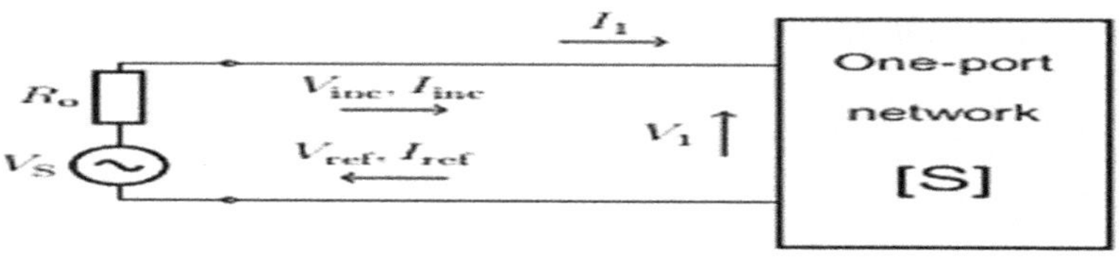

down pointedly at the frequency band bounds. On the other hand, both high pass or low pass filters can be deployed to match a given load impedance. High pass filter structures can provide a wideband impedance transformation for RF/Microwave impedance that can be approximately designed by using a resistance and capacitance in series, or by a conductance and an inductance in parallel. Uniformly, in order to improve the bandwidth, low pass filter networks can be approximately synthesized by using a resistance in series with an inductance, or by a conductance in parallel with a capacitance. However, both high pass and low pass filters are appropriate for restricted applications, contrasted with band pass filters owing to the fact that shows some inherent weaknesses.

Nevertheless, when the matching circuit design is established by using the network theory, the active device is represented by a two-port impedance (Z), admittance (Y), or scattering parameters matrix. As a result, in various cases, the coupling of the input and output networks out of the active device, increase the complication of the matching networks synthesis based on a two-port device.

Basically, microwave circuits or active devices, which demand impedance match, can be totally characterized by using the measured parameters at the network ports regardless of the network contents. In spite of an active device can be a multi-port network - more than 2 terminals -, the two-port theory is applicable only between the output and input terminals or conversely, whereas all the other terminals are terminated in known loads. For example, for a three port FET active device, the two-port theory can be used in order to characterize the transistor behavior between the input terminal (the gate port) and the output port (the drain port), while the source terminal is supposed RF and DC supplied.

The Real Frequency Technique (RFT)

Generally, in the design process of narrow band power amplifiers, the main idea is to match the output and input impedances of the power amplifier to a given standard resistive load and source impedances over a narrow frequency band (Jarry & Beneat, 2016). In this case, two or three elements are chosen in order to synthesize the input and output matching circuits. Thereafter, the impedance values of the synthesized elements are obtained by using the nonlinear optimization techniques.

However, when we treat the broadband power amplifiers matching problems, the challenge becomes more difficult. Firstly, we have no outline about the optimum selection for the network topology. Moreover, even we successfully determine the optimum circuit topology for the matching circuits, this topology necessitate more than two or three elements to reach the broadband designs as well as the transfer function to be optimized is extremely nonlinear owing to the components values and necessitate good initials. Ultimately, we begin the nonlinear optimization with improvised components values. At the end of the optimization process, mostly we will either find the right values of the optimized elements, or end up with non-divergent endless iterations.

Figure 4. General block diagram of a single stage power amplifier.

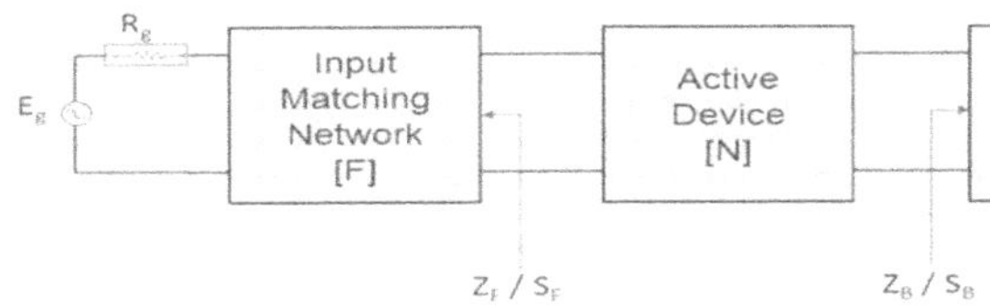

Consequently, in this subsection, we introduce novel matching technique, which is called the Real Frequency Technique (RFT), in order to design broadband power amplifiers. In the principle, the RFT is a semi-analytic broadband design technique that can be used to solve the broadband matching problems as well as realize lossless matching circuits with optimum network topologies.

Referring to Figure 4, a single stage RF/Microwave power amplifier consists of input matching network (Front-end equalizer [F]), active device and output matching network (Back-end equalizer [B]). The power amplifier can be driven either with large signals or small signals.

In the Real Frequency Technique, the input and output matching networks are completely characterized by the so-called Darlington driving point network functions among which impedance (Z_F and Z_B), reflectance (S_F , S_B), short emittance functions or equivalently admittance ($Y_F = 1 / Z_F$, $Y_B = 1 / Z_B$). in order to optimize performance parameters of the power amplifier, such as noise figure (NF), transducer power gain (TPG), voltage standing wave ratios (VSWR), … the positive real impedance functions $Z_F\left(p\right)$ and $Z_B\left(p\right)$ are deployed. The "*p*" represents the complex Laplace domain variables defined as: $p = \sigma + j\omega$.

The impedance functions $Z_F(p)$ and $Z_B(p)$ which are defined as rational functions of the complex Laplace domain variables are specified as:

$$Z_F\left(p\right) = \frac{a_F\left(p\right)}{b_F\left(p\right)} = \frac{a_{1,F}P^n + a_{2,F}P^{n-1} + \ldots + a_{n,F}P + a_{(n+1),F}}{b_{1,F}P^n + b_{2,F}P^{n-1} + \ldots + b_{n,F}P + b_{(n+1),F}} \quad (5)$$

$$Z_B\left(p\right) = \frac{a_B\left(p\right)}{b_B\left(p\right)} = \frac{a_{1,B}P^n + a_{2,B}P^{n-1} + \ldots + a_{n,B}P + a_{(n+1),B}}{b_{1,B}P^n + b_{2,B}P^{n-1} + \ldots + b_{n,B}P + b_{(n+1),B}} \quad (6)$$

Microstrip Broadband Matching Techniques

At high frequencies, the active devices are mismatched in the nature with the source and load impedances, and this leads to the loss of the RF signal transmitted through the power amplifier. As a result, to resolve this problem, an appropriate matching networks should be designed at the interface between the active device and the source and the load impedances respectively, or at the interface between any two sections having mismatched impedances. In other words, to ensure a maximum transfer of the RF power available from the input port to the input port of the active device, from the later to the output port, or between devices, a matching network must be synthesized in order to transform the mismatched impedances into desired impedances (Bahl, 2009).

Practically, there are broad variety of matching techniques that can be used for this aim, but the selection of the best approach might not always be clear (Ribate, et al., 2018). However, the following

Figure 5. Multi-section quarter wave transformer

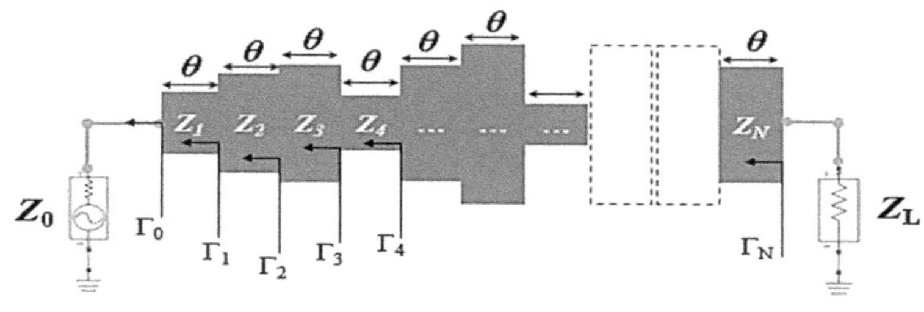

parameters must be properly taken into the account when selecting the matching technique: frequency bandwidth, ease of implementation and circuitry simplicity.

In this work, we combine several matching approaches in order to achieve the broadband performance. The first one is the Binomial transformer, which is a multi-section quarter wave transformer based on discrete sections of transmission lines having different characteristic impedances, but the same electrical length $\beta l = \theta$ with the aim of transforming the load impedance Z_L into the characteristic impedance of the feed line Z_0. A multi-section quarter wave transformer schematic is reported in figure 5.

Referring to the four sections binomial transformer illustrated in figure 6, the transmission lines sections impedances can be approximately calculated by using the following expressions:

$$ln\frac{Z_{n+1}}{Z_n} = 2\rho_n = 2^{-N} C_n^N ln\frac{Z_L}{Z_0} \text{Where: } C_n^N = \frac{N!}{\left(N-n\right)!n!} \quad (7)$$

Z_n and Z_{n+1} are the impedances of the nth and $\left(n+1\right)th$ transmission lines sections respectively, and ρ_n is the return loss coefficient at the junction between Z_n and Z_{n+1}, while C_n^N are the binomial coefficients.

However, at microwave frequencies, excluding an MMIC, it is generally tedious and difficult to implement matching networks synthesized with lumped elements (Hollister, 2007). As a result, the second technique deployed in this work is an approximate transformation of predesigned matching circuits based on lumped elements into matching networks based on transmission lines, which are more easy to implement on the PCB substrate.

Figure 6. Four sections quarter wave binomial transformer

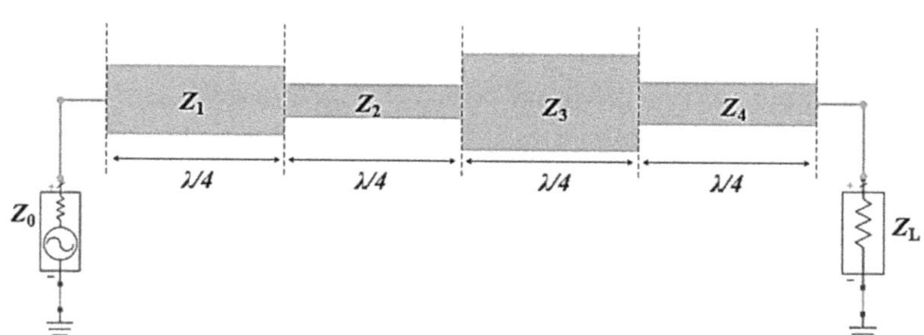

Figure 7. Transformation of lumped elements into transmission lines

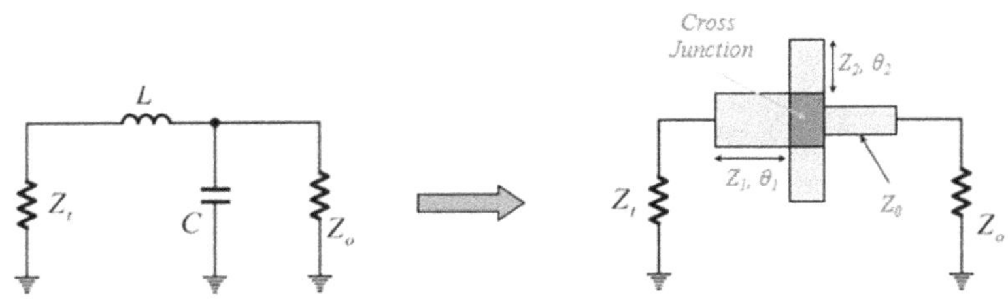

Referring to figure 7, the approximated values of Z_1, θ_1, Z_2 and θ_2 can be therefore calculated as:

$$X_1 = Z_1 \tan\left(\theta_1\right) \approx \omega L \tag{8}$$

$$\mathrm{X}_2 = \frac{Z_2}{2\tan\left(\theta_2\right)} \approx \omega C \tag{9}$$

Broadband Biasing Circuits Design

At microwave frequencies, the active device performance relies considerably on the DC biasing conditions. As a result, biasing circuits are essential parts, play a key role in the design and implementation of power amplifiers. They provide for a given active device the efficient bias conditions in terms of current and voltage and must be designed neatly to avoid low frequency oscillations, owing to the high capacitance, high device gain and low resistivity of the substrate as well as to provide a great deal to the stability and the effectiveness of the overall power amplifier network (Bahl, 2009). However, even though the transistor technology driven the main biasing parameters, the design principles are ultimately the same for all biasing methods.

As matter of fact, there are various biasing schemes available for power amplifiers and can be found as an integrated part of the matching circuits or could be stand-alone. In addition, the reactive effect of biasing circuits is either absorbed in the matching networks or simply negligible. In other words, if the RF power leak out through the bias terminal when the active device is supplied with a given DC bias, degradation of RF performance in terms of output power, gain, noise figure, IP3 and PAE can be takes places.

As a result, the bias voltage applied to the active devices must be carefully isolated by the helps of the biasing circuits which their role is to prevent the RF signal from leaking into the bias terminal. In this subsection, the design considerations of the biasing circuits of broadband power amplifiers will be described.

Figure 8. (a) device current versus gate voltage. (b) device I-V characteristics

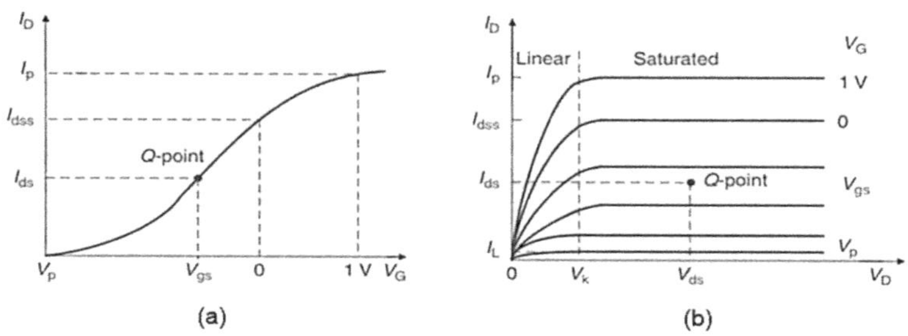

Biasing of Active Devices

Basically, the active devices biasing includes two important parts: the biasing circuit design and the selection of an adequate quiescent bias point in order to ensure an optimum performance of the overall circuit in terms of output power, gain, power added efficiency and noise figure. Both are essential for the power amplifiers operation, particularly dealing with high power levels and low noise.

Indeed, the selection of the proper quiescent bias point (Q-point) and remain constant current irrespective of the active device parameter instabilities owing to temperature fluctuations and process is one of the challenging tasks in the design of the DC biasing networks. However, this is linked to the active device's current – voltage known as DC I-V characteristics. Figure 8 depicts the I-V characteristics of an FET.

As we can see in figure 8, the broken line subdivides the I-V characteristics into two domains: the linear domain and the saturated domain. When the active device operates in the linear region, the signal loss takes place because of the resistive nature of the linear region, whereas the saturated region provides power and gain. As a result, all microwave transistor amplifiers are operated in this region.

Biasing Structures

Basically, the biasing structures described in this subsection, their task is to provide an adequate quiescent bias point for the power amplifier. Essentially, a biasing circuit includes an RF choke (in short RFC) and a DC block. Figure 9 illustrates a generic DC bias structure. The intent of the RFC is to prevent the RF signal from leaking through the bias port. In other words, the RFC are usually deployed in order to suppress the AC signals while overtaking the DC signals.

However, when the biasing circuit present a very high impedance to the matching circuits, its effect becomes negligible. In practice, the DC bias blocks are generally implemented by using either MIM capacitors in MIMICs or chip capacitors in hybrid circuits. Theoretically, the capacitor should be presents a short circuit at the operating band of frequencies. The selection of the capacitor value must be done in such a way that the condition $\omega C = 2\pi f C$ is very high. As a result, $Z_C = -j\left(\frac{1}{\omega C}\right)$ is very small, therefore, the signal will be transmitted through the capacitor with minimum reflection or loss. Also, the capacitor type or value should be selected in a such a way that there is no simultaneous resonance in the operating frequency band, otherwise, a very high impedance take place at the capacitor that is susceptible to instabilities of the power amplifier as reported in the previous subsection.

Figure 9. A universal biasing configuration

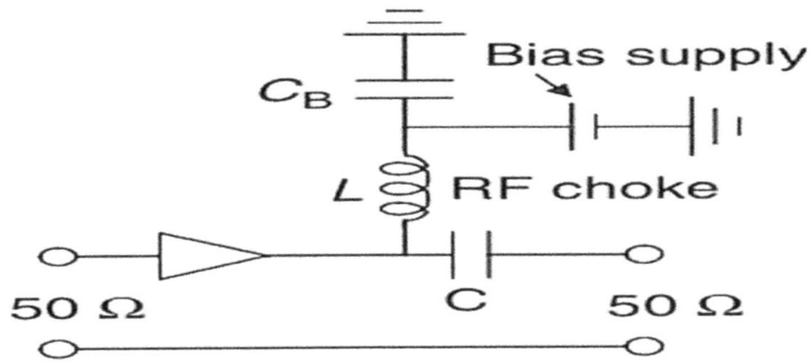

Similarly, in order to prevent the RF signal, the RFC acting as a DC short. At low frequency, lumped inductors are generally deployed as RFC. Also, its values must be selected in a such way that its impedance is very high at the operating frequency band.

In fact, at high frequencies the active devices are potentially unstable, the matching circuits become open circuited owing to low values of RFC and DC capacitors. In a such situation, in order to guarantee the unconditionally stability of power amplifiers, the biasing circuits provide the needed terminations.

From a practical standpoint, the gate base current for a given transistor is negligible where is compared to the drain collector current. Consequently, the drain collectors are generally biased through RFC, whereas the gate base can be supplied through RFC or simply by using resistors having high values.

Because in fact, that the BJT and HBT devices demand a positive base emitter voltage as well as the LDMOS and MOSFET devices need positive gate source voltage, a single power supply is deployed. On the other hand, FET and HEMT devices require a positive voltage for drains and negative voltage for gates, as a result, a dual power supplies are used. Figure 10 illustrates simple biasing configurations where a resistive voltage divider is deployed for base gate biasing.

The R1 and R2 values are determined by using the following equations:

$$V_B = \frac{R_2}{R_1 + R_2} V_C,\ V_{gs} = \frac{R_2}{R_1 + R_2} V_D,\ V_{gs} = \frac{R_2}{R_1 + R_2} V_G \tag{10}$$

Figure 10. DC bias networks: (a) BJT, (b) FET and (c) MOSFET

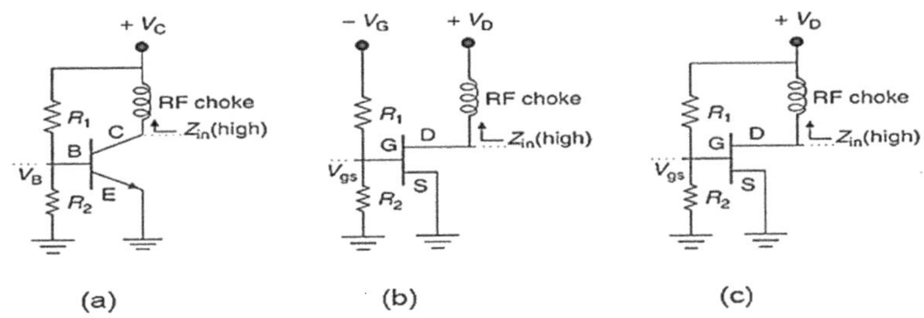

Figure 11. A number of biasing schemes for FET devices for low power applications

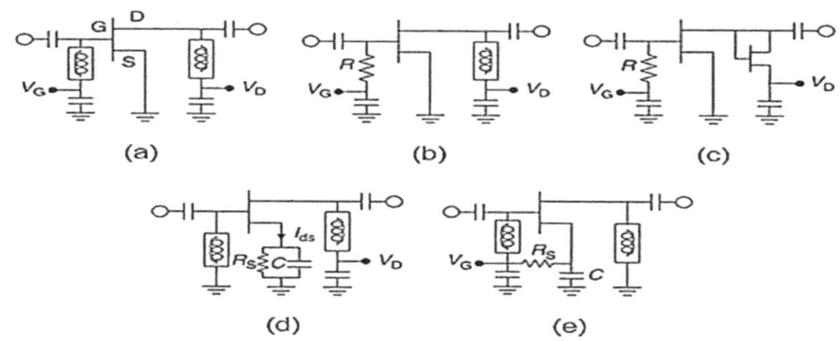

Note that the R1 and R2 values must be selected in a such way that the bias current is minimal as well as does not impact the Power Added Efficiency (PAE) of the power amplifier.

Five practical biasing structures for FET and HEMT devices are stated in figure 11. The (a), (b), and (c) configurations demand dual power supplies with the smallest possible source inductance. The (d) and (e) biasing networks with source resistors need only a single power supply, consequently, there are widely deployed for small and medium power applications.

Indeed, every scheme has some characteristics. The structure (a) is suitable for low noise/high gain applications and provides a high efficiency, while the configuration (b) gives high output power, high gain, high efficiency with a moderate low noise. However, the circuit (c) gives high gain, moderate noise, medium power and low efficiency, whereas permits a broad bandwidth at low frequencies. The (d) and (e) schemes provides moderate noise, low efficiency and low power with high gain. The RS resistor deployed in those schemes allows an automatic transient protection. Note that both (a) and (b) circuits are part of matching networks.

Microstrip Biasing Schemes

At microwave frequencies, the RFC is usually realized by using a shunt stub, also known as a high-impedance λ/4 line, terminated by an RF bypass capacitor C_B, while the DC block can be implemented by deploying either a 3-dB backward wave coupler or a capacitor described in figure 12.a. (Bozanic & Sinha, 2016)

In order to increase the bandwidth, two sections of λ/4 transmission line are used. In figure 12.b, λ/4 high impedance transmission line followed by λ/4 low impedance transmission line are used in order to guarantee an open circuit across the main RF line when it is needed. For the purpose to provide a broad bandwidth, a radial stub is used instead to the low impedance line section as depicted in figure 12.c.

Scattering Parameters

At low frequencies, the passive and active networks elements are commonly characterized by using the linear immittance parameters. As the operating frequency increases, the linear immittance parameters, which are defined as the ratios of the measured voltage and current, show more and more limitations. The most conspicuous limitation of immittance parameters owing to the fact that are measured under

Figure 12. Microstrip biasing circuit: (a) λ/4 impedance transformer with a coupled line DC block. (b) two λ/4 impedance transformers and a chip capacitor DC block. (c) λ/4 impedance transformer in series with a radial line stub and a chip capacitor DC block.

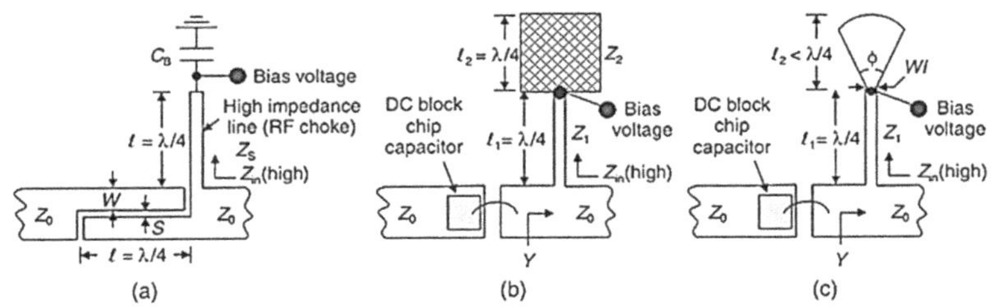

short and open circuits loads, which becomes more and more complicated to implement minutely at high frequencies.

In other hand, in the transmission line theory, the reflection coefficient is defined as the ratio of the reflected voltage wave to the incident voltage wave at any transmission line discontinuity.

It turns out that at microwave frequencies, the reflection coefficients are considerably easier to measure than static currents and voltages, consequently, would it not be better if we had a method of describing the behavior of a multi-port circuit by using some reflection coefficients as parameters?

In this subsection, we introduce an alternative method of characterizing two-port networks resulting in the so-called "S-Parameters", which are become the bedrock of microwave active network synthesize (Poole & Darwazeh, 2016). The "S" in S-parameters stands of "Scattering". Such parameters, which are linear complex coefficients, can characterize the behavior of a given two-port circuit in terms of reflected and incident power waves at the circuit terminals. In other words, the S-parameters depict what happens when an incident power wave propagating through a transmission line locates a discontinuity. In such situation, a portion of the incident energy reflected toward the source, while the remaining part of the incident power will be transmitted across the discontinuity. This is correct except two particular situations. The first take place when the discontinuity is a short circuit or open circuit. In this case, the incident power wave is completely reflected toward the source load. The second situation where the load impedance and the transmission line system are successfully matched –there is no efficaciously no discontinuity). In a such situation, the incident power wave is totally absorbed at the discontinuity.

Definition of S-Parameters

The scattering matrix or parameters of a two-port system can be defined in terms of the reflected and incident voltages and currents respectively. In other words, the S-parameters determines the relationships between the input and output incident waves as well as the input and the output reflected waves, when the network is ended on its characteristic impedance Z0. Typically, the reference characteristic impedance of the load and source used in commercial network analyzers is 50 Ω.

The scattering parameters can be determined with the help of figure 13, in which the load and source reflection coefficients are substituted with reflected and incident waves.

In general, a traveling wave can be characterized as follows:

Figure 13. Two port circuit with arbitrary source and load

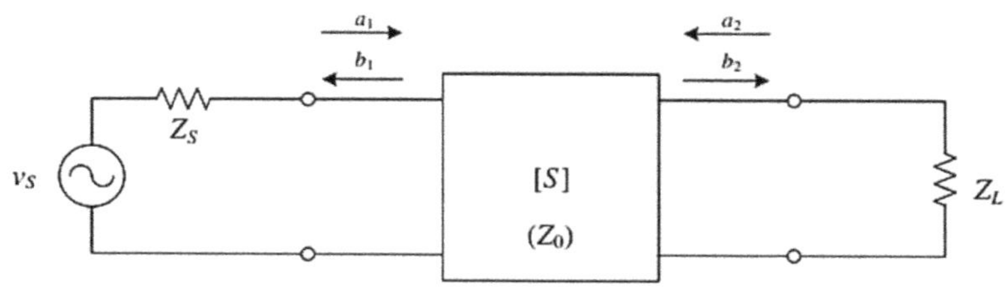

- A portion of the traveling wave emanating from the source and incident at the two port device (a_1) will be reflected as b_1, while another portion will be transmitted over the two port device.
- A portion of the transmitted wave is reflected from the load turn into incident at the output port of the two port device (a_2).
- A portion of the traveling signal (a_2) is reflected from the output port of the two port device backward to the load as b_2, while another portion is transmitted across the two port device toward the source.

Such characteristics indicate a call of two refection coefficients known as input return loss coefficient (S_{11}) and output return loss coefficient (S_{22}), as well as two gain coefficients known as revers isolation coefficient (S_{12}) and power gain coefficient (S_{21}).

As a result, the input port return loss S_{11} is defined as the measured ratio between b_1 and a_1 with a_2 set to zero. In the same way, the output return loss coefficient S_{22} is defined as the measured ratio between b_2 and a_2 with a_1 set to zero, and so on.

For a given two-port device, the scattering parameters are measured as follows:

$$\left.\begin{aligned} S_{11} &= \left.\frac{b_1}{a_1}\right|_{a_2=0} \\ S_{22} &= \left.\frac{b_2}{a_2}\right|_{a_1=0} \end{aligned}\right\} \tag{11}$$

$$\left.\begin{aligned} S_{12} &= \left.\frac{b_1}{a_2}\right|_{a_1=0} \\ S_{21} &= \left.\frac{b_2}{a_1}\right|_{a_2=0} \end{aligned}\right\} \tag{12}$$

Finally, the following two equations can therefore be inferred:

$b_1 = S_{11}a_1 + S_{12}a_2$

Figure 14. Two port sing stage power amplifier

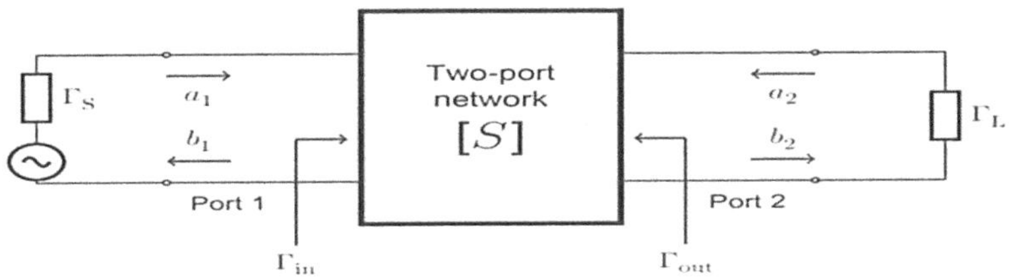

$b_2 = S_{21}a_1 + S_{22}a_2$ (13)

In the general case, the scattering parameters representation can be extended to describe the behavior of any network with an arbitrary n-pair of terminals by considering the reflected and incident power waves upon each port.

Expressions of Input and Output Impedances of a Two-Port Network in Terms of the Scattering Parameters

Consider the two-port circuit illustrated in figure 14, which is driven by a source having the return loss coefficient Γ_S and ended with an arbitrary impedance having the return loss coefficient Γ_L.

Typically, in a two port network, where the forward gain coefficient $S_{21} \neq 0$ and the reverse isolation coefficient $S_{21} \neq 0$, the input return loss Γ_{in} will be dependent on the load impedance value and reciprocally. Moreover, the immittance expressions of a two-port system driven by an RF generator and terminated with an arbitrary load, can be therefore derived is the scattering parameters form.

In order to derive the input return loss Γ_{in} of the two port system which is terminated with a load having the reflection coefficient Γ_L to the scattering parameters, we first starting by state the following relationship of the power wave upon the load impedance:

$$a_2 = b_{2\ L} \quad (14)$$

By substituting equation (13) into equation (12), the reflected wave at the output port b2 becomes:

$$b_2 = S_{21}a_1 + S_{22}b_{2L} \quad (15)$$

By developing equation (14) we can obtain the following expression of b2:

$$b_2 = \frac{S_{21}a_1}{1 - S_{22L}} \quad (16)$$

Since the return loss coefficient of a two port system is defined as:

$$_{in} = \frac{b_1}{a_1} \tag{17}$$

And by applying equation (13) and equation (17), the input return loss can be written as:

$$\Gamma_{in} = S_{11} + \frac{S_{12} b_2 \Gamma_L}{a_1} \tag{18}$$

Simply combining equation (16) and equation (18), we can obtain the following expression of the input return loss:

$$\Gamma_{in} = S_{11} + \frac{S_{12} S_{21} \Gamma_L}{1 - S_{22} \Gamma_L} \tag{19}$$

Assuming that $\Delta = S_{11} S_{22} - S_{12} S_{21}$ the reader can easily remark that equation (19) becomes:

$$\Gamma_{in} = \frac{S_{11} - \Delta \Gamma_L}{1 - S_{22} \Gamma_L} \tag{20}$$

Note that both form of the input return loss expression (equation (19) and (20)) are applicable in general use.

In the same manner, the output terminal yields output return loss coefficient with an arbitrary source impedance:

$$\Gamma_{out} = S_{22} + \frac{S_{12} S_{21} \Gamma_S}{1 - S_{11} \Gamma_S} \tag{21}$$

Which can in the same way be written as:

$$\Gamma_{out} = \frac{S_{22} - \Delta \Gamma_S}{1 - S_{11} \Gamma_S} \tag{22}$$

These important expressions of the input and output return loss (equation (19) and (21)) we will afterwards show another derivation of the signal flow graph technique section.

S-Matrix Classification: Reciprocal and Lossless Networks

When the power losses are identical between any pair of terminals regardless of the direction of propagation, in this case we call this type of network a reciprocal network. In terms of the scattering parameters, this means $S_{ij} = S_{ji}$ for any $i \neq j$. Typically, various passive networks such as attenuators, cables, couplers and power dividers are therefore reciprocal. In fact, the only case where a passive circuit is not reciprocal take place when it includes anisotropic materials which have several electrical characteristic (for example the relative dielectric constant) depending on the direction of the propagating waves. Ferrite is an example of an anisotropic material, which is deployed in the isolators and circulators fabrication in order to make these nonreciprocal circuits.

However, transistors and many other active devices are basically nonreciprocal, because in the most cases in fact, the power gain and the reverse transmission coefficients are different ($S_{12} \neq S_{21}$).

The conclusion that is directly inferred from the above would be that any reciprocal device has a symmetric s-parameter matrix. This means that all values over the upper right to the lower left diagonal are equal.

In other words, the scattering parameters matrix of a reciprocal device must fulfill the condition $\left[S\right]^{\dagger}\left[S\right]$, where $\left[M\right]^{\dagger}$ represents the conjugate transpose of a specified matrix (Hermitian Matrix). A typical scattering parameter matrix for a given two port reciprocal network can therefore be expressed as:

$$\begin{bmatrix} S_{11} & S_{12} \\ S_{21} & S_{22} \end{bmatrix} = \begin{bmatrix} x & y \\ y & x \end{bmatrix}$$

However, the lossless network is defined as a network where the quantity of power signals incoming in the all network terminals is equal to quantity of power signals leaving all network terminals. In other words, there is no power dissipation inside the circuit itself, $P_{in} = P_{out}$.

Assuming that P_{in} and P_{out} represent entire input and output powers, respectively. For an n-port network, the input and output powers are defined as:

$$P_{in} = \sum_{i=1}^{n} \left|a_i\right|^2 = \left[a\right]^{\dagger}\left[a\right] \quad (23)$$

$$P_{out} = \sum_{i=1}^{n} \left|b_i\right|^2 = \left[b\right]^{\dagger}\left[b\right] \quad (24)$$

By definition, for a lossless network, $P_{in} = P_{out}$, so by combining equation (23) and equation (24) we obtain:

$$\left[a\right]^{\dagger}\left[a\right] = \left[b\right]^{\dagger}\left[b\right] \quad (25)$$

Since, $b = \left[S\right]\ a$, equation (24) can therefore be written as:

$$\left[a\right]^{\dagger}\left[I\right]\left[a\right] = \left(\left[S\right]\left[a\right]^{\dagger}\right)\left(\left[S\right]\left[a\right]\right) \tag{26}$$

Where [I] represent the identity matrix. Then we can rewrite equation (26) as:

$$\left[a\right]^{\dagger}\left[I\right]\left[a\right] = \left[a\right]^{\dagger}\left[S\right]^{\dagger}\left[S\right]\left[a\right] \tag{27}$$

However, $\left[S\right]^{\dagger}\left[S\right] = \left[I\right]$ (28)As a result of equation (28), for an n x m port scattering matrix, for all m we obtain:

$$\sum_{i=1}^{n}\left|S_{im}\right|^{2} = 1 \tag{29}$$

The Signal Flow Graphs

The signal flow graph is a useful technique that serve to analyze and visualize the power flow inside a microwave network characterized by S-parameters. By definition, a signal flow graph is a pictorial modeling of a matrix of simultaneous equations. As a useful tool was primarily developed for control theory purposes, thereafter was applied by Kuhn [] in order to analyze RF networks described by means of S-parameters. Once the microwave network has been characterized by using the signal flow graph technique, the simplification and analysis of this network can be done by applying some basic rules. Comparing the use of algebraic technique and the signal flow graph technique, the latter is often faster even though both approaches return some results.

Typically, a signal flow graph is constituted with Branches and Nodes, which are determined as follows:

- Each terminal, i, of a given RF network has two nodes, a_i and b_i.
- Node a_i is characterized with a signal wave coming up port i, while node b_i is described with a signal wave reflected from terminal i.
- The voltage value of a node is equivalent to the sum of all signal waves entering this node.
- The directed path between two nodes is called Branch, describing the signal flow emanating from a given node towards another.
- Each branch is associated with coefficient. The latter can be a reflection coefficient or a scattering parameters coefficient.

The scattering parameters of a given two port system can be modeled using the signal flow graph representation as illustrated in figure 15.

Figure 15. Signal flow graph representation of a two port circuit.

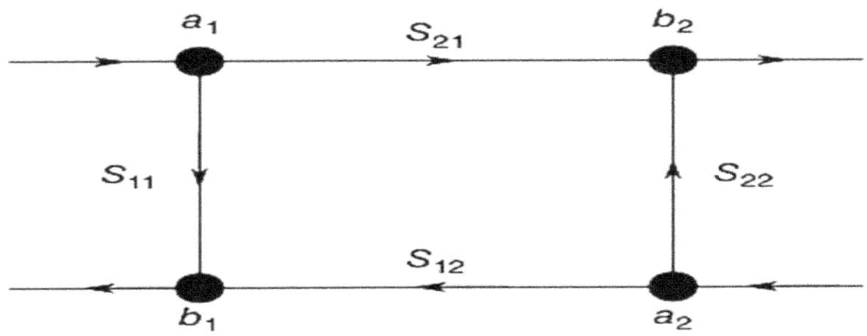

Stability Analysis of A Two Port Networks

In general, any power amplifier at microwave frequencies can be made to oscillate if an external positive feedback is applied, for instance, a high gain MMIC chip in a ceramic or plastic package with bad isolation (Bahl, 2009). As a result, the unavoidable parasitic effects are frequently enough to induce oscillations at the operating frequency or band of frequencies, of interest, if care is not taken in the synthesize and implementation of the designed power amplifier. It is worthwhile to note that the unwanted oscillations can be detected by the help of various indications among which any unexpected variation in the DC parameters of the power amplifier, networks that are very sensitive of their environment and output power without any input signals represent just a few examples. In general, almost all microwave power amplifiers are able to oscillate anywhere from 1 MHz to 40 GHz or higher.

Clearly, the overall stability of a given power amplifier must be fulfilled not only in the operating frequency ranges, but for all frequencies: specifically, for the low frequency band where the higher gain of the device could induce oscillations owing to the feedback elements that are inherently exist in the network or in the device itself (Ribate, et al., 2017).

As already outlined, any active device can be transformed into an oscillator by applying a positive feedback with an appropriate phase across this device. However, this feedback may be studied including inter-stage biasing circuits, source induction and closed circuits owing to the devices in parallel, or unintentional such as poor grounding and poor isolation. In many cases, once the stability problem has been detected, it is mainly easy to correct it. For various microwave applications, conditional stability is commonly tolerable over the operating frequencies where the terminal impedances are defined, whereas unconditional stability is more suited out of the frequency band where there is no idea about the system specifications.

In practice, various configurations and structures are deployed in order to protect power amplifiers from unknown impedances, among which balanced structures and ferrite isolators. However, if care is not taken in the synthesis of power amplifier isolator circuits, the stability problems may turn up outside the operating frequency of the designed isolator, where this latter is neither unilateral nor matched.in other hand, the stabilization of networks becomes very critical in the case of power amplifiers compared to small-signal amplifiers.

From large signal conditions point of view, the active device characteristics may be ultimately go through large fluctuations. In this respect, the power amplifiers design requires accurate stability analysis techniques, an appropriate nonlinear active device models and a suitable CAD tools. Moreover, the

Figure 16. Tow port circuit with an arbitrary source and load impedances

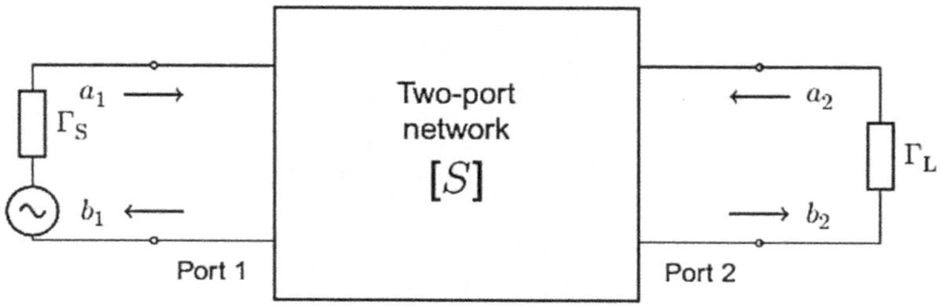

stability circuit of power amplifiers becomes an integral part of the design, because the system mandate that the power amplifiers must be unconditionally stable, while most active devices that are deployed in the design process are only conditionally stable.

Basically, there are two perspectives to examine a two-port stability. The first address the problem of stability in terms of immittance parameters, while the second investigates the stability in terms of scattering parameters.

In this subsection, we treat the problem of stability from the scattering parameters point of view. As already depicted, there are two categories of stability known as unconditionally stability and conditionally stability. A two port network is said to be unconditionally stable or inherently stable if the amplitude of input power P_{in} and output power P_{out} are both less than unity, for all the real parts of source impedance Z_G and load impedance Z_L. In other words, at a given frequency or band of frequencies, there is no combination of load and source impedances that can be made this two-port device to oscillate at that frequency or band of frequencies.

However, a two port is said to be conditionally stable or potentially unstable at a given frequency band, if a combination of source and load impedances can be existing and make it to oscillate at this frequency band.

Stability Circles

As stated before, a two-port device is conditionally stable because there is some combination of load and source impedances that make it unstable. In other words, there is a passive value of the load reflection coefficient Γ_L that will drive the input return loss Γ_{in} to be greater than unity ($\left|\Gamma_{in}\right| > 1$), and vice versa. As we will show in this subsection, the boundaries between unstable and stable regions on the load and source planes are in the form of circle, well-known as a stability circle (Pozar, 2009). The radius and centers of the stability circles are defined in terms of the scattering parameters of the active device.

Referring to figure 16, and from the load return loss plane point of view, the load impedance values that induce instability are defined as those which drive the input return loss magnitude of the active device to be greater than or equal to unity.

Referring to the simplified expressions reported in the previously subsection, the input return loss of a two port circuit which is terminated on an arbitrary load characterized by Γ_L, is defined as:

$$\Gamma_{in} = S_{11} + \frac{S_{12}S_{21}\Gamma_L}{1 - S_{22}\Gamma_L} = \frac{S_{11} - \Delta\Gamma_L}{1 - S_{22}\Gamma_L} \quad (30)$$

Considering the Γ_L plan, the boundary that separates unstable from stable regions is obtained by setting $|\Gamma_{in}| = 1$ in the previously equation. From the latter equation, we can clearly remark that all value of Γ_L that remain on his outlines fulfill the following:

$$|S_{11} - \Delta\Gamma_L| = |1 - S_{22}\Gamma_L| \quad (31)$$

Especially, for simplification purposes, we define the following variables:

$$C_2 = S_{22} - \Delta S_{11}^* \quad (32)$$

$$D_2 = |S_{22}|^2 - |\Delta|^2 \quad (33)$$

By squaring and assembling both sides of equation (31), we obtain:

$$|S_{11}|^2 + |\Gamma_L|^2\left(|\Delta|^2 - |S_{22}|^2\right) + 2\Re e(\Gamma_L)\Re e(C_2^*) + 2\,\mathrm{Im}(\Gamma_L)\,\mathrm{Im}(C_2^*) = 1 \quad (34)$$

Equation (34) can be simplified as:

$$|S_{11}| + |\Gamma_L|^2\left(|\Delta|^2 - |S_{22}|^2\right) + \Gamma_L C_2^* + \Gamma_2^* C_2^* = 1 \quad (35)$$

Consequently,

$$|\Gamma_L|^2 - \Gamma_L \frac{C_2^*}{D_2^2} - \Gamma_L^* \frac{C_2}{D_2^2} = \frac{|S_{11}|^2 - 1}{D_2^2} \quad (36)$$

By adding $|C_2|^2 / D_2^2$ to both sides of equation (36), we obtain:

$$|\Gamma_L|^2 + \frac{|C_2|^2}{D_2^2} - \Gamma_L \frac{C_2^*}{D_2^2} - \Gamma_L^* \frac{C_2}{D_2^2} = \frac{|S_{11}|^2 + |C_2|^2 - 1}{D_2^2} \quad (37)$$

Therefore, we can easily infer that equation (37) is in the form of a circle in the Γ_L plan, defined as:

$$\left|\Gamma_L - C_{SL}\right|^2 = \left|\gamma_{SL}\right|^2 \tag{38}$$

Where the center of the load plan stability circle being at:

$$C_{SL} = \frac{C_2^*}{\left|S_{22}\right|^2 - \left|\Delta\right|^2} \tag{39}$$

And the radius being at:

$$\gamma_{SL} = \frac{\sqrt{\left(\left|S_{11}\right|^2 - 1\right) - \left(\left|S_{22}\right|^2 - \left|\Delta\right|^2\right) + \left|C_2\right|^2}}{\left|S_{22}\right|^2 - \left|\Delta\right|^2} \tag{40}$$

This equation can be simplified to yield:

$$\gamma_{SL} = \frac{\left|S_{12}S_{21}\right|}{\left|S_{22}\right|^2 - \left|\Delta\right|^2} \tag{41}$$

In the same manner, we can determine the stability circles of the source plane by taking into consideration the boundary of the load plane stability which is given by $\left|\Gamma_{Out}\right| = 1$, where, this latter is defined as:

$$\Gamma_{out} = S_{22} + \frac{S_{12}S_{21}\Gamma_S}{1 - S_{11}\Gamma_S} = \frac{S_{22} - \Delta\Gamma_S}{1 - S_{11}\Gamma_S} \tag{42}$$

In the same way, by following a similar analysis as reported above, we start with equation (42), then we can determine the stability circles of the source plan in terms of center and radius as follows:

$$C_{SS} = \frac{C_1^2}{\left|S_{11}\right|^2 - \left|{}''\right|^2} \tag{43}$$

And

Figure 17. Stability circles of source plane (shaded area represent stable region)

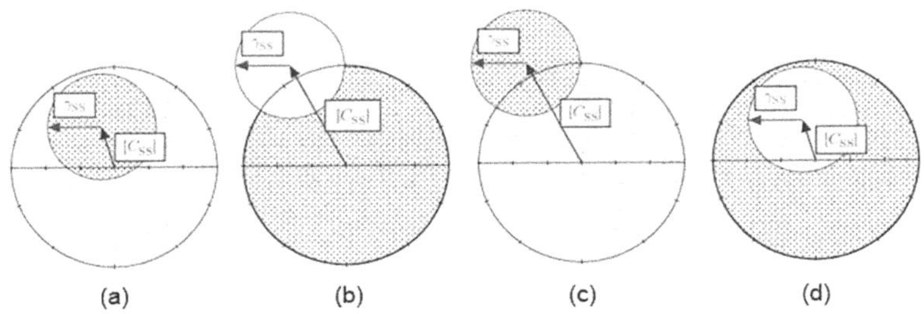

$$\gamma_{SS} = \frac{\left|S_{12}S_{21}\right|}{\left|S_{11}\right|^2 - \left|\Delta\right|^2} \tag{44}$$

Where

$$C_1 = S_{11} - \Delta S_{22}^* \tag{45}$$

The subscripts S (in C_{SL}, C_{SS} *and* γ_{SS}) denote stability, while the subscripts S and L symbolize source and load respectively.

Since the system mandate that the power amplifier must be operate under stable conditions, the stability circles serves to determine the outlines of the authorized impedances in terms of their respective return loss planes. Her an essential question appears; the stability region it is represented by the interior or exterior area of the stability circles? This question can be answered by considering the case when we choose $\Gamma_S = 0$, which will result in $\left|\Gamma_{out}\right| = \left|S_{22}\right|$ (from equation 41). Therefore, it is possible to infer some effective and simple considerations. For a given two-port network, if $\left|S_{22}\right| < 1$, the source plane origin of the smith chart must reside inside the stability region. In other words, if the stability circle of the source plane includes the smith chart origin, the interior area of the stability circle will symbolize the stable region as illustrated in figure 17.a. however, if the stability plan does not include the Smith Chart origin as well as $\left|S_{22}\right| < 1$, the unstable region of the two-port network will be represented by the inside area of the stability circle (figure 17.b).

As a result, the necessary condition that allows the stability circle of the source plan to include the Smith Chart origin is given by:

$$\gamma_{SS} > \left|C_{SS}\right| \tag{46}$$

Once again, in the case when the stability circle of the source plane does not contain the Smith Chart origin and the $\left|S_{22}\right| > 1$, the stabile region will be represented by the interior area of the stability circle as illustrated in figure 17.c. Finally, if the stability circle of the source plane does include the origin and $\left|S_{22}\right| > 1$, the stable area will modeled by outside region of the stability circle as depicted in figure 17.d.

Figure 18. An unconditional stable two-port network stability circle

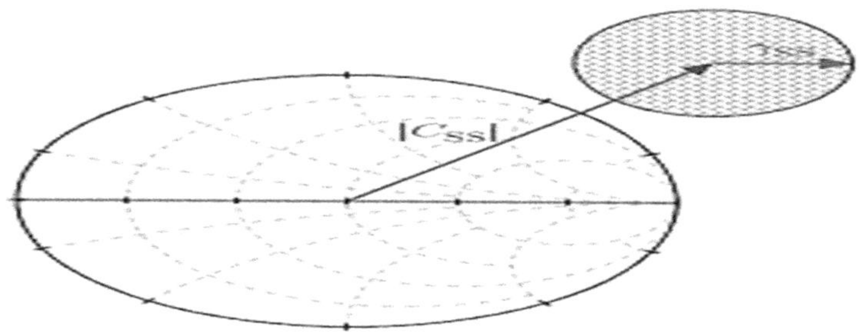

In the same manner, by applying the similar logic as reported above, we obtain the same results in the case of the load plan, thus we will not repeat it here.

Stability Analysis in Terms of the Scattering Parameters: The Rollet Stability Factor

From a practical point of view, it is so advantageous to have a simple formulation that permits us to analyses the stability conditions of a given two-port active device by means of a quick calculation, without the necessity to draw the stability circles. However, several stability criteria have been developed and proposed, but the most broadly deployed is the so-called Rollet stability factor, that was primarily introduced by John Rollet, who was the first to suggest a single stability factor in terms of the scattering parameters (Gonzalez, 1997).

As already stated in the previously subsection, the stability factor introduced by Rollet is frugally a simple scattering parameters form of a few earlier stability factors that had been developed in terms of immittance parameters. Even though the Rollet stability factor has some weaknesses, its historical importance and ubiquitous use impose use to allocate a little time here in order to clarify how it is developed and applied. The scattering parameters form of the Rollet stability factor is generally symbolized by the letter "K" (in upper case form), in order to differentiate it from the immittance parameters form, which is usually denoted by "k" (in lower case form).

On the other hand, the Rollet stability factor for an unconditionally stable two-port network, can be therefore obtained by considering its stability circles. An example of a source stability circle which is placed outside the unit circle in the Γ_S plan is illustrated in figure 18. The latter shows that in the shaded area (unstable region), there are no circuit terminations that can be implemented by using only the passive circuit elements.

Referring to figure 18, the stability criterion for the source plan can be defined as:

$$\gamma_{SS} - \left|C_{SS}\right| > 1 \tag{47}$$

From equation (43) et (44), equation (47) can expressed as:

$$\frac{\left|S_{11}-\Delta S_{22}^{*}\right|-\left|S_{12}S_{21}\right|}{\left|S_{11}\right|^{2}-\left|\Delta\right|^{2}} \tag{48}$$

Therefore, equation (48) can be simplified to yield the following:

$$2\left|S_{12}S_{21}\right|<1-\left|S_{11}\right|^{2}-\left|S_{22}\right|^{2}+\left|\Delta\right|^{2} \tag{49}$$

As a result, the Rollet stability factor can be therefore expressed as:

$$k=\frac{1-\left|S_{11}\right|^{2}-\left|S_{22}\right|^{2}+\left|\Delta\right|^{2}}{2\left|S_{12}S_{21}\right|} \tag{50}$$

And the Rollet stability criterion will be:

$$k>1 \tag{51}$$

In the same way, by applying a similar analysis for the load plan, we can therefore obtain identical result to equation (49).

However, the full stability criterion in the scattering parameters form as published by Rollet includes the input and output reflection coefficients S_{11} and S_{22} respectively. The criterion states simultaneous conditions for the unconditional stability as follows:

$$\begin{cases} k>1 \\ \left|S_{11}\right|<1 \\ \left|S_{22}\right|<1 \end{cases} \tag{52}$$

On the other hand, the Rollet stability factor K by itself is not enough to prove stability as exhibited by a number of researchers among of whom is Woods, who introduces a new condition of the stability criteria expressed as follows:

$$\begin{cases} k>1 \\ and \\ \left|\Delta\right|<1 \end{cases} \tag{53}$$

Where $\Delta = S_{11}S_{22} - S_{12}S_{21}$.

Many deficiencies of the basic Rollet stability criteria have been pointed out by many other researchers, who have proposed other conditions that must be considered besides the Rollet stability factor. One of those researchers is Bodway, who introduces other conditions that must be taken into account in the stability analysis, where expressed as follows:

$$\begin{cases} k > 1 \\ B_1 > 0 \\ B_2 > 0 \end{cases} \tag{54}$$

And

$$B_1 = 1 + \left|S_{11}\right|^2 - \left|S_{22}\right|^2 - \left|\Delta\right|^2 \tag{55}$$

$$B_2 = 1 - \left|S_{11}\right|^2 + \left|S_{22}\right|^2 - \left|\Delta\right|^2 \tag{56}$$

In the past few decades, the Rollet stability factor has come under improved investigation. As a result, new stability criterion has appeared in order to perform the stability analysis of two-port networks.

One of those stability criterion is a geometrical method which is firstly introduced by Edwards and Sinsky in 1992. As already mentioned, the Edwards and Sinsky criterion use a geometrical approach to reach two stability factors, namely μ1 and μ2 respectively, which are defined as follows:

$$\mu_1 = \frac{1 - \left|S_{11}\right|^2}{\left|S_{22} - \Delta S_{11}^*\right| + \left|S_{12}S_{21}\right|} \tag{57}$$

$$\mu_2 = \frac{1 - \left|S_{22}\right|^2}{\left|S_{11} - \Delta S_{22}^*\right| + \left|S_{12}S_{21}\right|} \tag{58}$$

Where μ1 and μ2 measure the radius at the output and the input planes respectively from the nearest unstable point to the center of the Smith Chart (Z_0).

In order to ensure that the unstable area lies outside the unit circle, the Edwards and Sinsky stability criteria must fulfill simultaneously the following conditions:

$$\begin{cases} \mu_1 > 1 \\ and \\ \mu_2 > 1 \end{cases} \tag{59}$$

Figure 19. Single stage RF power amplifier.

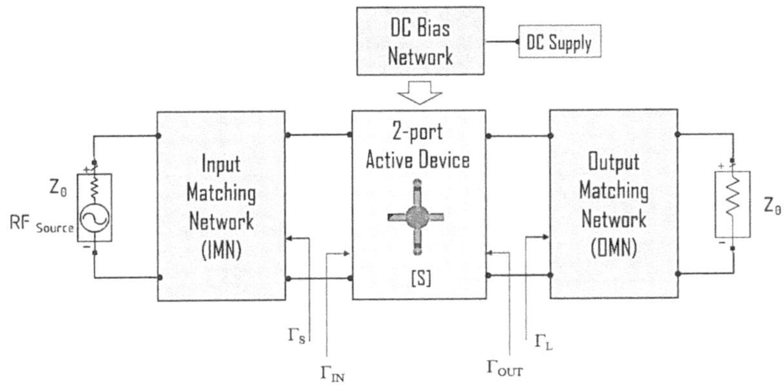

The Proposed Broadband Power Amplifier Circuit Design

Regardless of its physical implementation, the function of a power amplifier is to provide a finite positive power gain over the operation frequency band. In other words, the PA task is to increase the power level of the RF signal presents at its input port up to a predetermined level at its output port over the operating frequency band. On the other hand, the broadband power amplifier synthesizes results in a trade-off among various contradictory requirements such as power gain and wider bandwidth, low distortion and output power or efficiency vs linearity.

Referring to figure 19, the simplest single stage RF power amplifier includes a single active device DC supplied through a DC bias network and connected to a source and load having the system characteristic impedance Z_0, through the input and the output matching networks respectively.

By using the concepts mentioned above, the overall schematic circuit of the proposed Single Stage Broadband Power Amplifier including the biasing network is illustrated in figure 20. Figure 21 and 22 shows the proposed broadband input and output matching networks respectively.

Stability Analysis of the Proposed BPA

As already outlined, any power amplifier providing a positive power gain can be made to oscillate if an external positive feedback is applied. These oscillations are a result of unavoidable parasitic effects that are generally sufficient to provide oscillation if the power amplifier is not neatly designed. For this reason, design techniques can be deployed in order to ensure the stability requirements. For the proposed BPA, the unconditionally stability is more demanded out of the frequency band where the component impedance systems are not specified. As mentioned in the previous sections, the unconditionally stability is reached when the following equations are simultaneously validated:

$$\begin{cases} k > 1 \\ B_1 > 0 \\ B_2 > 0 \end{cases} \tag{60}$$

Figure 20. The proposed BPA schematic

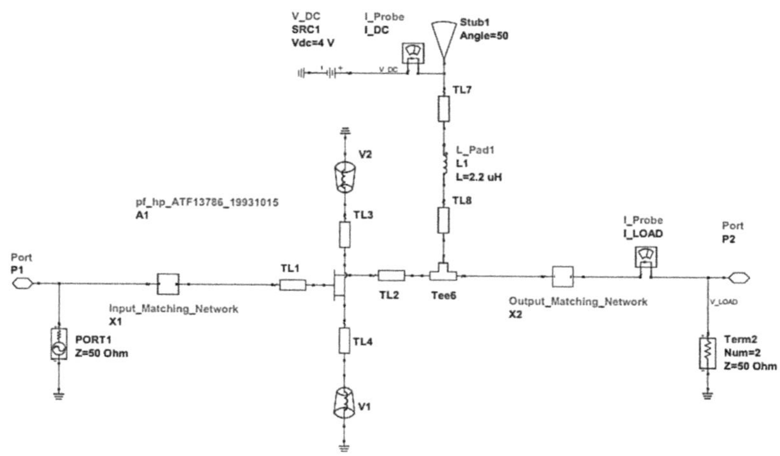

Where,

$$k = \frac{1 - \left|S_{11}\right|^2 - \left|S_{22}\right|^2 + \left|\Delta\right|^2}{2\left|S_{12}S_{21}\right|} > 1 \tag{61}$$

$$B_1 = 1 + \left|S_{11}\right|^2 - \left|S_{22}\right|^2 - \left|\Delta\right|^2 \tag{62}$$

Figure 21. The proposed broadband input matching network

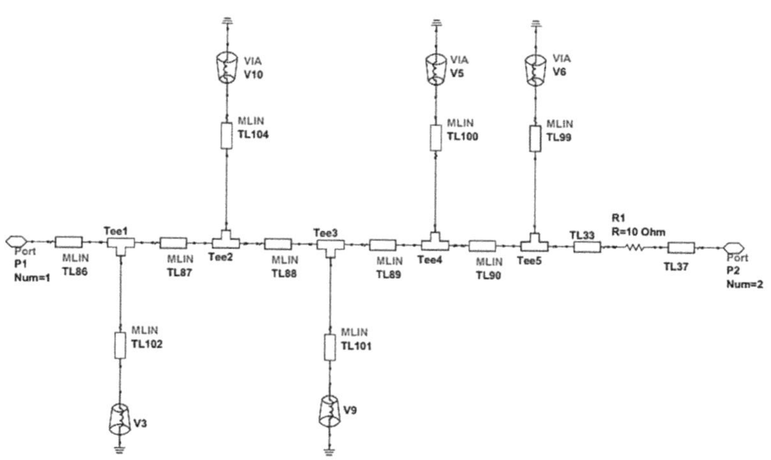

Figure 22. The proposed broadband output matching network

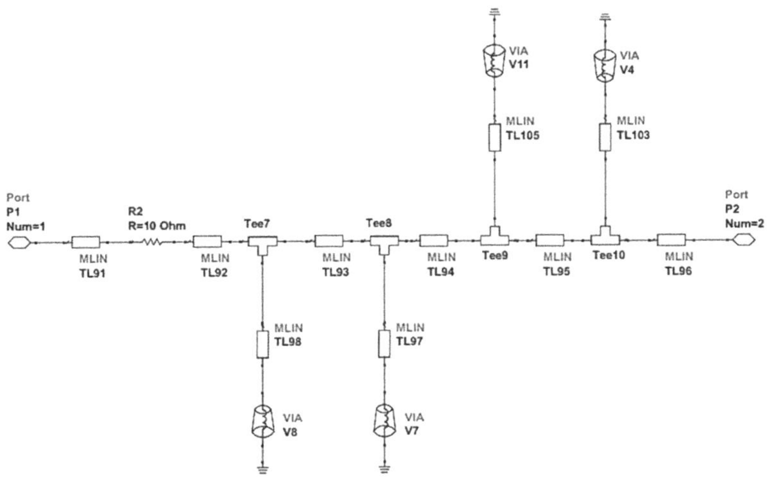

$$B_2 = 1 - |S_{11}|^2 + |S_{22}|^2 - |\Delta|^2 \quad (63)$$

From figure 23, we can easily remark, that the Rollet stability factor K is greater than 1.2 while the Bodway stability factor B_1 is also greater than 0.85. As a result, the proposed BPA fulfills the unconditionally stability conditions.

Small Signal Analysis of the Proposed BPA

As underlined in the earlier subsections, at microwave frequencies, it is so difficult to measure directly the currents and voltages. For this reason, the active devices characterization is performed in terms of the scattering parameters which are based on the measurement of the incident and reflected waves.

Figure 23. Curves of Rollet and Bodway stability factors versus frequency

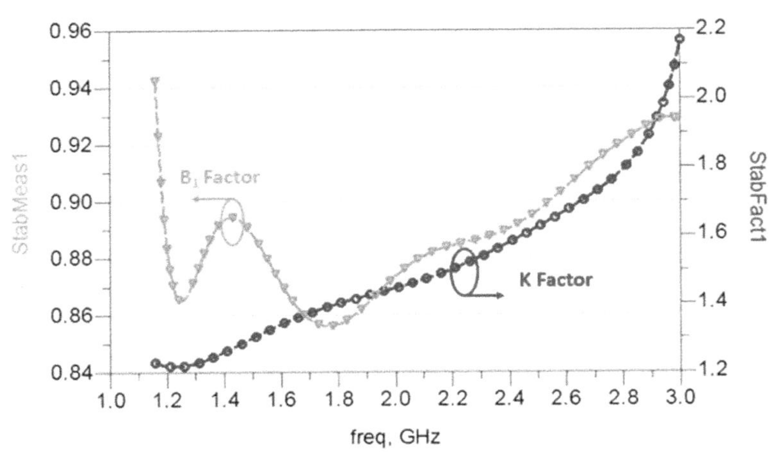

Figure 24. Curves of input and output reflection coefficients versus frequency

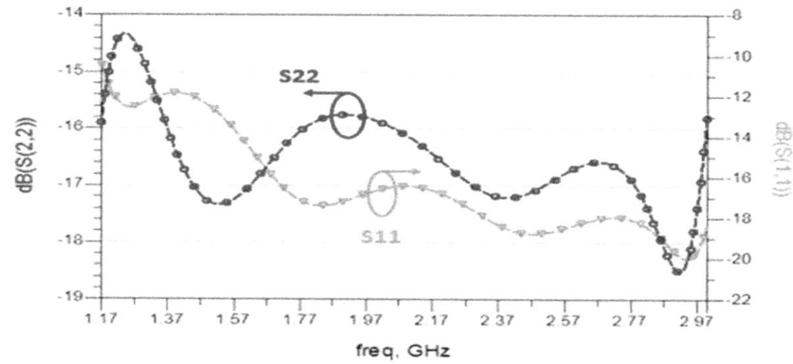

The simulated small signal scattering parameters of the proposed BPA are reported in figures 24 and 25. As can be noted from figure 24, the input return loss S_{11} of the proposed circuit varies between -11 dB as a maximum value and – 20 dB as a minimum value, while the output reflection coefficient S_{22} changes between a maximum value of -14.3 dB and a minimum value of -18.5 dB.

On the other hand, since the internal feedback of a given two port circuit from the output port to the input port is represented by S_{12} parameter, the smaller value of this parameter the greater is the degree of isolation between the output terminal and the input port, consequently, the greater is the degree of stability of the whole circuit. However, the nonlinear behavior of the active device, and intrinsically the two port network, from the input port to the output port is characterized by S_{21}, known as the power gain which is depends on the input signal level.

As can easily be noted from figure 25, over the operating frequency ranging from 1.17 GHz to 3 GHz, the revers isolation coefficient S_{12} is less than -20 dB while the S_{21} parameter which is the power gain varies between a maximum value of 17.3 dB and a minimum value of 8 dB.

Therefore, the small signal scattering parameters simulation of the proposed BPA provides satisfactory performance in terms of matching networks, power gain and revers isolation over the entire bandwidth and under an unconditionally stability.

Large Signal Performance of the Proposed BPA

The simulated large signal performances of the proposed BPA including the output power, the 1-dB compression point and the Power Added Efficiency (PAE), have been done at 2.25 GHz and under 50 Ω input and output impedances.

By definition, at a given frequency f or band of frequencies $B = \left[f_{Low}\ ;f_{High}\right]$, the output power of a given power amplifier is expressed as:

$$P_{out} = P_{out}\left(f\right) = \frac{1}{2}\Re\left\{V_{out} * I_{out}^{*}\right\},\ f \in \left[f_{Low}\ ;f_{High}\right] \tag{64}$$

Figure 25. Curves of power gain and revers isolation coefficients versus frequency

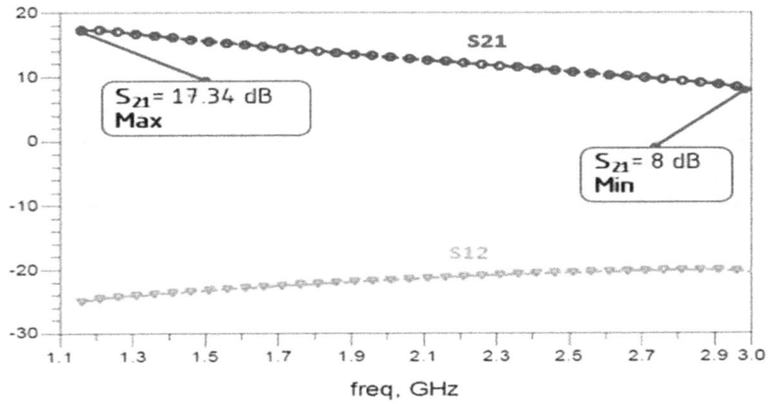

Note that, at microwave frequencies, the power quantities are commonly expressed in logarithmic units due to the wide dynamic range of the signals involved in a given power amplifier. Particularly, assuming 1 mW as a reference, power levels are therefore expressed in dBm (decibel over 1 mW) as:

$$P_{dBm} = 10.log_{10}\left(\frac{P}{1mW}\right) = 10.log_{10}\left(P_{mW}\right) = 10.log_{10}\left(P_{W}\right) + 30 \tag{65}$$

On the other hand, while the input power is augmented, we eventually reach a point where the output power can no keep increasing linearly with the input power. Such point is named 1-dB compression point, in which, the output power of the two port network deviates with 1 dB from the linear region. Such point may be useful while performing a BPA to estimate the range on the input levels for which the output power is linearly proportional with them. In other words, the 1-dB compression point is a useful tool that allows to determine the dynamic range of a given BPA.

The output power and the 1-dB compression point for the proposed BPA are reported in figure 26. The simulated results show that the proposed circuit achieves a maximum output power of 17 dBm, this leading to 50.11 mW, while the 1-dB compression point is marked at 4 dBm of the input power.

From an energetic standpoint, a power amplifier can be regarded as a device which is able to transform the DC energy from supplies into RF energy at a given frequency band. The effectiveness of this transformation is usually evaluated by means of Power Added Efficiency (PAE), which is the ratio between the added power (P_{add}) and the DC power (P_{DC}) emanating from the bias supply. The PAE is therefore calculated by the help of the following equations:

$$\eta_{add} = \frac{P_{add}}{P_{DC}} = \frac{P_{out} - P_{in}}{P_{dc}} \text{ Where } P_{DC} = V_{bias}\frac{1}{T}\int I_{bias}\left(t\right)dt \tag{66}$$

The simulated PAE of the proposed BPA is depicted in figure 27. At 2.25 GHz, the proposed BPA achieves a maximum PAE of 8.55%.

Figure 26. Output power curve versus input power & 1-dB compression point at PIN=4 dBm

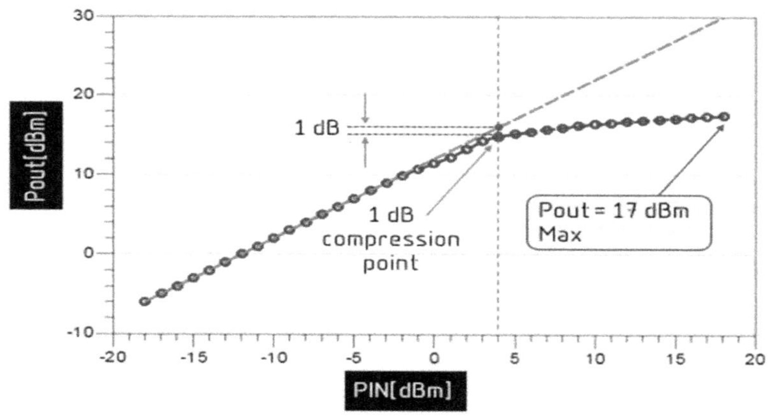

Referring to table 1, to author's best acknowledge, the matching techniques deployed in this work are clearly very sufficient and allows to achieve the broad bandwidth. By comparing the simulated performance of the proposed BPA with the contemporary state-of –the art BPAs, the proposed circuit fulfils good output power, good power gain and excellent impedance matching over a broad bandwidth.

CONCLUSION

In this chapter, a single stage solid state broadband power amplifier operating in the frequency band ranging from 1.16 GHz to 3 GHz which covers the mainstream communication standards running in L and S band has been presented. The chapter started with a theoretical background of broadband power amplifier design considerations including broadband matching techniques, broadband biasing networks design, scattering parameters and stability analysis. With carefully designed bias network, and input/ output matching circuits, good matching is fulfilled over the wider bandwidth. The small signal analysis

Figure 27. Power added efficiency curve versus input power

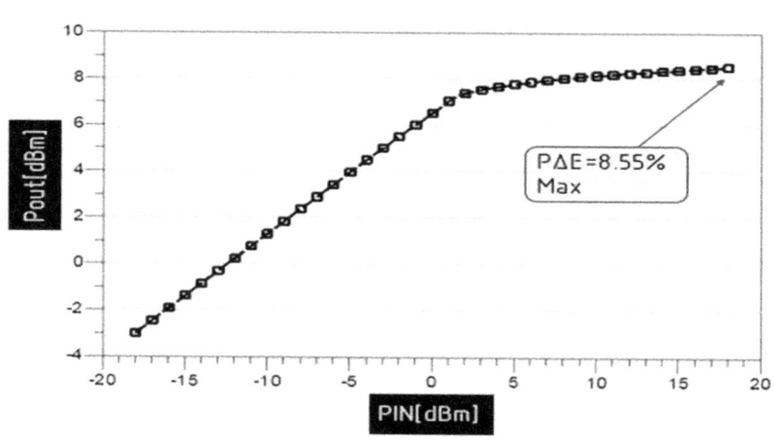

Table 1. Performance comparison of the proposed BPA and the contemporary state-of-the art BPAs

	François, et al., (2015)	**Chen, et al., (2016)**	**Rachakh, El Abdelaoui, et al., (2017)**	**Ribate, Mandry, et al., (2018)**	**This Work**
Freq [GHz]	1.9 – 2.7	1.8 – 2.8	1.75 – 2.15	1.1 – 3	1.16 – 3
Gain [dB]	11	28	11.7	14.9	17.34
Psat [dBm]	28.1	25	8	17	17
PAE [%]	13.7	6.1	11.7	14.9	8.5
S11 [dB]	-	-25	-22	-35	-20
S22 [dB]	-	-19	-19	-25	-18.5
Supply [V]	2.5	5	12	3	4

shows that the input return loss varies between -10.50 dB and -20 dB, while the output return loss varies between -14.34 dB and – 18.5 dB. The large signal simulation report that the proposed BPA exhibits the saturated output power of 17 dBm (50.11 mW) with a maximum PAE of 8.55%. The proposed BPA is unconditionally stable over the whole operating bandwidth from 1.16 GHz to 3 GHz. Except the PAE, a power amplifier with those features is more suitable for many microwave applications operating in L and S bands such as avionics, medical microwave applications (microwave imagining, radiometry, hyperthermia, …), cellular communications (Single Radio Access Network (SRAN), BTS, …) digital broadcasting, and many others applications.

REFERENCES

Bahl, I., & Bhartia, P. (2003). *Microwave Solid State Circuit Design*. Hoboken, NJ: John Wiley & Sons.

Bahl, I. J. (2009). *Fundamentals of RF and Microwave Transistor Amplifiers*. Hoboken, NJ: John Wiley & Sons. doi:10.1002/9780470462348

Bozanic, M., & Sinha, S. (2016). *Power Amplifier for the S-, C-, X- and ku bands*. Switzerland: Springer International. doi:10.1007/978-3-319-28376-0

Chen, C. Q., Hao, M. L., Li, Z. Q., Du, Z. B., & Yang, H. (2016). A 1.8–2.8 GHz Highly Linear Broadband Power Amplifier for LTE-A Application. *Progress in Electromagnetics Research C.*, *66*, 47–54. doi:10.2528/PIERC16050503

Colantonio, P., Giannini, F., & Limiti, E. (2009). *High Efficiency RF and Microwave Solid State Power Amplifiers*. UK: John Wiley & Sons. doi:10.1002/9780470746547

Francois, B., & Reynaert, P. (2015). Highly linear fully integrated wideband RF PA for LTE advanced in 180-nm SOI. *IEEE Transactions on Microwave Theory and Techniques*, *63*(3), 649–658. doi:10.1109/TMTT.2014.2380319

Gonzalez, G. (1997). Microwave Transistor Amplifiers Analysis and Design (2nd ed.). Upper Saddle River, NJ: Prentice Hall.

IEEE Std 521-1984. (1984). *IEEE Standard Letter Designations for Radar-Frequency Bands.*

Jarry, P., & Beneat, J. N. (2016). *Microwave Amplifier and Active Device Design Using the Real Frequency Technique*. Hoboken, NJ: John Wiley & Sons. doi:10.1002/9781119073093

Kazimierczuk, M. K. (2015). *RF Power Amplifiers* (2nd ed.). Hoboken, NJ: John Wiley & Sons.

Kumar, N., & Grebennikov, A. (2015). Distributed Power Amplifiers for RF and Microwave Communications. Norwood, MA: Artech House.

Poole, C., & Darwazeh, I. (2016). Microwave Active Circuit Analysis and Design. Amsterdam, The Netherlands: Elsevier Ltd.

Pozar, D. M. (2012). *Microwave engineering* (4th ed.). Hoboken, NJ: John Wiley & Sons.

Rachakh, A., Abdellaoui, L., Zbitou, J., & Errkik, A., Tajmouati, A., & Mohamed, L. (2018). A Novel Configuration of a Microstrip Microwave Wideband Power Amplifier for Wireless Application. *Telkomnika*, *16*(1), 2014–2031. doi:10.12928/telkomnika.v16i1.7369

Ribate, M., Mandry, R., Latrach, M., Errkik, A., & El Abdellaoui, L. (2017). GaAs FET Broadband Power Amplifier for L and S Bands Applications. In *Proceedings of the 2nd International Conference on Computing and Wireless Communication Systems* (p.42). Academic Press. 10.1145/3167486.3167530

Ribate, M., Zbitou, J., Mandry, R., Errkik, A., & Latrach, M. (2018). Broadband GaAs FET Power Amplifier for L and S Bands Applications. *International Journal of Intelligent Engineering and Systems*, *11*(5), 96–105. doi:10.22266/ijies2018.1031.09

Yeom, K. W. (2015). Microwave Circuit Design, a Practical Approach Using ADS. Upper Saddle River, NJ: Prentice Hall.

Chapter 6

A New Compact Coplanar Waveguide Stop Band Filter Based on Circular Shaped DGS for Microwave and Millimeters Wave Applications:
Coplanar Waveguide Technology

Elmahjouby Sghir
University of Hassan 1st, Morocco

Ahmed Errkik
University of Hassan 1st, Morocco

Mohamed Latrach
ESEO Group, France

ABSTRACT

This chapter introduces an overview of coplanar technology and the general techniques and process to improve the response and characteristics of microwave components. A new circular defected ground structure (DGS) with shaped coplanar line is investigated for compact stopband filter (SBF) for microwave and millimeter wave applications. With this structure, the response of resonant element in 20 GHz exhibits the bandstop function. The proposed DGS is also modified by introducing four symmetrical slots with L-configuration in conductor line of a coplanar circuit to improve separately the stopband and passband performances. An equivalent circuit model derived for the proposed structures will be provided.

DOI: 10.4018/978-1-7998-0117-7.ch006

INTRODUCTION

In this chapter, an overview of coplanar waveguide technology is presented. Varieties of defected ground structure with the recent DGS units which developed to replace the Electromagnetic Band Gap (EBG) circuits in the goal to improve response and characteristics of microwave components such as filters are introduced. Finally, a new stopband CPW filter using Circular Defected Ground Structures (C-DGS) is proposed with good performances, a coplanar transmission line connected to circular DGS resonator form the first stop band filter with a center frequency of 20 GHz which can be tuned easily to millimeter-wave frequency for more application domain.

The growing demand and need for micro wave and millimeters wave circuits for wireless applications reveals new requirements for the transmission and reception channels in telecommunication, which should give excellent performance in terms of response as bandwidth, quality factor and the parasite rejection, also the level of integration as the size, the development and integration of other components and their interconnection. The Defected Ground Structure with Coplanar transmission line (CPWDGS) is one of the attractive solutions to above problems and employed to give a good solution.

In this chapter, a circular defected ground structure (DGS) with shaped coplanar line is investigated for compact stopband filter (SBF) for microwave and millimeter wave applications. With this structure, the first proposed response of resonant element in 20 GHz exhibits the bandstop function. The proposed DGS is also modified by introducing four symmetrical slots with L-configuration in conductor line of a coplanar circuit to improve separately the stopband and passband performances. The insertion loss can be reduced by introducing four symmetrical slots with L-configuration in conductor line of a coplanar circuit to improve separately the stopband and passband performances. Additionally, the operating frequency ranges are extended to millimeter-waves with little increase in radius of circular defected ground. It combined between those requirements: Simple structure and design, easy fabrication and good performance.

OVERVIEW OF COPLANAR TECHNOLOGY

In recent years, communications systems have been continuously expanding. The gradual saturation of the frequency bands allocated, by the extraordinary explosion of mobile telecommunications market and the continuing increase in communications services, leading to the development of higher-frequency applications, particularly in the frequency domain microwaves and millimeters. This is why mainstream applications, such as satellite communications, guidance, intelligent systems ..., require permanent improvements in performance for their low realization costs, low weight and size.

The micro strip technology, which remains the most widespread, however, has some limitations for applications at millimeter wavelengths. Firstly, the additional technological steps to achieve the metalized holes are expensive. Secondly, these holes induce parasitic effects which degrade the performance of the circuits at these frequencies. Coplanar technology is another approach that allows the use of uniplanar structures with several advantages and reliability. This coplanar technology becomes attractive with increasing production volumes and its compatibility with other new technologies. This approach consists in using the uniplanar structures. However, its advantages compared to micro strip technology, such as:

- Simplifies fabrication and lower production cost

- It facilitates easy shunt as well as series surface mounting of active and passive devices, and
- It eliminates the need for wraparound and via holes and (Alsanoursi et al., 2013; Aytouna et al., 2016; Boutejdar et al., 2008).
- It reduces radiation loss (Boutejdar et al., 2016)
- Less electromagnetic coupling between two adjacent lines and low dispersion
- The reduction of parasitic effects (Boutejdar, 2016).
- A small seize (Ahn et al., 2001; Browne, 1987)

Transmission Line

A transmission line is a set of one, two or more conductors carrying an electrical signal from a transmitter (a source) to a receiver (a charge). The most practical lines are Coaxial cable, Balanced lines(Twisted pair, Twin-lead, Lecher lines…), Single-wire line and Planar lines.

Planar transmission lines can be used for interconnecting components as well as making them like filters, antennas, impedance transformers, couplers,.... The dielectric medium that separates the conductors (air, ceramic, Teflon,...) plays an important role in the speed of propagation of the signal and the electrical losses (Browne, 1992; Browne, 1987).

A transmission line is characterized by its characteristic impedance, its attenuation constant (which specifies the losses in the line), and the speed of propagation of the signals, which depends on the dielectric used.

Coplanar Transmission Line

Structure of the Line

The coplanar transmission line (CPW), is a key element in the design of based microwave integrated circuits (MICs) as well as monolithic microwave integrated circuits (MMICs) because these circuits can be manufactured on the same existing substrate with the same processes as the rest of structure simply by applying patterns as hole, slots or DGS…, All these gives the coplanar technology a great advantage over other types (Browne, 1989; Browne, 1990).

It has been proposed for the first time as an alternative to the microstrip line by C. P. Wen in 1969 (Chen, 2006). It consists of three metal surfaces placed on the same plane at the same distance from the dielectric substrate. The center strip conductor leads the microwave signal. The two other side surfaces serve as ground plane and are separated from the signal by two slots coplanar, as shown in Figure 1.

Figure 1 shows the schematic of a coplanar transmission line on a dielectric substrate. ‘W’ represents the width of the center conductor, ‘S’ the width of the coplanar slot, ‘Wg’ the width of the two ground planes, ‘Hs’ the thickness of the substrate and ‘t’ the thickness of the metal conductors.

The Different Propagation Modes

In specific frequency range, a transmission line filled with uniform dielectric can support a simple and defined propagation mode: TEM for coaxial lines, TE for waveguides...

An ideal coplanar line consists of a central conductor and two semi-infinite mass planes deposited on the same surface of a semi-insulating substrate of infinite thickness (Elmahjouby et al., 2017). Due

Figure 1. Schema of a coplanar line

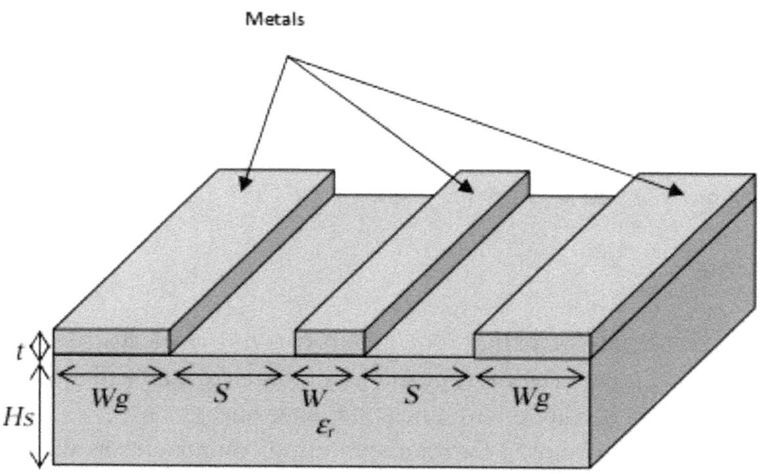

to its symmetry and geometry, a coplanar line supports two fundamental propagation modes without cutoff frequency (Elmahjouby et al., 2014).

- Quasi-TEM mode, often called the odd mode or coplanar mode: Where the fields in the two slots are 180° out of phases, as shown in Fig 1(a).
- Quasi-TE pair mode or non TEM mode, urealso called the even mode or slotline mode: Where the fields are in phases, as shown in Figure 1(a).

The propagation mode desired in line is the coplanar mode (odd mode) (Jackson, 1989). In this mode, the RF signal propagates in the central conductor and the ground planes are equipotential in structure.

However, the quasi-slot line mode (even mode) can be excited in the presence of a discontinuity such as a junction. Its results, a zero potential on the central conductor and opposite potentials on the two ground planes. This mode is a highly dispersive (James et al., 2011; Karim et al., 2005).

Passive Elements for RF and Microwave Application in Coplanar Technology

It is possible to realize passive elements on a coplanar line in order to constitute circuits as antennas and a planar filters. Therefore, to integrate these passive elements on the same substrate with other active elements such as transistors, which can be used in MIC's and MMIC's circuits for microwave and millimeters applications.

Therefore, we have presented a technique that allowing for the realization of the different passive elements constituting the circuit such as (Koo et al., 2007; Koster et al., 1989).

- Inductors (series and parallel),
- capacities (series and parallel),

An open circuit is capacitive in nature and the short circuit is inductive in nature.

Figure 2. Field patterns modes: quasi-TEM mode or coplanar mode (a), quasi-slotline mode (b)

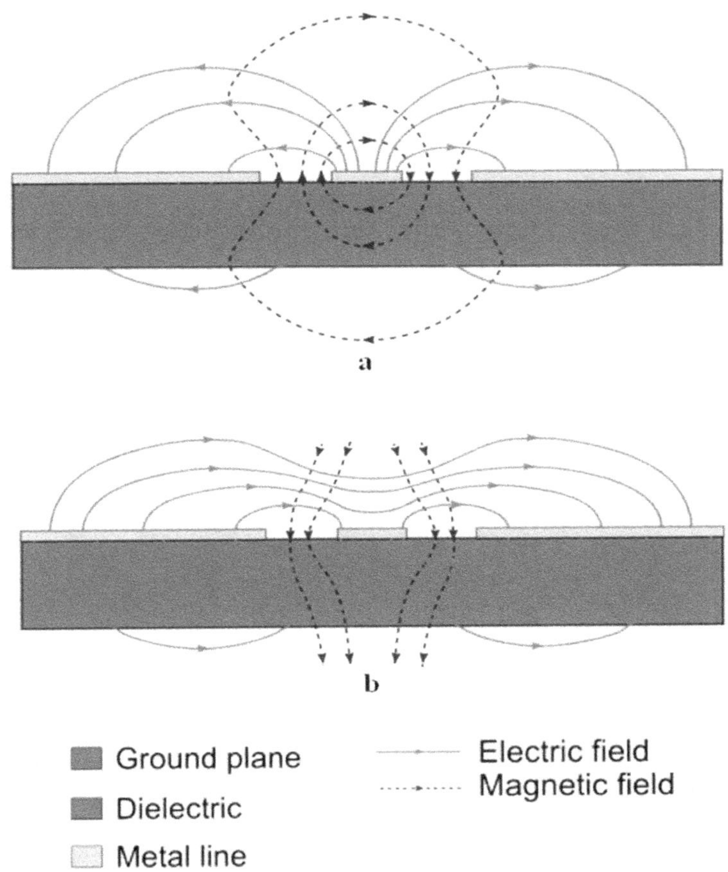

Improvement of the Response and Characteristics of Microwave and Millimeters Components

There are different techniques and configurations to improve the response and characteristics of microwave and millimeters components in coplanar technology.

Periodic Structures

Periodic structure is the repetition of numbers of physics cells keeping the same distance that separates two successive units in the same structure. The response and characteristics of microwave and millimeter components depends on number of periodic cells. By cascading resonator cells or DGS units in the ground plane or in the conductor line, the cutoff frequency, level of rejection and the bandwidth are essential parameters of filters witch improved with periodic cells number, as in (Lim et al., 2002; Mukesh et al., 2017; Park et al., 1999).

Table 1. Different configuration of series elements of a CPW Line and its equivalent circuits

	Circuit element	Realization using CPW
Coplanar capacity	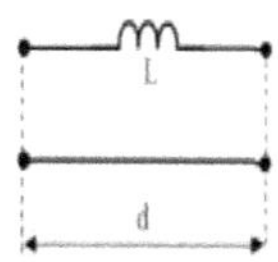	 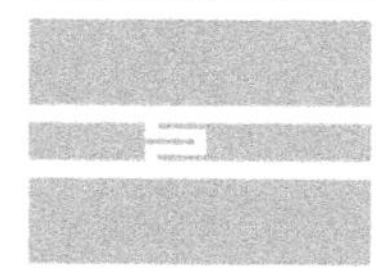
Coplanar Inductance	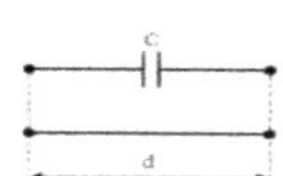	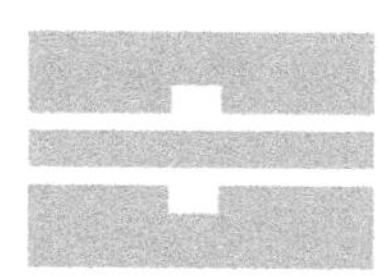 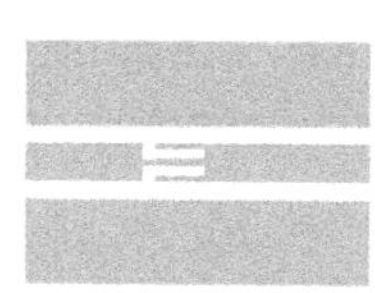

DGS Cells

Table 2. Different configuration of Parallel elements of a CPW Line and its equivalent circuits

	Circuit element	Realization using CPW
Coplanar capacity	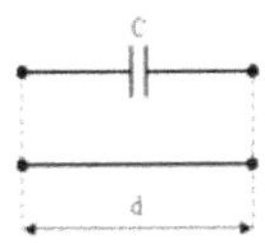	
Coplanar Inductance	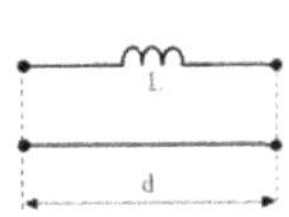	

Introduce of DGS units especially in the ground of structure of coplanar line. DGS units are very useful in the design of planar filters, it is a good candidate technique for reducing the size, and improving the response and characteristics of filters as the cutoff frequency, level of rejection and the bandwidth. However, some DGSs configuration can not improve characteristics and the response frequency of filter without adding of a periodic units for structures (Park et al., 1999). A lot off works using DGS structure are designed, fabricated, and measured with good performance as in (Radisic et al., 1998; Riaziat et al., 1990). Also, in this chapter a new stopband CPW filter using Circular Defected Ground Structures (C-DGS) is proposed with good performances, a coplanar transmission line connected to circular DGS resonator form.

Loading the Holes Slots by Identical Geometry Units

In the research, there were two possibilities for efficient use of performance of DGS structures witch are modified defected ground structure DGS as in (Radisic et al., 1998), and periodic DGS cells as in (Lim et al., 2002; Mukesh et al., 2017). There are a lot off slot geometries etched in the microwave coplanar line and in the ground plane of CPW, in order to realize structure with a good performance of the previous presented filters. But in some cases, these structures are not always satisfying in response, especially in terms of rejection level and in the separation between the insertion loss and return loss. The proposed idea to resolve this problem, is Loading the these etched geometries by similar geometry units to the first proposed unit with the difference that the new structure consists of small seize. This technique has been reported in the literature as in, (Sporkmann, 1998; Stegens et al., 1987).

Introduce of Passive Elements as Shaped, Slots and Interdigitated http://www.sonnetsoftware.com/products/lite/interdigitated-capacitor.html*Capacity in Structure to Improve the Response*

Some configuration as slots resonator and DGSs cells structure of microwave and millimeter components can not improve and provide the characteristics and the response frequency of circuits, for that, it is necessary to find the optimized size of the unit to improve the frequency characteristics of the circuit, but the change in the size of structure cause a shift in the level response in the desired frequency. The question here, is how to improve the characteristics and the response frequency of circuit, while maintaining the desired frequency, and kept the reduce seize of structure. In order to illustrate this objective with this method, by introduce a passive elements to provide response frequency, a wide band stop filters are designed, simulated, improved by introducing four symmetrical slots with L-configuration in conductor line of a coplanar circuit to provide separately the stopband and passband performances.

Overview on Defected Ground Structure DGS

The history of the Defected Ground Structure was firstly with Korean scholar (Stegens et al., 1987). Which is etched in the ground plane of planar circuit, this disturbance will change the response of circuit, to give a new characteristics of transmission line such as capacitance and inductance effect, and it provides a promising approach to reduce size of passive components and to increase reject band, (Thakur et al., 2006).

The concept of the defected ground structure (DGS) has been derived from the idea of photonic bandgap structure (PBGs) (Walker et al., 1993).

Figure 3. Various DGSs configuration: (a) dumbbell-shaped, (b) arrowhead-slot, (c) cross-shaped, (d) square open-loop with a slot in middle section, (e) spiral-shaped, (f) H-shaped, (g) square slots connected with narrow slot at edge, (h) U-shaped, (i) inter-digital slots, (j) split-ring resonators, (k) concentric ring shaped, (l) L-shaped, (m) V-shaped, (n) U-head dumbbell, (o) double equilateral U, (p) Fractal etched, (q) open loop dumbbell

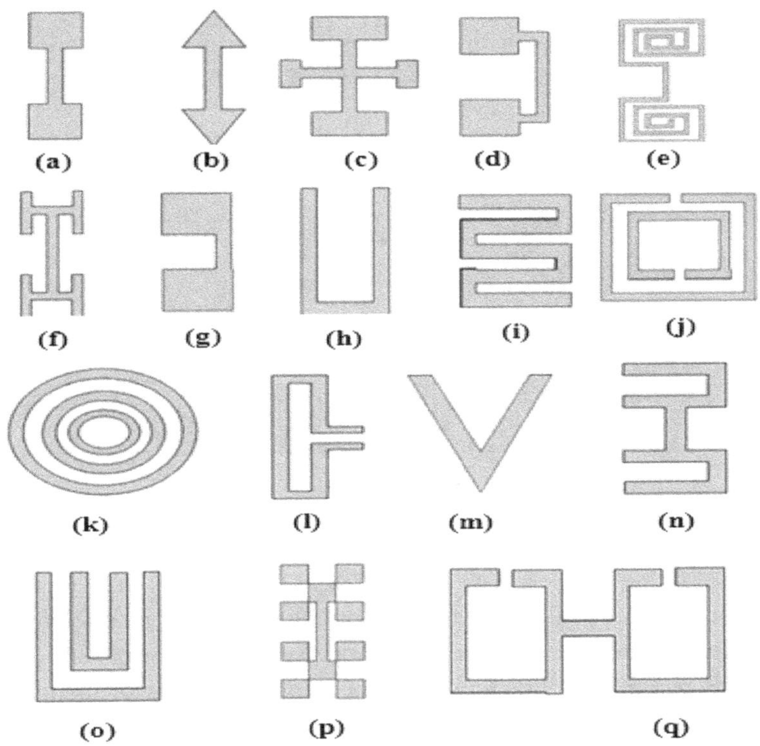

Although this later structure (PBG) applied at optical frequencies, and it has been derived for using planar circuits in electromagnetic microwave frequency with the name of electromagnetic bandgap (EBG) (Wang et al., 2008). Also the DGS is utilized to design many microwave circuits type, such as microstrip antenna, dividers, couplers, amplifiers, oscillators, (Wen et al., 1969; Weng et al., 2008; Wolff, 2006). and microwave filters as this chapter.

Initial Proposed DGS Stopband Filter Design

A Stopband filter with DGS circular structure is designed, Figure 1 depicts the schematic of the proposed structure. It consists of a 50Ω conventional coplanar transmission line, with a signal line width of W=108μm and the gap width, G=60μm.The substrate is with dielectric permittivity of er=11.9 and thickness of H=200 μm.

The DGS configuration in this structure is a circular slot etched in the ground plane with a diameter d. This circular DGS is connected to the line gap by a rectangle transverse slot with length of ls and width of ws. The Momentum (an EM solver integrated in ADS Agilent) is used for deriving and studying the filter's electrical performances.

Figure 4. CPW line with Circular DGS structure

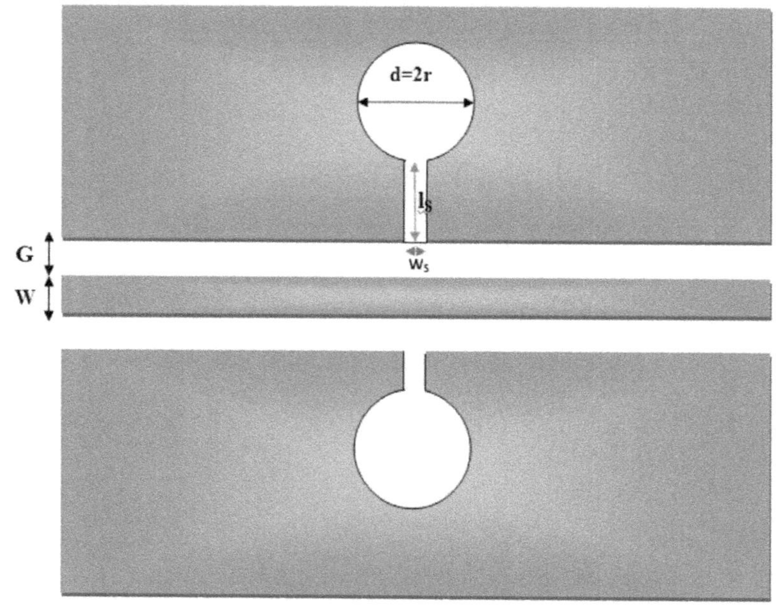

Response of Initial Circular DGS Filter

It is widely known that the DGS circular slot etched plays the role of resonator element and its frequency resonance depend on circular geometry parameters. Figure 5 depicts the response of the CPW DGS Filter.

As noted from the S-parameter studies, we have a stop band behavior of the circuit with a good bandstop around 20 GHz.

Moreover, we analyzed the dependence of the rectangle slot on the S-parameters response, the ray of circular etched is kept constant at r = 600µm, also the width of the rectangle slot is fixed at ws = 60µm.

Figure 5. Simulated S-parameters for the CPW DGS Circular structure

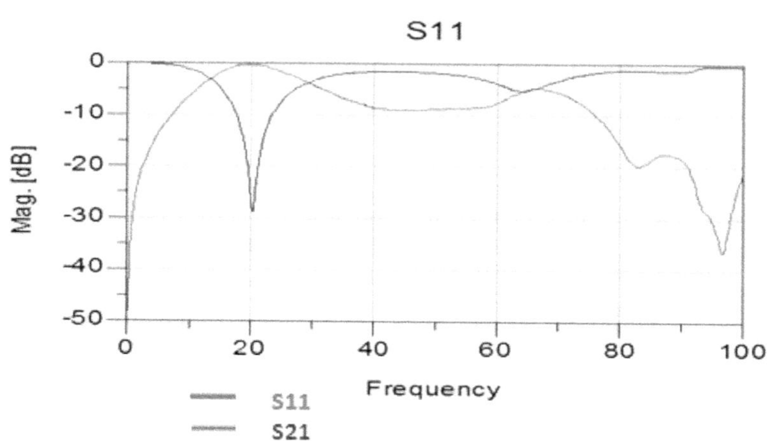

The length ls was optimized for two different value of ls = 400μm and 500μm. The simulation results obtained with ls = 500μm are shown in Figure 6.

As a result, there is a significant difference in S-parameter between the initial response structure and this later with the new length ls. We see that the problem is resolved and insertion loss (S21) and return loss (S11) are affecting in the passband, S21 is better than -10dB and S11 is better than -3 dB in stopband with the new configuration.

Modeling of the Circular DGS Dtopband Filter

Circular DGS combined with coplanar line causes a resonant response of the structure transmission, and the resonant frequency dependants by circular etched parameters and size of the slot. Its RLC-equivalent circuit of the cell is proposed, and the different parameters can be extracted as follows.

The equivalent impedance equation of the single resonant model may be expressed as:

$$Z = \left(jwc + \frac{1}{jwL} \right)^{-1} \quad (1)$$

The resonant frequency of the parallel circuit is defined as:

(2)

Figure 6. Simulated S-parameters for the CPW DGS Circular with ls = 500μm

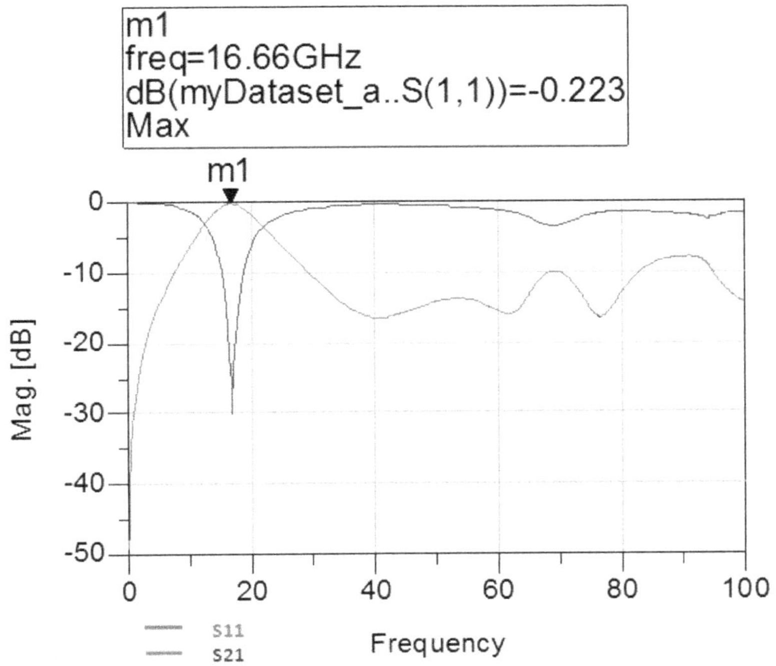

Figure 7. RLC-equivalent circuit for CPW DGS Filter

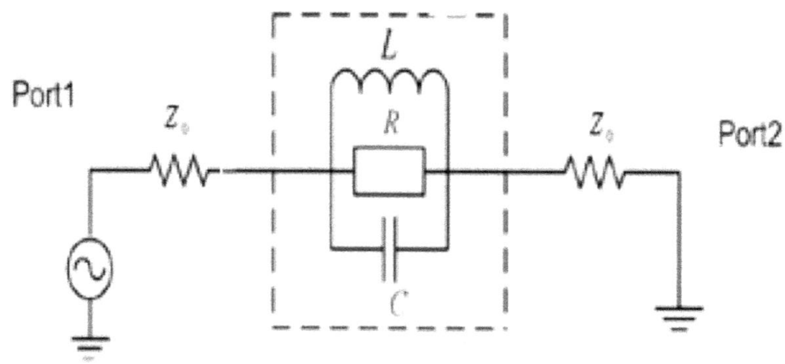

The -3dB cut-off angular frequency, w_c, can be determined by:

$$\left|S_{21}\right| = \left|\frac{2Z_0}{2Z_0 + Z}\right| \tag{3}$$

Substituting (1) into (3), the capacitance is obtained as:

$$C = \frac{w_c}{2Z_0\left(w_0^2 - w_C^2\right)} \quad C = \frac{w_c}{2Z_0\left(w_0^2 - w_C^2\right)} \tag{4}$$

The inductance can be determined by:

$$L = \frac{1}{4\left(\grave{A}f_0\right)^2 C} \tag{5}$$

The resistances R of the circuit model are determined around the resonant frequency ω_0 as:

$$R = \frac{2Z_0}{\sqrt{\frac{1}{\left|S_{11}\left(w_0\right)\right|^2} - \left(2Z_0\left(w_0C - \frac{1}{w_0L}\right)\right)^2 - 1}} \tag{6}$$

Where, Z_0 is the characteristic impedance of the coplanar wave guide line, w_0 is the angular resonance frequency, w_c is the -3dB cutoff angular frequency, and S11(w) is the input reflection coefficient of the equivalent circuit, which can be obtained from EM simulation results. The resistance R is added to the LC circuit to model the radiation, conductor and dielectric losses.

Table 3. parameters of proposed RLC-Equivalent circuit for CPW DGS filter

Parameter	Value
f01	20 GHz
R	2800 Ω
C	0. 115 pF
L	0. 534 nH

In order to show the validity of the equivalent circuit and the extracted parameters for the proposed used DGS structure, the results of EM simulation and equivalent circuit are all shown in Figure 3. An inductive capacitive and resistance RLC equivalent circuit can be used to model the used circular DGS.

Simulation results are depicted in Figure 8, which shows the characteristics of a stop band filter. The DGS cell is simulated using Momentum Agilent. The values of the cut-off frequency fc and resonance frequency f0 can be found from the transmission characteristics of the circular DGS, with the values of f0 = 20 GHz and fc = 13.11 GHz.

As noted from the S-parameter studies in the Figure 4, there is a good agreement between the EM and circuit simulated S-parameters.

The good agreement between the EM and circuit simulated S-parameters ensure the validity of the proposed equivalent circuit model of circular DGS filter. The values of resonant circuit: resistance R, inductive L and capacitive C are shown in table 3.

To investigate the effect of dimensions on the response frequency. Especially the length of rectangular etched slot. Two other value of length ls are chosen, respectively at 200 μm and 500 μm. All other dimensions are kept constant.

The simulation results are shown in Figure. 9, we can note that the modification of the length ls of the slot have a big influence on the frequency response. The resonance frequency moves up to higher frequency, while the length ls of rectangular etched slot decreases. There is a significant decalate of resonance frequency of 5 GHz when the dimension of length changed with only 300 μm.

Tuning of Circular DGS CPW Stopband Filter Characteristics

To investigate the frequency characteristics of the proposed DGS, the ADS Agilent momentum has been used to design and to optimize the proposed circular DGS filter. The proposed circular DGS cell can provide attenuation poles at certain frequencies without requiring other cells or resonator. The change of dimensions of circular DGS unit can involves the response of resonance frequency. This effect is due with the increase or decrease of both the length ls of rectangular slot and the value of radium of circular DGS unit of the coplanar filter.

In order to verify the tunability of the proposed stopband filter in other frequency response, the methods of reconfigurability in this study based on the effect of the change of dimensions of rectangular etched slot, and the variation of the radium of circular DGS cell. Firstly, the length ls is fixed in the value of 200 μm and the radium optimized in four values, the other changed dimensions are kept constant. Secondly, the same process with other value of length of the rectangular etched slot ls=500 μm. The simulation results are shown in Figure 10.

Figure 8. Comparison on parameter extraction of equivalent circuit and EM simulation results for the circular DGS structure: (a) return loss (S11) and (b) insertion loss (S21)

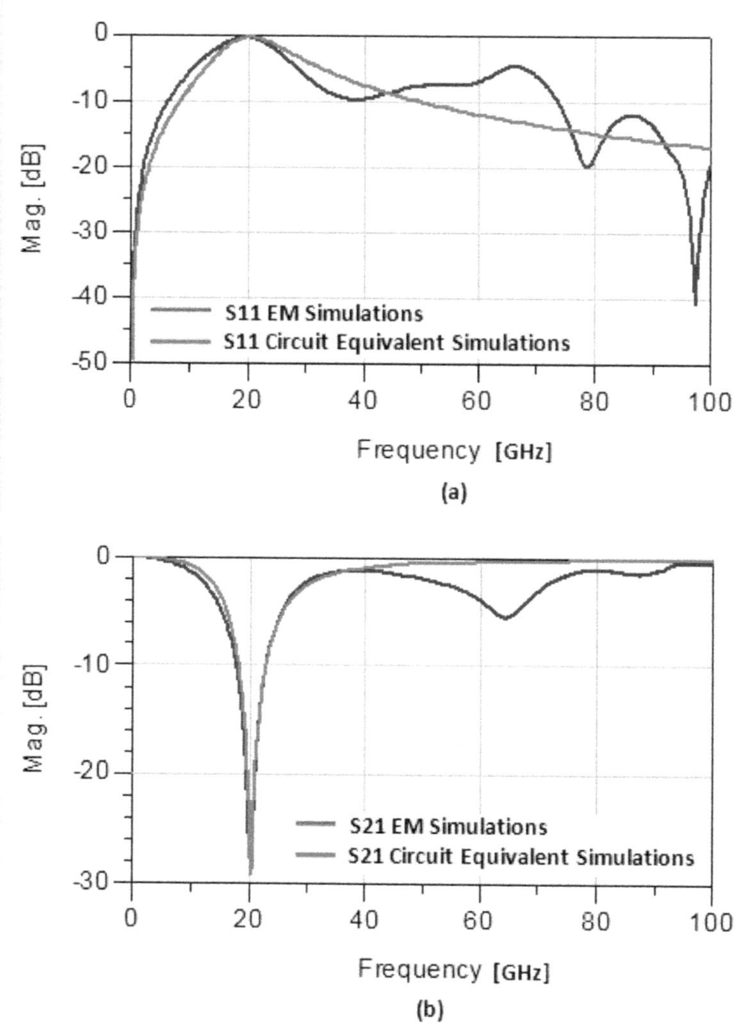

The simulation results are illustrated in Figure 10 and Figure 11. As well explain in the ex paragraph writing, the resonance frequency of the proposed circular DGS cell can be generated by a combination of resistance, inductive and capacitive elements. Thus, to explain the relationship between the variation of dimensions and the simulated frequency response of the proposed circular DGS cell. As the radium of the proposed circular DGS decrease, both the characteristic impedance and the parallel resistance, inductive and capacitive elements of the equivalent circuit decrease, while the cutoff and resonance frequency increased. In addition, the attenuation pole location moves up to a higher frequency.

Adjusting the value of the radium r shifts the cut-off frequency and the attenuation pole of the proposed stop band filter in the frequency domain. It means that, we could obtain other response frequency of the proposed stopband filter, and the designed filter can be used in others applications domain, and

Figure 9. EM simulation results of insertion loss S21 and S12 for two value of length ls of the circular DGS structure

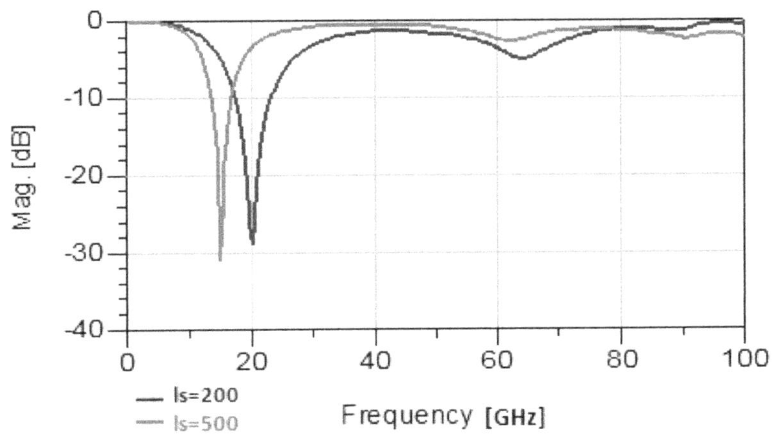

other several microwave and millimeter applications, by adjusting just the radium r of the circular DGS cell. These results prove the advantages of the proposed structure, compared to the conventional filter.

PROPOSED DEVICE OF CIRCULAR DGS STOPBAND FILTER

The initial DGS stopband filter design is modified by introducing four symmetrical slots with L-configuration to improve the insertion loss (S21) and return loss (S11) performance in stopband and bandpass of the circular DGS CPW.

The parameters of the DGS StopBand filter are kept unchanged while dimensions of the four L-slots are introduced, as shown in Figure 12.

The dimensions of the various elements of every L-slot are as length (l=220µm), gap (g=20µm) and height (a=40µm), and x the distance separation between two slots(x=260µm).

Figure 13 shows the simulated S-parameters of the proposed device of stopband filter with four L-slots.

As a result, the stopband resonance is around of the frequency desired (20 GHz), it is approximately centered between 16 GHz and 23 GHz, and it exhibits more than 10 dB stopband with minimum |S21| of −29 dB around 20 GHz, and offering a drastically improved response performance of stopband filter. That slots inserted in the conductor line of a cpw structure can modify the insertion loss (S21) and return loss (S11) behavior, and improved more performance, also can modify its bandwidth to get an ultra wide bandpass after the desired stopband rejection of the CPW-SBF, approximately between 26 GHz and 85 GHz The insertion loss (S21) is better than −3 dB and return loss (S11) is better than −10dB within this band.

Figure 10. Effect of variation of the radium r with the length ls=500 µm on EM parameters of circular DGS stop band filter: (a) EM simulation results of return loss S11. (b) EM simulation results of insertion loss S21

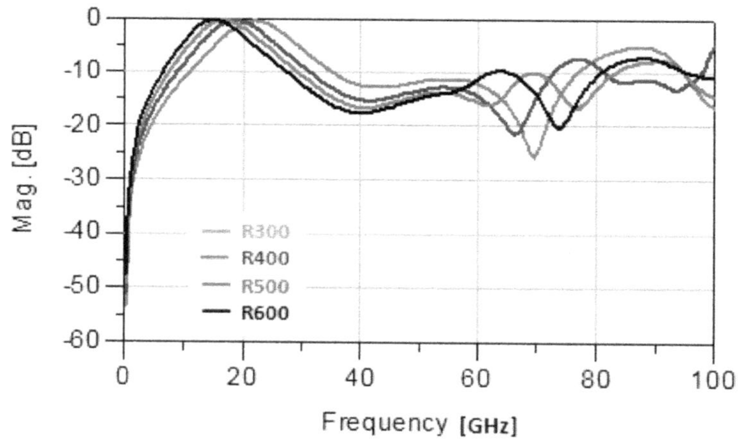

(a)

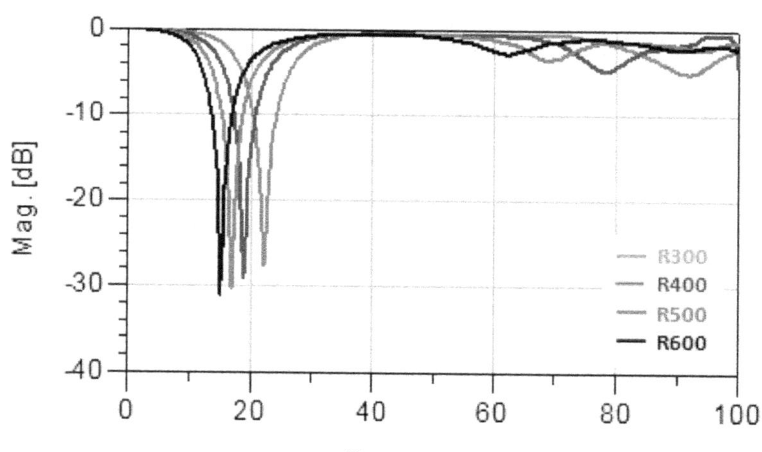

(b)

DISTRIBUTION OF SURFACE CURRENT AT PASS-BAND AND STOP-BAND FREQUENCY RANGES

In order to investigate and demonstrate the relationship between the EM-simulation results and the surface current density distribution of the proposed circular DGS stopband filter. The iam of this study basing on behaviors of surface current density energies. Figure 14 shows the current distribution simulated at two different frequencies respectively in the passband Figure 14 (a) and stopband Figure 14 (b). The first EM field distribution is at frequency 40 GHz and the second one at frequency 20 GHz.

Figure 11. Effect of variation of the radium r with the length ls=200 µm on EM parameters of circular DGS stop band filter: (a) EM simulation results of return loss S11. (b) EM simulation results of insertion loss S21

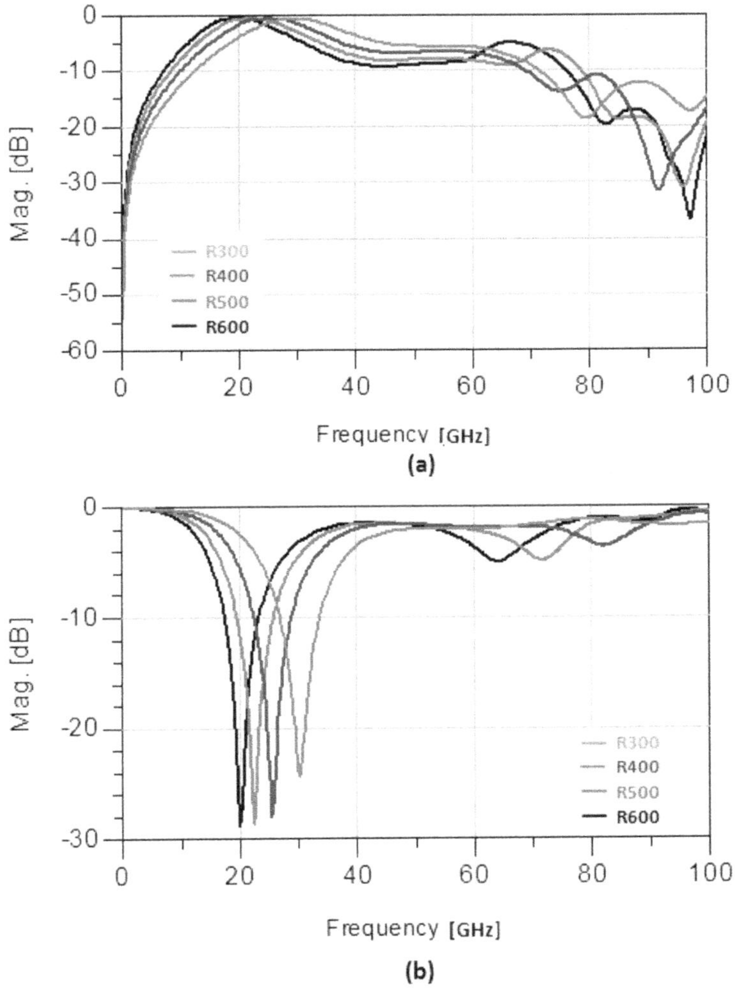

As a result, in Figure 14(a) we can observe that the current energies was transmitted along the proposed filter structure from the input to the output and it is important around the four L-slots and the circular DGS, at the frequency of 40 GHz, which means that the filter is in the passband state. In the other side, as figure 14(b), shows, at the 20 GHz frequency, there is a high current distribution energies in the first part of structure, while blocked around the port 1, the the four L-slots and the circular DGS cell, and no current near to the output port 2, therefore no a transmission of RF power from the input port to the output port, which means that the filter is in the stopband state.

We can note from this experiment proves, that proposed filter with circular DGS cell has similar behaviors between the EM-simulation results and the energies distribution.

Figure 12. The proposed device of stopband filter with four slots L-configuration

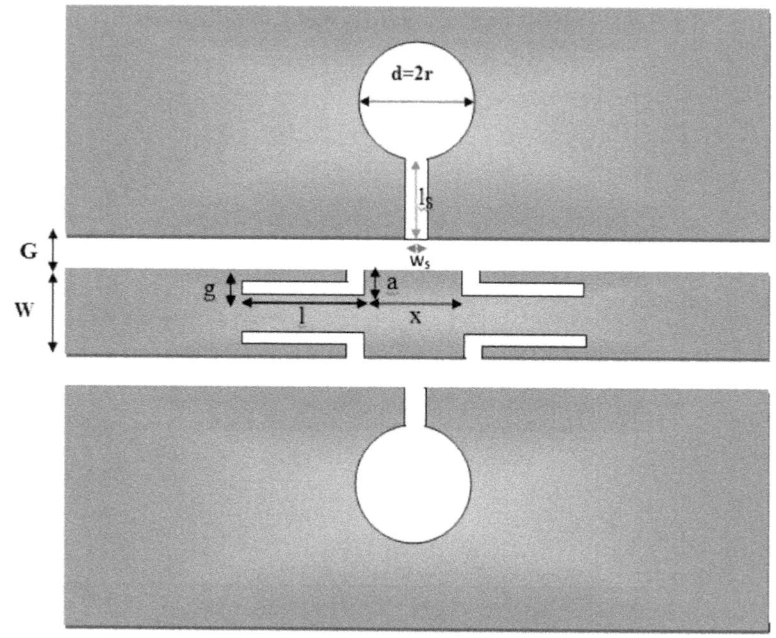

Figures 15 and 16 shows respectively the phase simulation of return loss (S11) and insertion loss (S21) of the designed circular DGS stopband filter between 0.1 GHz and 100 GHz, we can note that throughout the pass and stop bands are acceptably linear for high frequency response and UWB applications.

Figure 13. Simulated S-parameter of the proposed device

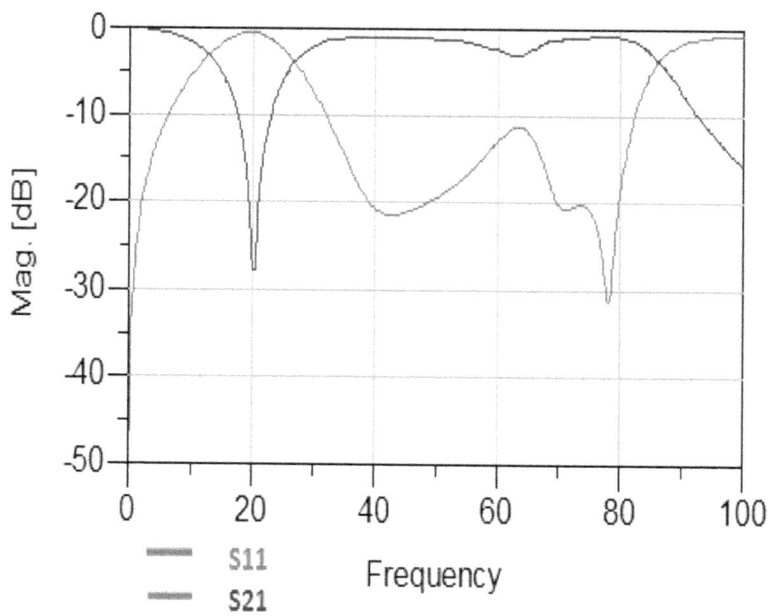

Figure 14. Simulated surface current density: (a) At f = 40GHz, (b) At f = 20GHz

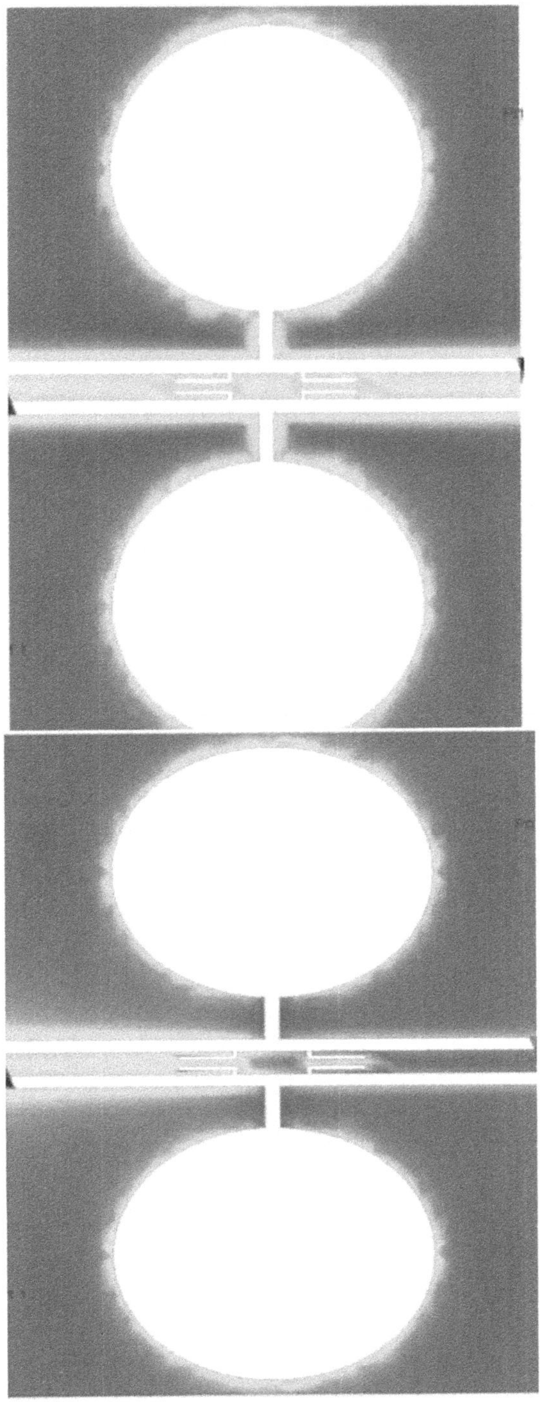

(b)

Figure 15. S11 simulation phase response of the proposed CPW stopband filter

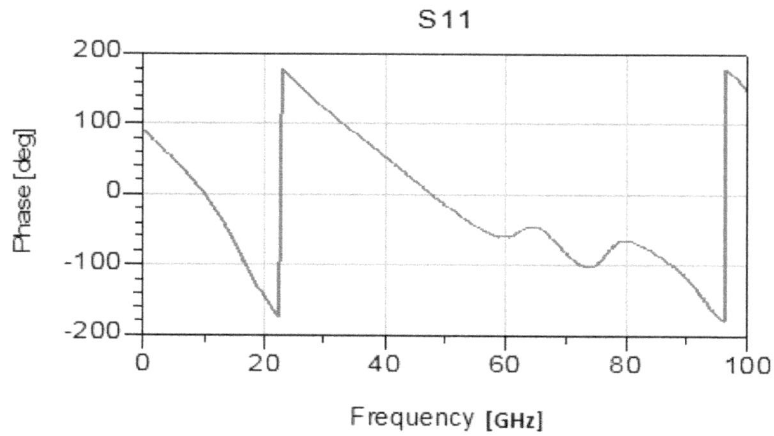

The simulation result of the proposed circular DGS stopband filter group delay is shown in Figure 17. Group delay is the rate of change of transmission phase angle with respect to frequency. It is clear that the variations in the group delay have good linearity for the pass band frequency, with flat group delay variation around 20 GHz in the stopband. The simulation results are shown in Figure 17.

CONCLUSION

In this chapter, an overview of coplanar technology and the general techniques, process to improve the response and characteristics of microwave components. Thereafter, a novel compact coplanar stopband filter using one circular DGS cell, with excellent suppressed, sharp rejection slope and wide stopband rejection is presented. Different techniques for adjusting the insertion loss and return loss behaviors are

Figure 16. S21 simulation phase response of the proposed CPW stopband filter

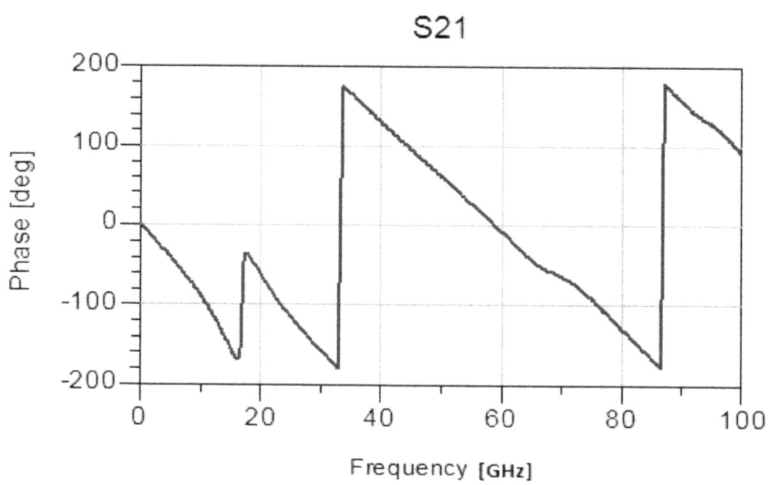

Figure 17. The simulated group delay of the proposed circular DGS stopband filter

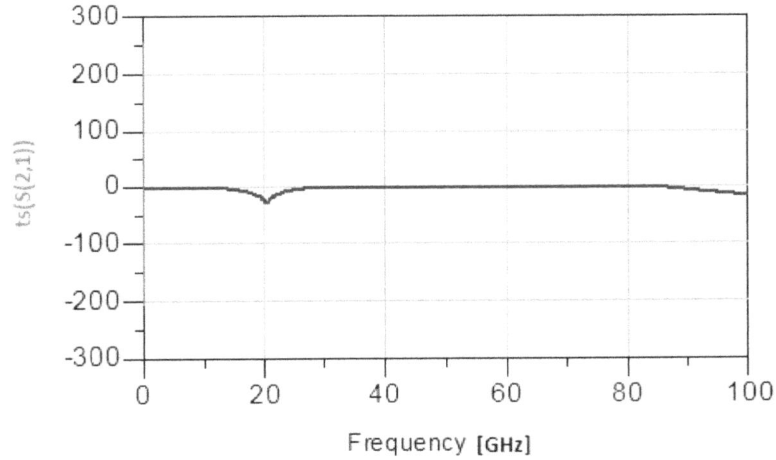

discussed. The slots with L-configuration play a key role in adjusting the stopband center frequency and to improve separately the stopband and passband performances. It is an effective technique to enhance the stopband as well as passband performances. The equivalent circuit models of varieties of DGS unit are also presented.

REFERENCES

Ahn, D., Park, J. S., Kim, C. S., Kim, J. Y., Qian, X., & Itoh, T. (2001). A design of the low-pass filter using the novel microstrip defected ground structure. *IEEE Transactions on Microwave Theory and Techniques*, *49*(1), 86–93. doi:10.1109/22.899965

Alsanousi, A. A., Ashoka, A., & Jeberson, W. (2013). A Compact Microstrip Low Pass Filter Based on DGS with Shaped Microstrip Line. *IACSIT International Journal of Engineering and Technology*, *3*(6).

Aytouna, F., Zbitou, J., Aghoutane, M., Touhami, N., Tribak, A., & Latrach, M. (2016). A Novel CPW Low Cost Lowpass Filter Integrating Periodic Structures. [IJECE]. *Iranian Journal of Electrical and Computer Engineering*, *6*(3), 1106–1111.

Boutejdar, A. (2016). Design of 5 GHz-compact reconfigurable DGS-bandpass filter using varactor-diode device and coupling matrix technique. *Microwave and Optical Technology Letters*, *58*(2), 304–309. doi:10.1002/mop.29561

Boutejdar, A., & Abd Ellatif, W. (2016). Improvement of Compactness of Low Pass Filter Using New Quasi-Yagi-DGS-Resonator and Multilayer-Technique. *Progress In Electromagnetics Research C*, *69*, 115–124. doi:10.2528/PIERC16073003

Boutejdar, A., Makkey, M., Elsherbini, A., & Omar, A. (2008). Design of compact stop band extended microstrip low pass filters by employing mutual-coupled square shaped defected ground structures. *Microwave and Optical Technology Letters*, *50*(4), 1107–1111. doi:10.1002/mop.23273

Browne, J. (1987). Broadband Amps Sport Coplanar Waveguide. *Microwaves & RF*, *26*(2), 131–134.

Browne, J. (1987). Coplanar MIC Amplifier Bridges 0.5 To 18.0 GHz. *Microwaves & RF*, *26*(6), 194–195.

Browne, J. (1989). Coplanar Waveguide Supports Integrated Multiplier Systems. *Microwaves & RF*, *28*(3), 137–138.

Browne, J. (1990). Coplanar Circuits Arm Limiting Amp with 100-dB Gain. *Microwaves & RF*, *29*(4), 213–220.

Browne, J. (1992). Broadband Amp Drops through Noise Floor. *Microwaves & RF*, *31*(2), 141–144.

Chen, H. J., Huang, T. H., & Chang, C. S. (2006). A novel cross-shape DGS applied to design ultra-wide stop-band low-pass filters. *IEEE Microwave and Wireless Components Letters*, *16*(5), 252–254. doi:10.1109/LMWC.2006.873594

Close-Up, T. (1988, April). *Microwaves & RF*, *27*(4), 79.

Elmahjouby, S., Errkik, A., Mandry, R., Tajmouati, A., & Latrach, M. (2016). A New CPW Compact Stopband Filter Using DGS Loaded. In *proceedings of International conference on computing wireless and communication systems.* (pp 33-35). Settat, Morocco: University of Hassan 1st.

Elmahjouby, S., Errkik, A., Zbitou, J., Tajmouati, A., El Abdellaoui, L., & Latrach, M. (2017). A Novel Compact CPW LowPass Filter Integrating Periodic Triangle DGS Cells. *Transactions on Machine Learning and Artificial Intelligence*, *5*(4), 457–462.

Jackson, R. W. (1989). Mode conversion at discontinuities in finite-width conductor-backed coplanar waveguide. *IEEE Transactions on Microwave Theory and Techniques*, *37*(10), 1582–1589. doi:10.1109/22.41005

Karim, M. F., Liu, A. Q., Alphones, A., Zhang, X. J., & Yu, A. B. (2005). CPW band-stop filter using unloaded and loaded EBG structures. *IEE Proceedings, Microwaves, Antennas and Propagation*, *152*(6), 434. doi:10.1049/ip-map:20050096

Koo, J. J., Oh, S. M., & Hwang, M. S., Park, C., Jeong, Y., Lim, J., … Ahn, D. (2007, October). A new DGS unequal power divider. In *Proceedings of European Microwave Conference*. (pp. 556-559). IEEE.

Koster, N. H. L., Koslowski, S., Bertenburg, R., Heinen, S., & Wolff, I. (1989). Investigations on Air-bridges used for MMICs in CPW technique. In Proceedings of 19th EuMC, (pp. 666-671).

Lim, J., Kim, S. C. S., & Lee, Y. T. (2002). A spiral-shaped defected ground structure for coplanar waveguide. *IEEE Microwave and Wireless Components Letters*, *12*(9), 330–332. doi:10.1109/LMWC.2002.803208

Mukesh, K. K., Binod, K. K., & Sachin, K. (2017). Defected Ground Structure: Fundamentals, Analysis, and Applications in Modern Wireless Trends. *International Journal (Toronto, Ont.)* of Antennas and Propagation, Volume 2017, Article ID 2018527. p 22.

Park, J. I., Kim, C. S., Kim, J., Park, J. S., Qian, Y., Ahn, D., & Itoh, T. (1999, November). Modeling of a photonic bandgap and its application for the low-pass filter design. In 1999 Asia Pacific Microwave Conference. Microwaves Enter the 21st Century. Conference Proceedings (Vol. 2, pp. 331-334). IEEE.

Radisic, V., Qian, Y., Coccioli, R., & Itoh, T. (1998). A novel 2-D photonic bandgap Structure for microstrip lines. *IEEE Trans. Microwave and Guided Wave Lett*, *8*(2), 69–71. doi:10.1109/75.658644

Riaziat, M., Majidi-Ahy, R., & Jaung-Feng, I. (1990). Propagation modes and dispersion characteristics of coplanar waveguides. *IEEE Transactions on Microwave Theory and Techniques*, *38*(3), 245–251. doi:10.1109/22.45333

Sharma, A. K., & Itoh, T. (1993). Special Issue on Modeling and Design of Coplanar Monolithic Microwave and Millimeter-Wave Integrated Circuits. *IEEE Transactions on Microwave Theory and Techniques*, *41*(9).

Sor, J., Quin, Y., & Itoh, T. (2011). Miniature Low-Loss CPW Periodic Structures for Filter Applications. *IEEE Transactions on Microwave theory and techniques, 49*(12).

Sporkmann, T. (1998). The Current State of the Art in Coplanar MMICs. *Microwave Journal*, *41*(8), 60–74.

Sporkmann, T. (1998). The Evolution of Coplanar MMICs over the past 30 Years. *Microwave Journal*, *41*(7), 96–111.

Stegens, R. E., & Alliss, D. N. (1987). Coplanar Microwave Integrated Circuit for Integrated Subsystems. *Microwave Sys. News Comm. Tech.*, *17*(11), 84–96.

Thakur, J. P., & Jun-Seok, P. (2006). A new design approach for circular polarize antenna with DGS under the unbalanced feed-lines. *36th European Microwave Conference*, (pp. 1483-1485). 10.1109/EUMC.2006.281339

Walker, J. L. B. (1993). A Survey of European Activity on Coplanar Waveguide: Atlanta, Georgia. *IEEE MTT-S International Microwave Symposium Digest*, *2*, 693–696.

Wang, J. F., Qu, S. B., Xu, Z., Zhang, J. Q., Yang, Y. M., Ma, H., & Gu, C. (2008). A candidate three-dimensional GHz left-handed metamaterial composed of coplanar magnetic and electric resonators. *Photonics and Nanostructures-fundamentals and Applications, 6*(3-4), 183-187.

Wen, C. P. (1969). Coplanar waveguide: A surface strip transmission line suitable for non reciprocal gyromagnetic device application. *IEEE Transactions on Microwave Theory and Techniques*, *17*(12), 1087–1090. doi:10.1109/TMTT.1969.1127105

Weng, L. H., Guo, Y. C., Shi, X. W., & Chen, X.-Q. (2008). An overview on defencted ground structure. *Progress in Electromagnetics Research B*, *7*, 173–189. doi:10.2528/PIERB08031401

Wolff, I. (2006). *Coplanar Microwave Integrated Circuit*. Hoboken, NJ: John Wiley & Sons.

Chapter 7
Autoencoders in Deep Neural Network Architecture for Real Work Applications:
Convolutional Denoising Autoencoders

Houda Abouzid
Abdelmalek Essaadi University, Morocco

Otman Chakkor
Abdelmalek Essaadi University, Morocco

ABSTRACT

The most heard sound exists as a mixture of several audio sources. All human beings have the ability to concentrate on a single source of their interest and ignore the other sources as disturbing background noise. To apply this powerful gift to a machine, it must obligatory pass through the source separation process. If there is not enough information about the process of mixture of those sources and their nature as well, the problem is known by Blind Source Separation BSS. This thesis is dedicated to study the BSS as a solution for human machine interaction. The objective consists in recovering one or several source signals from a given mixture signal. Recently, the science research is towards artificial intelligence and machine learning applications. The proposed approach for the separation will be to apply a Deep Neural Network method based on Keras. Extracting features from the audio with signal processing techniques and machine learning to learn a representation from the audio for the compression tasks and the suppression of the noise will improve the state-of-the-art.

INTRODUCTION

The Blind Source Separation known as the problem of BSS consists to find statistically independent signals from of their mixtures (observations) and this is done without any prior knowledge of the structure of mixtures or source signals.

DOI: 10.4018/978-1-7998-0117-7.ch007

Source separation occurs in a variety of applications such as locating and tracking targets in radar and sonar, the separation of speakers (problem called "Cocktail party"), detection and separation in systems of multiple access communication, Independent Component Analysis (ICA) of biomedical signals (e.g., EEG or ECG), etc. This problem has been intensively studied in the literature and many effective solutions have already been proposed (Belouchrani, Abed-Meraim, Cardoso, & Moulines, 1997).

In addition to the separation problem, there is another challenge that should be not ignored during the separation process which is the denoising of the unuseful sources aiming at a high-quality restitution. To do so, there are different techniques used for this purpose and between them the autoencoders.

Autoencoders are unsupervised learning techniques because there is no really need for explicit labels to train the model. The algorithm takes the input data which is in our case represented by audio signals, and try to reconstruct it with only fewer number of bits from the latent space. This operation is done with compression of the data during the time training of the neural network. As the Principle Component Analysis (PCA) (Abouzid & Chakkor, 2018) first goal is to reduce the dimension space, the autoencoders play the same role. In general, the idea is to project the dataset in a smaller space with removing some unuseful parts. As most of the researchers in signal processing field know that the PCA uses linear transformation, but the autoencoder uses the non-linear transformation. This is the big difference between them.

Blind Source Separation

Problem Formulation

The major goal of source separation is to estimate various signals from the observation of their mixture. The mixture is denoted $X(t)$ throughout this manuscript, is usually represented as a multi-channel (or multi-channel) time signal. So, it is commonly represented as a vector function of a time variable t such that:

$$\forall t, X(t) = \left[x_1(t), \ldots, x_M(t)\right]^T \tag{1.1}$$

Where $X(t)$ is the matrix of mixtures according to the time and $x_i(t)$ is the ith parameter of this matrix.

The model of source separation supposes the existence of N independent signals $s_1(t), \ldots, s_N(t)$ and M observations denoted $x_1(t), \ldots, x_M(t)$ which represent the mixtures. If those mixtures are supposed to be instantaneous, then the equation will be:

$$x_i(t) = \sum_{j=1}^{N} a_{ij} s_j(t) \tag{1.2}$$

This last equation can be denoted by in another writing:

$$x(t) = As(t) \tag{1.3}$$

Where $s(t)=\left[s_1(t),\ldots,s_N(t)\right]^T$ is the column vector of dimension $N\times 1$ which gathers the source signals, the vector $x(t)$ groups together in the same manner the M signals observed, and $A=\left[a_1,\ldots,a_N\right]$ is the matrix of mixture of dimension $M\times N$ where $a_{i=}\left[a_{1i},..,a_{Mi}\right]^T$ contains the coefficients of the mixture.

We will assume that for any pair (i,j) the vectors a_i and a_j are linearly independent (see the figure 1).

In another hand, and in a real environment, there are delays, repetitions and reverberations. A more general way to rephrase the problem is to add a delay term resulting from the convolution of the sources with the associated impulse response (the propagation filters) denoted as $h_j(n)$. The formulation of the problem becomes in convolutional mixtures as follows:

$$x(n)=\sum_{j=1}^{N}h_j(n)*\left\{a_j s_j(n)\right\} \tag{1.4}$$

According to the properties of the convolution, the equation (1.4) becomes:

$$x(n)=\sum_{i=1}^{M}\sum_{j=1}^{N}\sum_{k=1}^{L} a_{kij}\cdot s_j(n-\delta_{kij}) \tag{1.5}$$

Where a_{ij} represents the coefficients corresponding to the mixing filler, and δ_{ij} represents the order of the RIF filler. The figure 2 shows an example of a convolutive mixture.

Figure 1. instantannuous mixture of audio sources by an XY stereo configuration

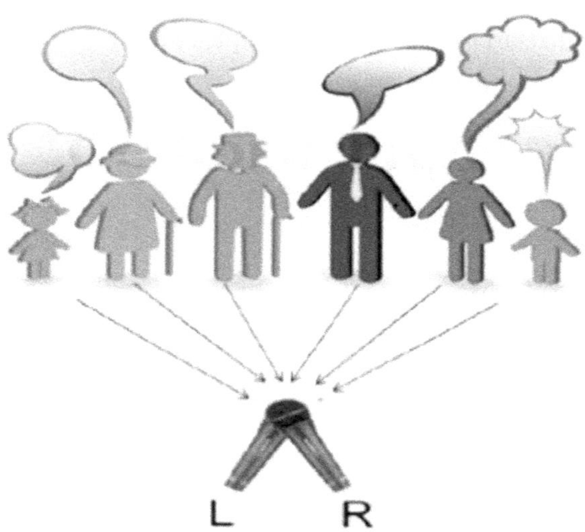

The Main Assumptions for Separation

Separation of sources from a blind viewpoint usually requires the necessary assumptions in order to achieve the separation process. The following remarks are the assumptions that are required:

Hypothesis 1: The components of the source vector $s(t)$ are correlated. This can be clearly seen in Equation 1.2 This hypothesis is elementary for blind separation using the matrix decomposition.

- **Hypothesis 2:** the simplest case to study is always the one when the number of sources is less than or equal to the number of sensors, this is called successively the mixing system over-determined or determined. Several algorithms deal with these two situations and thus, the ICA method is applied on this type of mixture only, however it remains limited to solve the case of under-determined mixture when the number of sources is much greater than the number of microphones (the most complex case and close to reality).
- **Hypothesis3:** the independence of the sources is a fundamental condition to be assumed for the BSS problem. When the source signals are assumed mutually statistically independent, this gives from a mathematical point of view, the joint probability density of all the signals is equal to the product of their marginal densities. The following equation summarizes this sentence mathematically:

$$p(s_1(t-\delta 1),\ldots,p(s_N(t-\delta_N)) = \prod_{j=1}^{N} p(s_j(t-\delta_j)) \tag{1.6}$$

for any moment t and for any offset δ_j with $j = 1,\ldots,N$. This independence condition is simplified in the case of instantaneous linear mixtures by setting $\delta_j = 0$.

Figure 2. Convolutive mixture of audio sources

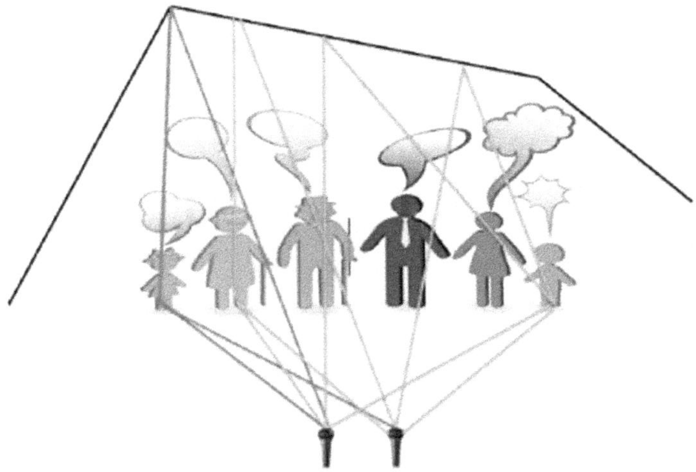

This assumption is valid for Higher Order Statistics (HOS) methods, however for methods using Second Order Statistics (SOS) it is reduced to non-correlation of sources for signals colored stationary or for non-stationary signals.

- **Hypothesis 4:** Each source is assumed to be mutually statistically independent from the other sources and the samples from each source are independents and identically distributed. Which means that at every moment, each source is an independent statistical distribution for all its present and future moments. This assumption is essential and important for the separation and will be presented by default for all the state of the art of this chapter.

From the hypothesis 3, at most there is one source signal that supposed to be gaussian.

In the case of instantaneous linear mixing, the signals $y = Bx$ where y is the estimated sources and B is the separation matrix, are supposed to be independent if and only if the separation matrix is denoted by this form of writing: $B = DPA^{-1}$. Here D represents the diagonal matrix and P is the permutation matrix. This means that independence is equivalent to the separation to a scale factor and to a close permutation. In this context, if there is more than one Gaussian source, the property of independence does not necessarily lead to separation.

- **Hypothesis 5:** Most **BSS** methods assume linearity of mixing system. In the case of the ICA, this assumption is primary. However, there are other algorithms that treat the type of non-linear mixture (Deville and Hosseini, 2009; Hosseini and Deville, 2013).
  Hypothesis 6: The mixing matrix A is assumed to be of full rank, which implies the existence of the separation matrix B. This assumption is always verified for the ICA method (whether for an instant or convolutive mixture).
- **Hypothesis 7:** All signals are ergodic, which means that the time averages are the same as the statistical averages. Practically, the latter is reduced to a recording of limited duration.

Presentation of the Various Tools

Neural Network Principal

Artificial neural networks (ANNs) is a function that can be trained on a set of data. Given their general nature, ANNs would seem useful tool for nonlinear transformation systems. ANN are classifiers widely used intensively in many domains to accomplish different computational tasks like signal separation classification (Amari et al., 1996), the acquisition, image processing and analysis of multiple images (Egmont-Petersen et al., 2002), the recognition of outlines that are applied in computer vision and automatic recognition of the word (Mao et al., 2014) the segmentation of objects in images, like to segment microstructures from metallographic (de Albuquerque et al., 2009) and many other useful scientific domains.

ANN are inspired from a biological concept which means the natural neural network existing in the human brain. The smallest unit of a neural network is called "neuron" which is a mathematical representation of a biological neuron that forms the layer and build the system.

Biological Neuron

A natural neuron receives many information called **neurotransmitters** from his **dendrites**. These neurotransmitters are released by their synapses. When the amount information exceeds a certain threshold, the neuron will be activated and then will send an electric signal into its **axon** which allows to this last to transmit neurotransmitters using his synapses (see figure 3 for more details).

Simple Neuron

Neural networks are typically structed in '**layers**. Layers are made up of a number of interconnected '**nodes'** which apply an **'activation function'**. The neural model is introduced to the network via the '**input layers'**, that communicate with the **'hidden layers'** where the actual processing is done via a system of **'weighted connections.** The resulted responses are linked with the hidden layers to the end of the neural network which are known as the **'output layers.** The figure bellow shows an example of a neural network that uses two inputs, five hidden layers and two outputs.

A simple neuron can be seen as a function transmitting a prediction $\tilde{y}$ according to a variable input x .

Mathematically, the neuron formal is a combination of a simple operation with an activation function f (see the figure 6). To do the analogy to the biological neuron, the weight w represents the synapses, z the dendrites and y the axon. The equations are:

$$y = f(z) = f(\sum_{i=1}^{n} w_i x_i - b) \quad (1.7)$$

$$W^T = (-b, w_1, \cdots, w_n) \quad (1.8)$$

Figure 3. A biological neuron, source (Anon, 2019)

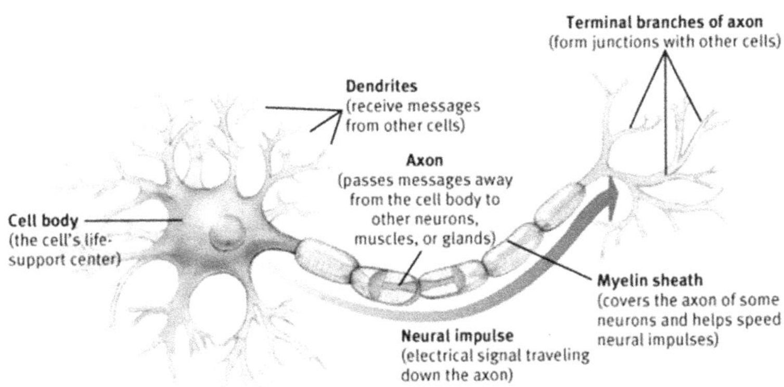

Figure 4. Example of a neural network, source (WildML, 2019)

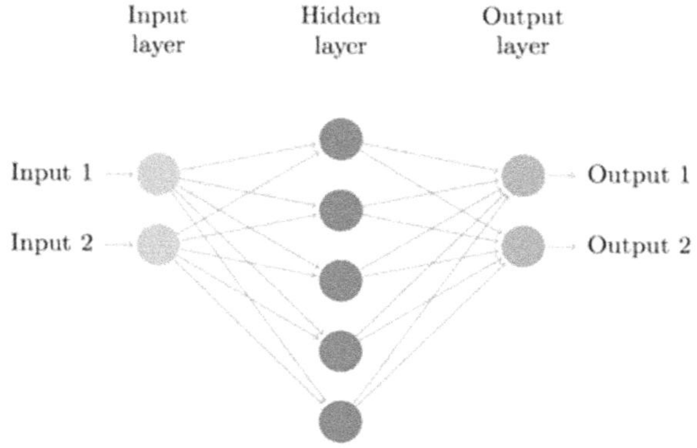

Figure 5: Principle of neural network

Figure 6. Neuron formal, source (Anon, 2019)

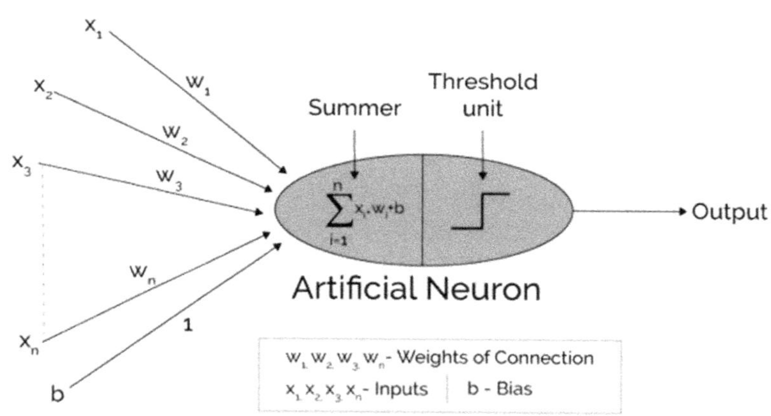

$$x = \begin{pmatrix} x_1 \\ \vdots \\ x_n \end{pmatrix} \quad (1.9)$$

To simplify, we usually put the **biais** in the matrix of weight and increasing the size of the entry by 1. About the activation function there are many and the most famous between them are for example the **ReLU**, **tanh** and **sigmoide**. It depends about the result that is searched for to be obtained after training the model. The goal that should be into account is to find the good weights and the biais so that the Neural Network respond to a certain criterion.

Multilayer Perceptron (MLP)

The architecture of a multilayer perceptron **MLP** has N layers with the same principle and a big number of neurons in each layer. The column vector of weights becomes a matrix because in this case there are more neurons in input and output (see figure 7).

The layers who are between the input layer and the output layer are called Hidden layers. The number of hidden layers is free as the number of neurons according to the classification task.

For a network of L layers, the matrix of weights between the layers l and $l+1$ denoted $W^{(l,l+1)}$ and has the dimension of $n_{l+1} * (n_l + 1)$ with n_l represents the number of neurons in the layer l.

Machine Learning

The Learning

The learning is the time when the Neural Network optimizes the matrices weights $W^{(l,l+1)}$ of the network so that it learns the data.

There are 3 different types of the learning: supervised learning, unsupervised learning and semi supervised learning.

1. Supervised learning

In the supervised setting, the data $z_i(x_i, y_i) \in \mathbb{R}^m \times \mathbb{R}^n$ are formed from in input $x_i \in \mathbb{R}^m$ and a target, or a label $y_i \in \mathbb{R}^n$. The goal is then to learn a function $f(x_i) \in \mathbb{R}^n$ able to predict the target. The nature of the target defines the kind of problem to solve.

In the case where the target is discrete, we then speak of a problem of **classification**. The simplest case is the binary classification, where there are only two classes to separate.

In this case, we can represent the two classes by 0 and 1. If there are more than two classes, a good way to represent the target is by a one-hot vector (i.e. a vector where all the elements are zero except

Figure 7. A NN with L layers, source (Pavlovsky, 2019)

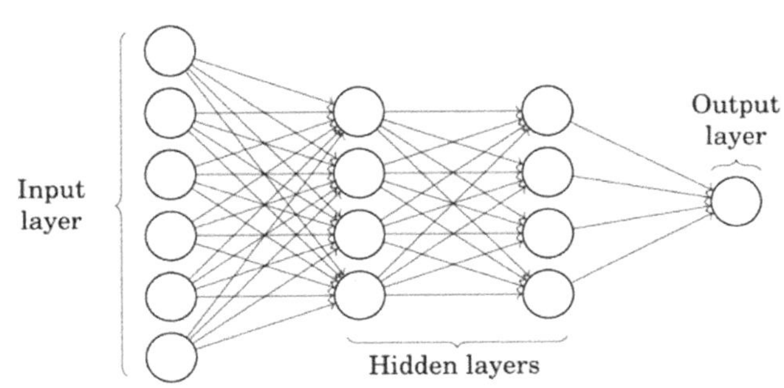

one which is equal to 1). It is also possible that an example of an entry can belongs to several classes at the same time, in this case we will talk about labeling rather than classification.

If the target is continuous, then we are talking about a **regression** problem. In this case, the goal of learning is to find a continuous function that predicts y given the x entries.

Unsupervised Learning

In the unsupervised learning, the data $z_i = x_i \in \mathbb{R}^m$ have no target. Unsupervised learning can allow us to estimate the probability density of inputs, to reduce dimensionality, to find a partitioning of data, or to find a new representation of data.

Dimensionality Reduction

Dimensionality reduction, is useful in many cases. Among other things, it can be used to visualize high-dimensional data, extract features, speed up learning, or reduce the amount of memory required.

Representational Learning

Representational learning is the case that interests us the most in this document.

The learned representation may be smaller, larger, or equal to the input dimension. Dimensionality reduction is therefore a special case of representation learning. A goal of representation learning can be to find a space from which it will be easier to solve a given task. the representation learning is also seen as a feature extraction.

Semi-Supervised Learning

Sometimes when there are a lot of Input data, but it's expensive to get targets for that data, or there are only a few targets for just a subsample of the data. This happens often in the case of musical audio where it is easy to find huge databases of examples non-labeled in large quantities. On the other hand, according to the task to be solved, it may be that the labeling requires the work of an expert who will annotate each example by hand.

It is possible to take advantage of this situation thanks to the semi-supervised framework. In this case, it is assumed that additional information can be derived from unlabeled data, which will improve the performance of the supervised task.

Some Machine Learning Models

Here, we will describe some famous learning machine algorithms.

Classification

Support Vector Machine (SVM)

The classification is a method of machine learning that distinguish between two or more different classes. When the data is linearly separable (i.e., can be separated by linear decision surfaces), a simple linear classifier can easily perform the classification work. Perceptron and logistic regression are two examples of linear classifiers.

However, in the case where classes are not linearly separable in the input space, it may be preferable to use a nonlinear classifier. Support vector machines (**SVMs**) (Abouzid & Chakkor, 2017) is an example of a very widely used nonlinear classifier. There are some non-linear SVM that uses the kernel trick to project data into a high-dimensional space before performing a linear classification.

The SVM presents a quadratic convex optimization problem, which guarantees to find the global minimum, but at the cost of a computation time of the order of N^2, which becomes prohibitive for large datasets.

Multi-Layer Perceptron (MLP)

The MLP is an approximator. It can approximate any non-linear function. The optimization of the MLP is done by gradient backpropagation algorithm as a supervised learning technique. This optimization is not convex, which brings the risk of falling into a minimum local. However, optimizing an MLP is faster (in real time computation) than an SVM for large data sets.

A multilayer perceptron is a neural network connecting multiple layers in a directed graph, which means that the signal path through the nodes only goes one way. Since there are multiple layers of neurons, MLP is a deep learning technique.

MLP is widely used for solving problems that require supervised learning as well as research into computational neuroscience and parallel distributed processing. Applications such as speech recognition, image recognition and machine translation.

Figure 8. SVM classification, source (Abouzid & Chakkor, 2017)
a: SVM classification of 448000 samples b: SVM classification of 55 of mixtures samples of mixtures

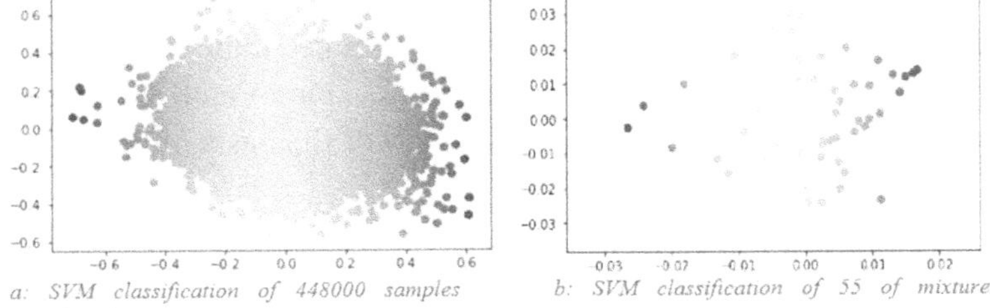

a: SVM classification of 448000 samples samples of mixtures *b: SVM classification of 55 of mixtures*

Unsupervised Partitioning

With a non-tagged dataset, one can use the empirical density distribution of the data to conclude partitioning. The K-means method is probably the simplest unsupervised partitioning technique. We start by initializing K points representing the random K means in the input space. Then, in a first phase, we assign to each data of the set the closest mean. In a second phase, the mean K is recalculated according to the data points assigned to it. This alternates the two phases to convergence. This convergence method is called expectation-maximization (EM).

The Gaussian Mixture Model (GMM) is a more advanced version of K-means. Here, instead of using only means, Gaussians are used to describe the data distribution. The GMM therefore has more parameters to optimize than K-means. However, it allows for a more refined modeling of the data density distribution. EM can also be used to train a GMM.

High Dimensional Data and Dimensionality Reduction

When the data is of type audio, it is important to consider the fact that "audio" is a very large representation. For example, if we consider 10 seconds of recorded music at a sampling frequency of 44100 Hz, we need $44100 \times 10 = 441000$ samples to represent this part of audio. It is a big problematic to work with audio extracts the least long if one does not try to reduce the dimensionality of the representation. With such high-dimensional data, there will be of course limited computational resource problems.

To solve the problem of the dimensionality, there are several techniques of dimensionality reduction that project the data into a subspace of the input space in which the data is better distributed. If we project data in 2 or 3 dimensions, we can then talk about visualization. Here are some of them dimensionality reduction techniques.

Principle Component Analysis (PCA)

Principal Component Analysis (PCA) is a dimensionality reduction method that is used in a multitude of areas.

The PCA algorithm represents a set of data in a new axis system in which each dimension is decorrelated. This decorrelation of the variables allows us to reduce the representation to the desired dimension by choosing the axes for which the variance of the data is greatest. Therefore, the PCA gives us a linear compression that best explains the variance in the data.

The PCA principle is that for a large data, it is possible to find a similarity between them and find a way easier and less expensive to describe them in a new basis. This is done by taking the covariance matrix of the data and diagonalize it to get then the diagonalization matrix. This matrix is composed of proper vectors with their associated proper values.

By changing the landmark by a rotation matrix U, the obtained data is presented in a new reference. It is just enough to choose only the vectors having the strong values. Once the data is described in a new basis of proper vectors, it is too easy to isolate the most part of the basis of the other party that contains that few information.

Independent Component Analysis (ICA)

The first work about the source separation problem has been studied by the researchers Herault, Jutten and Ans (Hérault, Jutten, & Ans, 1985) to separate the signals carried by nerve fibers. They applied an algorithm that uses unsupervised learning architecture that its idea was inspired from nerve cell. ICA leads to solve an optimization problem. It requires the measurement of the independence of random variables. This is a first obligatory condition that have to be supposed while dealing with the ICA problem. The whitening of the signals to be processed makes it possible to simplify the formulation of the problem.

The objective is to estimate N source signals, supposed stationary, ergodic and independent, from the measurement of M observed signals resulting from a mixture of the source signals. The simplest case is when it is assumed that the mixture is linear and instantaneous, that the observation noise can be neglected and that $M = N$.

Deep Learning

According to the researcher Yoshua Bengio, a specialist in artificial intelligence, he defined the Deep learning as is a set of automatic learning methods attempting to model with a high level of abstraction of data through articulated architectures of different nonlinear transformations. These techniques have led to significant and rapid progress in the areas of sound or visual signal analysis including facial recognition, speech recognition, computer vision, automated language processing (Le Cun, Bengio, & Hinton, 2015)

Recently, the vast majority of scientific research focused on deep learning architectures. Moreover, in theory, with enough data, capacity, and computation time, some shallow models can represent any complex function. However, these architectures still have shortcomings in the power of representation and non-local generalization. The limited power of representation of shallow architectures results in using more resources than necessary to represent a given function.

Deep learning techniques are a class of machine learning algorithms that:

- Use different layers of non-linear processing unit for feature extraction and transformation; each layer takes as input the output of the previous one; algorithms may be supervised or unsupervised, and their applications include model recognition and statistical classification.
- Operate with multi-level learning of detail or representation of data; across the different layers, we move from low-level parameters to higher-level parameters, where the different levels correspond to different levels of abstraction of the data.
- Deep learning uses hidden layers of artificial neural networks and sets of complex propositional computations. Deep learning algorithms oppose shallow learning algorithms because of the number of transformations performed on the data between the input layer and the output layer, where a transformation corresponds to a processing unit defined by weights and thresholds.

To resume all the things that have been mentioned above, the figure 9 shows the relation between deep learning, machine learning and artificial network. Different applications are developed according to each type of learning (figure 9).

Autoencoders

Deep Neural Network (DNN)

Researchers working in neural network domain have shown that DNN including Recurrent Neural Network (RNN) and Convolutional Neural Network (CNN) can model complex functions and perform good results in different tasks like audio signal processing and image recognition.

A neural network is one type of machine learning models. In the mid-1980s and early 1990s, many applications used architectural advancements based on neural network. With the demand of big data, required amount of time and the quality of the results. Deep Neural Network appears in many areas of applications such as:

- Image recognition
- Signal separation
- Automatic translation
- Autonomous cars
- Medical diagnosis
- Automatic moderation of social networks
- Financial prediction and automated trading
- Identification of defective parts
- Detection of malware or fraud
- Smart robots

About how DNN works, the computer programs that use deep learning use algorithms to run the process of training. Each algorithm in the hierarchy applies a nonlinear transformation on its input and uses what it learns to create a statistical model as output. Iterations continue until the output has reached

Figure 9. Applications related to machine learning algorithms, source ("KissPNG - HD png images and illustrations. Free unlimited download.,")

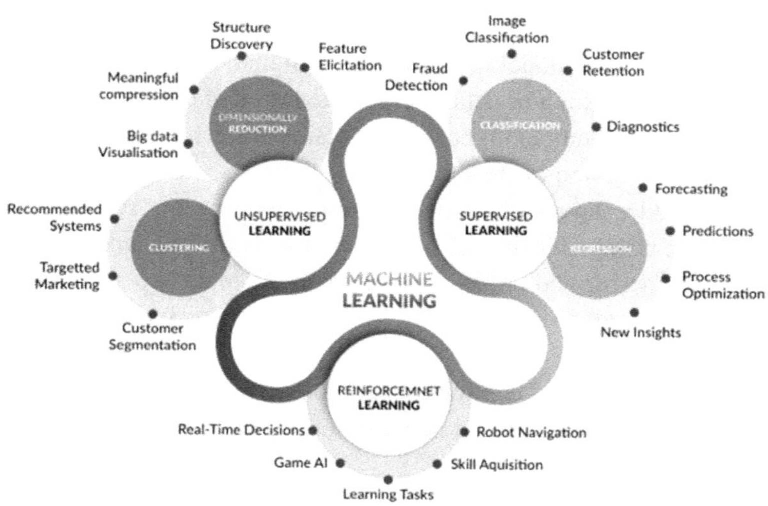

an acceptable level of accuracy. The number of processing layers through which data must pass is what inspired the label deep.

The machine learning process passes through some specifics steps that are shown in the Figure 10.

Network (CNN)

The Convolutional networks also called **convnet** (for "Convolutional Network"), or CNN (for "Convolutional Neural Network") is generally represented in the following form:

The one can distinguish two parts, a first part which is called the **convolutive** part of the model and the second part, which is called the **classification** part of the model which corresponds to a model MLP (Multi Layers Perceptron). The purpose of the convolution is the dimension reduction. The convolution has the advantage of having only a few weights to calculate (those of the filter) and that it reuses them for the whole data (image, audio file,), while the MLPs will have a unique weight to calculate for each neuron. This further reduces the number of calculations considerably.

The CNN is the architecture in which adjacent network layers are fully connected to one another. That is, every neuron in the network is connected to every neuron in adjacent layers (see figure 11):

Convolutional Neural Network (CNN) is a type of network of acyclic artificial neurons (**feed-forward**), in which the connection pattern between neurons is inspired by the animal's visual cortex. The neurons of this region of the brain are arranged so that they correspond to regions that overlap during tiling of the visual field. Their functioning is inspired by biological processes, they consist of a multilayer stack of perceptron, whose purpose is to pretreat small amounts of information. Convolutional neural networks

Figure 10.

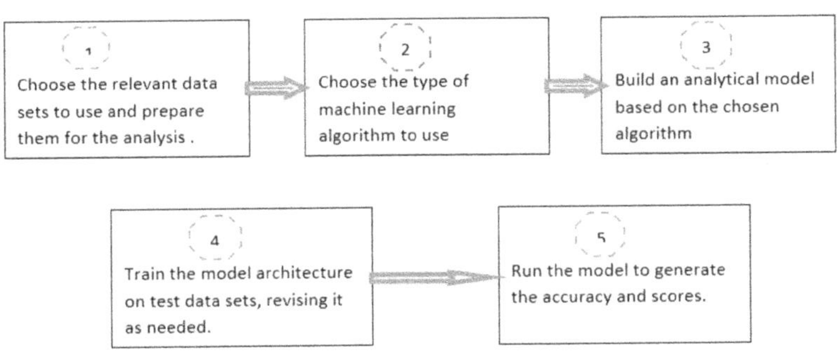

Figure 11. Architecture of a convolutional neural network, source ("https://www.natural-solutions.eu/blog/la-reconnaissance-dimage-avec-les-rseaux-de-neurones-convolutifs.," n.d.)

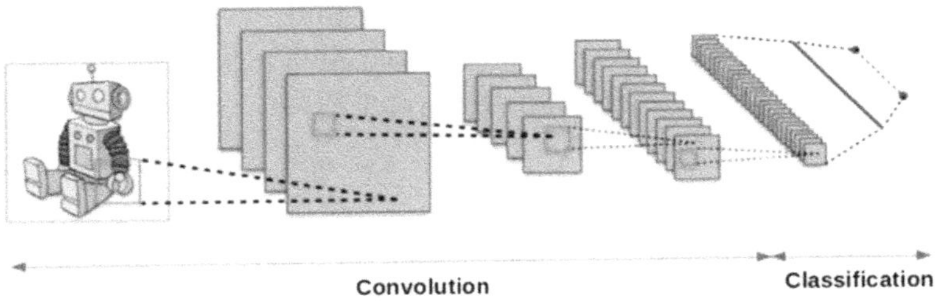

Figure 12. Convolutional NEURAL network, source ("http://neuralnetworksanddeeplearning.com/chap6.html.," n.d.)

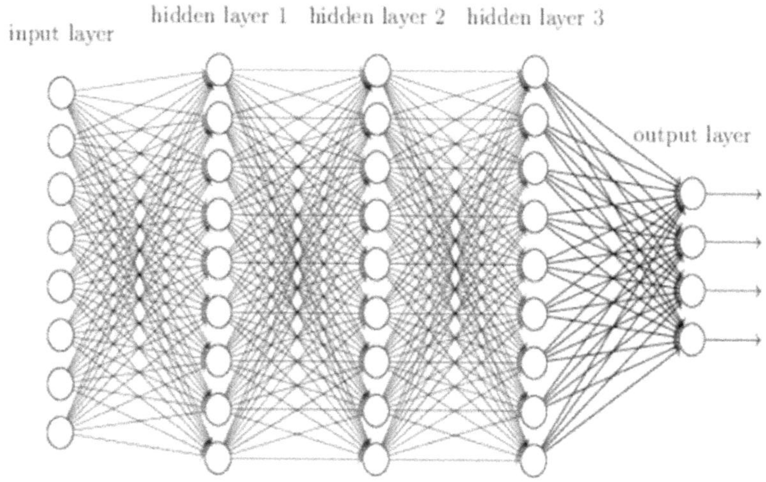

have wide applications in image and video recognition, recommendation systems and natural language processing.

So, the CNN is the network of neurons that use the convolution in two dimensions in the place of linear operation for the different layers. Why the use of the convolution on two dimensions? Simply that is allow to extract the spatial features of the dataset. Then the one can get the interesting results about the network for the treatment of images and audio files.

Audio Data

At the level of operation of audio data using CNN, this type of data has already been used in onset detection. Here the author used the CNN to detect the appearance of notes in a very interesting way. It is in fact a technical important way that has been applied for audio and that improved its well-done work for the detection appearance of music. The CNN has demonstrated its performance in comparison with a lot of systems. It considered also more appropriate for this type of data than the other networks such as the Recurrent Neural Networks (RNN) and the Superflux (the algorithm off low spectral Improved (Böck & Widmer, 2013).

The Neural Network can be used also for the classification task for the data as that is mentioned before, as an example of that the classification of the electroencephalograms (EEG). Here (Stober, S., Cameron, D. J., & Grahn, J. A, 2014) the author used the convolutional neural networks (CNNs), and he demonstrated that will allow to recognize individual rhythms with a mean accuracy of 22.9% over all subjects by just looking at the EEG recorded during the silence between the stimuli.

Knowing that the CNN can carry the training on both CPUs, a single GPU, multiple graphics processors in parallel.

Table 1. Precision results of the detection appearance between RNN and CNN, source (Schlüter & Böck, 2014)

	Precision
RNN [10,5]	0.892
CNN [1]	0.905
+ Dropout	0.909
+ Fuzziness	0.914
+ ReLU	**0.917**
SuperFlux [5]	0.883

Recurrent Neural Network (RNN)

Contrary to the MLP, the Recurrent Neural Network has not a propagation in only one direction but in which information can spread in both directions, including deep layers in the first layers. In this, they are closer to the true functioning of the nervous system, which is not one-way. These networks have recurrent connections in the sense that they retain information in memory: they can take into account at a time t a number of past states. Indeed, they are able to memorize only the so-called near past, and begin to "forget" after about 50 iterations. A layer may be connected to itself through intermediary the other layers of the network. This two-way transfer of information makes training much more complicated, and only recently have effective methods been developed such as **Long Short Term Memory (LSTM)**. These broad-based "short-term memory" networks have revolutionized speech recognition by machines (Speech Recognition) or understanding and text generation (Natural Language Processing). The interest of the looping is to help the network to recognize the data by knowing its characteristics in the treatment of the current given data. That will be useful to treat the sequential data such as an audio data because the sound is correlated temporally. RNN are used specially in automatic speech recognition or handwriting.

Convolutional Autoencoder (CAE)

Autoencoder is an unsupervised learning algorithm that has the input value as the output value that aim to reconstruct the input data in the output with the possible minimum distortion. The goal behind using the autoencoders is to reduce the dimension reduction and compression (Hughes & Correll, 2016), requiring fast time computation, improving performance by removing redundant variables (Abouzid, Chakkor, Reyes, & Ventura, 2019), visualizing high dimensional data and suppression noise from the original data. Recently, the computer science field has demonstrated a great performance on extracting features of images using a deep learning architecture of convolutional neural network and especially the convolutional autoencoders.

The convolutional autoencoder has an architecture containing convolutional layers (Kingma & Welling, 2013) that outperforms a simple autoencoder for compression data.

There are three types of autoencoders which are: Convolutional, denoising and sparse autoencoders. Let's begin by the first type:

The convolutional autoencoder is an autoencoder with convolutional and deconvolutional layers. In general, the CAE is used to learn audio features from the input samples. Convolutional autoencoders

Figure 13. Feed forward and recurrent neural networks, source (Hughes, & Correll, 2016)

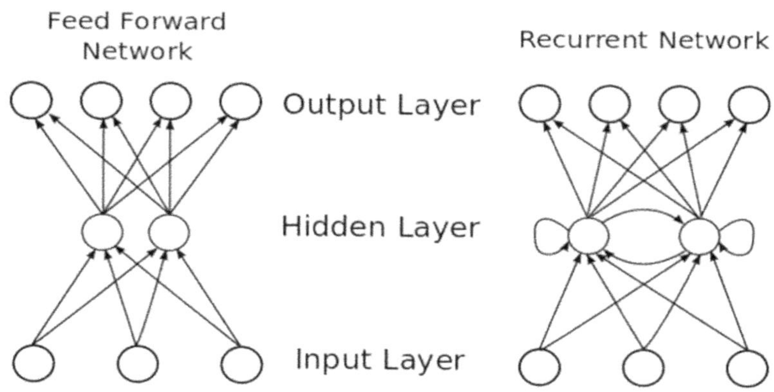

are often used for feature extraction in images (Mao, Shen, & Yang 2016, Guo, Liu, Zhu, & Yin, 2017), but not often used in audio signal treatment. In (Abouzid, Chakkor, Reyes, & Ventura, 2019) the authors applied two types of autoencoders: convolutional and denoising to extract audio features and compress the data to reconstruct it then at the output layer. In (Elhami & Weber, 2019) the authors used the CAE as an application to voice conversion, they explored the extraction features that contain both the vocal characteristics of the speaker and the content of the speech. The application is working via the proposed short-time discrete cosine transform (STDCT). The authors introduced a deep neural mapping at the encoding bottleneck to enable converting a source speaker's speech to a target speaker's speech while preserving the source-speech content.

To deal with the audio separation, in (Grais & Plumbley 2017) the authors proposed to use deep fully convolutional denoising autoencoders (CDAEs) for monaural audio source separation. They used as many CDAES as the number of mixed sources. The idea behind this is to separate one source for each trainable CDAE and ignore the other mixed signals as a disturbing background noise. This proposed method has shown its importance results with better accuracy then the deep feedforward neural networks (FNNs) even with fewer parameters than FNNs.

1. Denoising Autoencoder (DAE)

A Denoising Autoencoder (DAE) takes a noisy spectrum, and then outputs its cleaned-up version (Liu, Smaragdis, & Kim, 2014). The first goal of the DAE is to suppress the noise existing in the data,

Figure 14. Feature Extraction Network, source (Elhami, & Weber, 2019)

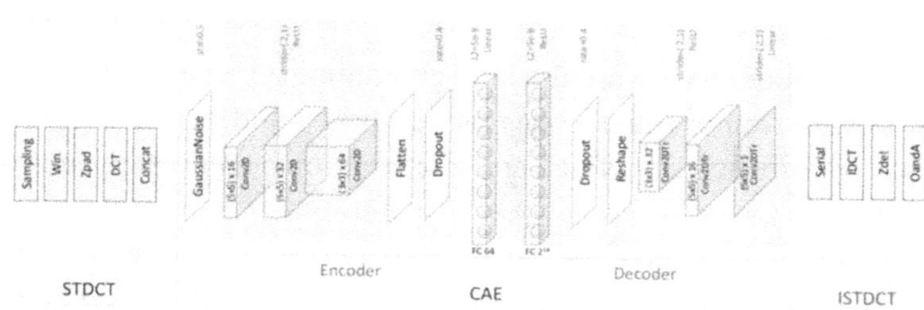

and especially if this data represents an audio file (Xu, Du, Dai, & Lee, 2014). A Denoising autoencoder is a feature learning technique using some perturbed input with stationary noise that need to be removed. The first fundamental purpose of a Denoising Autoencoder is taking a noisy spectrum, and then outputs its cleaned-up version (Abouzid, Chakkor, Reyes, & Ventura, 2019). To use the DAE for the source separation task, different types of noise can be applied to test the application.

In (Liu, & Yang, 2018) the authors applied a denoising autoencoder to separate music sources using 1D convolutions instead of 2D convolutions that has the benefit of having the possibility to add recurrent layers right after the convolution layers. This approach has demonstrated its effectiveness that achieved about 5.74 signal-to-distortion ratio (SDR) in vocals with MUSDB in SiSEC 2018. The figure 16 shows the proposed method. The separation model in this approach is a fully-convolutional network (FCN) with all convolution layers use 1D convolutions. This model has a specific name which is the ARC model that means applying a denoising autoencoder with recurrent skip connections. In this model the shape of the output will be (channels, temporal points).

Some others researchers have applied CNN with symmetric skip connections for singing voice separation like in (Jansson, Humphrey, Montecchio, Bittner, Kumar, & Weyde, 2017). The difference here is that they used 2D convolutions in their Convolutional Neural Networks (CNN). The shape of the output will be as (channels, frequency bins, temporal points). In this case applying a recurrent network to that tensor will cause a problem, this is why the authors in (Liu, & Yang, 2018) thought about using only 1D convolutions.

Sparse Autoencoder (SAE)

Another type of autoencoder has the name of Sparse Autoencoder or (SAE), this one represents another learning algorithm aiming to learn features from unlabeled data. This approach is usually used for computer vision applications, audio processing and natural language processing (Ng, (2011). It is also used in speech emotion recognition (Deng, Zhang, Marchi, & Schuller, 2013), here this method applied the SAE for the feature transfer learning for the signal emotion recognition. The authors' idea was a common emotion-specific mapping rule that is learnt from a small set of labelled data in a target domain. So then, newly reconstructed data are obtained by applying this rule on the emotion-specific data in a different field. The algorithm is described below. *Sparse Autoencoder algorithm used for feature transfer learning, source* (Deng, Zhang, Marchi, & Schuller, 2013)

Algorithm: Sparse Autoencoder Feature Transfer Learning
Input: The two labelled data sets T_t and T_s, and the corresponding class labels $C_1, \ldots, C_L$.
Output: Learnt classifier H for the target task.
1: Initialize reconstructed data $\tilde{T}_s = \varnothing$.
2: **for** $l = 1$ to L **do**
3: Initialize a single layer autoencoder $SA^l(W, b)$.
4: Choose class-specific examples $T_t^{C_l}$ from T_t.
5: Train $SA^l(W, b)$ using $T_t^{C_l}$.
6: Choose class-specific examples $T_t^{C_l}$ from T_t.
7: Reconstruct data $\tilde{T}_t^{C_t} = SA^l_{\mathrm{Recon}}(T_s^{C_t})$.

Figure 15. Diagram of the proposed separation model, ARC. Each tuple in the figure represents (output channels, filter size, stride) of the corresponding convolution layer. An STFT window size 2,048 is used and a spectrogram is symmetric in the frequency dimension, so the effective dimension of a spectrogram is (2, 048/2 + 1) = 1025. T is the number of temporal frames, source (Liu, & Yang, 2018).

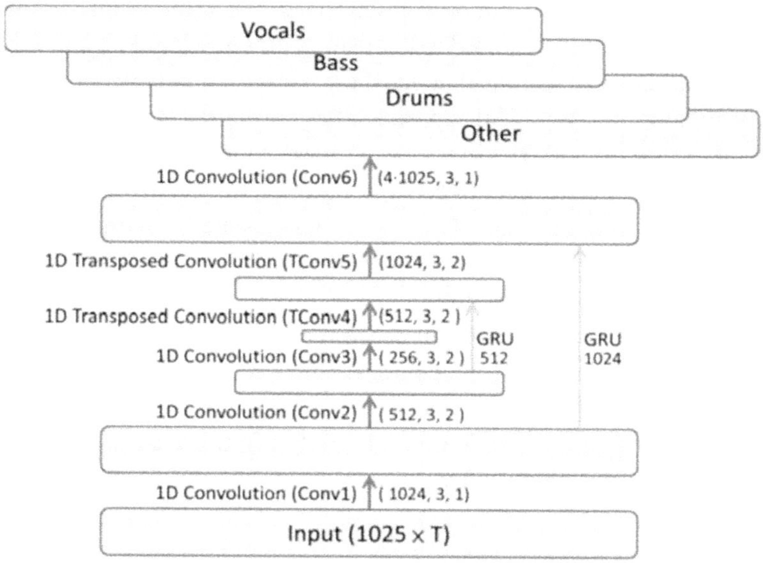

8: Update the reconstructed data $T_s = \tilde{T}_s \cup \tilde{T}_s^{C_t}$

9: **end for**

10: Learn a classifier H by applying supervised learning algorithms (e.g., SVM) to the reconstructed data $\tilde{T}_s$.

11: **return** the learnt classifier H.

The specific use of the Sparse Autoencoder for audio source separation in one hand that the audio signal is sparse by default and the speech can be segmented into units of analysis in the other hand.

Sparse autoencoders offers method that provides information without requiring to reduce the number of nodes in the hidden layers. The chosen *activation* function is the penalized in the layers of the network to reconstruct the *loss* function. The two parts of the network: the encoding and the decoding are the learned by activating only a small number of the neurons. This is having no relation with the regulation method as we normally regularize the weights of a network, not the activation. The figure below presents a generic sparse autoencoder.

To clarify the difference between the undercomplete autoencoder and the sparse autoencoder is that the first one uses the entire network for leaning from every observation, while the second one will be forced to selectively activate regions of the network depending on the input data. The advantage of using a Sparse autoencoder is then limiting the network's capacity to memorize the input data without limiting the networks capability to extract features from the data. In this way, the encoding dimensionality can be chosen and applying the regularization by the sparsity constraint.

To perform the regularization there are two principal ways to choose between them:

Figure 16. An example of a generic sparse autoencoder, source (Jeremy Jordan, 2019).

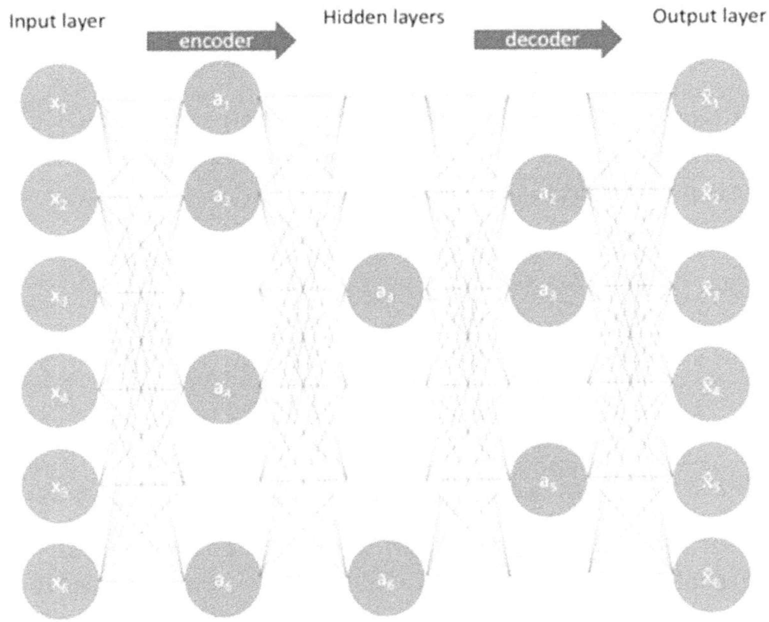

1- L1 Regularization: By adding a term to that penalizes the absolute value of the vector of activations a in layer h for observation *I scaled by a tuning parameter λ.*

$$\Im(x,\hat{x}) + \lambda \sum_i \left| a_i^{(h)} \right| \tag{1.10}$$

2- **KL-Divergence:** KL-divergence is a measure of the difference between two probability distributions.

The activation of a neuron can be determined by the sparsity parameter defined in the equation:

$$\hat{\rho}_j = \frac{1}{m} \sum_i \left[a_i^{(h)}(x) \right] \tag{1.11}$$

Where j denotes the specific neuron in layer h, m is the number of the training of the observation denoted as x. This operation allows neurons to fire only for a subset (collection of samples) of the observation. The description of the parameter ρ can be seen as a *Bernoulli* random variable distribution to make the comparison between the ideal distribution and the observed distributions over all hidden layer nodes $\hat{\rho}$. So, to perform this idea, the KL divergence is presented in the equation denoted (1.12):

$$\Im(x,\hat{x}) + \sum_j KL(\rho \parallel \hat{\rho}_j) \tag{1.12}$$

To know what is a Bernoulli distribution, a simple definition is given bellow:

Definition: *A Bernoulli distribution is "the probability distribution of a random variable which takes the value 1 with probability pp and the value 0 with probability "q=1−p". This corresponds quite well with establishing the probability a neuron will fire (Jeremy Jordan, 2019).*

When there are two *Bernoulli* distributions, another term is born which is known by the *loss* function. This last determines the average activation of the hidden unit $\hat{\rho}$ according to the KL divergence. This difference could be writen in the equation bellow:

$$\sum_{j=1}^{l^{(h)}} \rho \log \frac{\rho}{\hat{\rho}_j} + (1-\rho) \log \frac{1-\rho}{1-\hat{\rho}_j} \quad (1.13)$$

CONCLUSION

To summarize, in this work we have presented different method used for the signal processing separation task used for many real applications such that blind audio source separation, image reconstruction, speech reconstruction and noise removal (data denoising (ex. images, audio), anomaly detection, information retrieval, speech recognition and so many other signals time processing real applications.

In this works we have presented also a specific new approach used for the same purpose, that deal with deep neural network. This chapter has been focused about the tree types of autoencoders used for discovering the structure of the data. The main result behind the choice of autoencoders is the compression of the data based on correlations between the input feature vector and the suppression of unwanted information like the noise mixed with the audio signals. The autoencoder architecture ensures the goal of presenting the original data input as a meaningful attribute's presentation. Working with a specific type of autoencoders or mixing between two or more of them in a neural network model presents the challenge of letting the network learns meaningful and generalizable latent space representation.

REFERENCES

Abouzid, H., & Chakkor, O. (2018, April). Dimension Reduction Techniques for Signal Separation Algorithms. In *International Conference on Big Data, Cloud and Applications* (pp. 326-340). Cham, Switzerland: Springer.

Abouzid, H., Chakkor, O., Reyes, O. G., & Ventura, S. (n.d.). Signal speech reconstruction and noise removal using convolutional denoising audioencoders with neural deep learning. *Analog Integrated Circuits and Signal Processing,* 1-12.

Abouzid, H., & Chakkor, O. (2017, November). Blind source separation approach for audio signals based on support vector machine classification. In *Proceedings of the 2nd international conference on computing and wireless communication systems* (p. 39). ACM.

Amari, S. I., Cichocki, A., & Yang, H. H. (1996). A new learning algorithm for blind signal separation. In Advances in neural information processing systems (pp. 757-763).

Anon. (2019). [online] Available at https://sites.google.com/site/mrstevensonstechclassroom/hl-topics-only/4a-robotics-ai/neural-networks-computational-intelligence.

Anon. (2019). [online] Available at https://www.quora.com/How-do-artificial-neural-networks-work

Belouchrani, A., Abed-Meraim, K., Cardoso, J., & Moulines, E. (1997). A blind source separation technique using second-order statistics. *IEEE Transactions on Signal Processing*, *45*(2), 434–444. doi:10.1109/78.554307

Böck, S., & Widmer, G. (2013, September). Maximum filter vibrato suppression for onset detection. In *Proc. of the 16th Int. Conf. on Digital Audio Effects (DAFx)*. Maynooth, Ireland.

de Albuquerque, V. H. C., de Alexandria, A. R., Cortez, P. C., & Tavares, J. M. R. (2009). Evaluation of multilayer perceptron and self-organizing map neural network topologies applied on microstructure segmentation from metallographic images. *NDT & E International, 42*(7), 644-651.

Deng, J., Zhang, Z., Marchi, E., & Schuller, B. (2013, September). Sparse autoencoder-based feature transfer learning for speech emotion recognition. In *2013 Humaine Association Conference on Affective Computing and Intelligent Interaction* (pp. 511-516). IEEE. 10.1109/ACII.2013.90

Deville, Y., & Hosseini, S. (2009). Recurrent networks for separating extractable-target nonlinear mixtures. Part I: Non-blind configurations. *Signal Processing*, *89*(4), 378–393. doi:10.1016/j.sigpro.2008.09.016

Egmont-Petersen, M., de Ridder, D., & Handels, H. (2002). Image processing with neural networks—a review. *Pattern recognition, 35*(10), 2279-2301.

Elhami, G., & Weber, R. M. (2019). *Audio Feature Extraction with Convolutional Neural Autoencoders with Application to Voice Conversion*. No. CONF.

Grais, E. M., & Plumbley, M. D. (2017, November). Single channel audio source separation using convolutional denoising autoencoders. In *2017 IEEE Global Conference on Signal and Information Processing (GlobalSIP)* (pp. 1265-1269). IEEE. 10.1109/GlobalSIP.2017.8309164

Guo, X., Liu, X., Zhu, E., & Yin, J. (2017, November). Deep clustering with convolutional autoencoders. In *International Conference on Neural Information Processing* (pp. 373-382). Cham, Switzerland: Springer.

Hérault, J., Jutten, C., & Ans, B. (1985). Détection de grandeurs primitives dans un message composite par une architecture de calcul neuromimétique en apprentissage non supervisé. In 10 Colloque sur le traitement du signal et des images, FRA, 1985. GRETSI, Groupe d'Etudes du Traitement du Signal et des Images.

Hosseini, S., & Deville, Y. (2013). Recurrent networks for separating extractable-target nonlinear mixtures. Part II. Blind configurations. *Signal Processing*, *93*(4), 671–683. doi:10.1016/j.sigpro.2012.08.027

http://neuralnetworksanddeeplearning.com/chap6.html. (n.d.). Retrieved from https://www.kisspng.com/png-deep-learning-machine-learning-artificial-intellig-1401501/preview.html

https://www.natural-solutions.eu/blog/la-reconnaissance-dimage-avec-les-rseaux-de-neurones-convolutifs. (n.d.). Retrieved from https://www.kisspng.com/png-deep-learning-machine-learning-artificial-intellig-1401501/preview.html

Hughes, D., & Correll, N. (2016). Distributed machine learning in materials that couple sensing, actuation, computation and communication. *arXiv preprint arXiv:1606.03508.*

Jansson, A., Humphrey, E., Montecchio, N., Bittner, R., Kumar, A., & Weyde, T. (2017). Singing voice separation with deep U-net convolutional networks. In *Proceeds of 18th International Society for Music Information Retrieval Conference* (pp. 23-27).

Jeremy Jordan. (2019). *Introduction to autoencoders.* Available at https://www.jeremyjordan.me/autoencoders/

Kingma, D. P., & Welling, M. (2013). Auto-encoding variational bayes. *arXiv preprint arXiv:1312.6114.*

KissPNG - HD png images and illustrations. Free unlimited download. (n.d.). Retrieved from https://www.kisspng.com/png-deep-learning-machine-learning-artificial-intellig-1401501/preview.html

LeCun, Y., Bengio, Y., & Hinton, G. (2015). Deep learning. nature, 521(7553), 436.

Liu, D., Smaragdis, P., & Kim, M. (2014). Experiments on deep learning for speech denoising. In *Fifteenth Annual Conference of the International Speech Communication Association.*

Liu, J. Y., & Yang, Y. H. (2018, December). Denoising Auto-encoder with Recurrent Skip Connections and Residual Regression for Music Source Separation. In *2018 17th IEEE International Conference on Machine Learning and Applications (ICMLA)* (pp. 773-778). IEEE. 10.1109/ICMLA.2018.00123

Mao, J., Xu, W., Yang, Y., Wang, J., & Yuille, A. L. (2014). Explain images with multimodal recurrent neural networks. *arXiv preprint arXiv:1410.1090.*

Mao, X., Shen, C., & Yang, Y. B. (2016). Image restoration using very deep convolutional encoder-decoder networks with symmetric skip connections. In Advances in neural information processing systems (pp. 2802-2810).

Ng, A. (2011). Sparse autoencoder. CS294A Lecture notes, 72(2011), 1-19.

Pavlovsky, V. (2019). Introduction To Artificial Neural Networks. [online] Vojtech Pavlovsky. Available at https://www.vaetas.cz/posts/introduction-artificial-neural-networks/

Schlüter, J., & Böck, S. (2014, May). Improved musical onset detection with convolutional neural networks. In *2014 Ieee International Conference on Acoustics, Speech and Signal Processing* (pp. 6979-6983). IEEE.

Stober, S., Cameron, D. J., & Grahn, J. A. (2014, October). Does the beat go on?: Identifying rhythms from brain waves recorded after their auditory presentation. In *Proceedings of the 9th Audio Mostly: A Conference on Interaction with Sound* (p. 23). ACM.

WildML. (2019). Implementing a Neural Network from Scratch in Python – An Introduction. [online]

Xu, Y., Du, J., Dai, L. R., & Lee, C. H. (2014). An experimental study on speech enhancement based on deep neural networks. *IEEE Signal Processing Letters, 21*(1), 65–68. doi:10.1109/LSP.2013.2291240

Chapter 8
Performance Analysis of Hierarchical Routing Protocols in Heterogenous WSNs

Yassine Yazid
National School of Applied Sciences of Tangier, Morocco & Abdelmalek Essaadi University, Morocco

Imad Ezzazi
National School of Applied Sciences of Fes, Morocco & Université Sidi Mohamed Ben Abdellah, Morocco

Mounir Arioua
National School of Applied Sciences of Tetouan, Morocco & Abdelmalek Essaadi University, Morocco

Ahmed El Oualkadi
https://orcid.org/0000-0002-4953-1000
National School of Applied Sciences of Tangier, Morocco & Abdelmalek Essaadi University, Morocco

ABSTRACT

Since the appearance of WSN, the energy efficiency has been widely considered as a critical issue due to the limited battery-powered nodes. In this regard, communication process is the most energy demanding in sensor nodes. Subsequently, using energy-aware routing protocols in order to decrease the communications costs as much as possible and increase the network lifetime is of paramount importance. In this chapter, we have mainly focused on the most recent-based clustered routing algorithms for heterogeneous WSNs, namely Selected Election Protocol (SEP), and Distributed Energy Efficient Clustering Protocol (DEEC). In addition, we have proposed an efficient clustered routing protocol based on Zonal SEP algorithm. Indeed, we have evaluated the performance of the proposed protocol according to different scenarios in order to guarantee the best distribution of heterogeneous nodes in the network. The results have shown that the proposed clustered routing approach outperforms the existed Z-SEP protocol in terms of energy efficiency and stability.

DOI: 10.4018/978-1-7998-0117-7.ch008

INTRODUCTION

Over the past decades, wireless sensor networks (WSNs) have become an emerging technology through their increasing use in various fields. This is mainly due to their countless advantages, including the ease of network deployment, simplified management and fast communications, that have attracted significant attention in many applications, i.g. medical monitoring, environmental, military and surveillance applications. Generally, a typical WSN combines a group of tiny autonomous devices called sensor nodes which are absolutely dispersed in different ways, randomly by planes or drones. For instance, nodes can be distributed randomly in a forest fire zone in order to create a suitable temperature map to facilitate its tracking and extinguishing, or they are manually installed in specific locations in a building or in a human body to monitor environmental and physical conditions such as temperature and humidity. After sensing process, each concerned sensor node transmits the data through the network to a collection data base called a base station or sink. Generally, WSN suffers from major restrictions that make it difficult to achieve higher levels of performance and efficiency. In particular, there are many common constraints such as the node's weak storage capacity, low processing power, unstable network topology, interference and energy stress. However, energy is classified as a major problem in WSN since sensor nodes rely on small batteries with limited power supply. The autonomous and limited batteries directly affect the energy resources of the sensors, when nodes process and communicate the data as well as retransmit data due to channel impairments. This results in a significant transmission energy consumption with a delay in the response time. Overall, ensuring effective fault tolerance techniques can mitigate these problems. However, the fact that sensor nodes are limited in terms of power supply and bandwidth makes designing energy efficient techniques for WSN a challenging task for the research community.

Sensor nodes can exhaust their energy when processing and transmitting data in a wireless environment. Particularly, the radio communication is the main responsible of high energy dissipation and short network lifetime. Therein, the cost of a jump in terms of energy is measured by the distance between two nodes involved in the communication (transmitter and receiver). Hence, the destination can be reached either with many small jumps called multiple jumps, or with a small number of large jumps called single jumps. Whereas, the overall cost of routing is the sum of the energies consumed at all jumps, hence multi-hop routing has been deemed to be more efficient than single-hop routing (Fedor & Collier, 2007). Therefore, the amount of the consumed energy is proportional to the length of jumps. Hence, the longer the jumps between transceivers are, the higher the amount of energy consumed will be. Thus, to overcome these constraints, several methods and techniques have been used to obtain better results in term of energy efficiency. Accordingly, to control the energy of nodes it is highly required to resort to energy efficient methods to relay data from the sender node to the receiver one (i.e. Relay node or Base station) which guarantees a prolonged network lifetime. For this reason, routing approaches have recently been considered as one of the most promising energy efficiency techniques employed in WSN networks to reduce the energy burdens of communications. Principally, the routing techniques are categorized into three classes depending on network structure: flat, location-based and hierarchical architectures. The hierarchical (i.e. clustering) routing architectures have been largely used in WSN due to their energy efficiency and load balancing in the network compared to other techniques (Al-Karaki & Kamal, 2004). This method plays an important role to remedy the energy constraints in WSN. It is a technique of partitioning for high-density networks into subgroups of nodes called clusters. Thereby, the cluster formation allows a better management of the routing on the network. Each subset (cluster) consists of many sensor nodes grouped around a leader node which is considered as the collector node of the cluster. In addition, the collector nodes named cluster-heads (CHs) coordinate and order the activities in the cluster. The transmissions power is adjusted by allowing certain nodes to set on sleep mode, or by organizing the transmission instances to avoid collisions.

In this chapter, we have provided a comparative study of clustering-based routing protocols in terms of their energy efficiency, network lifetime, throughput and stability performances. We have focused mainly on the recent based routing algorithms belonging to two different clustered routing protocols families that are dedicated to heterogeneous WSNs namely respectively Selected Election Protocol (SEP) and Distributed Energy Efficient Clustering Protocol (DEEC). We have considered the most efficient protocols named SEP, Enhanced SEP, Threshold SEP, Zonal SEP, DEEC, Enhanced DEEC, Developed DEEC and Threshold DEEC. Moreover, we have proposed a new routing protocol based on Z-SEP architecture in order to enhance the WSN performance. The proposed routing approach enables to provide better performances that the Z-ESP protocol. In the end, a comparative study between our proposed method and the mentioned protocols is carried out based on Matlab simulation results.

The chapter is organized as follows. Section 2 discusses the main related works the related works. Section 3 describes the WSN system model and the radio communication energy model adopted by sensor nodes. Section 4 presents an overview of the main characteristics of hierarchical architectures and their characteristics. Section 5 provides a detailed description of heterogenous hierarchical routing. Section 6 describes the proposed routing scheme in detail. Section 7 discloses the study of the selected routing protocols belonging to SEP and DEEC and the modified zonal stable election protocol. Section 8 concludes the chapter.

RELATED WORKS

Since its appearance before two decades ago, WSN technology has been attracted by different communities including the scientific and industrial ones. This extreme and intense focus has resulted in extensive research that has contributed to the remarkable development of this wireless technology along with its various strands. Meanwhile, many research projects have been carried out to improve WSN's fusibility and robustness by focusing on each of its constraints including power consumption, network topology, routing, coverage and others. In this section we discuss the main related works in the literature that have been dealing with clustered routing techniques. In recent works, various clustered routing protocols have been developed for both homogenous and heterogenous networks in order to overcome some constraints or to reduce the influence of certain vulnerable aspects on the network functioning. This is in order to extend the lifetime of the network as well as increase the scalability and the energy consumption balance. For Instance, several techniques are proposed for homogeneous networks, the most well-known is LEACH algorithm (Low Energy Adaptive Clustering Hierarchy)(W. R. Heinzelman, Chandrakasan, & Balakrishnan, 2000). In order to reduce the energy consumption of sensor nodes, this hierarchical algorithm uses the random rotation approach for cluster head selection. A few variants of the LEACH protocol have been proposed in WSN (Tyagi & Kumar, 2013). The operating phases of LEACH are the installation phase and the balancing phase. In (Lindsey & Raghavendra, 2002) the authors have presented the protocol PEGASIS (Power-Efficient Gathering in Sensor Information Systems) which is an extension of the LEACH protocol. This algorithm eliminates the dynamic clustering overflow created by LEACH, where each node communicates only with a nearby neighbor, then transmits to the CH, which in turn transmits rotatably the selected data to the base station. The PEGASIS protocol saves more energy and is more robust in the event of node failure than the LEACH protocol. In addition, HEED (Hybrid Energy-Efficient Distributed Clustering) is one of homogeneous routing protocols in which the CHs election probabilities are based on residual energy and the degree of the node (Younis & Fahmy, 2004). Moreover, the threshold sensitive Energy Efficient sensor Network protocol (TEEN) is also one of hierarchical protocol dedicated especially for time-critical applications. The network architecture of the sensors is based on a hierarchical grouping where cluster-heads broadcasts their data to two of its members standing in the Hard Threshold (HT) and the Soft Threshold (ST) (Manjeshwar & Agrawal, 2001). The Adaptive Threshold Sensitive Energy Efficient Sensor Network (APTEEN) which is an enhancement of TEEN that is intended to both the collection of periodic data and the responsiveness of time sensitive incidents. Ultimately, the cluster heads diffuse attributes, threshold values and transmission schedules to all nodes. Subsequently, the cluster gathers the aggregated data, which leads to significant energy saving (Manjeshwar & Agrawal, 2001). The Virtual Grid Architecture (VGA) routing protocol also combines the data gathering and the network processing to attain both better energy efficiency and high network lifetime (Al-Karaki, Ul-Mustafa, & Kamal, 2004). Lately, various homogeneous routing schemes have been appeared as discovered in the work of the authors in (Singh & Sharma, 2015) and (Yan, Zhou, & Ding, 2016).

On other side, to fulfill the need of heterogeneous routing protocols, numerous protocols have been designed such as the protocol SEP (Smaragdakis, Matta, & Bestavros, 2004) that introduces two heterogeneous types of nodes, normal nodes and advanced nodes. Advanced nodes have more energy than normal nodes. The two nodes have a moderate probability of becoming CHs. Advanced nodes are more likely to become CHs than normal nodes. Another scheme called Enhanced SEP enhances SEP protocol by adding a third level of heterogeneity by defining extra nodes with intermediate level of energy called intermediate nodes which lies between normal node and advance node (Faisal et al., 2013). T-SEP (Threshold SEP) is also one of the proactive protocols that uses three different levels of heterogeneity where nodes are classed into normal, intermediate and advanced (Aderohunmu & Deng, 2009). Moreover, the election probability of nodes to take the role of CH is the same as it used in E-SEP. To ensure better selection of CHs, T-SEP takes into consideration soft and hard thresholds for such type of nodes to become a cluster head when each node in the network generates randomly a number between 0 and 1. Z-SEP is another SEP based routing protocols that follows a hybrid approach by using two transmission techniques, direct transmission and cluster head transmission. The protocol DEEC (Qing, Zhu, & Wang, 2006) is a distributed multilevel clustering algorithm for heterogeneous WSN. The cluster head selection process of this protocol is based on the ratio of energy between energy of each node and the network's energy average. D-DEEC is a developed version of the protocol DEEC. Even more, E-DEEC (Saini & Sharma, 2010a) is actually DEEC protocol but with exactly three levels of heterogeneity.The protocol Threshold-DEEC(Saini & Sharma, 2010b) uses the same technique for CH selection proposed in DEEC. However, nodes select randomly in each round a number between 0 and 1 to decide the nodes which will become a CH. Furthermore, the authors in (Tanwar, Kumar, & Rodrigues, 2015a) have given a systematic review about heterogeneous WSN protocols by providing a relative performance comparison between the existing protocols respectively in terms of cluster head selection, energy efficiency, and security for some specified applications. Moreover, these authors (Mahboub, En-Naimi, Arioua, Ez-Zazi, & El Oualkadi, 2016) have proposed a new technique for heterogenous networks called multi-zonal selected election protocol (MPSEP) that divides the network into multiple triangle zone allocations. In (Mehra, Doja, & Alam, 2017) the authors have proved that the proposed type of zonal SEP approach shows better performances in terms of lifetime and provides more reliability in terms of successfully received packets by base station compared to DEEC, SEP, Z SEP and LEACH. Besides, to overcome the problem of fast energy consumption at the level of clustered head since they play an additional role contrary to other member nodes, authors in (Naranjo, Shojafar, Mostafaei, Pooranian, & Baccarelli, 2017) have proposed a new approach to deal with this problem which consists of a new protocol called Prolong selected election protocol (P-SEP). To assess the performance of heterogeneous routing techniques in WSN, these authors (Yazid et al., 2019) have presented a comparative study between the most efficient heterogeneous routing protocols SEP, E-SEP, Z-SEP and T-SEP . In order to reduce the amount of energy spent by sensor nodes in wireless sensor network, a proposed technique has been presented in (Rani, Kakkar, Kakkar, & Raman, 2019). Regarding to these authors this proposed method has shown satisfactory results in term of energy efficiency by taking into account the distances separating the nodes and the base station. Moreover, the authors in (Wang et al., 2019) have presented a new clustering scheme using a new CH selection called An Energy Centers Searching using Particle Swarm Optimization (EC-PSO) to avoid the energy holes among the cluster head in the network. Recently, a new scheme have been proposed in (Dutt, Agrawal, & Vig, 2018). called Cluster-head Restricted Energy Efficient Protocol (CREEP). This method was inspired from the existing DEEC protocol which aims to improve the lifetime of the network depending on the use of two heterogeneity levels of CH selection.

SYSTEM MODEL

In WSN, the wireless communication between nodes and the base station is responsible for energy draining activities. Thus, to study the sensors energy variation in the network, we adopt the model proposed by authors in (W. Heinzelman, 2000). It is a first-rate radio model that provides an evaluation of the energy consumed during the transmissions and the receptions between a transmitter node and a receiver node.

According to the radio-frequency dissipation model illustrated in Figure 1 above, in order to obtain an acceptable signal-to-noise ratio (SNR) when transmitting a *k*-bit message over the distance *d* between two the transmitter and the receiver. The energy consumed by the radio communication is estimated using the following equation:

$$E_{Tx}(k,d) = \begin{cases} k \times E_{elec} + \varepsilon_{fs} \times d^2 , & if\, d < d_0 \\ k \times E_{elec} + \varepsilon_{mp} \times d^4 , & if\, d \geq d_0 \end{cases}$$

Where E_{elec} is the energy required to run the transmitter and receiver circuits, ε_{fs} and ε_{mp} are the bit energies used for the free space transmission amplifier ($\approx d^2$) and Two multipath beams ($\approx d^4$), respectively, and d is the distance between the transmitter and receiver. Indeed, when the inter-node distance

Figure 1. Sensor nodes radio enerrgy dissipation model

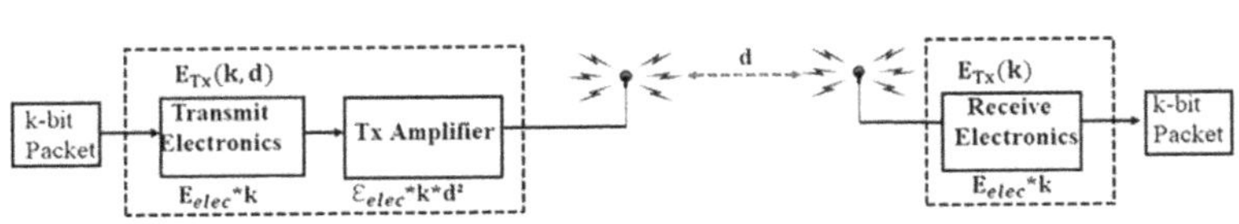

$d < d_0$, the transmission energy is consumed by the free space model, while when $d \geq d_0$ the transmission energy is used in the two-radius ground model. Thus, the transition from free space propagation to two-beam ground propagation is made at a threshold distance, which is then expressed as a function of the free space and the two-beam amplifiers as follows:

$$d_0 = \sqrt{\frac{\varepsilon_{fs}}{\varepsilon_{mp}}}$$

Where ε_{fs} and ε_{mp} are respectively the power amplifier when using free space (fs) model and the multipath (mp) model.

The energy consumption required for reception depends on the circuitry energy E_{elec} and received packet size k, which is given by:

$$E_{Rx}(k) = E_{elec} \times k$$

Therefore, in this model the reception of a message is not an inexpensive operation, thus the protocols must try to minimize not only the transmission distances but also the number of transmission and reception operations for each message. In this model the radio channel is assumed to be symmetrical, so that the energy required to transmit a message from node X to node Y is the same as that required to transmit a message from node Y to node X for a given SNR.

HIERARCHICAL ARCHITECTURE FOR WSN

In view of the limited energy resources of sensor nodes in WSN, the direct and multi-hop communications of sensor nodes to base station became unpractical since those nodes lose their energy very fast due to the high distances between those sensors and their BS especially for those sensors in extremity of the network. Thereupon, a hierarchical approach was adopted to overcome the problem of energy consumption in order to rise network lifespan.

Furthermore, in hierarchical approaches, the network's sensor nodes are divided into groups called clusters (See Figure 2). In each cluster there are two type of nodes, normal ones and a CH which has a high level of energy. Generally, normal nodes are responsible for sensing and forwarding collected data to the CH which is responsible for treating, processing and transmitting those aggregated data directly to the base station or through other CHs and then to BS. Typically, CHs nodes lose more energy than normal nodes since they transmit the entire collected data to the BS which is located far from the cluster location. Furthermore, in order to keep a balancing amount of energy between sensor nodes in the network, the role of being leader (CH) is periodically changed. Additionally to its efficient results in term of energy consumption, this clustering method shows numerous advantages such as collision avoidance by occurring transmissions to BS only by CHs, scalability since it allows to add nodes easily in the cluster, deliverance of high level of QoS and fault tolerance.

Figure 2. Clustering method in WSN

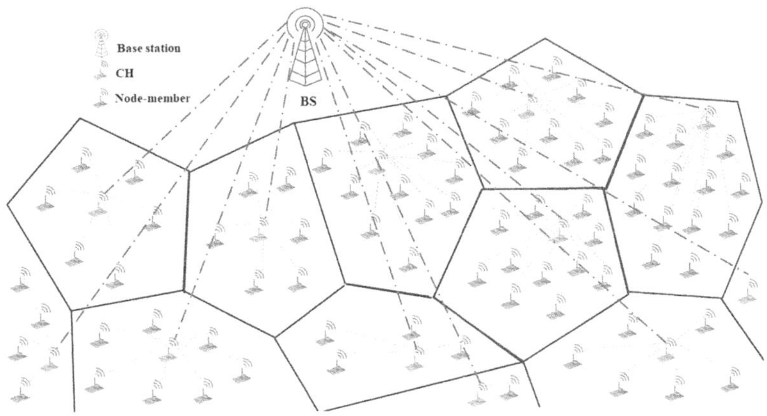

HETEROGENEOUS HIERARCHICAL PROTOCOLS

In this section, we evaluate the performances of each cited protocol in previous sections and our proposed technique in order to pick the efficient one standing in many evaluated parameters like number of transmitted packets to base station and to cluster head, the variation of dead nodes eventually in the time (rounds). In the light of this study a clustered WSN model is implemented in Matlab simulator to evaluate the routing schemes. We have considered 100 heterogeneous sensor nodes randomly deployed in a square field of dimension 100 m x 100 m and communicate with a BS which is located in the middle of the network. In all simulation the initial energy average is assumed to be 0.5 J. in other words, the total energy average of the all nodes in the network before occurrence of any transmission. whereas the energy of nodes is different according the type of this nodes, there were advanced nodes, super nodes, intermediate nodes and normal nodes. Moreover, the percentage of each node type is variated in accordance of each specified protocol. Overall, the used parameters in our simulations are listed in Table 1. In either case we did not take into consideration the energy loss caused by signal collision and interference in wireless channel.

Distributed Energy Efficient Clustering

In this subsection we will take a closer look at the four different protocol selected to carry out our study. Correspondingly, it is a detailed description of the DEEC, Enhanced DEEC, DDEEC and Threshold DEEC protocols.

Distributed Energy Efficient Clustering Protocol (DEEC)

DEEC is one of the clustering-based protocols that are designed to deal with nodes for multi-level communication in heterogenous WSNs. For creating cluster-heads, it improves the function of the election probability of nodes in the cluster by taking in consideration levels of energy initial and residuals whenever nodes with high energy are more probable to become CH than nodes with low energy. Effectively, nodes in the network do not have the same residual energy, consequently the nodes with low energy

Table 1. Simulation parameters.

Parameter	Value
Network's size	100 m x 100 m
Total number of nodes	100
Initial energy E_o	0.5 J
Message size P_{opt} E_{elec} E_{fs} E_{amp} E_{da} Soft Threshold Hard Threshold	4000 bits 0.1 50 nJ/bit 10 nJ/bit/m^2 0.0013 pJ/bit/m^2 5 nJ/bit/signal 100 2
α μ β Base station position	0.3 0.15 0.6 (x= 50 m, y=50 m)

will die quickly in the fixed rotating epoch. In addition, DEEC can be used only when the base station is located near the sensor nodes (Qing et al., 2006).

The network energy average $\overline{E}(r)$ that defines the probability of nodes S_i in term of the radio initial energy $E_i(r)$ during round r is calculated by:

$$\overline{E}(r) = \frac{1}{N}\sum_{i=1}^{n} E_i(r)$$

The probability of each node to become a cluster head in a selection can be given as:

$$P_i = P_{opt}\left(1 - \frac{\overline{E}(r) - E_i(r)}{\overline{E}(r)}\right) = P_{opt}\left(\frac{E_i(r)}{\overline{E}(r)}\right)$$

Consequently,

$$\sum_{i=1}^{n} P_i = \sum_{i=1}^{n} P_{opt}\left(\frac{E_i(r)}{\overline{E}(r)}\right) = N * P_{opt}$$

Where N is the total number of nodes in the network and $N * P_{opt}$ defines the optimal number of CHs. The value of P_{opt} in two levels of heterogenous network is given by the equations (4) and (5). Then,

$$P_i\left(S_i\right)=\begin{cases}\dfrac{P_{opt}.E_i(r)}{\left(1+\alpha.m\right).\overline{E}(r)} & ,if\ the\ node\ S_i\ is\ normal\\ \dfrac{P_{opt}.\left(1+\alpha\right).E_i(r)}{\left(1+\alpha.m\right).\overline{E}(r)} & ,if\ the\ node\ S_i\ is\ advanced\end{cases}$$

Although, we can use the expression below if there are more than two required heterogneity levels in the network:

$$P_{multi}=\frac{P_{opt}.N.\left(1+\alpha\right)}{N+\sum_{i=1}^{N}\alpha_i}$$

Enhanced Distributed Energy Efficient Clustering Protocol (E-DEEC)

E-DEEC (Saini & Sharma, 2010a) presents an enhanced version of the protocol DEEC with exactly three levels of heterogeneity in the network. Three types of nodes are distributed, normal, advanced and super nodes. The probability used for CH selection is given by:

$$P_i\left(S_i\right)=\begin{cases}\dfrac{P_{opt}.E_i(r)}{\left(1+\left(\alpha+\beta.m_0\right).m\right).\overline{E}(r)} & ,\ if\ node\ S_i\ is\ normal\\ \dfrac{P_{opt}.\left(1+\alpha\right).E_i(r)}{\left(1+\left(\alpha+\beta.m_0\right).m\right).\overline{E}(r)} & ,\ if\ node\ S_i\ is\ advanced\\ \dfrac{P_{opt}.\left(1+\beta\right).E_i(r)}{\left(1+\left(\alpha+\beta.m_0\right).m\right).\overline{E}(r)} & ,\ if\ node\ S_i\ is\ super\end{cases}$$

Where K_{opt} defines the optimal number of clusters in the network in each round.

Threshold Distributed Energy Efficient Clustering Protocol (T-DEEC)

T-DEEC uses the same technique for CH selection proposed in DEEC. Precisely, nodes choose in each round a number between 0 and 1 randomly to decide the nodes which will become a CH. For more, a threshold value is adjusted depending on nodes probabilities to become CHs. Hence, if the value is less than the threshold, nodes are selected to be CHs (Saini & Sharma, 2010b). The threshold value is given as follow:

$$T(S) = P_{opt}.\left(\frac{p}{1 - p.\left(1 \bmod \frac{1}{p}\right)}\right).\frac{Node's\ residual\,energy}{Network's\,energy\,average}.K_{opt}$$

Developed Distributed Energy Efficient Clustering Protocol (D-DEEC)

D-DEEC is one of energy efficient clustering protocols inspired by the protocol DEEC. D-DEEC uses the same technique as DEEC for estimation of average energy in the network and CH selection algorithm based as well on residual energy. Furthermore, the difference is in the expression that defines the probabilities associated for normal and advanced nodes to become CHs. Since advanced nodes get more residual energy than normal nodes, advanced nodes are selected as CH even in the rounds when the energy is equal to that of the normal nodes which push them to die quickly. To avoid this problem, D-DEEC introduces a threshold condition for residual energy to prevent the missing of advanced node quickly and to balance the role of being CH with normal nodes after crossing rounds when energy is nearly equal between normal and advanced nodes (Elbhiri, Saadane, & Aboutajdine, 2010). The expression of residual energy threshold Th_{rev} is given approximately as below:

$$Th_{rev} = \left(\frac{7}{10}\right).E_0$$

Then the expression of probability defined by D-DEEC is:

$$P_i\left(S_i\right) = \begin{cases} \dfrac{P_{opt}.E_i(r)}{\left(1+\alpha.m\right).\overline{E}(r)}, for\ normal\,nodes\,S_i\ when\ E_i(r) < Th_{rev} \\ \dfrac{P_{opt}.\left(1+\alpha\right).E_i(r)}{\left(1+\alpha.m\right).\overline{E}(r)}, for\ advanced\,nodes\,S_i\ when\ E_i(r) > Th_{rev} \\ c.\dfrac{P_{opt}.\left(1+\alpha\right).E_i(r)}{\left(1+\alpha.m\right).\overline{E}(r)}, for\ normal\,and\,advanced\,nodes\,S_i\ when\ E_i(r) \le Th_{rev} \end{cases}$$

Where, C is a real positive number related to the number of cluster heads.

Selected Election Protocol of Shortlisted Protocols

In this subsection we present the mathematic side in detail of selected protocols belonging to Selection Election Protocol it is about SEP itself, Enhanced SEP, Zonal SEP and Threshold SEP.

Selected Election Protocol

SEP algorithm (Smaragdakis et al., 2004), improves the stable region of the hierarchy process by using the characteristic parameters of the heterogeneity, namely the fraction of the advanced nodes *m* and the additional energy factor between the advanced and normal nodes (α). In order to extend the stable region, SEP tries to maintain the constraint of a well-balanced energy consumption. Intuitively, advanced nodes must become cluster leaders more often than normal nodes, which equate to a fairness constraint on energy consumption. The total energy of the new heterogenization is equal to:

$$E_{Total} = n \times (1-m) \times E_0 + n \times m \times E_0 \times (1+\alpha) = n \times E_0 \times (1+\alpha \times m)$$

Where n is total number of nodes, m is the fraction of advanced nodes, E_0 is the supposed initial energy of nodes and α is the energy of advanced nodes.

And the probabilities of normal and advanced nodes are respectively given by

$$P_{nrm} = \frac{P_{opt}}{1+\alpha \times m} \quad \text{and } P_{adv} = \frac{P_{opt} \times \left(1+\alpha\right)}{1+\alpha \times m}$$

Where P_{opt} is the optimal probability of each node to become a CH.

Furthermore, those must satisfy the conditions below to be CHs in the network:

$$T(s_{nrm}) = \begin{cases} \dfrac{P_{nrm}}{1-P_{nrm} \times (r \bmod \frac{1}{P_{nrm}})}, & if\ s_{nrm} \in G' \\ 0, & Otherwise \end{cases}$$

$$T(s_{adv}) = \begin{cases} \dfrac{P_{adv}}{1-P_{adv} \times (r \bmod \frac{1}{P_{adv}})}, & if\ s_{adv} \in G'' \\ 0, & Otherwise \end{cases}$$

where G' and G'' are the sets of nodes that have not been CH in last $\frac{1}{P_{nrm}}$ and $\frac{1}{P_{adv}}$ rounds respectively for normal and advanced nodes. r is the associated round.

Enhanced SEP

E-SEP is an enhanced type of the protocol SEP. Unlike SEP, this protocol deals with multi-level of sensor nodes in the network: normal, intermediate and advanced nodes (Faisal et al., 2013). The total energy in the network is given by

$$\begin{aligned} E_{Total} &= n \times (1 - m - \lambda) \times E_0 + n \times m \times E_0 \times (1 + \alpha) + n \times \lambda \times E_0 \times (1 + \mu) \\ &= n \times E_0 \times (1 + \alpha \times m + \lambda \times \mu) \end{aligned}$$

Where, α is the energy of advanced nodes, λ is the proportion of intermediate nodes that have μ times more energy than normal nodes where $\mu = \frac{\alpha}{2}$, n is total number of nodes, m is the fraction of advanced nodes, E_0 is the supposed initial energy of nodes .In the same way as SEP, in E-SEP CHs are selected depending on probability of each type of node. Probabilities of becoming CH for normal, intermediate and advanced nodes, respectively are given by

$$P_{nrm} = \frac{P_{opt}}{(1 + \alpha \times m + \lambda \times \mu)}$$

$$P_{\mathrm{int}} = \frac{P_{opt} \times (1 + \alpha)}{(1 + \mu)(1 + \alpha \times m + \lambda \times \mu)}$$

$$P_{adv} = \frac{P_{opt} \times (1 + \alpha)}{(1 + \lambda)(1 + \alpha \times m + \lambda \times \mu)}$$

Threshold Selected Election Protocol (T-SEP)

T-SEP is also one of proactive protocols that uses three different levels of heterogeneity where nodes are classed into normal, intermediate and advanced (Kashaf, Javaid, Khan, & Khan, 2012). Moreover, the probabilities of being CH is the same as E-SEP. Whereas to ensure better selection of CHs, T-SEP takes into consideration a threshold for such type of nodes to become a cluster head when each node in the network generates randomly a number between 0 and 1, if the generated value is less than threshold then this node becomes CHs. For each type of nodes, we got these different formulas for the calculation of threshold in function of their probabilities:

$$T(s_{nrm}) = \begin{cases} \dfrac{P_{nrm}}{1 - P_{nrm} \times (r \bmod \dfrac{1}{P_{nrm}})} & , \ if\ s_{nrm} \in G' \\ 0 & , \ Otherwise \end{cases}$$

$$T(s_{\text{int}}) = \begin{cases} \dfrac{P_{adv}}{1 - P_{\text{int}} \times (r \bmod \dfrac{1}{P_{\text{int}}})} & , \ if\ s_{\text{int}} \in G'' \\ 0 & , \ Otherwise \end{cases}$$

$$T(s_{adv}) = \begin{cases} \dfrac{P_{adv}}{1 - P_{adv} \times (r \bmod \dfrac{1}{P_{adv}})} & , \ if\ s_{adv} \in G''' \\ 0 & , \ Otherwise \end{cases}$$

Where G', G'' and G''' are the sets of nodes that have not been CH in last $\dfrac{1}{P_{nrm}}$, $\dfrac{1}{P_{\text{int}}}$ and $\dfrac{1}{P_{adv}}$ rounds respectively for normal, intermediate and advanced nodes.

Zonal Threshold Selected Election Protocol (Z-SEP)

Z-SEP shows two levels heterogeneity of nodes in the network as in SEP (Faisal et al., 2013). But the difference is that in Z-SEP, normal nodes that are distributed near the BS use the direct transmission technique whereas advanced nodes use transmission via cluster heads. Generally, the network when using this technique, the network is divided into three parties, zone 0, zone 1 and zone 2. In the zone 0, which located in the center of the network, nodes are assumed to transmit their aggregated data in direct manner to base station. In both zone 1 and zone 2, nodes use the clustering technique to send the collected data to the BS.

PROPOSED SCHEMES

In our proposed method we have added some modification on the existed routing clustered protocol Z-SEP in order to extend the network lifetime and the number of transmitted packets. We have tried to develop two new versions of Zonal-SEP protocol. In first version, we kept the same network architecture as Z-SEP the network remains with three zones (zone 0, zone 1 and zone 2) as shown in Figure 3. Whereas the distribution of nodes is modified hence we selected only 40% of nodes to be distributed in zone 0 where transmission is forced to be realized between those nodes and the base station in direct manner. However, the rest 60% nodes are distributed equally in zone 1 and zone 2 under the assumption that the transmission follows the clustering.

Figure 3. The distribution of nodes for the Z-SEP protocol

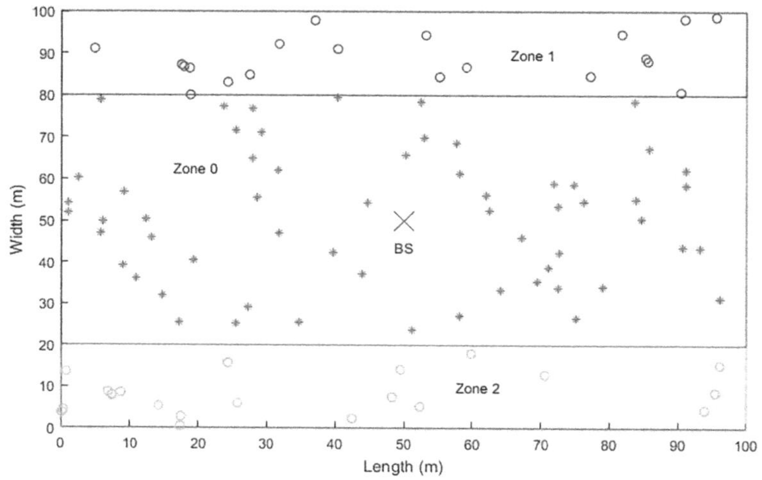

Notably, we noticed in existed ZSEP that the distances of normal nodes and their base station in the zone 0 are very significant, thus we decided to shorten these distances in order to reduce the energy consumption. In addition, we know that radio communication with ZigBee standard can assure transmissions with emitter and receiver in a threshold distance of 100 meters. Since the zone 0 is a 60 *m* of large and 60 *m* of length thus direct communication can be affected in this case. Whereas in the network's extremities zones, nodes are very far from the BS which requires indirect transmission methods as clustering.

Otherwise, in second version we sub-divided the network exactly into five zones called respectively Zone 0, Zone 1, Zone 2, Zone 3 and Zone 4. In the zone 0 the communications are performed directly between sensor nodes and the base station. Whereas, in Zone 1, Zone 2, Zone 3 and Zone 4 the transmissions are carried out using the clustering technique by electing the cluster head which acts as an intermediate collector between its attached nodes numbers and the base station. By using these techniques, we aim to observe the network lifetime and transmitted number of packets variation using two and three levels of heterogeneity, respectively.

We distributed two type of nodes (advanced and normal) nodes on a network of 100 m of length and 100 m in width. The normal nodes in red stars are distributed on the zone 0 which is localized in the center of the network with 40 *m* in length and 40 *m* in width as it ullustraed in Figure 5 . Though, the zone 1, zone 2, zone 3 and zone 4 are fixed with 30 *m* of width and 70 *m* of length. Likewise, each zone is illustrated in different colour in Figure 5. To emphasize, we supposed that the nodes in each zone have no interaction with nodes in other different zone, precisely nodes in zone 1 can never interact with those in zone 0, zone 2, zone 3 or zone 4. Generally, it is a sort of cluster of clustering method of totally separated zones in the same network area. In addition, we keep the same mathematical expressions used for SEP and original Z-SEP.

In summary, we present three version of modified Z-SEP protocol. In first one, we keep the same network architecture as in original Z-SEP and we tried to find the efficient distribution of nodes by their type in three different zones and modulating the size of zone 1 and zone 2 by taking into consideration the fraction of nodes distributed in each zone as it illustrated in the Figure 4. The second one, we change radically the existed Z-SEP network structure and nodes distribution. In third and last one, we kept the

Figure 4.: Modified Z-SEP protocol version 1

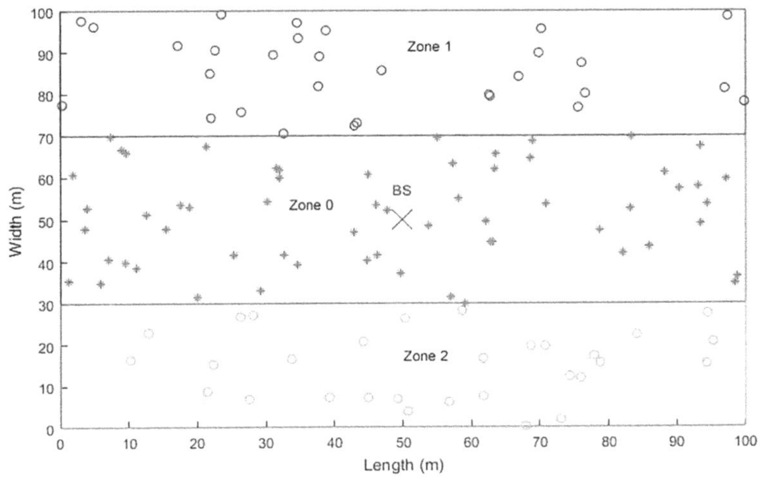

same network form though we have changed the zones dimensions with a specified fraction number of advanced nodes and normal nodes in each zone as depicted in Figure 5.

SIMULATION AND COMPARISON

In this subsection, we exhibit the diverse results of realized simulation to emphasize the analytics and comparative studies in order to pick the efficient routing protocols based on clustering technique in heterogeneous networks. In this end, we arranged all simulation results in these follow figures and tables.

Figure 5. Modified Z-SEP version 3

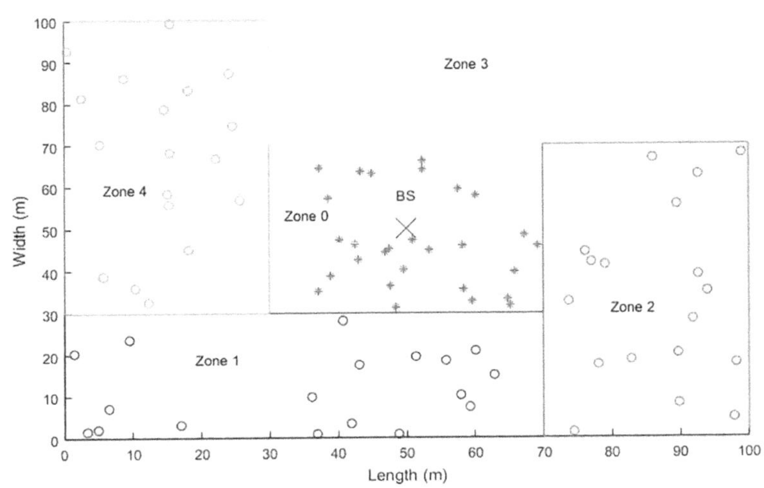

Each simulation result represents correspondingly the feedback of each foreseen protocol under specific circumstances.

In Figure 6, we plot the number of dead nodes in the network versus the network lifetime (i.e. transmission rounds). The simulation results show that T-SEP offers the best result in term of network lifespan and energy efficiency compared to other SEP based protocols as it is indicated in table 2. under the same circumstances, on the network of 100 m length and 100 m of width sensor nodes that adopt T-SEP technique to communicate with their base station resisted in total for 5504 rounds more respectively than our modified Z-SEP V3 (i.e. 3315), E-DEEC (i.e. 2648), our modified Z-SEP V2 (i.e. 2590), our modified Z-SEP V1 (i.e. 2433), existed Z-SEP (i.e. 2104), E-SEP (i.e. 2061), T-DEEC (i.e. 2045), SEP (i.e. 2005), DEEC (i.e. 1898) and at end of the list the protocol D-DEEC with only 1898 rounds in total.

In Figure 7, we plot the average energy versus transmission rounds respectively. This figure compares the performance of each studied routing algorithms in terms of residual energy average consumed by nodes in the network in function of the runoff of each round. Moreover, it depicts the last standing rounds of nodes in the network by using each protocol.

Therefore, in order to understand the stability and instability periods of the network of 100 sensor nodes. The Table 2 exhibits the rounds when the first node of the network is dead (i.e. 1% of dead nodes), the half of network's nodes (i.e. 50%) and when the network is empty of alive nodes, in other words 100% of nodes are dead. Rather, the obtained results show that T-SEP performs better than the other protocols in terms of energy and network lifetime as it mentioned in Figure 6 and Figure 7. When using T-SEP protocol, the percentages of 1%, 50% and 100% of dead nodes occur at 2026, 2604 and 5504 rounds, respectively. Thus, 469, 1618 and 3315 for preferment our proposed protocol. Consequently, the lifetime of T-SEP protocol is over others with increase of 39%, 51%, 52%, 55%, 61%, 62%, 63%, 65% and 81% respectively for our modified Z-SEP V3, E-DEEC, our modified Z-SEP V2, our modified Z-SEP V1, Existed Z-SEP, E-SEP, T-DEEC, SEP, DEEC and D-DEEC. Even though, T-SEP protocol exceeds all other protocol especially our modified Z-SEP V3 which classes in last of the queue of better stability feedback. Nevertheless, our modified Z-SEP V2 is classed second place after T-SEP. Correspondingly Figure 8 and Figure 9 display the number of packets transmitted successfully from nodes members to

Figure 6. The variation of dead nodes in the network in function of transmission rounds

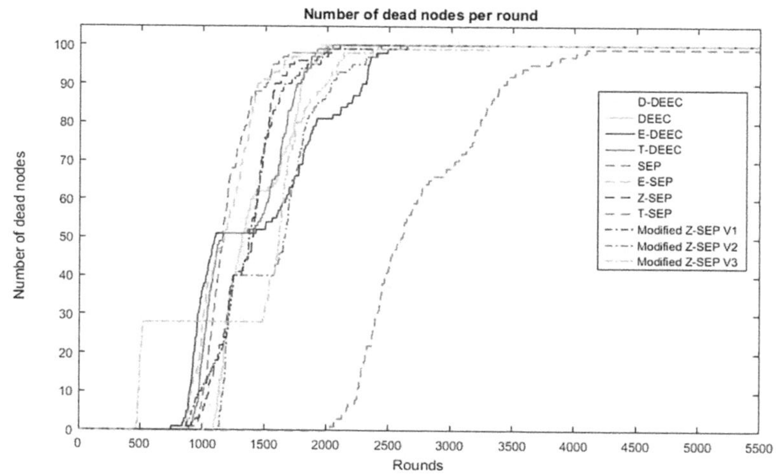

Figure 7. The energy variation of energy average per rounds

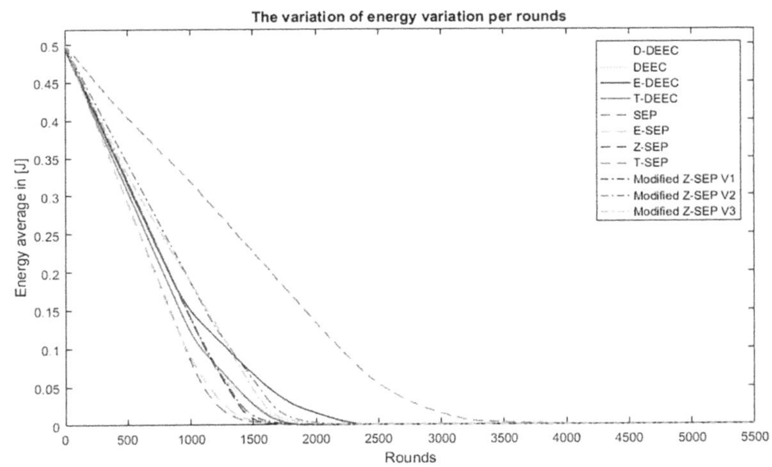

their CH in concerned cluster and from those CHs directly to BS whenever in zones that requires clustered method or peer to peer transmissions (i.e. zone 0). Table 3 provides a clear comparison between studied algorithm in term of the maximum throughput received by CH and BS for each protocol. The results depict that SEP provides the best throughput at intra clusters (i.e. 117408 packets), flowed by our modified Z-SEP V3 (i.e. 107488 packets), T-SEP (i.e. 104767 packets), E-SEP (i.e. 103275 packets), our modified Z-SEP V2 (i.e. 93510 packets), our modified Z-SEP V1 (i.e. 80757 packets), D-DEEC (i.e. 80595 packets), Z-SEP (i.e. 79793 packets), T-DEEC (i.e. 77415 packets), DEEC (i.e. 67122 packets), and then E-DEEC (i.e. 63747 packets). Accordingly, this difference in term of received packet by CH is explained by the fact that some of the studied algorithms like SEP use a high number of clusters unlike others.

Furthermore, concerning the throughput to the BS, E-DEEC. DEEC provides the best performance presents the best performance (i.e. 80028 packets) compared to DEEC (i.e. 75160), our modified Z-SEP V2 (i.e. 59715), T-DEEC (i.e. 56210), D-DEEC (i.e. 55524), Z-SEP (i.e. 54673), our modified Z-SEP V2 (i.e. 54294), T-SEP (i.e. 33328), our modified Z-SEP V3 (i.e. 27248), E-SEP (i.e. 14834), and then SEP (i.e. 14360). Above all, for our simulations we supposed that there were no interferences in network, thus no packet loss. Typically, the power consumption of the nodes is inversely proportional to the number of laps exhausted. In addition, on the first round, all nodes in the network can communicate, but just after some rounds, some of them become unable to complete any radio communication operation due to insufficient residual energy in their batteries.

Accordingly, the distribution of nodes in the network area plays a primeval role on the energy consumption of the all nodes. For this reason, deploying normal nodes to monitor near the base station and advanced nodes to monitor other parts of the network is a ruse way to achieve this purpose. Furthermore, the number of selected nodes for each type is also an important factor to increase or decrease the lifetime of the network. For instance, for the proposed scheme version 1 where the network was repartitioned onto two nodes type heterogeneity accurately with 40% of normal nodes on zone 0 and 60% of advanced nodes are dispersed evenly on both zones 1 and 2 had shown better results than the distribution with 60% of normal nodes on zone 0 and 60% of advanced nodes shared equally between zone 1 and zone 2.

Figure 8. Total number of packets sent to BS per round

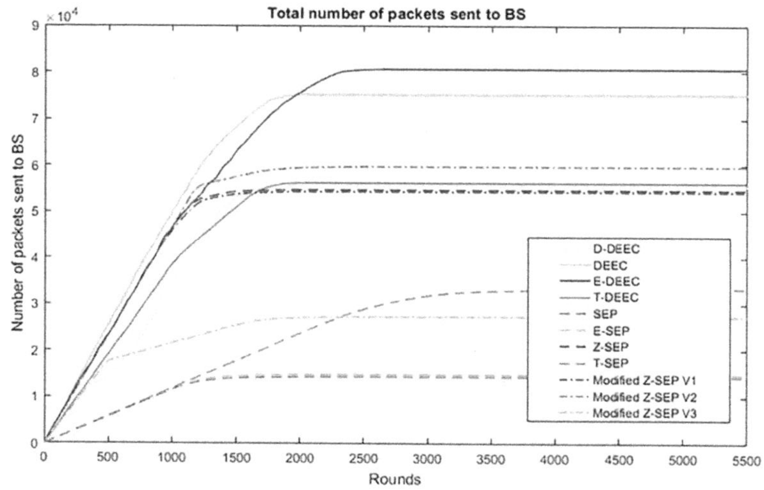

Moreover, the number of allocated number cluster per rounds influence also the performance offered by the adopted protocols which explains the differences in the number of transmitted packets to base station and to CH.

The protocol T-SEP have proved to be the efficient algorithm in terms of network lifetime which consequently allows to offer higher instability period and important number of successfully transmitted packets to both the base station and CHs. Whereas, the proposed routing method provides better performances in terms of number of transmitted packets with 59715 packets to base station and 93510 to CHs. Thus, The proposed approach is deemed to be affective than T-SEP protocol as this approach adopts the hard and soft threshold which means the transmissions are occurred only if the value of those thresholds

Figure 9: Total number of packets sent to CH per round

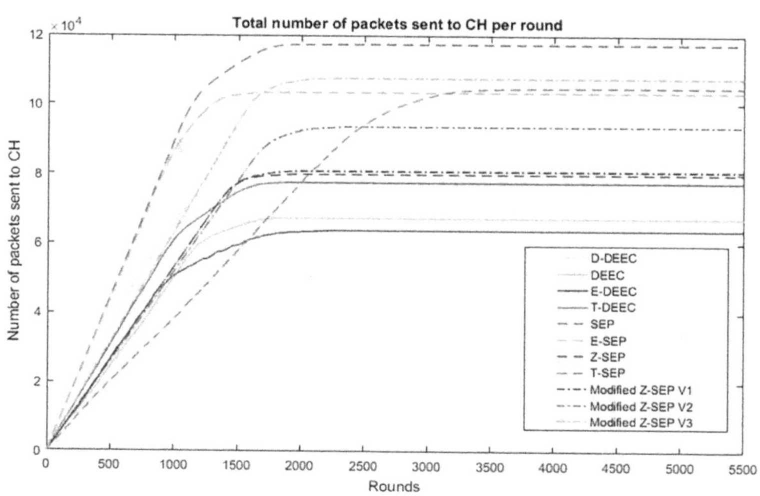

Table 2. Performance comparison of SEP based routing protocols

	First dead nodes	50% of dead nodes	100% of dead nodes	Instability period
DEEC	1095	1346	1898	903
E-DEEC	750	1097	2648	1898
D-DEEC	1009	1346	1889	880
T-DEEC	894	1162	2045	1151
SEP	879	1162	2005	897
E-SEP	859	1178	2061	859
Z-SEP	870	1404	2104	1234
T-SEP	2026	2604	5504	3478
Modified Z-SEP version 1	875	1375	2433	1558
Modified Z-SEP version 2	1136	1661	2590	1454
Modified Z-SEP version 3	469	1618	3315	2846

is reached which thus make it constrained for certain applications that require frequent transmissions in particularly when the aggregation data in the real time is needed.

CONCLUSION

In this chapter, a comparative study between efficient routing protocols based on clustering techniques dedicated for heterogeneous networks is presented. Thereby, this work has not focused only on existed clustered protocol but also proposed a modified version of Z-SEP. Notably, this work has mainly focused on two different types of clustered routing protocols, i.g., the SEP based routing (i.e. SEP, T-SEP, Z-SEP)

Table 3. Number of transmitted packets to cluster heads and to the BS

	Max Nu. of packet sent to CH	Max Nu. of packet sent to BS
DEEC	67122	75160
E-DEEC	63747	80028
D-DEEC	80595	55524
T-DEEC	77415	56210
SEP	117408	14360
E-SEP	103275	14838
Z-SEP	79793	54673
T-SEP	104767	33328
Modified Z-SEP version 1	80757	54294
Modified Z-SEP version 2	93510	59715
Modified Z-SEP version 3	107488	27248

and the DEEC based-routing (i.e. DEEC, DDEC, TDEEC, EDEEC). Accordingly, we have provided a performance comparison of the studied routing protocols regarding their energy efficiency, network lifetime, throughput and stability. The simulation results have shown that the proposed M-SEP protocol outperforms the exited Z-SEP, SEP, E-SEP, E-SEP, DEEC, D-DEEC, D-DEEC and TDEEC in terms of energy efficiency, network lifetime and stability. However, only T-SEP protocol can perform better than the proposed routing approach in some cases. Hence, this one uses three levels of nodes heterogeneity accordingly the normal, intermediate and advanced nodes with fixed energy threshold. For this reason, we put as perspective to enhance the performance of our proposed scheme by inducing three levels of heterogeneity rather than two. Thereby, T-SEP is proved to be the suitable choice for energy constrained heterogeneous WSN in order to save energy and prolong the network lifespan. This is nevertheless, only for applications that do not require monitoring and collection of aggregated data within specific time periods. However, our proposed technique is deemed to be practical for all applications as they exhibit satisfactory results in terms of throughput and network lifetime.

REFERENCES

Aderohunmu, F. A., & Deng, J. D. (2009). *An enhanced stable election protocol (sep) for clustered heterogeneous wsn*. New Zealand: Department of Information Science, University of Otago.

Al-Karaki, J. N., & Kamal, A. E. (2004). Routing techniques in wireless sensor networks: A survey. *IEEE Wireless Communications, 11*(6), 6–28. doi:10.1109/MWC.2004.1368893

Al-Karaki, J. N., Ul-Mustafa, R., & Kamal, A. E. (2004). Data aggregation in wireless sensor networks-exact and approximate algorithms. In *HPSR. 2004 Workshop on High Performance Switching and Routing, 2004.* (pp. 241–245). IEEE. 10.1109/HPSR.2004.1303478

Dutt, S., Agrawal, S., & Vig, R. (2018). Cluster-Head Restricted Energy Efficient Protocol (CREEP) for Routing in Heterogeneous Wireless Sensor Networks. *Wireless Personal Communications, 100*(4), 1477–1497. doi:10.100711277-018-5649-x

Elbhiri, B., Saadane, R., & Aboutajdine, D. (2010). Developed Distributed Energy-Efficient Clustering (DDEEC) for heterogeneous wireless sensor networks. In *2010 5th International Symposium on I/V Communications and Mobile Network (ISVC),* (pp. 1–4). Citeseer.

Faisal, S., Javaid, N., Javaid, A., Khan, M. A., Bouk, S. H., & Khan, Z. A. (2013). Z-SEP: Zonal-stable election protocol for wireless sensor networks. *ArXiv Preprint ArXiv:1303.5364.*

Fedor, S., & Collier, M. (2007). On the problem of energy efficiency of multi-hop vs one-hop routing in wireless sensor networks. In *21st International Conference on Advanced Information Networking and Applications Workshops (AINAW'07)* (Vol. 2, pp. 380–385). IEEE. 10.1109/AINAW.2007.272

Heinzelman, W. (2000). *Application-Specific protocol architectures for wireless networks* (Doctoral dissertation, Massachusetts Institute of Technology).

Heinzelman, W. R., Chandrakasan, A., & Balakrishnan, H. (2000). Energy-efficient communication protocol for wireless microsensor networks. In *Proceedings of the 33rd annual Hawaii international conference on System sciences, 2000.* (pp. 10-pp). IEEE. 10.1109/HICSS.2000.926982

Kashaf, A., Javaid, N., Khan, Z. A., & Khan, I. A. (2012). TSEP: Threshold-sensitive stable election protocol for WSNs. In *2012 10th International Conference on Frontiers of Information Technology (FIT)* (pp. 164–168). IEEE.

Lindsey, S., & Raghavendra, C. S. (2002). PEGASIS: Power-efficient gathering in sensor information systems. In *Aerospace conference proceedings, 2002. IEEE* (Vol. 3, p. 3). Citeseer.

Manjeshwar, A., & Agrawal, D. P. (2001). TEEN: a routing protocol for enhanced efficiency in wireless sensor networks. In null (p. 30189a). IEEE. doi:10.1109/IPDPS.2001.925197

Mehra, P. S., Doja, M. N., & Alam, B. (2017). Zonal based approach for clustering in heterogeneous WSN. *International Journal of Information Technology*, 1–9. doi:10.100741870-017-0071-2

Qing, L., Zhu, Q., & Wang, M. (2006). Design of a distributed energy-efficient clustering algorithm for heterogeneous wireless sensor networks. *Computer Communications*, *29*(12), 2230–2237. doi:10.1016/j.comcom.2006.02.017

Rani, R., Kakkar, D., Kakkar, P., & Raman, A. (2019). *Distance Based Enhanced Threshold Sensitive Stable Election Routing Protocol for Heterogeneous Wireless Sensor Network* (pp. 101–122). Berlin, Germany: Springer. doi:10.1007/978-3-662-57277-1_5

Saini, P., & Sharma, A. K. (2010a). E-DEEC-enhanced distributed energy efficient clustering scheme for heterogeneous WSN. In *2010 1st International Conference on Parallel Distributed and Grid Computing (PDGC)* (pp. 205–210). IEEE. 10.1109/PDGC.2010.5679898

Saini, P., & Sharma, A. K. (2010b). Energy efficient scheme for clustering protocol prolonging the lifetime of heterogeneous wireless sensor networks. *International Journal of Computers and Applications*, *6*(2).

Singh, S. P., & Sharma, S. C. (2015). A Survey on Cluster Based Routing Protocols in Wireless Sensor Networks. *Procedia Computer Science*, *45*, 687–695. doi:10.1016/j.procs.2015.03.133

Smaragdakis, G., Matta, I., & Bestavros, A. (2004). SEP: A stable election protocol for clustered heterogeneous wireless sensor networks. Boston, MA: Boston University Computer Science Department.

Tyagi, S., & Kumar, N. (2013). A systematic review on clustering and routing techniques based upon LEACH protocol for wireless sensor networks. *Journal of Network and Computer Applications*, *36*(2), 623–645. doi:10.1016/j.jnca.2012.12.001

Wang, J., Gao, Y., Liu, W., Sangaiah, A., Kim, H.-J., Wang, J., ... Kim, H.-J. (2019). An Improved Routing Schema with Special Clustering Using PSO Algorithm for Heterogeneous Wireless Sensor Network. *Sensors (Basel)*, *19*(3), 671. doi:10.339019030671 PMID:30736392

Yan, J., Zhou, M., & Ding, Z. (2016). Recent Advances in Energy-Efficient Routing Protocols for Wireless Sensor Networks: A Review. *IEEE Access: Practical Innovations, Open Solutions*, *4*, 5673–5686. doi:10.1109/ACCESS.2016.2598719

Yazid, Y., Ez-zazi, I., Salhaoui, M., Arioua, M., Ahmed, E. O., & González, A. (2019). Extensive Analysis of Clustered Routing Protocols For Heteregeneous Sensor Networks. In *Proceedings of the Third International Conference on Computing and Wireless Communication Systems, ICCWCS 2019, Faculty of Sciences, Ibn Tofaïl University -Kénitra- Morocco*. EAI. 10.4108/eai.24-4-2019.2284208

Younis, O., & Fahmy, S. (2004). HEED: A hybrid, energy-efficient, distributed clustering approach for ad hoc sensor networks. *IEEE Transactions on Mobile Computing*, *3*(4), 366–379. doi:10.1109/TMC.2004.41

Chapter 9
Non-Negative Matrix Factorization for Blind Source Separation

Nabila Aoulass
University Abdelmalek Essaadi, Morocco

Otman Chakkour
University Abdelmalek Essaadi, Morocco

ABSTRACT

NMF method aim to factorize a non-negative observation matrix X as the product X =G.F between two non-negative matrices G and F, respectively the matrix of contributions and profiles. Although these approaches are studied with great interest by the scientific community, they often suffer from a lack of robustness with regard to data and initial conditions and can present multiple solutions. The work of this chapter aims to examine the different approaches of NMF, thus introducing the constraint of sparsity in order to avoid local minima. The NMF can be informed by introducing desired constraints on the matrix F (resp G) such as the sum of 1 of each of its lines. Applications on images made it possible to test the interest of many algorithms in terms of precision and speed.

INTRODUCTION

Provide The separation of sources is the operation which, from the observations, makes it possible to obtain a set of signals proportional to the sources and to identify the contribution of each of the sources within the observed mixture. Thus, we distinguish two subproblems:

1. the identification of the mixture.
2. the reconstruction of the sources.

DOI: 10.4018/978-1-7998-0117-7.ch009

This opposite problem is badly posed because without any information on the sources and on the mixture, an infinity of solutions would be admissible. It is then necessary to formulate additional hypotheses and to take into account additional information on mixing and sources. The problem of separation sources can be approached from two points of view. The first is that the decomposition of observations on a basis of elementary signals to eliminate the redundancy of information between the different observations. So, the first methods were proposed by C. Jutten and J. Hérault who realized a nonlinear (ACP) in which we can diagonalize the covariance matrix by the decomposition in eigenvalues (EVD). Due to the limitation of diagonalizable matrices, singular value decomposition (SVD) makes (PCA) always possible based on the orthogonality constraint. It offers the least error (with respect to some measures) with the same reduced complexity, compared to other models. But it is not the only. The NMF is used in place of other low rank factorizations, such as the (SVD) because of its two primary advantages: storage and interpretability. Due to the non-negativity constraints, the NMF produces a so-called "additive parts-based" representation of the data. One consequence of this is that the factors of decomposition matrix are generally naturally sparse, thereby saving a great dea of storage when compared with the (SVD)'s dense factors. But is not for free. On the one hand, the decomposition of the SVD is known to have a polynomial complexity On the other hand, it has been recently demonstrated that the factorization of NMF has a non-deterministic polynomial computation complexity (NP). for which the existence of a optimal algorithm of a polynomial time is unknown. Moreover, non-orthogonal factors do not allow representation as in (PCA) but are used as a basis for unsupervised or prior modeling for supervised learning. A second, more recent approach is that of the Independent Component Analysis (ICA), it will be necessary to wait for the work of P.Comon to generalize this concept. The latter demonstrates, in the case of linear mixtures, that if the source signals are assumed to be mutually independent and non-Gaussian (except for at most one source), it is possible to separate these signals to a scale factor and a permutation by seeking to minimize the dependence measurements between the estimated signals at the output of the separation system. The implicit objective of the (ICA) is often to find physically significant components. However, in some field of environmental science, and using data that has the property of non-negativity, the solutions estimated by the methods based on the (ICA) lack of physical interpretability. In addition, the (ICA) cannot determine the variances (energies) of the independent components as well as the order of the independent sources because the basic functions are classified by non-Gaussianities . In NMF, the non-negativity constraint leads to the representation based on parts of the input mixture that helps to develop structural constraints on the source signals. NMF does not require independent evaluation and is not limited to the length of the data. It provides more important basic vectors for the reconstruction of the underlying signal than the activation vectors. Among the difficulties of matrix factorization in the area of blind separation, the ratio between the number of observations and the number of sources is a problem of interest for a large number of applications and has allowed the taxonomy that we recall below. In most applications and relying on instantaneous linear mixtures, the number m of samples in X is much larger than the numbers n of observations and p of sources. We then separate the determined case p = min (n, m), over-determined p < n, finally underdetermined such that p > min(n,m). When a single-channel source separation problem is considered under-determined, it cannot usually be solved without prior knowledge of the underlying sources in the mixture. For this reason, the problem of estimating multiple overlapping sources from an input mixture is unclear and complex in the (BSS) environment. But (NMF) provides a solution to this single-channel source separation problem by using its non-negativity constraint as well as a supervised mode of operation for source separation.

Background

NMF is defined as:

$$X \approx F \cdot G \tag{1}$$

Where $X \in \mathbb{R}^{n \times m+}$ is the spectograme, $F \in \mathbb{R}^{n \times p+}$ is matrix of basis vectors (columns), $G \in \mathbb{R}^{p \times m+}$ is the matrix of activations (rows) of the input mixture. In NMF when the spectrogram of mixture X is given, the matrices G and F can be computed via an optimization problem by:

$$\min_{F,G} D(X \,||\, F.G) \tag{2}$$

where D denotes the divergence. The reduced dimension p depends on the application and is imposed by the problem that we seek to solve. It is the same for the content of the product matrices that vary also depending on the application and the processed data and can have different physical meanings. Then NMF is applied in different domains for example we note: Single-Channel Speech Separation using Sparse Non-Negative Matrix Factorization, deals with the separation of multiple speech sources from a single microphone recording. The approach is based on a sparse non-negative matrix factorization, that is used to learn speaker models from a speech corpus. These models are then used to separate the audio stream into its components. Wind Noise Reduction using Non-negative Sparse Coding, introduces a speaker independent method for reducing wind noise in single channel recordings of noisy speech. The method is based on sparse nonnegative matrix factorization and relies on a noise model that is estimated from isolated noise recordings. Linear Regression on Sparse Features for Single-Channel Speech Separation, addresses the problem of separating multiple speakers from a single microphone recording by the formulation of a linear regression model, that estimates each speaker based on features derived from the mixture. separate useful information from superimposed biomedical data corrupted by a large level of noise and interference, for example, by using non-invasive recordings of human brain activities targeted at understanding the ability of the brain to sense, recognize, store and recall patterns and comprehending crucial elements of learning: association, abstraction and generalization.

Algorithms for Solving the NMF Problem

NMF Method Based on the Frobenius Norm

The multiplicative methods can be obtained in two different ways, either by a heuristic approach, or by a Maximization-Minimization (MM) approach.

Heuristic Approach

Multiplicative methods based on the Frobenius norm solve the problem (1) by rewriting it as a matrix trace, i.e.

$$J(G,F) = Tr\left(\left(X - G.F\right)^T \left(X - G.F\right)\right) \tag{3}$$

By developing this expression, we can show 3 functions in the expression of J(G, F), that is,

$$J(G,F) = Tr\left(X^T.X\right) - Tr\left(F^T.G^T.X\right) - Tr\left(X^T.G.F\right) + Tr\left(F^T.G^T.G\right) = J_1 - J_2 + J_3 \tag{4}$$

The calculation of the gradient of each of these three functions is carried out here with respect to a matrix F, while noting that the calculation of that with respect to G is similar.

$$\frac{\partial J}{\partial F} = 2G^T.\left(G.F - X\right) \tag{5}$$

we can identify the two non-negative functions appearing in the writing of the gradient,

$$\nabla_F^+ = 2GT.G.F \tag{6}$$

$$\nabla_F^- = 2G^T.X \tag{7}$$

The heuristic approach consists in using these two non-negative functions in the update rules of F, and transpose the results for the update of G:

$$F \rightarrow F \circ \frac{\nabla_F^- J\left(G,F\right)}{\nabla_F^+ J\left(G,F\right)} \tag{8}$$

$$G \rightarrow G \circ \frac{\nabla_G^- J\left(G,F\right)}{\nabla_G^+ J\left(G,F\right)} \tag{9}$$

the multiplicative update rules for the Frobenius NMF are available in:

$$F \rightarrow F \circ \frac{\left(G^T.X\right)}{\left(G^T.G.F\right)} \tag{10}$$

$$G \rightarrow G \circ \frac{\left(X.F^T\right)}{\left(G.F.F^T\right)} \tag{11}$$

The Maximization-Minimization (MM) Approach

Is based on two steps to obtain the update rules. The principle of this method consists of: first look for increasing the cost function by a function called auxiliary function. Then in a second step, to perform the minimization of the auxiliary function.

In the problematic of the classical NMF, we can take the function J (G, F) reduced to its vector formulation and rewrite it in the form of a Taylor development in the second order, considering that, the column current of the matrix F, is the only variable of the problem.

$$J(\theta)=J\left(\theta^{k}\right)+\nabla J\left(\theta^{k}\right)\left(\theta-\theta^{k}\right)+\frac{1}{2}\left(\theta-\theta^{k}\right)^{T}.G^{T}-G\left(\theta-\theta^{k}\right) \tag{12}$$

where $J\left(\theta^{k}\right)=-G^{T}\left(X-G.\theta^{k}\right)$ is the quadratic general form of the auxiliary function, $H\left(\theta,\theta^{k}\right)$ is given as a function of a positive $A\left(\theta^{k}\right)$ matrix, Lee and Seung propose to choose the matrix A in the form:

$$A\left(\theta^{k}\right)=diag(\frac{G^{T}.G.\theta^{k}}{\theta^{k}}) \tag{13}$$

The cancellation of the gradient of $H\left(\theta,\theta^{k}\right)$ and the Hadamard product leads to an exact relationship of the minimizer,

$$\theta^{k+1}=\frac{\theta^{k}oG^{T}x}{G^{T}.G.\theta^{k}} \tag{14}$$

The matrix formulation can be obtained by collecting the columns of the matrix F, which gives rise to the multiplicative updating rules for the Frobenius NMF problem:

$$F\rightarrow F^{\circ}\frac{\left(G^{T}.X\right)}{\left(G^{T}.G.F\right)} \tag{15}$$

$$G\rightarrow G^{\circ}\frac{\left(X.F^{T}\right)}{\left(G.F.F^{T}\right)} \tag{16}$$

Regularized ISRA Algorithm

- Using $D_f(X\,||F.G) = D_f(X^T\,||\,F^T.G^T)$, it obtain a similar update for F: Now just iterate between.
    - Updating F.
    - Updating G.
    - Checking$||\mathrm{X}-\mathrm{F.G}||$. if the change since the last iteration is small, then declare convergence.
        - Initialize F, G
        - repeat.

$$F \to F^\circ \frac{\left(G^T.X\right)}{\left(G^T.G.F\right)} \tag{17}$$

$$G \to G^\circ \frac{\left(X.F^T\right)}{\left(G.F.F^T\right)} \tag{18}$$

- until convergence return F, G.

NMF Methods Based on Fullback Libeler Divergence

The strategy is of type MM and the auxiliary function obtained is based on the concavity of the logarithmic function. Here we develop only update rules for the NMF type KL. The cost function to be minimized is expressed as,

$$D_{KL}(X\,||F.G) = \sum_{i,j}\left[X\log\frac{X}{G.F}\right] - X + G.F]_{i,j} \tag{19}$$

We adopt the notation f, and x as the respective current column vectors of F and X,

$$J(f) = D_{KL}(X\,||F.G) = \sum_i x_i \log x_i - x_i + \sum_j G_{i,j}f_j - x_i \log G_{i,j}f_j \tag{20}$$

Using the concavity of the logarithmic function and Jensen's inequality, the previous cost can be increased by the following auxiliary function:

$$H\left(f,f^{k}\right)=\sum_{i}x_{i}\log x_{i}+\sum_{j}G_{i,j}f_{j}-x_{i}\sum_{j}\frac{G_{i,j}f_{j}^{k}}{\sum_{l}G_{i,j}f_{l}^{k}}(\log G_{i,j}f_{j}-log\frac{G_{i,j}f_{j}^{k}}{\sum_{l}G_{i,j}f_{l}^{k}}) \quad (21)$$

The Maximization-Minimization Theorem implies that f^{k+1} verifies:

$$J\left(f^{k}\right)>\min\left(H\left(f,f^{k}\right)\right)=H\left(f^{K+1},f^{k}\right))\geq J\left(f^{K+1}\right) \quad (22)$$

Minimize H(., f^{k}) from f:

$$\frac{\partial H}{\partial f_{j}}=\sum_{i}G_{i,j}-\frac{f_{j}^{k}}{f_{j}}\sum_{i}x_{i}\frac{G_{i,j}f_{j}^{k}}{\sum_{l}G_{i,j}f_{l}^{k}}=0 \quad (23)$$

The minimum is then given by:

$$f^{k+1}=\frac{f^{k+1}}{G.1}\circ\left(G^{T}\frac{x^{\circ}1}{G.f^{k}}\right)$$

By grouping these vectors, we get the update expression of F:

$$F=\frac{F}{G^{T}.1}\left(G^{T}\frac{x^{\circ}1}{G.F}\right) \quad (24)$$

By transposition, the rule for updating the matrix G is written:

$$G=\frac{G}{1.F^{1}}\circ\left(\frac{1^{\circ}x}{G.F}F^{T}\right) \quad (25)$$

Regularized EMML Algorithm

- Using $D_{KL}(X\,||F.G)=D_{KL}(X^{T}\,||\,F^{T}.G^{T})$, it obtain a similar update for F: Now just iterate between.
  - Updating G.
  - Updating F.
  - Checking $D_{KL}(X\,||F.G)$. if the change since the last iteration is small, then declare convergence.
    - Initialize F, G
    - repeat.

$$F \rightarrow \frac{F}{G^T.1}\left(G^T \frac{x^{\circ}1}{G.F}\right) \quad (26)$$

$$G \rightarrow \frac{G}{1.F^1} \circ \left(\frac{1 \circ x}{G, F} F^T\right) \quad (27)$$

- until convergence return F,G.

UNIQUENESS OF NMF SOLUTION

Given the formulation of the NMF problem, it is clearly to be feared the existence of invertible square matrices T such that the pairs (WT, T^{-1} H) are also solutions of the problem, since F.G =(F.S)(S^{-1} .H) and that the cost of reconstruction depends only on the product FG. Given a solution (F^0 , G^0), the pair(F^0 .S, S^1 .G) is the solution of the problem if and only if the two matrices F^0 .S and (S^{-1} .G) have positive or zero coefficients. At least two types of cases can be exhibited where this is possible:

Trivial Invariances: If we insist that S and its inverse S^{-1} have positive or zero coefficients, the F.S and (S^{-1} .G) products are also positive and we are in the presence of a new solution to the initial problem. In this case, it is easy to prove that such a matrix S is necessarily the product of a permutation matrix and a diagonal matrix with positive coefficients . The permutation matrix introduces K degrees of invariance, but does not really change the solution; moreover, the uniqueness of a quantity is often defined at a close permutation. With regard to scale factors (diagonal matrix), the question can be solved by imposing a standardization on one of the factors F or G (in practice, one will often choose to standardize the columns of F in norm L_2). These invariances are therefore not a real obstacle to a possible uniqueness of the solution, which will be defined by a permutation and a change of scale.

Local invariances: The product ($F^0 . G^0$) can also remain invariant on points in the vicinity of the couple that makes it. Suppose that F^0 and G^0 have strictly positive coefficients. Given a square matrix U, not necessarily with positive coefficients, we can find ε sufficiently small such that (I +εU) is invertible. One can make the limited development: (I +εU))≈(I −εU). The matrices (I +εU) and (I +εU)−1 perform local transformations around (F0.G0); provided that this point is not situated on the edges of the positive quadrant, and that ε is chosen sufficiently small, the points (F^0 (I +εU)) and ($(I+\varepsilon U)^{-1}$) G^0 remain in this quadrant. Thus, we obtain a new solution to the problem of NMF, the pair (F^0 (I +εU), $(I+\varepsilon U)^{-1} G^0$).

Non-Uniqueness of the General Case

We have considered only pairs of solutions expressing themselves relative to each other via a linear transformation: (F, G) and (G.S, S^{-1} .F). In reality, there is nothing to require that two solutions producing the same product F.G be connected in this way.

$$X=\begin{matrix} 1 & 1 & 0 & 0 \\ 1 & 0 & 1 & 0 \\ 0 & 1 & 0 & 1 \\ 0 & 0 & 1 & 1 \end{matrix} = X.I = I.X$$

where I denotes the identity matrix. The matrix V is of rank 3.We can choose as factorizations ($F^0 =$ X, G^0 = I)or(F^1 = I, G^1 = X) There is no invertible matrix S such that $F^0 S = F^1$, for reasons of rank. However, if we choose K = rg (X), we can show that such counterexamples are impossible. In this case, all solutions are connected to each other by linear transformations.

SPARSE NMF

The concept of sparse coding refers to a representational scheme where only a few units (out of a large population) are effectively used to represent typical data vectors. In effect, this implies most units taking values close to zero while only few take significantly non-zero values. In this paper, we use a sparseness measure based on the relationship between the L_1 norm and the L_2 norm:

$$sparseness(x) = \frac{\sqrt{n} - \sum|x_i| / \sqrt{\sum x_i^2}}{\sqrt{n} - 1} \tag{24}$$

where n is the dimensionality of x. Our aim is to constrain (NMF) to find solutions with desired degrees of sparseness. The first question to answer is then: what exactly should be sparse? The basis vectors F or the coefficients G? This is a question that cannot be given a general answer; it all depends on the specific application in question. Further, just transposing the data matrix switches the role of the two, so it is easy to see that the choice of which to constrain (or both, or none) must be made by the experimenter. The sparse NMF problem can be formulated as

$$E(F,G) = X - F.G^2 \tag{25}$$

is minimized, under optional constraints

$$\text{sparseness}(f_i) = S_f, \ \forall \text{i} \tag{26}$$

$$\text{sparseness}(g_i) = \text{Sg}, \ \forall \text{i} \tag{27}$$

where f_i is the i:th column of F and g_i is the i:th row of G. Here, M denotes the number of components, and S_f and S_g are the desired sparsenesses of F and G (respectively). These three parameters are set by the user.

Algorithm with Sparseness Constraint

- Initialize F and G to random positive matrices
- If sparseness constraints on G apply, then project each column of G to be non-negative, have unchanged L_2 norm, but L_1 norm set to achieve desired sparseness.
    - Iterate.
    - if Sparseness constraints on G apply,then.
    - Set G:= G−μF(FG−X) convergence.
    - Project each row of H to be non-negative, have unit norm, L_2 and L_1 norm set to achieve desired sparseness.
        - Else Take standard multiplicative step.
        - Until convergence return F, G.

EXISTENCE OF INFORMED METHODS OF NMF WITH SET VALUES AND SUM TO 1 VARIABLES

The non-negativity constraint can sometimes be explicitly associated with a sum-1 constraint Sum-to-1 variables can be found in several domains and in different ways.

Normalization of G lines or F columns

In a first case, we can assume that the normalization concerns the lines of the first factor in the expression (1.5). This one is written:

$$F_{P\times m}.1_{m\times m} = 1_{p\times m} \tag{28}$$

The normalized matrix - also named stochastic matrix row and denoted tilde F below - is obtained from any non-standard version of F as follows:

$$F' = \frac{F}{F.1_{m*m}} \tag{29}$$

where the fraction represents the term-term division of the elements of the matrices operands. Similarly, a new matrix G can be displalyed as:

F’.G’

From expressions (3.3) and (3.4), we deduce the expression of which depends on F^T

$$G' = G \circ \left(1_{n*m}.F\right) \tag{30}$$

In particular, the optimization problem consists in estimating the matrices G and F such that:

$$\begin{cases} X \approx G.F \\ S.CF_{\mathrm{m\times n}}.1_{m\times m} = 1_{p\times m} \end{cases} \tag{31}$$

By multiplying on the right the equation X = G.F by the matrix 1m timesm and by combining this expression with the constraints (3.3), we obtain:

$$X_{\mathrm{m\times m}}.1_{\mathrm{m\times n}} = G_{\mathrm{n\times p}}.F_{\mathrm{p\times n}}.1_{\mathrm{m\times n}} = G_{\mathrm{n\times p}}.1_{\mathrm{p\times n}} \tag{32}$$

ie the sum of each line of G is equal to the sum of the lines of X corresponding. However, in practice, we do not have an exact but approximate relation between X, G and F, because of the possible presence of noise, outliers or approximation of the model. Then the relation (31) becomes:

$$X_{\mathrm{m\times m}}.1_{\mathrm{m\times n}} \approx G_{\mathrm{n\times p}}.1_{\mathrm{p\times n}} \tag{33}$$

We introduce auxiliary variables Z and T of appropriate dimensions which must be no negative and are respectively bound to matrices F and G

$$F = \frac{Z}{Z.1_{\mathrm{m\times n}}} \tag{34}$$

$$G = \frac{\mathrm{T}^\circ\left(\mathrm{X}.1_{\mathrm{m\times n}}\right)}{\mathrm{T}.1_{\mathrm{p\times p}}} \tag{35}$$

Resolution the Problem Informed NMF

Among the parameterizations to solve the approximate NMF problem, we have:

$$argminD(X\,||F.G)S.C \begin{cases} F_{\mathrm{m\times n}}.1_{m\times m} = 1_{p\times m} \\ X_{\mathrm{m\times m}}.1_{\mathrm{m\times n}} \approx G_{\mathrm{n\times p}}.1_{\mathrm{p\times n}} \end{cases} \tag{36}$$

Problem could be formulated as an optimization problem with Z:

$$\{T',Z\} = argminD(X|\frac{T^\circ(X.1_{m\times n})}{T.1_{p\times p}}.\frac{Z}{Z.1_{m\times n}}) \quad (37)$$

The Split Gradient Method (SGM) is an attractive approach to solving the non-negative constraint for the matrix factorization problem (NMF). SGM can be classified in gradient-based approaches.

$$\frac{\partial D}{\partial Z_{ik}} = \sum_{m}^{j=1} \frac{\partial D}{\partial F_{ij}} \cdot \frac{\partial F}{\partial Z_{ik}} \quad (38)$$

The matrix difference with respect to Z_{ik} is derived:

$$\frac{\partial D}{\partial Z_{ik}} = \frac{1}{\sum_l [Z]_{il}} \cdot \left(\delta_{jk} - [F]_{ij}\right) \quad (39)$$

And δ_{jk} is delta function, so:

$$-\frac{\partial D}{\partial Z_{ik}} = \frac{1}{\sum_l [Z]_{il}} \left(\left(\frac{\partial D}{\partial F_{ik}} \right) - \sum_j [F]_{ij} \left(-\frac{\partial D}{\partial Z_{ik}} \right) \right) \quad (40)$$

From $-\partial D/\partial Z = P - Q$ such that:

$$P = (-\partial D/\partial F)_s \quad (41)$$

$$Q = (F^\circ(-\partial D/\partial F)_s.1_{m\times m}) \quad (42)$$

where the index represents a shift applied to the partial differentiation $-\partial D/\partial F$ The SGM from allows to perform a gradient descent on Z according to

$$z^{K+1} = z^K + \cdot {}_F^k \circ z^K \circ \left[\frac{(-\partial D/\partial F)_s}{F^\circ(-\partial D/\partial F)_s.1_{m\times m}} - 1_{p\times m} \right] \quad (42)$$

using this definition:

$$\left(-\frac{\partial D}{\partial F^k}\right)_s = \left(-\frac{\partial D}{\partial F^k}\right) - \min\left(-\frac{\partial D}{\partial F_{ij}^k}\right).1_{p\times n} \quad (43)$$

so:

$$(-\partial D/\partial F_{ij})_s \leftarrow -(-\partial D/\partial F_{ij}) + \eta),\ \forall(i,j) \tag{44}$$

$$(-\partial D/\partial G_{ij})_s \leftarrow -(-\partial D/\partial G_{ij} + \mu),\ \forall(i,j) \tag{45}$$

with:

$$\eta = \min(-\partial D/\partial F_{ij}).1p\times n \tag{46}$$

$$\mu = \min(-\partial D/\partial G)._{ij}1m\times p \tag{47}$$

. Algorithm SGM avec Variable de Somme-at -1

- Repeat
    - Read X.
    - Initialization de G et F.
    - while the stop rule is not filled, do:
    - Calculated F and fixed G by:

$$♣\ F^{K+1} = F^{K} + \cdot{}_F^k \circ F^{K} \circ \left[\frac{(-\partial D/\partial F)_s}{F^\circ(-\partial D/\partial F)_s.1_{m\times m}} - 1_{p\times m} \right] \tag{48}$$

$$G^{K+1} = G^{K} + \cdot{}_F^k \circ F^{K} \circ \left[\frac{(-\partial D/\partial F)_s}{\frac{G}{X.1_{m\times p}}\circ(\partial D/\partial F)_s.1_{m\times m}} - 1_{p\times m} \right] \tag{49}$$

♣ end.

PROJECTED GRADIENT APPROACHES

The projection of NMF in the iterations on the constraints is here called a projective NMF. Thus, a special parameterization that takes into account this knowledge is introduced. So, new algorithms for this parameterization are planned.

Parameterization of the Profile Matrix

In several applications, the values of some entries may be provided by experts. The corresponding parameterization takes into account this knowledge (missing values). Let $\Omega_{i,j}$ be a binary matrix p times m which informs of the presence or absence of constraints on each element F_{ij} of the matrix F, so:

$$\Omega_{i,j} = \begin{cases} 1 & if \quad \mathrm{F_{ij}} \text{ is known} \\ 0 & if \quad \text{not} \end{cases} \tag{50}$$

Then the binary matrix p×m ω is defined as $\omega = 1_{\mathrm{p \times m}} - \Omega$. And φ is the sparse matrix:

$$\varphi = \mathrm{F} \circ \omega \tag{51}$$

By construction φ_{ij} the (i, j) th element of φ is zero when $\Omega_{i,j}$=0.

We can easily prove that:

$$\varphi \circ \Omega = \varphi,\ \varphi \circ \Omega = 0 \tag{52}$$

in [10] ΔF is defined as the free part of the matrix profile in the form:

$$\Delta \mathrm{F} = \mathrm{F} - \varphi \circ \Omega. \tag{53}$$

Following the general procedure in [10] so, the matrix F is obtained in the following form:

$$\mathrm{F} = \Omega \circ \varphi + \Omega \circ \Delta \mathrm{F} \tag{54}$$

In addition, it may be noted that this parameterization relates only to non-negative matrices. So, F satisfies the following inequality,

$$\mathrm{F} \geq \varphi \tag{55}$$

The φ matrix is fixed and known in advance while ΔF represents the non-negative free matrix whose structure is imposed and which must be found.

INTRODUCTION TO BAYESIAN NMF

The Bayesian NMF offers the problem a theoretical statistical framework and allow to expand the capabilities of systems modeling. In these approaches, the signal is described in a unified and systematic way by random variables representing each element to be modeled: distribution of frequencies, amplitudes, noise, evolution of partials, notes, instruments, etc... These interdependent random variables thus form a network, characterized by the choice of their probabilistic distributions. The problem is then solved via the theory of Bayesian inference, so we estimate the optimal models and parameters from the observed signal. Two probabilistic quantities intervene: the a priori laws, which translate the knowledge on the parameters which one chooses to inject into the system, and the likelihood, which establishes the relation between the parameters and the signal. The methodology of statistical signal analysis specifically considered in this part is based on various general principles:

- The observed signal X is considered as the realization of a random variable x.
- It is modeled via a model of law of probability, or through other random variables, called latent whose distribution is modeled.
- The developments are aimed at obtaining a method for estimating θ parameters of the model.
- This estimation can be obtained by the method of maximum likelihood (**MV**), which consists to maximize, the likelihood p(X|θ) (probability of the observations knowing the model).
- It can also be obtained by the maximum posteriori method (**MAP**), related to Bayes theorem. It consists to maximize the posterior distribution p(θ|X) of the model. This so-called Bayesian inversion also makes it possible to include a priori knowledge of the parameters to be estimated in the modeling, via the prior distribution p(θ).

PROBABILISTIC INTERPRETATIONS OF THE NMF

The relation between the statistical framework and NMF is truly established in the literature when a probabilistic model of non-negative observation is set, and formal equivalence is established between the minimization of the function of cost in the NMF problem and estimation of model parameters in the statistical problem. For example, the author poses in [12] the following model on a signal S:

$$\mathrm{F} \sim \mathrm{P}(\mathrm{F}, \Theta^{f}) \tag{56}$$

$$\mathrm{G} \sim \mathrm{P}(\mathrm{G}, \Theta^{g}) \tag{57}$$

$$s_{\mathrm{v,i,Ä}} \sim P(s_{\mathrm{v,i,Ä}}, f_{v,i}, g_{\mathrm{i,Ä}}) \tag{58}$$

And,

$$x_{\mathrm{v,Ä}} = \sum_{i} s_{\mathrm{v,i,Ä}} \tag{59}$$

such that each latent variable is distributed according to a generalized Poisson distribution (denoted P), with the non-negative intensity parameter λ, where:

$$\mathrm{P}(\mathrm{s}, \Gamma) = \exp(\mathrm{s}.\log(\Gamma) - \Gamma - \log(\Gamma(\mathrm{s}+1)) \tag{60}$$

with " (s+1)= s! is the function of Gamma, The probability priors P(G|ů) and P(F|ů) will be specified later. We call the variables $S_i = s_{v,i,Ä}$the latent sources.

$$\text{Log } P(X|G,F) = \log \sum_s P(X \mid S)P(S \mid G,F) \tag{61}$$

$$= \log \prod_{v,Ä} P\left(x_{v,Ä}; g_{v,Ä} f_{i,Ä}\right)$$

$$= \sum_v \sum_Ä x_{v,Ä} \log[G.F]_{v,\tau} - \log \Gamma(x_{v,Ä}) + 1)$$

PROBABILISTIC MODEL FOR NON-NEGATIVE MATRIX FACTORIZATION

the non-negative matrix decomposition is formulated like that:

$$X = G.F + E \tag{62}$$

such as: X ∈R R Rn×m and G and F two latent matrices such as G ∈R R Rnxk et G ∈R R Rk×m, whose values are positive. where E is the noise captured by the matrix E ∈R R Rn×m. The indices of observed inputs can be represented by: $\Omega = \{(i,j) | x_{i,j} \text{ observed}\}$.

The matrices F and G are considered as independent variables. The main challenge is to design appropriate prior information on G and F by defining a prior distribution P(G) and P(F).To ensure the non-negativity of the factors, it is necessary to choose probability densities with positive supports. Using Bayes' theorem, the common posterior distribution of unknown factors is given by:

$$P(F,G|X) = \frac{P(X \mid F,G) \times P(F) \times P(G)}{P(X)} \tag{63}$$

such as P(X|G,F) is the likelihood of the observations and p(X) is the probability density of the observations. It should be noted that the two matrices G and F are independent, the joint distribution is reduced to a simple product:

$$P(G,F) = P(G)P(F) \tag{64}$$

and P (X) appears as a normalization term. Then equation (62) can be written as:

$$P(F,G|X) \propto P(X|F,G) \times P(F) \times P(G) \tag{65}$$

and the estimate **MAP** is usually performed taking the maximum equation (71) or its logarithm. The two matrices G and F can be calculated from equation (64) and using several Bayesian estimators that perform essentially approximate calculations except in the case of special conjugate distributions. The independence property between the two estimated matrices G and F makes it possible to express the common law a posteriori P(G,F|X), thus all Bayesian algorithms use the Markov Chain Monte Carlo (**MCMC**) to generate samples distributed according to later marginal laws. The markov chain is a sequence of random variables such that the next value of the sequence depends only on the previous value We consider the state theta(t), then the new state $\theta(t+1)$ depends only on the current state. There are two major algorithms, the Metropolitan Hasting Algorithm and the Gibbs

Sampling, which can be implemented to calculate the next set of values that will form the Markov chain from which the elements of matrices G and F are estimated. After initializing the algorithm, the Metropolitan Hasting can be summarized as follows:

1. We generate a random variable (the candidate variable) by sampling from a proposal distribution that depends solely on the previous state.
2. We check if the generated candidate variable satisfies a defined condition. If satisfied, the generated variable will be the new state. Otherwise, the previous value will be affected. This procedure will be repeated for a large number of iterations.

GIBBS SAMPLING

Gibbs sampling operates by sampling new values for each parameter θ_i from its marginal distribution taking into account the current values of the other parameters θ_{-i} and the observed data D. So, if we are sampling new values for each θ_i parameter of $p(\theta_i, \theta_{-i}, D)$, we will eventually converge on samples from the posterior, which can be used to approach the posterior $p(\theta_i, |D)$.

Gibbs Sampling Algorithm

- Suppose that $\theta_1,\theta_2 \sim P(\theta_1,\theta_2)$ and we can sample according to $P(\theta_1,\theta_2)$ et $P(\theta_2,\theta_1)$.
- Initialization by $P(\theta_1^{(0)}, \theta_2^{(0)})$ the sampling Gibbs:
    - Sample $\theta_1 \sim P(\theta_1, \theta_2^{j-1})$
    - Sample $\theta_2 \sim P(\theta_2, \theta_1^{j})$
        - $P(\theta_2,\theta_1)= P(\theta_1,\theta_2)/ P(\theta_1)$ (66)
        - $P(\theta_2,\theta_1)= P(\theta_2,\theta_1)/ P(\theta_1)$ (67)
        - $P(\theta_1) = \int P(,_1,,_2)\, d\theta_2$ (68)
        - $P(\theta_2) = \int P(,_1,,_2)\, d\theta_1$ (69)

We take a probabilistic approach to this problem. We express a likelihood function for the observed data and treat the latent matrices as random variables. Each value of X comes from the product of G and F, with some added Gaussian noise,

$$X_{i,j} \approx N\{X_{i,j}, G_i.F_j, \tau^{-1}\} \quad (70)$$

So, G_i, F_j denotes the i-th and j-th lines of G and F, with:

$$N\{x,\mu,\tau^{-1}\}= \tau^{1/2} (2\pi)^{1/2} \exp\{(-\tau/2).(x-\mu)^2\} \quad (71)$$

The set of parameters of our model is denoted θ = G,F,τ. In the Bayesian approach to inference, we want to find the distributions on θ parameters after observing the data D = $X_{i,j}$.

We can use Bayes theorem,

$$P(, | D) \propto P(D | ,).P(,) \quad (72)$$

We cannot normally calculate the exact $P(, | D)$, but some choices of priors allow us to get a good approximation Then Schmidt et al choose an exponential a priori on G and F, so that each element in G and F is supposed to be distributed exponentially independently with rate parameters $\lambda_{i,k}^{G}$, $\lambda_{k,j}^{F}$ > 0.

$$G_{i,k} \sim \varepsilon(G_{i,k} | \lambda_{i,k}^{G}) \quad (73)$$

$$\lambda_{i,k}^{F} \sim \varepsilon(G_{i,k} | \lambda_{k,j}^{F}) \quad (74)$$

such that $\mu(x|») = »\exp(-»xu(x))$ is the density of the exponential distribution, and u(x) is the unit pitch function. For precision τ we use a distribution λ of form $\alpha_\tau > 0$ and $\beta_\tau > 0$,

$$P(Ä) \sim G(Ä\alpha_{Ä}, {}^2{}_{Ä}) = \frac{\beta_{Ä}^{\pm Ä}}{“(\pm_{Ä})} X^{\pm Ä-1} \exp(-^2{}_{Ä} X) \quad (75)$$

such that $\Gamma(x)$ is gamma function:

$$\Gamma(x)= \int_0^\infty x^{t-1}\, e^{-xt}\, dt \quad (76)$$

SOLUTIONS AND RECOMMENDATIONS

Classification of Hyperspectral Data by NMF

The information on the surface of the Earth is traditionally an image that is to say a signal provided by a sensor that is used for the reconstruction of truth-terrain, it is brought to an observation system in the vast majority of applications to the electromagnetic radiation, like light, which is the visible manifestation of this radiation. The latter is an electromagnetic wave with the main properties of emission, absorption,

Figure 1. The color image

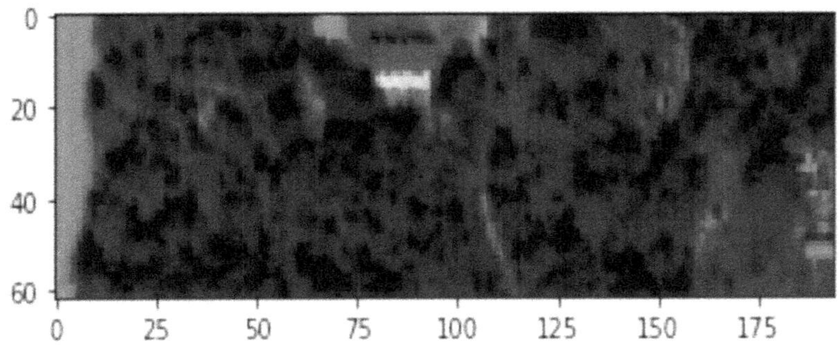

reflection and transmission. The information on the surface of the Earth is traditionally an image that is to say a signal provided by a sensor that is used for the reconstruction of truth-terrain, it is brought to an observation system in the vast majority of applications to the electromagnetic radiation, like light, which is the visible manifestation of this radiation. The latter is an electromagnetic wave with the main properties of emission, absorption, reflection and transmission. The phenomenon of reflection of light by surfaces forms the basis of most applications of remote sensing. Indeed, the various surfaces react differently with electromagnetic radiation by the reflection property, which is the basis of spectral signatures.

The main objective of the spatial remote sensing image analysis is the interpretation of the latter. This interpretation is generally performed by the classification process. The methods of classifying spatial remote sensing images are generally based on the spectral response of the sensors. This response makes it possible to associate the spectral properties of the terrestrial surface with certain properties of coverage or use of the soil. So, in this section we want to classify a hyperspectral image from its Normalized Difference Vegetation Index (NDVI). We must note that the purpose of this work is not to obtain the best classifier. However, with this current work, we want to show that it is possible to define a classification using the NMF which can present NDVI by the intensity of the spectra and its spatial positions.

Database

We use NEOP AOP data are distributed provided by NEON (National Ecological Observatory Networks field sites) Teaching Data Subset:Data Institute 2017 .The OSBSTinyIm.mat file contains the spectra of 346 bands, 194 columns and 62 lines. Now, we can display a color image we can display a color image

The Calculation of NDVI

The normalized difference vegetation index is constructed from the red (R) and near infrared (NIR) channels. The normalized vegetation index highlights the difference between the visible band of red the near infrared:

$$NDV\ \ I = (NIR - V\ \ IS)/(NIR + V\ \ IS)$$

Figure 2. The vegetation index

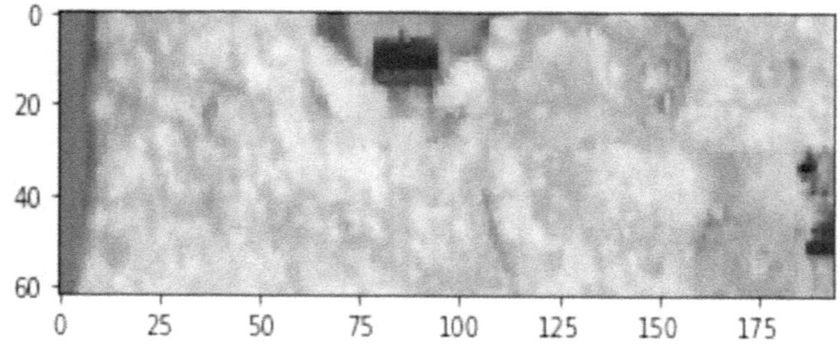

424/5000 NDVI values range from -1 to +1, with negative values corresponding to surfaces other than plant cover, such as snow, water or clouds for which the reflectance in the red is higher than that of the near infrared. For bare floors, reflectances being about the same order of magnitude in the red and the near infrared, the NDVI has values close to 0.

Classification by NMF

After having calculated NDVI we apply the different approaches of NMF: classical determinist, informed and the Bayesian approach.

The components have three categories of NDVI values between 0.1 and 0.7, less than 0.1 and greater than 0.7. The detection involves identifying the presence and locating a target from its spectral intensity.

In order to measure the convergence speed of different NMF algoritms, the maximum number of iterations is fixed at 100 iteration. In effect, the results obtained in the figures (5.19) and (5.20) show us that the convergence of classical NMF (respectively informed NMF, Bayesian NMF) is provided from 30 iteration (respectively less than 20, equal to 10 iteration).

Figure 3. The vegetation index greater than 0.

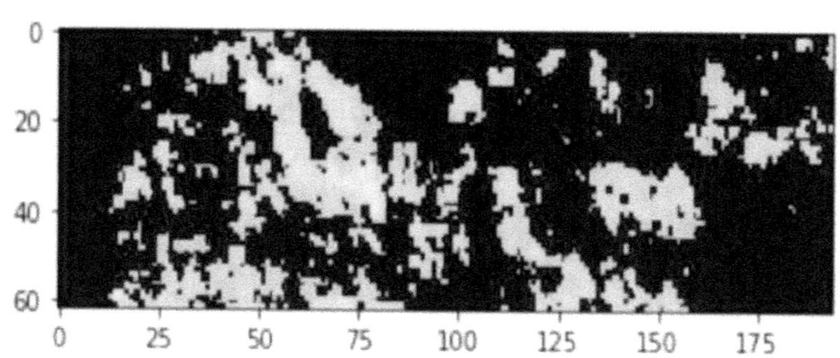

Figure 4. The vegetation index between 0.1 and 0.8

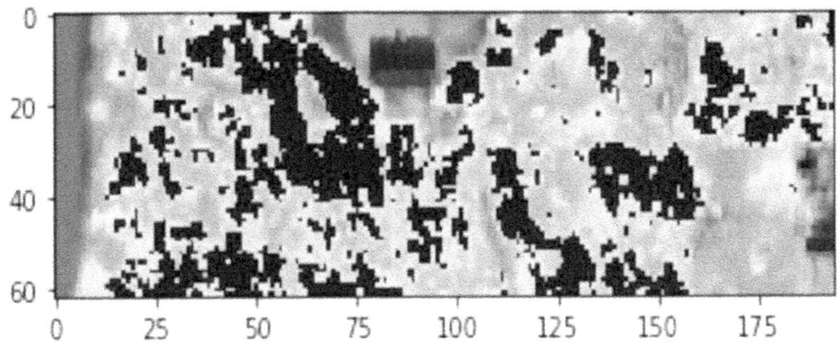

Figure 5. The vegetation index less than 0.1

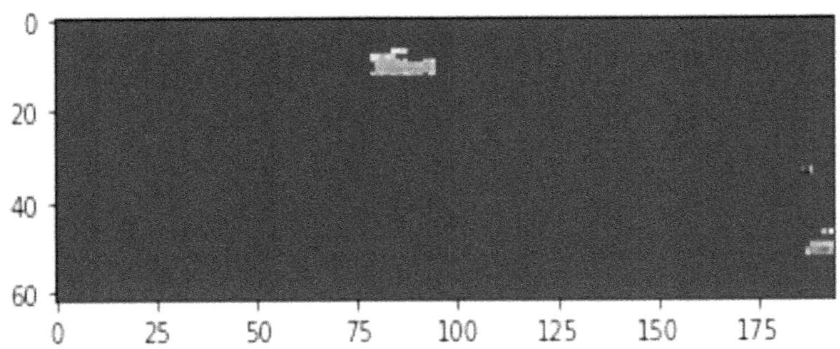

CONCLUSION

Geometrically, the (NMF) consists in finding a cone belonging to the positive orthant which includes the components of the vectors of the observed data. From this point of view, the cone is not always unique without additional constraints. From this geometrical interpretation, on the one hand, it appears that the (NMF) is not unique, which poses a problem for the BSS. On the other hand, in (NMF), the criteria can be convex only according to one of the two matrices produced but not for both. The algorithms therefore

Figure 6. The three components of classic NMF

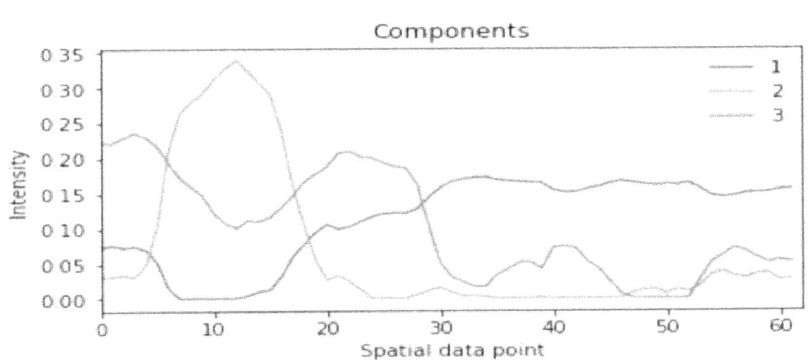

Figure 7. The three components of informed NMF

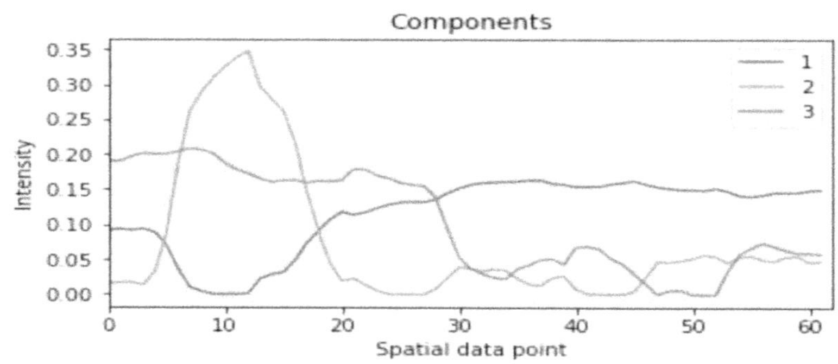

Figure 8. The three components of Bayesian NMF

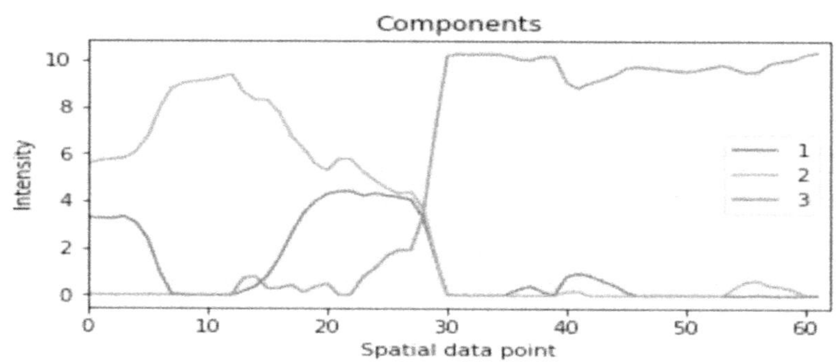

only allow to converge towards a local minimum. Therefore, the convergence result strongly depends on the initialization of the algorithm. The NMF is used in place of other low rank factorizations, because of its two primary advantages: storage and interpretability. Due to the non-negativity constraints, the NMF produces a so-called "additive parts-based" representation of the data. One consequence of this is that the factors F and G are generally naturally sparse, thereby saving a great deal of storage. The NMF also has impressive benefits in terms of interpretation of its factors. So, the basis vectors naturally correspond to conceptual properties of the data.

Figure 9. The loss of the objective function of the classic NMF

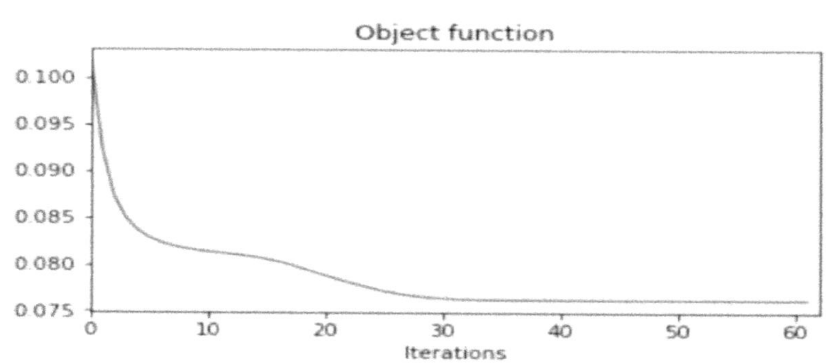

Figure 10. The loss of the objective function of informed NMF

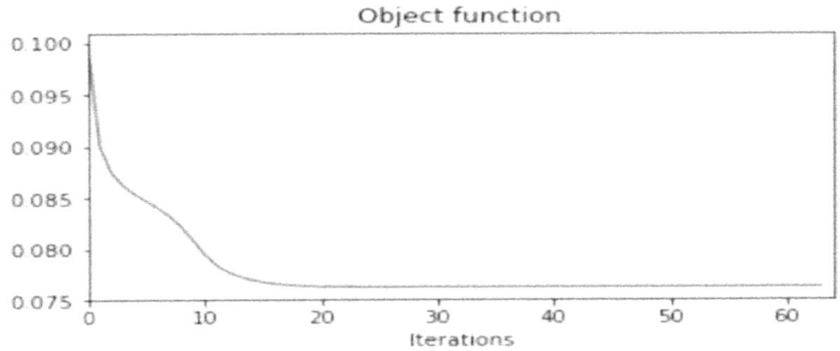

FUTURE RESEARCH DIRECTIONS

(NMF) provides a solution to this single-channel source separation problem by using its non-negativity constraint as well as a supervised mode of operation for source separation.

In future work, we will look at application of single channel separation of speech, we will present the optimal parameters found for the experiments as well discuss how parameters affect performance.

REFERENCES

Database https://www.neonscience.org/calc-ndvi-py

ADDITIONAL READING

7. Nancy Bertin,2010. Les factorisations en matrices non-négatives. Approches contraintes et probabilistes, application à la transcription automatique de musique polyphonique .

8. Patrik O. Hoyer,2004. Non-negative Matrix Factorization with Sparseness Constraints.

Figure 11. Loss of objective function for Bayesian NMF

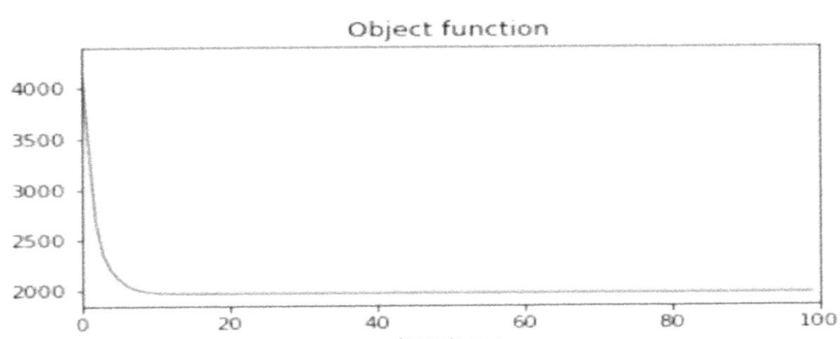

5. Moussa Sofiane Karoui,2012. Méthodes de séparation aveugle de sources et application à la télédétection spatiale.

3. Langville, A. N., Meyer, C. D., Albright, R., Cox, J., & Duling, D. (2014). *Algorithms.* Initializations, and Convergence for the Nonnegative Matrix Factorization.

1. Abdelhakim Limem,(2017). Méthodes informées de factorisation matricielle non négative .

2. Limem, A., Delmaire, G., Puigt, M., Roussel, G., & Courcot, D. 2014 Non-negative matrix factorization under equality constraints—a study of industrial source identification. In: Applied Numerical Mathematics.

6. Mona Nandakumar M and Edet Bijoy K,2014. An Experimental Survey on Non-Negative Matrix Factorization for Single Channel Blind Source Separation.

4. Mohamed Aziz Sbai,2012. Traitement des signaux parcimonieux et application

Chapter 10
Solar Micro-Inverter

Sivaraman P.
TECH Engineering, India

Sharmeela C.
Anna University, India

ABSTRACT

A solar micro inverter is a small-size inverter designed for single solar PV module instead of group of solar PV modules. Each module is equipped with a micro inverter to convert the DC electricity into AC electricity and the micro inverter is placed/installed below the module. The advantages of micro inverters are: reduced effect of shading losses, module degradation and soiling losses, enabled module independence, different rating of micro inverter can be connected in parallel to achieve the desired capacity, additional modules can be included at time which allows the good scalability, string design and sizing are avoided, failure of any micro inverter does not affect the overall power generation, individual MPPT controller for each module increases the power generation, any orientation and tilt angle allows higher design flexibility, lower DC voltage increasing the safety, easy to design, handle and install, requires less maintenance, draws attention of design engineers, contractors, etc.

INTRODUCTION TO SOLAR PV SYSTEM

Solar energy is the one of the major reliable renewable sources. Solar photovoltaic system is used to convert the light energy (photons) into electrical energy from the sunlight. A simple solar PV system consists of solar PV panel and inverter to convert the sunlight into electricity. Solar panel is a device that used to convert the photons in the sun light into DC electricity and inverter is used to convert the DC electricity into AC electricity (Chetan et al., 2012). The simple solar PV system is shown in Figure 1.

As per IEEE standard 929 – 2000, solar PV system applications are classified into small application up to 10 kW, intermediate (medium) application from 10 kW to 500 kW and large application more than 500 kW (Patel et al., 2014). Solar PV systems are classified into two types based on the utility power supply, i.e, is standalone system and other one is grid connected system (Chetan et al., 2012).

DOI: 10.4018/978-1-7998-0117-7.ch010

Figure 1. Simple solar PV system

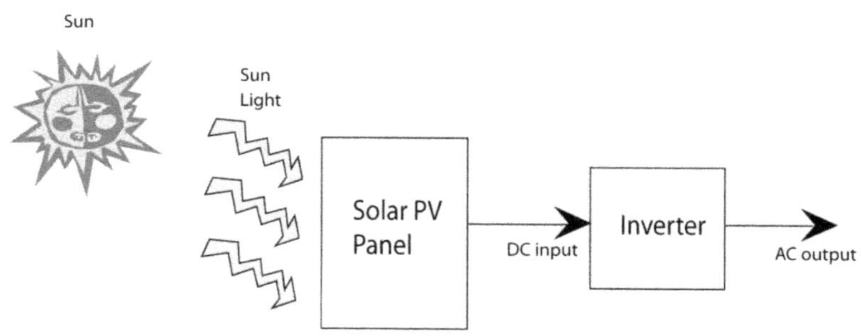

Grid connected solar PV system is operating in parallel with the utility power supply (Chetan et al., 2012). This system export the power from solar PV system into utility power grid and importing the power from utility power grid to the system. The typical grid connected solar PV system is shown in Figure 2.

Solar PV systems are classified into two types based on no of phases, they are single phase system and three phase system. The energy consumption of residential systems are very less and residential systems are mostly requires lesser rating solar PV for their demand. Small application of less than 10 kW solar PV systems are mostly used for residential systems. Commercial and small scale industrial systems have the power demand more than 10 kW to several hundred kW. These systems require medium application of solar PV system to cater the loads.

In most of the cases, the grid connected PV systems are preferably used in small scale commercial and industrial systems as compared to standalone PV systems (Chetan et al., 2012). In grid connected system, solar power is used to power the own plant loads and excess amount of power generation from the solar PV is exported into the utility power grid. Large solar PV systems are more than 500 kW rating primarily used for exporting the power to the utility power grid. The three phase solar PV systems are mostly preferred when compared to three numbers of single phase PV systems.

Figure 2. Grid connected solar PV system

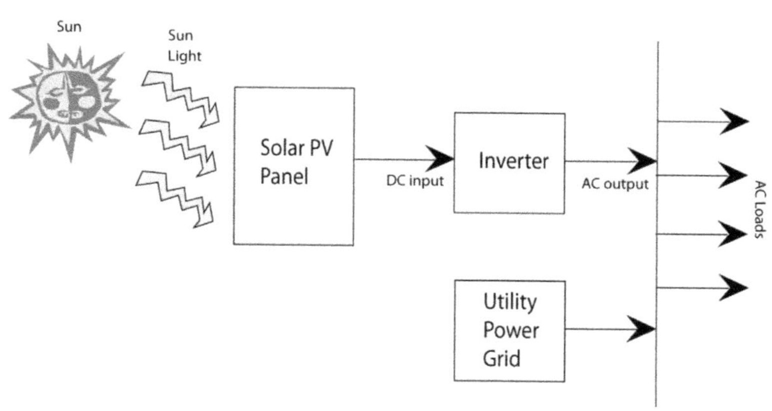

CLASSIFICATION OF SOLAR INVERTERS

Earlier days, grid connected solar PV inverters are classified into string inverter and central inverter. These string inverter and central inverters are working (MPPT and control) based on the string connection. Central inverters are used in utility scale power generation (ground mounted system) normally in Mega Watt power range and string inverters are majorly used in rooftop solar PV application in kilo Watt power range (Chetan et al., 2012). In many places in rooftop PV application, the roofs are facing many obstrucles and it create shadow in the roof area. So the becomes not suitable for solar PV application with string inverter because of stringing design. Also string inverter is working based on connected string, monitoring and control is possible at stringing level only. Individual level monitoring and control is not possible with string inverters. In order to overcome the problems with string inverter design, the concept of micro inverter is came into the picture. It enables the individual module level monitoring and control.

Introduction to Micro Inverter

A solar micro inverter is small size inverter designed for single solar PV module instead of group of solar PV modules. Each module are equipped with a micro inverter to convert the DC electricity into AC electricity and the micro inverter is placed/installed below the module. The Figure 3 shows the typical schematic diagram of micro inverter provided per solar PV module and figure 4 shows actual micro inverter provided per solar PV module installed in site.

Figure 4 clearly shows the location of micro inverter installed below the solar PV module.

Now a day's one micro inverter shall be used for two solar PV modules and four solar PV modules also. The Figure 5 shows the typical schematic diagram of one micro inverter provided for two solar PV modules.

The Figure 6 shows the typical schematic diagram of one micro inverter for four PV modules.

Demerits and De-merits of Micro Inverter

The main advantages of micro inverters are

- Reduced effect of shading losses, module degradation and soiling losses
- Enables module independence
- Different rating of micro inverter can be connected in parallel to achieve the desired capacity
- Additional modules can be included at time which allows the good scalability
- String design and sizing are avoided
- Failure of any micro inverter does not affect the overall power generation
- Individual MPPT controller for each module increases the power generation
- Any orientation and tilt angle allows higher design flexibility
- Lower DC voltage increasing the safety
- Easy to design, handle and installation, requires less maintenance

The main disadvantages of micro inverters are

- Higher capital investment

Figure 3. Typical schematic of micro inverter provided per solar PV module

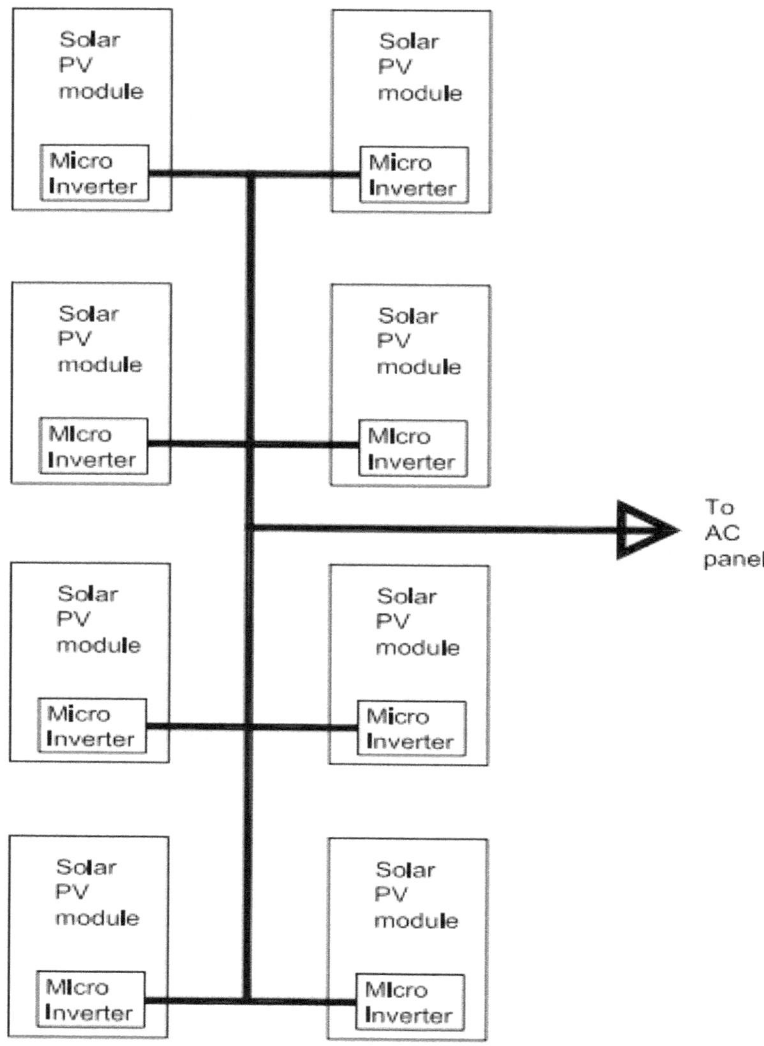

Figure 4. Micro inverter provided per solar PV module

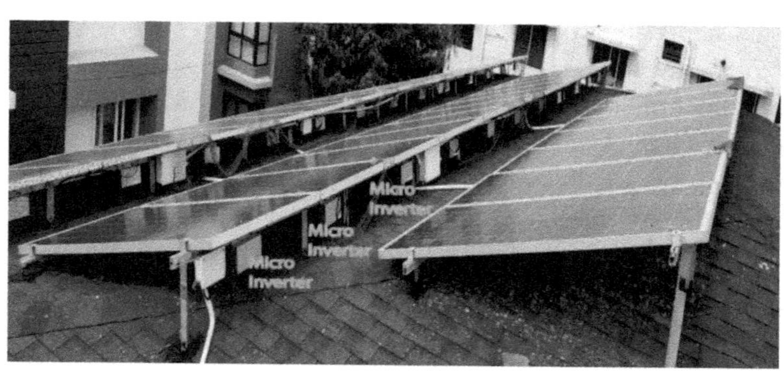

Figure 5. Typical schematic of one micro inverter provided for two PV modules

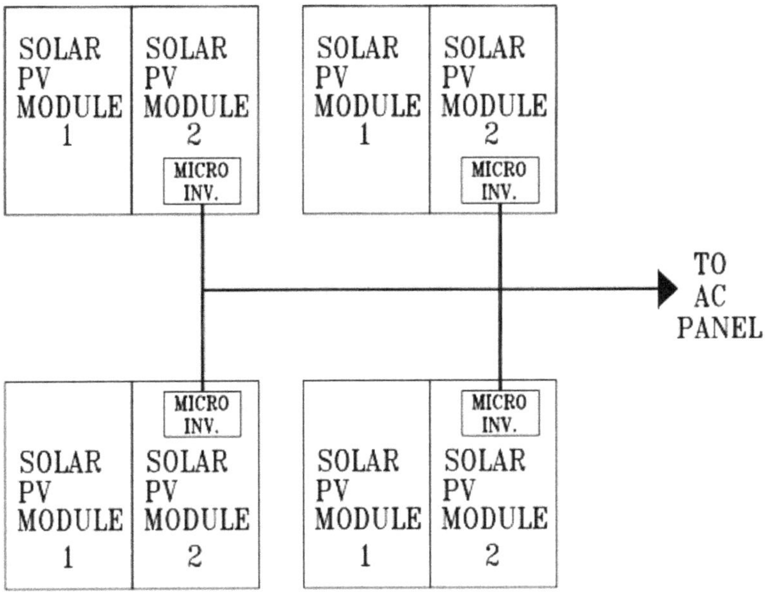

- Compared to string inverter, micro inverter efficiency is less

Suitable Places Where Micro Inverters are Preferred

The micro inverters are mostly preferred for:

- Places requiring multiple tilt angle and multiple orientation
- Places where concern for more shading
- Places where higher system availability is required
- Distributed systems and Building integrated PV systems

Figure 6. Typical schematic one micro inverter for four PV modules

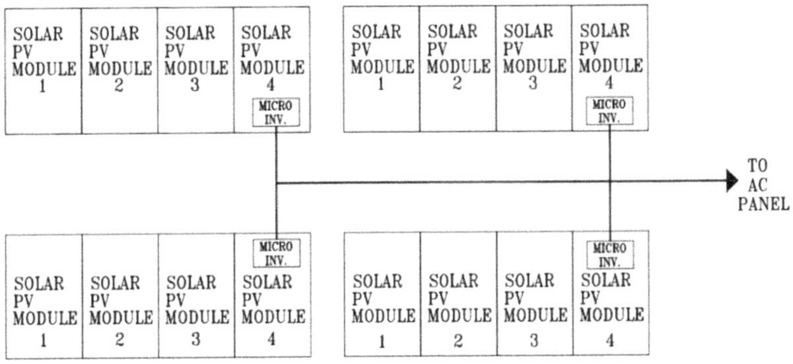

Features of Micro Inverter

The solar micro inverter has the following features

- The module monitoring features for micro inverters are available at both DC and AC side. The following main parameters are monitored in the micro inverter
  a) DC voltage and DC current
  b) AC voltage and AC current
  c) AC output power
  d) Produced energy
  e) Frequency
  f) Module temperature
  g) Inverter temperature
- Web based monitoring using IoT s also possible for micro inverters.
- PLCC communication protocol
- For control and networking using Wi-Fi, Ethernet and mobile network

Ratings of Micro Inverter

Now a days, solar micro inverters are available up to 1200 W AC power rating. Typical power rating and configurations of solar micro inverters are (1in1, 2in1 and 4in1) as follows (Krishnaswami et al., 2011).

- 250W – 1 module in 1 inverter
- 300W – 1 module in 1 inverter
- 500W – 2 modules in 1 inverter
- 600W – 2 modules in 1 inverter
- 700W – 2 modules in 1 inverter
- 1000W – 4 modules in 1 inverter
- 1200W – 4 modules in 1 inverter

The typical name plate details for 250 W and 300 W micro inverter for single solar PV module is listed in Table 1.

The typical name plate details for 500 W, 600 W and 700 W micro inverter for two solar PV module is listed in Table 2.

The typical name plate details for 1000 W and 1200 W micro inverter for two solar PV module is listed in Table 3.

Table 1. Name plate details of 250 W and 300 W micro inverter

S. No	Description	Rating	
1	IP rating	IP 67	IP 67
2	Ambient temperature range (°C)	`-40 to +65	`-40 to +65
3	Operating temperature range (°C)	`-40 to +85	`-40 to +85
4	No of cells in PV modules	60 or 72	60 or 72
5	DC max input power (W)	310	380
6	DC min start up voltage (V)	22	22
7	DC max input voltage (V)	60	60
8	DC max input current (A)	11.5	11.5
9	MPPT voltage (V)	27-48	29-48
10	Operating range (V)	16-60	16-60
11	AC maximum continuous output power (W)	250	300
12	Nominal voltage (L-N) (V)	230	230
13	AC operating voltage range (V)	180-275	180-275
14	AC continuous output current (A)	1.09	1.3
15	AC continuous maximum output current (A)	1.4	1.7
16	Nominal frequency (Hz)	50	50
17	Power factor at rated power	>0.99	>0.99
18	Efficiency (%)	96.5	96.5

MODELLING AND ANALYSIS OF MICRO INVERTERS VS STRING INVERTER USING PVSYST

Introduction

The selection of type of inverter is based on actual site condition and it will vary with site to site. Before selecting the type of inverter for any, site survey and feasibility analysis is essential for inverter selection. The simulation analysis for micro inverter is performed in PVsyst V6.78 simulation software and result of the analysis is explained and the same is compared with string inverter.

Example 1: Comparative Analysis of 10 kW Micro Inverter and 10 kW String Inverter

Location: Chennai, Tamilnadu, India
Latitude: 13.16° N
Longitude: 79.96° E
Time zone: UT+5.5
Resource data: Meteonorm 7.2

Table 2. Name plate details of 500 W, 600 W and 700 W micro inverter

S. No	Description	Rating		
1	IP rating	IP 67	IP 67	IP 67
2	Ambient temperature range (°C)	`-40 to +65	`-40 to +65	`-40 to +65
3	Operating temperature range (°C)	`-40 to +85	`-40 to +85	`-40 to +85
4	No of cells in PV modules	60 or 72	60 or 72	60 or 72
5	DC max input power (W)	620	760	800
6	DC min start up voltage (V)	22	22	22
7	DC max input voltage (V)	60	60	60
8	DC max input current (A)	10.5	11.5	11.5
9	MPPT voltage (V)	27-48	29-48	33-48
10	Operating range (V)	16-60	16-60	16-60
11	AC maximum continuous output power (W)	500	600	700
12	Nominal voltage (L-N) (V)	230	230	230
13	AC operating voltage range (V)	180-275	180-275	180-275
14	AC continuous output current (A)	2.17	2.61	3.04
15	AC continuous maximum output current (A)	2.8	3.3	3.9
16	Nominal frequency (Hz)	50	50	50
17	Power factor at rated power	>0.99	>0.99	>0.99
18	Efficiency (%)	96.50%	96.50%	96.50%

Case 1: Analysis of 10 kW Micro Inverter

The solar PV module name plate details are listed in Table 4.

The no of modules used is 32, no of modules in series is 1 and no of modules in parallel is 32. Total module area is 62.8 m^2 and cell area is 56.1 m^2.

The solar PV micro inverter name plate details are listed in Table 5.

The system configuration is listed in Table 6.

The PV array loss factors considered for the analysis is listed in Table 7.

The system loss factors considered for the analysis is listed in Table 8:

Results

The summary of PVsyst simulation is listed in Table 9.

The month wise energy production is shown in figure 7.

The month wise performance ratio is shown in figure 8.

The energy production is 15.63 MWh/Year

Performance ratio is 77.87%

Table 3. Name plate details of 1000 W and 1200 W micro inverter

S. No	Description	Rating	
1	IP rating	IP 67	IP 67
2	Ambient temperature range (°C)	`-40 to +65	`-40 to +65
3	Operating temperature range (°C)	`-40 to +85	`-40 to +85
4	No of cells in PV modules	60/72	60/72
5	DC max input power (W)	1240	1520
6	DC min start up voltage (V)	22	22
7	DC max input voltage (V)	60	60
8	DC max input current (A)	10.5	10.5
9	MPPT voltage (V)	27-48	32-48
10	Operating range (V)	16-60	16-60
11	AC maximum continuous output power (W)	1000	1200
12	Nominal voltage (L-N) (V)	230	230
13	AC operating voltage range (V)	180-275	180-275
14	AC continuous output current (A)	4.35	5.22
15	AC continuous maximum output current (A)	5.6	6.7
16	Nominal frequency (Hz)	50	50
17	Power factor at rated power	>0.99	>0.99
18	Efficiency (%)	96.00%	96.00%

Case 2: Analysis of 10 kW String Inverter

Table 4. Name plate details of solar PV module

S. No	Description	Rating
1	Manufacturer	Trina solar
2	Module power rating (W_p)	320
3	V_{mpp} (V)	33
4	I_{mpp} (A)	8.62

Table 5. Name plate details of micro inverter

S. No	Description	Rating
1	Manufacturer	Enphase
2	Model	IQ7X-96-x-INT
3	Unit nominal power rating (W)	315
4	Maximum power rating (W)	320
5	Operating voltage (V)	30 - 64

Table 6. System configuration

S. No	Description	Rating
1	Tilt	10°
2	Azimuth	0°
3	No of modules	32

Table 7. PV array loss factor

S. No	Description	Rating
1	LID	2%
2	Module quality loss	0%
3	Module mismatch losses	1% at MPP
4	Wiring Ohmic loss	1.4 mΩ
5	Thermal loss factor	29 W/ m²K
6	Array soiling losses	2%

Table 8. System loss factors

S. No	Description	Rating
1	Wiring Ohmic loss	1%@STC
2	Unavailability of the system	7.3 days, 3 periods 2%

Table 9. Summary of PVsyst simulation

S. No	Month	Global Horizontal Irradiance kW/m2	Global Tilted Irradiance kW/m2	Ambient Temperature °C	Energy produced by Array (MWh)	Energy exported to grid (MWh)	PR (%)
1	January	149.3	164	25.1	1.4	1.1244	74.1
2	February	163.9	176.5	26.19	1.482	1.438	79.6
3	March	195.8	202.1	28.06	1.662	1.613	77.9
4	April	193.7	192.4	29.68	1.587	1.54	78.2
5	May	192.4	185.5	31.73	1.526	1.481	78
6	June	168.9	161.2	30.72	1.35	1.215	73.6
7	July	161.1	154.7	30.44	1.296	1.258	79.4
8	August	161.9	159.1	29.49	1.335	1.295	79.5
9	September	160.5	163	28.8	1.35	1.31	78.5
10	October	137.4	143	27.43	1.22	1.184	80.9
11	November	115.1	123.3	25.56	1.05	0.917	72.6
12	December	123.6	135.1	24.89	1.165	1.131	81.8

Figure 7. Month wise energy production

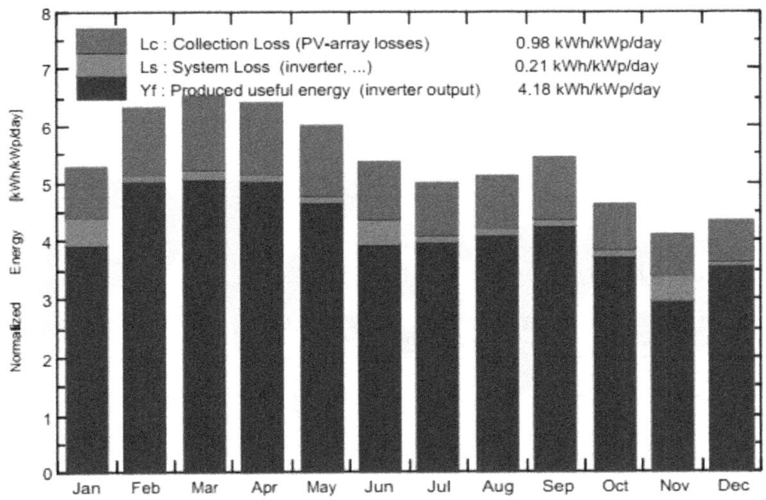

The solar PV module name plate details are listed in Table 10.

The no of modules used is 32, no of modules in series is 16 and no of modules in parallel is 2. Total module area is 62.8 m^2 and cell area is 56.1 m^2

The solar PV string inverter name plate details are listed in Table 11.

The system configuration is listed in Table 12.

The PV array loss factors considered for the analysis is listed in Table 13.

The system loss factors considered for the analysis is listed in Table 14.

Figure 8. Moth wise performance ratio

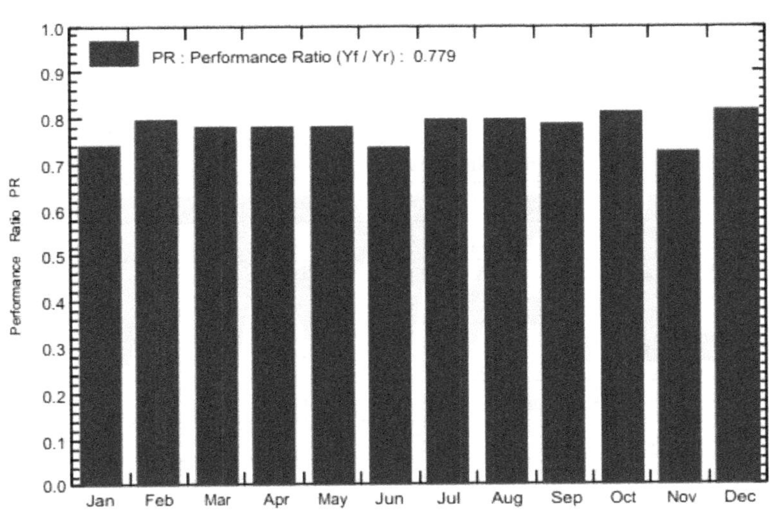

Table 10. Name plate details of solar PV module

S. No	Description	Rating
1	Manufacturer	Trina solar
2	Module power rating (W_p)	320
3	V_{mpp} (V)	33
4	I_{mpp} (A)	8.62

Table 11. Name plate details of string inverter

S. No	Description	Rating
1	Manufacturer	Delta
2	Model	RPI M10A
3	Unit nominal power rating (kW)	10
4	Operating voltage (V)	200 - 800

Table 12. System configuration

S. No	Description	Rating
1	Tilt	10°
2	Azimuth	0°
3	No of modules	32

Results

Table 13. PV array loss factor

S. No	Description	Rating
1	LID	2%
2	Module quality loss	0%
3	Module mismatch losses	1% at MPP
4	Wiring Ohmic loss	522 mΩ
5	Thermal loss factor	29 W/ m^2K
6	Array soiling losses	2%

Table 14. System loss factors

S. No	Description	Rating
1	Wiring Ohmic loss	1%@STC
2	Unavailability of the system	7.3 days, 3 periods 2%

Table 15. Summary of PVsyst simulation

S. No	Month	Global Horizontal Irradiance kW/m2	Global Tilted Irradiance kW/m2	Ambient Temperature °C	Energy produced by Array (MWh)	Energy exported to grid (MWh)	PR (%)
1	January	149.3	164	25.1	1.41	1.259	75.5
2	February	163.9	176.5	26.19	1.499	1.461	80.8
3	March	195.8	202.1	28.06	1.697	1.653	79.8
4	April	193.7	192.4	29.68	1.607	1.565	79.4
5	May	192.4	185.5	31.73	1.535	1.495	78.7
6	June	168.9	161.2	30.72	1.349	1.218	73.8
7	July	161.1	154.7	30.44	1.296	1.262	79.7
8	August	161.9	159.1	29.49	1.338	1.302	80
9	September	160.5	163	28.8	1.368	1.332	79.8
10	October	137.4	143	27.43	1.221	1.189	81.2
11	November	115.1	123.3	25.56	1.057	0.926	73.4
12	December	123.6	135.1	24.89	1.166	1.137	82.2

The summary of PVsyst simulation is listed in Table 15.

The month wise energy production is shown in figure 9.

The month wise Performance Ratio (PR) is shown in figure 10.

The energy production is 15.8 MWh/Year

Performance ratio is 78.72%

Figure 9. Month wise energy production

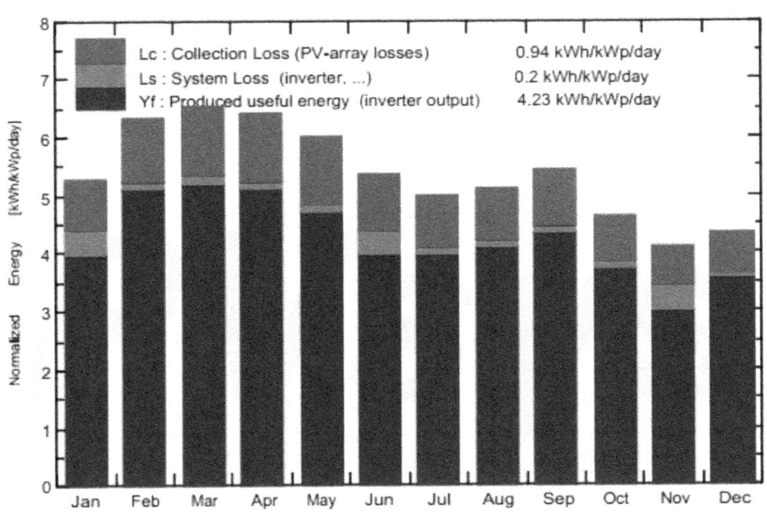

Figure 10. Month wise performance ratio

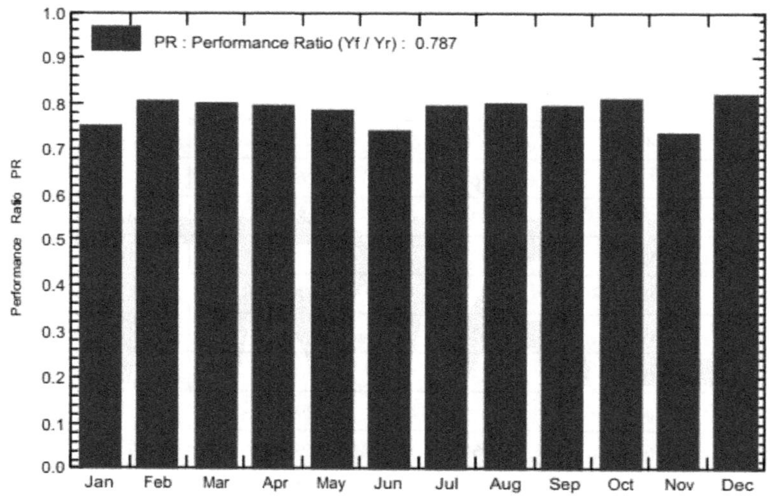

Comparison of Case 1 and Case 2

The comparison of energy production and performance ratio for micro inverter and string inverter is listed in Table 16.

From Table 16, the energy difference is 0.17 MWh/year and PR is 0.85%. It means, micro inverter is giving extra energy of 0.17MWh/year and PR is 0.85% as compared with string inverter.

Example 2: Comparative Analysis of 20 kW Micro Inverter and 20 kW String Inverter

Location: Chennai, Tamilnadu, India
Latitude: 13.16° N
Longitude: 79.96° E
Time zone: UT+5.5
Resource data: Meteonorm 7.2

Case 1: Analysis of 20 kW Micro Inverter

The solar PV module name plate details are listed in Table 17.

Table 16. Comparison of case 1 and case 2

S. No	Case	Type of inverter	Energy production (MWh/year)	PR (%)	Difference of energy (MWh/year)	Difference of PR (%)
1	Case 1	Micro Inverter	15.63	77.87	0.17	0.85
2	Case 2	String Inverter	15.8	78.72		

Table 17. Name plate details of solar PV module

S. No	Description	Rating
1	Manufacturer	Trina solar
2	Module power rating (W_p)	320
3	V_{mpp} (V)	33
4	I_{mpp} (A)	8.62

Table 18. Name plate details of micro inverter

S. No	Description	Rating
1	Manufacturer	Enphase
2	Model	IQ7X-96-x-INT
3	Unit nominal power rating (W)	315
4	Maximum power rating (W)	320
5	Operating voltage (V)	30 - 64

The no of modules used is 63, no of modules in series is 1 and no of modules in parallel is 63. Total module area is 124 m^2 and cell area is 110 m^2.

The solar PV micro inverter name plate details are listed in Table 18.

The system configuration is listed in Table 19.

The PV array loss factors considered for the analysis is listed in Table 20.

The system loss factors considered for the analysis is listed in Table 21:

Table 19. System configuration

S. No	Description	Rating
1	Tilt	10°
2	Azimuth	0°
3	No of modules	63

Table 20. PV array loss factor

S. No	Description	Rating
1	LID	2%
2	Module quality loss	0%
3	Module mismatch losses	1% at MPP
4	Wiring ohmic loss	0.69 mΩ
5	Thermal loss factor	29 W/m2K
6	Array soiling losses	2%

Table 21. System loss factors

S. No	Description	Rating
1	Wiring Ohmic loss	2%@STC
2	Unavailability of the system	7.3 days, 3 periods 2%

Table 22. Summary of PVsyst simulation

S. No	Month	Global Horizontal Irradiance kW/m2	Global Tilted Irradiance kW/m2	Ambient Temperature °C	Energy produced by Array (MWh)	Energy exported to grid (MWh)	PR (%)
1	January	149.3	164	25.1	2.757	2.437	73.7
2	February	163.9	176.5	26.19	2.919	2.816	79.1
3	March	195.8	202.1	28.06	3.272	3.157	77.5
4	April	193.7	192.4	29.68	3.125	3.016	77.8
5	May	192.4	185.5	31.73	3.004	2.901	77.6
6	June	168.9	161.2	30.72	2.659	2.381	73.2
7	July	161.1	154.7	30.44	2.551	2.465	79.1
8	August	161.9	159.1	29.49	2.628	2.538	79.1
9	September	160.5	163	28.8	2.658	2.566	78.1
10	October	137.4	143	27.43	2.402	2.321	80.5
11	November	115.1	123.3	25.56	2.068	1.796	72.3
12	December	123.6	135.1	24.89	2.294	2.216	81.4

Results

The summary of PVsyst simulation is listed in Table 22.

The month wise energy production is shown in Figure 11.

The month wise performance ratio is shown in Figure 12.

The energy production is 30.61 MWh/Year

Performance ratio is 77.47%

Case 2: Analysis of 20 kW String Inverter

The solar PV module name plate details are listed in Table 23.

The no of modules used is 64, no of modules in series is 16 and no of modules in parallel is 4. Total module area is 126 m^2 and cell area is 112 m^2.

The solar PV string inverter name plate details are listed in Table 24.

The system configuration is listed in Table 25.

The PV array loss factors considered for the analysis is listed in Table 26.

The system loss factors considered for the analysis is listed in Table 27.

Figure 11. Moth wise energy production

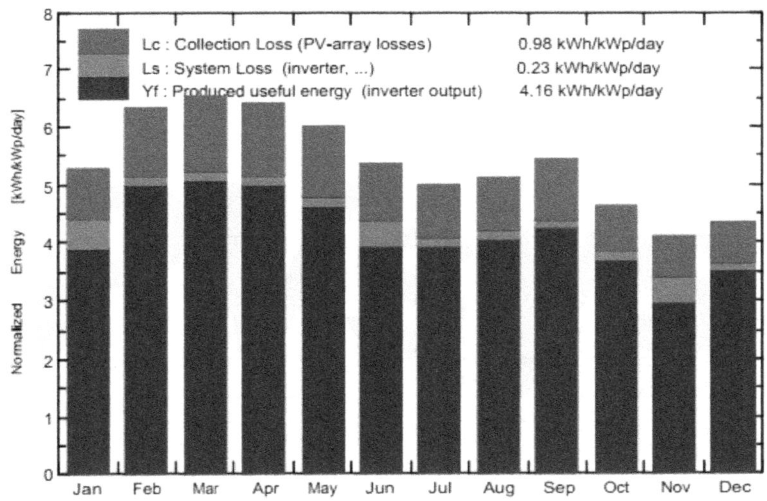

Results:

Figure 12. Moth wise performance ratio

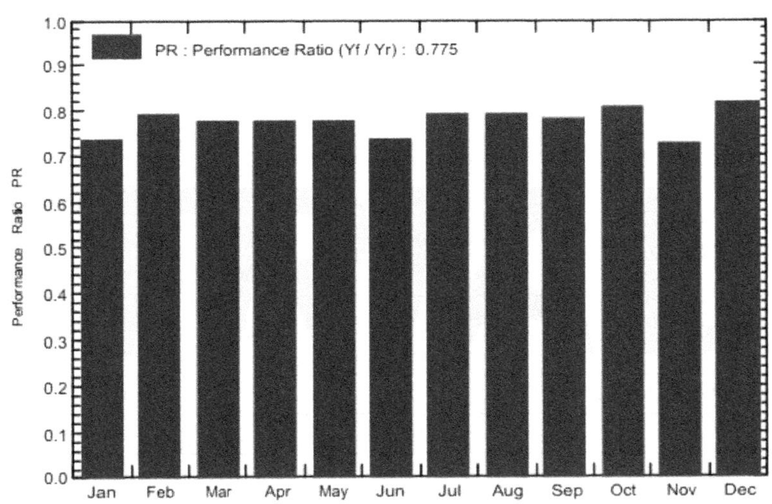

Table 23. Name plate details of solar PV module

S. No	Description	Rating
1	Manufacturer	Trina solar
2	Module power rating (W_p)	320
3	V_{mpp} (V)	33
4	I_{mpp} (A)	8.62

Table 24. Name plate details of string inverter

S. No	Description	Rating
1	Manufacturer	Delta
2	Model	RPI M20A
3	Unit nominal power rating (kW)	20
4	Maximum power (kW)	21
5	Operating voltage (V)	200 - 820

Table 25. System configuration

S. No	Description	Rating
1	Tilt	10°
2	Azimuth	0°
3	No of modules	64

The summary of PVsyst simulation is listed in Table 28.

The month wise energy production is shown in Figure 13.

The month wise performance ratio is shown in Figure 14.

The energy production is 31.36 MWh/Year

Performance ratio is 78.14%

Table 26. PV array loss factor

S. No	Description	Rating
1	LID	2%
2	Module quality loss	0%
3	Module mismatch losses	1% at MPP
4	Wiring Ohmic loss	261 mΩ
5	Thermal loss factor	29 W/m2K
6	Array soiling losses	2%

Table 27. System loss factors

S. No	Description	Rating
1	Wiring Ohmic loss	2%@STC
2	Unavailability of the system	7.3 days, 3 periods 2%

Table 28. Summary of PVsyst simulation

S. No	Month	Global Horizontal Irradiance kW/m2	Global Tilted Irradiance kW/m2	Ambient Temperarture °C	Energy produced by Array (MWh)	Energy exported to grid (MWh)	PR (%)
1	January	149.3	164	25.1	2.82	2.498	74.4
2	February	163.9	176.5	26.19	2.998	2.896	80.1
3	March	195.8	202.1	28.06	3.394	3.276	79.1
4	April	193.7	192.4	29.68	3.214	3.104	78.8
5	May	192.4	185.5	31.73	3.069	2.967	78.1
6	June	168.9	161.2	30.72	2.699	2.42	73.3
7	July	161.1	154.7	30.44	2.592	2.509	79.2
8	August	161.9	159.1	29.49	2.675	2.587	79.4
9	September	160.5	163	28.8	2.736	2.644	79.2
10	October	137.4	143	27.43	2.441	2.363	80.7
11	November	115.1	123.3	25.56	2.115	1.841	72.9
12	December	123.6	135.1	24.89	2.333	2.258	81.6

Comparison of Case 1 and Case 2

The comparison of energy production and PR for micro inverter and string inverter is listed in Table 29.

From Table 29, the energy difference is 0.75 MWh/year and PR is 0.67%. It means, micro inverter is giving extra energy of 0.75MWh/year and PR is 0.67% as compared with string inverter.

Figure 13. Month wise energy production

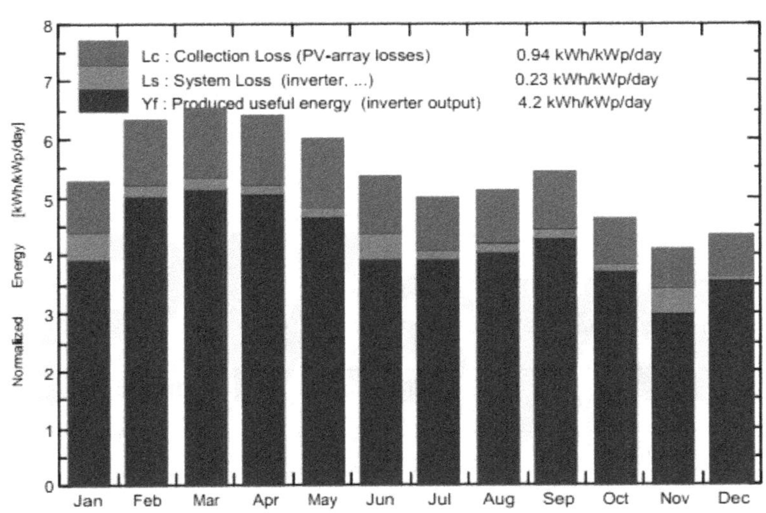

Figure 14. Month wise performance ratio

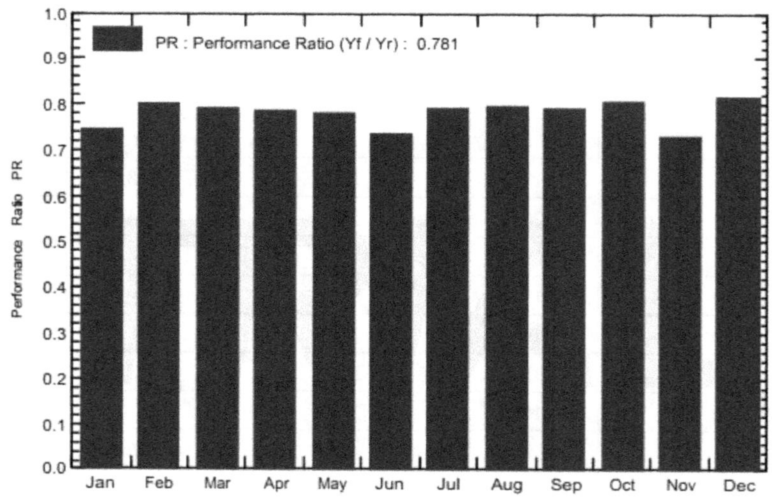

COMPARISON OF STRING INVERTER VS MICRO INVERTER

The micro inverter has more advantages over string inverter and comparison of string inverter vs micro inverters are listed in table 30 (Harb et al., 2013).

CONCLUSION

Micro inverters are DC to AC converters like string inverter and mounted below the solar PV module. These inverters are more suitable in areas where more shadings and multiple orientation of PV module and multiple tilt angles are required. Micro inverters enables the module independence and individual module parameters can be monitored. Presently micro inverters are commercially available up to 1200 W (AC) and 1520 W (DC) in the market.

Table 29. Comparison of case 1 and case 2

<table>
<tr><th>S. No</th><th>Case</th><th>Type of inverter</th><th>Energy production (MWh/year)</th><th>PR (%)</th><th>Difference of energy (MWh/year)</th><th>Difference of PR (%)</th></tr>
<tr><td>1</td><td>Case 1</td><td>Micro Inverter</td><td>30.61</td><td>77.47</td><td rowspan="2">0.75</td><td rowspan="2">0.67</td></tr>
<tr><td>2</td><td>Case 2</td><td>String Inverter</td><td>31.36</td><td>78.14</td></tr>
</table>

Table 30. Comparison of String inverter vs Micro inverter

S. No	Description of parameter	String inverter	Micro inverter
1	Module level monitoring	Not possible. Only string level monitoring is possible	Possible
2	String design sizing	Required	Not required
3	MPPT	Available	Available
4	Individual MPPT per PV module	Not possible	Possible
5	Module independence	Not possible	Possible
6	Any orientation of PV module	Not possible	Possible
7	Any tilt of PV module	Not possible	Possible
8	Efficiency	High compared with micro inverter	Low compared with string inverter
9	Capital cost	Low compared with micro inverter	High compared with string inverter
10	Shading losses	High compared with micro inverter	Low compared with string inverter
11	Module degradation	High compared with micro inverter	Low compared with string inverter
12	Soiling losses	High compared with micro inverter	Low compared with string inverter
13	Addition of module in future	Limited based on string design and sizing	Limited based on current carrying capacity of the cable and switch gear
14	LVRT/HVRT	Available	Available
15	Remote monitoring	Available	Available

REFERENCES

Harb, S., Kedia, M., Zhang, H., & Balog, R. S. (2013). Microinverter and string inverter grid connected photovoltaic system A comprehensive study, Photovoltaic Specialists Conference (PVSC), pp. 2885-2890, June 2013. 10.1109/PVSC.2013.6745072

Krishnaswami, H. (2011). Photovoltaic Microinverter using Single-stage Isolated High-frequency link Series Resonant Topology, IEEE Energy Conversion and Congress Exposition (ECCE), pp. 495-500, Sep. 2011.

Patel, J. A., & Solanki, P. D. (2014). Comparative Analysis of String Inverter and Micro Inverter for Solar Based Power System, *International Journal of Advanced Research in Electrical, Electronics and Instrumentation Engineering,* April.

Solanki, C. S. (2012). *Solar photovoltaics fundamentals, technologies and applications* (2nd ed.).

Chapter 11
LEDs for Solid-State Lighting:
State of the Art and Challenges

Muhammad Wasif Umar
Universiti Teknologi PETRONAS, Malaysia

NorZaihar Yahaya
Universiti Teknologi PETRONAS, Malaysia

ABSTRACT

Solid-state lighting technology is rapidly gaining acceptance in lighting industry street lighting, traffic lighting, decorative lighting, projection displays, display backlighting, automotive lighting, and so on. Differing from conventional light sources that use tungsten filament, plasma, or gases to generate light, solid-state lighting is based on organic or inorganic light emitting diodes (LEDs), and has the potential to generate light with almost 100% efficiency. LED luminaires have a long lifetime and are environmentally friendly with no toxic mercury contained. However, the success of these luminaires depends on system design, which comprises an understanding of several factors such as performance and control. In this chapter, we shall touch upon some technological advancements in the field of solid-state lighting technologies and challenges that limit their market penetration for consumer lighting.

INTRODUCTION

Light emitting diode (LED) based solid-state lighting (SSL) solutions have made tremendous progress over the last several years, with the potential to make much more over the coming years. The four major factors supporting their popularity are: energy efficiency, long lifetime, mercury free structure and design flexibility. Since the origin of first red LED in 1962, the technology has grown rapidly. In the 1970s and 1980s seven-segment LED displays were in widespread but falling off due to liquid crystal displays with their lower power consumption and greater display flexibility at the time, for a while it seemed like LEDs would not be a popular technology. Due to defects in crystal structure and poor substrate creation, the light outputs of LEDs were not anywhere near its potential. However, in the 1990s high-brightness

DOI: 10.4018/978-1-7998-0117-7.ch011

gallium nitride (GaN) LEDs were discovered and it was a short step to the 2000s, when bright white LEDs became a lighting revolution (Bender, Marchesan, & Alonso, 2015).

LED application areas include LCD backlights, displays, transportation equipment lighting and general lighting (see Table 1). LEDs are used as a light source for LCD backlights in products such as TVs, mobile phones, monitors and cameras (Ahn, Hong, & Kwon, 2016; Moon & Oh, 2015). Display applications include outdoor billboards, electronic scoreboards and signage lighting (e.g., LED strips and lighting bars). Examples of transportation equipment lighting areas are passenger vehicle/train lighting (e.g., meter backlights, tail and brake lights) and ship/airplane lighting (e.g., flight control and searchlights) (Long et al., 2015). General lighting applications are divided into indoor lighting (e.g., LED lighting bulbs, surface and desk lighting), outdoor lighting (e.g., street lighting, decorative lighting and flood lighting) and special lighting (e.g., appliances and elevator lighting) (Bender et al., 2015). LED-based light sources (lamps, modules) and luminaires for general lighting are rapidly gaining acceptance with a growing list of applications, such as, street lighting, commercial/business lighting and consumer applications. In this chapter, we will focus on some of recent technological developments in the field of SSL technologies for consumer lighting applications and challenges that limit their market penetration.

BARRIERS TO ADOPTION

In previous section, a few key historical advancements and profound energy, economic, performance and application benefits of the LED technology has been mentioned. However, in spite of the numerous advantages, LEDs have their set of challenges as well, which have been holding back their market growth to an extent. The following lists some of the technical and market barriers to LED technology. Overcoming these barriers is essential to their rapid market deployment.

- **Cost**
- **Thermal management**
- **Electrical Drives**
- **Global standardization**

Table 1. Application areas of LEDs

Applications	Examples
LCD backlighting	• TVs • Mobile phone • Monitors • Cameras
Displays	• Billboards • Electric scoreboards • Signage lighting
Transportation	• Vehicle/Train lighting • Ships/Airplane lighting
General lighting	• Indoor lighting • Outdoor lighting • Special lighting

Let us delve further into these challenges and explore how efforts are being made on various fronts to expedite the market penetration of LED-based SSL products.

COST

Cost is probably the most important factor limiting the widespread adoption of LED luminaires. Although the cost of LED bulbs is getting much cheaper, it is still considered relatively expensive compared to compact fluorescent lamp (see Table 2). For example in the United States, price of an LED lamp is currently about seven times higher than the price of a halogen bulb and double the cost of a dimmable compact fluorescent lamp (Gerke, Ngo, Alstone, & Fisseha, 2014; Nardelli, Deuschle, de Azevedo, Pessoa, & Ghisi, 2017).

However, there is still significant room for improvement in terms of price. The U.S. Energy Information Administration (eia) has released a report predicting the drastic price reduction of LED lamps in the near future (see Fig. 1).

Although this trend appears to be levelling off in 2025, improvements continue. As consumers begin to understand the full energy and performance benefits of LED based fixtures, a market for high efficiency, high fidelity products is expected to merge.

Table 2. Comparison of typical market prices for several light sources in the US

Light Source	Price (US$/klm)
Halogen Bulb (A 19, 20 lm/W)	$2.50
Compact Fluorescent lamp (A 19, 70 lm/W)	$2
Dimmable Compact Fluorescent lamp (A 19, 70 lm/W)	$7
Linear Fluorescent System (108 lm/W)	$4
High-Watt HID System (115 lm/W)	$3
Low-Watt HID System (104 lm/W)	$4
LED light bulb (A 19, 100 lm/W dimmable)	$14

Figure 1. Average lighting efficacy (light output per unit of energy consumed) and cost per bulb

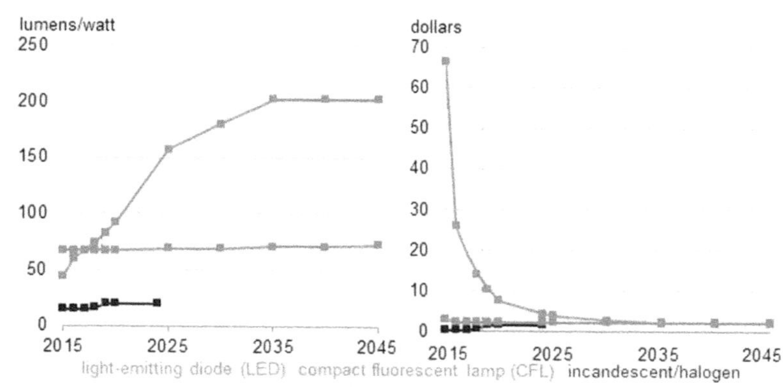

THERMAL MANAGEMENT

Heat and improper thermal management are also important failure accelerators in LED luminaires. From LED chip to related products such as LED bulbs, modules and fixtures, all are sensitive to temperature changes. Prolonged heat can cause a significant drop in performance or even premature failure of the whole system. Therefore, an efficient thermal management mechanism is an important factor in ensuring that LED systems deliver the expected level of performance during their lifetime (Lasance & Poppe, 2014). This section offers a brief account of prominent technologies currently available to manipulate the heat flow in LED luminaires.

Depending on their applications, thermal management techniques are broadly divided into:

- **Active solutions**
- **Passive solutions**
- **Thermal interface-based solutions**

Active Solutions

Active thermal management requires the inclusion of a sub-system that forces air out of the LED lighting system to keep it cool. This technology includes conventional electric motorized fans and diaphragm-based forced air cooling (synJet) (see Fig. 2). SynJet is a novel cooling device having an oscillating diaphragm to perform the periodic suction of hot air, which is then ejected out of an opening.

Figure 2. Active cooling systems for LED lighting
a) [2]Active cooling module b) [3] Downlight synJet cooler

a) active colling module

b) downlight synJet coller

The greatest advantage of using active cooling solutions is that they can result in remarkable size, weight and cost reductions in high power LED based products.

Passive Solutions

This technique involves metal heat-sinks to draw heat out from LEDs (see Fig. 3). A heat-sink is a system that transfers heat from any solid material to a fluid medium like air or water. Since the operation of passive cooling solutions is orientation-dependent, modules that use these tend to be bulky and heavy. Being noiseless, this cooling method does not suffer performance degradation in the long run. However, these systems cannot operate reliably in hot climates and enclosed fixtures.

Thermal Interface Material (TIM) Based Solutions

Thermal interfacial materials (TIMs) are thermally conductive materials which are applied to establish an effective thermal path between a heat source such as LED chip and cooling system such as heat sink (see Fig. 4). An effective TIM replaces the gaps created by the non-smoothing mating surfaces with a material whose thermal conductivity is much greater than that of air. Several terms refer to this category of TIM, including adhesives, pastes and rubber-like compounds (Raypah, Dheepan, Devarajan, & Sulaiman, 2016).

These materials are broadly divided into two parts, i.e. silicon-based and silicon-free. Silicon-based TIMs are widely used for outdoor LED fixtures installed in harsh climatic conditions. Silicon offers unmatchable thermal and optical stability. Whereas, Silicon-free TIMs eliminates the problem of silicon outgassing.

The basic concept detailed in this section shows that whatever thermal management system is chosen, it must be able to offer efficient heat transfer under the range of conditions the unit may be exposed to.

Figure 3. Metal heat-sinks for LED modules

Figure 4. Typical Thermal interfacial material (TIM) representation

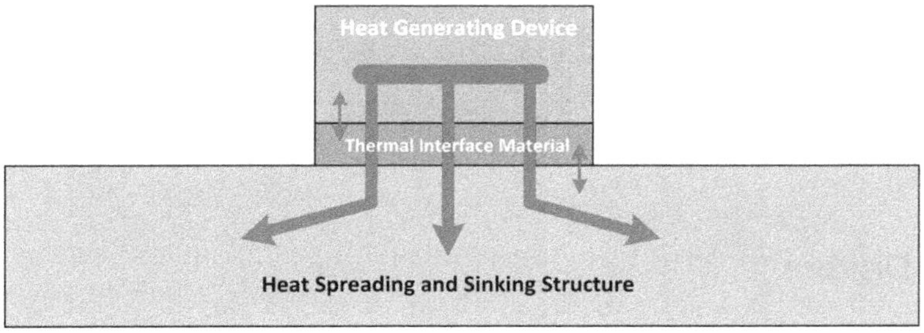

ELECTRICAL DRIVES

Operating an LED module always requires control gear (a ballast), which in the field of LEDs is called a driver (Chen & Chung, 2011; Li, Tan, Hui, & Tse, 2013). Divers are essentially a self-contained power supply that regulates power to an LED or a string of LEDs (Li et al., 2016). However unfortunately, LED drivers are the weakest links in the solid-state lighting reliability chain. A recent study showed that approximately 59% of the registered failures in LED luminaires were due to the power supply itself: 52% were power circuitry failures and 7% were associated with electronic control circuitry (Pedro Santos Almeida, Camponogara, Dalla Costa, Braga, & Alonso, 2015). This is mostly because the driver sometimes contains internal components having incompatible life span with that of the LEDs being driven, which later leads to premature failure of the whole fixture. Hence, driver topology, functionality and cost dictates the exact performance of an LED luminaire.

Basically, LED drivers can be classified into two major groups:

- **Single-stage drivers**
- **Multiple-stage drivers**

Single-Stage Drivers

Single-stage drivers are composed of a PFC converter where the bus voltage is applied directly to the LEDs (see Fig. 5). The single-stage configuration is the best choice when space is a critical factor. Because of its inherent reduced component count, its circuit can easily fit, e.g., in LED replacement bulbs (B. Wang, Ruan, Yao, & Xu, 2010).

However, in general, the LED driver must comply with many requirements, such as output current control, PFC, reduced output ripple, long life and high efficiency; a single converter cannot be optimized to perform all these tasks.

Figure 5. A driver with one stage of energy processing

Figure 6. A driver with independent stages

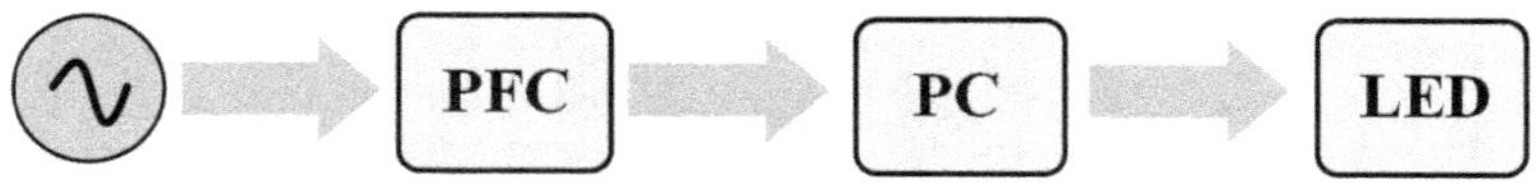

Multiple-Stage Drivers

As discussed earlier, it is necessary for the LED driver to meet various requirements. To include all those functions in a single-stage converter is sometimes a challenging task. Because of that, a possible solution that has been extensively presented in literature is the use of multiple converters to process the energy from the grid to the fixture.

The simplest solution to obtain a multiple-stage driver is to use independent stages, i.e. a PFC stage followed by a power-control (PC) stage (see Fig. 6).

This solution can easily overcome the task of reducing the bus capacitance. However, the efficiency is reduced because of energy reprocessing. Besides that, the cost is usually high due to the increase in the number of active switches and gate drivers (Pedro S Almeida, Dalla Costa, Alonso, & Braga, 2013; Ma et al., 2012).

To overcome the problem of high cost of multiple independent-stage drivers, another possible solution is to integrate the PFC and PC (see Fig. 7).

The main reason of this integration is to reduce the number of controllable switches, which implies both cost reduction and simplification of control gate driving circuitry (Alonso et al., 2013; Sichirollo, Alonso, & Spiazzi, 2015).

The main problem with integrated drivers is their low efficiency caused by reprocessing of the power. So, one of the logical solutions to solve the efficiency problem is to reduce the total amount of the power being reprocessed (Hu & Zane, 2011; S. Wang et al., 2012)(see Fig. 8).

Although single-stage topologies are more efficient, compact and easier to implement but multiple-stage in general are very effective in reducing overall capacitance. Table 3 presents a qualitative comparison of all the topological solutions discussed in this section.

Figure 7. A driver with integrated stages

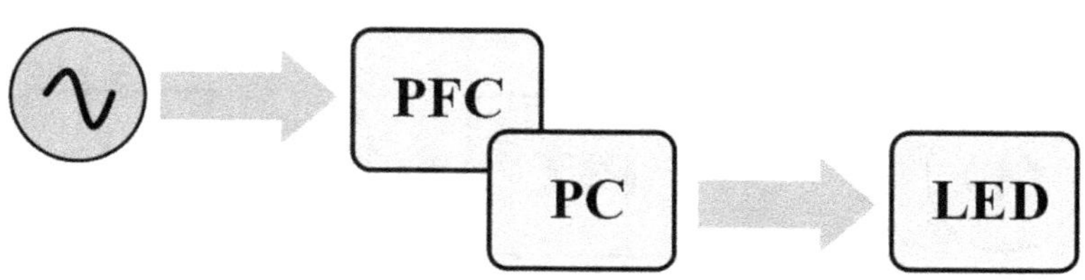

Figure 8. A driver with a reduction in power processing

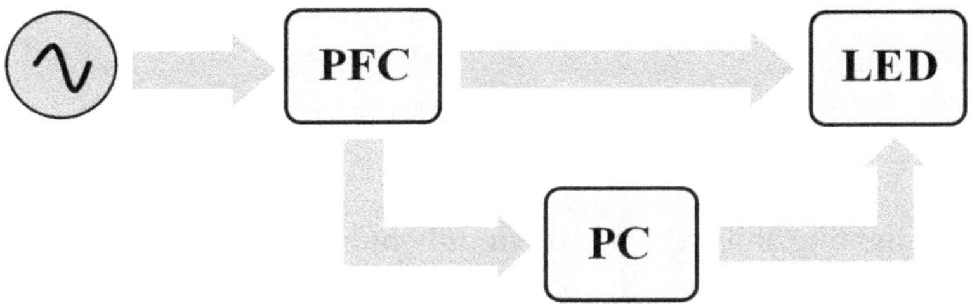

GLOBAL STANDARDS

Standards are an essential feature of an emerging technology such as solid-state lighting, helping to establish a framework for product introduction and effective market penetration. Several regulatory agencies around the world have been very busy over the last few years in the field of SSL, with a number of new standards introduced and many more in development. Currently, regulatory and standards organizations that are impacting SSL products include:

- American National Standards Institute (ANSI)/ National Electrical Manufacturers association (NEMA) efforts include industry norms, guidelines and standards for SSL products.
- Illuminating Engineering Society of North America (IESNA) is the recognised technical authority on illumination and currently publishes LM-79 and LM-80 SSL standards. IESNA is also developing a specification (EN-21) for long term reliability of the fixtures.
- Electrotechnical Commission (IEC), also known as IEC62861, creates so called guide to principal component reliability testing for LED light sources and luminaires.
- International Commission on Illumination (CIE) has several standards for lighting performance including SSL.
- Federal Communication commission (FCC) is involved with radio frequency aspects of SSL.
- Federal Trade Commission (FTC) is pursuing more accurate labelling for light bulbs.
- Department of Energy (DOE) CALiPER program tests and reports on available SSL products.
- Environmental Protection Agency (EPA) has the ENERGY STAR qualified LED lighting.

This is actually a rather short summary of what is occurring in the area of SSL regulations and standards. The pace at which the process for developing SSL standards is going, clearly shows that in a few

Table 2. Comparison of LED driver configurations

Topology	Cost/Component count	Conversion efficiency	Control complexity	Design flexibility
Single-stage	Low	High to medium	Low	Little
Multiple-stage independent	High	Medium to low	Medium	Good
Multiple-stage independent	Medium	Low	Low	Little
Multiple-stage with reduced power processing	High	High	High	Good

years, the list will be completely different. Moreover, complying with all of them, especially in a coordinated, harmonized manner could be a daunting task and provide a considerable delay to market entry.

FINAL REMARKS

LED based solid-state lighting shifts the paradigm in lighting technology. Though, we are still in the midst of the early phase of the solid-state lighting revolution, where lighting performance is on par with conventional products and there are substantial benefits in terms of efficacy, lifetime and cost of ownership, but products are still expensive in terms of initial cost. Current adoption trends indicate that these barriers will not be fundamental limitations to solid-state lighting technology, but rather are a normal disruption of a large, entrenched market. Researchers believe that with continuous advances in efficiency and reductions cost, the significance of these barriers will be further reduced. So, we can safely assume that the future of solid-state lighting is certainly bright and efficient.

REFERENCES

Ahn, H.-A., Hong, S.-K., & Kwon, O.-K. (2016). A fast switching current regulator using slewing time reduction method for high dimming ratio of LED backlight drivers. *IEEE Transactions on Circuits and Systems. II, Express Briefs*, *63*(11), 1014–1018. doi:10.1109/TCSII.2016.2548158

Almeida, P. S., Camponogara, D., Dalla Costa, M., Braga, H., & Alonso, J. M. (2015). Matching LED and driver life spans: A review of different techniques. *IEEE Industrial Electronics Magazine*, *9*(2), 36–47. doi:10.1109/MIE.2014.2352861

Almeida, P. S., Dalla Costa, M. A., Alonso, J., & Braga, H. A. C. (2013). Application of series resonant converters to reduce ripple transmission to LED arrays in offline drivers. *Electronics Letters*, *49*(6), 414–415. doi:10.1049/el.2012.4412

Alonso, J. M., Calleja, A. J., Gacio, D., Cardesín, J., Lopez, E., Dalla Costa, M. A., ... Do Prado, R. N. (2013). High-power-factor light-emitting diode lamp power supply without electrolytic capacitors for high-pressure-sodium lamp retrofit applications. *IET Power Electronics*, *6*(8), 1502–1515. doi:10.1049/iet-pel.2012.0142

Bender, V. C., Marchesan, T. B., & Alonso, J. M. (2015). Solid-state lighting: A concise review of the state of the art on LED and OLED modeling. *IEEE Industrial Electronics Magazine*, *9*(2), 6–16. doi:10.1109/MIE.2014.2360324

Chen, N., & Chung, H. S.-H. (2011). A driving technology for retrofit LED lamp for fluorescent lighting fixtures with electronic ballasts. *IEEE Transactions on Power Electronics*, *26*(2), 588–601. doi:10.1109/TPEL.2010.2066579

Gerke, B. F., Ngo, A. T., Alstone, A. L., & Fisseha, K. S. (2014). *The evolving price of household LED lamps: Recent trends and historical comparisons for the US market: Lawrence Berkeley National Lab.* Berkeley, CA: LBNL.

Hu, Q., & Zane, R. (2011). Minimizing required energy storage in off-line LED drivers based on series-input converter modules. *IEEE Transactions on Power Electronics*, *26*(10), 2887–2895. doi:10.1109/TPEL.2010.2087772

Lasance, C. J., & Poppe, A. (2014). Thermal management for LED applications. New York, NY: Springer. doi:10.1007/978-1-4614-5091-7

Li, S., Tan, S.-C., Hui, S., & Tse, C. (2013). *A review and classification of LED ballasts*. Paper presented at the 2013 IEEE Energy Conversion Congress and Exposition. 10.1109/ECCE.2013.6647106

Li, S., Tan, S.-C., Lee, C. K., Waffenschmidt, E., Hui, S. R., & Chi, K. T. (2016). A survey, classification, and critical review of light-emitting diode drivers. *IEEE Transactions on Power Electronics*, *31*(2), 1503–1516. doi:10.1109/TPEL.2015.2417563

Long, X., He, J., Zhou, J., Fang, L., Zhou, X., Ren, F., & Xu, T. (2015). A review on light-emitting diode based automotive headlamps. *Renewable & Sustainable Energy Reviews*, *41*, 29–41. doi:10.1016/j.rser.2014.08.028

Ma, H., Lai, J.-S., Feng, Q., Yu, W., Zheng, C., & Zhao, Z. (2012). A novel valley-fill SEPIC-derived power supply without electrolytic capacitor for LED lighting application. *IEEE Transactions on Power Electronics*, *27*(6), 3057–3071. doi:10.1109/TPEL.2011.2174446

Moon, J., & Oh, K. (2015). Effects of light-emitting diode (LED) configuration on luminance and color of an edge-lit backlight unit. *Journal of Display Technology*, *11*(9), 768–775. doi:10.1109/JDT.2015.2443374

Nardelli, A., Deuschle, E., de Azevedo, L. D., Pessoa, J. L. N., & Ghisi, E. (2017). Assessment of Light Emitting Diodes technology for general lighting: A critical review. *Renewable & Sustainable Energy Reviews*, *75*, 368–379. doi:10.1016/j.rser.2016.11.002

Raypah, M. E., Dheepan, M., Devarajan, M., & Sulaiman, F. (2016). Investigation on thermal characterization of low power SMD LED mounted on different substrate packages. *Applied Thermal Engineering*, *101*, 19–29. doi:10.1016/j.applthermaleng.2016.02.092

Sichirollo, F., Alonso, J. M., & Spiazzi, G. (2015). A novel double integrated buck offline power supply for solid-state lighting applications. *IEEE Transactions on Industry Applications*, *51*(2), 1268–1276. doi:10.1109/TIA.2014.2350071

Wang, B., Ruan, X., Yao, K., & Xu, M. (2010). A method of reducing the peak-to-average ratio of LED current for electrolytic capacitor-less AC–DC drivers. *IEEE Transactions on Power Electronics*, *25*(3), 592–601. doi:10.1109/TPEL.2009.2031319

Wang, S., Ruan, X., Yao, K., Tan, S.-C., Yang, Y., & Ye, Z. (2012). A flicker-free electrolytic capacitor-less AC–DC LED driver. *IEEE Transactions on Power Electronics*, *27*(11), 4540–4548. doi:10.1109/TPEL.2011.2180026

ENDNOTES

1 www.eia.gov
2 www.sunon.com
3 www.boydcorp.com
4 www.fischerelektronik.de

Chapter 12
Modelling of Lamb Waves Propagation in Orthotropic Plate

Salah Nissabouri
https://orcid.org/0000-0001-7702-3030
FST Settat, Morocco

Mhammed El Allami
CRMEF Settat, Morocco & FST Settat, Morocco

El Hassan Boutyour
FST Settat, Morocco

ABSTRACT

In this chapter, we model by Finite Element Method (FEM) the Lamb waves' propagation and their interactions with symmetric and asymmetric delamination in sandwich skin. Firstly, a theoretical model is established to obtain the equation of lamb modes propagation. Secondly, dispersion curves are plotted using Matlab program for the laminate [0]4. The simulations were then carried out using ABAQUS CAE by exciting the fundamental A0 Lamb mode in the frequency 300 kHz. The delamination was then estimated by analyzing the signal picked up at two sensors using two techniques: Two Dimensional Fast Fourier Transform (2D-FFT) to identify the propagating and converted modes, and Wavelet Transform (WT) to measure the arrival times. The results showed that the mode A0 is sensible to symmetric and asymmetric delamination. Besides, based on signal changes with the delamination edges, a localization method is proposed to estimate the position and the length of the delamination. In the last section, an experimental FEM verification is provided to validate the proposed method.

INTRODUCTION

The detection of defects in isotropic and composite structures is becoming a priority, particularly in the aeronautical industries. Indeed, most parts are subjected to different constraints during their manufacturing process and life cycle. These mechanical or thermal constraints cause internal or surface defects. So,

DOI: 10.4018/978-1-7998-0117-7.ch012

reliable methods must then be found to diagnose the state of pieces health. NDT (Non Destructive Testing) techniques are used to assess the health of the part without destruction. Some NDT methods use: x-rays, laser holography, or microscopy to locate defects. These methods are expensive and require specific conditions of use. For that reason, Lamb waves are an attractive tool used to control long distances such as pipes. However, they are dispersive, which means that they change frequency by propagating through the structure. So, before performing the control, the authors have to choose an adequate frequency based on dispersion curves. Moreover, Multilayer structures, or multilayer composites, may contain critical internal defects such as: delamination facilitating rapid alteration and leading to further problems.

Nondestructive Testing of Composite Material

A composite material is an assembly of at least two materials that has different properties. The first which is called fiber that ensures the mechanical strength. The second named matrix consisting of plastic, metal or ceramic material. The figure 1 shows the component of unidirectional composite.

Composites are designed to improve mechanical proprieties of structures. They are widely used in different fields especially in aeronautic industry. Their characteristics are influenced by the proportions of the matrix and the reinforcements. There are other parameters that also affect the properties of a composite like: size, orientation and distribution of the fiber. The heterogeneity of composites structures lead to their weakness and facilitates the appearance of internal and external damages such as: fiber breakage, matrix cracking, through-thickness hole, local delamination.

Composite structures require careful monitoring and inspection to identify damage and take corrective action to ensure safe and continuous operation. In metal structures, the control mainly concerns corrosion, tensile stress cracking, and stress corrosion. In composite structures, control is also concerned with delamination growth, compression fatigue, manufacturing defects, fiber degradation and failure. These types of damage usually occur below the surface and cannot be easily detected.

One of composite structures is Sandwich materiel made from two thin skins bonded to a thick core (see figure 2). The skin studied in this paper is an orthotropic plate [0]4 with three mutually perpendicular planes of symmetry.

Types of Defects in Composite Structures

Composite structures are affected by several potential damages as described in figure 3 such as: delamination, fiber breakage, matrix cracking, crushing of the core (in the case of a honeycomb material).

Figure 1. Unidirectional composite

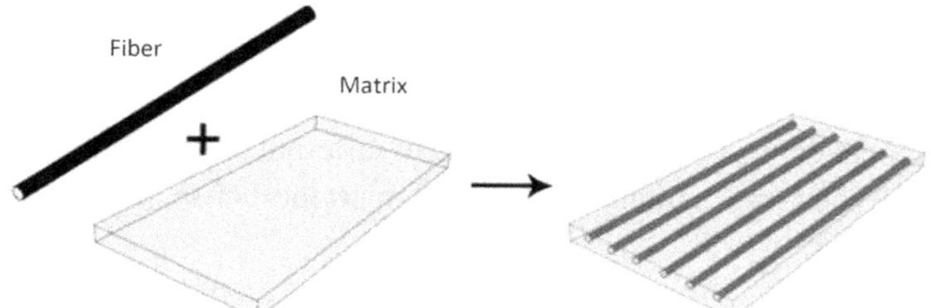

Figure 2. Sandwich structure

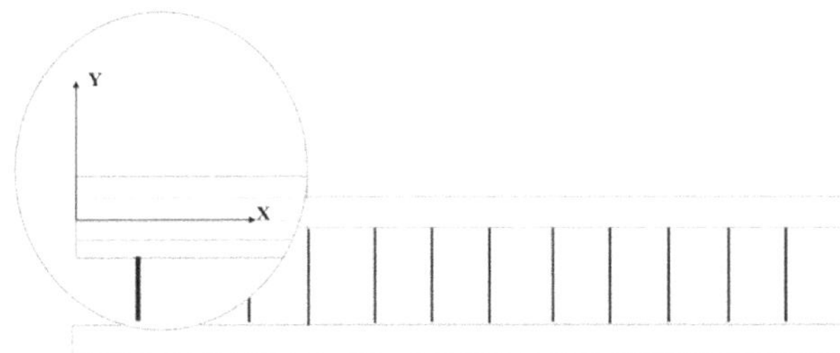

Delamination

Delamination is the separation of two adjacent layers; it is one of the most common defects. This type of damage is due to the heterogeneous properties of the matrix and the fibers. The thermal stresses may result in residual / inter-laminar stresses that may be sufficient to cause delamination due to the gap between the properties of two adjacent layers, see Vladimir (1994).

Damaged Fibers

Since the fibers withstand the majority of tensile shear forces applied in the plane, damage to the fibers can have adverse effects on the overall strength of the composite structure. Typical damages associated with fibers are: torsion and fracture.

Cracking of the Matrix

Matrix cracking occurs in the transverse directions and can respectively perpendicular and parallel to the fibers. Matrix cracking is generally the first damage to occur when a composite laminate is subjected

Figure 3. Defects types in composite structures see Shashank (2014)

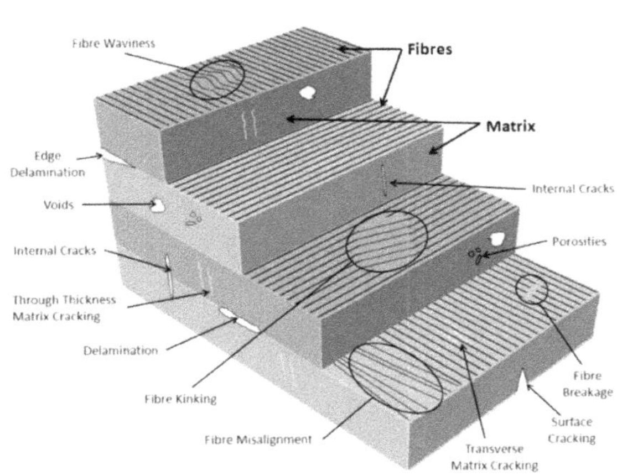

to quasi-static / cyclic tensile loading, see Gayathri (2010). Cracks across the thickness are due to the significant decrease in strength / stiffness of the matrix compared to reinforcement. This type of damage usually occurs during service due to traction, fatigue and impact load and is affected by the polymer matrix, see Hu (2007). Although it is one of the most common forms of damage in composites, cracking of the matrix itself, does not cause structural failure; however, it can initiate other critical failure modes such as delamination and fiber breakage.

Lamb Waves Control Technics

Lamb waves are most used for many reasons, they can propagate long distances along plates and shells so they permit quick inspection of large structures, and also, they are sensitive for the small variations either in material proprieties or in structure of the plate. However, they are dispersive which means that the interpretation of received signals can be complicated. So the key is to choose one pure mode to excite and to analyze its reflection, conversion and transmission.

At low frequency, two modes, A0 and S0, can propagate. As the frequency increases, more modes are possible to propagate. Among the contact technics to generate and to receive the signal, the authors find two methods: pitch catch and pulse echo. These two technics are simple but they require the coupling medium which limits the transmission of acoustic power. In the pulse-echo method, the reflected signal permits to evaluate the damage. In the pitch catch technic the transducer placed in the transmission side permits to evaluate the defect. In this paper two methods have been investigated to evaluate the delamination. Besides, the authors are considering the study of first anti-symmetric mode A0 (figure 4).

Post Process Tools

2D-FFT Analysis

Lamb waves' propagation is sinusoidal in the frequency and spatial domains. For that reason, a temporal Fourier Transform is applied to go from the time to frequency domain, after that, a spatial Fourier Transform is computed to obtain the frequency wave-number domain, see Cawely (1991). In practice, carrying out spatial Fourier methods to data obtained experimentally or numerically requires applying a 2D-FFT, using:

Figure 4. Lamb modes types: (a) symmetrical mode, (b) mode anti-symmetrical mode

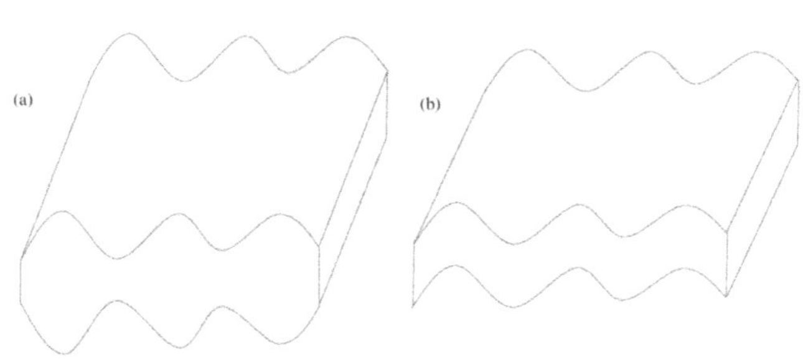

$$H(k,f) = \int_{-\infty}^{+\infty}\int u(x,t)e^{-i(kx+\omega t)}dxdt \tag{1}$$

Wavelet Analysis

Wavelet transform is an important tool in the time frequency domain of transient signals. The continuous wavelet transform of signal u(t) is defined as follows:

$$cwt(a,b) = \frac{1}{\sqrt{|a|}}\int_{-\infty}^{+\infty} u(t)\psi^*\left(\frac{t-b}{a}\right)dt \tag{2}$$

The obtained signal is a function of a and b, the translation and scale parameters, respectively. The parameter psi (t) is the transforming function named the mother wavelet. In this paper the authors use wavelet transform by 'Gaus1' to locate the peak which permits to determine the arrival time of the wave at specific frequency. Moreover, for each frequency f, the localization of maximal value of the wavelet coefficients cwt (a,b) allows identifying the arrival times t1 and t2. Knowing the group velocity and the delay (delta) Δt =t2-t1 between the two modes with the same nature, the authors can calculate the distant by the following equation:

$$\Delta x = V_g \Delta t \tag{3}$$

Knowing the delay between two different modes (symmetric and anti-symmetric) the authors calculate the distant by the following equation:

$$\Delta x = \Delta V_g \Delta t \tag{4}$$

With: $\Delta V_g = \frac{V_S V_A}{V_S - V_A}$.

Where: VS and VA are the group velocities respectively of S0 and A0 modes. Arrival time method was used by many authors to localize the delamination, see Ip (2004).

Previous Work

The damage causes wave scattering, mode conversion, and multiple reflections. To understand these mechanisms, theoretical, numerical and experimental studies are conducted.

Feng, Ribeiro, and Ramos (2017) analyzed the interaction of symmetric S0 and anti-symmetric A0 mode with the delamination using finite element simulations. The Lamb waves propagation in a 4-layer [0/90]s laminate were compared with propagations obtained in 1-layer [0] and in 3-layer [90/90/0] sub-laminates. Chiua, Roseb and Nadarajaha (2017) investigated the scattering of the S0 mode by a delamination in quasi-isotropic fibre-composite laminate. Guo and Cawley (1993) studied by finite element analysis and by experiment the interaction of the S0 Lamb mode with delamination. Nadarajah, Vien and Chiu (2015) presented results for the scatter field for various angles of incidence, and for varying defect sizes. Hayashi and Kawashima (2002) studied the reflections of Lamb waves at a delamination by semi-analytical finite element method. Ching-Tai and Veidta (2012) investigated the scattering characteristics of A0 mode Lamb wave at a delamination in a quasi-isotropic composite laminate. Bin (2017) suggested an algorithm to localize and identify the damage in Woven Glass Fiber reinforced epoxy (WGF/epoxy). Mustapha (2015) characterized fundamental symmetric and anti-symmetric Lamb modes in terms of their velocity and magnitude variation as they change gradually in the thickness of a composite sandwich plate with a high density foam core. Ng (2012) presented a theoretical and Finite Element (FE) investigation of the scattering characteristics of A0 at delaminations in a quasi-isotropic composite laminate. Veidt and Ng (2011) studied the influence of stacking sequence on fundamental anti-symmetric Lamb wave (A0) scattering characteristics through holes in composite laminates. Luca (2016) developed a valid Finite Element Model to simulate Lamb waves' propagation in a Carbon Fiber Reinforced Plastic (CFRP) laminate for damage detection purpose and investigated the effects of the wave interaction with respect to damage parameters such as size and orientation. Yang (2006) investigated some aspects of numerical simulation of excitation and detection of Lamb waves using piezoelectric disks in plate like composite laminates. Hanfei (2019) presents a new methodology for detecting and quantifying delamination in composite plates based on the high-frequency local vibration under the excitation of piezoelectric wafer active sensors. Michalcová and Hron (2018) deal with delamination length determination in double cantilever beam specimens using Lamb waves based on the wave velocity determination method.

Objectives

In this chapter we study the lamb wave's propagation in orthotropic plate. In the first part, we establish the mathematical equation describing the lamb wave's propagation along the main axis in single-layered orthotropic plate. The second part of this chapter is the numerical model where we model the lamb wave propagation using FEM based on ABAQUS. The application of 2D-FFT over displacements extracted from the numerical model permits to identify the propagating and the converted modes. In the third part of this chapter we propose a method of delamination localization. The application of Wavelet Transform allows calculating the arrival times of propagating modes. The delamination position and length are estimated on two sensors: before and after the defect. The results are different at the two sensors.

THEORETICAL MODEL

Establishment of Propagation Equation

Theoretical model of wave propagation has been presented by several authors Harsh (2012), Ameneh (2014), and Mark (2001). The study of the propagation of Lamb waves in anisotropic structures is complex and authors such as Nayfeh and Rose have greatly contributed Nayfeh (1989).

The plate is assumed to be infinite along the axes x1 and x2, the x3 axis being normal to the plate (Figure 5). In this configuration, the waves propagate along the axis x1. One can however reduce to this case by performing a rotation of axis x3 of the stiffness matrix. The waves propagating in the x1

Figure 5. Plate Model

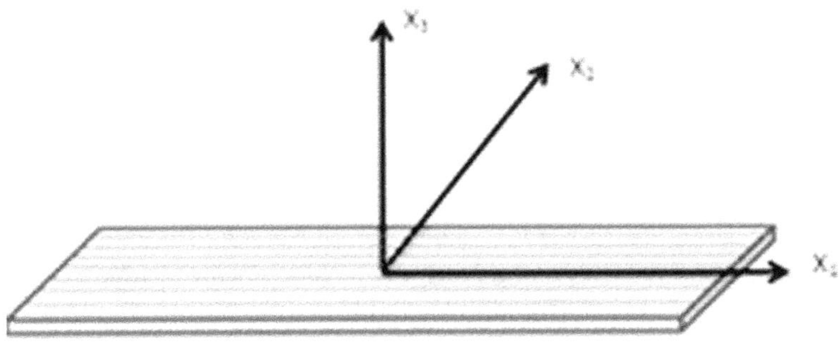

direction and the plate being infinite along x2, the components of the displacement vector are therefore independent of Coordinates along the x2 axis.

To establish the propagation equation of lamb waves the authors applied the fundamental principle of dynamics besides the authors neglect the forces of inertia and gravity.

$$p\frac{\partial^2 u_i}{\partial t^2} = \frac{\partial T_{ij}}{\partial x_j} \tag{5}$$

With ρ is the density of the material. Taking into account the Hooke's law:

$$T_{ij} = C_{ijkl}\overrightarrow{\frac{\partial u_l}{\partial x_k}} \tag{6}$$

With Cijkl is the stiffness tensor written in the form:

$$C = \begin{bmatrix} C_{11} & C_{12} & C_{13} & C_{14} & C_{15} & C_{16} \\ C_{12} & C_{22} & C_{23} & C_{24} & C_{25} & C_{26} \\ C_{13} & C_{23} & C_{33} & C_{34} & C_{35} & C_{36} \\ C_{14} & C_{24} & C_{34} & C_{44} & C_{45} & C_{46} \\ C_{15} & C_{25} & C_{35} & C_{45} & C_{55} & C_{56} \\ C_{16} & C_{26} & C_{36} & C_{46} & C_{56} & C_{66} \end{bmatrix}$$

The propagation relation (5) then takes the form:

$$\rho\frac{\partial^2 u_i}{\partial t^2} = C_{ijkl}\frac{\partial^2 u_l}{\partial x_j \partial x_k} \tag{7}$$

Wave Propagation in a Single-Layered Orthotropic Plate

The displacement vector can be written as:

$$u_l = U_l e^{jk(x_1+\alpha x_3-ct)} = U_l e^{j(kx_1+qx_3-\omega t)} \quad (8)$$

With k is the wave number along the axis x1, c = ω/k the phase velocity and α = q / k coefficient to be determined knowing that the unknown q is the component of the wave number along the axis x3.

The propagation equation could be written under the form:

$$[K_{mn}][U_n]=0 \quad (9)$$

Where K is a symmetric matrix:

$$K=\begin{bmatrix} C_{11}-\rho c^2+2C_{15}\alpha+C_{55}\alpha^2 & C_{16}+(C_{14}+C_{56})\alpha+C_{45}\alpha^2 & C_{15}+(C_{13}+C_{55})\alpha+C_{35}\alpha^2 \\ C_{16}+(C_{14}+C_{56})\alpha+C_{45}\alpha^2 & C_{16}-\rho c^2+2C_{46}\alpha+C_{44}\alpha^2 & C_{15}+(C_{13}+C_{55})\alpha+C_{35}\alpha^2 \\ C_{15}+(C_{13}+C_{55})\alpha+C_{35}\alpha^2 & C_{15}+(C_{13}+C_{55})\alpha+C_{35}\alpha^2 & C_{15}+(C_{13}+C_{55})\alpha+C_{35}\alpha^2 \end{bmatrix}$$

The system thus defined admits a non-trivial solution if the determinant of the matrix K is null. It is then possible to solve the characteristic equation which relies α with c. Thus, for each value of c, this equation admits 6 roots noted. For each root, the system of equations permits to obtain the following ratios:

$$V_r=\frac{U_{2r}}{U_{1r}}=\frac{K_{11}(\alpha_r)K_{23}(\alpha_r)-K_{13}(\alpha_r)K_{12}(\alpha_r)}{K_{13}(\alpha_r)K_{22}(\alpha_r)-K_{12}(\alpha_r)K_{23}(\alpha_r)},\quad r\in\{1,2,3,4,5,6\} \quad (10)$$

$$W_r=\frac{U_{3r}}{U_{1r}}=\frac{K_{11}(\alpha_r)K_{23}(\alpha_r)-K_{12}(\alpha_r)K_{13}(\alpha_r)}{K_{12}(\alpha_r)K_{33}(\alpha_r)-K_{23}(\alpha_r)K_{13}(\alpha_r)},\quad r\in\{1,2,3,4,5,6\}$$

According to Hooke's law and superposition theorem, it is possible to obtain the values of the displacements and the constraints according to these ratios:

$$(u_1,u_1,u_1)=\sum_{r=1}^{6}(1,V_r,W_r)U_{1r}e^{jk(x_1+\alpha_r x_3-ct)}$$

$$(T_{13},T_{23},T_{33})=\sum_{r=1}^{6}jk(D_{1r},D_{2r},D_{3r})U_{1r}e^{jk(x_1+\alpha_r x_3-ct)}$$

With:

$$\begin{bmatrix} D_{1r} \\ D_{2r} \\ D_{3r} \end{bmatrix} = \begin{bmatrix} C_{13} + \alpha_r C_{35} + (C_{36} + \alpha_r C_{34}) V_r + (C_{35} + \alpha_r C_{33}) W_r \\ C_{15} + \alpha_r C_{55} + (C_{56} + \alpha_r C_{45}) V_r + (C_{55} + \alpha_r C_{35}) W_r \\ C_{14} + \alpha_r C_{45} + (C_{46} + \alpha_r C_{44}) V_r + (C_{45} + \alpha_r C_{34}) W_r \end{bmatrix} \tag{11}$$

By writing the boundary conditions at x3 = ± h / 2 the authors obtain two systems that can be grouped in a single matrix form [D] [U] = 0 such as:

$$\begin{bmatrix} D_{11}E_1 & D_{12}E_2 & D_{13}E_3 & D_{14}E_4 & D_{15}E_5 & D_{16}E_6 \\ D_{21}E_1 & D_{22}E_2 & D_{23}E_3 & D_{24}E_4 & D_{25}E_5 & D_{26}E_6 \\ D_{31}E_1 & D_{32}E_2 & D_{33}E_3 & D_{34}E_4 & D_{35}E_5 & D_{36}E_6 \\ D_{11}E_1^* & D_{12}E_2^* & D_{13}E_3^* & D_{14}E_4^* & D_{15}E_5^* & D_{16}E_6^* \\ D_{21}E_1^* & D_{22}E_2^* & D_{23}E_3^* & D_{24}E_4^* & D_{25}E_5^* & D_{26}E_6^* \\ D_{31}E_1^* & D_{32}E_2^* & D_{33}E_3^* & D_{34}E_4^* & D_{35}E_5^* & D_{36}E_6^* \end{bmatrix} \begin{bmatrix} U_{11} \\ U_{12} \\ U_{13} \\ U_{14} \\ U_{15} \\ U_{16} \end{bmatrix} = \begin{bmatrix} 0 \\ 0 \\ 0 \\ 0 \\ 0 \\ 0 \end{bmatrix} \tag{12}$$

With:

$$E_r = e^{j\frac{k\alpha_r h}{2}} \text{ and } E_r^* = e^{-j\frac{k\alpha_r h}{2}}, \; r \in \{1,2,3,4,5,6\}$$

In general case, due to the anisotropy (21 independent coefficients), the resolution seems to be difficult, however the authors can calculate for the materials which present symmetric conditions like: monoclinic and orthotropic materials. A material having three planes of symmetry orthogonal two by two is called orthotropic. In this case, the constants C16, C26, C36 and C45 are nulls and the stiffness matrix comprises only nine independent coefficients.

Propagation Along the Main Axis

By assimilating the direction of propagation and the principal axis (of symmetry) of the material, the stiffness matrix can then be given in the form:

$$C = \begin{bmatrix} C_{11} & C_{12} & C_{13} & 0 & 0 & 0 \\ C_{12} & C_{22} & C_{23} & 0 & 0 & 0 \\ C_{13} & C_{23} & C_{33} & 0 & 0 & 0 \\ 0 & 0 & 0 & C_{44} & 0 & 0 \\ 0 & 0 & 0 & 0 & C_{55} & 0 \\ 0 & 0 & 0 & 0 & 0 & C_{66} \end{bmatrix} \tag{13}$$

With these new conditions, the matrix K (10) becomes:

$$K = \begin{bmatrix} C_{11} - \rho c^2 + C_{55}\alpha^2 & 0 & (C_{13} + C_{55})\alpha \\ 0 & C_{66} - \rho c^2 + C_{44}\alpha^2 & 0 \\ (C_{13} + C_{55})\alpha & 0 & C_{55} - \rho c^2 + C_{33}\alpha^2 \end{bmatrix} \tag{14}$$

And the values of the coefficients can be calculated:

$$\begin{cases} \alpha_1 = -\alpha_2 = \dfrac{-B - \sqrt{B^2 - 4AC}}{2A} \\ \alpha_3 = -\alpha_4 = \dfrac{-B + \sqrt{B^2 - 4AC}}{2A} \\ \alpha_5 = -\alpha_6 = \sqrt{\dfrac{\rho c^2 - C_{66}}{C_{44}}} \end{cases} \tag{15}$$

$$\begin{cases} A = C_{33}C_{55} \\ B = C_{33}\left(C_{11} - \rho c^2\right) + C_{55}\left(C_{55} - \rho c^2\right) - \left(C_{13} + C_{55}\right)^2 \\ C = \left(C_{11} - \rho c^2\right)\left(C_{55} - \rho c^2\right) \end{cases} \tag{16}$$

The Dr Coefficients then become:

$$\begin{cases} D_{1r} = \left(C_{13} + C_{33}\alpha_r W_r\right) \\ D_{2r} = C_{55}(\alpha_r + W_r) \end{cases} \tag{17}$$

For $r \in \{1,2,3,4\}$, with:

$$W_r = \frac{\rho c^2 - C_{11} - C_{55}\alpha_r^{\ 2}}{\left(C_{13} + C_{55}\right)\alpha_r}$$

At this stage, the determinant of the system (13) is written in the form:

$$|D| = \begin{vmatrix} D_{11}C_1 & D_{13}C_3 & 0 & 0 & 0 & 0 \\ D_{21}S_1 & D_{23}S_3 & 0 & 0 & 0 & 0 \\ 0 & 0 & D_{11}S_1 & D_{13}S_3 & 0 & 0 \\ 0 & 0 & D_{21}C_1 & D_{23}C_3 & 0 & 0 \\ 0 & 0 & 0 & 0 & D_{35}S_5 & 0 \\ 0 & 0 & 0 & 0 & 0 & D_{35}C_5 \end{vmatrix}$$

With:

$$C_r = \cos\left(k\alpha_r\left(\frac{h}{2}\right)\right) \text{ and} S_r = \sin\left(k\alpha_r\left(\frac{h}{2}\right)\right),\ r\in\{1,2,3,4\}.$$

The authors can now write the characteristic equations of the symmetrical, anti-symmetric waves:

$$\begin{cases} D_{11}D_{23}cotan\left(\frac{k\alpha_1 h}{2}\right) - D_{13}D_{21}cotan\left(\frac{k\alpha_3 h}{2}\right) = 0 \\ D_{11}D_{23}tan\left(\frac{k\alpha_1 h}{2}\right) - D_{13}D_{21}tan\left(\frac{k\alpha_3 h}{2}\right) = 0 \end{cases} \tag{18}$$

Where h is the skin thickness, k is the wavenumber corresponding to the X direction.

Dispersion Curves

The Lamb waves propagating modes depend on the product frequency × thickness f.h. The result allowing the analytical description of the propagation is the establishment of dispersion curves (phase velocity, group or wave number versus the product frequency × thickness f.h). Many researchers have widely contributed in plotting dispersion curves of anisotropic material. Harsh Kumar Baid (2012) and Mark E. Orwat (2001) studied the dispersion in the case of transversely isotropic composites. Nayfeh (1989) has developed a transfer matrix method to plot dispersion curves of lamb waves propagating in multilayered anisotropic media. Demcenko (2010) has presented global matrix method to obtain dispersion curves of multilayered isotropic plate. Ameneh and Abdolreza (2014) have extended the relations of dispersion curves for multilayered composite-metal plates using the transfer matrix method. Others calculated dispersion curves using semi analytical or numerical approaches. Leger and Deschamps (2009) presented semi analytical methods. The authors have adopted the Matlab program developed by Elhadji (2014). Numerical resolution of the equation (18) permits to plot dispersion curves. A program was developed using Mathematica that allows calculating accurately the solutions (ω,Vp) by Newton-Raphson method. FORTRAN code was established in order to get the points to start the algorithm of Newton-Raphson.

The dispersion curves for the material that characteristics presented in Table 1, with the thickness 1.6 mm, presented in the Figure 6 are plotted in term of wave number versus frequency. The bleu curves for symmetric modes and the red for the anti-symmetric ones.

NUMERICAL MODEL USING FEM

Previous Work

Numerical simulations were carried out using finite element software ABAQUS to predict the A0 Lamb wave propagation behavior in a [0]4 plate. A number of modeling methods that had been used by previous

Table 1. Elastic coefficients Cij and density ρ of the materiel

ρ (kg/m3)	1500
C_{11} (GPa)	57
C_{22}=C_{33} (GPa)	15
C_{13}= C_{12}=C_{23} (GPa)	10
C_{55}=C_{66} (GPa)	4

Figure 6. Dispersion curves for orthotropic plate, h=1.6 mm (wave number vs frequency)

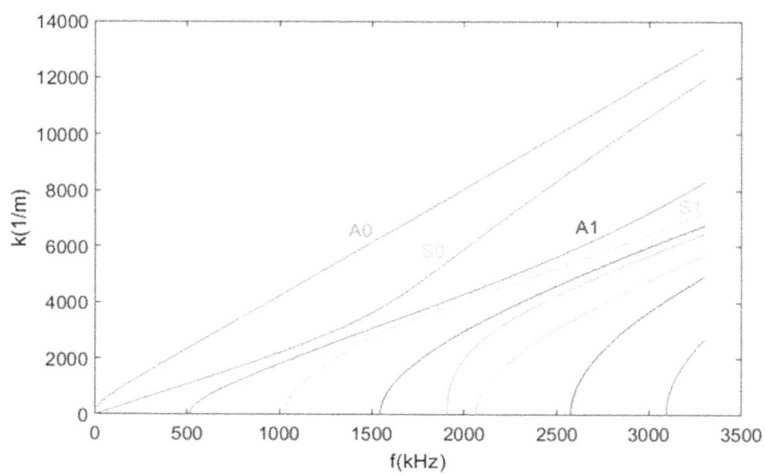

researchers to model the propagating wave: finite difference, boundary elements, strip elements, hybrid and finite element. The finite element method has been extensively and successfully used to model the Lamb waves and study their interactions with defects in structures.

Many researchers have used the numerical simulation to study lamb waves. Chiua, Roseb and Nadarajaha (2017) built a 3D model in ABAQUS of the 8 ply laminate [45/-45/0/90]S. Guo and Cawley (1993) investigated the influence of delamination position through the thickness and its interaction with the S0 mode using finite element analysis. The composite laminate modeled was the eight-layer cross ply. Nadarajah, Vien and Chiu (2015) studied the scattering of a zero-order anti-symmetric (A0) Lamb wave mode by semi-circular mid-plane edge delamination using the commercial FE package ABAQUS. Hayashi and Kawashima (2002) studied Lamb waves' propagation in laminated plates with delamination using the Strip Element Method and discussed the reflection and transmission for the case S0 and A0 excitation. Ng (2012) modeled the delamination as a volume split in quasi isotropic plate. The FE results were carried out by the explicit FE code LS-DYNA. Ng and Veidta (2012) used a three-dimensional FE method to simulate an eight-ply [45/-45/0/90]S quasi-isotropic composite laminate with a delamination. The simulations were computed by ANSYS software. Panda (2013) presented 3D FE simulations that were carried out to visualize the wave propagation and their interaction with the defect at various depths of an 8-layered Glass Fibre Reinforced Polymer (GFRP) by finite element software ABAQUS/Explicit.

Ramadas (2009) investigated the interaction of the primary anti-symmetric mode, A0 with symmetric delamination type defects in a quasi-isotropic laminated composite using 2D model ANSYS. Soleimanpour and Ng (2015) used 3D explicit finite element method to study cross-ply laminated composite beams [0/90/0/90]S, they investigated the mode conversion and scattering characteristics of guided waves at delamination. Gudimetla and Kharidi (2009) showed a procedure to simulate the propagation of Lamb waves in 8-layered Carbon Reinforced Fibre Plastic (CRFP) using a 2D model in ANSYS. The authors simulate the propagating the Lamb waves along the plane of the structure in the form of a time dependent force excitation. Basri and Chiu (2004) investigated how Lamb waves respond to the presence of material degradation in a plate-like structure using a series of finite element analyses. The propagation of these guided waves was interpreted with the dispersion characteristics and displacement profiles were analyzed in the frequency and wave number domain.

In the next section, the authors try to predict the interaction of A0 mode in orthotropic plate. The 2D numerical simulations were carried out using ABAQUS CAE. The authors consider the lamb waves propagation only in the skin to simplify the propagating modes identification. This assumption has been made in many papers Elhadji (2012), Bourasseau (2000) and, Diamanti (2004). They consider only the propagation in skin as it acoustic impedance is more than the acoustic impedance of the core.

Numerical Model

The skin has a length of L=400 mm and a thickness of h = 1.6 mm. The mechanical properties of each lamina are shown in Table 1. Once the geometry of the plate has been achieved, it remains to mesh and to define a sampling sufficient time.

Meshing and Time Sampling

To satisfy an accurate solution, the model has been meshed using the equation 17, and the time step is calculated by equation 18:

$$Max(\Delta x, \Delta y) < \frac{\lambda_{\min}}{10} \tag{19}$$

$$\Delta t < 0.7\frac{Max(\Delta x, \Delta y)}{V_L} \tag{20}$$

Excitation

In order to identify accurately the propagating modes, the simulation is performed on the range of frequencies to prevent dispersive phenomenon. For that reason the authors choose 300 kHz as frequency of excitation (see Figure 7).

Localization Method

The finite element method with Abaqus CAE code is used to compute the temporal displacement by 2D-FFT technic (equation 1) so as to identify the propagating and converted modes. Once the modes are identified, the next step consists in calculating the arrival times by WT (equation 2).

So there are two cases: in the first case where there is no modes conversion, the time delay permits to calculate the delamination position by equation 3, in the second case where the mode converts to another (A0 to S0 or S0 to A0), the equation 4 is used.

RESULTS AND DISCUSSION

Model: [0]4 without Defect

The A0 mode was excited from the left edge of the plate by concentrated force at the actuator C (see Figure 8). The A0 Lamb wave is the focus in this study because of its sensitivity to the small defects. To verify the excitability of pure mode A0, 2D- FFT was performed at the sensor A. A matrix was calculated at uniformly spaced points.

The figure 9 shows that there is only one mode propagating which is A0.

Model: Symmetric Delamination 60 x 0.1

Figure 10 is the delaminated model. The figure 11 is a snapshot of FE out-of plane displacement. The Lamb waves split into two parts and travel independently with different velocities in the upper and the lower sub-laminates. Mode conversions happen at the two edges of a delamination due to the change of boundary conditions. To characterize the delamination, the authors analyzed the signals picked up at the sensors A, and B.

Figure 7. Excitation windowed by Hann window

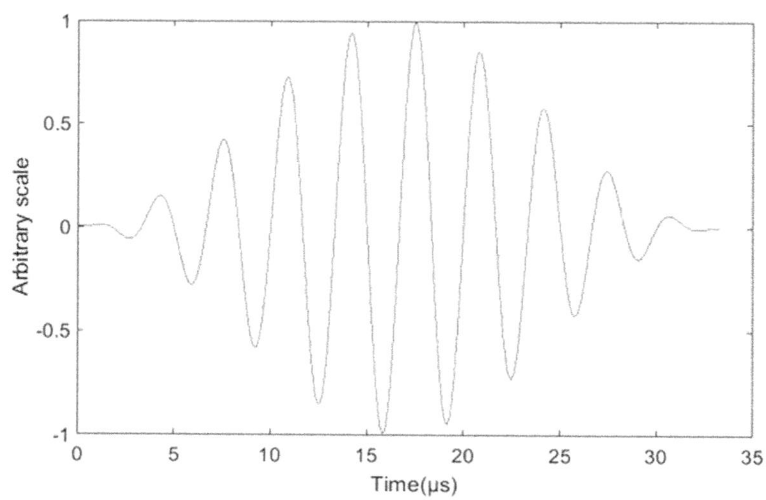

Figure 8. Numerical Model without defect, C: actuator, A, and B: sensors

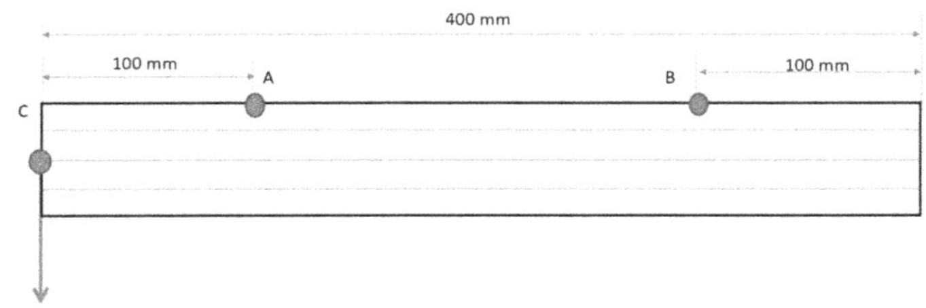

Figure 9. Superposition of 2D- FFT over analytical dispersion curves

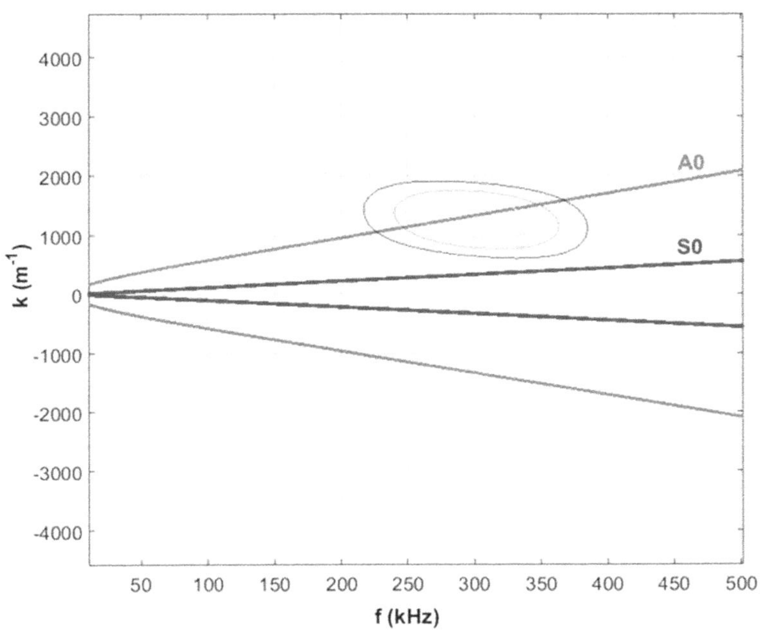

Modes Conversion

In the figure 12, the first wave group is a forward traveling wave group, which is an excited A0 mode at x = 0. When this interacted with the entrance of the delamination, there was a low amplitude reflected wave group. This reflected wave group appears as the second wave group A0A0. Delamination separates the plate into two sub-laminates: upper and lower. The transmitted wave converted to the symmetric and anti-symmetric mode A0S0 and A0A0 which travelled within the sub-laminates with different velocities. The first wave A0S0U - U, L to mean that the mode travel respectively in the upper and lower sub laminate- reached first the exit of delamination this is justified by the fact that symmetric mode is faster than the anti-symmetric mode. The backward mode A0S0US0U passes through the upper and then transmitted to the main laminate as the second wave group A0S0US0UA0. By the same manner, in the

Figure 10. Numerical model of symmetric delamination

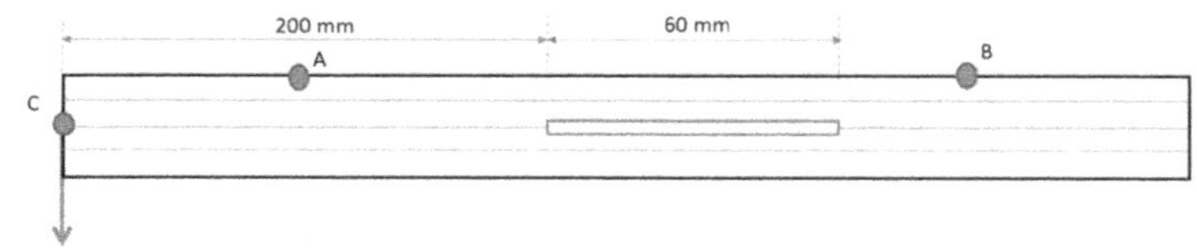

Figure 11. Snapshot of FE simulated out-of plane displacement in the upper and the lower sub-laminates separated by the delamination

Figure 12. A0 reflection at the entrance and at the end of asymmetric delamination (length 60), case A0 incident, R denotes Reflected mode and T denotes Transmitted mode

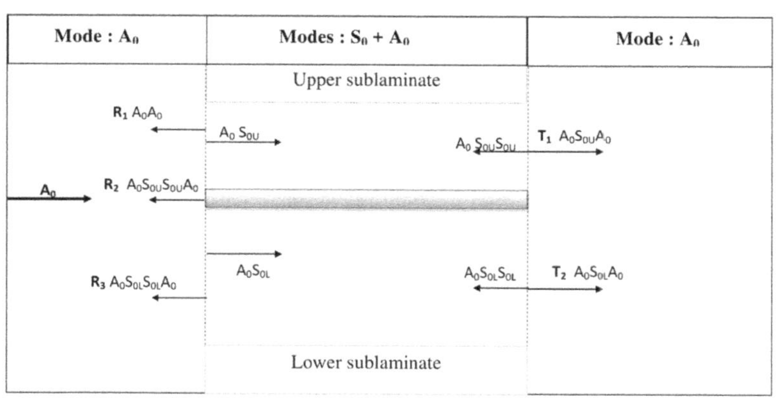

lower sub laminate the mode A0S0LS0L transmitted to the main laminate as A0S0LS0LA0. In the case of symmetric delamination the two modes A0S0US0UA0 and A0S0LS0LA0 reflected at the same time.

Results: Sensor A

Figure 13 is A-scan image from the numerical simulations, taken at 100 mm for a delamination length of 60 mm. As explained in the figure 12, the authors conclude that the first arrived wave is the incident

Figure 13. Out-of-plane displacement, taken on damaged composite [0]4 at x = 100 mm, case symmetric delamination length 60

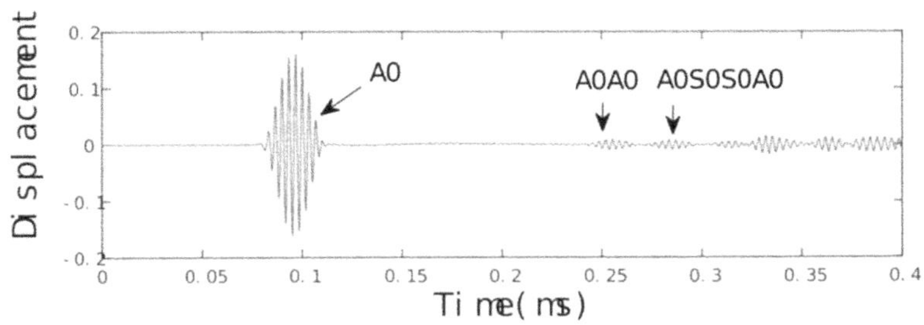

mode A0 and the second is the first reflected wave A0A0. The delamination is symmetric so two waves arrived at the same time as one packet which is A0S0US0UA0+ A0S0LS0LA0.

Based on the 2D-FFT computed using Matlab program applied to the surface of delaminated skin (before the delamination), the figure 14 shows that the incident and the reflected modes are anti-symmetric which confirms that the first reflected mode is A0A0.

Delamination Characterization from Data Acquired at Sensor A

Figure 15,below, presents wavelet coefficients versus time at sensor A. The time delay between the peaks permits to estimate the position and the length of delamination.

Knowing the group velocity of the modes A0 and S0: Vg(A0,300kHz)=1268 m/s, Vg(S0,300kHz)=5649 m/s, the authors can calculate the delamination position from the sensor A noted PD by Δt1 (Delta 1), and delamination length noted LD by Δt2 (Delta 2) as presented in table below:

Figure 14.

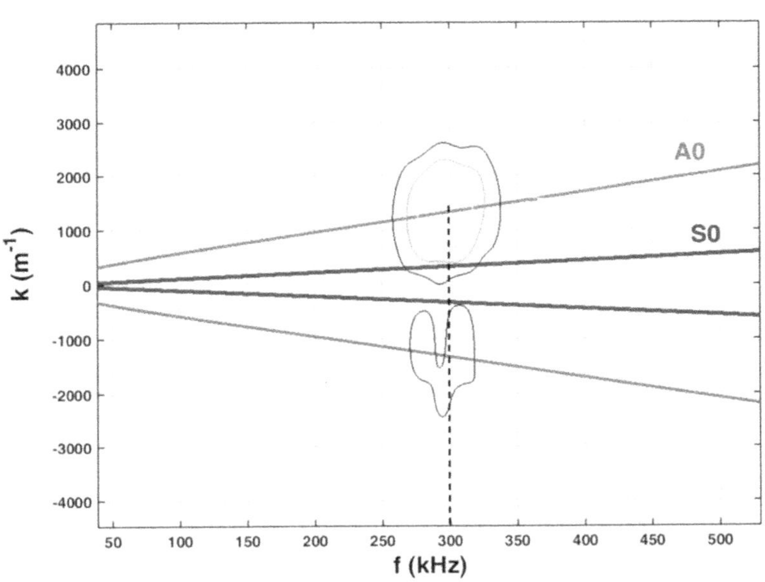

Figure 15. Coefficients of wavelet gaus1, case A0 incident, delamination length 60, between layer 2 and layer 3

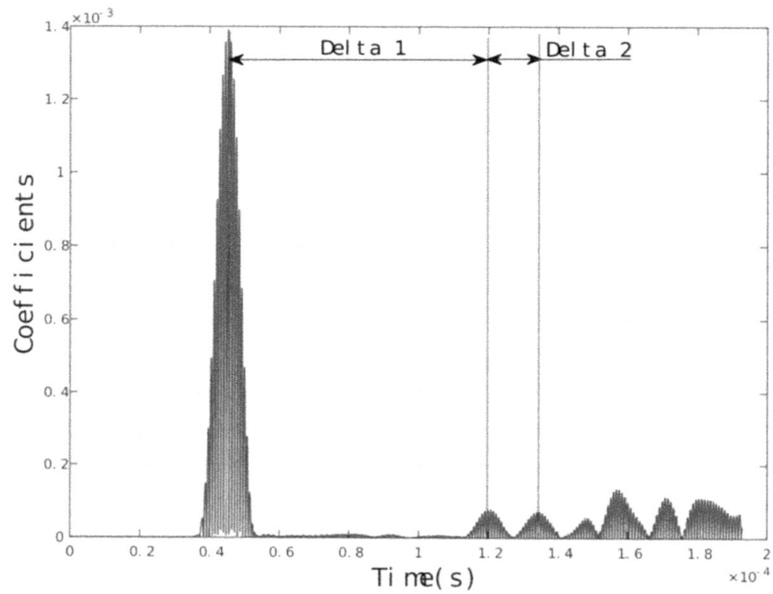

The sensor A permits to evaluate the position and the length of symmetric and asymmetric delamination. The Errors associated with the predicted symmetric delamination position range from 3.1% to 6.6%. However, in the case of asymmetric delamination, the error is less than 3.8%. As far as the length prediction is concerned, the errors range from 3.6% to 6.9% in symmetric delamination and from 2.9% to 5.9% in asymmetric delamination. So the authors conclude that more the delamination is long more the estimation is high.

Results: Sensor B

Figure 16, below, represents different wave modes in the case of A0 mode is incident on the transmission side of the main laminate at sensor B.

Table 2. Delamination localization from data acquired at sensor A

Delamination Position	Length	PD	PDTH	Error %	LD	Error %
1 and 2	60	103,8	100	3,8	62,0	3,3
	90	103,9	100	3,9	93,3	3,7
	120	103,8	100	3,8	123,4	2,9
2 and 3	60	103,2	100	3,2	62,2	3,7
	90	106,6	100	6,6	94,1	4,5
	120	103,1	100	3,1	120,9	0,8

Figure 16. Out-of-plane displacement, taken on damaged composite [0]4 at x = 300 mm. case symmetric delamination length 60

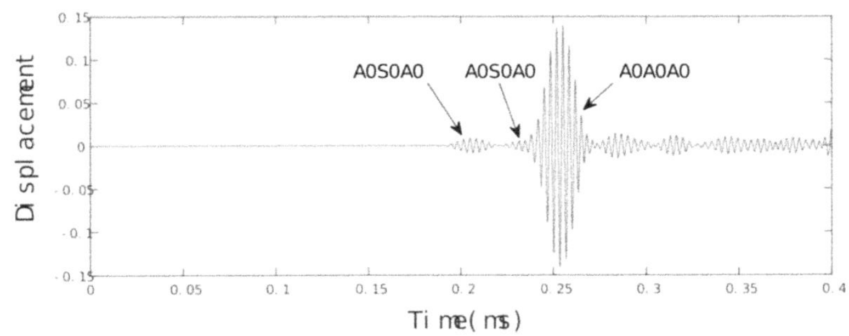

This figure represents different wave modes in the case of A0 mode is incident on the transmission side of the main laminate at sensor B. A new mode S0 was generated when the A0 interacted with the entrance of delamination as showed in the figure 12. This new mode travelled within the upper sub-laminates along the length of delamination. Then it reached the exit of delamination, interacted with the exit and generated a new mode, A0 in the forward direction. This new mode propagated in the main laminate, called the A0S0UA0 mode. In the lower sub laminate the mode is called A0S0LA0. Two modes arrived at the same time in the case of symmetric delamination.

Delamination Characterization from Data Acquired at Sensor B

Figure 17 presents wavelet coefficients versus time at sensor B. The Δt1 is the time delay between two transmitted symmetric modes. The Δt2 is the time delay between the first transmitted symmetric mode and the third anti-symmetric mode, so knowing the group velocity of the mode S0: Vg (S0, 300 kHz) =5649 m/s, the authors can calculate the delamination length noted LD1 by equation 3 and LD2 by equation 4. The Table 3 resumes the results found:

Figure 17. Coefficients of wavelet gaus1, case symmetric delamination length 60

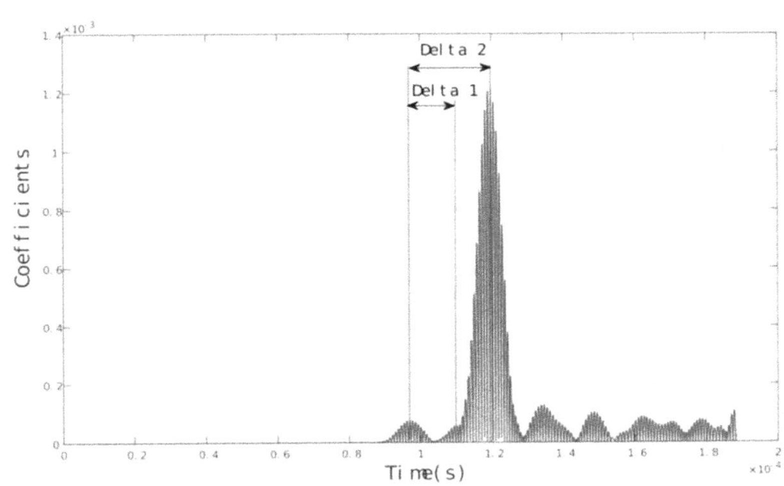

The sensor B permits to evaluate only the length of symmetric and asymmetric delamination. The Errors associated with the predicted symmetric delamination length range from 1.2 to 4.9.

In the case of asymmetric delamination the errors range from 3.5 to 5.9. So we conclude that more the delamination is long more the estimation is high. Also, the sensor B gives accurate estimation in case of symmetric delamination.

The table above compared the delamination length calculated at the two sensors A and B. It shows that sensor A gives more accurate estimation than the sensor B. Moreover, sensor A permits also to estimation the start of the delamination. So the optimal position to evaluate the position and the length of delamination is between the actuator and the defect.

EXPERIMENTAL VERIFICATION

To validate our model, the authors have based on Elhadji (2012) results; the author studied by FEM the A0 lamb mode interaction with delamination in sandwich skin. Based on Fourier Diamond see Bonnet (1983), he has estimated the delamination length. The obtained results are validated by experimental study. These studies have been done by the propagation of Lamb waves on air by using a Laser Doppler velocimetry.

To validate our model and our method we followed the same methodology, as presented in the figure 18.

The S(x,t) is the measured time-displacement matrix at the skin surface. IFFT is the Inverse Fast Fourier Transform.

Model: Delamination 24 x 0.1

The B-scan illustrated in the figure 20, represents the S(x,t): time–displacements matrix picked up at the surface of limited zone (see figure 19).

Figure 20 indicates all the interactions of A0 lamb mode with delamination in the skin. To evaluate the defect length, and following the Diamond method, the authors have first of all separated the incident and the reflected waves.

Then the separated incident waves get filtered to remove the undesirable reflections.

Figure 21 shows the filtered incident waves. The authors can easily see the mode perturbation in the studied zone. This is due to the delamination presence.

Table 3. Delamination localization from data acquired at sensor B

Delamination Position	Length	LD1	Error	LD2	Error	Min Error
1 and 2	60	49,1	18,1	62,1	3,5	3,5
	90	93,5	3,9	95,2	5,7	3,9
	120	127,1	5,9	127,7	6,4	5,9
2 and 3	60	45,5	24,2	62,9	4,9	4,9
	90	98,9	9,8	94,0	4,5	4,5
	120	121,5	1,2	122,5	2,1	1,2

Table 4. Comparison between results at sensors A and B

Delamination Position	Length	Sensor A		Sensor B	
		LD	Error	LD	Error
1 and 2	60	62,0	3,3	62,1	3,5
	90	93,3	3,7	93,5	3,9
	120	123,4	2,9	127,1	5,9
2 and 3	60	62,2	3,7	62,9	4,9
	90	94,1	4,5	94,0	4,5
	120	120,9	0,8	121,5	1,2
3 and 4	60	62,0	3,3	62,1	3,5
	90	93,3	3,7	93,5	3,9
	120	123,4	2,9	127,1	5,9

Figure 18. Fourier Diamond

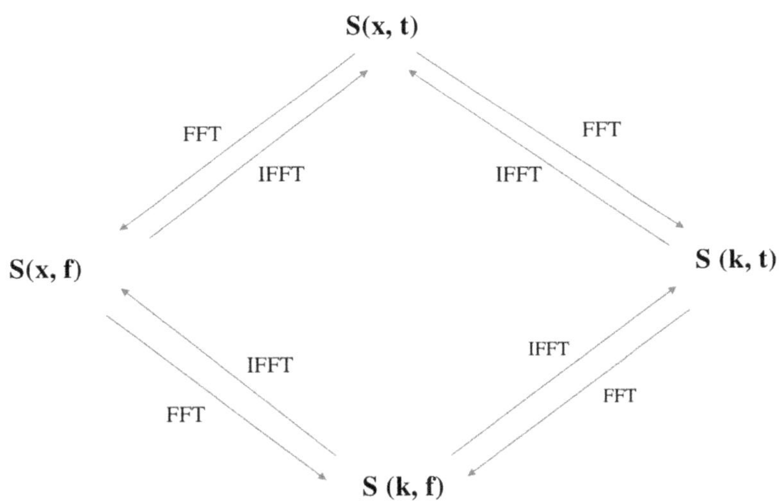

Figure 19. Model geometry

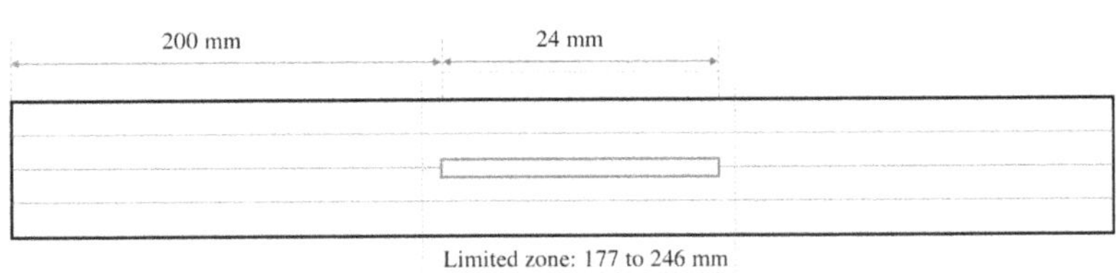

To evaluate the defect length the authors computed the temporal FFT to the filtered signal as presented in the figure 22.

Figure 20. Wavefields of A0 mode interaction with delamination

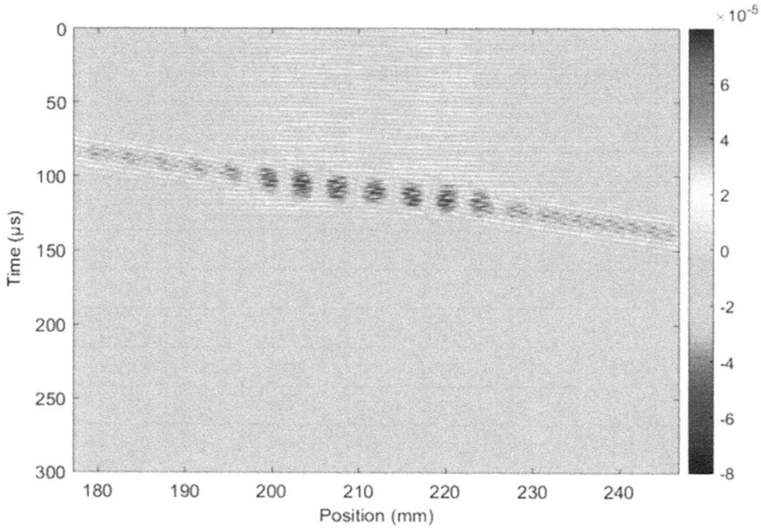

Figure 21. Filtered wavefields of incident waves

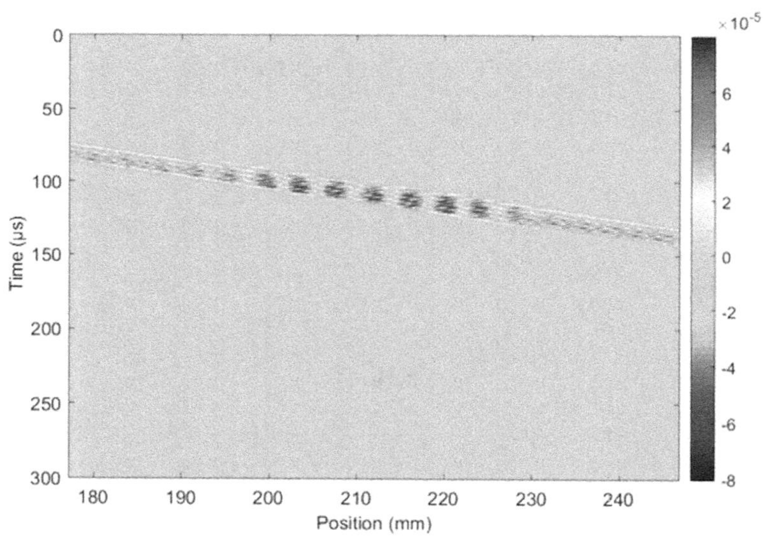

Figure 22 shows the principal lobe localized at the frequency: 300 kHz which verifies the excitation frequency. To measure the delamination length, a cut of the temporal FFT at the frequency 300 kHz must be done to show clearly the peaks of the lobes.

Figure 23 permits to measure the delamination length. The authors measure between the peaks of the lobes. The delamination length is 22 mm. To conclude, the model gives an accurate result with an error of 8.3%. The errors are due to the approximated numerical model.

Figure 22. S(x,f) temporal FFT applied to the filtered incident waves

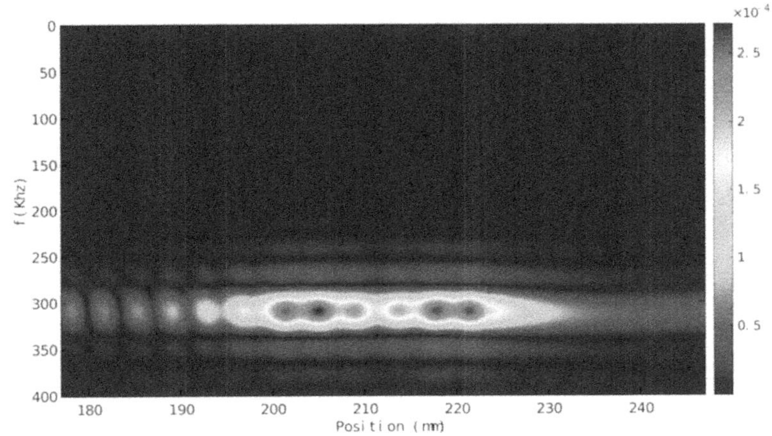

Figure 23. Cut of S(x,f) at the frequency 300 kHz

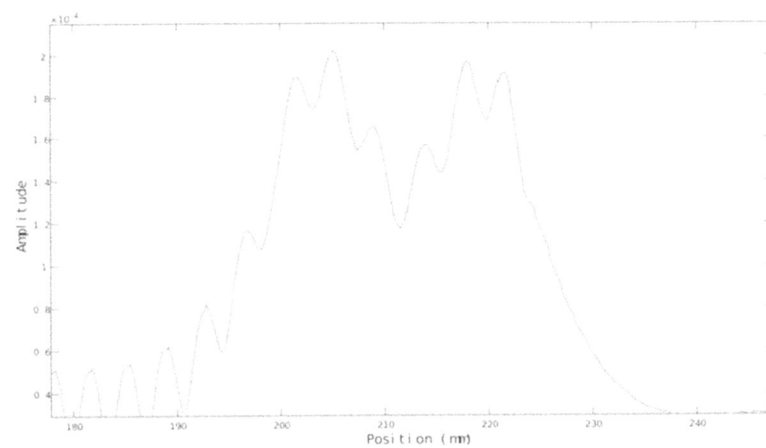

CONCLUSION

In this work, the authors have studied the numerical model describing the propagation of anti-symmetric mode A0 within a sandwich skin. Simulations based on Finite Element Method were carried out using ABAQUS Software in order to investigate the interaction of Lamb wave mode by symmetric and asymmetric through thickness delamination. The results showed the conversion phenomenon. To quantify the conversion and to characterize the delamination, the authors have computed the 2D-FFT in order to identify the propagating and the converted modes; after that, the WT was calculated at sensors A and B to measure the arrival times.

Moreover, comparing the signals obtained at two sensors: A, and B, it was found that the mode conversions happen at the edges of a delamination for A0 wave. In fact, this mode is sensitive to the position and to the length of delamination; this makes it become an effective tool to evaluate the delamination in orthotropic materiel. Besides, based on signal changes with the delamination edges, a localization method is proposed to estimate the position and the length of the delamination. The proposed method was validated based on previous experimental work.

These are the main results:

- The estimation is accurate when the delamination length increases,
- The optimal position to evaluate the beginning and the length of delamination is between the actuator and the defect which corresponds to the pulse-echo method.

As perspective, the authors model the whole sandwich structure and characterize the defect by reflection and transmission coefficients.

REFERENCES

Basri, R. & Chiu, W. K. (2004). Numerical analysis on the interaction of guided Lamb waves with a local elastic stiffness reduction in quasi-isotropic composite plate structures. Composite Structures, 87–99.

Bonnet, G. (1983). Au-delà d'une vitesse de groupe: vitesse d'onde et vitesse de signal. Première partie: déformation de l'amplitude et influence de l'affaiblissement. Annales des Télécommunications, 1–17.

Bonnet, G. (1983). Au-delà d'une vitesse de groupe: vitesse d'onde et vitesse de signal. Première partie: l'opérateur vitesse de groupe en l'absence d'affaiblissement. Annales des Télécommunications, 1–22.

Bourasseau, N., Moulin, E., Delabarre, C., & Bonniau, P. (2000). Radome health monitoring with Lamb waves: Experimental approach. NDT & E International, 33(6), 393–400. doi:10.1016/S0963-8695(00)00007-4 doi:10.1016/S0963-8695(00)00007-4

Cawley, P., & Alleyne, D. (1991). A two-dimensional Fourier transform method for the measurement of propagating multimode signals. J. Acoust. Soc.1159-1168.

Chiua, W. K., Roseb, L. R. F., & Nadarajaha, N. (2017). Scattering of the fundamental anti –symmetric Lamb wave by a midplane edge delamination in a fiber-composite laminate. Procedia Engineering, 188, 317–324. doi:10.1016/j.proeng.2017.04.490 doi:10.1016/j.proeng.2017.04.490

Chunhui, Y. (2006). Some aspects of numerical simulation for Lamb wave propagation in composite laminates. Composite Structures, 267–275.

Demcenko, A., & Mazeika, L. (2010). Calculation of lamb waves dispersion curves in multi-layered planar structures. Journal of Sound and Vibration, 1435–1449.

Diamanti,K., Hodgkinson, J. M. & Soutis, C. (2004). Application of a Lamb Wave Technique for the Non Destructive Inspection of Composite Structures.

Elhadji, B. N., & Hugues, D. (2012). Non Destructive Testing of sandwich composites: adhesion defects evaluation; Experimental and Finite Element Method simulation comparison. Proceedings of the Acoustics Nantes Conference.

Elhadji, B. N. (2014). Contrôle santé de structures sandwichs composites, caractérisation et évaluation non destructives de l'adhésion et du vieillissement – Simulations par éléments finis et expérimentation par mesure d'impédance et par ultrasons multiéléments. Université du Havre, École Doctorale S.P.M.I.I.

Feng, B., Ribeiro, A. L., & Ramos, H. G. (2017). Interaction of Lamb waves with the edges of a delamination in CFRP composites and a reference-free localization method for delamination. IEEE.

Gayathri, P., Umesh, K., & Ganguli, R. (2010). Effect of matrix cracking and material uncertainty on composite plates. Reliability Engineering & System Safety, 95(7), 716–728. doi:10.1016/j.ress.2010.02.004 doi:10.1016/j.ress.2010.02.004

Gudimetla, P., & Kharidi, A. (2009). Simulation of delaminations in composite laminates. International Conference on Precision, Meso, Micro and Nano Engineering, 11 -12.

Guo, N., & Cawley, P. (1993). The interaction of Lamb waves with delaminations in composite laminates. The Journal of the Acoustical Society of America, 94(4), 2240–2246. doi:10.1121/1.407495 doi:10.1121/1.407495

Hanfei, M., Asaad, M., Mohammad Faisal, H., Roshan, J., Md Yeasin, B., & Victor, G. (2019). Vibration-Based In-Situ Detection and Quantification of Delamination in Composite Plates. Sensors (Basel), 19(7), 1734. doi:10.3390/s19071734 PubMed doi:10.339019071734 PMID:30978968

Harsh, K. B. (2012). Detection of Damage in a Composite Structure Using Guided Waves. Los Angeles, CA: University Of California.

Hayashi, T., & Kawashima, K. (2002). Multiple reflections of Lamb waves at a delamination. Ultrasonics, 44(1-8), 193–197. doi:10.1016/S0041-624X(02)00136-1 PubMed doi:10.1016/S0041-624X(02)00136-1 PMID:12159931

Hu, H. W. (2007). Physical Aging in Long Term Creep of Polymeric Composite Laminates. Journal of Mechanisms, 23(3), 245–252. doi:10.1017/S1727719100001283 doi:10.1017/S1727719100001283

Ip, K. H., & Mai, Y. W. (2004). Delamination detection in smart composite beam using Lamb waves. Smart Materials and Structures, 13(3), 544–551. doi:10.1088/0964-1726/13/3/013 doi:10.1088/0964-1726/13/3/013

Leger, A., & Deschamps, M. (2009). Ultrasonic wave propagation in non-homogeneous media. Springer Science & Business Media. doi:10.1007/978-3-540-89105-5 doi:10.1007/978-3-540-89105-5

Luca, A. D. (2016). Numerical simulation of the Lamb wave propagation in impacted CFRP laminate. Procedia Engineering, 167, 109–115. doi:10.1016/j.proeng.2016.11.676 doi:10.1016/j.proeng.2016.11.676

Maghsoodi, A., Ohadi, A., & Sadighi, M. (2014). Calculation of Wave Dispersion Curves in Multilayered Composite-Metal Plates. Shock and Vibration. Hindawi Publishing.

Mark, E. O. (2001). Experimental Investigation of Lamb Waves in Transversely Isotropic Composite Plates. Cambridge, MA: Massachusetts Institute of Technology.

Michalcová, L., & Hron, R. (2018). IOP Conf. Series: Materials Science and Engineering 326, doi:10.1088/1757-899X/326/1/012006

Mustapha, S., & Ye, L. (2015). Propagation behavior of guided waves in tapered sandwich structures and debonding identification using time reversal. Wave Motion, 57, 154–170. doi:10.1016/j.wavemoti.2015.03.010 doi:10.1016/j.wavemoti.2015.03.010

Nadarajah, N., Vien, B. S. & Chiu, W. K. (2015) Computational Study of the A0 Scattered Field Due to an Edge Delamination. Journal of Mechanics Engineering and Automation. 229-236.

Nayfeh, A., & Chimenti, D. (1989). Free wave propagation in plate of general anisotropic media. Journal of Applied Mechanics, 56(4), 881–886. doi:10.1115/1.3176186 doi:10.1115/1.3176186

Ng, C. T., & Veidta, M. (2012). Scattering of the fundamental antisymmetric Lamb wave at delaminations in composite laminates. Acoustical Society of America, 115.

Ng, C. T., Veidt, M., Rose, L. R. F., & Wang, C. H. (2012). Analytical and finite element prediction of Lamb wave scattering at delaminations in quasi-isotropic composite laminates. Journal of Sound and Vibration, 331(22), 4870–4883. doi:10.1016/j.jsv.2012.06.002 doi:10.1016/j.jsv.2012.06.002

Panda, R. S. (2013). Lamb wave interactions with delaminations in composite laminates using air-coupled ultrasonic visualization. 14th Asia-Pacific Conference on Non-Destructive Testing.

Ramadas, C. (2009). Interaction of the primary anti-symmetric Lamb mode (A0) with symmetric delaminations: numerical and experimental studies. Smart Materials and Structures.

Shashank, P. (2014). Lamb Wave Propagation and Material Characterization of Metallic and Composite Aerospace Structures for Improved Structural Health Monitoring (SHM). Ontario, Canada: Carleton University Ottawa.

Soleimanpour, R. & Ng, C. T. (2015). Mode conversion and scattering analysis of guided waves at delaminations in laminated composite beams. Structural Monitoring and Maintenance, 213-236.

Veidt, M., & Ng, C. T. (2011). Influence of stacking sequence on scattering characteristics of the fundamental anti-symmetric Lamb wave at through holes in composite laminates. J. Acoust, 129.

Vladimir, V. B. (1994). Delaminations in composite structures: its origin, buckling, growth and stability. First International Conference on Composites Engineering (ICCE/1), New Orleans, LA. 28-31.

Yang, B., Xuan, F.-Z., Chen, S., Zhou, S., Gao, Y., & Xiao, B. (2017). Damage localization and identification in WGF/epoxy composite laminates by using Lamb waves: Experiment and simulation. Composite Structures, 165, 138–147. doi:10.1016/j.compstruct.2017.01.015 doi:10.1016/j.compstruct.2017.01.015

Chapter 13

Fault Analysis and Protection of Low-Voltage DC Microgrid Equipped by Renewable Energy Resources

Navid Bayati
Aalborg University, Denmark

Amin Hajizadeh
Aalborg University, Denmark

Mohsen Soltani
Aalborg University, Denmark

ABSTRACT

This chapter consists of two sections, 'Modelling of DC Microgrids' and 'Protection of DC Microgrids'. In the first section, the new developments in DC Microgrids are discussed. Then, the Modelling of renewable energy resources-based DC Microgrid using characteristics and mathematics equations of each component are presented and then they are simulated by MATLAB. Afterward, the fault analysis and fault current behavior of the studied DC Microgrid are investigated. In the second section, a method of protecting the DC Microgrid and locating the fault in different parts of the system is proposed. The proposed method protects DC Microgrid using localized protection devices. And, the effectiveness of the proposed protection method is validated in a DC Microgrid with ring configuration.

GENERAL INTRODUCTION

This chapter is summarized as follows: in the first section, the new developments on DC Microgrids are discussed and future trends of DC Microgrids in power grid, aircraft and vehicles are introduced. Then, a DC Microgrid including Photovoltaic (PV), Wind Turbine (WT), battery, DC motor, and power

DOI: 10.4018/978-1-7998-0117-7.ch013

electronic converters is simulated in MATLAB for fault studies. In addition, the behavior of these components during fault is investigated by mathematical and simulation analysis. The protection challenges and constraints are also discussed in this section. In the second section, the new protection methods for DC Microgrids are presented and compared, then, a protection method based on the characteristics of the DC Microgrids and DC current nature is proposed. The protection method uses communication links and localized protection devices to locate and detect fault. In addition, this method uses the equation of current and voltage variation to find the location of fault. Finally, the proposed protection method isolates the faulty zone and insures the normal operation of other zones. The presented method uses the characteristics of current to detect the location of faults in the DC Microgrid. As a final point, the results of the proposed methods are shown, and the effectiveness of these methods is demonstrated by simulation in MATLAB environment.

MODELING OF DC MICROGRIDS

Introduction

Using Distributed Energy Resources (DERs) in the power grids opens new opportunities. A microgrid is a distribution system with different DERs and loads which can operate as an isolated power system (Hatziargyriou et al., 2007). A Microgrid is operated as an islanded or grid-connected mode. During the normal operation situations, the Microgrid is connected to the grid, and local DERs supply the loads, and, if the power of loads increases, the grid also can supply the loads. If the power of loads decreases to a value less than the power of DERs, the excess power will be injected into the grid. Typically, the DERs in Microgrids are based on renewable energy, for instance, WT and PV, and these DERs have different voltage with the main buses of Microgrid. Therefore, DERs require to be connected to the system by power converters (Kalesar et al., 2018). The best solution to protect sensitive loads from disturbances and power outages is Microgrids (Venkataramanan et al., 2002). The main benefits of a Microgrid are high reliability to make power balance between demand and generation by using storage devices, less transmission loss, and improving the stability of the system (Gunasekaran et al., 2018).

Microgrids are divided into two different types, AC and DC. The presenting of grid control by rectifiers made a rise in the application of DC power in power grids by HVDC transmission lines. Therefore, the development of the conversion technologies caused to emerge DC loads and power converters for different applications. Today, a DC Microgrid is normally planned for developing small scale facilities (Sannino et al., 2002), aircraft (Magne et al., 2013) or electric vehicles (Hu et al., 2016). DC Microgrids have several advantages over AC Microgrids. Majority of DERs, storage system, and also loads like telecommunication applications, vehicle charging stations and computers require or generate DC power. The loss of energy conversion to supplying DC loads from AC power is approximately 20% (Gao et al., 2017; Patterson et al., 2012). Therefore, due to the using DC power in DC Microgrid, loads are directly supplied by DC power, and thus, the power conversion loss of the system will reduce. On the other hand, for supplying remote communities, DC Microgrid is the most preferred solution, because due to the absence of the transmission line, the remote communities cannot access the main grid. Other advantages of DC Microgrids over AC Microgrids are:

- DERs, batteries and most loads are DC (Bayati et al., 2018).

- The DC cable loss is 15% less than AC cables (de Oliveira et al., 2014).
- The absence of frequency and phase make control of DC Microgrids more simple (Bayati et al., 2018).
- DC Microgrids have higher power flow capacity than AC Microgrids (Amamra et al., 2017).
- Less number of converters reduce heat and losses of the system (Elsayed et al., 2015).

In addition to the aforementioned advantages of DC Microgrids, the power converter-based DC Microgrids is improving for protection requirements. Due to the lack of comprehensive protection standards for DC Microgrids, these systems are faced with challenges in this issue. The primary concern of the protection studies in DC Microgrids is detecting and locating faults in the system, and isolating the faulty section. Most possible faults in the DC Microgrids are Pole to Pole (PP), and Pole to Ground (PG) faults, which the faults between negative and positive lines are PP, and fault between line and ground is PG. The most complex type of fault is determining of PG, which occurs by a resistance ground path, and therefore, due to the low surge fault current, the detection of PG faults is a challenging problem. Also, using controllers for power converters increases the difficulties of fault detection in DC Microgrids (Ericsen et al., 2006).

The first step of the fault protection in DC Microgrids is the modeling of the system. Modeling of the system must be accurate and considers the dynamic of the different components. Therefore, fault analysis can be investigated, and the results will be used for designing a protection scheme. Modeling of the electrical elements during fault are investigated by three different methods:

- Equivalent circuit modeling: model of different components by electric circuits in terms of voltage, current or impedance behavior of the component (Ammerman et al., 2010).
- Equation-based model: this model is relying on the mathematical equations of each component by combining dynamic and differential equations (Ziani et al., 2009).
- Heuristic models: modeling the system based on the experimental data, which can be added to the equation based model or equivalent circuit for improving the accuracy of the model (Andrea et al., 2010).

The nature of the DC current is different from AC systems. Thus, the behavior of the fault current in DC Microgrids is different from AC Microgrids. Therefore, to analyze the performance of the system during the fault, and proposing a protection scheme, in the first stage, a dynamic model of the components is required (Bayati et al., 2018). DERs such as PV, WT, and fuel cells are the main components of the DC Microgrids, and the dynamic modeling of these resources is the essential part of the DC Microgrid modeling. The dynamic model of PV during the fault is shown in (Yang et al., 2010), however, for increasing the speed of the simulations, this model is simplified. On the other hand, the accurate model of the PV by dividing the model into three different sub-model, optic, heat and electrical model, is proposed in (Sahli et al., 2018). Moreover, the simplified and accurate dynamic model of WT is proposed in (Zhao et al., 2015; Conroy et al., 2009), respectively. Also, the modeling of other components of DC Microgrids, such as Battery (Farrokhabadi et al., 2018), cables (Wei et al., 2006), loads (Arif et al., 2018), and power converters (Frances et al., 2018) are presented in previous researches.

In addition, one of the important issues in the operation of DC Microgrids is protection of these systems. And, DC Microgrids faced several challenges for designing a protection method, which motivate researchers to design a scheme to solve all challenges. As brief the challenges of protecting of DC Microgrids are:

- Lack of suitable standards for DC systems
- Lack of zero-crossing point
- Most of the protection methods are designed for AC system, thus, they cannot implement in the DC systems
- High rise of fault current

However, in recent years, several standards for DC Microgrid protection have been released (Augustine et al., 2018), but these standards only cover the DC Microgrid protection with a nominal voltage less than 120 V. Therefore, still more protection standards require for LVDC DC Microgrids, with a nominal voltage between 120 to 1500 V. In Table 1 the recent developments of the protection standards are summarized.

On the other hand, in recent years, several DC Microgrid protection methods have been proposed. One of the most used protection methods in power and AC systems are overcurrent protection (Bayati et al., 2017). Due to the post-fault challenges in DC Microgrids, the overcurrent protection needs several improvements. In (Saleh et al., 2017), A LC circuit is paralleled with the overcurrent relay, and this makes an explicit frequency during DC fault, and overcurrent relay detects the fault. However, due to the lack of selectivity in this method, the protection device cannot coordinate with other protection devices. In addition, (Cairoli et al., 2013) developed a protection scheme for DC Microgrids with ring configurations based on the coordinated control of power converters and mechanical switches, and also, using converters for reconfiguration of switches decreases the fault current. However, this method increases the complexity and cost of the protection scheme. In summary, the advantages of overcurrent protection of DC Microgrids are simplicity and applicable in fault interrupt approaches. And the disadvantages are cannot detect the HIF, should be used with other methods to provide selectivity.

Another widely used method for protection of DC Microgrids is differential protection method. In this method, the fault is detected by calculating the difference of the current in both sides of the protected component. First, the relay measures these differences, and if this amount rises higher than a threshold, the relay sends the signal of the trip to the C.Bs. In (Park et al., 2013), the differential protection is used in a DC Microgrid to detect and locate the fault. The fault is isolated by Intelligent Electronic Device (IED), and this device controls C.Bs, which connected to nodes. The IED calculates the differential current between each node, and during the fault, IEDs isolate the components quickly. In case of C.Bs failure, IEDs send the signals to the down or upstream IEDs to operate their C.Bs for backup protection. Therefore, the faulted component is completely isolated, and the healthy sections operate normally. The

Table 1. Recent standards for DC Microgrid protection

Name of Standard	Scope
IEEE 1547-2018	The requirements of safety, security, maintenance for Hybrid AC/DC Microgrids
IEEE 1547	Smart inverter protection, modeling, and function
IEEE 946-2004	The guidelines for selecting the type and quantity of the equipment, and protection equipment.
IEC 60364	Protection against electric shock, overcurrent, voltage disturbances

results of the differential based protection methods in DC Microgrids shows that this method can detect and isolate the fault section quick and accurate. But, due to the need for communication instructors, this method is expensive and able to have communication failure, decrease accuracy in case of noisy environments. Another protection device for protection Microgrids is Fault Current Limiters (FCLs), which highly used in AC Microgrids. A FCL is a power electronic protection device that limits the current value of the system during faults (Khodadoost et al., 2018). But, due to the high rise current in DC Microgrids during the fault, using of FCLs in DC Microgrids faced with several restrictions such as temperature limitation and cost issues.

Moreover, the impedance-based protection methods are used to locate the fault in the DC Microgrids. In (Yang et al., 2012), a simple method based on distance protection by using only two DC measurements on the lines is proposed. In this method, one more reference voltage to avoid communication requirements is used to estimates the location of the fault. The problem of distance protection methods is the high sensitivity to fault resistance. In addition, very small cable resistance in the DC systems makes a challenge for this technique to use is the DC Microgrids. A solution for these challenges is that the distance estimation method injects a voltage or current spectrum to the system to estimate the fault distance. Current or voltage signals with a random magnitude are generated by these protection devices which could be a full bridge inverter or large capacitor coupled to a source. In addition, signal processing methods, for instance, Fast Fourier Transform (FFT) is used to calculate the impedance of the faulted system and distance. Due to the developments of artificial intelligence (AI) in recent years, AI-based protection methods have been developed in DC Microgrids. Among these methods, due to the less cost and complexity, Artificial Neural Network (ANN) is the best technique for locating and detecting faults in DC Microgrids. But, directly sampling data of the fault in the system, increase the calculation time of the ANN, and also the signal noise increase the chance of the protection scheme failure. Other methods such as wavelet transform and FFT extract the characteristics of the fault signal, which can be linked to the ANN by sending extracted features to the ANN. The wavelet transform protection method suggested in (Li et al., 2012); the required signals in the time-frequency are demonstrated by wavelet transform. This method calculates the transient signals as the frequency changes during the fault. However, wavelet transform based protection methods needs data acquisition units with a high-sampling rate.

After a summarized introduction to the recent trends and motivation of DC Microgrid protection, to address the presented challenges regarding protection of DC Microgrids, in this chapter, a protection method for DC Microgrids with ring configuration is proposed. The proposed protection scheme uses different zones to isolate the faulty section and also detect and classify the fault in the shortest time. This method uses the transient behavior of the fault current to detects the fault and identify the faulty zone. This method is tested using MATLAB environment to proves the effectiveness of the proposed method, and also compared with other research papers.

In this section, first, the new developments of implementation and application of DC Microgrids are presented. Afterward, the equations for modeling of DC Microgrid components are discussed, and an accurate model for studying of DC Microgrids is proposed. Finally, the fault analysis in DC Microgrids is presented to find the fault current behavior in different situations. On the other hand, the impact of fault impedance on fault current behavior in different locations are studied.

RECENT DEVELOPMENT OF DC MICROGRIDS

Massive developments achieved in the past century, developments of high voltage transmission lines, PV, WT, and other power system issues. All of these developments caused the appearance of large scale centralized power grids which consumers are connected to the grid by transmission lines. In recent years, the concept of DC Microgrids made a new trend for developing power grids. This trend is shown in Figure 1 (a). And, in Figure 1 (b), the total installed power of Microgrids between 2015 and 2019 is depicted. Based on this figure, the total power of Microgrids is increased by 123% in 5 years.

Today, the various Microgrids are implemented for different purposes, and they are introduced here for better understanding of the Microgrid operations. Some of the established Microgrids are only used for islanded grids in different applications. Therefore, there is a wide area of the concept and application of the Microgrids.

In recent years, Japan has been increased the usage of renewable energy resources, but, this reduces the power quality of the power system in Japan. In Japan, mainly renewable resources are PV and WT, which both of these DERs have several obstacles. Therefore, the usage of Microgrids in Japan solves these problems, and thus, Japan implemented most Microgrids in the world (Morozumi et al., 2007). The project of Aichi in 2006 is designed by PV, fuel cells and battery energy storage system (BESS). The first test of the project showed the difficulties in the voltage and frequency of the AC Microgrid. The AC Microgrid of Kyotango is another Microgrid in Japan, which includes biogas, two PV sources and a

Figure 1. (a) The usage of Microgrids and traditional power grids between 1910 and 2010 (Shaver et al., 2017) (b) Total power of Microgrids

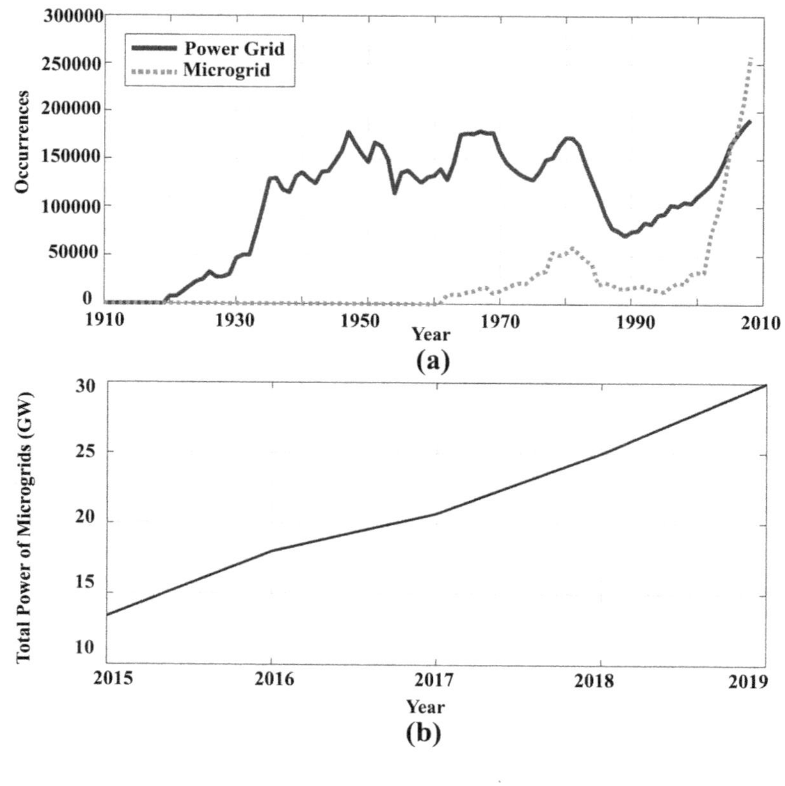

WT. And also, this Microgrid operates as a virtual power plant in Japan. In 2017, an AC Microgrid for supplying 117 homes in Shioashiya was implemented by using PVs. However, all of these traditional Microgrids are AC, which faced several protection and control problems. Therefore, a DC Microgrid is implemented in Nushima Island, which has PV, WT, and BESS. Due to the installation of the DC Microgrid in the island, this system is operated as islanded mode. Thus, implementing the DC Microgrids in recent years has been dramatically increased. The Hefei University of Technology Microgrid, Bosch DC Microgrid at California Honda Facility, Stone Edge Farm Winery Microgrid, Burlington DC Microgrid, and Xiamen University DC Microgrid are some of the new DC Microgrid developments in recent years.

Also, the aforementioned advantages of DC Microgrids over AC Microgrids are summarized in the Table 2. This table shows the reasons of the motivation for using DC Microgrids.

Feasibility of DC Microgrids

Due to the major impacts of changing from AC to DC grid in the traditional power system structure, this transition will be complex. An accurate evaluation of the feasibility of using DC Microgrids in the AC systems must consider many technical factors. The lack of standards for DC systems increases the difficulty of implementing DC Microgrids. DC Microgrids require standards in terms of several risks: damage to loads, faults, electrocution, arcing. And, the key parameters of a protection method are simplicity, cost, speed, and reliability. In the DC Microgrid protection, solid state switches, no load switches fuses, circuit breakers (C.B), and fuses are the components which are lacking standards. In addition to the protection components, design paradigms, such as condition-based maintenance, transient operation, load shedding, and fault isolation, also require the standards.

Table 2. The comparison of AC and DC Microgrids

Under studied factor	AC Microgrid	DC Microgrid
Line impedance	High line resistance	Low line resistance
Frequency	Requiring a frequency monitoring	Don't need to a frequency monitoring
Stability	Lower stability, due to sensitive to external disturbances	More stable
Power transmitted	Due to the high line losses, the power transmitted is limited	Due to the lack of skin effect, ad lower losses, the power transmitted capability is higher than AC Microgrid.
Control strategies	• The frequency and phase should be controlled. • Reactive power controller is essential • Synchronization in essential	Due to the lack of frequency and phase in DC systems, the control schemes are more simple, and do not need any reactive power controlling.
Protection strategies	• Most protection methods are designed for AC systems • Clearing fault at zero crossing point • Several standards for AC Microgrid protection is existed.	• Due to the differences in nature of DC and AC systems, the AC protection methods cannot apply on DC systems • High rise of fault current • Due to the lack of zero crossing point, clearing fault is more challenging. • Lack of standards for DC Microgrid protection.
Losses	Due to the high line resistance and more converters, loss is too high	Due to the lower number of converters, and lack of skin effect, the losses in DC Microgrids are low.

On the other hand, the massive growth in the distributed generation (DG), increasing the feasibility of using DERs, and the importance of implementing more efficient systems increases the effectiveness of DC Microgrids. The factors which must be taken into consideration for implementation of DC Microgrids are:

- The return in investment
- The size of the DC Microgrid
- The type of available DER

Evaluation of DC Microgrid Development

Today, a large number of DC Microgrid testing and developments have concentrated on large scale DC Microgrids, such as electric vehicles, laboratories, aircraft, and marine industries. Moreover, many laboratories in universities have DC Microgrids for research in these areas. The DC Microgrids are designed to reach to a number of aims: increasing the usage of renewable energy resources, improving the grid stability, reducing outage time of loads and emissions, the protecting system from supply outages, and financial motivation such as reduction in power losses, and improved returns on renewable energy resources. The developments in the DC Microgrid technologies increase the motivation of the implementation of these systems.

Several papers reviewed the results of grid extension, solar systems, and Microgrids (Levin et al., 2012; Levin et al., 2016; Chaurey et al., 2010), and (Deichmann et al., 2011) compared the Microgrids, grid extension, and renewable based homes. These comparisons are based on analyzing the combination of distribution, fuel, and cost for renewable energies. And the results show that while grid extension has the minimum cost for rural electrification, decentralized systems have more cost efficiency in remote areas. Technically, (Levin et al., 2012) suggested that in remote areas, over half of the consumptions can be supplied by islanded power systems. Moreover, reducing the cost of installation PVs in recent years indicates that the usage of PV based DC Microgrids are the best option for remote areas.

Current Applications

Due to the limitation in the modeling of DC Microgrids, it would be best served to introduce some of the applications of DC Microgrids.

- *Data Centers*

Data centers have a high energy demand and entirely DC loads, therefore, the DC Microgrid is the best option for this system. Lawrence Berkeley National Laboratory (LBNL) data centers have transited to the 380 VDC from traditional 480 VAC. The results of this modifying show 10% to 28% saves in energy consummation. Therefore, it shows data centers can reduce their power demand by almost 20% using DC Microgrids.

- *Commercial*

A DC Microgrid was implemented on USA which most of the loads in this DC Microgrid is DC lighting (George et al., 2006), and these loads are supplied by 21kW PV system. The transition from AC to DC in this system causes a large reduction in lighting density. Low voltage control system and lower lighting density save approximately 20% of energy consumption.

- *Residential*

Another usage of DC Microgrids is residential buildings. In (Kakigano et al., 2010), a DC Microgrid for the residential complex has been implemented; consist of residential loads, PV and natural gas engine. In this system, total conversion losses were 6710 kWh before using the DC system, and by using DC Microgrid, the conversion losses reduced to 1030 kWh. This study showed that using DC Microgrid can save 15% of the total consumption of buildings.

- *Aircraft*

One of the recent developments of DC Microgrids is the usage of this system in more electric aircrafts (MEAs) (Gao et al., 2016). Due to reduced cost, higher efficiency, and the lower total weight, the using of DC Microgrids in MEAs is significantly increased. The nominal voltage range of these DC systems is between 200 and 600V, and in this application, the weight and volume are very important. Normally due to the cable size and lower equipment, the weight of DC Microgrids is lower than AC systems.

Plug-in electric vehicle (PEV) charging stations

Utilizing PEVs is an important step to make a low carbon transport system (Hajimiragha et al., 2010). However, PEV charging rises the consumption of power systems. Due to the remote usage and high current required for this station, the connecting of these stations to the grid has several constraints. Therefore, one of the solutions for these limitations is DC Microgrid (Sechilariu et al., 2014). Thus, the limitations of PEV charging stations are cost, and high current of lines. In this context, due to the cable characteristics, lack of skin effect and reactive power, the maximum current flow of DC Microgrids is higher than the AC system.

- *Maritime*

Maritime power systems are one of the high-power transportation systems which have high power loads require a high power supply. In recent years, one of the important concern of governments become emissions and cost of the maritime systems. Thus, a DC Microgrid removes bulky low-frequency transformers, and decrease the rating of the switchgear. Therefore, it reduces the weight, and size of the system. It will result in 15% fuel saving and more space for cargo (Jin et al., 2016).

DC MICROGRID MODELING

Many studies have been proposed for modeling of distributed generation and Microgrids, but, small amount of research has been carried out for DC Microgrid modeling. By increasing the penetration of DC Microgrids, the detailed modeling of DC Microgrids should not be ignored in the power system simulations. However, the detailed model and dynamic modeling of DC Microgrids increases the complexity and time of simulation, therefore, using a simplified or dynamic model for each part of DC Microgrids depends on the aim of the study. In this section, detailed and simplified models of each part of DC Microgrids are explained, and the suitability of each model for fault and protection study are discussed.

Photovoltaic

PVs are mainly modeled by an equivalent circuit model, and the main advantage of this modeling is the simplicity of analyzing of electrical models in standard electrical software, which a PV model can be easily integrated with other components.

The Ideal Model

The p-n junction absorbs photons during the presence of irradiation and generates a difference in potential to cause a current flow in the external circuit. The resulting current is photocurrent, which depends on the irradiation and absorption capacity of the semiconductor. The output current of the cell is shown by equation (1), and the equivalent circuit is shown in figure 2. In the ideal single diode, the model includes a current source connected to a parallel diode (Chin et al., 2015).

$$I = I_{PV} - I_D = I_{PV} - I_0[\exp(\frac{qV}{akT}) - 1] \tag{1}$$

Where I_{PV} is photocurrent, I_D is diode current, I_0 is saturation current, k is Boltzmann's constant, T is the temperature of the junction, α is ideality factor of the diode, V is voltage of PV, and q is charge of electrons.

R_S-Model With a Single Diode

Due to simplicity and neglecting many elements, the ideal model of PV cannot be used in the PV simulation. For increasing the accuracy of the model, a series of resistance, R_S, is used to represent the losses between the electrode and silicon surface, and the resistance of electrodes (Lasnier et al., 2017). The equation of current based on the R_S-model with the single diode is given by equation (2), and the model is depicted in figure 3.

$$I = I_{PV} - I_0[\exp(\frac{q(V + IR_S)}{akT}) - 1] \tag{2}$$

R_P-Model With a Single Diode

Due to the lack of temperature parameters, such as open circuit voltage, and temperature coefficient in R_S-model, the temperature variations decrease the accuracy of this model. Therefore, to solve this problem, a resistance paralleled with a diode to improve the temperature sensitivity by modeling the leakage current in a p-n junction. Thus, the expression of current in R_P-model with single diode is given as follows:

Figure 2. Ideal model of PV

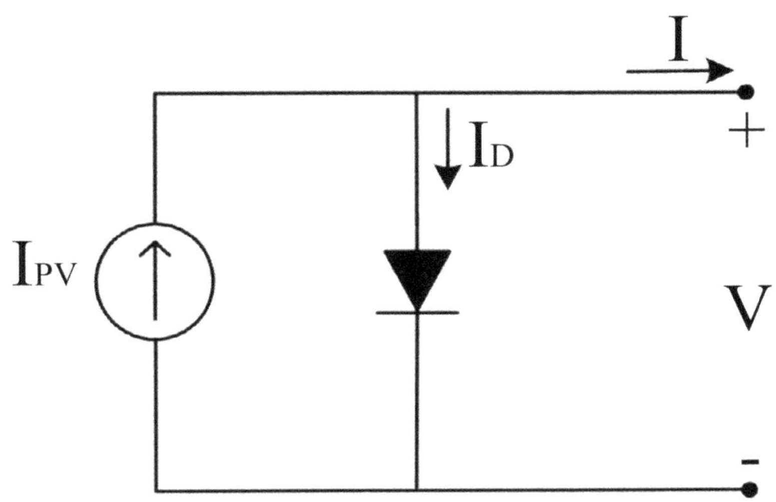

$$I = I_{PV} - I_0[\exp(\frac{q(V + IR_S)}{akT}) - 1] - \frac{V + IR_S}{R_P} \qquad (3)$$

Despite the taken into account the temperature variations, the accuracy reduces at low irradiances. The diagram of this model is shown in Figure 4.

Figure 3. R_S-model with a single diode

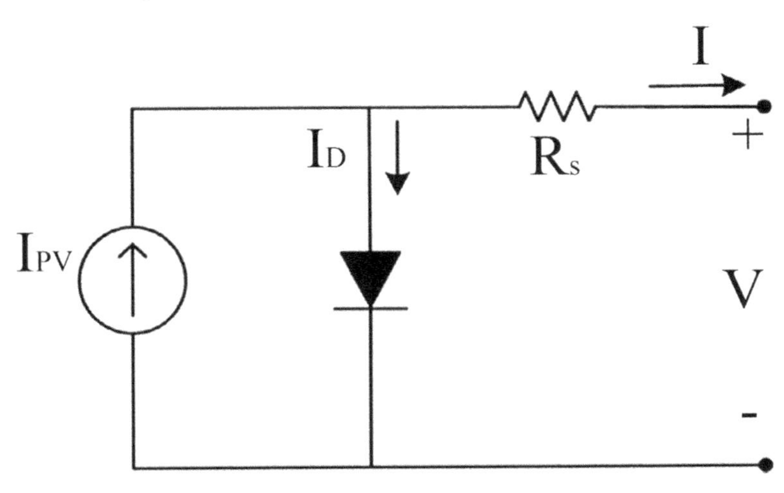

Figure 4. R_P-model with a single diode

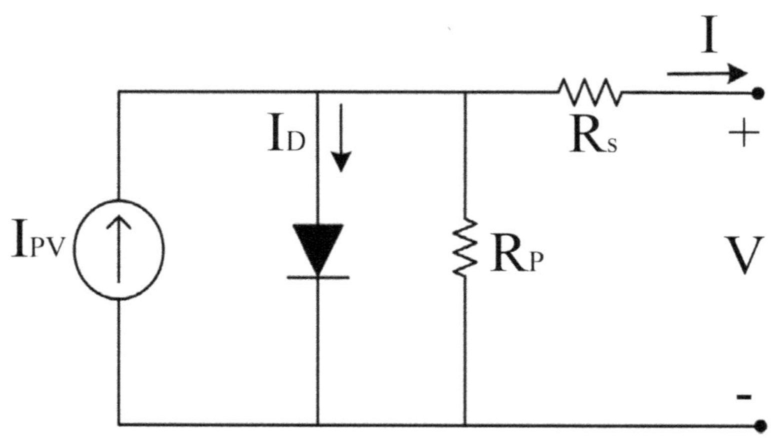

Two-Diode Model

Another loss factor in PVs is recombination current loss, which is modeled by a parallel diode. The two-diode model increases the accuracy of a model particularly in low voltage systems, and this model is depicted in Figure 5. The equation of current behavior is shown by:

$$I = I_{PV} - I_{01}[\exp(\frac{q(V + IR_S)}{a_1 kT}) - 1] - I_{02}[\exp(\frac{q(V + IR_S)}{a_2 kT}) - 1] - \frac{V + IR_S}{R_P} \quad (4)$$

Where all variables are the same to the other models, and the subscript 1 and 2 show the number of the diode. In the low irradiations and low voltage, the two-diode model has the highest accuracy. However, the presence of two diodes in this model increases the simulation time.

Figure 5. Two-diode model

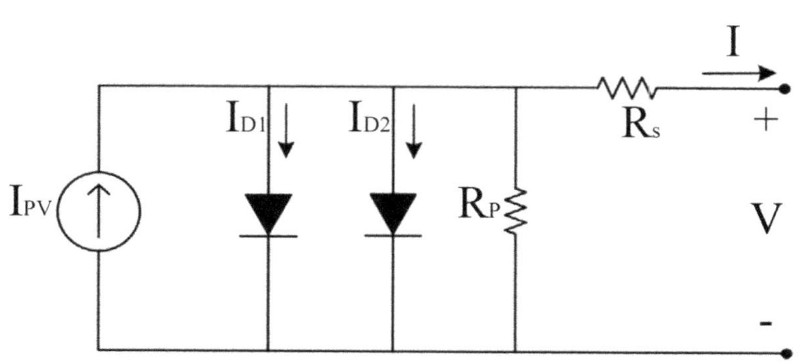

PV System During the Fault

In the DC Microgrid with high penetration of PV systems, from the PV side, faults can be occurred in two different locations, into the PV system or the outside of the PV system. During the fault, PVs inject the fault current into the DC Microgrid. The behavior of the PV system during the fault and the value of injected fault current depends on the converter design. When the PV converter pushes maximum power of PV into the DC Microgrid, during the fault and voltage drop, it will be the same with a constant power source. However, if the fault current rises higher than the maximum current rating of the converter, it limits the fault current. In the protection studies, the limited current is assumed to be twice the maximum current rating. Therefore, PV systems with a typical voltage source converter are modeled by a controlled current source (Hooshyar et al., 2013), which is controlled by the PV power and terminal voltage.

The PV panels for faults in the PV systems can be modeled by electrical equivalent models. During the fault, the voltage of the PV panel is shown by:

$$V = R_f I \tag{5}$$

Which the R_f is the value of fault resistance, and the value of PV current is shown by equations (6) to (9) for different PV models:

$$\text{Ideal Model: } I = I_{PV} - I_0[\exp(\frac{qR_f I}{akT}) - 1] \tag{6}$$

$$R_s\text{-Model with single diode: } I = I_{PV} - I_0[\exp(\frac{q(R_f I + IR_S)}{akT}) - 1] \tag{7}$$

$$R_P\text{-Model with single diode: } I = I_{PV} - I_0[\exp(\frac{q(R_f I + IR_S)}{akT}) - 1] - \frac{R_f I + IR_S}{R_P} \tag{8}$$

$$\text{Two-diode Model: } I = I_{PV} - I_{01}[\exp(\frac{q(R_f I + IR_S)}{a_1 kT}) - 1] - I_{02}[\exp(\frac{q(R_f I + IR_S)}{a_2 kT}) - 1] - \frac{R_f I + IR_S}{R_P} \tag{9}$$

Which, based on the value of fault resistance, the value of current will be different. The variation of fault current in terms of fault resistance for BP solarMSX-60 with data of Table 3 is shown in figure 6.

Based on Figure 6 and experimental values for BP solarMSX-60 in (Ishaque et al., 2011), by increasing the fault resistance and voltage, the value of fault current must be reduced approximately from 3.8 A to 3.72 A. Therefore, for fault studies, the ideal and R_S-model is not suitable, and two-diode model

Table 3. Parameters for four different models

Parameters	Ideal Model	R_S-Model with a single diode	R_P-Model with a single diode	Two-diode Model
q	$-1.60217646\times10^{-19}$	$-1.60217646\times10^{-19}$	$-1.60217646\times10^{-19}$	$-1.60217646\times10^{-19}$
K	1.3806503×10^{-23}	1.3806503×10^{-23}	1.3806503×10^{-23}	1.3806503×10^{-23}
α	1.2	1.2	1.2	1.2
T	300	300	300	300
I_0	9.094×10^{-8}	9.094×10^{-8}	9.094×10^{-8}	$I_{01}=I_{02}= 4.704\times10^{-10}$
I_{PV}	3.8	3.8	3.8	3.8
R_S	-	79	0.2	0.35
R_P	-	-	304.83	176.4

has the highest accuracy. However, using the two-diode model increases the simulation time. Therefore R_P-model with single diode provides a suitable balance between accuracy and simulation time.

Wind Turbine

WTs are one of the main DERs in DC Microgrids. Thus, modeling of WTs in DC Microgrid studies is an essential part. Several sections of a WT must be modeled as shown in Figure 7, which start from the modeling of wind speed. The input of the rotor model is wind, rotor speed, and pitch angle, and this model converts the kinetic energy of wind in the mechanical part then applies it to the generator. At the next part, generator and power converter transform the mechanical power in the electrical power and supply the DC Microgrids with the current. The voltage, rotor speed, and pitch angle controller control its control parameters by connecting to the different models. The protection scheme limits the converter current detects the fault and switch off the WT during disturbances. However, increasing the detail of the model increases the simulation time. Thus, a model for protection studies does not require several sections of the modeling. Therefore, WT can be modeled by a simplified model for protection studies.

Figure 6 The fault current variation for four different models

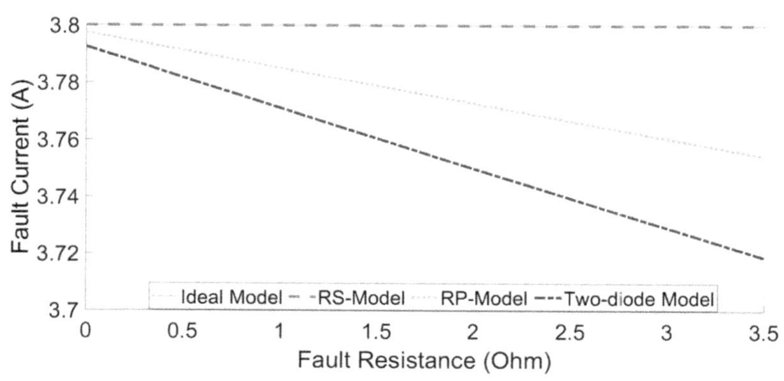

Figure 7. Different sections of a WT

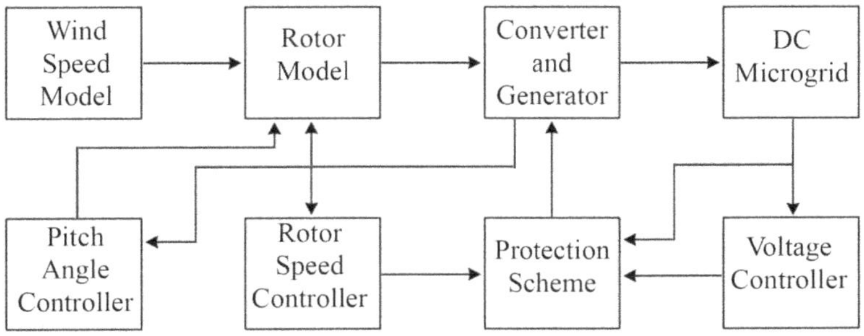

In (Erlich et al., 2008), a simplified order of WT is proposed, as shown in Figure 8. Therefore, it is an interesting model for protection and fault studies, due to the increasing simulation time. In figure 8, the transient impedance of the WT is explained by equation (10):

$$Z' = r_s + jx' = r_s + j\omega_0(L_h + L_S - \frac{L_h^2}{L_h + i_R}) \tag{10}$$

Where, r, x, L are resistance, reactance, and inductance, respectively. And the subscripts of R and S show rotor and stator part, respectively. The subscripts of h show the main field, Z′ is the transient impedance, and ω is the angular speed.

The value of the voltage source is defined by:

$$u' = j\omega_0 k_R \psi_R \tag{11}$$

Figure 8. Simplified model for WT

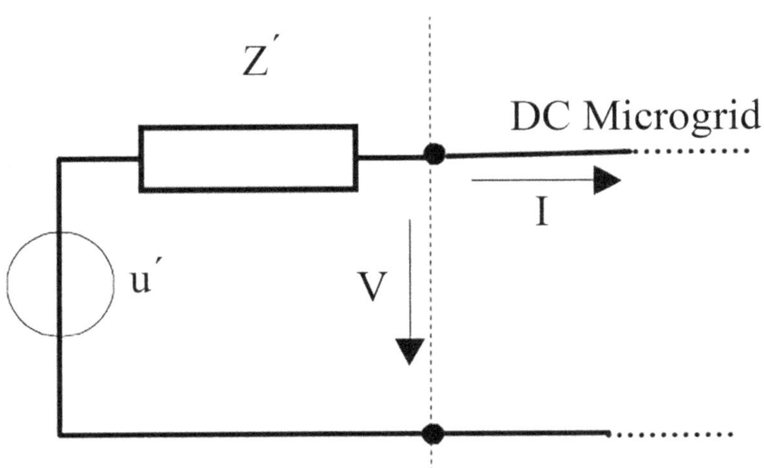

Where, the value of k is an arbitrary reference frame, and ψ is complex flux-linkage. To calculate the unknown parameters, the following differential equations must be solved:

$$\frac{d\psi_{Rd}}{dt} = -\frac{r_R}{L_R}\psi_{Rd} - (\omega_R - \omega_0)\psi_{Rq} + k_R r_R i_{Sd} + u_{Rd} \tag{12}$$

$$\frac{d\psi_{Rq}}{dt} = (\omega_R - \omega_0)\psi_{Rd} - \frac{r_R}{L_R}\psi_{Rq} + k_R r_R i_{Sd} + u_{Rq} \tag{13}$$

$$\frac{d\omega_R}{dt} = \frac{1}{\theta_m}(k_R(\psi_{Rd} i_{Sq} - \psi_{Rq} i_{Sd}) + t_m) \tag{14}$$

Where, the subscripts of q and d is quadrature and direct axis component, respectively. And θ is the inertia of complete rotor shaft.

DC Line

The equivalent model of a transmission line is shown in Figure 9, which, R, L, and C are resistance, inductance, and reactance, respectively. In the DC Microgrids, due to the short length of lines, the capacitive effects of lines can be neglected. Therefore, a π equivalent of line with only resistance and inductance is suitable for protection studies.

On the other hand, the high power of future DC Microgrids, such as aerospace and marine systems, causes increase in fault current, thus, taking into account the skin effect is an essential part of modelling of DC lines in DC Microgrids, and also, the impact of skin effect in this high-frequency current can impact on the protection scheme design. The skin effect is a phenomenon which AC current flows in the surface of the line, but, in the normal operation of DC systems, the lines do not have any skin effect.

Figure 9. Equivalent model of DC line

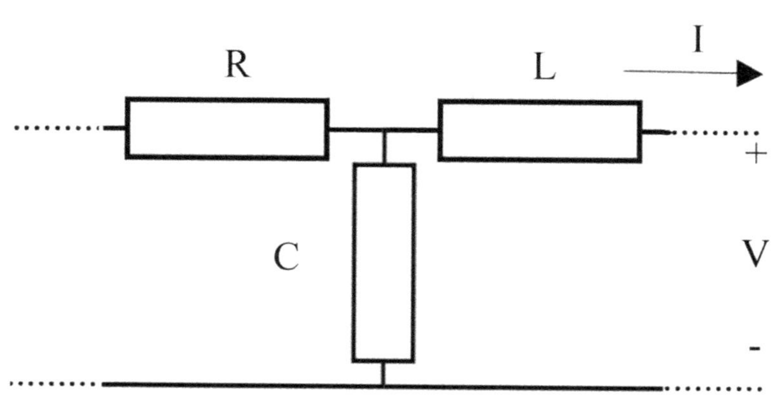

However, the impact of the skin effect increases by increasing the frequency, then, it reduces the effective area of the cable. Then, the resistance of the line increases for higher frequencies. Considering the impact of skin effect on the modeling of the DC Microgrid behavior during fault is a key for increasing the accuracy of protection schemes.

An approach of skin effect modeling is the use of an equivalent ladder circuit (Fletcher et al., 2010), consist of several resistance and inductances. This model is shown in Figure 10, which Figure 10 (a) shows the line structure in four different segments, and figure 10 (b) shows the equivalent ladder circuit of the line. Value of each resistance and inductance in figure 10 (b) is related to its number in figure 10 (a). Section number four shows the largest circle in the center of the line, and by increasing the area, the value of inductance and resistance will reduce. due to the independent behavior of the external segment with frequency, it modeled by only a resistance. Therefore, due to the lower resistance and inductance in the skin effect model, the value of the fault current will be higher than the simplified model.

Power Converter

In a DC Microgrid, the behavior of the power converters during fault divided into an uncontrolled fault current by resources and a passive discharge current by capacitor discharges. The modular multilevel converter is most attractive converter in the Microgrids. This type of converter is designed by controllable submodules, which are two-level half-bridge converters. For the modular multilevel converter, the fault current behavior is influenced by two different aspects. The capacitor is distributed in different series submodules, and IGBTs of sub-modules blocks the discharge of the capacitors. Second, converters have series reactors, which limit the rising rate of fault current during the capacitor discharge.

For fault studies, a simplified model of converter increases the simulation speed and also, it must consider the capacitor discharge during fault. Thus, an equivalent RLC model is proposed in (Leterme et al., 2014), as shown in Figure 11. A method for modeling the impact of the capacitor of the converters is to consider the decreasing trend of the voltage of a converter. A modular multilevel converter has several submodules and arms. Therefore, the maximum magnitude of the injected voltage by one arm

Figure 10. Ladder circuit (a) structure of line (b) equivalent model

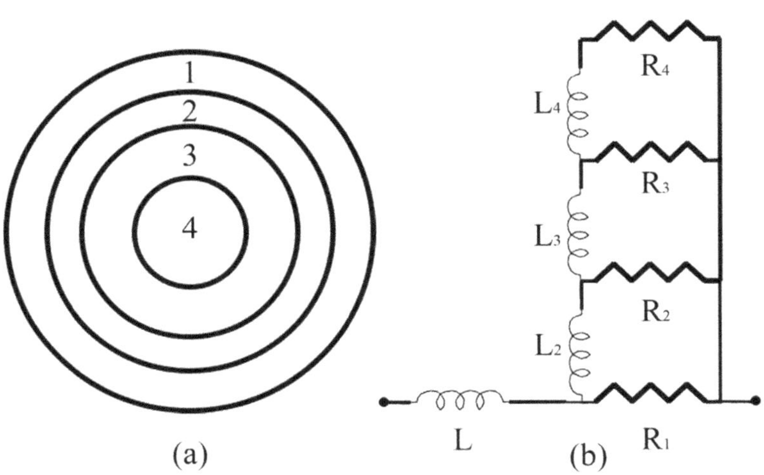

Figure 11. RLC model of the converter during fault

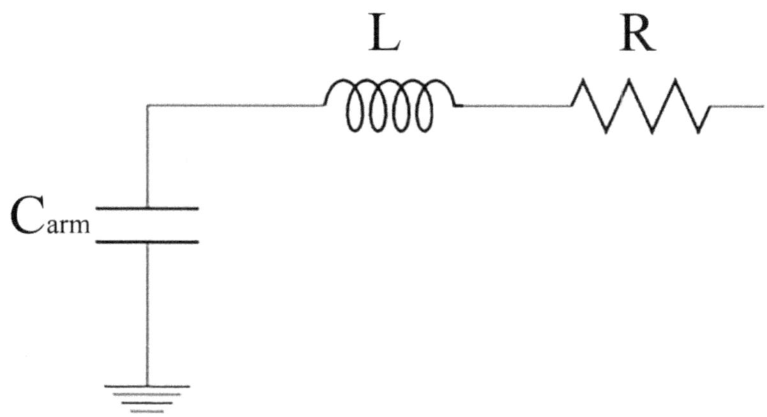

of the converter is the total of the available voltage in that arm. The voltage variations of each converter arm in terms of the current are as follows:

$$\begin{cases} \dfrac{du_u}{dt} = \dfrac{n_u i_u}{C_{arm}} \\ \dfrac{du_L}{dt} = \dfrac{n_L i_L}{C_{arm}} \end{cases} \tag{15}$$

Where, u_u and u_L are the voltage of the top and bottom of the arm, respectively. C_{arm} is equivalent series capacitance, n_L and n_u are the numbers of submodules in the lower and upper of the arm, i_L and i_u is the current of bottom and top of the arm, respectively. The voltage of the one module is calculated by summation of the upper and lower arm voltage. In addition, during the fault, the current of the upper and lower arm will be the same. Therefore, the equation of calculating the capacitor is obtained by:

$$\frac{du_{leg}}{dt} = \frac{i_{leg}}{C_{arm}} \tag{16}$$

Where, i_{leg} and u_{leg} are the voltage and current of the arm during fault, respectively. The value of resistance of the model can be calculated by:

$$R = n_u.R_{IGBT} + (N - n_u).R_d + n_L.R_{IGBT} + (N - n_L).R_d \tag{17}$$

Where, N is the number of the legs. The equivalent resistance is the sum of the resistance of all components during the discharge. The resistance of the lower diode is R_d, and resistance of the IGBT is R_{IGBT}.

Battery

In a DC Microgrid, due to the uncertainties of renewable energy resources, batteries are the main component of each DC Microgrid. The equivalent circuit of a battery consists of resistors, capacitors, and a DC voltage source to provide an accurate dynamic model for this component. This model is shown in Figure 12, which has suitable accuracy and complexity for fault studies.

The paralleled RC is the response of battery nonlinear polarization. The value of V_{DC} is the open circuit voltage of the battery, R_1 and R_2 are polarization resistances, R_0 is the connection resistance of the battery, and C_1 is polarization capacitance. During the first moment of fault, due to the presence of the two capacitors in this model, these capacitors participate in injecting current as capacitor discharge. Also, the voltage resource of the model will supply a constant fault current to the system.

ANALYZING OF DC MICROGRID DURING THE FAULT

Based on the previous sections, an equivalent model by modeling each component can be designed and analyzed by using simulation tools. The case study of this model is an islanded 600 V DC Microgrid with ring configuration. The case study is analyzed in different fault situations to investigate the challenges of protection of a DC Microgrid and behavior of the fault current in different stages. As shown in Figure 13, the studied DC Microgrid consists of PV, WT, battery, and loads. In this system, the fault can occur in three different sections in the 5s. The fault current behaviors for three different sections are shown in Figure 14. The fault resistance for all faults is 0.1 Ohm. As shown in Figure 14, fault 3 has the highest magnitude, due to the high fault current of DC motor and battery.

A fault current in DC Microgrid can be analyzed in two different stages. The first stage is capacitor discharge, and it starts immediately after fault, and continue until the voltage reaches to its minimum. For example, for a fault at F_3 the fault current and voltage behaviors are depicted in Figure 15. During the fault, the capacitor begins discharging in the cable, and the fault current behavior can be given by (Bayati et al., 2018):

Figure 12. Equivalent model of battery

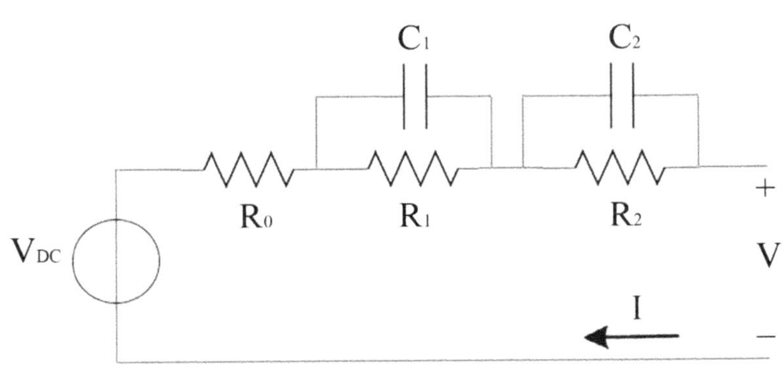

Figure 13. Islanded 600 V DC Microgrid

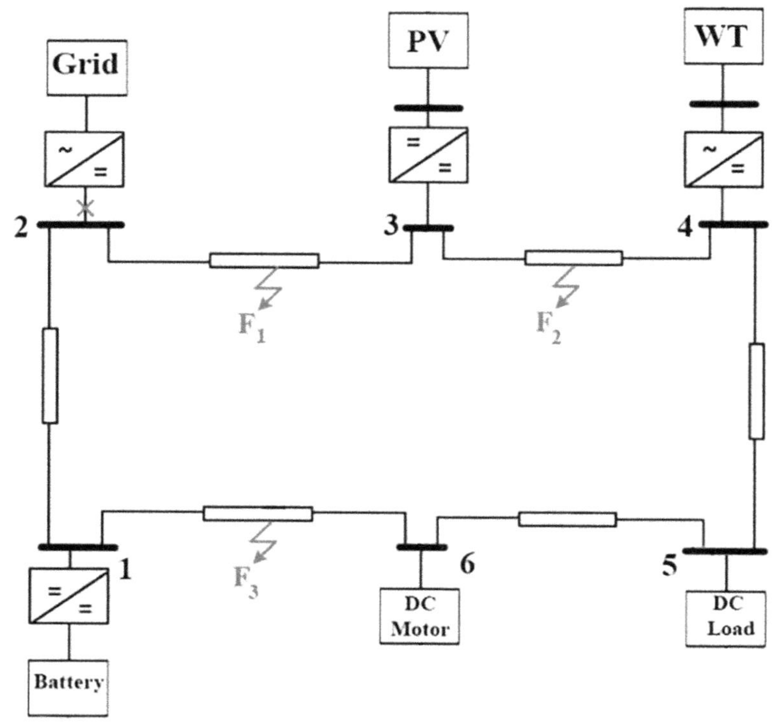

Figure 14. Fault current for three different place

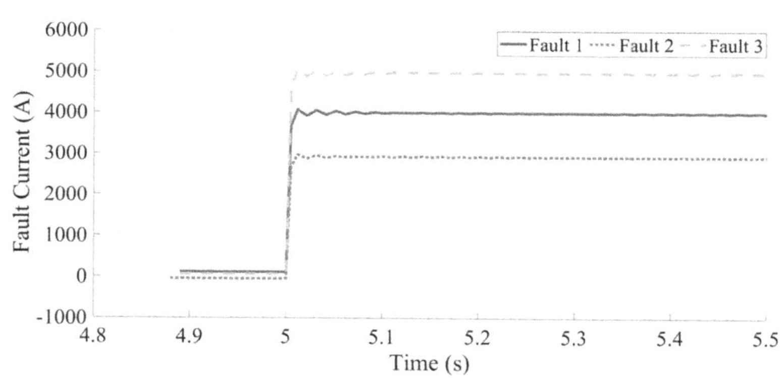

Figure 15. Fault behavior of the DC Microgrid

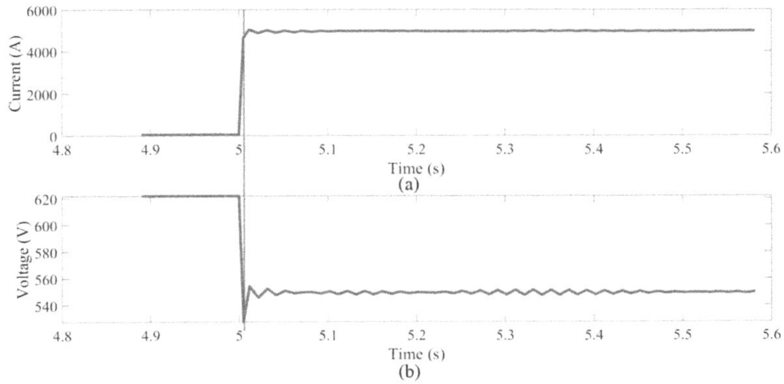

$$
\begin{cases}
i(t) = \dfrac{V_0}{\omega L} e^{-at} \sin(\omega t) - \dfrac{I_0 \omega_0}{\omega} e^{-at} \sin(\omega t - B) \\
a = \dfrac{R}{2L} \\
\mathrm{B} = \arctan(\omega \,/\, \alpha) \\
\omega = \sqrt{\dfrac{1}{LC} - (\dfrac{R}{2L})^2} \\
\omega_0 = \sqrt{a^2 + \omega^2}
\end{cases}
\tag{18}
$$

Where, R, L, C, V_0, and I_0 are equivalent resistance, inductance, capacitance, initial voltage and current, respectively.

The second stage starts when the voltage reaches to the minimum value, and this impacts on freewheel diodes to conducts. Thus, this provides a current path independent to the IGBT state, and the current ten times of the nominal value will flow through freewheel diodes.

One of the challenges of protection methods is detecting High Impedance Faults (HIFs). Figure 16 shows the magnitude of fault current in the DC Microgrid for F2 in terms of different fault resistances. The magnitude of fault current decreases by increasing the amount of fault resistance, therefore, for HIFs, distinguish between fault and overload is a challenge for protection schemes. In this system, the value of the nominal current of the line between buses 3 and 4 is 52 A. As shown in figure 16, the fault current with fault resistance of 3 Ohms is 68 A, which is difficult for protection devices to distinguish between this and 20% overload.

Another challenge of the DC Microgrid protections is the bidirectional fault current in the ring configurations. Therefore, the fault current cannot be measured from one side. As shown in figure 17, the fault current of F1 is injected by two sides of the line. Hence, a protection scheme must measure the fault current from two sides of the line to locate the fault. In addition, using measurement and protection devices in each line increases the cost of the protection scheme, thus, minimizing the number of measurement devices and data decrease the cost.

Figure 16. fault current reduction in terms of fault resistance

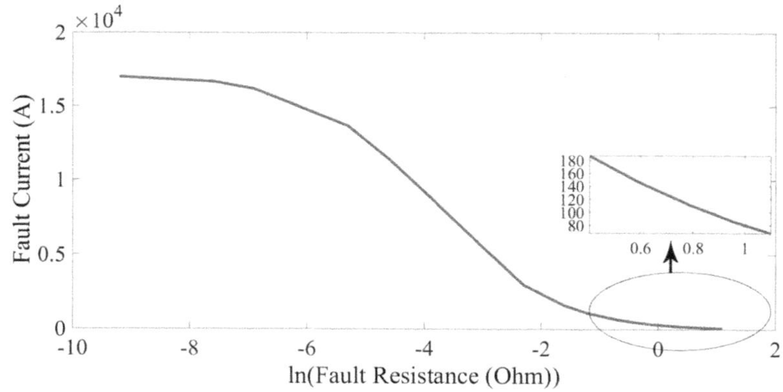

PROPOSED DC MICROGRID PROTECTION SCHEME

In this research, a protection method for detection, classification, and isolation of the fault in the DC Microgrids is proposed. The first step of the proposed protection method is fault detection. Then, based on the direction of the current of each relay, the fault is classified into PP and PG faults. Therefore, the DC Microgrid is divided into three different zones, which can operate independently to other zones. After detection of a fault and defining the faulty zone, the faulty zone is isolated and other zones are operated by the normal condition. In this scheme, the number of relays and protection devices are minimized to reduce the cost, complexity, and communication links. The structure of the protection scheme and the studied DC Microgrid are shown in Figure 18. Which, three relays and four C.Bs are connected to the Central Protection Unit (CPU) to detect and classify fault, and then send the trip signal to necessary C.Bs.

In this scheme, the DC Microgrid is divided into three different zones, zone 1 is connected to the grid, and zones 2 and 3 have a battery. Therefore, each zone can disconnect from other zones and operate with new conditions. Due to the isolating the fault by three zones, this scheme only requires three relays and four C.Bs.

Figure 17. Bidirectional fault current for F1

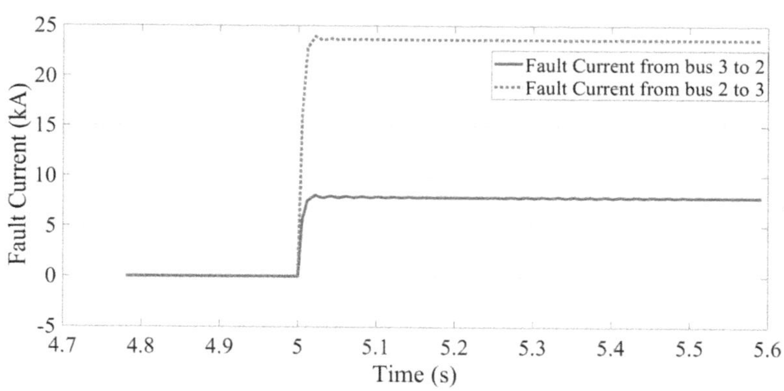

Figure 18. Case study and zones of a DC Microgrid

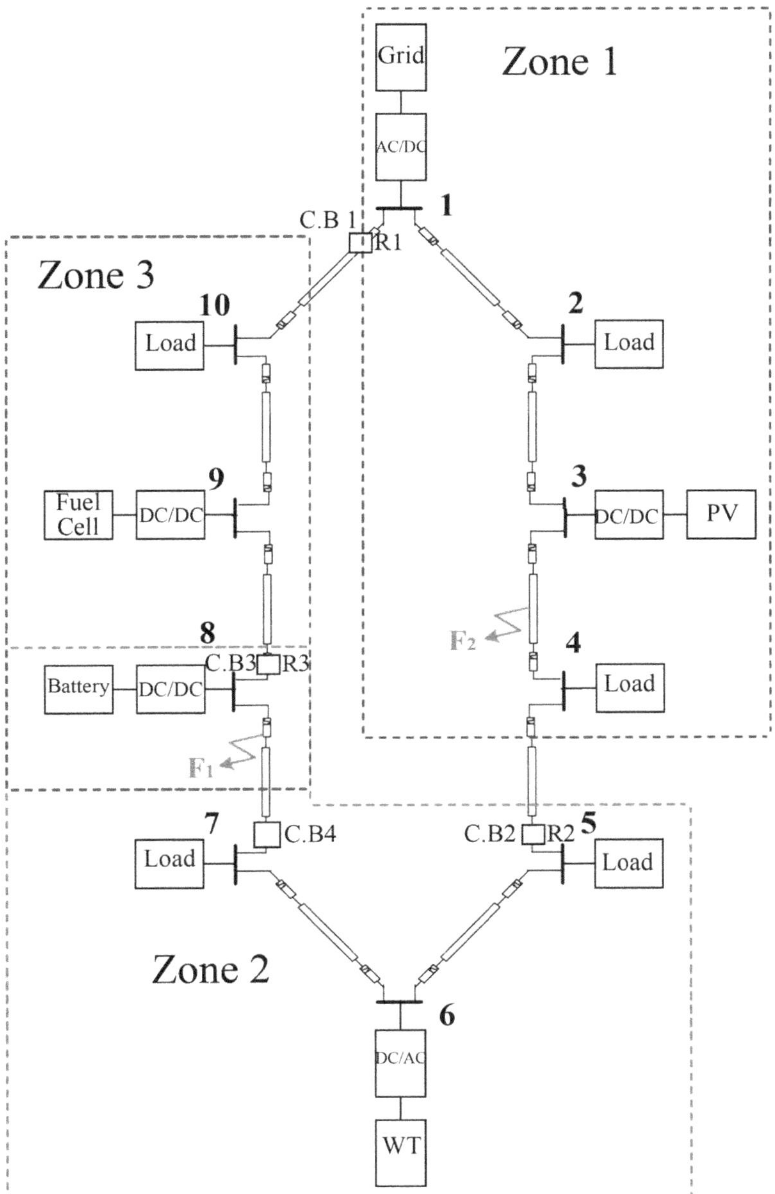

Fault Detection

During the normal operation of the DC Microgrid, the value of current is approximately constant. However, during the fault, the value of current will be increased to a higher value in a short time. Accordingly, the value of the derivative of current is calculated by $\Delta I/\Delta t$, which during the fault by increasing the value of current in a short period of time, the derivative of current will dramatically increase. Thus, the fault can be detected by using the variation of derivative of current. Also, during the load increasing or over current situations, the value of the derivative of current will increase. Therefore, determining the value of a threshold for ΔI/Δt is a pivotal problem.

The settings for the ΔI/Δt are classified as a fixed threshold for the protected DC Microgrid. The value of the threshold for ΔI/Δt depends on the fault type and location, and it is defined here by Th. Thus, the threshold must be calculated for both sides of the zones, and the value of Th for fault at the end of the zone is T_{min} and for close location is T_{max}. Therefore, after increasing the value of ΔI/Δt between T_{min} and T_{max}, the relay will go to the classification stage.

Based on the equation (18), the value of ΔI/Δt can be determined by:

$$\frac{dI}{dt} = -\alpha \frac{v_0}{\omega L} e^{-\alpha t} \sin \omega t + \frac{I_0 \omega_0}{\omega} \alpha e^{-\alpha t} \sin(\omega t - B) + \frac{V_0}{L} e^{-\alpha t} \cos \omega \,\mathrm{t} - \mathrm{I}_0\, \omega_0 e^{-\alpha t} \cos(\omega t - B) \tag{19}$$

Then, this value should be more than a threshold, ε, and then in can be rewrite as follows:

$$\frac{V_0}{L} e^{-\alpha t} \cos \omega \,\mathrm{t} - \mathrm{I}_0\, \omega_0 e^{-\alpha t} \cos(\omega t - B) \geq \alpha I(t) + \varepsilon \tag{20}$$

Where the value of the ε is calculated by:

$$\varepsilon = \frac{I_{OL} - I_n}{\Delta t} \tag{21}$$

Where, I_{OL} is the maximum value of overload current, and I_n is the nominal current, and Δt is the rise time for overload, which can be assumed as 10 ms. Therefore, using the above equations proves that if a fault occurs in the system, the fault detection method detects the fault by the lowest operation time.

Fault Classification

The second stage of the proposed DC Microgrid protection scheme is the classification of the fault the determine the type and location of the fault based on the direction of the fault current. During the normal operation of the DC Microgrid, the current direction is indicated by N.C, and the sign of the current for clockwise and anticlockwise currents are indicated by 1 and -1, respectively. In the DC Microgrids, lines are divided into positive and negative lines. During the PG faults, the direction of the unfaulty lines will be N.C, and the direction of the faulty line will change to inject to the fault location. But, in the PP faults, both lines inject the current into the fault location. Thus, based on the DC Microgrid of figure 18, and faults in different locations, the direction of the fault current in different locations and types are determined by Table 4.

Where, the superscripts P and N show the polarity of the line, positive and negative, respectively. And subscripts 1, 2 and 3 for I indicate the relay number. Therefore, by comparing the states of the system with Table 4, the zone of fault is determined, and the relay will go to the isolation stage.

Fault Isolation

The most important task of a protection scheme is isolating the fault location with the lowest interruption in other sections in the quickest time. Therefore, in the proposed protection scheme, after fault detection and classification, the faulty zone will be isolated by two C.Bs at two sides of the zone. For this aim, the CPU sends the trip signal to the desired C.Bs. Due to the short length of the lines in the DC Microgrids, all C.Bs and relays are connected to the CPU by Wi-Fi to reduce the communication link cost and failure and increase the speed of the sending data. In this research, the value of the delays for C.Bs, Wi-Fi, and measuring are assumed 6 ms.

After isolation, three different situations will happen for the DC Microgrids, as shown in figure 19. The isolated zones 1, 2 and 3 are shown in figure 19 (a), (b), and (c), respectively. Therefore, after isolation, in the case of (a), the DC Microgrid operates in islanded mode, (b) and (c) operates in grid-connected mode but with a radial configuration. It is undeniable note that the number of zones depends on the configuration of the protected DC Microgrid.

On the other hand, one of the impacts of isolation in the DC Microgrids is high rise current of the system after reconnection of zones. And, this surge current can damage to the converters and critical loads, therefore, installation a Fault Current Limiter (FCL) series with loads limits the value of the current during fault. an FCL is made by a inductance and reactance with a high value to restrict the value of current during the fault. For this reason, an FCL is serried with loads and the FCLs are activated after detection of the fault, and after isolation of faulty zone FCLs will deactivate.

SIMULATION AND RESULTS

To demonstrate the protection scheme, a DC Microgrid with 500 V and a PV, fuel cell, WT, battery, and 5 loads are simulated in Simulink/MATLAB. The data of the DC Microgrid is shown in Table 5, and the structure of the case study is shown in figure 18. The proposed method is tested in different situations and fault detection, classification and isolation for different situations are evaluated in this section. Also, for considering the delay in communication and measuring, a 10 ms delay is applied to the measured

Table 4. Fault current direction in different zones and types

Zone	Type of fault	Faulted Line Polarity	I_1^P	I_1^N	I_2^P	I_2^N	I_3^P	I_3^N
1	PP	-	1	1	-1	-1	1	1
2	PP	-	-1	-1	1	1	-1	-1
3	PP	-	-1	-1	-1	-1	1	1
1	PG	Positive	1	N.C	-1	N.C	1	N.C
1	PG	Negative	N.C	1	N.C	-1	N.C	1
2	PG	Positive	-1	N.C	1	N.C	-1	N.C
2	PG	Negative	N.C	-1	N.C	1	N.C	-1
3	PG	Positive	-1	N.C	-1	N.C	1	N.C
3	PG	Negative	N.C	-1	N.C	-1	N.C	1

Figure 19. The isolation topologies of the DC Microgrid (a) fault in zone 1 (b) fault in zone 2 (c) fault in zone 3

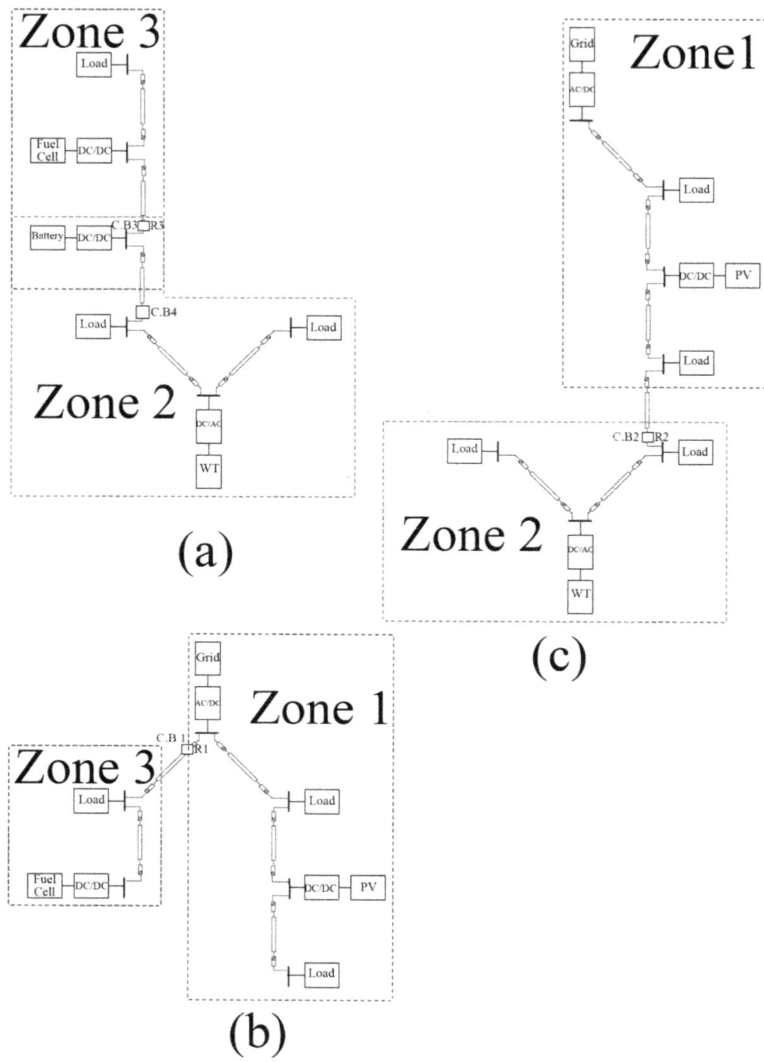

data. The system is designed in the Simulink of MATALB using Simulink ode3 with fixed step time of 20 microseconds. And the summarized block diagram of fault protection scheme in the MATLAB is shown in Figure 20. This method only uses the maximum two current data for each zone, then, the measurements and the computational time are minimized. Figure 20 (a) is the fault detection, and figure 20 (b) is the fault classification and isolation part.

Fault Detection

A PP fault is occurred in the F1 with a fault resistance of 0.001 Ohm at the t=2.5 s, as shown in figure 18, and the fault current is shown in Figure 21 (a). The fault reaches to the maximum value in 18 ms,

Figure 20. The brief block diagram of the protection method (a) fault detection (b) fault isolation

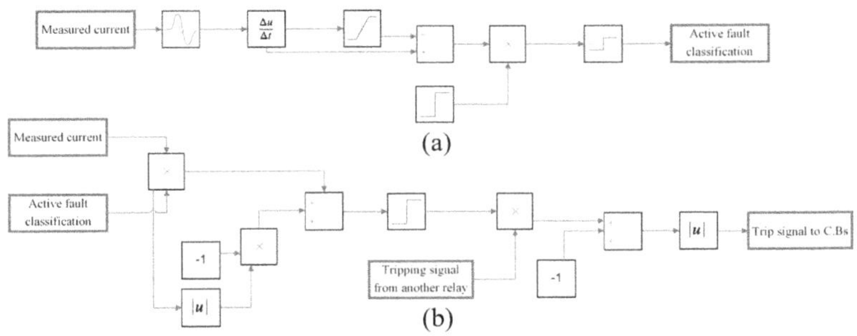

Table 5. The data for the protected DC Microgrid

Component	Data
WT	Power= 275 kW
PV	PV model= SunPower SPR-305-WHT Number of series module=25 Number of parallel strings =66 V_{OC}=64.2 I_{OC}=5.96
Fuel cell	Number of cells in series= 1000 Universal gas constant= 8314 Absolute temperature= 1273
Battery	Rated capacity= 200 Ah Type= Nickel-Metal-Hydride
Line	R for each segment= 0.002 Ohm L for each segment= 0.01 mH
Load	Power= 82 kW

and the maximum fault current is 135 kA. In Figure 21 (b), the signal of the fault detection method is

Figure 21. The fault detection method (a) fault current for fault resistance 0.001 Ohm (b) fault detection signal (c) fault current for fault resistance 0.1 Ohm (d) fault detection signal

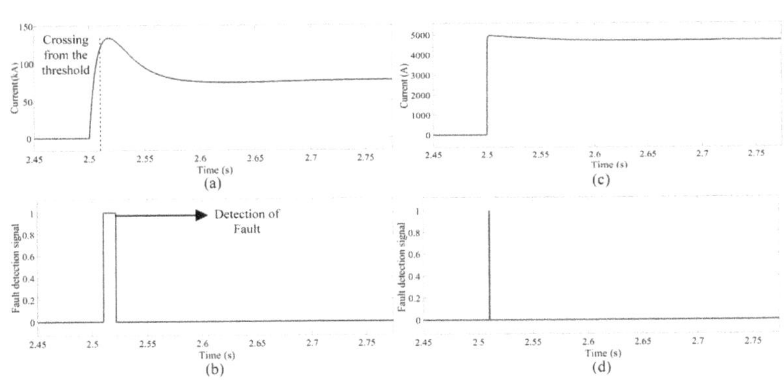

Figure 22. The fault current of (a) F1 (b) F2

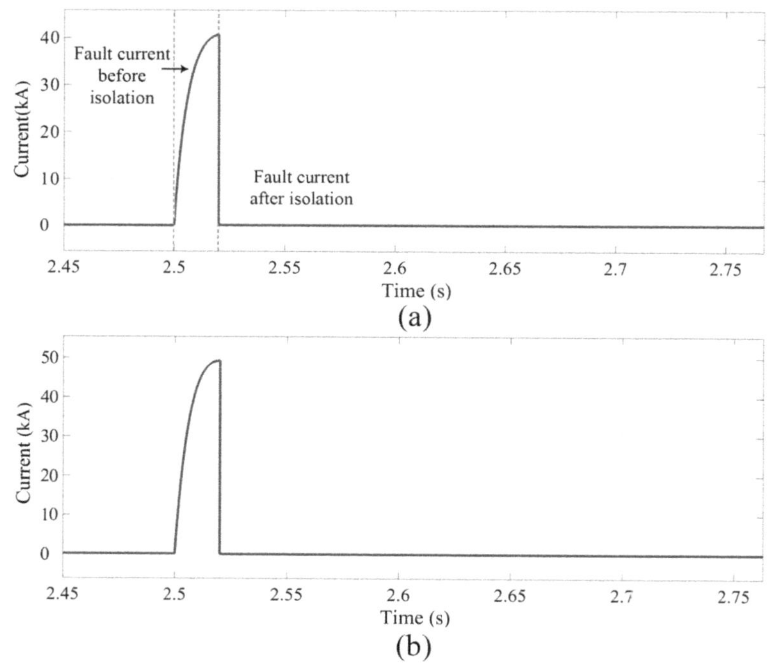

shown, in this issue, the difference between fault time and the detection time is the error. The signal of the fault detection method indicates the trip signal for the classification stage, and for F1 the fault is detected after 10 ms. Also, in Figure 21 (c), the fault current with a fault resistance of 0.1 Ohm at F1 is shown, and figure 21 (d) is the fault detection signal of this fault, which the fault is detected after 10 ms. Therefore, this method is independent of the fault resistance, and in both situations, faults are detected in 10 ms. The activation of the second stage is the output signal of the fault detection stage, and as shown in Figures 21 (b) and (c), it is a 0 and 1 signal, which 1 shows the detection of the fault.

Fault Classification and Isolation

Based on the previous section, the protection area of the DC Microgrid is divided into three different zones, which each zone can operate separately after fault. therefore, after fault detection, the fault classification classifies the fault zone and isolate the faulted zone. By using the settings in the Table 4, the classification stage is calculated the faulty zone and send the trip signal to the C.Bs. Due to the simplicity of the proposed method, the computational time of the classification stage is assumed to be zero. Thus, a fault occurs at the F1, and F2 and the protection method by using two C.Bs isolate the faulty zone, as shown in Figure 22. Therefore, as shown in Figure 22 (a), fault F1 is isolated after 20 ms, which a fast protection method is essential for DC Microgrids. In figure 22 (b), the F2 occurs, and the fault current is shown in figure 22 (b), and the fault is isolated in 20 ms.

On the other hand, one of the main goals of the proposed method is isolating the faulted zone and normally operate of other zones after fault detection. In Figure 23, the power consumption of loads

Figure 23. The load currents, load at (a) Bus 10 (b) Bus 5 (c) Bus 7 (d) Bus 4 (e) Bus 2

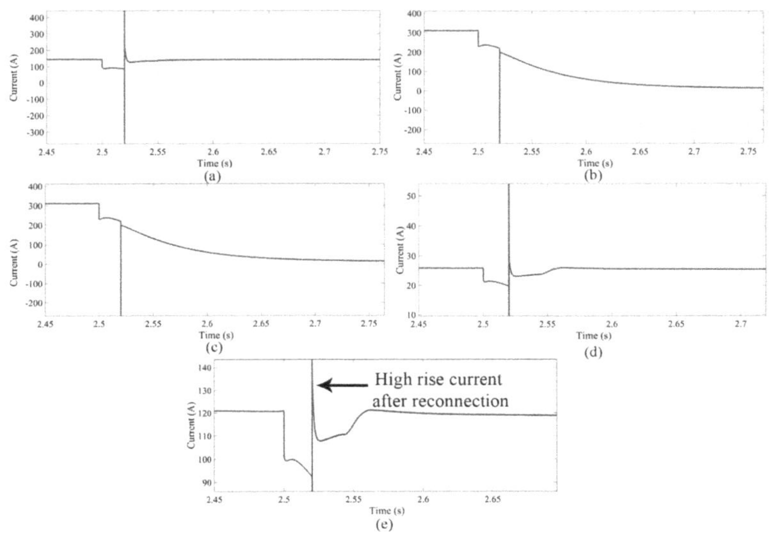

during and after a fault at F1 is presented. Based on the results, loads are affected during the fault and fault detection process, but, after isolation of the fault, loads in other zones operate normally. During the fault at F1, due to the location of the fault, the zone number 2 is isolated, and the structure of the DC Microgrid changes to figure 19 (b). After a fault, due to the outage of the zone 2, the current of loads at buses 5 and 7 reduce to zero as shown in figures 22 (b) and (c). However, the loads at the buses 2, 4, and 10 are still operated in normal operation after isolation of the faulty zone. The results show that after fault detection and isolation, the faulty zone is detected and isolated in 20 ms, and other zones are operated normally without any interruption in load consumption.

Due to the capacitors and inertia of WT after fault, the current of the loads have a time constant to reach zero. Also, loads in the zones 1 and 3 are restored to the normal operation, after 50 ms. Therefore, during 50 ms, the fault is detected, classified, and isolated, and the loads are restored to the normal operation modes. Thus, the main goal of this method is isolating the fault and supplying unfaulted loads after fault isolation. As shown in the Figure 23, loads in the Bus 10, 4, and 2 are supplied as normal operation

Figure 24. The load current after operation of protection scheme

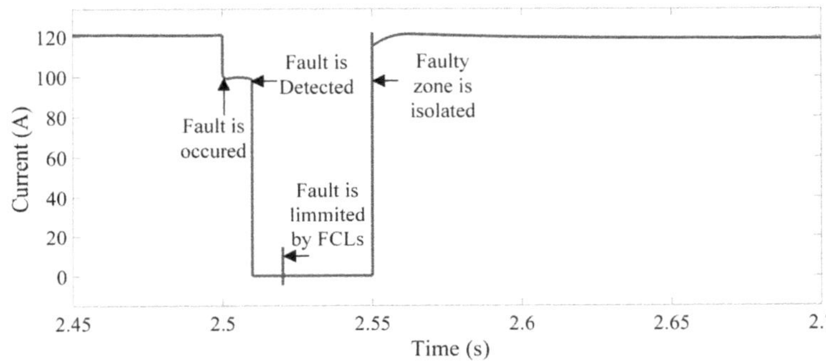

after fault isolation. However, due to the 1% voltage drop after fault isolation, as shown in the figure 23, the current of loads are decreased approximately 1% after fault isolation. Consequently, all loads are operated normally, and the faulty zone is detected and isolated less than 20 ms, and DC Microgrid restores to the normal operation by 50 ms. The fault current of the DC Microgrid goes higher than the threshold, as evident from figure 23. The performances of the proposed method are determined for the cases and the load current of the unfaulty zones are restored to the normal values.

However, after isolation the fault, and high-rise current is injected to the loads, as shown in the figure 23. This current can damage to the converters and other devices in the DC Microgrid. Therefore, an FCL is serried with each load to limit the value of fault current. The fault current after installation of FCLs for load at bus 2 is shown in the figure 24, and operation of the protection scheme in different stages of the fault is depicted. Therefore, finally, the proposed protection scheme detects, locates, isolates, and limits the fault in approximately 50ms, without any damage and interruption to other devices and zones.

In addition, the qualitative and quantitative results of the proposed DC Microgrid protection method are compared with other techniques. Table 6 presents the comparison of three methods with the proposed method based on the cost, reliability, accuracy, and requiring communication facilities. Due to uses the limited number of PDs in the proposed method, the cost of this method is lower than other compared methods. The suggested method in the (Dhar et al., 2018) uses two PDs for each line without determining any zones; therefore after fault, a large part of the systems will be deenergized. In (Salehi et al., 2019), a protection strategy by using communication-assisted fault detection is presented, and it uses communication links between every adjacent PD. Thus, the base of this method is communication links, which increase the cost and also the probability of communication failure. The suggested method in (Sztykiel et al., 2018) uses differential relays to design an adaptive protection method. In this method, a new frequency filtering Kernel based Discrete Time Frequency Transform is proposed. But, the speed of the suggested method is too low.

Based on Table 6, the qualitative comparison of the proposed method with other methods shows the advantages of the proposed method over other related works. Therefore, using the proposed method increases the reliability and accuracy as well as decreases the cost and require communication links of the protection scheme in a DC Microgrid. Also, one of the main factors for evaluating a protection method is clearing time, which the proposed method has the lowest clearing time, therefore, the effectiveness of the proposed method is proven by clearing the fault at the lowest speed.

Table 6. Comparison of three different methods with the proposed method

Performance	Proposed method	(Dhar et al., 2017)	(Monadi et al., 2016)	(Naik et al., 2019)
Cost	Low	High	High	Moderate
Reliability	High	Moderate	Low	Low
Accuracy	High	High	Moderate	Moderate
Requiring communication links	Moderate	High	High	Low
Clearing time	50 ms	85 ms	97 ms	100 ms

CONCLUSION AND DISCUSSION

Due to the bidirectional power flow in the DC Microgrids with ring configuration, protecting these systems is a challenging task. In addition, if the DC Microgrid determined a fault in one zone, it may face a black-out in the whole DC Microgrid. On the other hand, the accurate model of a DC Microgrid for fault studies is an essential part of designing a protection scheme. Due to the increasing the penetration of power electronic devices, the sensitivity of the DC Microgrids to fault current is increased, and the first step of designing a protection scheme for a system is accurate modeling of the system. Therefore, in this chapter, the model each component of DC Microgrid is accurately modeled, and the behavior of all components are accurately studied during the fault. Also, the behavior of different DC Microgrid modeling methods is studied during the fault, and the mathematical equations of these models are proposed. Then, in the second section, a protection method, which protects the DC Microgrid in three different stages by isolating the DC Microgrid in different zones, was proposed. There are many methods for protecting the Microgrids, but there is still a challenge of DC Microgrid protection by considering the operation time and selectivity. Therefore, in this chapter, a fault detection method based on the transient behavior of the fault current using localized relays is proposed based on the derivative of the current, where the fault is detected in 10 ms. Then, a fault classification is proposed to identify the faulted zone, and it is connected to a fault isolation stage. This stage causes a normal operation for healthy zones after the fault. The simulation results show that the proposed method isolates the faulty zone in less than 50 ms. Thus, the operation of the protection method on the DC Microgrid was investigated for different faults, and the analysis proves the effectiveness of the proposed method on the DC Microgrid protection. In addition, due to the high surge current during reconnection of zones to the grid, an FCL is connected to loads and protects loads against high currents decreasing the probability of damage to power electronic devices. The results show that the healthy zones of the DC Microgrid are restored to the normal operation mode in 50 ms, which is very low compared with other studies. Moreover, the proposed protection method causes a fast restoring after fault occurrence for the DC Microgrid. The comparison of this method with other methods shows the superiority of the proposed method in terms of cost, speed, reliability, and resiliency.

REFERENCES

Amamra, S. A., Ahmed, H., & El-Sehiemy, R. A. (2017). Firefly algorithm optimized robust protection scheme for DC microgrid. *Electric Power Components and Systems*, *45*(10), 1141–1151.

Ammerman, R. F., Gammon, T., Sen, P. K., & Nelson, J. P. (2010). DC-arc models and incident-energy calculations. *IEEE Transactions on Industry Applications*, *46*(5), 1810–1819.

Andrea, J., Schweitzer, P., & Tisserand, E. (2010, October). A new DC and AC arc fault electrical model. In *2010 Proceedings of the 56th IEEE Holm Conference on Electrical Contacts* (pp. 1-6). IEEE.

Arif, A., Wang, Z., Wang, J., Mather, B., Bashualdo, H., & Zhao, D. (2018). Load modeling—A review. *IEEE Transactions on Smart Grid*, *9*(6), 5986–5999.

Augustine, S., Quiroz, J. E., Reno, M. J., & Brahma, S. (2018). *DC Microgrid Protection: Review and Challenges (No. SAND2018-8853). Sandia National Lab.* Albuquerque, NM: SNL-NM.

Bayati, N., Dadkhah, A., Sadeghi, S. H. H., Vahidi, B., & Milani, A. E. (2017, June). Considering variations of network topology in optimal relay coordination using time-current-voltage characteristic. In *2017 IEEE International Conference on Environment and Electrical Engineering and 2017 IEEE Industrial and Commercial Power Systems Europe (EEEIC/I&CPS Europe)* (pp. 1-5). IEEE.

Bayati, N., Hajizadeh, A., & Soltani, M. (2018). Protection in DC microgrids: A comparative review. *IET Smart Grid*, *1*(3), 66–75.

Bayati, N., Hajizadeh, A., & Soltani, M. (2018, June). Impact of Faults and Protection Methods on DC Microgrids Operation. *In 2018 IEEE International Conference on Environment and Electrical Engineering and 2018 IEEE Industrial and Commercial Power Systems Europe* (EEEIC/I&CPS Europe) (pp. 1-6). IEEE.

Bayati, N., Hajizadeh, A., & Soltani, M. (2018, September). Accurate Modeling of DC Microgrid for Fault and Protection Studies. *In 2018 International Conference on Smart Energy Systems and Technologies (SEST)* (pp. 1-6). IEEE.

Cairoli, P., Kondratiev, I., & Dougal, R. A. (2013). Coordinated control of the bus tie switches and power supply converters for fault protection in DC microgrids. *IEEE Transactions on Power Electronics*, *28*(4), 2037–2047.

Chaurey, A., & Kandpal, T. C. (2010). A techno-economic comparison of rural electrification based on solar home systems and PV microgrids. *Energy Policy*, *38*(6), 3118–3129.

Chin, V. J., Salam, Z., & Ishaque, K. (2015). Cell modelling and model parameters estimation techniques for photovoltaic simulator application: A review. *Applied Energy*, *154*, 500–519.

Conroy, J., & Watson, R. (2009). Aggregate modelling of wind farms containing full-converter wind turbine generators with permanent magnet synchronous machines: Transient stability studies. *IET Renewable Power Generation*, *3*(1), 39–52.

de Oliveira, T. R., Bolzon, A. S., & Donoso-Garcia, P. F. (2014, October). Grounding and safety considerations for residential DC microgrids. *In IECON 2014-40th Annual Conference of the IEEE Industrial Electronics Society* (pp. 5526-5532). IEEE.

Deichmann, U., Meisner, C., Murray, S., & Wheeler, D. (2011). The economics of renewable energy expansion in rural Sub-Saharan Africa. *Energy Policy*, *39*(1), 215–227.

Dhar, S., Patnaik, R. K., & Dash, P. K. (2017). Fault detection and location of photovoltaic based DC microgrid using differential protection strategy. *IEEE Transactions on Smart Grid*, *9*(5), 4303–4312.

Elsayed, A. T., Mohamed, A. A., & Mohammed, O. A. (2015). DC microgrids and distribution systems: An overview. *Electric Power Systems Research*, *119*, 407–417.

Ericsen, T., Hingorani, N., & Khersonsky, Y. (2006). Power electronics and future marine electrical systems. *IEEE Transactions on Industry Applications*, *42*(1), 155–163.

Erlich, I., Kretschmann, J., Mueller-Engelhardt, S., Koch, F., & Fortmann, J. (2008, July). Modeling of wind turbines based on doubly-fed induction generators for power system stability studies. In 2008 IEEE Power and Energy Society General Meeting-Conversion and Delivery of Electrical Energy in the 21st Century (pp. 1-8). IEEE.

Farrokhabadi, M., König, S., Cañizares, C. A., Bhattacharya, K., & Leibfried, T. (2018). Battery energy storage system models for microgrid stability analysis and dynamic simulation. *IEEE Transactions on Power Systems*, *33*(2), 2301–2312.

Fletcher, S. D. A., Elders, I. M., Norman, P. J., Galloway, S. J., Booth, C. D., Burt, G. M., ... Hill, J. E. (2010). The impact of incorporating skin effect on the fault analysis and protection system performance of DC marine and aerospace power systems.

Frances, A., Asensi, R., García, Ó., Prieto, R., & Uceda, J. (2018). Modeling electronic power converters in smart DC microgrids—An overview. *IEEE Transactions on Smart Grid*, *9*(6), 6274–6287.

Gao, F., Bozhko, S., Asher, G., Wheeler, P., & Patel, C. (2016). An improved voltage compensation approach in a droop-controlled DC power system for the more electric aircraft. *IEEE Transactions on Power Electronics*, *31*(10), 7369–7383.

Gao, L., Liu, Y., Ren, H., & Guerrero, J. (2017). A DC microgrid coordinated control strategy based on integrator current-sharing. *Energies*, *10*(8), 1116.

George, K. (2006). DC power production, delivery and utilization. *Electric Power Research Institute White Paper*.

Gunasekaran, M., Mohamed Ismail, H., Chokkalingam, B., Mihet-Popa, L., & Padmanaban, S. (2018). Energy management strategy for rural communities' DC Micro Grid power system structure with maximum penetration of renewable energy sources. *Applied Sciences*, *8*(4), 585.

Hajimiragha, A., Canizares, C. A., Fowler, M. W., & Elkamel, A. (2010). Optimal transition to plug-in hybrid electric vehicles in Ontario, Canada, considering the electricity-grid limitations. *IEEE Transactions on Industrial Electronics*, *57*(2), 690–701.

Hatziargyriou, N., Asano, H., Iravani, R., & Marnay, C. (2007). Microgrids. *IEEE Power & Energy Magazine*, *5*(4), 78–94.

Hooshyar, H., & Baran, M. E. (2013). Fault analysis on distribution feeders with high penetration of PV systems. *IEEE Transactions on Power Systems*, *28*(3), 2890–2896.

Hu, K. W., & Liaw, C. M. (2016). Incorporated operation control of DC microgrid and electric vehicle. *IEEE Transactions on Industrial Electronics*, *63*(1), 202–215.

Ishaque, K., Salam, Z., & Taheri, H. (2011). Simple, fast and accurate two-diode model for photovoltaic modules. *Solar Energy Materials and Solar Cells*, *95*(2), 586–594.

Jin, Z., Sulligoi, G., Cuzner, R., Meng, L., Vasquez, J. C., & Guerrero, J. M. (2016). Next-generation shipboard dc power system: Introduction smart grid and dc microgrid technologies into maritime electrical networks. *IEEE Electrification Magazine*, *4*(2), 45–57.

Kakigano, H., Nomura, M., & Ise, T. (2010, June). Loss evaluation of DC distribution for residential houses compared with AC system. In The 2010 International Power Electronics Conference-ECCE ASIA- (pp. 480-486). IEEE.

Kalesar, B. M., Noshahr, J. B., Kermani, M., Bavandsavadkoohi, H., & Ahbab, F. (2018, June). Effect of Angles of Harmonic Components of Back to Back Converter of Distributed Generation Resources on Current Behavior of Distribution Networks. In *2018 IEEE International Conference on Environment and Electrical Engineering and 2018 IEEE Industrial and Commercial Power Systems Europe (EEEIC/I&CPS Europe)* (pp. 1-4). IEEE.

Khodadoost, A. A. A., Bayati, N., Reza, M., Gharehpetian, G. B., & Sadeghi, S. H. (2018). Fault Current Limiter optimal sizing considering different Microgrid operational modes using Bat and Cuckoo Search Algorithm. *Archives of Electrical Engineering*.

Lasnier, F. (2017). *Photovoltaic engineering handbook*. Abingdon, UK: Routledge.

Leterme, W., & Van Hertem, D. (2014). Reduced modular multilevel converter model to evaluate fault transients in DC grids.

Levin, T., & Thomas, V. M. (2012). Least-cost network evaluation of centralized and decentralized contributions to global electrification. *Energy Policy*, *41*, 286–302.

Levin, T., & Thomas, V. M. (2016). Can developing countries leapfrog the centralized electrification paradigm? *Energy for Sustainable Development*, *31*, 97–107.

Li, W., Luo, M., Monti, A., & Ponci, F. (2012, May). Wavelet based method for fault detection in medium voltage DC shipboard power systems. In *2012 IEEE International Instrumentation and Measurement Technology Conference Proceedings* (pp. 2155-2160). IEEE.

Magne, P., Nahid-Mobarakeh, B., & Pierfederici, S. (2013). Active stabilization of dc microgrids without remote sensors for more electric aircraft. *IEEE Transactions on Industry Applications*, *49*(5), 2352–2360.

Monadi, M., Gavriluta, C., Luna, A., Candela, J. I., & Rodriguez, P. (2016). Centralized protection strategy for medium voltage DC microgrids. *IEEE Transactions on Power Delivery*, *32*(1), 430–440.

Morozumi, S. (2007, April). Micro-grid demonstration projects in Japan. In 2007 Power Conversion Conference-Nagoya (pp. 635-642). IEEE.

Naik, J., Dhar, S., & Dash, P. K. (2019). Adaptive differential relay coordination for PV DC microgrid using a new kernel based time-frequency transform. *International Journal of Electrical Power & Energy Systems*, *106*, 56–67.

Park, J. D., Candelaria, J., Ma, L., & Dunn, K. (2013). DC ring-bus microgrid fault protection and identification of fault location. *IEEE Transactions on Power Delivery*, *28*(4), 2574–2584.

Patterson, B. T. (2012). Dc, come home: Dc microgrids and the birth of the enernet. *IEEE Power & Energy Magazine*, *10*(6), 60–69.

Sahli, M., Correia, J. P. M., Ahzi, S., & Touchal, S. (2018). Multi-physics modeling and simulation of heat and electrical yield generation in photovoltaics. *Solar Energy Materials and Solar Cells*, *180*, 358–372.

Saleh, K. A., Hooshyar, A., & El-Saadany, E. F. (2017). Hybrid passive-overcurrent relay for detection of faults in low-voltage DC grids. *IEEE Transactions on Smart Grid*, *8*(3), 1129–1138.

Sannino, A., Postiglione, G., & Bollen, M. H. (2002, October). Feasibility of a DC network for commercial facilities. In *37th IAS Annual Meeting Record of the 2002 IEEE Industry Applications Conference.* (Vol. 3, pp. 1710-1717). IEEE.

Sechilariu, M., Wang, B. C., Locment, F., & Jouglet, A. (2014). DC microgrid power flow optimization by multi-layer supervision control. Design and experimental validation. *Energy Conversion and Management*, *82*, 1–10.

Shaver, L. (2017). *Implementation of a DC Microgrid*. University of Wisconsin-Madison.

Venkataramanan, G., & Illindala, M. (2002). Microgrids and sensitive loads. In *Conference Proceedings of 2002 IEEE Power Engineering Society Winter Meeting.* (Vol. 1, pp. 315-322). IEEE.

Wei, X. X., & Li, Y. (2006, May). A high-frequency transmission line model for the low voltage power distribution network. In *2006 1ST IEEE Conference on Industrial Electronics and Applications* (pp. 1-7). IEEE.

Yang, B., Li, W., Zhao, Y., & He, X. (2010). Design and analysis of a grid-connected photovoltaic power system. *IEEE Transactions on Power Electronics*, *25*(4), 992–1000.

Yang, J., Fletcher, J. E., & O'Reilly, J. (2012). Short-circuit and ground fault analyses and location in VSC-based DC network cables. *IEEE Transactions on Industrial Electronics*, *59*(10), 3827–3837.

Zhao, H., Wu, Q., Guo, Q., Sun, H., & Xue, Y. (2015). Distributed model predictive control of a wind farm for optimal active power control part I: Clustering-based wind turbine model linearization. *IEEE transactions on sustainable energy, 6*(3), 831-839.

Ziani, A., & Moulai, H. (2009). Extinction properties of electric arcs in high voltage circuit breakers. *Journal of Physics. D, Applied Physics*, *42*(10), 105205.

Chapter 14
Anti-Plane Shear Wave in Microstructural Media:
A Case Wise Study of Micropolarity, Irregular, and Non-Perfect Interface

Mriganka Shekhar Chaki
https://orcid.org/0000-0001-5093-4432
Indian Institute of Technology Dhanbad, India

Abhishek Kumar Singh
Indian Institute of Technology Dhanbad, India

ABSTRACT

The present chapter encapsulates the characteristic behavior of anti-plane shear wave propagation in a micropolar layer/semi-infinite structural media. Two types of interfacial complexity have been considered at the common interface which give rise to two distinct mathematical models: (1) Model I: Anti-plane shear wave in a micropolar layer/semi-infinite structure with rectangular irregular interface and (2) Model II: Anti-plane shear wave in a micropolar layer/semi-infinite structure with non-perfect interface. For both models, dispersion equations have been deduced in algebraic-form and in particular, the dispersion equation of new type of surface wave resulted due to micropolarity has been obtained. The deduced results have been validated with classical cases analytically. The effects of micropolarity, irregularity, and non-perfect interface on anti-plane shear wave have been demonstrated through numerical study in the present chapter.

INTRODUCTION

Micropolar theory suggested by Eringen (1966) differs from the classical continuum due to its involvement of microrotations of the material rigid particles. Platelet composites, aluminium epoxy, soils, liquid crystals with rigid molecules, concrete with sand, animal bloods and many other materials do not exhibit properties of classical continuum accurately due to their granular or molecular structure of the

DOI: 10.4018/978-1-7998-0117-7.ch014

body. This inaccuracy, hence, micromechanics has been ignored in engineering fields for many years, but, in recent decades, micropolar theory has attracted attentions of many researchers and engineers.

Askar (1972) suggested that if shear wave is induced at one end of a micropolar crystal, and by strain gauges, if the incoming waves are detected at the other end, then some of the energy will be found missing. The reason for this experiment was that some of the translational energy has been transformed into rotational energy through the coupling between these two modes and strain gauge cannot detect rotations. This clears the very necessity of the question that why the consideration of micropolar theory is needed in engineering applications. For any engineering applications involving macrostructures, information on microstructural properties are essential for stability and non-destructive testing. Eringen (1999) stated that to deal with the measurement of microscopic-level quantities in high frequency and short-wavelength regions, doing experiments with micropolar constants requires precision and more elaborate instrumentation. As a result, the development of engineering and pragmatic applications regarding microstructures is very humble. Hence, any theoretical development on microstructures will be of great concern now-a-days, especially, in civil, mechanical and seismic engineering applications.

In aspect of wave propagation phenomena, it has been observed that if the wavelength and the average dimension of the microelements are of same order of magnitude, the fundamental motion of the microelements gets affected significantly (Eringen, 1966). If wave propagation gets affected by the spin inertia of micropolar body, it gives rise to new-type of surface waves not reported in the classical elasticity. Hence, the study of wave propagation in micropolar continuum is very much essential in the vicinity of finding these new-type of surface waves. Although several models can be found consisting of microstructural continuum in the literature, but, in aspect of wave propagation phenomena, very few works can be mentioned. In the vicinity of our present work, a great attention has been given on the work of Midya (2004). Midya (2004) have studied Love wave propagation in micropolar layer/half-space configuration. This layer/half-space configuration is very popular and essential for the propagation of surface waves. It is to be noted that within classical elasticity, anti-plane surface waves do not exist in an elastic semi-infinite space (Achenbach, 1973), but in a layer/semi-infinite configuration, anti-plane shear horizontal waves (viz. Love wave) can propagate provided the phase velocity should be greater than bulk shear velocity of upper layer and lesser than that of lower semi-infinite space (Love, 2013). However, in micropolar elasticity, wave propagation is possible in half-space due to microrotations, but authors are more interested in dispersive nature of waves. Hence, in the present chapter, emphasis has been given more on the layer/half-space configuration due to anti-plane shear waves.

Interfacial complexity in man-made structures is based on real world scenario and could arise due to aging of glue at the interface of two solids, microdefects, diffusion impurities, and other forms of damages in several engineering problems. The study of non-parallel boundaries in context of wave propagation in elastic media is also essential to geophysicists in order to make analysis and prediction for the seismic behavior at continental margins and mountain roots. Due to abrupt thickening, many shapes and sizes of irregularity can be found at the interface between two media, viz. rectangular, parabolic, triangular notch, corrugation etc. On the other hand, non-perfect bonding occurs due to "slip" condition between two media at the interface where the interfacial shearing stress can be balanced by the slip (Murty, 1976). However, considering such complexity in mathematical model creates challenges.

The main objective of the present chapter is to study propagation characteristics of anti-plane shear horizontal surface wave in a homogeneous micropolar layered structure by introducing interfacial complexity. Two distinct models can be considered having different types of interfacial complexity (i.e. irregular boundary (Model I) and non-perfect boundary (Model II)) along with source of disturbance and studied in comparative manner. For the solution of these types of mathematical modeling problems, emphasis needs to be given on analytical methods instead of numerical ones, since analytical methods are consistent, free from numerical approximations and the obtained results can easily be validated analytically with pre-established results. However, numerical calculation and graphical illustrations will be executed for the obtained analytical results in order to bring out more exquisite outcomes of the present work.

Background

Conventional linear elasticity theory successfully describe the man-made and naturally occurring materials for homogeneous macro-continuum. But, the theory fails when underlying heterogeneity of materials at small scale (or micro-continumm) is considered. The classical theory of elasticity doesn't reflect the inherent complexity of material properties mainly due to Hook's law being too ideal case of elasticity. Love (2013) stated that cast iron, building stone, cement and many materials used in engineering structures do not obey Hooke's law for any strains that are large enough to be observed, but, on the other hand, those materials which follow the Hooke's law for small strains do not obey it for larger ones. Some materials, e.g., steel, aluminium, concrete etc. are observed to be agreeing with experimental results within the elastic limits. However, materials like fibrous, polymers, asphalts, etc. exhibit significant inconsistency in the results between the experiments and theory of classical elasticity. These disagreements are due to the dominance of atomic structures of the material which are neglected in classical elasticity. Hence, the classical elasticity theory needs to be developed by considering the existence of a microstructure at a scale which is small compared to the characteristic dimension (macroscale) of the material specimen (Sobczyk and Kirkner, 2012).

For the in-depth knowledge of historical development of micropolar theory, authors would like to emphasize on some theories associated with rotational properties of material point existed before micropolar theory. "Couple stress theory" given by Voigt (1887) assumed that force stresses along with the couple stresses act on the faces of an elementary parallelepiped and are asymmetric in nature. In "Cosserat theory" (Cosserat and Cosserat, 1909), the concept of rotation of a geometrical points in the continuum body was first introduced which gave rise to the additional degrees of freedom due to rotation at each point and revealed asymmetry of strain and stresses. The behavior of materials like Heterogeneous materials with granular structure, composites under loading that causes rotation of grains (superplastic materials, acoustic metamaterials etc.) and many others can be explained by this theory. But, Cosserat theory was ignored for many years. After almost 50 years, researchers have understood the importance of this theory and focused on developing it for which many new microstructural models came into picture, viz. Couple stress theory (Mindlin and Tiersten, 1962; Toupin, 1962), Cosserat pseudo-continuum theory (Nowacki, 1974) etc. Later, Eringen (1966) gave micropolar theory and generalized this theory by extending it to Microstretch and Micromorphic continuum which is also called Eringen's 3M theory (Eringen, 1999). Linear Micropolar elasticity theory suggested by Eringen (1966) is chosen as the main topic of the present chapter.

Eringen's (1966) suggested linear theory of micropolar elasticity which articulates more generalized continuum by adding an independent microrotation vector to the classical continuum mechanics. Micropolar theory bears tractions with couple stresses affected by spin tensors where the particle orientation is defined by orthonormal trihedron whose vectors are called "Directors". The deformation of micropolar body can be understood by Figure 1 where for the unit basis vector $\mathbf{e}_i$, the position vector

Figure 1. Geometry of Micropolar deformation

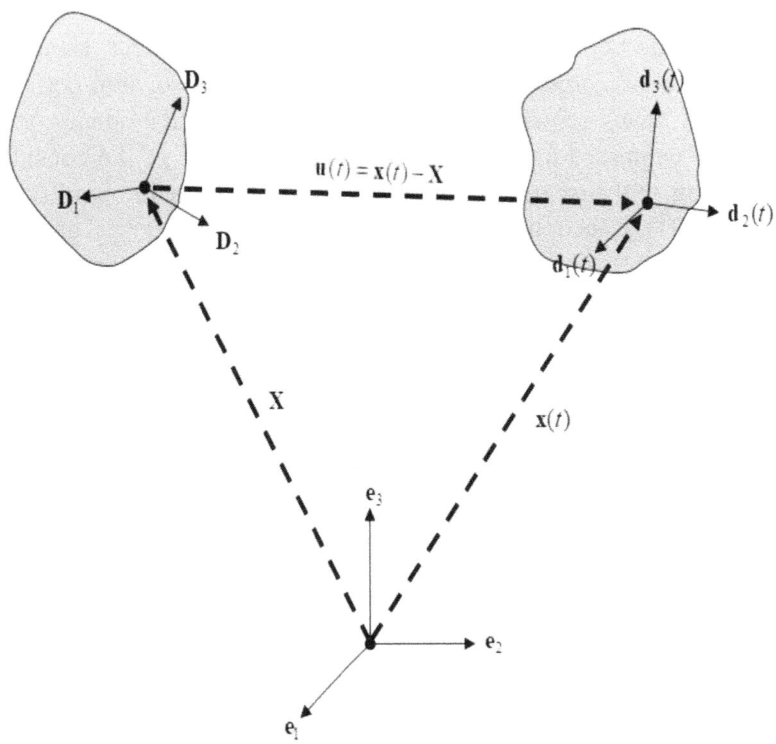

of particle is $\mathbf{X}$ in undeformed body with directors $\mathbf{D}_i$, position vector of particles is $\mathbf{x}(t)$ in the deformed body with directors $\mathbf{d}_i(t)$ and $\mathbf{u}(t)$ represents relative position vector. According to Figure 1, the micropolar continuum deformation can be described by the relations $\mathbf{x} = \mathbf{x}(\mathbf{X},\ t)$, $\mathbf{H} = \mathbf{H}(\mathbf{X},\ t)$ where microrotation tensor $\mathbf{H} = \mathbf{d}_i \otimes \mathbf{D}_i$. For details, readers may prefer the book of Eremeyev et al. (2012) and PhD thesis of Dilbag (2008).

The application of micropolar theory in several fields has been discovered gradually since it wasn't popular until recent years. Bažant and Christensen (1972) worked with micropolar elasticity numerically for steel-concrete grid structures in civil engineering. Yang and Lakes (1982) applied the micropolar theory to determine elastic properties of human compact bone. Gauthier (1982) have identified aluminum-epoxy as micropolar material for the first time and this material is also used for numerical analysis in the present study of the chapter. Li and Xie (2004) applied FEM to linear micropolar elasticity and studied some scale effects phenomena in MEMS. Altenbach and Eremeyev (2012) gave detail description of 3-D Cosserat model and micro-inhomogeneous materials. Recently, Abreu et al. (2017) studied micropolar modelling of rotational waves in seismology suggesting the implementation of theory of micropolar media in global seismological problems. Although, in context of shear wave propagation in massive layered structures with microrotations, very few relevant papers can be found in the literature. Ariman (1972) first discussed wave propagation in a micropolar elastic half-space. Later, Eringen (1984) have studied plane waves in nonlocal micropolar elasticity. Midya (2004) theoretically deduced the dispersion equation for Love wave propagation in micropolar layer/semi-infinite structure. Recently,

Kaur et al. (2017) studied shear wave propagation in vertically heterogeneous viscoelastic layer over a micropolar elastic half-space.

In real world scenario, the bonding between layers may not be perfect. Hence, many researchers have tried to develop models incorporating non-parallel boundaries or non-perfect boundaries. Due to mathematical simplicity, many researchers have considered particular shapes or sizes of interfacial complexity which can be expressed as a function of spatial variable of the considered model. Some analytical methods dealing with irregular boundary can be found in the works of Slavin and Wolf (1970), Fung and Eom (1981), Bouchon et al. (1989) whereas some numerical approaches can be found in Koketsu (2000), Li and Ito (2006), Ge and Chen (2008). In the vicinity of the solution technique, the authors would like to point out that researchers have always preferred the analytical methods to solve a mathematical modeling problem.

It was Mal (1962) who studied the frequency equation for Love waves due to abrupt thickening of the crustal layer where the solution technique to tackle rectangular shaped irregular interface has also been adopted as the solution methodology in the present chapter. This solution technique has also been considered by Bhattacharya (1962) to study the irregularity in the thickness of the transversely isotropic crustal layer and also by Chattopadhyay et al. (2008) to investigate propagation of SH waves in an irregular monoclinic crustal layer. Recently, Singh et al. (2018) has considered irregularity on SH-type wave propagation in micropolar elastic structure and the findings are also included in the present chapter. On the other hand, Murty (1976) gave a mathematical model to introduce linear slip condition which depicts the imperfect bonding between two solids. Termonia (1990), Nagy (1992), Kumar and Singh (1997), Kumar and Sharma (2008) and many others have discussed linear slip interfaces in their works. Recently, Kaur et al. (2016) discussed influence of imperfectly bonded micropolar elastic half-space with non-homogeneous viscoelastic layer on propagation behavior of shear wave.

Early publications regarding wave propagation in micropolar solid doesn't provide any insight on new-type of wave generation due to micropolar structural irregularity. Although, due to the involvement of microrotations in micropolar elasticity, new-types of surface waves have been already reported, but yet not extensively studied in aspect of structural interfacial complexity. Interfacial irregularity or non-perfect bonding may effect significantly on SH wave and new-type of dispersive wave generated due to micropolarity. Hence, in the present chapter, in order to explore the propagation characteristics of anti-plane Shear wave, two distinct microstructural layered models have been considered having irregularity at the interface (Model I) and non-perfect bonding at the interface (Model II) and studied in distinct cases. The obtained dispersion relations for Model I and II consist of irregularity parameter and non-perfect bonding parameter respectively whose impact have been studied through analytical and numerical discussions. Hence, the present chapter reveals some exquisite outcomes which certainly enrich the literature and provide a framework for future experiments or simulations in context of wave propagation in microstructures.

ANTI-PLANE SHEAR WAVE IN MICROSTRUCTURAL MEDIA

The main focus of the present chapter is to construct a layer/semi-infinite configuration with interfacial complexity in aspect of anti-plane Shear wave propagation and to study its propagation characteristics. Anti-plane Shear Horizontal wave or Love wave is a well-known surface wave which is widely used in studying material characteristics through wave propagation (Ewing et al., 1957). In order to study any complex medium, researchers always favor surface waves due to their dispersive characteristics and mathematical simplicity. Since, the consideration of rotation of geometrical points in elastic continua gives rise to microstructural continua, hence, to study material characteristics of such complex medium, anti-plane shear wave propagation has been chosen for the present study.

In order to explore the effect of interfacial complexity, a particular geometrical shaped irregular interface has been considered in a micropolar layer/semi-infinite configuration, on the other hand, non-perfect interface has been considered in another layer/semi-infinite configuration. These two configurations have been studied in two distinct cases as **Model I**: an upper homogeneous micropolar elastic layer overlying a homogeneous micropolar elastic semi-infinite medium with rectangular shaped irregularity at the interface, and **Model II**: an upper homogeneous micropolar elastic layer non-perfectly bonded to a lower homogeneous micropolar elastic semi-infinite. A comparative study of these two models can certainly reveal some significant results in aspect of wave propagation. Dispersion relations for both models can be deduced analytically in algebraic form for both SH wave and new-type of dispersive wave generated due to microrotations. Moreover, numerical analysis can be done based on these relations which may divulge the effect of various influencing parameters, viz. wave number, irregularity parameter, non-perfect bonding parameter, micropolarity etc. on both SH wave and new-type wave. Authors have chosen aluminum-epoxy micropolar material for numerical study.

Like any Elastodynamic problem, governing equations of assumed material structure in aspect of Euclidean frame of reference needs to be established first. Hence, governing equations consisting of constitutive equations having stress-strain relations, equations of motion and strain energy density functions has been explored in details in the next section in aspect of displacement and rotational components of micropolar continuum.

Governing Equations of Micropolar Elasticity

Let a volume V in $\mathbb{R}_3$ be occupied by a micropolar elastic solid with the boundary $A = \partial V$ where $\mathbb{R}_3$ manifests the three-dimensional space. Infinitesimal deformations of a micropolar elastic solid is considered with displacement and microrotation field

$$\mathrm{u}=\mathrm{u}\left(\mathrm{x},\ t\right),\theta=\theta\left(\mathrm{x},\ t\right) \tag{1}$$

where **u** and **θ** are twice differentiable vector functions of the displacement fields and microrotations along with $\|\mathbf{u}\| \ll 1,\ \ \|\nabla\mathbf{u}\| \ll 1,\ \ \|\theta\| \ll 1,\ \ \|\nabla\theta\| \ll 1$ where **X** is the position vector and t is time.

The non-negative strain energy density function **W** for micropolar media in quadratic form of **ε and χ is**

$$\mathbf{W} = \frac{\lambda}{2}\mathrm{tr}^2\varepsilon + \frac{\mu+\kappa}{2}\mathrm{tr}(\varepsilon\,.\,\varepsilon^T) + \frac{\mu}{2}\mathrm{tr}(\varepsilon\,.\,\varepsilon) + \frac{\beta_1}{2}\mathrm{tr}^2\chi + \frac{\beta_2}{2}\mathrm{tr}(\chi\,.\,\chi) + \frac{\beta_3}{2}\mathrm{tr}(\chi\,.\,\chi^T). \tag{2}$$

On using Cauchy theorem, Cauchy stress tensor and Couple stress tensor are

$$\sigma = \mathbf{n} \cdot \mathbf{T} \text{ and } \mathbf{m} = \mathbf{n} \cdot \mathbf{M} \tag{3}$$

where **n** is the outer normal to ∂V and

$$\left.\begin{aligned} \mathbf{T} &= \frac{\partial \mathbf{W}}{\partial \varepsilon} = \lambda \mathbf{I} \operatorname{tr} \varepsilon + (\mu + \kappa)\varepsilon + \mu \varepsilon^T, \\ \mathbf{M} &= \frac{\partial \mathbf{W}}{\partial \chi} = \beta_1 \mathbf{I} \operatorname{tr} \chi + \beta_2 \chi^T + \beta_3 \chi, \end{aligned}\right\} \tag{4}$$

Here, the strain measures are $\varepsilon = \nabla \mathbf{u} - \mathbf{I} \times \theta$, $\chi = \nabla \theta$ along with the conditions $3\lambda + 2\mu + \kappa \geq 0$, $2\mu + \kappa \geq 0$, $\kappa \geq 0$, $3\beta_1 + \beta_2 + \beta_3 \geq 0$, $\beta_2 + \beta_3 \geq 0$, $\beta_3 - \beta_2 \geq 0$.

From balance laws of equations, in particular, balance law of linear momentum, the equations of motion for micropolar media are obtained as

$$\left.\begin{aligned} &\nabla.\mathbf{T} = \rho \ddot{\mathbf{u}}, \\ &\nabla.\mathbf{M} - \mathbf{T}_\times = \rho \ddot{\theta}, \end{aligned}\right\} \tag{5}$$

On using constitutive equations (4) in balance law of linear momentum (5), the equations of motion can be deduced for the micropolar media (in tensor form) as follows (Eringen, 1966):

$$(\alpha + \beta)\theta_{l,lk} + \gamma\theta_{k,ll} + \kappa \in_{klm} u_{m,l} - 2\kappa\theta_k = \rho j \ddot{\theta}_k, \tag{6}$$

$$(\lambda + \mu)u_{l.lk} + (\mu + \kappa)u_{k,ll} + \kappa \in_{klm} \theta_{m,l} = \rho \ddot{u}_k, \tag{7}$$

Equations (6) and (7) serve as equations of motion for micropolar elastic body in any Euclidean frame of reference.

Issues and Formulation of the Problems

It is to be noted that research on micropolar theory in aspect of several engineering fields, especially, implementing micropolar theory in massive layered structures has been done in very limited way. Now-a-days, several scientists are still exploring the possibilities of implementing micropolar theory in massive macrostructures. In context of macrostructures, layer/semi-infinite configuration is very much popular and appropriate to study shear wave propagation phenomena. Hence, this configuration has been chosen to study Shear wave propagation characteristics in the present chapter. The emphasis has been given on the work of Midya (2004) where he has not only theoretically deduced the dispersion equation for Love wave propagation in micropolar layer/half-space configuration, but also the dispersion equation of new-type of dispersive surface wave generated due to the presence of micropolarity. However, the aforementioned paper lacks the structural (interfacial) complexity of real-world scenario (geophysical/man-made).

In aspect of geophysical phenomena as well as civil and mechanical engineering applications, dealing with massive layered structures have many discrepancies. Most of these discrepancies are due to lack of implementation of complexity generated due to real world scenario. Although, many researchers have preferred numerical methods to deal with such complexity (FEM, BEM, XFEM etc.), authors are more interested in analytical methods, since analytical methods can provide algebraic expressions of displacements and stress components and are also free from imprecisions due to numerical approximations (no error estimation required). Very few works can be found in literature considering complex interface, since dealing with such interface creates mathematical challenges to establish analytical solutions. Till date, no works have been found in aspect of wave propagation in micropolar layer/semi-infinite space configuration having non-perfect interface

Singh et al. (2018) have considered Love wave propagation in micropolar layered structure with irregular interface, but imperfect bonding has not yet been studied for this considered model. Despite the content of the present chapter being influenced by the work of Singh et al. (2018), the comparison between the model of Singh et al. (2018) and a micropolar layered model with imperfect bonding in context of Shear wave propagation brings novelty to the literature.

Now, mathematical modelling problems need to be established in order to study the anti-plane surface wave propagation characteristics in micropolar continuum for which consideration of suitable structural configuration of micropolar body is essential. Hence, two layer/semi-infinite configurations with interfacial complexity have been considered in distinct cases which are dealt with analytical methods to solve the mathematical problems by keeping mathematical simplicity. Two types of mathematical modelling problems has been formulated by deriving equations of motion from established Governing equations (6) and (7) for Rectangular Cartesian Coordinate system and by assuming suitable boundary conditions in order to study the impact of distinct types of interfacial complexity of layered micropolar structures on propagation characteristics of anti-plane Shear wave.

In this section, first mathematical modelling problem (Model I) consists of anti-plane Shear horizontal wave propagation in an upper homogeneous micropolar elastic layer overlying a homogeneous micropolar elastic semi-infinite medium with rectangular shaped irregularity at the common interface, and second mathematical problem (Model II) comprised of anti-plane Shear horizontal wave propagation in an upper homogeneous micropolar elastic layer non-perfectly bonded to a lower homogeneous micropolar elastic semi-infinite. Detailed geometric configurations of these models have been discussed in the next sections.

Mathematical formulation of Model I: Anti-plane shear wave in micropolar layer/semi-infinite structure with rectangular irregular interface

Anti-plane SH wave propagation has been considered in a model consisting of a homogeneous micropolar layer (M_1) of finite width H overlying a micropolar semi-infinite medium (M_2) such that the irregularity residing at the interface is considered in the form of rectangle as shown in Figure 2. In Figure 2, the span and depth of the irregularity are considered to be s and H' respectively. A Cartesian coordinate system has been imposed such that the origin O is at the middle of the span of the irregularity; z-axis is in the direction of the wave propagation; z-axis is in downward towards depth of semi-infinite medium; and y-axis is normal to the plane. A source of disturbance O' is assumed on the z-axis at depth $d > H'$.

The equation of interface for rectangular irregularity is

Figure 2. Geometry of Mathematical Modelling Problem (Model I)

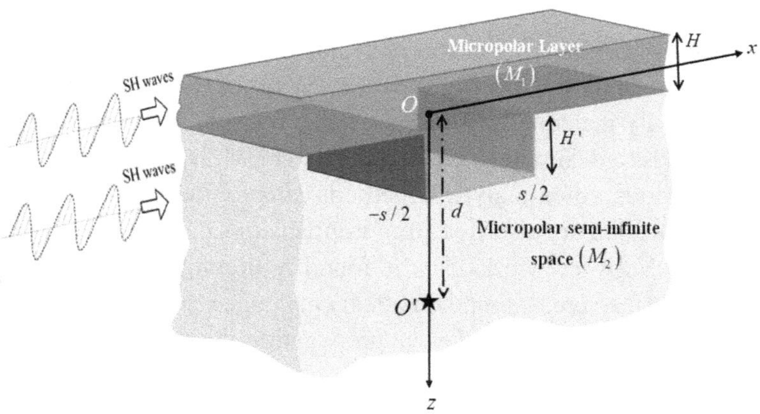

$$z = \varepsilon h(x) \text{ with } h(x) = \begin{cases} 0 & \text{for } |x| > s/2, \\ s & \text{for } |x| \geq s/2, \end{cases} \tag{8}$$

where $\varepsilon = H'/s \ll 1$ is the perturbation parameter.

Due to anti-plane Shear Horizontal (SH) wave, displacement components and microrotational components are considered as $\left(0,\ u_2^{(n)}(x,z,t),\ 0\right)$ and $\left(\theta_1^{(n)}(x,z,t),\ 0,\ \theta_3^{(n)}(x,z,t)\right)$ for the medium M_n, n=1,2.

Hence, the equations of motion (6) and (7) reduce to

$$\left.\begin{aligned} &\gamma_n \nabla^2 \theta_1^{(n)} + (\alpha_n + \beta_n)\frac{\partial}{\partial x}\left(\frac{\partial \theta_1^{(n)}}{\partial x} + \frac{\partial \theta_3^{(n)}}{\partial z}\right) - 2\kappa\theta_1^{(n)} - k_n \frac{\partial u_2^{(n)}}{\partial z} = \rho_n j_n \frac{\partial^2 \theta_1^{(n)}}{\partial t^2}, \\ &\gamma_n \nabla^2 \theta_3^{(n)} + (\alpha_n + \beta_n)\frac{\partial}{\partial z}\left(\frac{\partial \theta_1^{(n)}}{\partial x} + \frac{\partial \theta_3^{(n)}}{\partial z}\right) - 2\kappa\theta_3^{(n)} + k_n \frac{\partial u_2^{(n)}}{\partial x} = \rho_n j_n \frac{\partial^2 \theta_3^{(n)}}{\partial t^2}, \\ &(\mu_n + \kappa_n)\nabla^2 u_2^{(n)} + \kappa_n\left(\frac{\partial \theta_1^{(n)}}{\partial z} - \frac{\partial \theta_3^{(n)}}{\partial x}\right) = \rho_n \frac{\partial^2 u_2^{(n)}}{\partial t^2}, \end{aligned}\right\} \tag{9}$$

where $\mu_n, \kappa_n, \alpha_n, \beta_n, \gamma_n$ are the elastic moduli due to micropolar media, ρ_n is the material density and j_n is the microinertia for the medium M_n, n=1,2.

Boundary Conditions for Model I

For the present model (Model I), the boundary conditions are:

1. For stress-free upper layer:

$$T_{zy}^{(1)} = 0 \text{ at } z = -H \tag{10.1}$$

2. For couple stress free upper layer:

$$M_{zz}^{(1)} = 0 \quad \text{and} \quad M_{zx}^{(1)} = 0 \quad \text{at} \quad z = -H \tag{10.2}$$

3. For the continuity of displacement component at the interface:

$$u_2^{(1)} = u_2^{(2)} \quad \text{at} \quad z = \varepsilon h(x) \tag{10.3}$$

4. For the continuity of microrotational components at the interface:

$$\theta_1^{(1)} = \theta_1^{(2)} \quad \text{and} \quad \theta_3^{(1)} = \theta_3^{(2)} \quad \text{at} \quad z = \varepsilon h(x) \tag{10.4}$$

5. For the continuity of stresses at the interface:

$$T_{zy}^{(1)} - \varepsilon h'(x) T_{xy}^{(1)} = T_{zy}^{(2)} - \varepsilon h'(x) T_{xy}^{(2)} \quad \text{at} \quad z = \varepsilon h(x) \tag{10.5}$$

6. For the continuity of the couple stresses at the interface:

$$M_{zz}^{(1)} = M_{zz}^{(2)} \text{ and } M_{zx}^{(1)} = M_{zx}^{(2)} \quad \text{at} \quad z = \varepsilon h(x) \tag{10.6}$$

It is to be noted that the irregularity effect has not been considered in couple stress for mathematical simplicity. Hence, Equations of motion (9) together with the boundary conditions (10.1)-(10.6) provide the complete mathematical modelling problem for Model I.

Mathematical formulation of Model II: Anti-plane shear wave in micropolar layer/semi-infinite structure with non-perfect interface

The second model consisting of a homogeneous micropolar layer (M_1) of finite width H lying over a micropolar semi-infinite medium (M_2) has been considered such that the interface is considered to be non-perfect as shown in Figure 3. The non-perfect bonding at the interface implies that the layer and semi-infinite medium move independently. The Cartesian coordinate system is considered to be similar as of Model I. The equations of motion will also be similar to Equation (9) (in Model I) for both layer and semi-infinite medium.

Boundary Conditions for Model II

The boundary conditions for this particular model will be almost similar to Model I except the boundary condition associated to irregularity, i.e. equations (10.3)-(10.6) need to be changed. Due to the presence of non-perfect bonding at the interface, the boundary condition (10.3) will be changed to

Figure 3. Geometry of Mathematical modelling problem (Model II)

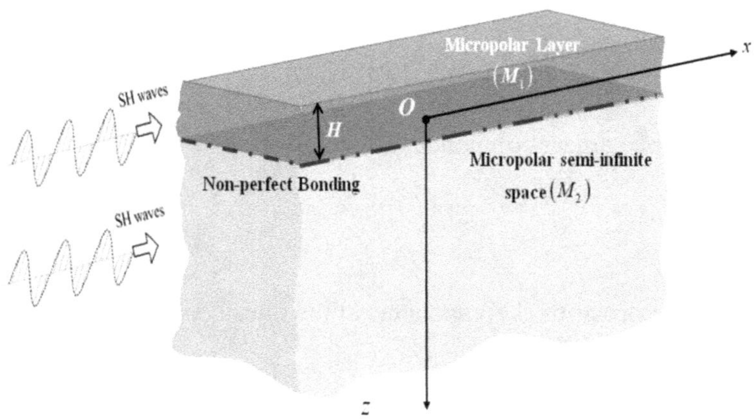

$$T_{yz}^{(1)} = \Gamma\left(u_2^{(2)} - u_2^{(1)}\right) \text{ at } z=0, \tag{27}$$

where Γ represents the non-perfect bonding parameter. Γ=0 is the case of ideally smooth interface whereas Γ→∞ is the case of wielded interface (Murty, 1976). In this case also, the effect of non-perfect bonding has not been considered for couple stress for mathematical simplicity.

Hence, Equations of motion (9) along boundary conditions (10.1)-(10.2), (27) and (10.4)-(10.6) at z=0 represent the complete mathematical modelling problem of Model II.

SOLUTIONS AND RECOMMENDATIONS

In this section, two different types of solution methodology have been adopted for mathematical modelling problems of Model I and II. For Model I, Fourier and inverse Fourier transforms along with perturbation and integral approximations have been considered whereas for Model II, variable separable method has been employed in order to deduce dispersion relations. However, before applying the aforementioned methods, the equations of motion needs to simplified.

Now, on using Helmotz decomposition, the following has been obtained:

$$\left.\begin{aligned} \theta_1^{(n)} &= \frac{\partial \phi^{(n)}}{\partial x} + \frac{\partial \psi^{(n)}}{\partial z}, \\ \theta_3^{(n)} &= \frac{\partial \phi^{(n)}}{\partial z} - \frac{\partial \psi^{(n)}}{\partial x}, \end{aligned}\right\} \tag{28}$$

Employing Equation (28) in Equation (9), the following has been deduced:

$$\left.\begin{aligned}
&\nabla^2\phi^{(n)}-\frac{2(\omega_0^{(n)})^2}{(c_4^{(n)})^2+(c_5^{(n)})^2}\phi^{(n)}=\frac{1}{(c_4^{(n)})^2+(c_5^{(n)})^2}\frac{\partial^2\phi^{(n)}}{\partial t^2},\\
&\nabla^2\psi^{(n)}-\frac{2(\omega_0^{(n)})^2}{(c_4^{(n)})^2}\psi^{(n)}-\frac{(\omega_0^{(n)})^2}{(c_4^{(n)})^2}u_2^{(n)}=\frac{1}{(c_4^{(n)})^2}\frac{\partial^2\psi^{(n)}}{\partial t^2},\\
&\nabla^2u_2^{(n)}-p_0^{(n)}\nabla^2\psi^{(n)}=\frac{1}{(c_2^{(n)})^2}\frac{\partial^2u_2^{(n)}}{\partial t^2},
\end{aligned}\right\} \tag{29}$$

where

$$(c_4^{(n)})^2=\frac{\gamma_n}{\rho_n j_n},\ (c_5^{(n)})^2=\frac{\alpha_n+\beta_n}{\rho_n j_n},\ (\omega_0^{(n)})^2=\frac{\kappa_n}{\rho_n j_n},\ p_0^{(n)}=\frac{\kappa_n}{\mu_n+\kappa_n}\ \text{and}\ (c_2^{(n)})^2=\frac{\mu_n+\kappa_n}{\rho_n}.$$

The following assumption has been made for the harmonic wave propagation

$$\left\{\phi^{(n)}(z,x,t),\ \psi^{(n)}(z,x,t),\ u_2^{(n)}(z,x,t)\right\}=\left\{\phi^{(n)}(z,x),\ \psi^{(n)}(z,x),\ u_2^{(n)}(z,x)\right\}e^{i\omega t}, \tag{30}$$

where ω is angular frequency.

Solution Methodology for Model I

Employing Fourier and Inverse Fourier transformation on displacement fields as well as rotational components replacing spatial variable *x* by Fourier transform parameter η, Equation (29) reduces to

$$\left.\begin{aligned}
&\left[\frac{\partial^2}{\partial z^2}-\eta^2-\frac{2(\omega_0^{(n)})^2}{(c_4^{(n)})^2+(c_5^{(n)})^2}+\frac{\omega^2}{(c_4^{(n)})^2+(c_5^{(n)})^2}\right]\bar{\phi}^{(n)}=0,\\
&\left(\frac{\partial^2}{\partial z^2}-\eta^2+a_n\right)\left(\frac{\partial^2}{\partial z^2}-\eta^2+b_n\right)(\bar{\psi}^{(n)},\bar{u}_2^{(n)})=0,
\end{aligned}\right\} \tag{31}$$

where $(\omega_0^{(n)})^2\neq 0$ and

$$a_n,b_n=\frac{1}{2}\left[\omega^2\left(\frac{1}{(c_2^{(n)})^2}+\frac{1}{(c_4^{(n)})^2}\right)-(2-p_0^{(n)})\frac{(\omega_0^{(n)})^2}{(c_4^{(n)})^2}\pm\left[\left\{\omega^2\left(\frac{1}{(c_2^{(n)})^2}+\frac{1}{(c_4^{(n)})^2}\right)-(2-p_0^{(n)})\frac{(\omega_0^{(n)})^2}{(c_4^{(n)})^2}\right\}^2-\frac{4\omega^2}{(c_2^{(n)})^2(c_4^{(n)})^2}(\omega^2-2(\omega_0^{(n)})^2)\right]^{1/2}\right]$$

For small values of κ_n, a_n and b_n, it is assumed that

$$\left.\begin{aligned} a_n &\simeq \frac{\omega^2 - 2(\omega_0^{(n)})^2}{(c_4^{(n)})^2} + \mathcal{O}(\kappa_n^2), \\ b_n &\simeq \frac{\omega^2}{(c_2^{(n)})^2} + \mathcal{O}(\kappa_n^2). \end{aligned}\right\} \tag{32}$$

In view of above, the mcrorotational components and displacement field for medium M_1 and medium M_2 are

$$\left.\begin{aligned} \phi^{(1)} &= \frac{(\omega_0^{(1)})^2}{2\pi}\int_{-\infty}^{\infty}\Big(A\cos rz + B\sin rz\Big)e^{-i\eta x}d\eta, \\ \psi^{(1)} &= \frac{(\omega_0^{(1)})^2}{2\pi}\int_{-\infty}^{\infty}\Big(C\cos sz + D\sin sz + E\cos pz + F\sin pz\Big)e^{-i\eta x}d\eta, \\ u_2^{(1)} &= \frac{\omega^2 - 2(\omega_0^{(1)})^2 - a_1(c_4^{(1)})^2}{2\pi}\int_{-\infty}^{\infty}\Big(C\cos sz + D\sin sz\Big)e^{-i\eta x}d\eta \\ &\quad + \frac{\omega^2 - 2(\omega_0^{(1)})^2 - b_1(c_4^{(1)})^2}{2\pi}\int_{-\infty}^{\infty}\Big(E\cos pz + F\sin pz\Big)e^{-i\eta x}d\eta, \end{aligned}\right\} \tag{33}$$

and

$$\left.\begin{aligned} \phi^{(2)} &= \frac{(\omega_0^{(2)})^2}{2\pi}\int_{-\infty}^{\infty}\Big(A'e^{-r_1 z}\Big)e^{-i\eta x}d\eta, \\ \psi^{(2)} &= \frac{(\omega_0^{(2)})^2}{2\pi}\int_{-\infty}^{\infty}\Big(C'e^{-s_1 z} + E'e^{-p_1 z}\Big)e^{-i\eta x}d\eta, \\ u_2^{(2)} &= \frac{\omega^2 - 2(\omega_0^{(2)})^2 - a_2(c_4^{(2)})^2}{2\pi}\int_{-\infty}^{\infty}(C'e^{-s_1 z} + \frac{2}{s_1}e^{s_1 z}e^{-s_1 d})e^{-i\eta x}d\eta \\ &\quad + \frac{\omega^2 - 2(\omega_0^{(2)})^2 - b_2(c_4^{(2)})^2}{2\pi}\int_{-\infty}^{\infty}(E'e^{-p_1 z} + \frac{2}{p_1}e^{p_1 z}e^{-p_1 d})e^{-i\eta x}d\eta, \end{aligned}\right\} \tag{34}$$

where

$$r = \sqrt{\frac{\omega^2 - (\omega_0^{(1)})^2}{(c_4^{(1)})^2 + (c_5^{(1)})^2} - \eta^2},\quad s = \sqrt{\frac{\omega^2 - 2(\omega_0^{(1)})^2}{(c_4^{(1)})^2} - \eta^2} + \mathcal{O}(\kappa_1^2),\quad p = \sqrt{\frac{\omega^2}{(c_2^{(1)})^2} - \eta^2} + \mathcal{O}(\kappa_1^2),$$
$$r_1 = \sqrt{\eta^2 - \frac{\omega^2 - (\omega_0^{(2)})^2}{(c_4^{(2)})^2 + (c_5^{(2)})^2}},\quad s_1 = \sqrt{\eta^2 - \frac{\omega^2 - 2(\omega_0^{(2)})^2}{(c_4^{(2)})^2}} + \mathcal{O}(\kappa_2^2),\quad p_1 = \sqrt{\eta^2 - \frac{\omega^2}{(c_2^{(2)})^2}} + \mathcal{O}(\kappa_2^2),$$

$\kappa_1 \neq 0 \text{ and } \kappa_2 \neq 0$ along with the conditions:

$\mathrm{Re}(r_1) > 0,\ \mathrm{Re}(s_1) > 0, \mathrm{Re}(s_1) > 0 \text{ and } \omega > (\sqrt{2}\omega_0^{(1)}, \sqrt{2}\omega_0^{(2)}).$

In this case, the range of $\omega > (\sqrt{2}\omega_0^{(1)}, \sqrt{2}\omega_0^{(2)})$ can be considered as cut-off frequency of SH wave for the considered layered structure.

Now, the subsequent steps need to be followed:

Step 1: Due to irregular interface, perturbation method has been applied to the arbitrary constants considered in the assumed solutions, i.e.

$$A \cong A_0 + \varepsilon A_1, B \cong B_0 + \varepsilon B_1, C \cong C_0 + \varepsilon C_1, D \cong D_0 + \varepsilon D_1, E \cong E_0 + \varepsilon E_1, F \cong F_0 + \varepsilon F_1,$$
$$A' \cong A_0' + \varepsilon A_1', C' \cong C_0' + \varepsilon C_1', E' \cong E_0' + \varepsilon E_1'$$

where ε is perturbation parameter[1].

Step 2: Using Equations (33) and (34) in the boundary conditions, a system of integral equations are obtained. This system are simplified by putting η+λ=*k* in the inner integral where *k* is wave number and λ is considered as a constant which will imply *d*η=*dk*. Further, on using inverse Fourier transforms, much more simplified form of system of integral equations has been obtained.

Step 3: The obtained system of integral equations contains nine (9) integral equations. This system can be divided into two subsystem of integral equations each containing nine (9) integral equations by carefully comparing the absolute terms (terms not containing ε) and coefficients of ε.

At the end of Step 3, a system of eighteen (18) integral equations can be achieved where it is required to solve for the 18 arbitrary constants $A_i, B_i, C_i, D_i, E_i, F_i, A_i', C_i', E_i'$ (for $i = 0, 1$) by tedious mathematical calculations and later, the constant's expressions will be used to find the dispersion relation.

Dispersion Relation for Anti-plane Shear Wave in Model I

On following the above steps, the displacement component in the layer M_1 from Equation (33) can be rewritten as

$$u_2^{(1)} = \frac{1}{2\pi}\int_{-\infty}^{\infty} \frac{4e^{-p_1 d}\left(\omega^2 (c_2^{(2)})^2 - \omega^2 (c_4^{(2)})^2 - 2(c_2^{(2)})^2(\omega_0^{(2)})^2\right)(\kappa_2 + \mu_2)}{(c_2^{(2)})^2 \Delta_1} \times \left[1 - \frac{\varepsilon (c_2^{(2)})^2 \left[p_1 R_1(k)(\kappa_2 + \mu_2) + R_2(k)\right] e^{p_1 d}}{4(\kappa_2 + \mu_2)\left(\omega^2 (c_2^{(2)})^2 - \omega^2 (c_4^{(2)})^2 - 2(c_2^{(2)})^2(\omega_0^{(2)})^2\right)}\right] \left(\cos pz - \sin pz \tan pH\right) e^{-ikx} dk. \tag{35}$$

where Δ_1, $R_1(k)$ and $R_2(k)$ are given in Appendix 1.

Within the integral of Equation (35), the following expression can be deduced as

$$p_1 R_1(k)(\kappa_2 + \mu_2) + R_2(k)$$
$$= \frac{2(\kappa_2+\mu_2)s}{\pi}\int_{-\infty}^{\infty}\Bigg[\left(\omega^2 - 2(\omega_0^{(2)})^2 - \omega^2\frac{(c_4^{(2)})^2}{(c_2^{(2)})^2}\right)\Big(4(\kappa_2+\mu_2)pp_1^2 + 2p_1^2(\kappa_2 p_1 + \mu_2 p_1 + p\kappa_1 \tan pH$$
$$+ p\mu_1 \tan pH) - 2p_1^2\Delta_1 - 4(\kappa_1+\mu_1)(p^2 p_1 - k\lambda p_1) - 2p_1^2(\kappa_2 p_1 + \mu_2 p_1 + p\kappa_1 \tan pH + p\mu_1 \tan pH)$$
$$- 2p_1^2\Delta_1 - 2\lambda k(\kappa_2 p_1 + \mu_2 p_1 + p\kappa_1 \tan pH + p\mu_1 \tan pH) - 2\lambda k\Delta_1\Big)\frac{e^{-p_1 d}}{p_1\Delta_1}\Bigg]^{\eta = k-\lambda}\frac{1}{\lambda}\sin\frac{\lambda s}{2}\,d\lambda. \tag{36}$$

where $\bar{h}(\lambda) = \frac{2s}{\lambda}\sin\frac{\lambda s}{2}$, which can be calculated Equation (8) using inverse Fourier transformation.

Adopting asymptotic formulas described in the works of Willis (1948) and Tranter (1966) and for large value of *s*, neglecting the terms containing 2/*s* and higher powers of 2/*s*, Equation (36) can be approximated linearly as

$$p_1 R_1(k)(\kappa_2 + \mu_2) + R_2(k) \simeq 2(\kappa_2+\mu_2)\frac{H'}{\varepsilon}\varphi(k), \tag{37}$$

where

$$\varphi(k) = \frac{4e^{-p_1 d}}{\Delta_1}\left(\omega^2 - 2(\omega_0^{(2)})^2 - \omega^2\frac{(c_4^{(2)})^2}{(c_2^{(2)})^2}\right)$$
$$\left[-pp_1(\kappa_2+\mu_2)\tan pH + pp_1(\kappa_1+\mu_1)\tan pH - (\kappa_1+\mu_1)p^2 - (\kappa_2+\mu_2)p_1^2\right].$$

In view of Equations (37), Equation (35) reduces to

$$u_2^{(1)} = \frac{1}{2\pi}\int_{-\infty}^{\infty}\frac{4e^{-p_1 d}\left(\omega^2(c_2^{(2)})^2 - \omega^2(c_4^{(2)})^2 - 2(c_2^{(2)})^2(\omega_0^{(2)})^2\right)(\cos pz - \sin pz \tan pH)}{(c_2^{(2)})^2\Delta_1\left[1 + \frac{H'}{2}\frac{(c_2^{(2)})^2\varphi(k)e^{p_1 d}}{\left(\omega^2(c_2^{(2)})^2 - \omega^2(c_4^{(2)})^2 - 2(c_2^{(2)})^2(\omega_0^{(2)})^2\right)}\right]}e^{-ikx}dk. \tag{38}$$

From the integral expression of (38), the poles can be calculated by equating the denominator to zero, which on simplification, provides the required dispersion relation

$$\tan pH = \frac{\kappa_2 p_1 + \mu_2 p_1 - 2(\kappa_1+\mu_1)p^2 H' - 2(\kappa_2+\mu_2)p_1^2 H'}{p\kappa_1 + p\mu_1 + 2(\kappa_2+\mu_2)pp_1 H' - 2(\kappa_1+\mu_1)pp_1 H'}, \tag{39}$$

which exists under the condition $c_2^{(1)} < c < c_2^{(2)}$ and is very much dependent on the irregularity parameter H'.

Dispersion Relation for New-type Surface Wave in Model I

The microrotational component $\psi^{(1)}$ for the layer M_1 from Equation (33), can be expressed as

$$\psi^{(1)} = I_1 + I_2, \tag{40}$$

where

$$I_1 = \frac{(\omega_0^{(1)})^2}{2\pi} \int_{-\infty}^{\infty} \left[(C_0 + \varepsilon C_1)\cos sz + (D_0 + \varepsilon D_1)\sin sz\right] e^{-ikx} dk,$$
$$I_2 = \frac{(\omega_0^{(1)})^2}{2\pi} \int_{-\infty}^{\infty} \left[(E_0 + \varepsilon E_1)\cos pz + (F_0 + \varepsilon F_1)\sin pz\right] e^{-ikx} dk.$$

Now following the similar process as mentioned before, the required dispersion relations have been deduced as

$$(L_1 \cos rH + L_2 \cos sH - L_3 \sin rH + L_4 \sin sH)(L_5 \cos rH - L_6 \cos sH + L_7 \sin rH - L_8 \sin sH) = 0 \tag{41}$$

and

$$\tan pH = \frac{\kappa_2 p_1 + \mu_2 p_1 - 2(\kappa_1 + \mu_1)p^2 H' - 2(\kappa_2 + \mu_2)p_1^2 H'}{p\kappa_1 + p\mu_1 + 2(\kappa_2 + \mu_2)pp_1 H' - 2(\kappa_1 + \mu_1)pp_1 H'}. \tag{42}$$

where L_i's are provided in Appendix 1.

Here, equation (42) describes the dispersion relation for new-type of dispersive wave with the condition

$$\sqrt{\frac{(c_4^{(1)})^2 + (c_5^{(1)})^2}{\left(1 - \frac{2(\omega_0^{(1)})^2}{\omega^2}\right)}} < c < \sqrt{\frac{(c_4^{(2)})^2 + (c_5^{(2)})^2}{\left(1 - \frac{2(\omega_0^{(2)})^2}{\omega^2}\right)}}$$

from which the cut-off frequency can be found for this new-type wave and the phase velocity of new-type dispersive wave is found more than the velocity of longitudinal-microrotational wave of micropolar layer and less than that of half-space. It is to be noticed that Equation (41) doesn't contain any irregularity parameter H'. Moreover, Equation (42) is found to be similar to Equation (39). However, the absence of irregularity parameter in the new-type wave dispersion equation can easily be explained by the fact that the irregularity has not been considered in couple stresses.

Figure 4. Geometry of propagation of new-type dispersive wave in elastic layer with micropolar semi-infinite medium.

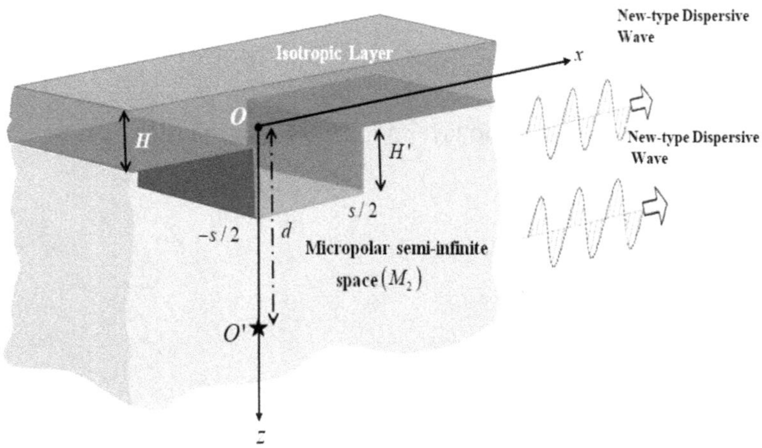

Special Case

When the layered structure is comprised of a elastic layer without microrotations (i.e. $|s|,|r| \to \infty$, $\alpha_1 = \beta_1 = \gamma_1 \to 0$ and $k_1 \to 0$) and a micropolar elastic semi-infinite medium M_2 with irregularity at the common interface (see Figure 4), the dispersion relation for new-type surface wave, i.e. Equation (41) reduces to

$$[r_1^2(\alpha_2 + \beta_2 + \gamma_2) - \alpha_2 k^2](\beta_2 k^2 + \gamma_2 s_1^2) - k^2 r_1 s_1 (\beta_2 + \gamma_2)^2 = 0 \tag{43}$$

where $c_2^{(1)} < c < \sqrt{\dfrac{(c_4^{(2)})^2 + (c_5^{(2)})^2}{1 - 2(\omega_0^{(2)})^2 / \omega^2}}$

It is to be noted that dispersion relation (43) for new-type wave exists only due to micropolar semi-infinite medium and it does not depend on the irregularity parameter. This dispersion relation (43) can also be achieved by considering the anti-plane shear wave in a micropolar semi-infinite medium with boundary conditions $\tau_{zy}^{(2)} = 0$, $m_{zz}^{(2)} = 0$, $m_{zx}^{(2)} = 0$ at $z=0$ which further, on considering $\alpha_2 = \beta_2 = \gamma_2 \to 0$ and $\kappa_2 = 0$ will reduce to non-dispersive equation of Rayleigh wave propagating in classical elastic semi-infinite medium (see Figure 5). This also suffices the fact that SH wave doesn't propagate in an elastic semi-infinite medium.

Solution Methodology for Model II

Following similar solution technique as before in Model I, the solutions for this model (Model II) can be considered as

Figure 5. Geometry of propagation of Rayleigh wave in elastic semi-infinite medium

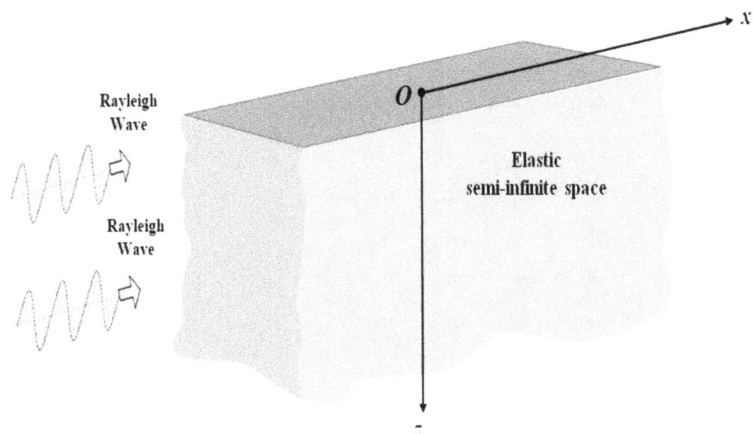

$$\left.\begin{aligned}\phi^{(1)} &= (\omega_0^{(1)})^2\left(A\cos rz + B\sin rz\right)e^{i(kx-\omega t)},\\ \psi^{(1)} &= (\omega_0^{(1)})^2\left(C\cos sz + D\sin sz + E\cos pz + F\sin pz\right)e^{i(kx-\omega t)},\\ u_2^{(1)} &= \left(\omega^2 - 2(\omega_0^{(1)})^2 - a_1(c_4^{(1)})^2\right)\left(C\cos sz + D\sin sz\right)e^{i(kx-\omega t)}\\ &\quad + \left(\omega^2 - 2(\omega_0^{(1)})^2 - b_1(c_4^{(1)})^2\right)\left(E\cos pz + F\sin pz\right)e^{i(kx-\omega t)},\end{aligned}\right\} \quad (44)$$

and

$$\left.\begin{aligned}\phi^{(2)} &= (\omega_0^{(2)})^2 A' e^{-r_1 z} e^{i(kx-\omega t)},\\ \psi^{(2)} &= (\omega_0^{(2)})^2\left(C' e^{-s_1 z} + E' e^{-p_1 z}\right)e^{i(kx-\omega t)},\\ u_2^{(2)} &= \left(\omega^2 - 2(\omega_0^{(2)})^2 - a_2(c_4^{(2)})^2\right)C' e^{-s_1 z} e^{i(kx-\omega t)}\\ &\quad + \left(\omega^2 - 2(\omega_0^{(2)})^2 - b_2(c_4^{(2)})^2\right)E' e^{-p_1 z} e^{i(kx-\omega t)},\end{aligned}\right\} \quad (45)$$

For solution methodology, variable separable method can be used, but perturbation or Fourier transformations are not necessary for the solution of this mathematical modelling problem.

Dispersion Relation for Anti-plane Shear Wave in Model II

In view of (44) and (45) and on using boundary conditions for Model II i.e. equations (10.1)-(10.2), (27) and (10.4)-(10.6), the dispersion relation is obtained by eliminating the arbitrary constants, i.e.

$$\begin{vmatrix} m_{11} & m_{12} & 0 & 0 & 0 & 0 & 0 & 0 & 0 \\ m_{21} & m_{22} & m_{23} & 0 & 0 & 0 & 0 & 0 & 0 \\ m_{31} & 0 & m_{33} & 0 & 0 & 0 & 0 & 0 & 0 \\ m_{41} & m_{42} & 0 & m_{44} & m_{45} & m_{46} & m_{47} & 0 & 0 \\ m_{51} & m_{52} & 0 & m_{54} & m_{55} & m_{56} & m_{57} & 0 & 0 \\ m_{61} & 0 & m_{63} & m_{64} & 0 & 0 & m_{67} & m_{68} & m_{69} \\ 0 & m_{72} & m_{73} & 0 & m_{75} & m_{76} & 0 & m_{78} & m_{79} \\ m_{81} & 0 & m_{83} & m_{84} & 0 & 0 & m_{87} & m_{88} & m_{89} \\ 0 & m_{92} & m_{93} & 0 & m_{95} & m_{96} & 0 & m_{98} & m_{99} \end{vmatrix} = \delta_1 \times \delta_2 = 0 \tag{46}$$

where

$$\delta_1 = \begin{vmatrix} m_{11} & m_{12} & 0 \\ m_{21} & m_{22} & m_{23} \\ m_{31} & 0 & m_{33} \end{vmatrix}, \ \delta_2 = \begin{vmatrix} m_{44} & m_{45} & m_{46} & m_{47} & 0 & 0 \\ m_{54} & m_{55} & m_{56} & m_{57} & 0 & 0 \\ m_{64} & 0 & 0 & m_{67} & m_{68} & m_{69} \\ 0 & m_{75} & m_{76} & 0 & m_{78} & m_{79} \\ m_{84} & 0 & 0 & m_{87} & m_{88} & m_{89} \\ 0 & m_{95} & m_{96} & 0 & m_{98} & m_{99} \end{vmatrix}$$

and m_{ij}'s are provided in Appendix 2.

Now, Equation (46) gives rise to two situations, i.e. $\delta_1=0$ gives the dispersion relation for anti-plane shear wave propagation in Model II:

$$\tan pH = \frac{\Gamma\, p_1(\mu_2 + \kappa_2)}{\left(\Gamma + (\mu_2 + \kappa_2)p_1\right)p(\mu_1 + \kappa_1)} \tag{47}$$

and $\delta_2=0$ gives the dispersion relation for new-type surface wave generated due to micropolar media which is also similar to Equation (41). In this case also non-perfect bonding parameter doesn't appear in the new-type wave dispersion equation due to the fact that the non-perfect bonding has not been considered in couple stresses.

Validation

If Model I is comprised of regular common interface $\left(H' = 0\right)$ and if the layer and semi-infinite medium of Model II is having perfectly common interface $\left(\Gamma \to \infty\right)$, then dispersion equations (39) and (47) matches with the results of Midya (2004). The dispersion equation (41) of new-type wave can also be validated analytically with Midya (2004).

Further, for Model I, when both the layer and semi-infinite medium are without microrotations and having regular common interface, i.e.

$$H' = 0,\ \alpha_1 = \beta_1 = 0,\ \alpha_2 = \beta_2 = 0,\ \gamma_1 \to 0,\ \gamma_2 \to 0,\ \kappa_1 \to 0,\ \kappa_2 \to 0$$

and also for Model II, when the layer and semi-infinite medium are without microrotations and having perfectly common interface, i.e.

$$\Gamma \to \infty,\ \alpha_1 = \beta_1 = 0,\ \alpha_2 = \beta_2 = 0,\ \gamma_1 \to 0,\ \gamma_2 \to 0,\ \kappa_1 \to 0,\ \kappa_2 \to 0,$$

the dispersion equations (39) and (47) reduce to

$$\tan kH \sqrt{\frac{c^2}{\left(\beta^{(0)}\right)^2} - 1} = \frac{\mu_2 \sqrt{1 - \frac{c^2}{\left(\beta^{(1)}\right)^2}}}{\mu_1 \sqrt{\frac{c^2}{\left(\beta^{(0)}\right)^2} - 1}}, \tag{48}$$

where $\beta^{(0)} = \sqrt{\frac{\mu_1}{\rho_1}}$ and $\beta^{(1)} = \sqrt{\frac{\mu_2}{\rho_2}}$

Eq. (48) represents classical Love wave dispersion equation proposed by Love (2013) which also validates the results obtained in the present work.

In this particular case, Eq. (41) completely vanishes which can also be validated by the fact that new-type waves are being generated due to the presence of microrotation of material particles.

Numerical Analysis

In order to reveal the effect of wave number, micropolarity, irregular interface (Model I) and non-perfect bonding of the interface (Model II) on the phase velocity of SH wave as well as new-type dispersive wave, Figure 6 to Figure 10 have been drawn for which the data from Table 1 have been considered along with the assumptions: $\omega^2 / \left(\omega_0^{(n)}\right)^2 = 10$ (low frequency) and $\alpha_n = \beta_n = \gamma_n$.

The Effect of Wave Number

In Figure 6 to Figure 10, it can be easily observed that the wave number $\left(kH,\ k / \sqrt{\mu_1 / \gamma_1},\ k / \sqrt{\mu_2 / \gamma_2}\right)$ has a decreasing effect on phase velocity of shear wave $\left(c / c_2^{(1)}\right)$ as well new-type wave $\left(c / c_4^{(1)},\ c / c_4^{(2)}\right)$, i.e. the increase in wave number tends to decrease phase velocity of shear wave and new-type surface wave in both fundamental as well as 1st order vibrational mode.

Table 1. Numerical data for micropolar layer and semi-infinite medium (Khurana and Tomar, 2009)

Upper layer Medium	Lower semi-infinite Medium
$\lambda_1 = 7.85\times10^{11}$ dyne/cm^2	$\lambda_1 = 7.59\times10^{11}$ dyne/cm^2
$\mu_1 = 1.37\times10^{11}$ dyne/cm^2	$\mu_2 = 1.35\times10^{11}$ dyne/cm^2
$\kappa_1 = 0.0139\times10^{11}$ dyne/cm^2	$\kappa_2 = 0.0149\times10^{11}$ dyne/cm^2
$\rho_1 = 2.7$ gm/cm^3	$\rho_1 = 2.2$ gm/cm^3
$j_1 = 0.212$ cm^2	$j_2 = 0.196$ cm^2
$\gamma_1 = 0.288\times10^{11}$ dyne	$\gamma_2 = 0.268\times10^{11}$ dyne

The Effect of Irregularity

In Figure 6, non-dimensional phase velocity $\left(c/c_2^{(1)}\right)$ of SH wave against non-dimensional wave number (*kH*) has been plotted for a constant value of irregularity parameter $\left(H'/H=0.3\right)$ on dispersion curve of SH-type wave whereas Figure 7 has been plotted for no irregularity $\left(H'/H=0\right)$. A comparative study of Figure 6 and Figure 7 unfolds the effect of rectangular irregularity on SH wave which is decreasing in nature, i.e. with the increase of rectangular irregularity at the interface, the phase velocity will decrease in both fundamental and higher order modes. Moreover, it has been already proved analytically that in the dispersion relation (Equation (24)) of new-type wave, no term of irregularity appears. Hence, there is no point of drawing graphs in this case.

The Effect of Non-Perfect Bonding

For the numerical analysis, non-perfect bonding parameter Γ of Model II has been non-dimensionalized by assuming $R=\frac{\left(\mu_1+\kappa_1\right)k}{\Gamma}$, where *R=0* represents perfect interface and *R*→∞ represents ideally smooth interface. In Figure 10, non-dimensional phase velocity $\left(c/c_2^{(1)}\right)$ of SH wave against non-dimensional wave-number (*kH*) has been plotted for a distinct values of non-dimensional non-perfect bonding parameter for Model II. Solid line represents the case of perfectly bonded interface (*R*=0) whereas dotted line associated with the case of ideally smooth interface (*R*=50), i.e. layer and semi-infinite medium move independently. Figure 10 reveals that with the increase of non-perfect bonding at the interface, the phase velocity gets decreased in all vibrational modes. It has also been proved analytically in the dispersion relation of new-type wave for Model II, no term of non-perfect bonding appears in the equation.

Figure 6. The variation of phase velocity $\left(c / c_2^{(1)}\right)$ *against wave number (kH) of SH wave revealing the effect of micropolarity in the medium with interface having rectangular irregularity* $\left(H'/ H = 0.3\right)$.

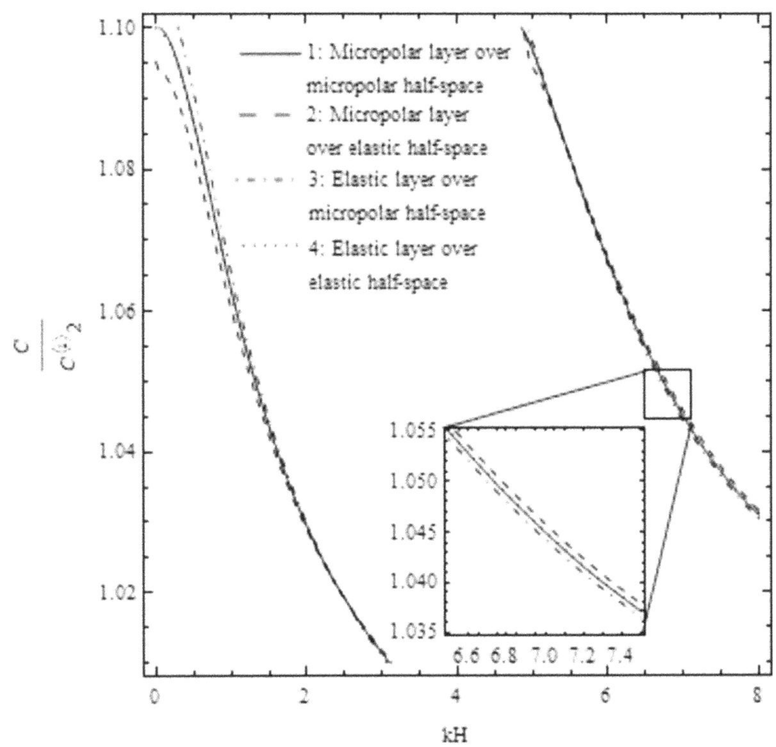

The Effect of Micropolarity

In Figure 6 to Figure 9, the effect of the presence and absence of micropolarity in the medium has been revealed through dispersion curves. In fundamental vibrational mode, it has been observed that irrespective of the presence/absence of irregularity in Model I, the phase velocity $\left(c / c_2^{(1)}\right)$ is maximum for the case when microrotations are absent in the layer (elastic layer, micropolar semi-infinite medium) whereas it is minimum when microrotations are absent in semi-infinite medium (micropolar layer, elastic semi-infinite medium). For the case when both layer and semi-infinite medium are either micropolar or without microrotations (elastic), the change in phase velocity is found to be moderate. More specifically, the phase velocity of shear wave is observed to be more for micropolar layer/semi-infinite structure as compared to elastic layer/semi-infinite structure. But, in higher order vibrational modes, this particular behavior gets reversed in the presence of irregular interface (Figure 6) whereas it remains unchanged in absence of irregular interface (Figure 7). The stated behavior suggests that in fundamental vibrational mode, phase velocity of SH wave gets increased when the microrotations are introduced in the lower elastic semi-infinite medium irrespective of whether the upper layer is micropolar or elastic. On the other hand, when microrotations are introduced in the elastic layer, the phase velocity of SH wave gets decreased irrespective of the semi-infinite medium being micropolar or elastic. In higher order vibrational mode for the case of irregular interface, the observed behavior in the fundamental mode gets

Figure 7. The variation of phase velocity $\left(c / c_2^{(1)}\right)$ *against wave number (kH) of SH wave revealing the effect of micropolarity in the medium with interface having no irregularity* $\left(H' / H = 0\right)$.

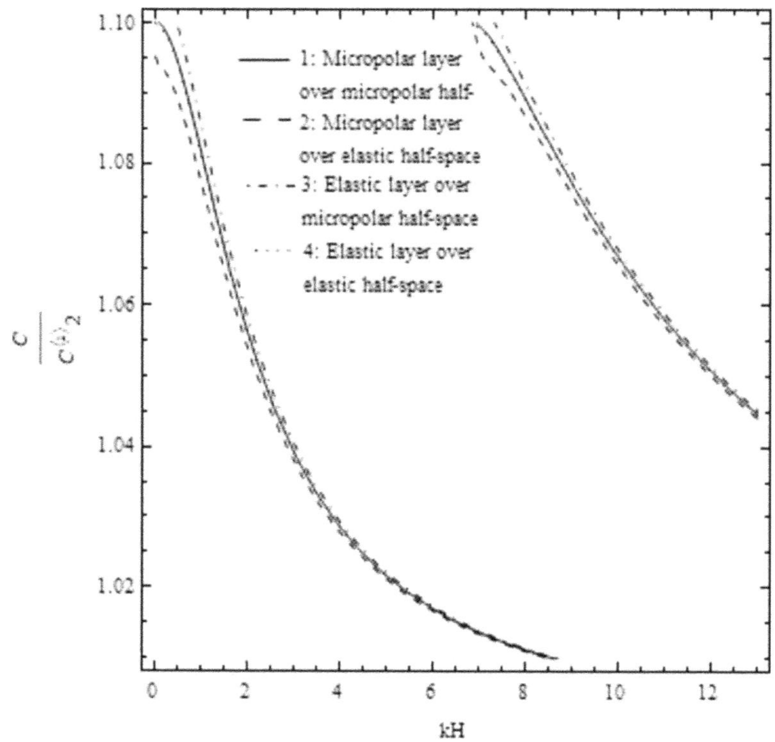

reversed, but, in case of perfect interface, the behavior remains unchanged for higher order vibrational modes. Further, Figure 5 and Figure 6 also reveal that the effect of microrotational components in semi-infinite medium dominates over that of layer in all vibrational mode irrespective of the presence/absence of irregular interface.

Figure 8 and Figure 9 reveal that the presence of microrotations in the layered structure encourages the phase velocities $\left(c / c_4^{(1)}\right)$ and $\left(c / c_4^{(2)}\right)$ of new-type surface wave which is also apparent from the fact that this new-type dispersive wave appears due to the presence of microrotations in the structure. It has been observed that the microrotations in the layer influence the phase velocity of new-type surface wave more as compared to the one in the semi-infinite medium.

FUTURE RESEARCH DIRECTIONS

In order to discuss some practical cases of the implementation of present study, some recent developments on microstructure in the area of engineering, geophysics and seismology need to be mentioned:

- It is well known that conventional sensors measure velocity or acceleration fields but they fail to measure rotation rates of material particles. Hence, to measure energy dissipation due to rotation

Figure 8. The variation of phase velocity $\left(c / c_4^{(1)}\right)$ against wave number $\left(k / \sqrt{\mu_1 / \gamma_1}\right)$ of new-type wave for the case when the micropolar layer is overlying a semi-infinite medium with or without micropolarity. (Singh, 2018)

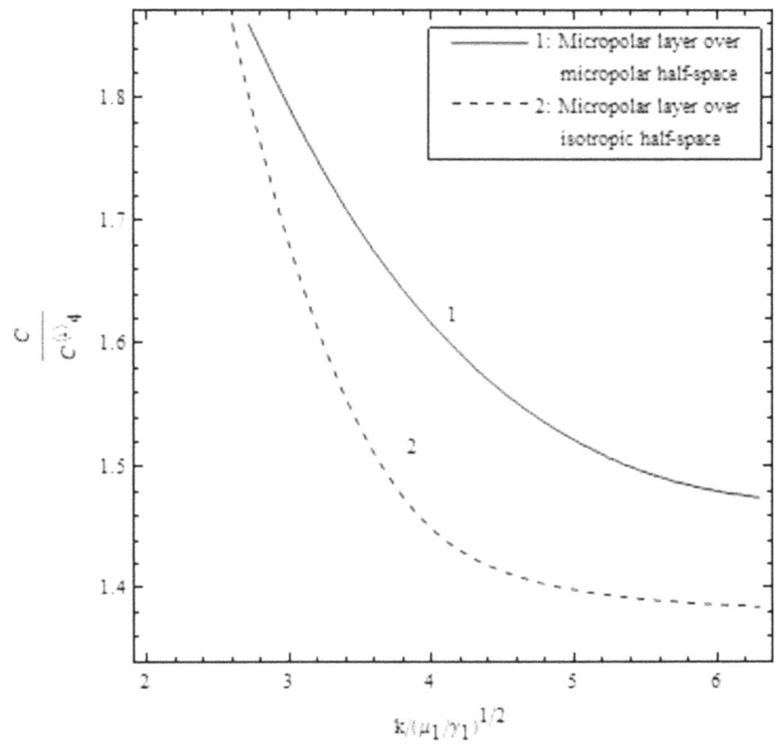

in microstructures, rotational sensors viz. mechanical sensors (Nigbor, 1994), laser sensors (Igel et al., 2005), rotational seismometers etc. are very much essential (Askar, 1972). However, despite measuring rotational energy, structural irregularity plays a significant role in energy dissipation, for which, the geometrical construction considered in the present study may provide some insight to physically interpret the results obtained by sensors and to achieve accuracy and enhance effectiveness.

- Many experimental results associated with granular structures are not well-explained. Rotations of particles and agglomerates in granular materials can be important in many processes, in particular, restructurisation of the medium and onset of instability (Merkel et al., 2011; Turco, 2018). Moreover, even a small rotation of particles can change the wave propagation characteristics and hence, the whole dynamic process (Sadovskii & Sadovskaya, 2015). In view of this, the outcome of the present study plays an important role in such experiments.
- Rotational seismology has now gained much attraction of seismologists and geophysicists due to the discovery of new-type of waves generated due to rotations. In vortex structures, the state of predestruction in fault zones, rotations of blocks of geomedium, the study of seismic wave (in particular, Love wave) propagation in microstructural layered configuration is very significant (Lee et

Figure 9. The variation of phase velocity $\left(c / c_4^{(2)}\right)$ *against wave number* $\left(k / \sqrt{\mu_2 / \gamma_2}\right)$ *of new-type wave for the case when a micropolar semi-infinite medium is overlaid by layer with or without micropolarity. (Singh, 2018)*

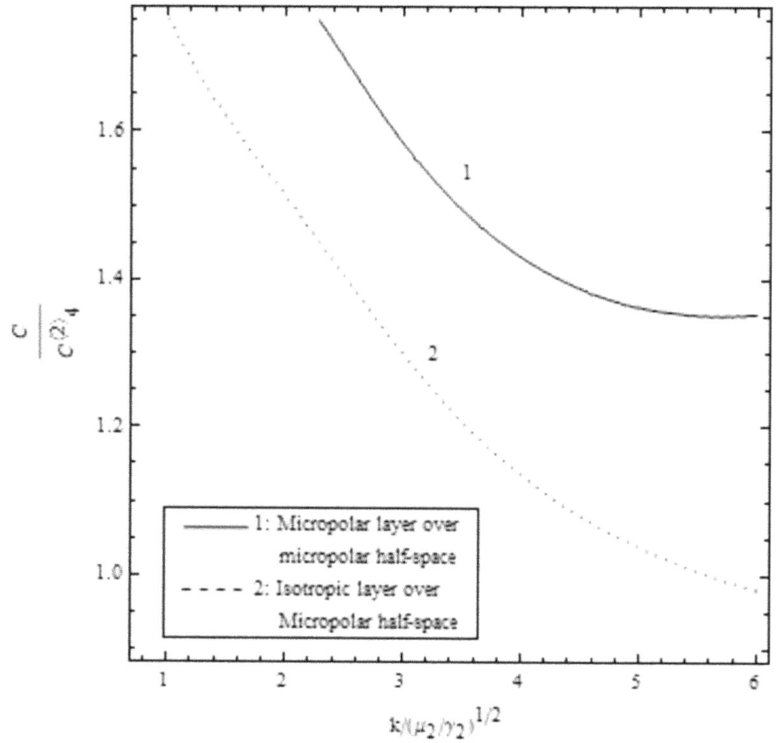

al., 2009; Abreu et al., 2017). Since, interfacial complexity is non-avoidable in layered structures, the present chapter provides two possible theoretical frameworks close to real world scenario.

- One of the possible applications of the present study can be found in acoustic metamaterials. Microstructural media has been recently used as metamaterials to study the wave propagation phenomena (Madeo, 2015; Khlopotin, 2015). Hence, the results obtained in present study having irregular and non-perfect bonding may contribute significantly in studying metamaterials for large scale which has great concern in acoustical engineering. Moreover, the present study may also be applied to metasurfaces, i.e. two-dimensional counterparts of metamaterials which are now being suggested to control seismic wave destructions (Palermo and Marzani, 2018).

Considering engineering applications in microscopic level in aspect of wave propagation in the fields of civil engineering, mechanical engineering, earthquake engineering and many more, the theoretical development articulated in the present chapter has many possibilities for future extensions:

- The obtained displacement and microrotational components can be used to find stress components and strain energy function for micropolar media which are essential in the vicinity of engineering applications.

Figure 10. The variation of phase velocity $\left(c / c_2^{(1)}\right)$ against wave number (kH) of SH wave for distinct values of non-perfect bonding parameter R.

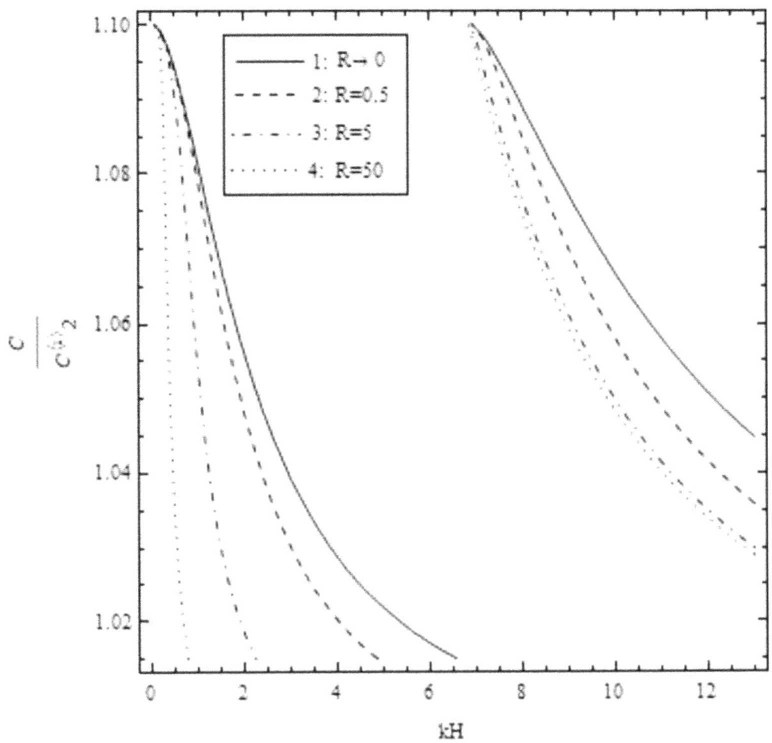

- The similar studies can be done for different materials instead of aluminium epoxy where material particles are having rigid directors (viz. platelet composites, liquid crystals, animal blood with rigid cells, foams, concrete with sand and muddy fluids etc.).
- Since no numerical simulations or experimental setup has been provided for the present study, there are always possibilities to validate the obtained results with suitable numerical or experimental methods.
- The analytical method as well as numerical analysis for present study can also be adopted for more generalized continua (microstretch or micromorphic materials). Further, the constitutive equations of microstructure considered in the present study can also be more generalized by assuming (i) posority of medium (Eringen, 2003a), (ii) thermoelasticity with microtemperatures (Iesan, 2007), (iii) micromorphic electromagnetic thermoelasticity (Eringen, 2003b), (iv) non-linear theory (Eringen and Suhubi, 1964) etc.
- The study can be further extended by considering distinct shapes and sizes of irregularity. Some of the functions associated with irregular interface are as follows:

$$h(x) = \begin{cases} 0 & \text{for} \quad |x| > s/2 \\ 1 + \dfrac{2x}{s} & \text{for} \quad -s/2 \le x \le 0 \\ 1 - \dfrac{2x}{s} & \text{for} \quad 0 \le x \le s/2 \end{cases} \quad \text{or} \quad h(x) = \begin{cases} 0 & \text{for} |x| > s/2 \\ \left(1 - \dfrac{4x^2}{s^2}\right) & \text{for} |x| \le s/2 \end{cases}$$

where $h(x)$ is a function associated with irregular boundary which can referred from the equation (8) of the present chapter. The first function represents a triangular notch shaped irregularity whereas the second one represents parabolic shaped irregular interface. In view of this, Model I can be studied again which may give some distinctive results.

- The chosen geometry of the problem is consisting of layered structure which can be extended to multilayered structure with more interfacial complexity and may also be studied for other types of surface waves to find a new-type of waves or leaky waves due to microrotations.

CONCLUSION

The outcome of the study in the current chapter can be encapsulated as the following:

1. Wave number of anti-plane shear wave and reported new type of dispersive surface has significant decreasing effect on their respective phase velocities.
2. For irregularity and non-perfect bonding in Model I and II respectively, the phase velocity of Shear wave decreases.
3. In fundamental vibrational mode, phase velocity of SH wave gets increased when the microrotations are introduced in the lower elastic semi-infinite medium irrespective the upper layer being micropolar/elastic and when microrotations are introduced in the elastic layer, the phase velocity of SH wave gets decreased irrespective of the semi-infinite medium being micropolar or elastic.
4. In higher order vibrational mode, for the case of irregular interface, the behavior found in the fundamental mode gets reversed, but, in case of perfect interface, it remains unchanged for higher order vibrational modes.
5. The effect of microrotational components in semi-infinite medium dominates over that of layer in all vibrational mode independent of irregular interface.
6. The phase velocity of the reported new dispersive wave gets influenced by the presence of microrotational components in both models. Moreover, the effect of microrotation in the layer dominates over that of semi-infinite medium.
7. The non-perfect bonding at the interface has decreasing effect on the phase velocity of SH wave in all vibrational modes.
8. Dispersion of new type of dispersive surface wave is independent of irregular as well as non-perfect common interface.

The above findings bestow a theoretical framework in context of wave propagation phenomena in solid elastic continuum having particles with rotation. In the present study, the assumed models contain interfaces which are considered mathematically with complexity, i.e. for irregular boundary, the irregularity depth has been considered very small as compared to layer width such that reflection/transmission phenomena of SH wave can be ignored, whereas, in non-perfect interface, shearing stress has been considered to be balanced by slip condition of displacement fields. Hence, it can be easily observed that such structures, although very essential for theoretical development, are hard to implement in numerical simulations and experimental works. Hence, numerical study has been done on the dispersion relations obtained from the models and extensively and comparatively analyzed for both Model I and II. The outcome of the study certainly has an impact in the field of mechanical, civil and earthquake engineering as well as in studying geological phenomena.

REFERENCES

Abreu, R., Kamm, J., & Reiß, A. S. (2017). Micropolar modelling of rotational waves in seismology. *Geophysical Journal International*, *210*(2), 1021–1046. doi:10.1093/gji/ggx211

Achenbach, J. D. (1973). *Wave propagation in elastic solids*. Amsterdam, The Netherlands: North Holland.

Altenbach, H., & Eremeyev, V. A. (Eds.). (2012). Generalized Continua-from the Theory to Engineering Applications (Vol. 541). Berlin, Germany: Springer Science & Business Media.

Ariman, T. (1972). Wave propagation in a micropolar elastic half-space. *Acta Mechanica*, *13*(1-2), 11–20. doi:10.1007/BF01179655

Askar, A. (1972). Molecular crystals and the polar theories of the continua experimental values of material coefficients for KNO3. *International Journal of Engineering Science*, *10*(3), 293–300. doi:10.1016/0020-7225(72)90044-4

Bažant, Z. P., & Christensen, M. (1972). Analogy between micropolar continuum and grid frameworks under initial stress. *International Journal of Solids and Structures*, *8*(3), 327–346. doi:10.1016/0020-7683(72)90093-5

Bhattacharya, J. (1962). On the dispersion curve for Love wave due to irregularity in the thickness of the transversely isotropic crustal layer. *Gerlands Beitrage zur Geophysik*, *6*, 324–334.

Bouchon, M., Campillo, M., & Gaffet, S. (1989). A boundary integral equation-discrete wavenumber representation method to study wave propagation in multilayered media having irregular interfaces. *Geophysics*, *54*(9), 1134–1140. doi:10.1190/1.1442748

Chattopadhyay, A., Gupta, S., Sharma, V. K., & Kumari, P. (2008). Propagation of SH waves in an irregular monoclinic crustal layer. *Archive of Applied Mechanics*, *78*(12), 989–999. doi:10.100700419-008-0209-6

Cosserat, E., & Cosserat, F. (1909). *Theorie des Corps Deformables*. Paris, France: Hermann et Fils.

Dilbag, S. (2008). *Some dynamical problems in micropolar elasticity*, (Doctoral dissertation, Department of Mathematics, Panjab University, India).

Eremeyev, V. A., Lebedev, L. P., & Altenbach, H. (2012). Foundations of micropolar mechanics. Berlin, Germany: Springer Science & Business Media.

Eringen, A. C. (1966). Linear theory of micropolar elasticity. *Journal of Mathematics and Mechanics*, 909-923.

Eringen, A. C. (1984). Plane waves in nonlocal micropolar elasticity. *International Journal of Engineering Science*, *22*(8-10), 1113–1121. doi:10.1016/0020-7225(84)90112-5

Eringen, A. C. (1999). *Microcontinuum Field Theories-I, Foundations and Solids*. New York, NY: Springer-Verlag. doi:10.1007/978-1-4612-0555-5

Eringen, A. C. (2003a). Micropolar mixture theory of porous media. *Journal of Applied Physics*, *94*(6), 4184–4190. doi:10.1063/1.1598640

Eringen, A. C. (2003b). Continuum theory of micromorphic electromagnetic thermoelastic solids. *International Journal of Engineering Science*, *41*(7), 653–665. doi:10.1016/S0020-7225(02)00274-4

Eringen, A. C., & Suhubi, E. S. (1964). Nonlinear theory of simple micro-elastic solids-I. *International Journal of Engineering Science*, *2*(2), 189–203. doi:10.1016/0020-7225(64)90004-7

Ewing, W. M., Zardetzky, W. S., & Press, F. (1957). *Elastic Waves in Layered Media*. New York, NY: McGraw-Hill Book Co. doi:10.1063/1.3060203

Fung, A., & Eom, H. (1981). A theory of wave scattering from an inhomogeneous layer with an irregular interface. *IEEE Transactions on Antennas and Propagation*, *29*(6), 899–910. doi:10.1109/TAP.1981.1142679

Fung, A., & Eom, H. (1981). A theory of wave scattering from an inhomogeneous layer with an irregular interface. *IEEE Transactions on Antennas and Propagation*, *29*(6), 899–910. doi:10.1109/TAP.1981.1142679

Gauthier, R. D. (1982). Experimental investigations on micropolar media. *Mechanics of micropolar media*, 395-463.

Ge, Z., & Chen, X. (2008). An efficient approach for simulating wave propagation with the boundary element method in multilayered media with irregular interfaces. *Bulletin of the Seismological Society of America*, *98*(6), 3007–3016. doi:10.1785/0120080920

Ieşan, D. (2007). Thermoelasticity of bodies with microstructure and microtemperatures. *International Journal of Solids and Structures*, *44*(25-26), 8648–8662. doi:10.1016/j.ijsolstr.2007.06.027

Igel, H., Schreiber, U., Flaws, A., Schuberth, B., Velikoseltsev, A., & Cochard, A. (2005). Rotational motions induced by the M8.1 Tokachioki earthquake, Sept. 25, 2003. *Geophysical Research Letters*, *32*(8), L08309. doi:10.1029/2004GL022336

Kaur, T., Sharma, S. K., & Singh, A. K. (2016). Influence of imperfectly bonded micropolar elastic half-space with non-homogeneous viscoelastic layer on propagation behavior of shear wave. *Waves in Random and Complex Media*, *26*(4), 650–670. doi:10.1080/17455030.2016.1185191

Kaur, T., Sharma, S. K., & Singh, A. K. (2017). Shear wave propagation in vertically heterogeneous viscoelastic layer over a micropolar elastic half-space. *Mechanics of Advanced Materials and Structures*, *24*(2), 149–156. doi:10.1080/15376494.2015.1124948

Khlopotin, A., Olsson, P., & Larsson, F. (2015). Transformational cloaking from seismic surface waves by micropolar metamaterials with finite couple stiffness. *Wave Motion*, *58*, 53–67. doi:10.1016/j.wavemoti.2015.07.002

Khurana, A., & Tomar, S. K. (2009). Longitudinal wave response of a chiral slab interposed between micropolar solid half-spaces. *International Journal of Solids and Structures*, *46*(1), 135–150. doi:10.1016/j.ijsolstr.2008.08.018

Koketsu, K. (2000). Finite difference traveltime calculation for head waves travelling along an irregular interface. *Geophysical Journal International*, *143*(3), 729–734. doi:10.1046/j.1365-246X.2000.00269.x

Kumar, R., & Sharma, N. (2008). Propagation of waves in micropolar viscoelastic generalized thermoelastic solids having interficial imperfections. *Theoretical and Applied Fracture Mechanics*, *50*(3), 226–234. doi:10.1016/j.tafmec.2008.07.010

Kumar, R., & Singh, B. (1997). Reflection and transmission of elastic waves at a loosely bonded interface between an elastic and micropolar elastic solid. *Indian Journal of Pure and Applied Mathematics*, *28*, 1133–115.

Lee, W. H., Igel, H., & Trifunac, M. D. (2009). Recent advances in rotational seismology. *Seismological Research Letters*, *80*(3), 479–490. doi:10.1785/gssrl.80.3.479

Li, L., & Xie, S. (2004). Finite element method for linear micropolar elasticity and numerical study of some scale effects phenomena in MEMS. *International Journal of Mechanical Sciences*, *46*(11), 1571–1587. doi:10.1016/j.ijmecsci.2004.10.004

Li, Z., & Ito, K. (2006). *The immersed interface method: numerical solutions of PDEs involving interfaces and irregular domains, 33*. SIAM. doi:10.1137/1.9780898717464

Love, A. E. H. (Ed.). (2013). *A treatise on the mathematical theory of elasticity*. Cambridge, UK: Cambridge University Press.

Madeo, A., Neff, P., Ghiba, I. D., Placidi, L., & Rosi, G. (2015). Wave propagation in relaxed micromorphic continua: Modeling metamaterials with frequency band-gaps. *Continuum Mechanics and Thermodynamics*, *27*(4-5), 551–570. doi:10.100700161-013-0329-2

Mal, A. K. (1962). On the frequency equation for Love waves due to abrupt thickening of the crustal layer. *Geofisica Pura e Applicata*, *52*(1), 59–68. doi:10.1007/BF01996000

Merkel, A., Tournat, V., & Gusev, V. (2011). Experimental evidence of rotational elastic waves in granular phononic crystals. *Physical Review Letters 107*(22), 225, 502.

Midya, G. K. (2004). On Love-type surface waves in homogeneous micropolar elastic media. *International Journal of Engineering Science*, *42*(11-12), 1275–1288. doi:10.1016/j.ijengsci.2004.03.002

Mindlin, R. D., & Tiersten, H. F. (1962). Effects of couple-stresses in linear elasticity. *Archive for Rational Mechanics and Analysis*, *11*(1), 415–448. doi:10.1007/BF00253946

Murty, G. S. (1976). Reflection, transmission and attenuation of elastic waves at a loosely-bonded interface of two half spaces. *Geophysical Journal International*, *44*(2), 389–404. doi:10.1111/j.1365-246X.1976.tb03663.x

Nagy, P. B. (1992). Ultrasonic classification of imperfect interfaces. *Journal of Nondestructive Evaluation, 11*(3-4), 127-139.

Nigbor, R. L. (1994). Six-degree-of-freedom ground-motion measurement. *Bulletin of the Seismological Society of America*, *84*(5), 1665–1669.

Nowacki, W. (1974). *Micropolar Elasticity, International Center for Mechanical Sciences, Courses and Lectures No. 151*. New York, NY: Springer-Verlag.

Palermo, A., & Marzani, A. (2018). Control of Love waves by resonant metasurfaces. *Scientific Reports*, *8*(1), 7234. doi:10.103841598-018-25503-8 PMID:29740063

Sadovskii, V., & Sadovskaya, O. (2015). Modeling of elastic waves in a blocky medium based on equations of the cosserat continuum. *Wave Motion*, *52*, 138–150. doi:10.1016/j.wavemoti.2014.09.008

Singh, A. K., Chaki, M. S., & Chattopadhyay, A. (2018). Remarks on impact of irregularity on SH-type wave propagation in micropolar elastic composite structure. *International Journal of Mechanical Sciences*, *135*, 325–341. doi:10.1016/j.ijmecsci.2017.11.032

Slavin, L. M., & Wolf, B. (1970). Scattering of Love waves in a surface layer with an irregular boundary for the case of a rigid underlying half-space. *Bulletin of the Seismological Society of America*, *60*(3), 859–877.

Sobczyk, K., & Kirkner, D. J. (2012). Stochastic modeling of microstructures. Berlin, Germany: Springer Science & Business Media.

Termonia, Y. (1990). Fibre coating as a means to compensate for poor adhesion in fibre-reinforced materials. *Journal of Materials Science*, *25*(1), 103–106. doi:10.1007/BF00544192

Toupin, R. A. (1962). Elastic materials with couple-stresses. *Archive for Rational Mechanics and Analysis*, *11*(1), 385–414. doi:10.1007/BF00253945

Tranter, C. J. (1966). *Integral transform in mathematical physics*. London, UK: Methuen and Company Limited.

Turco, E. (2018). In-plane shear loading of granular membranes modeled as a Lagrangian assembly of rotating elastic particles. *Mechanics Research Communications*, *92*, 61–66. doi:10.1016/j.mechrescom.2018.07.007

Voigt, W. (1887). Theoritiscke studien uber die elastizitats verhaltnisse der krystalle. *Abhandlungen der Braunschweigischen Wissenschaftlichen Gesellschaft*, *34*, 3–51.

Willis, H. F. (1948). LV. A formula for expanding an integral as a series. *The London, Edinburgh and Dublin Philosophical Magazine and Journal of Science*, *39*(293), 455–459. doi:10.1080/14786444808521694

Yang, J. F. C., & Lakes, R. S. (1982). Experimental study of micropolar and couple stress elasticity in compact bone in bending. *Journal of Biomechanics*, *15*(2), 91–98. doi:10.1016/0021-9290(82)90040-9 PMID:7076689

ADDITIONAL READING

Boehmer, C. G., Neff, P., & Seymenoğlu, B. (2016). Soliton-like solutions based on geometrically nonlinear Cosserat micropolar elasticity. *Wave Motion*, *60*, 158–165. doi:10.1016/j.wavemoti.2015.09.006

Boehmer, C. G., & Tamanini, N. (2015). Rotational elasticity and couplings to linear elasticity. *Mathematics and Mechanics of Solids*, *20*(8), 959–974. doi:10.1177/1081286513511093

Dehkordi, A. R. (2008). *3D finite element cosserat continuum simulation of layered geomaterials*, Doctoral dissertation.

Dell'Isola, F., & Eremeyev, V. A. (2018). Some Introductory and Historical Remarks on Mechanics of Microstructured Materials. In *Advances in Mechanics of Microstructured Media and Structures* (pp. 1–20). Cham: Springer. doi:10.1007/978-3-319-73694-5_1

Eremeyev, V. A. (2005). Acceleration waves in micropolar elastic media. [MAIK Nauka/Interperiodica.]. *Doklady Physics*, *50*(4), 204–206. doi:10.1134/1.1922562

Eremeyev, V. A. (2018). On the material symmetry group for micromorphic media with applications to granular materials. *Mechanics Research Communications*, *94*, 8–12. doi:10.1016/j.mechrescom.2018.08.017

Eringen, A. C. (2004). Electromagnetic theory of microstretch elasticity and bone modeling. *International Journal of Engineering Science*, *42*(3-4), 231–242. doi:10.1016/S0020-7225(03)00288-X

Ghosh, N. C., Nath, S., & Debnath, L. (2001). Propagation of waves in micropolar solid-solid semi-spaces in the presence of a compressional wave source in the upper solid substratum. *Mathematical and Computer Modelling*, *34*(5-6), 557–563. doi:10.1016/S0895-7177(01)00083-8

Grekova, E. F. (2016). Plane waves in the linear elastic reduced Cosserat medium with a finite axially symmetric coupling between volumetric and rotational strains. *Mathematics and Mechanics of Solids*, *21*(1), 73–93. doi:10.1177/1081286515577042

Grekova, E. F., Kulesh, M. A., & Herman, G. C. (2009). Waves in linear elastic media with microrotations, part 2: Isotropic reduced Cosserat model. *Bulletin of the Seismological Society of America*, *99*(2B), 1423–1428. doi:10.1785/0120080154

Hård af Segerstad, P., Toll, S., & Larsson, R. (2008). A micropolar theory for the finite elasticity of open-cell cellular solids. *Proceedings - Royal Society. Mathematical, Physical and Engineering Sciences*, *465*(2103), 843–865. doi:10.1098/rspa.2008.0267

Karttunen, A. T., Reddy, J. N., & Romanoff, J. (2018). Micropolar modeling approach for periodic sandwich beams. *Composite Structures*, *185*, 656–664. doi:10.1016/j.compstruct.2017.11.064

Kaur, T., Sharma, S. K., Singh, A. K., & Chaki, M. S. (2016). Moving load response on the stresses produced in an irregular microstretch substrate. *Structural Engineering and Mechanics*, *60*(2), 175–191. doi:10.12989em.2016.60.2.175

Manzari, M. T. (2004). Application of micropolar plasticity to post failure analysis in geomechanics. *International Journal for Numerical and Analytical Methods in Geomechanics*, *28*(10), 1011–1032. doi:10.1002/nag.356

Nowacki, W. (1986). *Theory of Micropolar Elasticity*. Pergamon Press.

Randow, C. L., & Gazonas, G. A. (2009). Transient stress wave propagation in one-dimensional micropolar bodies. *International Journal of Solids and Structures*, *46*(5), 1218–1228. doi:10.1016/j.ijsolstr.2008.10.024

Sachio, N., Benedict, R., & Lakes, R. (1984). Finite element method for orthotropic micropolar elasticity. *International Journal of Engineering Science*, *22*(3), 319–330. doi:10.1016/0020-7225(84)90013-2

Zhang, P., Wei, P., & Li, Y. (2018). The elastic wave propagation through the finite and infinite periodic laminated structure of micropolar elasticity. *Composite Structures*, *200*, 358–370. doi:10.1016/j.compstruct.2018.05.095

KEY TERMS AND DEFINITIONS

Dispersion Relation: A relation between the wavelength/wavenumber of a wave and its frequency.

Elastodynamics: The study of elastic waves and their propagation characteristics in linear elasticity with variation in time.

Irregular Interface: An interface between two media having non-parallel or irregular boundary, e.g. abrupt thickening, corrugation etc.

Layer/Semi-Infinite Structure: A structure comprised of a layer of finite width overlying a half-space. In reference to the considered models: Layer is $-H < z < 0, -\infty < x < \infty$ and semi-infinite medium is $z > 0, -\infty < x < \infty$.

Micropolar Elasticity: The elasticity which incorporates a local rotation of points along with the translation assumed in classical elasticity, couple stress and force stress. The theory was suggested by Eringen (1966).

New-Type Surface Waves: Surface waves which arise due to microrotations of material particles and are not encountered in classical elasticity.

Non-Perfect Interface: An interface where the two media are loosely bonded to each other and a linear relationship between slip and shear stress exits.

Perturbation Theory: A theory consists of mathematical methods for finding an approximate solution to a problem which requires addition of perturbation parameter.

SH Wave: A shear wave that is horizontally polarized where its particle motions are subjected to periodic shear and direction of propagation are contained in a horizontal plane.

ENDNOTE

1 Please note that due to mathematical simplicity, the perturbation parameter used for perturbation method has been considered similar to the perturbation parameter considered for structural irregularity in Eq. (7) and (8). [for more details, please see A. K. Mal (1962)].

APPENDIX 1

$$
\begin{aligned}
\Delta_1 &= (\kappa_2 p_1 + p_1\mu_2 - p\kappa_1 \tan pH - p\mu_1 \tan pH),\\
R_1(k) &= \frac{1}{2\pi}\int_{-\infty}^{\infty}\Big[\left(\omega^2 - 2(\omega_0^{(1)})^2 - a_1(c_4^{(1)})^2\right)D_0 s + \left(\omega^2 - 2(\omega_0^{(1)})^2 - b_1(c_4^{(1)})^2\right)F_0 p + \left(\omega^2 - 2(\omega_0^{(2)})^2 - a_2(c_4^{(2)})^2\right)\\
&\quad \times\left(C_0' s_1 - 2e^{-s_1 d}\right) + \left(\omega^2 - 2(\omega_0^{(2)})^2 - b_2(c_4^{(2)})^2\right)\left(E_0' p_1 - 2e^{-p_1 d}\right)\Big]^{\eta=k-\lambda}\bar{h}(\lambda)d\lambda,
\end{aligned}
$$

$$
\begin{aligned}
R_2(k) = \frac{1}{2\pi}\int_{-\infty}^{\infty}\Big[&(\mu_1+\kappa_1)(\omega^2 - 2(\omega_0^{(1)})^2 - a_1(c_4^{(1)})^2)(-C_0 s^2) + (\mu_1+\kappa_1)(\omega^2 - 2(\omega_0^{(1)})^2 - b_1(c_4^{(1)})^2)\\
&\times(-E_0 p^2) + \kappa_1(\omega_0^{(1)})^2(-ikB_0 r - C_0 s^2 - p^2 E_0) + (\mu_1+\kappa_1)(\omega^2 - 2(\omega_0^{(1)})^2 - a_1(c_4^{(1)})^2)(C_0 k\lambda)\\
&+ (\mu_1+\kappa_1)(\omega^2 - 2(\omega_0^{(1)})^2 - b_1(c_4^{(1)})^2)(k\lambda E_0) + \kappa_1(\omega_0^{(1)})^2(-i\lambda B_0 r - kC_0\lambda - E_0 k\lambda)\\
&- (\mu_2+\kappa_2)(\omega^2 - 2(\omega_0^{(2)})^2 - a_2(c_4^{(2)})^2)(s_1^2 C_0' + 2s_1 e^{-s_1 d}) - (\mu_2+\kappa_2)(\omega^2 - 2(\omega_0^{(2)})^2 - b_2(c_4^{(2)})^2)\\
&\times(p_1^2 E_0' + 2p_1 e^{-p_1 d}) - \kappa_2(\omega_0^{(2)})^2(-ikA_0' r_1 + p_1^2 E_0' + s_1^2 C_0') - (\mu_2+\kappa_2)(\omega^2 - 2(\omega_0^{(2)})^2 - a_2(c_4^{(2)})^2)\\
&\times\left(k\lambda C_0' + \frac{2}{s_1}e^{-s_1 d}\lambda ik\right) - (\mu_2+\kappa_2)(\omega^2 - 2(\omega_0^{(2)})^2 - b_2(c_4^{(2)})^2)\left(\lambda k E_0' + \frac{2}{p_1}e^{-p_1 d}\lambda k\right)\\
&- \kappa_2(\omega_0^{(2)})^2(-i\lambda A_0' r_1 - kC_0'\lambda - E_0'\lambda k)\Big]^{\eta=k-\lambda}\bar{h}(\lambda)d\lambda,
\end{aligned}
$$

$$
\begin{aligned}
L_1 &= [\alpha_1 k^2 + r^2(\alpha_1+\beta_1+\gamma_1)][k^2 r\gamma_1 + rs^2\gamma_1][s_1((k^2+r^2)\alpha_1 + r^2(\beta_1+\gamma_1)) + k^2 s_1(\beta_2+\gamma_2)]rs\\
&\quad - [\alpha_1 k^2 + r^2(\alpha_1+\beta_1+\gamma_1)][k^2 r\gamma_1 + rs^2\gamma_1][k^2 s(\beta_1+\gamma_1) + s((k^2+r^2)\alpha_1 + r^2(\beta_1+\gamma_1))]rs_1,\\
L_2 &= k^2 rs(\beta_1+\gamma_1)[k^2 r\gamma_1 + rs^2\gamma_1][s_1((k^2+r^2)\alpha_1 + r^2(\beta_1+\gamma_1)) + k^2 s_1(\beta_2+\gamma_2)],\\
L_3 &= [\alpha_1 k^2 + r^2(\alpha_1+\beta_1+\gamma_1)][k^2 r\gamma_1 + rs^2\gamma_1][k^2 s(\beta_1+\gamma_1) + s((k^2+r^2)\alpha_1 + r^2(\beta_1+\gamma_1))]k^2\\
&\quad + [\alpha_1 k^2 + r^2(\alpha_1+\beta_1+\gamma_1)]k^2[k^2 s(\beta_1+\gamma_1) + s((k^2+r^2)\alpha_1 + r^2(\beta_1+\gamma_1))][-k^2 r(\beta_1+\gamma_1) + r(k^2\beta_2 + s_1^2\gamma_2)],\\
L_4 &= k^2 rs[k^2 s(\beta_1+\gamma_1) + s((k^2+r^2)\alpha_1 + r^2(\beta_1+\gamma_1))][-k^2 r(\beta_1+\gamma_1) + r(k^2\beta_2 + s_1^2\gamma_2)](\beta_1+\gamma_1),\\
L_5 &= k^2 r[k^2 r\gamma_1 + rs^2\gamma_1][k^2 s(\beta_1+\gamma_1) + s((k^2+r^2)\alpha_1 + r^2(\beta_1+\gamma_1))](\beta_1+\gamma_1)r_1\\
&\quad + kr(\beta_1+\gamma_1)k^2[k^2 s(\beta_1+\gamma_1) + s((k^2+r^2)\alpha_1 + r^2(\beta_1+\gamma_1))][-krr_1(\beta_1+\gamma_1) + krr_1(\beta_2+\gamma_2)],\\
L_6 &= -kr[k^2 s(\beta_1+\gamma_1) + s((k^2+r^2)\alpha_1 + r^2(\beta_1+\gamma_1))][-krr_1(\beta_1+\gamma_1) + krr_1(\beta_2+\gamma_2)](-k^2\beta_1 + s^2\gamma_1),\\
L_7 &= -krk[k^2 r\gamma_1 + rs^2\gamma_1][k^2 s(\beta_1+\gamma_1) + s((k^2+r^2)\alpha_1 + r^2(\beta_1+\gamma_1))](\beta_1+\gamma_1)r\\
&\quad + kr(\beta_1+\gamma_1)[k^2 r\gamma_1 + rs^2\gamma_1][k((k^2+r^2)\alpha_1 + r^2(\beta_1+\gamma_1)) + k((-k^2+r_1^2)\alpha_2 + r_1^2(\beta_2+\gamma_2))]rs,\\
L_8 &= -kr[k^2 r\gamma_1 + rs^2\gamma_1][k((k^2+r^2)\alpha_1 + r^2(\beta_1+\gamma_1)) + k((-k^2+r_1^2)\alpha_2 + r_1^2(\beta_2+\gamma_2))](k^2\beta_1 - s^2\gamma_1).
\end{aligned}
$$

APPENDIX 2

$$
\begin{aligned}
&m_{11} = ipY\cos pH, a_{12} = pY\sin pH, m_{21} = ipY(\mu_1 + \kappa_1), m_{22} = \Gamma\left(Y - p_0^{(1)}(\omega_0^{(1)})^2\right), m_{23} = -\Gamma\left(Y_1 - p_0^{(2)}(\omega_0^{(2)})^2\right),\\
&m_{31} = m_{21}, m_{33} = p_1 Y_1(\mu_2 + \kappa_2), m_{41} = -kp(\beta_1 + \gamma_1)\cos pH, m_{42} = ikp(\beta_1 + \gamma_1)\sin pH, m_{44} = Z\cos rH,\\
&m_{45} = -iZ\sin rH, m_{46} = iks(\beta_1 + \gamma_1)\sin sH, m_{47} = -ks(\beta_1 + \gamma_1)\cos sH, m_{51} = i(\beta_1 k^2 - \gamma_1 p^2)\sin pH,\\
&m_{52} = -(\beta_1 k^2 - \gamma_1 p^2)\cos pH, m_{54} = -ikr(\beta_1 + \gamma_1)\sin rH, m_{55} = kr(\beta_1 + \gamma_1)\cos rH,\\
&m_{56} = -(\beta_1 k^2 - r_1 s^2)\cos sH, m_{57} = k(\beta_1 k^2 - r_1 s^2)\sin sH, m_{61} = ip(\omega_0^{(1)})^2, m_{63} = p_1(\omega_0^{(2)})^2,\\
&m_{64} = ik(\omega_0^{(1)})^2, m_{67} = is(\omega_0^{(1)})^2, m_{68} = s_1(\omega_0^{(2)})^2, m_{69} = -ik(\omega_0^{(2)})^2, m_{72} = -ik(\omega_0^{(1)})^2,\\
&m_{73} = ik(\omega_0^{(2)})^2, m_{75} = ir(\omega_0^{(1)})^2, m_{76} = -ik(\omega_0^{(1)})^2, m_{78} = ik(\omega_0^{(2)})^2, m_{79} = r_1(\omega_0^{(2)})^2,\\
&m_{81} = -kp(\beta_1 + \gamma_1)(\omega_0^{(1)})^2, m_{83} = ikp_1(\beta_2 + \gamma_2)(\omega_0^{(2)})^2, m_{84} = Z(\omega_0^{(1)})^2, m_{87} = -ks(\beta_1 + \gamma_1)(\omega_0^{(1)})^2,\\
&m_{88} = iks_1(\beta_2 + \gamma_2)(\omega_0^{(2)})^2, m_{89} = -Z_1(\omega_0^{(2)})^2, m_{92} = -(\omega_0^{(1)})^2(\beta_1 k^2 - \gamma_1 p^2),\\
&m_{93} = -(\omega_0^{(2)})^2(\beta_2 k^2 + \gamma_2 p_1^2), m_{95} = kr(\omega_0^{(1)})^2(\beta_1 + \gamma_1), m_{96} = -(\omega_0^{(1)})^2(\beta_1 k^2 - \gamma_1 s^2),\\
&m_{98} = (\omega_0^{(2)})^2(\beta_2 k^2 + \gamma_2 s_1^2), m_{99} = -ikr_1(\omega_0^{(2)})^2(\beta_2 + \gamma_2), Y = \omega^2 - 2(\omega_0^{(1)})^2 - b_1(C_4^{(1)})^2 + p_0^{(1)}(\omega_0^{(1)})^2,\\
&Y_1 = \omega^2 - 2(\omega_0^{(2)})^2 - b_2(C_4^{(2)})^2 + p_0^{(2)}(\omega_0^{(2)})^2, Z = \alpha_1 k^2 + r^2(\alpha_1 + \beta_1 + \gamma_1), Z_1 = \alpha_2 k^2 - r_1^2(\alpha_2 + \beta_2 + \gamma_2).
\end{aligned}
$$

Chapter 15
Antilock Braking System Formal Modeling

Abdessamad Jarrar
https://orcid.org/0000-0002-8792-7014
University Hassan 1st, Morocco

Youssef Balouki
University Hassan 1st, Morocco

ABSTRACT

Antilock Braking System (ABS) is one of the most critical systems in the context of vehicles' mechatronics. The main purpose of the ABS system is allowing the wheels to stop while preventing sliding. It is responsible for ensuring a secure stopping of the vehicle, a very critical factor in trip safety. Therefore, the process of its construction should be performed with high care, and this is why theoretical modeling is highly needed. In order to help engineers to develop and study such a critical system, we propose a standard model that includes the essence of Antilock Braking System using a formal method called Event-B. This model may be used to reveal some bugs during proving that may go otherwise undetected. At the same time, the model can be animated to observe the system behavior.

INTRODUCTION

The Antilock Braking system is an active safety system that aims to reduce the number of road accidents (Cao, Chen, & Zhou, 2018; Voevodin et al., 2018). ABS is an electronically controlled system that allows the driver to keep up control of the vehicle amid crisis braking while at the same time keeping the wheels under control. Besides, by keeping brake weight just underneath the point of making a wheel lock up, ABS guarantees that the most extreme braking power is utilized to stop the vehicle while minimizing the braking distance as much as possible (Mammosser, Boisveert, & Micheau, 2013). In an ordinary ABS, the measure of the angular velocity of the wheels and the linear acceleration are done by means of sensors. These data received from these sensors are utilized to settle on the choice, regardless of whether

DOI: 10.4018/978-1-7998-0117-7.ch015

the wheel is going to lock up. If a tendency toward wheel block is occurred, the brake cylinder pressure is reduced. As soon as the wheels disallowed from locking, the pressure will increase again.

The ABS system is very critical in the road security domain, thus its design of should be proceed carefully. Therefore, we used one of most sophisticated techniques in system design that are formal methods (Adamiec, Dziubiński, Siemionek, & Drozd, 2018). Formal methods are rigorous computer techniques that allow, with the help of specialized languages and logical rules, to ensure (ideally) that there is no lack of computer programs - which is particularly important when they involve human lives - software or mathematical demonstrations. By using this kind of method, we can improve our understanding of the system behavior and at the same time detect ambiguities, incompleteness as well as inconsistencies. In this paper, we use The Event-B formal method that adopts a mathematical notation based on set theory. Furthermore, Event-B use refinement to represent systems at different abstraction levels which facilitate system design (Jarrar & Balouki, 2018; Debbeh, Bon, & Collart-Dutilleul, 2019). It includes mathematical proofs that allow the verification of consistency between abstraction levels and the preservation of invariants (Ondruš & Kolla, 2018). This method has been effectively used in different systems where not bugs were detected. For instance, the driverless meteor in Paris worked correctly and no bugs were identified after the theoretical proofs, neither at the functional approval. This achievement encourages Alstom and Siemens Transportation Systems to use this method to develop future metros. Therefore, we also used Event-B to develop a standard antilock braking system model.

The main contribution of this paper is providing a sort of a standard model of the ABS system. This model is suitable for any ABS systems regardless of the technologies and techniques used during fabrication which enlarge its use to almost all ABS systems. It includes the essence of ABS and can be used as a first abstract model; afterward it can be refined by adding more typical requirement and parameters of the studied system. Alongside the development and refinement of models, proof obligations ensure the consistency and absence of failure in the system which highly guarantees the correctness of the system.

The rest of the paper is structured as follows. Section 2 gives some background on related works, ABS, and the used method. In section 3, we present the set of requirements considered in this paper. The main content of the paper is in Section 4 describing our approach to develop the ABS. Section 5 concludes the paper.

BACKGROUND AND LITERATURE REVIEW

Overview of the Antilock Braking System

The Antilock Braking System ABS is a brake assist system used on vehicles, limiting wheel lock during periods of heavy braking. A secondary function in the traction control system, it is used on aircraft (during landing) and on motor vehicles or motorcycles, where it is becoming more and more standard equipment (Glamočić, 2018). Based on the second law of Newton, the motion equation of vehicle can be simplified as follows:

$$mV = -F_t - F_a \tag{1}$$

m: the total mass of the vehicle.
V: vehicle velocity.
F_t: Road friction force.

$$F_t = \mu(\lambda, V) F_N \tag{2}$$

With F_N is the total normal load.
F_a: The aerodynamic force that depends on the size, the shape, size, and the instantaneous velocity.

$$F_a = \frac{1}{4}\left(\frac{\rho}{2} C_d A_f V^2\right) \tag{3}$$

C_d: Vehicle drag coefficient.
A_f: vehicle frontal area.

Nowadays, ABS is used in almost all vehicles due to its ability to prevent accident when you lose control of the steering as you apply the brakes. With ABS, the driver will be able to steer the vehicle properly and at the same time allows the reduction of the braking distance.

To properly understand ABS, we firstly need to understand the principle of steering and braking. The moment the driver presses the braking pedal, he activates the brake pads on the wheel disks which will press against the disks and this will cause the stop of the wheels rotation. When the wheels stopping, a slippage between the road and wheels will be caused and this friction will stop the vehicle stop. Regarding the steering mechanism, it is dependent on the simple principle of wheels rolling. The wheels will always try to avoid slipping on the road. In other words, the velocity of the wheel at the contact point with the road equal zero. The wheels have two kind of motion, they rotation on their axis and they move also along the direction of the car (Otremba, Romero Navarrete, & Lozano Guzmán, 2018). These two kinds of motions cause the existence of two type of velocity: the translational velocity and the rotational. The figure bellows illuminate the both kinds of velocity:

According to the second law of Newton, the motion equation can be simplified as follows:

$$J\omega = -T_b + F_t R \tag{4}$$

J: the inertia moment of wheels.
ω: the speed of wheels.
T_b: braking torque.
R: wheels radius.

In order to prevent the slipping, the sum of the two velocities -translational and rotational- should equal zero (Zulhilmi, Peeie, Eiman, Izhar, & Asyraf, 2019). For instance, in an automobile, to avoid slipping all its wheels should be rolling at all times. This is the reason why the car turns when the front wheels turn (see figure 2).

Figure 1. Kinds of velocity

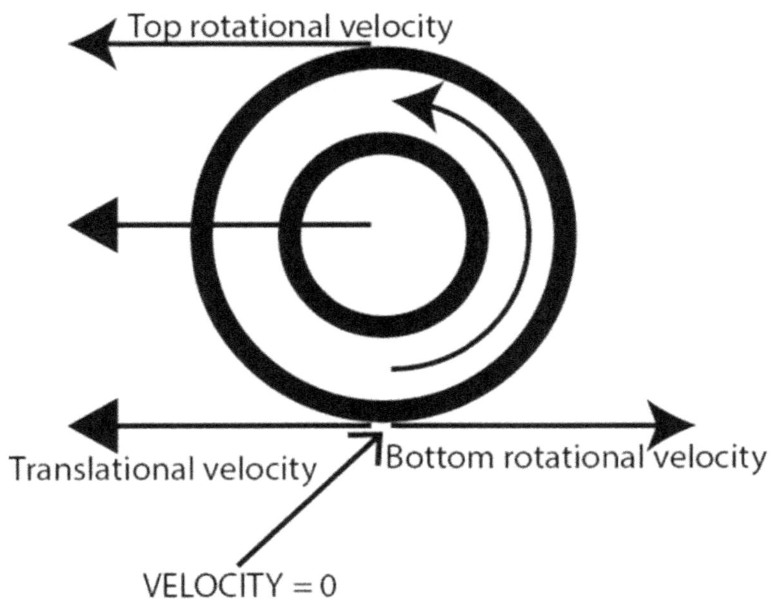

Assume a hypothetical condition in which even after turning the front wheels the car carried straight on, the rolling velocity will be inclined and the translational velocity remains straight. Therefore, if these two velocities been added up the resultant velocity will not be zero; this will cause the car to slip. Since slippage occurs between the road and the tire as the increase of force at wheels, the speed of wheel will be lower than the speed of vehicle. To specify the difference between these velocities, a parameter λ called wheels slip, this parameter is expressed as follows:

$$\lambda = \frac{V - \omega R}{V} \quad (5)$$

The only way to avoid car's slip is by making the translational velocity also inclined (D'Avico, 2019). This can be possible if the whole car makes turn. All four wheels may satisfy the rolling condition as the car makes a turn at an instantaneous center.

Due to the activation of braking pads, the wheels will stop spinning even though the wheels are turned. While the wheels are not rotating, there will be no rotational velocity. In this case, the vehicle does not need to turn as this has become like a normal case of braking. The vehicle will not react to the driver's commands of steering and will keep sliding in a straight line, and this may be able to cause an accident. Additionally, if a car does not include an ABS system, if will suffer from braking on surfaces with different levels of traction. In other words, if left and right wheels are on different traction surfaces, and the driver applies a brake, a different in frictional will be generated (Feng & Jun, 2019). This issue produces torque which will cause the car will go out of control into an uncontrollable spin.

What ABS does to solve these problems is simply prevent the wheels from completely locking up. The ABS system includes speed sensors attached to all of the wheels. When the sensor detects that a wheel

Figure 2. Car rotation

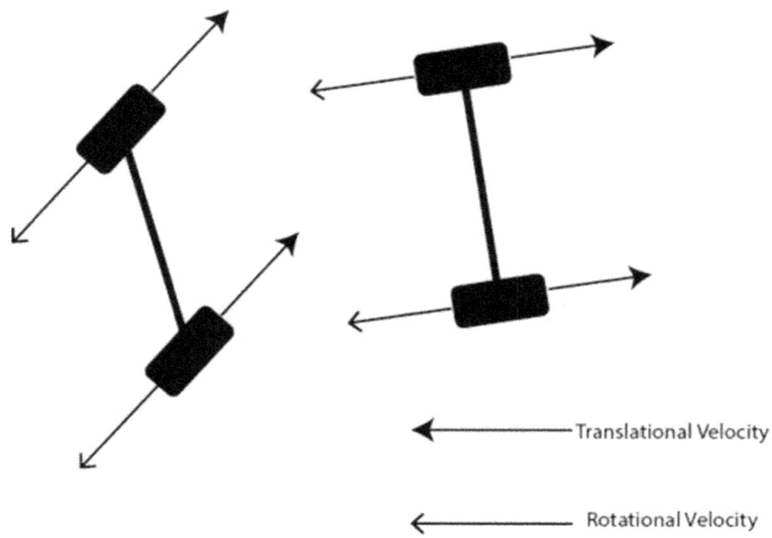

is about to lock up, a modulator unit partially releases the brake pad on that wheel. This way, the wheels will be allowed to spin intermittently during braking, and this affect can be seen immediately on steering.

When releasing intermittently brakes, the braking distance will be reduced. This is due to how the coefficient of friction between the tire and the road varies with respect to slip (Deshmukh, Shendge, & Phadke, 2018) (see figure 3).

As mentioned before, the frictional coefficient is approximately zero when wheels are rolling. When a wheel is slipping, sliding friction comes into play. The frictional values for other level of slippage can be expected to vary exponentially (Boere & da Silva, 2018). But since rubber is a complex material, which has a strange behavior during braking, the frictional coefficient of the tire and road varies as below with respect to slippage:

The frictional coefficient has a peak value at around 12% of slip ratio and below and above this the frictional coefficient reduces. During the braking in a car that does not include ABS, the frictional coefficient is predominantly sliding friction. The clever algorithm ABS system adjusts the pressure on wheels to maintain the slip ratio near to the peak value 12% as much as possible. This adjustment considerably minimizes the distance of braking.

By intelligent electronic braking distribution, which is a subsystem of modern ABS, one can solve the problem of cars instability during braking very easily (Onsay, Sharaf, Ashrey, & Eldemerdash, 2018). An EBD yaw rate of the car and the relative slip of each wheel are measured. The ABS system can reduce the frictional force that is normally produced by the wheels by reducing the pressure on the ones with maximum grip first. This will keep the yaw torque and the whole car under control.

The ABS has also its own limitations. This is the case with many other safety systems, to benefit from all the advantages of the ABS system, the driver must interact properly with it. Interacting properly with a safety system like ABS means continuing to drive carefully and carefully. Speeding, following too closely or driving when fatigued can negate the beneficial effects of ABS. For example, if you follow a vehicle too close, it may be impossible to maneuver to avoid it if the driver brakes suddenly (Tabone, Azzopardi, & Farrugia, 2018). Even if the ABS system intervenes in such a case, the driver simply will

Figure 3. Coefficient of friction / Slip ration variation

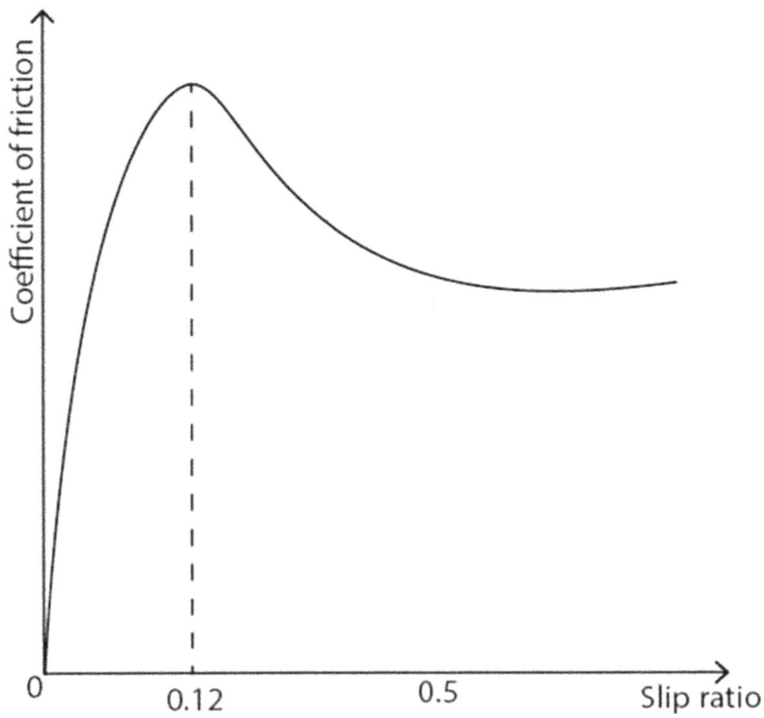

not have time to avoid the collision because the distance between the two vehicles is too short. Similarly, because it lengthens the driver's reaction time, fatigue or drowsiness can cause a driver to brake too late to avoid a collision.

The ABS system does not compensate for careless driving or poor road conditions. For the ABS system to provide the ideal performance, the driver must continue to exercise caution and good judgment while driving. Combined with safe driving practices, safety systems such as ABS have been shown to prevent or mitigate traffic accidents.

Modeling and Refinement in Event-B

In a few decades, the miniaturization of processors has allowed them to be introduced into equipment of all sizes and dedicated to all kinds of tasks, be the household appliances, nuclear power plants, medical devices, smart cards, bank machines, telecommunications systems, automobiles or airliners. All these devices are controlled by software programs, and this completely changes the situation as to the measures to be taken to avoid failures. To develop the analog part of these systems, the mathematical means to be implemented are the analysis, more precisely the differential calculus. Briefly, we are dealing with continuous functions, producing neighboring effects when the input deviations are small. Discrete systems, and especially programs, are fundamentally unstable. At the logic gates or machine instructions, the least bit change has potentially arbitrarily large effects. In a program expressed in a high-level language, a minimal change in a value, expression, or instruction just as easily leads to huge differences in behavior. Let us observe in passing the importance of this characteristic of the software, its flexibility: in comparison with a hardware system, it is very easy to intervene on a program to correct it, to change the functionalities, but also to introduce errors (including the ones we want to correct) (Clarke & Wing, 1996).

Due to the high risk in safety-critical systems, it is highly recommended to base their engineering on a certain theory. For example, electrical engineering is based on Maxwell's equations and Kirchhoff's laws; civil engineering is based on geometry and theory of material's strength. Similarly, software engineering has formal methods which are less considered. Therefore, there are engineers who do not know any theory building software, and the software often has bugs that may cost millions to fix. Now, formal methods can be used for verifying and specifying software in order to highly guarantee bugs' absence.

Event-B is a formal method that provides the correct-by-construct approach and formal verification by theorem proving (Abrial, 2010). Models in Event-B are presented based on abstract state machine notion, which presents the model states in term of a set of variables; these states are constrained by invariants. Invariants are the necessary properties that must be preserved during system function. Statuses transitions are described by events, which are a set of actions (Stankaitis et al., 2019). Each action changes the value of certain variable. Events may have some necessary conditions to be triggered; these conditions are called guards. Models in Event-B include sets, constants and axioms representing the static part of the model.

One of the main features of Event-B is refinement, which means starting modeling with an abstract model and then enriches it in successive steps by adding more details. This technique makes modeling easier than trying to model the whole system at once. We focus on a limited number of requirements in each step under the condition of cleverly choosing the refinement strategy. The figure below presents an outline of the process of modeling in Event-B:

Event-B models consistency, invariant preservation and the correctness refinement –refinements should not contradict- are ensured by discharging a number of verification conditions called proof obligations [17, 18]. For example, to prove that an invariant is preserved by an event, we prove that if an invariant is preserved before the event it will remain preserved after it. Mathematically speaking, let I be the model invariant, A are axioms, c are constants, s are set, v are variables before the event occurrence and v' are variables after the event. The following logical formula should be proved in order to prove invariant preservation:

$$A\left(c,s\right)\wedge I\left(v,c,s\right)\vdash I'\left(v',c,s\right) \quad (1)$$

Most of proof obligations are discharged automatically by means of a platform called Rodin. The remaining proofs may be dealt using an interactive prover included in Rodin.

With Rodin, the constants, abstract sets, and enumerated sets, as well as their properties, are grouped into components called contexts. Variables and events are grouped into components called templates. To create a context, select the project previously created in the Project Explorer view and clicks the Create new component button (Rodin, Jastram, & Butler, 2011). A window appears: it allows to enter the name of the component as well as its type: Machine (model) or Context. Machine is the value selected by default. Select Context and then click Finish to actually create the context. A context editor will appear (Abria et al., 2010).

Figure 4. Process of development in Event-B

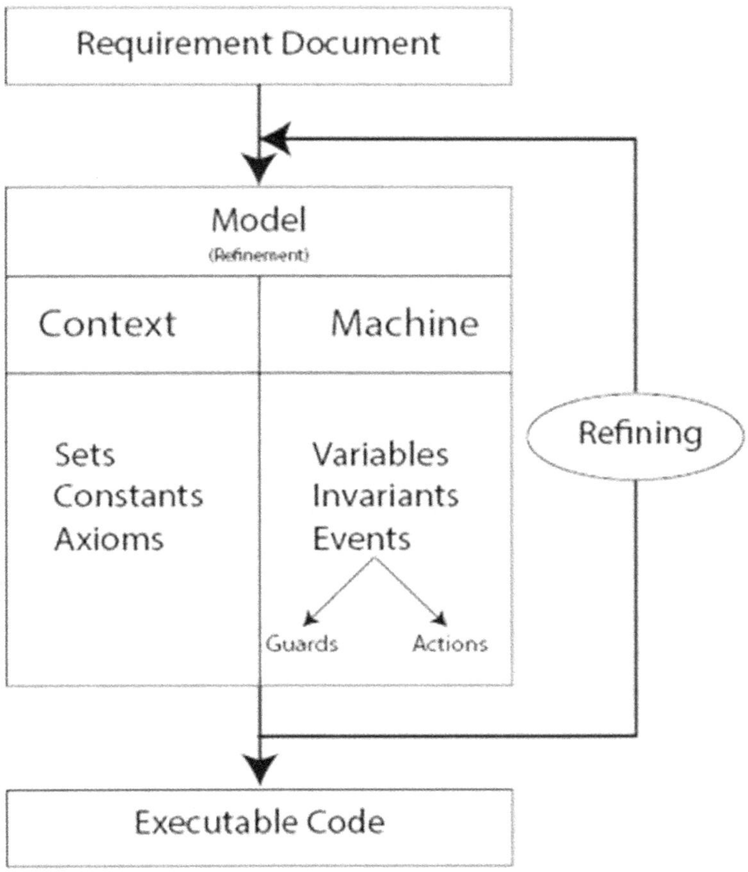

To create an enumerated set, select the context editor. Locate the New Enumerated Set Wizard button located above the context editor (it has the shape of a red uppercase S). Click on it. A window appears, allowing you to enter the name of the set and the different values. Click OK to create the enumerated set thus defined. Warning! The enumerated set is not one of Rodin's basic types: it is defined in terms of constants whose constituent elements are distinct two by two. These axioms are generated by the wizard. To modify the definition of an enumerated set, it is then necessary to manually modify the constants created by the wizard as well as the associated axioms.

To create the first model of a refinement chain, select the previously created project in the Project Explorer view, and then click the Create new component button. A window appears: it allows to enter the name of the component as well as its type: Machine (model) or Context. Click on Finish to actually create the context. A model editor appears. To use (have access to constants and sets) one or more contexts, click the button, at the bottom right of the model editor window, representing a downward arrow and located just to the left of the Add button. A drop-down list appears, containing the contexts defined in the project. Choose the context to use and click Add to add this context to the visibility list. To create a refinement of an existing model, select this model in the Project Explorer view and right-click to see a pop-up menu: select Refine. A window appears, inviting you to enter the name of the component to

Box 1.

The controller allows to operate the brake and display the ABS status.	FUN1

create. Click OK to actually create the refined model, containing all the variables and events defined in the abstract model. To create a variable, in a model, select the model editor. Locate the New Variable Wizard button located above the Model Editor (it has the shape of a yellow lowercase v). A window appears, allowing you to enter the name of the variable, its initialization and its invariant. Any modification of the name of the variable is reflected in the initialization and the invariant. To create an event, select the model editor. Locate the New Event Wizard button located above the Model Editor (it has the shape of a lowercase black e). A window will appear, allowing you to enter the name of the event, the list of local variables (3 fields labeled variable name (s)), guards and actions (substitutions).

With RODIN, the mathematical language has been simplified and clarified (Abrial, Butler, Hallerstede, & Voison, 2006):

- The + and - operators are arithmetic operators only. The set difference is noted and the Cartesian product is obtained by typing **.
- There are no more sequences in Rodin.
- The set of finite parts of a set is no longer part of the language. To indicate that a set S is finite, write finite (S).
- In a quantized predicate, the list of variables should no longer be parented. With Atelier B, we write! (X1, x2). (P (x1, x2)). With Rodin, we write! X1, x2.P (x1, x2).

REQUIREMENT DOCUMENT

As we are modelling formally a system, it is recommended to present requirements in a more formal way instead of a simple informal paragraph. We propose presenting the requirement document along two axis labelled and numbered: the first is "FUN" that stands for the specific task/Functional requirements of the system; the second is "ENV" which deals with the concerning environment and assumptions as for equipment situated around our intended system.

The requirement document for our proposed system is as follows (Bergmann et al., 2018; Batayneh, Jaradat, & Bataineh, 2018):

FORMAL DEVELOPMENT

When braking, it is important for safety not to block the wheels as this helps to maintain good grip conditions with the road and to avoid losing control of the vehicle in the event of a change of course or different conditions of contact of the wheels with the ground (one wheel on a puddle and the others on the dry bitumen). The antilock braking system Invented in 1920 by Gabriel Voisin is used as a safety

Box 2.

The controller receives wheel speed, linear speed, pedal status and engine condition	FUN2

Box 3.

The controller calculates and sends the control values	FUN3

Box 4.

When the pedal is engaged the indicator must display the ABS status	FUN4

system for vehicles such as airplanes, automobiles or motorcycles. Its principle is based on a braking assistance system which makes it possible to avoid the locking of the wheels during brutal and intense braking (Che et al., 2013). This system was developed to provide brake assistance by ensuring a better balance of the vehicle.

The antilock braking system configuration depends on four main parameters: wheel speed, the linear speed, the state of the pedal, the state of the engine. The wheel speed and the linear speed are the two main parameters needed for the ABS system to know either the wheels are locked or not by comparing the translational and rotational velocities. While, The state of the pedal and the the state of the engine represent the control parameters of the ABS. Therefore, the formalization of the proposed system should

Box 5.

The sensors make it possible to calculate the speed of the wheels, the linear speed and the state of the engine and also the state of the pedal	ENV1

Box 6.

The indicator shows the ABS status	ENV2

Box 7.

SETS Pedal_States Engine_States **CONSTANTS** Engage Disengage Ready Not_Ready **AXIOMS** axm1 : partition(PEDAL_STATES,{Engage},{Disengage}) axm2 : partition(ENGINE_STATES,{Ready},{Not_Ready})

begin with a simple definition of the various carried set defining the possible states of the pedal and engine. This can be modelled by the following context:

The context is presented in term of three parts: SETS, CONSTANTS and AXIOMS. The SETS part include the declaration of the PEDAL_STATES and ENGINE_STATES denoting repectivly the possible state of the pedal and the engine. In the CONSTANTS part, the possible values of states are declared interm of constants. Finaly, the AXIOMS part present two axioms modeling the PEDAL_STATES and ENGINE_STATES as partitions of Engage / Disengage and Ready / Not_Ready.

A partition of a set S is a set of non-empty parts of S two to two disjoint and whose union is X. The partition predicate is an easy way to enumerate sets. Mathematically, the partition predicate is defined as follows:

$$partition\left(S,\ x,\ y\right) \Leftrightarrow x \cup y = S \wedge \ x \cap y = \ \varnothing$$

Where x and y are two subsets of a set S.

As mentioned previously, The system state can be described by means of only four variables denoted in the dynamic part of the model (the machine) as follows:

The INVARIANTS part can present the different property that should be maintained during the system life time, however it may be used as we did for typing properties. Typing properties are predicate that define the type of each variable in the system.

The type chosen for speed in this formalization is ℕ referring to the fact that the speed is a positive numerical number (not necessarily a natural number).

Box 8.

VARIABLES engine_state pedal_state wheel_speed linear_speed **INVARIANTS** inv1: engine_state ∈ ENGINE_STATES inv2: pedal_state ∈ PEDAL_STATES inv3: wheel_speed ∈ ℕ inv4: linear_speed ∈ ℕ

Box 9.

```
INITIALISATION
THEN
act1: engine_state := Not_Ready
act2: pedal_state :=Disengage
act3: wheel_speed := 0
act4:linear_speed := 0
END
```

The initial state of the system is presented in an event called the INITIALISATION event that should not have guards while the initialization of the system should always be possible. In the case studied here, the initialization event can be presented as follows:

Initialy, the engine is supposed to be not ready and the pedal disengaged which mean the vehicule is stoping meaning that both the wheel and linear speed are null.

Beside the INITIALISATION event, the first event that trigger is obviously that event that start the engine (Potapov, Tarasova, Svirbutovich, & Fedyaev, 2018). Several factors affect the starting of the engine, for example: the power of the batteries, the performance of the starter, the viscosity of the lubricating oil, and also the installation of a Cold Start Device-Diesel is needed if the engine needs to start in very cold conditions. However, trying to enumerate all the factor that play a role in starting the engine is an impossible task, therefore, we focus on two main conditions (guards) to trigger the engine starting event: the first condition ensures that the engine is not already started; while the second prevent the engine from starting when the pedal is engaged. The Engine_Starting can be formalized as follows:

Even though the use of the brake pedal is usually associated with the concept of safety, it is important to keep in mind that braking does not mean pounding. In certain situations, braking can become potentially very dangerous, and the driver must have full control of his vehicle before making the decision to brake. The risk of collision due to too vigorous braking is very important, and it can have dramatic consequences, both material and human.

The usefulness of a brake pedal is, of course, more to demonstrate, but users should keep in mind that, as the braking force is proportional to the pressure exerted on the pedal, it is advisable to slow down slowly so as not to surprise other road users. The brake pedal is located in the middle of the pedal and it is handled with the right foot, alternating with the accelerator pedal.

For the manipulation of the pedal we present the events that make it possible to engage and disengage it:

In order to detect that the wheels are about to jam, each wheel is equipped with an electromagnetic rotation sensor. The process is very simple: on a disk integral with the axis of a wheel, and thus rotating with it, there is disposed at the periphery, a set of magnets, and a coil, placed on the hub of the wheel,

Box 10.

```
Engine_Starting
WHERE
grd1: engine_state = Not_Ready
grd2: pedal_state = Disengage
THEN
act1: engine_state := Ready
END
```

Box 11.

```
Engage_Pedal
WHERE
grd1: pedal_state = disengage
THEN
act1: pedal_state:= engage
END
Disengage_Pedal
WHERE
grd1: pedal_state= engage
THEN
act1: pedal_state:= disengage
END
```

near the disc. Thus, with each passage of a magnet in front of the coil - ie the sensor -, a pulse is created. The movement of the magnets in front of the sensor produces a variation of the magnetic field and causes the appearance of an analog alternating signal at the output of this sensor. This signal is then converted numerically by the analog / digital converter, before being transmitted to the computer, which constantly analyzes the information sent to it. If the sensor of a wheel no longer emits a signal, the computer immediately sends the order to the solenoid valves to suspend the pressure supply of the braking device of the locked wheel. When no current is sent into the solenoid valve, the brake circuit of the wheel is normal, and the pressure exerted by the master cylinder is directly transmitted to the wheels.

When a wheel is blocked, the computer sends a current to the solenoid valve. An electromagnet attracts the piston: the circuit is then opened and the pressure is evacuated towards the return circuit. The brake caliper then releases its pressure: the wheel is released. As soon as the wheel starts to turn again, the pressure is restored and braking continues, and so on, as long as braking is required.

After the calculations already made by the controller, it actuates the brake which implies a reduction of the speed of the wheels by the red_w value and a reduction of the linear speed by the value red_l. this can be formalized as belows:

In conjunction with the electronic anti-lock brake system computer, the warning light informs you that the function is switched off. Braking is then normal, without ABS. This light comes on when the ignition

Box 12.

```
Actuate_Brake
ANY
red_w
red_l
WHERE
grd1: engine_state = Ready
grd2: pedal_state = engage
grd3: wheel_speed ≠ 0
grd4: red_w ∈ ℕ
grd5: red_w ≤ wheel_speed
grd6: red_l ∈ ℕ
grd7: red_l ≤ linear_speed
THEN
act1: wheel_speed := wheel_speed − red_w
act2: linear_speed := linear_speed − red_l
END
```

Box 13.

Show_state_ABS **WHERE** grd1: pedal_state = engage **END**

Box 14.

End_Process **WHERE** grd1: wheel_speed = 0 grd2: linear_speed = 0 **THEN** act1: engine_state := Not_Ready **END**

is turned off and should turn off after a few seconds. The system is also equipped with an indicator that can show the state of the ABS. In case of brake engagement the indicator shows the status of the ABS:

And finally, we present the end of process event when the linear and wheel speed reaches a zero value as belows:

The RODIN platform: is the tool dedicated to development in event B, robustly supporting the refinement of models and mathematical proof. It was created in Zurich during the eponymous European project (FP7) between 2004 and 2007. It is an Open Source platform based on the Eclipse IDE. The ProB Plug-in is an animation tool for Event-B in the RODIN Platform. Animation is very important and can detect a series of problems, such as deadlocks or other unexpected behaviors of a model. This plug-in is based on a translation from Event-B to classic B.

The RODIN Platform along the ProB animator can ensure the correctness of the system. The table below presents the statistics proofs generated by RODIN:

This table measures the size of proofs generated including automatic and manual proofs. According to this report, we conclude that RODIN inference prover was able to establish 89% of proofs, which makes the task of modeling and proving easier. The combination of automatic and manual proofs ensures that the system developed here is correct by construction.

Table 1. Proofs statistics

Element name	Total	Auto	manual
Anti Braking System	73	65	8
Context	5	5	0
Machine	68	60	8

CONCLUSION

The ABS system is designed to serve when a driver brakes suddenly and tries to maneuver, especially when the road is slippery because wet or for some other reason. Under such conditions, the vehicle is more likely to slip or skid if the wheels block. The ABS system can benefit the driver in a variety of fairly common situations, including: A vehicle suddenly folds in front of yours, forcing you to brake and maneuver; an animal jumps on the road and you have to brake and maneuver to avoid it; bad weather makes the road slippery, which increases the possibility of wheel locks when you have to brake to stop.

The primary purpose of the ABS system is to allow the driver to maintain directional control after sudden braking. In support of the thesis of the effectiveness of the ABS system, he has been associated with: a 35% decrease in frontal collisions on wet roads, and; a 9% decrease in frontal collisions on a dry road.

This decrease in frontal collisions suggests that the ABS system allowed drivers to maneuver to avoid a collision. In addition, in controlled conditions trials: 58% of drivers of vehicles without ABS deviated from the desired course after braking; only 24% of drivers of vehicles with ABS did the same.

It is sometimes mistakenly thought that the primary role of ABS is to reduce stopping distances. In fact, the ABS system does not guarantee any reduction in stopping distance, which is only a secondary advantage of the system.

In a vehicle with ABS, turning the steering wheel too far can have serious consequences. If the driver of a vehicle without ABS suddenly turns the wheel under the effect of panic to avoid an accident, its action has no result since it lost directional control due to wheel lock. In a vehicle with ABS however, the same exaggerated rotation of the steering wheel will have an effect that could lead to other dangerous situations, such as a road trip, collision or rollover. The driver of a vehicle with ABS must therefore maneuver as calmly as possible.

The integration of formal methods in the development process of certain critical applications such as real-time systems and distributed systems has proved its worth in recent years. This has motivated their use in the development of software dealing with complex problems at the industrial level. A formal method is considered a software development approach based on mathematical notation and formal validation evidence. This approach uses a refinement process that starts from an abstract requirement specification to a refined and executable specification (or directly codable into a programming language). The main advantages of formal methods can be summarized in the following points: Precision and unambiguity: the use of a language based on formal and precise notations makes it possible to avoid ambiguity and redundancy in the specification.

The detection of conceptual errors as soon as possible: the application of proof of validation of the specification throughout the process of refinement of the latter, guaranteed the detection of design errors as early as possible in the development process of the product. 'application. In the absence of such validation, the design errors will be detected only after the implementation phase which will generate an additional cost. The satisfaction of the design (possibly the implementation) with the needs: it is guaranteed thanks to the process of refinement which starts from a specification of the needs and applies coherent rules of transformation to reach the final design. The control of data-processing consistency: which is directly supported by validation evidence. Reuse: the refinement of formal specifications and their successive decompositions make it possible to highlight interesting levels of abstraction for solving the problem and to promote reuse (specifications). These advantages encourage us to develop this proposed system by means of a formal method (Event-B) which highly guarantees the correctness of the system.

REFERENCES

Abrial, J.-R. (2010). Modeling in Event-B: System and Software Engineering. New York, NY: Cambridge University Press. doi:10.1017/CBO9781139195881

Abrial, J. R., Butler, M., Hallerstede, S., Hoang, T. S., Mehta, F., & Voisin, L. (2010). Rodin: An open toolset for modelling and reasoning in Event-B. *International Journal of Software Tools for Technology Transfer*, *12*(6), 447–466. doi:10.100710009-010-0145-y

Abrial, J. R., Butler, M., Hallerstede, S., & Voisin, L. (2006, November). An open extensible tool environment for Event-B. In *International Conference on Formal Engineering Methods* (pp. 588-605). Berlin, Germany: Springer. 10.1007/11901433_32

Adamiec, M., Dziubiński, M., Siemionek, E., & Drozd, A. (2018, September). Diagnostics of ABS and ESP systems. [IOP Publishing]. *IOP Conference Series. Materials Science and Engineering*, *421*(3).

Batayneh, W., Jaradat, M., & Bataineh, A. (2018, November). Intelligent Adaptive Control for Anti-Lock Braking System. In *ASME 2018 International Mechanical Engineering Congress and Exposition.* American Society of Mechanical Engineers. 10.1115/IMECE2018-87659

Bergmann, D., Fiege, L., Freitag, G., Gerlich, M., Grieser-schmitz, S., Lanze, E., & Robison, R. (2018). Vehicle with an anti-lock brake system and a method for braking a vehicle, U.S. Patent Application No. 14/345,904.

Boere, L. D., & da Silva, A. L. (2018). *Sizing of anti-lock braking system for Formula SAE vehicles (No. 2018-36-0201).* SAE Technical Paper.

Cao, F., Chen, C., & Zhou, L. (2018, May). Research on Simulation of Anti-lock Braking System Based on MATLAB. In *2018 3rd International Conference on Automation, Mechanical Control and Computational Engineering (AMCCE 2018).* Atlantis Press. 10.2991/amcce-18.2018.147

Chen, D., Han, S., Ye, A., Ren, X., Wang, W., & Wang, T. (2019). Prediction of tire–pavement friction based on asphalt mixture surface texture level and its distributions. *Road Materials and Pavement Design*, 1–20. doi:10.1080/14680629.2018.1560354

Clarke, E. M., & Wing, J. M. (1996). Formal methods: State of the art and future directions. [CSUR]. *ACM Computing Surveys*, *28*(4), 626–643. doi:10.1145/242223.242257

D'Avico, L. U. C. A. (2019). *Analysis and design of advanced anti-Lock braking systems.* (Doctoral dissertation, Italy).

Debbech, S., Bon, P., & Collart-Dutilleul, S. (2019). A model-based system engineering approach to manage railway safety-related decisions. *International Journal of Transport Development and Integration*, *3*(1), 30–43. doi:10.2495/TDI-V3-N1-30-43

Deshmukh, A., Shendge, P. D., & Phadke, S. B. (2018, June). Effect of Second Order Actuator on Performance of Anti-Lock Braking System with Multiple Sliding Surface. In *2018 Second International Conference on Intelligent Computing and Control Systems (ICICCS)* (pp. 313-317). IEEE. 10.1109/ICCONS.2018.8662886

Feng, Y. U., & Jun, X. I. E. (2019). Simulation analysis of efficiency and stability for vehicle's ABS control on combined steering and braking maneuvers. In *MATEC Web of Conferences* (*Vol. 272*, p. 01024). EDP Sciences. 10.1051/matecconf/201927201024

Fu, C., & Zheng, K. (2018, November). Patterns for Modeling Task-Level Timing Constraints with Event-B. In *2018 IEEE 9th International Conference on Software Engineering and Service Science (ICSESS)* (pp. 260-266). IEEE. 10.1109/ICSESS.2018.8663926

Glamočić, L. (2018). The Influence of Modern Technical Systems in Vehicles on Road Safety. *JTTTP-Journal of Traffic and Transport Theory and Practice, 3*(1-2).

Jarrar, A., & Balouki, Y. (2018). Formal modeling of a complex adaptive air traffic control system. *Complex Adaptive Systems Modeling*, *6*(1), 6. doi:10.118640294-018-0056-4

Jastram, M., & Butler, M. (2011). Rodin User's Handbook: Covers Rodin v. 2.8.

Kobayashi, T., Ishikawa, F., & Honiden, S. (2019). Consistency-preserving refactoring of refinement structures in Event-B models. *Formal Aspects of Computing*, 1–34.

Mammosser, D., Boisvert, M., & Micheau, P. (2013, December). Designing regenerative braking strategies for electric vehicles with an efficiency map. In Congrès français de mécanique. AFM, Maison de la Mécanique, 39/41 rue Louis Blanc, 92400 Courbevoie, France.

Ondruš, J., & Kolla, E. (2018, September). The impact of ABS system on the braking characteristics of the specified motorcycle on the dry road surface. [IOP Publishing]. *IOP Conference Series. Materials Science and Engineering*, *421*(2). doi:10.1088/1757-899X/421/2/022024

Onsy, A. M., Sharaf, A. M., Ashrey, M. M., & Eldemerdash, S. M. (2018). Influence of Intelligent Active Suspension System Controller Design Techniques on Vehicle Braking Characteristics. *SAE International Journal of Vehicle Dynamics, Stability, and NVH, 3*(10-03-01-0003).

Otremba, F., Romero Navarrete, J., & Lozano Guzmán, A. (2018, September). Modelling of a Partially Loaded Road Tanker during a Braking-in-a-Turn Maneuver. Actuators 7(3), (p. 45). Multidisciplinary Digital Publishing Institute. doi:10.3390/act7030045

Potapov, A. S., Tarasova, T. A., Svirbutovich, O. A., & Fedyaev, M. Y. (2018, May). Complex for experimental studies of anti-lock braking system operation. In *International Conference Aviamechanical engineering and transport (AVENT 2018).* Atlantis Press.

Stankaitis, P., Iliasov, A., Ait-Ameur, Y., Kobayashi, T., Ishikawa, F., & Romanovsky, A. (2019, January). A Refinement Based Method for Developing Distributed Protocols. In *2019 IEEE 19th International Symposium on High Assurance Systems Engineering (HASE)* (pp. 90-97). IEEE. 10.1109/HASE.2019.00023

Tabone, M., Azzopardi, J. P., & Farrugia, M. (2018, December). Emulation of Vehicle Speed and Impact on Vehicle Event Data Recorder. In *2018 18th International Conference on Mechatronics-Mechatronika (ME)* (pp. 1-7). IEEE.

Voevodin, E. S., Kashura, A. S., Fomin, E. V., Pulyanova, K. V., Askhabov, A. M., & Golub, N. V. (2018, May). Reduction of accidents on public roads by increasing informativeness of drivers. In International Conference Aviamechanical engineering and transport (AVENT 2018). Atlantis Press.

Zulhilmi, I. M., Peeie, M. H., Eiman, R. I. M., Izhar, I. M., & Asyraf, S. M. (2019). Investigation on Vehicle Dynamic Behaviour During Emergency Braking at Different Speed. *International Journal of Automotive and Mechanical Engineering*, *16*(1).

Chapter 16
Performance Improvement of Mechanical Components by Precision Coating

Korka Zoltan Iosif
University "Eftimie Murgu" From Resita, Romania

ABSTRACT

Precision coatings applied to mechanical components do not only provide different degrees of protection of their surfaces but can also contribute to significant improvement of performance. Given this fact, this chapter presents the main coating technologies and their practical applications. Beginning with the presentation of the main functions and types of coatings, the chapter progresses by offering information regarding the methods, the area of application, and the performance properties for each of the presented coating technologies. Furthermore, the quality and precision aspects, as well as the applicable standards, are discussed. Some practical examples regarding the performance improvement of different mechanical components by applying precision coatings are additionally presented. The final goal of this chapter is to enable the readers to choose the proper system for their application, by providing all the relevant information regarding various coating types.

INTRODUCTION

The terms "painting" and "coating" are often used interchangeably. However, a distinction is necessary: if "painting" generally describes a pigmented liquid or resin being applied to a surface as a thin layer that is converted to an opaque solid film after application, the term of "coating" is generally accepted to define the process of applying a liquid, gas, or solid layer on the surface of an object. The difference between the two terms further varies between countries. For example, in the USA, Canada, and the UK, a significant distinction between painting and other methods of surface processing, such as electroplating, anodizing or the application of a thin polymer layer over a surface has been made. Under these circumstances, this chapter will mostly use the term "coating".

DOI: 10.4018/978-1-7998-0117-7.ch016

Experts define the term of coating as the process of applying a layer on the surface of an object, commonly referred to as the substrate. This is done for either decorative or functional purposes. For example, jewelry makers apply coatings such as metal plating only for decorative purposes, while in industrial applications coating is utilized for improving the performance of mechanical components. Later on and following the development of innovative components in the fields of optics, electronics, sensors, and medical technology, the term of precision coating was also adopted in these fields, in order to meet product demands on uniformity, precision, high deposition rate, as well as stability.

Almost all known solid materials can be used as substrates. That is, metals, insulators, semiconductors, crystalline or amorphous materials, textile fabrics, films, and many more, can represent substrates. However, for optimal results, not all coating materials can be applied to any substrate materials. For example, consider that the stresses at the level of the coating layers may lower the adhesion of the deposition so that the coating can partially or completely detach. To prevent this from happening, in all the coating processes, the workpieces are usually mechanically and/or chemically treated, prior to deposition. This is done not only to achieve good adhesion of the layer to the substrate, but also to achieve a trouble-free process, a uniform application, and a good resistance of the coating against environmental influences.

If traditionally, the main functions of coatings were the protection and decoration of the substrates (Grainger & Blunt, 1998), more recently and because of years of research and development, new areas of interest emerged. These new functions can be broadly grouped as concerning (Korka, 2018):

- At the connection between coating and substrate;
- In the extent of the coating;
- At the interface between coating and the environment.

Figure 1 illustrates the major functions of coatings used in the main industrial applications in more detail, along with some common areas of use.

COATING PROCESSES

As already mentioned, the process of coating is concerned with the application of a thin layer of material on another base material, commonly referred to as the substrate. According to the German standard DIN 8580:2003, the coating is one of the main manufacturing processes in mechanical engineering.

Coating processes can be differentiated according to:

- Type of the substrate (metallic substrate or non- metallic substrate);
- Type of material applied on the substrate (metallic coating or non- metallic coating);
- Initial state of the coating material to be applied on the substrate (see Table 1 for the main coating methods used for mechanical components).

As Table 1 illustrates, there is a great number of coating processes, which were developed to cover various technical needs and economic considerations. Consequently, a wide range of coating technologies has become available - the most important processes used for mechanical components are next shortly described.

Figure 1. Main functions of coating and common illustration of use in industrial applications

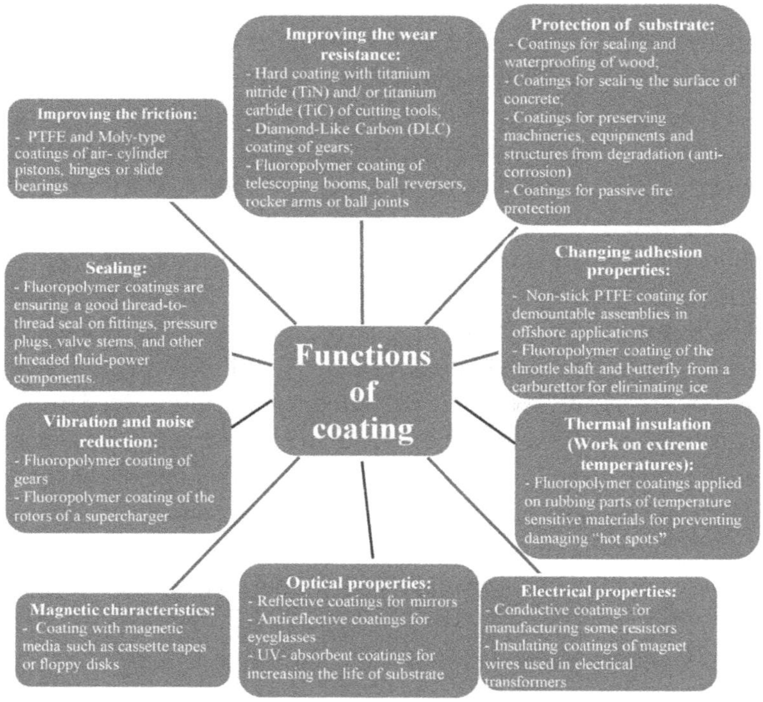

Chemical vapor deposition (CVD) is a coating process for the deposition of mostly thin layers from the gas phase on a solid substrate (Carlsson & Martin, 2010). The principle of the method is that reactive gases, which are often diluted in so-called carrier gases, are passed in a reaction chamber over a heated substrate, where they chemically decompose such that a new layer grows on the substrate surface The process involves the steps listed below, while a typical CVD reactor is shown in Figure 2:

- Formation of the reactive gas mixture;
- Mass transport of the reactant gas through a boundary layer to the substrate;
- Absorption of reactants in the substrate surface;
- The chemical reaction of the absorbents to build the coating layer.

Table 1. Coating processes according to the initial state of the coating material

gaseous or vapor	liquid	solved	solid
• chemical vapor deposition (CVD) • physical vapor deposition (PVD) (sputtering, evaporation, ion implantation, ion plating, and laser surface alloying)	• hot-dipping • organic coatings (lacquering, enameling, varnishing, rubber based coatings, Teflon coatings, bituminous coatings) • chemical solution deposition (CSD)	• electroplating • chromate conversion coating (to passivate steel, aluminum, zinc, cadmium, copper, silver, magnesium, or tin alloys) • galvanization • phosphate conversion coating • tinning; • electroless nickel plating	• thermal spraying (conventional flame spray, electric arc wire spray, plasma spray; high velocity oxy-fuel spray-HVOF) • powder coating • plasma-transferred-arc (PTA)

Figure 2. Typical CVD reactor

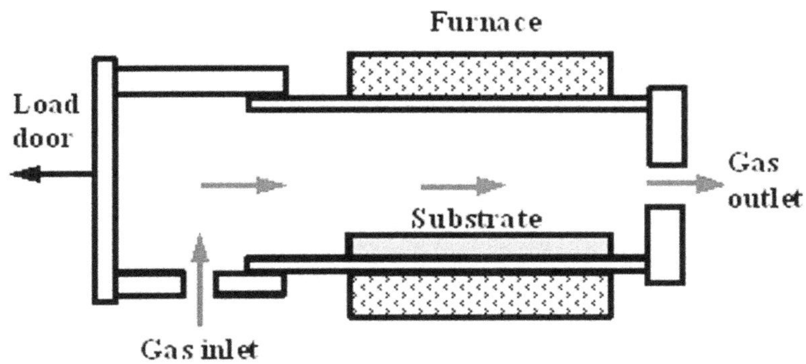

Depending on the operating conditions in which the chemical reactions are taking place, CVD is practiced at:

- Atmospheric pressure (APCD);
- Low (sub-atmospheric) pressure (LPCVD);
- At very low pressure (ultrahigh vacuum- typically below 10^{-6} Pa) (UHVCVD).

Upon on the physical characteristic of vapor, CVD can be:

Regarding the physical characteristics of the vapor, CVD is mainly used to deposit carbon, tungsten, titanium nitride, oxynitride, or carbide in ways that classic coating methods are not able to perform. Some practical examples of CVD applications for mechanical components are:

- coating of WC/Co carbide cutting tools with titanium nitride (TiN), titanium carbide (TiC), titanium carbonitride (TiCN) and others in order to improve the lifetime, hardness and wear resistance;
- PtAl or Cr layers in gas turbines to increase oxidation and corrosion protection;
- use of TiCN layers for erosion protection in gas turbines (e.g. in valves).

Physical vapor deposition (PVD) denotes a group of vacuum-based coating processes or thin-film technologies. Unlike chemical vapor deposition processes, the coating material is converted into the gas phase using physical processes. The gaseous material is then passed to the substrate, where it condenses and covers the entire area of the workpiece, rather than just certain surfaces. The PVD process flow diagram is shown in Figure 3. The group of PVD methods includes following technologies, as well as reactive variants of these processes (Chow, 2018):

- Laser surface alloying;
- Ion implantation;
- Ion plating;
- Evaporation;

Figure 3. PVD process flow diagram

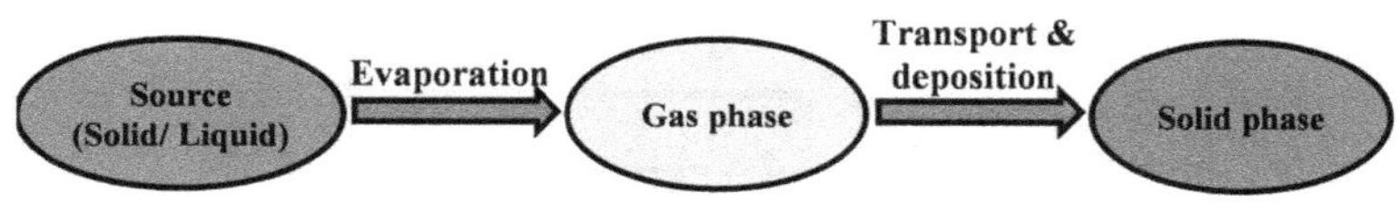

- Sputtering.

Layers of PVD are used in many areas of industry, especially in the field of machining. Most of the cutting tools are nowadays equipped with hard coatings based on the titanium nitride (TiN), titanium carbonitrides (TiCN), or titanium aluminum nitride (TiAlN). As early as the beginning of the 1990s, various research facilities developed further possible applications for the casting of aluminum and magnesium. In these applications, chromium-based layer systems such as chromium nitride (CrN), chromium vanadium nitride (CrVN), and chromium aluminum nitride (CrAlN), are particularly used. Furthermore, coatings made of CrN are often used for corrosion protection purposes.

Hot dipping is a group of coating methods in which a workpiece of a higher melting metal or alloy is completely immersed in a molten bath of a lower melting metal or alloy (Porter, 1994). When the piece is removed from the molten bath, the liquid metal adheres to the substrate and forms a solid metallic coating upon cooling. In order to apply this procedure, two basic conditions must be satisfied:

- The melting points of the two metals must not be too close to each other on the temperature scale - that is, the molten bath must be so hot that it is not nearly ready to solidify, and the workpiece should not be too hot when immersing that it already starts to melt.
- The molten metal must adhere to the substrate, meaning that the adhesion must be strong enough to wet the workpiece.

Additionally, other more practical conditions associated with this method need to be considered:

- Compared to the substrate, the coating material must have desirable, beneficial or value-enhancing properties;
- The substrate and the coating material must be available in sufficient quantity and at reasonable cost.
- The melting point of the coating metal must not be too high, since at very high temperatures, technological difficulties, costs and energy consumption increase as well.

While the most typical substrate materials suitable for hot-dip metallic coating are steel and cast iron, the commonly used coating materials are zinc (Zn), aluminum (Al), and Tin (Sn), which provide good corrosion protection under most atmospheric conditions. Figure 4 presents the operating principle of a hot-dip line used for zinc deposition on steel strips.

However, despite a very large number of known or theoretically possible alloys which could derive a significant amount of hot dipping processes, the number of actually implemented hot dipping processes remains relatively small because of the practical prerequisites.

Figure 4. Hot dipping of steel strip

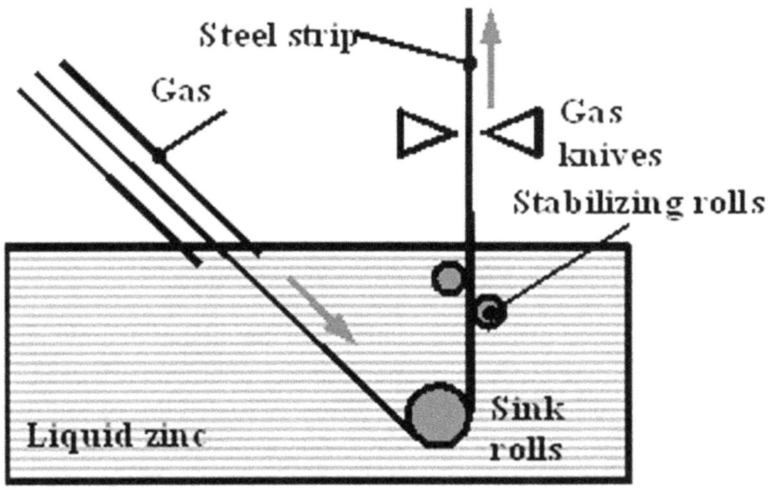

Organic coatings are complex combination of polymers, pigments, fluid carriers, additives and corrosion inhibitors (Popov, 2015). These represent the traditional and most common method used for delaying the corrosion outbreak. The main functions of organic coatings are

- Protection (in combination with one or more primer layers);
- Decoration (optical impact obtained by certain color effect);
- Special surface properties (modified electrical conductivity).

Depending on how the coating is applied, the process has following variants:

- Brushing or rolling;
- Spraying at low pressure (0.5- 0.7 bar), high pressure (1- 8 bar) or airless (60- 250 bar);
- Dip painting;
- Electrostatic spraying (using an electrostatic high voltage field of 80 to 150 kV).

Chemical solution deposition (CSD) denotes a group of (chemical) coating processes in which the substrate is first wet with a solution, after which the deposited layer is solidified. This process includes a wide range of procedures (Schneller T. et al., 2013):

- Metal-organic decomposition (MOD);
- Electro-chemical deposition (ECD);
- Electroless plating;
- Sol-gel process (method for producing solid materials from small molecules).

Unlike PVD or CVD, the process does not require a vacuum and is, therefore, faster and less expensive. With this procedure, it is also possible to coat large-surface substrates. The application of the solution to the substrate can be carried out through various methods, typically:

- *Spin coating* (after dosing the solution centered on a horizontal substrate surface, the substrate rotates, the axis of rotation is vertical to the surface, while the centrifugal forces are driving the solution outwards, thus forming a thin and even film on the substrate);
- *Dip coating* (the workpiece is repeatedly dipped in the coating solution and pulled out - during pulling out, a thin film of liquid remains on the substrate; temperature, ambient pressure, humidity, speed, and angle at which the substrate is pulled out of the coating solution are critical factors for coating thickness and quality) followed by drying and burning to ceramic;
- *Ink-jet printing* (the substrate is not coated by immersion in the solution, but by a special inkjet device like a printer).

Electroplating is the application of a metal coating to a metallic or another conductive surface by means of an electrochemical process. The article to be plated (the workpiece) is used as the cathode (negative electrode) of an electrolytic cell through which current is passed through, directly. The workpiece is dipped in an aqueous conducting solution (the bath) containing the corresponding metal in oxidized form, or as a complex ion. The anode is usually a billet of metal used for plating. During electrolysis, metal is deposited on the workpiece, while the metal of the billet dissolves. Following reactions are occurring at the two electrodes:

- ***At the cathode***, the metal ions from the bath (M^{n+}, where n is the metal valence) receive n electrodes (e), turning into metal (M) by a reduction reaction:

$$\mathrm{M}^{n+} + \mathrm{ne} \rightarrow \mathrm{M};$$

- ***At the anode***, the billet of metal (M) emits n electrodes (e) and dissolves the bath by an oxidation reaction, becoming metal ions (M^{n+}):

$$\mathrm{M} \rightarrow \mathrm{M}^{n+} + \mathrm{ne}.$$

The coating characteristics depend on various factors, such as chemical composition and temperature of the bath, or immersion time. Therefore, by adjusting these variables, coatings with different mechanical properties (ductile or brittle, soft or hard, dull or bright) can be obtained. Figure 5 exemplifies the principle of nickel electroplating. This coating process can be performed with a single metal, multiple layers or various metals, or even alloys (brass, for example).

While protection against corrosion, improvement of electrical conductivity, or aesthetics are the main characteristics which are improved by electroplating, the coating procedure can also be used to ameliorate friction, protect against abrasion and radiation, or, more generally, impart desired properties to a substrate that lacks those characteristics..

Chromate conversion coating refers to a group of surface engineering techniques where complex chromic acid salts are formed on metallic surfaces by the action of chromic acid. In this process, which is not to be confused with chrome plating or diffusion hardening, the dissolved metal ions of the base material are incorporated into the chromate deposition, forming passivation coatings of inorganic non-metallic protective layers. The procedure is mainly applied to passivize steel, zinc, aluminum, cadmium, copper, magnesium, silver and tin alloys.

Figure 5. Principle of nickel electroplating

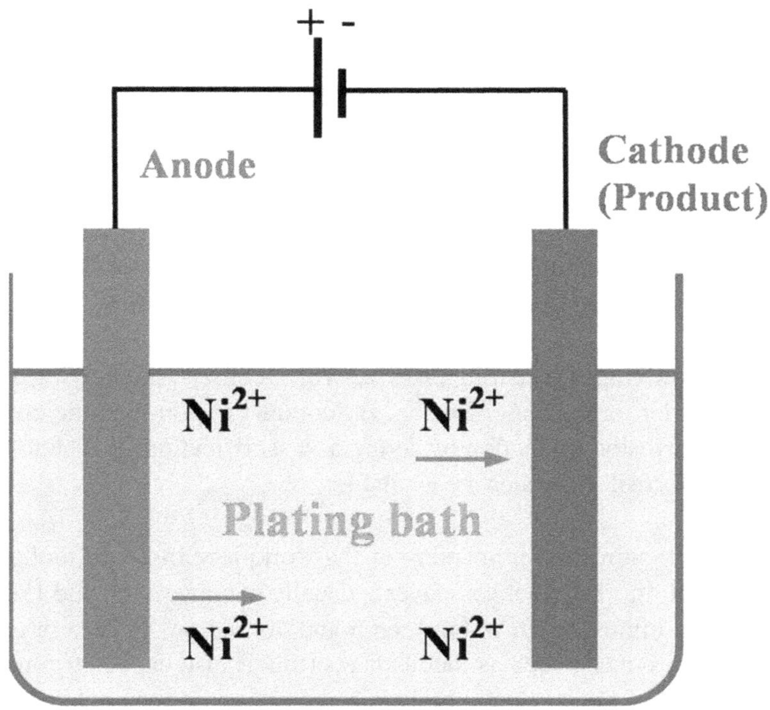

The most common function of chromate conversion coatings is corrosion protection, but they can also be used to create a primer for subsequent layers such as tarnish (for silver), to reduce the visibility of fingerprints, or to change the appearance (gloss, color). The types of chromate conversion coatings are differentiated according to their colors:

- ***Blue***: Under this shade, it is only a very slight blue color to be understood, the transition to transparent being often fluid. The metallic character of the surface is completely preserved, but the corrosion protection is relatively weak.
- ***Yellow***: The color can be compared with brass or a "dirty" gold tone. It is often somewhat inconsistent and sometimes it can even turn into red-like or green-like shades, this effect is called "iridescent". This method has been widely used because of its excellent corrosion protection and its economy.
- ***Olive***: The metal character of the surface is largely lost, but this shade is very strong and it ensures very good corrosion protection.
- ***Black***: The color is less strong than in a paint job and the metal character of the surface is partially preserved, corrosion protection is medium to good. The black chromate conversion coating should not be confused with the black chrome plating.

Due to new legal European rules, such as the RoHS (Restriction of Hazardous Substances) EU-Directive (2002/95/EC) and the REACh ("Registration, Evaluation, Authorization and Restriction of Chemicals") regulation (1970/2006), chromate conversion coating with carcinogenic chromium-containing substances is no longer permitted in many areas. This has created a great deal of interest in finding an equivalent or better substitute for the chromium-containing materials. In this respect, various alternatives have been tested:

- The changeover from chromate conversion coating to thick-film passivation;
- Reinforcement of the weaker coatings by subsequent sealing;
- Achieving the desired corrosion resistance by converting the zinc oxide pre-coated on a zinc alloy layer (ZnFe, ZnNi or ZnCo);
- Use of treatment baths containing complex titanium or zirconium fluorides that form a passivation layer of titanium or zirconium oxide, however, the development in this area is still ongoing.

Galvanization or **galvanizing** is a coating process, which consists of applying a protective layer of zinc on steel or iron, in order to prevent corrosion. By contrast to non-metallic coatings, the zinc deposition provides active corrosion protection by acting as a sacrificial layer. The following methods of applying a galvanization of steel substrates are available:

- ***Hot-dip galvanizing*** requires the immersion of the workpiece in liquid molten zinc with a temperature of around 450° C. The thickness layer is usually in between 50 and 150 microns when the parts are individually immersed, or in between 5 and 40 microns in case of continuous processing of semi-finished products, such as metal sheets (this is also called strip galvanizing). Due to the relatively large zinc layer thickness, galvanized components can achieve a life of more than 50 years. Hot dip galvanizing forms an alloy at the interface between steel and zinc. Due to the thermal effect, larger (sheet metal) parts can suffer deformations during the hot-dip galvanizing process. Therefore, a sheet thickness of at least 1.5 mm is recommended. Furthermore, the coating of hot dip galvanized surfaces requires special measures, since some organic coatings cannot prevent the formation of an oxide layer on the surface of the zinc, as the adhesion is difficult.
- ***Spray galvanizing*** is a variant of flame spraying in which zinc powder or a zinc wire is melted by an (oxygen-acetylene) flame or arc. This is atomized by compressed air and is applied to the workpiece, the still liquid zinc forming a porous layer. This method produces similarly good corrosion protection properties as those produced by hot-dip galvanizing. Due to the large inner surface of the porous layer, subsequent painting requires unusually large quantities of primer or filler. Compared to hot-dip galvanizing, this method has the disadvantage that it can be difficult to achieve a uniform layer of thickness, especially around cavities, but the thermal load on the workpiece is lower.
- ***Sherardizing*** or ***thermal diffusion galvanizing*** is a method where the parts to be coated are heated in batches with zinc powder in closed rotating drums. At temperatures between 320 and 500° C, the zinc evaporates and combines by means of diffusion with the substrate to an intermetallic phase, which grows out of the base material. This results in a uniform, temperature-resistant, hard and abrasion-resistant zinc-iron alloy layer. The method is suitable for parts with complicated geometries and cavities, as well as for low-cost coating of mass products. The dull-grey surface offers good paint adhesion and achieves excellent corrosion protection in combination with passivation or duplex systems.
- ***Zinc flake coating*** is a non- electrolytic process in which a mixture of zinc and aluminum flakes which are bound together by an inorganic matrix, is applied in layers with a thickness of 4 to 5

microns. Usually, the method is applied in two layers, after which it is additionally silicate-sealed or overcoated with an organic topcoat.

If it is applied in a right way, galvanization can be very beneficial for steel products in countless industries, because the zinc layer is protecting the metal substrate as it will be damaged first. On the other hand, if the processing is done incorrectly, for example if the cooling is too quick, the zinc layer may peel or chip off.

Phosphate conversion coating is a widely used process in surface technology in which a so-called conversion layer of firmly adhering metal phosphates is formed by chemical reactions of metallic surfaces with aqueous phosphate solutions. The method is mostly applied to steel, but can also be used on zinc, aluminum, cadmium, silver, and tin. Main areas of application are corrosion protection, adhesion, friction and wear reduction, as well as electrical insulation. The performance of this coating technology depends significantly on the crystal structure, as well as on the weight specifications. For example, a microcrystalline structure is the most favorable for corrosion resistance or subsequent coating. A rough grain structure impregnated with oil, nevertheless, may be more advantageous for wear resistance. These aspects are regulated by selecting the convenient phosphate solution, using diverse additives, as well as controlling the concentration or temperature of the bath and time of the process.

Tinning is a collective term for various surface finishing processes, which refers to depositing a tin coating on objects which are most often metallic. Even if tin has relatively good corrosion resistance, corrosion protection is not the main reason for choosing this metal as the coating material - the most important properties of tin coatings are food compatibility and the possibility of soldering. Items that are usually tinned consist of steel, copper or copper alloys. In the case of black plates, tinning can be performed by electroplating or hot-dipping.

Electroless nickel plating (ENP) is a chemical coating method which is usually used to improve wear properties and corrosion protection of metallic materials. The process requires dipping of the workpiece in a solution like hydrated sodium hypophosphite ($H_2O + NaPO_2H_2$), which reacts with the ions of the substrate and forms the nickel alloy layer. Unlike in the case of electroplating, the use an electric current for the deposition of the plating solution is not required. Therefore, by ENP non-conductive materials can be coated. Also, in contrast to electroplating, ENP creates a layer of uniform thickness, regardless of the surface profile. This coating method is useful when corrosion protection, wear resistance, and hardness is required. Main applications include drive shafts, rotors, oilfield valves, paper handling equipment, or other mechanical tools.

Thermal spraying methods represent surface coating technologies through which filler metals are melted inside or outside a spray burner, which then flow in the form of spray particles and spun onto the surface of the component to be coated. The workpiece surface is not melted but only subjected to a low thermal load. The coating formation takes primarily place by mechanical clamping because the spray particles flatten, more or less, when hitting the substrate. Therefore, the coating properties are significantly influenced by the temperature and the speed of the spray particles at the time of their impact on the surface to be coated. Quality features of thermally sprayed coatings are low porosity, good adhesion to the substrate, freedom from cracks, and homogeneous microstructure. Figure 6 presents a schematic representation of the thermal spraying process.

Any material that does not decompose during the melting process can be used for this coating technology. Frequently-used materials, in form of powder, wire, or rod are: pure metals (Al, Cu, Ni, Zn), self-fluxing alloys (FeNiBSi), oxides (Al_2O_3, Cr_2O_3), ceramics and plastic.

Figure 6. Schematic representation of the thermal spraying process

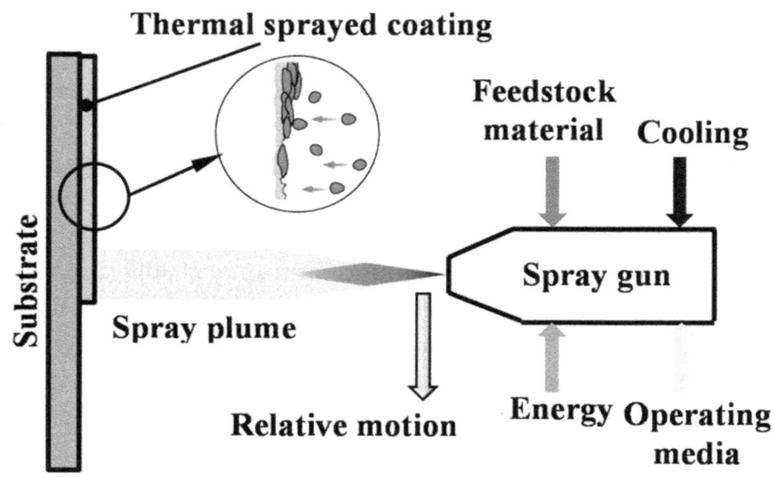

Methods utilized most often to perform a deposition by thermal spraying are:

- Conventional flame spray;
- Plasma spray
- Electric arc wire spray;
- High velocity oxy-fuel spray (HVOF).

Powder coating is a process in which a free- flow dry powder is applied on an electrically conductive substrate, which is processed under heat to allow the coating powder to flow and form a "skin" (McKeen, 2016). The main difference to the conventional liquid paint is that powder coating does not involve a solvent to keep the filler and binder together. The coating layers are typically in between 60 and 120 microns thick, while the substrates most often used are steel, galvanized steel, and aluminum.

Because the coating does not include a liquid component, it can be used to deposit thicker coatings, by comparison to typical liquid coatings. Therefore, there are no differences in terms of appearance, between horizontally and vertically coated surfaces. Furthermore, the coating process emits only a few volatile organic compounds (VOC), thus being more operator- and environment-friendly.

There are two main categories of powder coating: thermoplastics and thermosets. The thermoplastic coating method is characterized by the fact that the coating powder flows without taking any actions until the coating layer is formed, during the heating process. The thermosetting method uses a cross-linker which during the baking process, reacts with other chemical components, improving thus the performance properties.

Over other coating processes, powder coating offers following advantages:

- A significant faster curing time;
- Overspray can be recycled;
- A wide spectrum of specific effects are more easily accomplished.

Plasma-transferred-arc (PTA) is a thermal coating process for surface treatment, which is primarily used for the deposition of wear- and corrosion- resistant layers on a base material. In this respect, the surface of the workpiece is melted with a plasma arc generated by means of a plasma gas (argon or helium) through an arc between a cathode and an anode. The resulting high-density plasma arc serves as a heat source, while a metal powder is used as a coating material. The powder is heated in the plasma jet and applied against the workpiece surface, where it dissolves completely in the molten bath on the substrate. To prevent any external influence, the whole process takes place in the protective atmosphere of an inert gas (an argon-hydrogen mixture). The principle of the process is shown in Figure 7.

The PTA process allows a low dilution rate (5-10%), a small heat affected zone, a high application rate (up to 20 kg / h) and a strong metallurgical bond between the substrate and the coating layer. Due to the fact that the deposition material is in powder form, which is easier to produce than welding rods or wires, almost any materials and material combinations can be applied, at all hardness ranges. Moreover, the technology has a high degree of automation and therefore offers a good reproducibility of the welds.

Following this description of each of the coating processes used in the mechanical industry, Table 2 synthesizes their main advantages and disadvantages. Here, the readers are given all the relevant information regarding various coating types, in order to enable them to choose the proper system depending on their application.

QUALITY CONTROL OF PRECISION COATINGS

Generally, quality control involves a procedure, or a set of procedures developed to certify that a product or a service satisfies an agreed set of standard values or criteria, before, during, and after the manufacturing process. This is done in order to ensure the customers satisfaction or to comply with certain rules and regulations. Particularly, the functions of precision coatings, as described above in this chapter, can be provided only if the subsequent requirements are met:

- The materials used in the coating process have a verified quality;

Figure 7. Principle of Plasma-transferred-arc(PTA) process

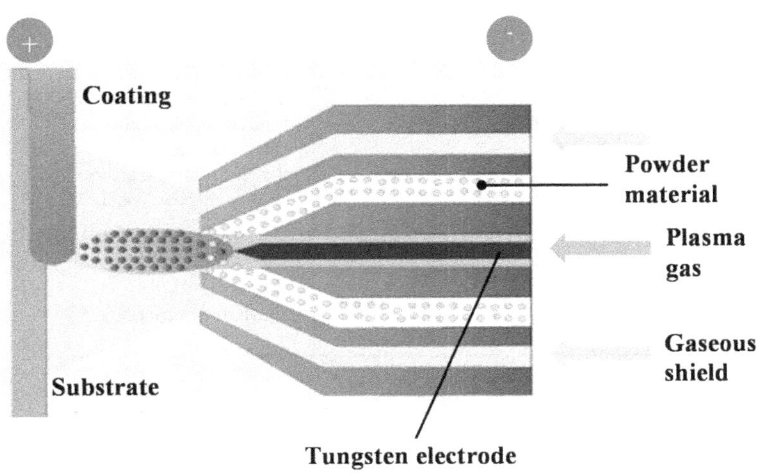

Table 2. Advantages, disadvantages and typical application fields of the coating processes which are used in mechanical industry

Coating process	Advantages	Disadvantages	Typical applications
Chemical vapor deposition (CVD)	- high purities and growth rates - economical - can deposit materials which are hard to evaporate - versatile - good reproducibility	- toxic and corrosive -involves high temperatures - needs high skilled operators - start-up costs are high	- forging tools - stamping tools - extrusion dies
Physical vapor deposition (PVD)	- more environmentally friendly than traditional coating technologies - ability to use a various range of organic and inorganic coatings on diverse substrates, using different finishing classes - a given film can be applied by more than one technique - more corrosion resistant and harder than the coatings obtained by electroplating methods - good strength at high temperatures and under impact conditions - excellent abrasion resistance and durability, so that protective topcoats are not necessary	- slow deposition rate - dissipates a large amount of heat - require skilled operating personnel - high capital costs	- cutting tools - automotive components - punching and forming tools - injection moulding tools
Hot dipping	- resistant to environmental impacts - can operate at high temperatures - high wear resistance - long warranty period	- high technological process - deformation of thin structures after heating - damaged coating cannot be repaired (only the whole structure)	- applications where long-time corrosion resistance is required (automotive parts, guide rails, heating equipment; furnace parts)
Organic coatings	- low treatment cost - can be applied in any colour specification - simplicity of the process	- are typically environmentally unfriendly to both the atmosphere and humans (due to VOC) - waste generated due to overspray	- corrosion protection of metallic parts
Chemical solution deposition (CSD)	- large thickness range (<10 nm- 10 μm) -covers large surfaces - simplicity of the process - requires only a solution container and a substrate mounting device	- waste of solution after every deposition - deposition rate depends on chemical and topographical nature of the substrate	- coating for corrosion protection - optical coating on glass - insulators
Electroplating	- versatile and inexpensive - low temperature operating conditions - good corrosion resistance - improves the mechanical properties of the substrate	- non- uniform plating - time consuming process - pollution potential - the electroplating solution has to be disposed safely after use	- car parts - wheel rims
Chromate conversion coating	- good corrosion protection - no foam and fumes during process - low product consumption	- very toxic - legislative restrictions	- automotive industry - interior parts in aerospace in military industry

continued on following page

Table 2. Continued

Coating process	Advantages	Disadvantages	Typical applications
Galvanization	- good and durable corrosive protection - does not require maintenance during the service life - high resistance to shock and vibration	- process requires high safety measures - parts can suffer deformations during the process	- corrosion protection of steel or iron pieces used in construction, industrial, marine and other industries
Phosphate conversion coating	- very good adhesion and electric isolation - good corrosion protection, adhesion and wear reduction - low friction - eco-friendly process	- chemical process which requires safety measures - existence of pores	- internal combustion engine parts
Tinning	- good corrosion protection - ductility - solderability	- high strength steel is sensitive to hydrogen embrittlement	- corrosion protection of parts manufactured from steel, copper or copper alloys
Electroless nickel plating (ENP)	- method is suitable for non-metallic substrates - process does not require a DC power source - can be coated deeply recessed areas and parts with a complex geometry - uniform thickness of the coating	-limited lifespan of the chemicals -high cost of waste treatment due to speedy chemical renewal - porous nature of the plating, leading to inferior material structure	- aeronautics - car parts - cutting tools - chemical and oil industry
Thermal spraying	- wide range of coating materials - high quality of the deposition - low heating of the substrate - high density and hardness - thicker coatings due to less residual stress	- high investment costs - requires qualified personnel - very difficult to achieve deposition on inner surfaces - high dust, fume and noise level - requires special health and safety measures	- repair procedures to restore components to their initial dimensions - gas turbine components - textile machine components - landing gears - jet engine components
Powder coating	- high durability - environmental friendly - nearly an unlimited selection of colours	- hard to achieve thin coating layers - high start-up costs - high sensitivity to UV rays	- refrigerators, washers, dryers - automotive parts (hubcaps, wheels, filters, radiators)
Plasma-transferred-arc (PTA)	- smooth deposits that significantly reduce the post weld machining - large variety of layer thickness - easily automated, ensuring a high degree of reproducibility - minimizes the thermal affection of the basic material	- high equipment costs - high dust, fume and noise level	- drilling hammers used in oil extraction - engine valves - seats and spindles of steam and water valves - turbine blades (deposition on leading edges)

- All the required substrate pretreatments were accurately implemented and controlled;
- The coating procedure was properly selected, executed and verified.

Consequently, in order to guarantee the customers satisfaction or to satisfy specific regulations related to precise coatings, following quality control-related actions should be carried out:

- Visual examination of the coating;
- Coating thickness measurement;
- Porosity check;
- Coating adhesion test;
- Hardness measurement;
- Checking the coating's wear resistance;
- Strength verification.

The above described tests may be restricted or extended depending on factors such as substrate, coating material and deposition procedure. However, it should be noted that hardness and thickness measurement are the most important topics of coating quality control, as they have an essential influence on optical aspect, life-time and the cost of the final product. Based on these arguments, the relevant aspects regarding coating hardness and thickness measurement are presented below.

Depending of the coating type (metallic or non-metallic), different methods to measure the coating hardness are available. For metallic coatings, the usual method to test the hardness is to use an indenter that enforces the test matter for a period of time. The shape of the indenter varies by the measured hardness including a ball (Brinell method), a cone (Rockwell method) or a pyramid (Vickers method). The testing methods are defined by the following standards:

- Brinell: ISO 6506 and ASTM E10;
- Rockwell: ISO 6508 and ASTM E18;
- Vickers: ISO 6507 and ASTM E92.

For non-metallic coatings the pencil hardness test, the pendulum test and the Buchholz indentation test are available. Following standards are outlining the testing methods:

- Pencil hardness test: ISO 15184 and ASTM D3363;
- Pendulum test: ISO 1522 and ASTM D4366;
- Buchholz indentation test: ISO 2815.

As the hardness testing can affect the appearance of the part which is evaluated, usually the measurement is performed on coating samples

Coating thickness measurement is an important process for both manufacturers and for customers receiving the finished product. Defective coatings do not fulfill their functions. Therefore, product liability requires rigorous quality control of coatings and the documentation of the obtained measurement results. Most of the testing methods are non-destructive, covering a wide range of coating procedures. Depending on the coating type and thickness, various methods and equipment are available to perform the necessary measurements. In the following section, the most commonly used testing methods and the applicable standards are presented briefly. These include:

1. *Magnetic induction (MI).* MI is a method employed for measuring both magnetic coatings over non-magnetic workpieces and non-magnetic coatings over ferrous workpieces, being applicable for thickness measurements in the range between 1 and 1000 μm. The method is non-destructive,

quick, precise, easy to operate, relatively low-cost and repeatable, being defined by the standards ISO 2178 and ASTM B499.

The measurement principle is based on the use of a coil placed inside the measurement probe, which generates a variable magnetic field. By placing the probe on the substrate, the density of this magnetic field changes. This alteration in the magnetic field is measured by a secondary coil, which is also placed in the probe. The output tension, which depends on the coating thickness is transferred to a microprocessor to be displayed as the coating thickness.

Typical applications of this method include plated finishes such as zinc, chrome, phosphate or cadmium over iron or steel workpieces, as well as powder or liquid coatings.

2. *Eddy current (EC).* EC is similar to the MI method, being used to measure the deposition of non-magnetic coatings like stainless steel, liquid or powder coatings over aluminum and copper. It is defined by the standards ISO 2360 and ASTM B244.

The testing method also uses a probe which contains a coil, however this coil has a double function: excitation and measurement. When the probe is placed near the substrate the impedance in the probe changes. This modification depends on the distance between the probe and the conductive substrate, i.e. the coating thickness.

3. *Beta backscatter (BB).* The BB method uses an isotope source for emission of beta radiation (electrons) which is directed through an aperture onto the coating. A portion of these particles is backscattered (reflected back) towards the source, being detected by a Geiger-Mueller counter tube, which translates this information into coating thickness. The coating thickness is proportional to the fre quency of the backscattered beta particles. The approach can be used to measure the thickness of silver, gold and tin coatings applied on electronic components, depositions on cutting tools or organic coatings over metals. The method is applicable only if the atomic numbers Z of the coating and the substrate differ by least Z=5. The standards ISO 3543 and ASTM B5610 define this testing method.
4. *X-ray fluorescence (XRF).* XRF is a versatile contactless method used to measure very thin multiple layers of alloy coatings applied on small components which have complex shapes. The standards ISO 3497 and ASTM B568 define this procedure. During measurement a part is exposed to X-radiation, which is focused by a collimator. This radiation causes characteristic X-ray emissions (fluorescence) from both the coating and the substrate. This emission is identified by an energy-dispersive detector. The wavelength or the energy level of the fluorescent X-ray is proportional to the atomic number of the material, and is specific for each singular coating or substrate. The quantity of the released energy depends on the coating thickness.

The method can detect a specific coating when intermediate layers are present, but cannot be implemented for measuring organic coatings. It is also moderately limited by the thickness of the coating, commonly not exceeding 0.5- 0.8 µm. Furthermore, unlike BB, the XRF method can measure coatings with similar atomic numbers, such as nickel applied on copper. The most common applications of XRF include the measurement of coatings applied to optical components, printed circuit boards, and electronic components.

EXEMPLES OF PRECISION COATINGS APPLIED FOR PERFORMANCE IMPROVEMENT OF MECHANICAL COMPONENTS

Precision coatings have been successfully employed in the industry in the last 50 years, allowing critical parts to gain higher protection against failure and, therefore, to improve reliability. Cutting tools can be taken as an example for how the implementation of innovative coating technologies is able to manage the tendency to use increasingly difficult materials to machine. Therefore, the challenge for innovative tooling concept is an on-going process which can be achieved by involving new materials, an adapted tool design, together with the implementation of high-performance coatings.

A recent VDI-GME study (2014) reveals that: "over 60% of all product innovations are based on new or improved materials as well as on methods in manufacturing and processing of those." This becomes obvious in traffic engineering where the weight reduction tendency is manifested using alloys such as those based on Al, Ti, and Mg. In addition, increasing requirements for new tools develop because of measures taken to develop the sustainability and productivity of production processes (Gupta et al., 2016; Sharma et al., 2016; Tai et al., 2014), such as:

- Hard machining;
- High-performance cutting;
- High-speed cutting;
- Minimization of lubricant consumption.

From the beginning, the development of precision coatings on tools has been focused on providing protection to adhesion and wear. With the increase of speed and cutting temperature research and development were focused on maintaining the coating characteristics under these harsh conditions. Furthermore, besides the requirement of providing a diffusion restraint between workpiece and tool. an additional function of the coatings is to ensure protection against oxidation. Moreover, the coating material needs to withstand to cyclic loads occurred during interrupted cuts. All these hard requirements are fulfilled only by a few coating materials or processes, respectively.

The first coating deposited by CVD on a cemented carbide tool was TiC, being introduced on the market in 1969 (Ettmayer et al., 20104). The fast success has contributed to a quick progress of the CVD technologies, together with alternative materials such as: TiN, TiCN and Al_2O_3 (Haubner et al., 2017; Liu et al., 2005). The year 1980 marks the first industrial application of TiN-coating through PVD process (Matthews, 1985). TiAlN followed after 9 years (Leyendecker et al., 1991a). Even if in 70s and 80s a lot of other coating materials were tested, TiC, TiN, TiCn, TiAlN and Al_2O_3 continue to be the most frequently used cutting materials for tools. Table 3 presents an evaluation of these coatings from different points of view. The listed hard materials are deposited by CVD, excepting TiAlN, which is coated by PVD. Other materials like CrN, CrAIN, TiB_2, ZrN, diamond and DLC (diamond like carbon) are deposited to a lesser extent.

The coating technology of tools is chosen, based on following general rule: at turning, thicker CVD coatings are suitable due to the larger wear volume, while for milling tools, thinner PVD coatings are preferred because of sharper cutting edges and residual compressive stress.

Presently, Al_2O_3 cemented carbides are the most commonly sold product for turning machining (Chien et al., 2013). This material is the best example of how performance improvements are possible by providing a cemented carbide tool with a proper geometry and a hard deposition. Therefore, the indexable

Table 3. Evaluation of hard coatings applied on cutting tools (Inspektor et al., 2014)

Evaluation	Hardness at ambient temperature	Hot hardness	Chemical stability	Oxidation stability
Excellent	TiC	Al_2O_3	Al_2O_3	Al_2O_3
Good	TiCN	TiAlN	TiAlN	TiAlN
Medium	Al_2O_3	TiN	TiN	TiN
Low	TiAlN	TiCN	TiCN	TiCN
Very low	TiN	TiC	TiC	TiC

inserts made from Al_2O_3 cemented carbides are now very sophisticated products, which are individually designed for each application in part.

Nowadays, due to modern sintering facilities the design of various cemented carbide is attainable. There is possible to vary not only the WC particle size, but also the content of cobalt (Co). Thus, a low Co content provides a harness improvement, an increase of the Young's modulus, and consequently a better compressive strength, while a higher Co content in the surface regions contributes to increased fracture toughness and therefore improved cutting edge stability. Thereby, the depositions get a suitable supporting effect which allows adequate hard coatings. In most of the applications, TiCN is implemented as a bond coat, while transitional layers of TiN, TiC or TiCN can be also applied (Connelly et al., 2005; Liu et al., 2005). Both approaches simultaneously contribute to an adhesion improvement and to a raised hardness acting like a buffer for the fragile Al2O3 layer and improving the mechanical properties.

Since recently, diamond deposition is another innovative coating procedure (Graham, 2014), being applied on industrial scale since the 1990s (Leyendecker et al., 1991b; Usuki et al., 1991). Nowadays diamond coated tools are used more and more often, as they allow processing of carbon fiber reinforced plastic (CFRP) (Uhlmann et al., 2014), graphite (Frank et al., 2007), ceramic, and nonferrous materials (Schubert & Nestler, 2009).

The most important discovery in the field of PVD coatings on tools dates back to the 1980s, when, using low-pressure plasmas, the synthesis of absolutely new types of material systems has been achieved: metastable, ternary nitrides in the MeAlN system, whereby Me stands for the refractory metals such as Ti, Zr, Hf, V, Ta, and Cr). TiAlN was brought on the market in 1989 as a result of vast researches (Münz, 1986; Knotek & Leyendecker, 1987; Holleck, 1988), being considered the most important PVD coating ever since (Veprek, 2008).

Currently, it is widely accepted that the expenses due to cutting tools represent about 3% of the total machining cots. Extending the life time of the tools by applying coatings is significantly decreasing the production costs. Thus, increasing the tools durability by 50% has the effect of reducing the machining costs by 1.5% per component. However, a significant part of the costs are also reduced because more performing tools allow higher cutting speeds, which is contributing to the decrease of production time and increase of productivity. Furthermore, the consumption of cooling lubricants, as well as the down times necessary for tool changes are both contributing to the decrease of machining costs by using more efficient tools. Additionally, the use of innovative cutting tools may further facilitate the machining of those workpiece materials which could not be cut until now. In sum, the rise of productivity in machining occurred due to coating systems and coating procedures.

Since the last decades, precision coatings are also increasingly implemented in the gear manufacturing industry, as a method of:

- Preventing pitting and micro-pitting;
- Improving wear resistance;
- Avoiding corrosion and fatigue;
- Decreasing friction;
- Reducing vibration and noise level.

Micro-pitting is a frequent gear failure mode, which is commonly illustrated by the initiation and evolution of micro-cracks on the gear teeth flanks, because of cyclic contact stress and sliding of the opposite gear flanks. The progress of these micro- cracks deteriorates the gear profile geometry, which contributes to the growth of stress applied along the gear flank. This leads to the formation of high contact stress areas, making the teeth engagement noisier, less efficient, and exposed to fatigue failure due to development of macro-pits and/ or flank bending cracks.

Besides high temperature because of poor lubrication and conditions of the gear meshing, damages occuring due to micro-pitting are often associated with the texture and roughness of the surface. Under these circumstances, the peaks and valleys from the gear flank surface are acting as local stress growers inducing micro-pitting under high sliding circumstances. Therefore, alternatives involving surface engineering of gear teeth flanks have gained interest. Recent studies have investigated the improvement of the surface quality (Britton et al., 2000; Kim & Olver, 1998; Oila & Bull, 2005; Zhang, 2005), the optimization of the micro-geometry in order to reduce the applied stress (Frazer et al., 2009), and the application of special coatings (Walton & Goodwin, 1998).

A recent study (Moorthy & Shaw, 2012) is claiming that Diamond-Like-Carbon (DLC) coating applied on gears is improving their contact fatigue performance. The research has evaluated the performance of helical gears provided with surface coatings either with or without Inorganic Fullerence Like Material (IFLM) nano-particles, such as tungsten disulfide (WS_2). Case-carburized and tempered helical gears manufactured from S156 steel were alternatively coated with Balint® C1000, Balint® C*, CrN + IFLM, C6+ IFLM and Nb-S coatings, after which their contact fatigue performance was tested on a back-to-back test rig.

Balint® C is a DLC type coating which contains metal–carbon (WC/C), while Balint® C* is a double coating involving layers of CrN and WC/C. Both coatings were applied with a thickness of about 2 μm by a Physical Vapour Deposition (PVD) process. The temperature during the process hasn't exceeded 200°C, in order not to affect the microstructure and hardness of the substrate (carburized and tempered gear steel).

WS_2 is a dry/solid lubricant powder, being one of the most lubricious materials ever discovered, with a very low coefficient of friction (0.03- dynamic and 0.07- static) and a temperature resistance from -270°C to +650°C. By incorporating it into the coating, it may improve the contact fatigue performance of gears by reducing mesh friction (Joly-Pottuz et al., 2005).

C6 and Nb–S are newly developed coatings by IonBond, UK, being deposited by PVD in layer with thicknesses of 2-3 μm. As this coating procedures are still under development stage, details regarding their composition and structure are not available yet.

The contact fatigue performance of uncoated gears and gears alternatively coated with Balint® C, Balint® C*, CrN + IFLM, C6 + IFLM and Nb–S were assessed at maximum contact stress stages varying from 1471 to 1898 MPa in Aeroshell oil at 100°C. The testing results revealed that the Balint® C coated gears have the lowest micropitting damage followed by C6 + IFLM and Nb–S. The Balint® C* coating show more micro-pitting contamination, but much lower than the uncoated and CrN + IFLM coated specimens. The Nb–S coated gears demonstrated the lowest gear flank profile deviation followed by Balint® C. Furthermore, the Nb–S coated gears clearly showed the best overall contact fatigue performance.

Because the operation of gears is accompanied by vibration and noise, the current legislative context requires special measures for reducing their associated intensity of the sound fields. In this context, complex coatings based on metal alloys or polymeric materials were developed.

Basiniuk et al. (2013) present their research results regarding the use of bronze-graphite Br05S20Gr0.5DMO0.5 coating. They suggested that, depending on the operation speed of the gears, a general noise decreasing level of 2 - 4 dBA can be obtained.

Korka et al. (2009, 2015, 2017) have studied the possibility of reducing the vibration and noise level by applying a fluoropolymer coating of Xylan 1052 (produced by Whitford Plastics Ltd, U.K.), on the gear teeth flanks. Their results have shown that this coating procedure is able to reduce the sound pressure level (SPL) of a gearbox, on average, by 7.7.6 dB, which is 11.25% from the initial noise level. They attributed this reduction to the additional damping during the meshing process ensured by the applied coating layers, and to the amelioration of the teeth flank roughness.

More recent research (Feng et al., 2019) presents the effect of applying a graphite-like carbon (GLC) coating followed by the addition of yttrium (Y), cerium (Ce) and tantalum (Ta), on the gears of a pump. The results show that the complex coating exhibited a significant reduction effect for vibration and gear noise. Compared with uncoated gears, the noise of the coated gears was decreased by 10 dB at 2000 rpm.

SUMMARY

The coating process refers to applying a layer on the surface of an object, commonly referred to as the substrate, either for decorative or functional purposes.

Almost all known solid materials can be used as substrates: metals, insulators, semiconductors, crystalline or amorphous materials, textile fabrics, films, and many more - for optimal results though, not all coating materials can be applied to any substrate materials.

The main functions of coatings include the protection and decoration of the substrates, the improvement of friction, the reduction of wear, vibration and noise, the change of adhesion properties, the thermal insulation, sealing and amelioration of electrical, optical or magnetic characteristics.

Coating processes can be differentiated according to the type of the substrate (metallic substrate or non- metallic substrate), the type of material applied on the substrate (metallic coating or non- metallic coating), and the initial state of the coating material to be applied on the substrate. Thus, a great number of coating processes have been developed to cover various technical needs and economic needs:

- **Chemical vapor deposition** (CVD) is a coating process for the deposition of mostly thin layers from the gas phase on a solid substrate.
- **Physical vapor deposition** (PVD) denotes a group of vacuum-based coating processes or thin-film technologies.
- **Hot dipping** refers to a group of coating methods in which a workpiece of a higher melting metal or alloy is completely immersed in a molten bath of a lower melting metal or alloy.

- **Organic coatings** are complex combination of polymers, pigments, fluid carriers, additives and corrosion inhibitors.
- **Chemical solution deposition** (CSD) denotes a group of (chemical) coating processes in which the substrate is first wet with a solution, after which the deposited layer is solidified.
- **Electroplating** represents the application of a metal coating to a metallic or another conductive surface by means of an electrochemical process
- **Galvanization** or **galvanizing** represents a class coating process, which consists of applying a protective layer of zinc on steel or iron, in order to prevent corrosion.
- **Phosphate conversion coating** is a widely used process in surface technology in which a so-called conversion layer of firmly adhering metal phosphates is formed by chemical reactions of metallic surfaces with aqueous phosphate solutions.
- **Tinning** is a collective term for various surface finishing processes, which refers to depositing a tin coating on objects which are most often metallic.
- **Electroless nickel plating** (ENP) is a chemical coating method which is usually used to improve wear properties and corrosion protection of metallic materials.
- **Thermal spraying** methods represent surface coating technologies through which filler metals are melted inside or outside a spray burner, which then flow in the form of spray particles and spun onto the surface of the component to be coated.
- **Powder coating** is a process in which a free- flow dry powder is applied on an electrically conductive substrate, which is processed under heat to allow the coating powder to flow and form a "skin".
- **Plasma-transferred-arc** (PTA) is a thermal coating process for surface treatment, which is primarily used for the deposition of wear- and corrosion- resistant layers on a base material.

Quality control procedures have been developed to certify that a product or a service satisfies an agreed set of standard values or criteria, before, during, and after the manufacturing process. Customers` satisfaction is also an important quality control criterion.

Hardness and thickness measurement are the most important topics regarding coating quality control, as they have an essential influence on optical aspect, life-time, and the cost of the final product.

The rise of productivity in machining occurred due to coating systems and coating procedures: for example, innovative coating technologies for cutting tools allow the use of increasingly difficult materials to machine. High temperature because of poor lubrication and conditions of the gear meshing, and damages occurring due to micro-pitting are other difficulties which can be overcome with the help of appropriate coating technologies.

Given that gears function is accompanied by vibration and noise, reducing the gears vibration and noise level can be achieved by applying complex coatings on the teeth flanks.

REFERENCES

Basiniuk, U. L., Levantsevich, M. A., Maksimchenko, N. N., & Mardasevich, A. I. (2013). Improvement of Triboengineering Properties and Noise Reduction of Tooth Gears by Cladding Functional Coatings on Working Surfaces of Interfaced Teeth. *Journal of Friction and Wear*, *34*(6), 575–582. doi:10.3103/S1068366613060044

Bobzin, K. (2017). High-performance coatings for cutting tools. *CIRP Journal of Manufacturing Science and Technology, 18*, 1–9. doi:10.1016/j.cirpj.2016.11.004

Britton, R. D., Elcoate, C. D., Alanou, M. P., Evans, H. P., & Snidle, R. W. (2000). Effect of surface finish on gear tooth friction. *Journal of Tribology, 122*(1), 354–360. doi:10.1115/1.555367

Carlsson, J. O., & Martin, P. M. (2010). Chemical Vapor Deposition. In P. M. Martin (Ed.), *Handbook of Deposition Technologies for Films and Coatings* (3rd ed.). Norwich, CT: William Andrew Publishing. doi:10.1016/B978-0-8155-2031-3.00007-7

Chien, H., Diaz-Jimenez, C., Rohrer, G. S., Ban, Z., Prichard, P., & Liu, Y. (2013). The Influence of Residual Thermal Stresses on the Mechanical Properties of Multilayer α-Al_2O_3/$TiCxN_{1-x}$ Coatings on WC/Co Cutting Tools. *Surface and Coatings Technology, 215*, 119–126. doi:10.1016/j.surfcoat.2012.07.088

Chow, L. A. (2018). Equipment and Manufacturability Issues in Chemical Vapor Deposition Processes. In K. Seshan, & D. Schepis (Eds.), *Handbook of Thin Film Deposition* (4th ed.). Norwich, CT: William Andrew Publishing. doi:10.1016/B978-0-12-812311-9.00009-8

Connelly, R., Pattanaik, A. K., & Sarin, V. K. (2005). Development of Moderate Temperature CVD Al_2O_3 Coatings. *International Journal of Refractory Metals & Hard Materials, 23*(4–6), 317–321. doi:10.1016/j.ijrmhm.2005.05.017

Directive 2002/95/EC of the European Parliament and of the Council of 27 January 2002 on *the restriction of the use of certain hazardous substances in electrical and electronic equipment.*

Ettmayer, P., Kolaska, H., & Ortner, H. M. (2014). History of Hardmetals. In V. K. Sarin (Ed.), Comprehensive Hard Materials. Amsterdam, The Netherlands: Elsevier Ltd. doi:10.1016/B978-0-08-096527-7.00001-5

European Commission. (2006). Regulation (EC) No 1907/2006 of the European Parliament and of the Council of 18 December 2006 concerning the Registration, Evaluation, Authorisation and Restriction of Chemicals (REACH), establishing a European Chemicals Agency.

Feng, L., Wang, Z., & Shen, W. (2019). Effects of complex graphite-like carbon coating on gear vibration noise. *Results in Physics, 12*, 1495–1499. doi:10.1016/j.rinp.2019.01.048

Frank, M., Breidt, D., & Cremer, R. (2007). Nanocrystalline Diamond Coatings for Machining. *Vakuum in Forschung und Praxis, 19*(S1), 34–38. doi:10.1002/vipr.200790041

Frazer, R. C., Shaw, B. A., Palmer, D., & Fish, M. Optimizing gear geometry for minimum transmission error, mesh friction, losses and scuffing risk through computer aided engineering. In: AGMA Technical Paper 09FTM07.

Graham, D. (2014). Diamonds Can Be A Tool's Best Friend. *Manufacturing Engineering*, (Supplement 4), 27.

Grainger, S., & Blunt, J. (1998). *Engineering Coatings: Design and Application* (2nd ed.). Cambridge, UK: Woodhead Publishing Ltd. doi:10.1533/9781845698577

Gupta, K., Laubscher, R. F., Davim, J. P., & Jain, N. K. (2016). Recent Developments in Sustainable Manufacturing of Gears: A Review. *Journal of Cleaner Production*, *112*(4), 3320–3330. doi:10.1016/j.jclepro.2015.09.133

Haubner, R., Lessiak, M., Pitonak, R., Köpf, A., & Weissenbacher, R. (2017). Evolution of Conventional Hard Coatings for its Use on Cutting Tools. *International Journal of Refractory Metals and Hard Materials*, *62*(B), 210-218.

Holleck, H. (1988). Metastable Layer Materials – Prediction of Constitution and Structure. *Surface and Coatings Technology*, *36*, 151–159. doi:10.1016/0257-8972(88)90145-4

Inspektor, A., & Salvador, P. A. (2014). Architecture of PVD Coatings for Metal cutting Applications: A Review. *Surface and Coatings Technology*, *257*, 138–153. doi:10.1016/j.surfcoat.2014.08.068

Joly-Pottuz, L., Dassenoy, F., Belin, M., Vacher, B., Martin, J. M., & Fleischer, N. (2005). Ultralow-Friction and Wear Properties of IF-WS2 under Boundary Lubrication. *Tribology Letters*, *18*(4), 477–485. doi:10.100711249-005-3607-8

Kim, T. H., & Olver, A. V. (1998). Stress history in rolling-sliding contact of rough surface. *Tribology International*, *31*(12), 727–736. doi:10.1016/S0301-679X(98)00085-1

Knotek, O., & Leyendecker, T. (1987). On the Structure of (Ti,Al)N-PVD-Coatings. *Journal of Solid State Chemistry*, *70*(2), 318–322. doi:10.1016/0022-4596(87)90071-5

Korka, Z. (2009). *Cercetari privind reducerea vibratiilor in functionarea reductoarelor cu roti dintate cilindrice (Research on reducing the vibration of cylindrical gearboxes).* (Unpublished doctoral dissertation). "Eftimie Murgu" University of Resita, Romania

Korka, Z. (2018). Precision Coatings. In K. Gupta (Ed.), *Micro and Precision Manufacturing. Engineering Materials*. Cham, Switzerland: Springer. doi:10.1007/978-3-319-68801-5_8

Korka, Z., Gillich, G. R., Mituletu, I. C., & Tufoi, M. (2015). Gearboxes noise reduction by applying a fluoropolymer coating procedure. *Environmental Engineering and Management Journal*, *14*(6), 1433–1439. doi:10.30638/eemj.2015.155

Korka, Z. I., Cojocaru, V., Gillich, N., & Miclosina, C. O. (2017). Comparative Assessment of Noise Reduction in Helical Gearboxes. *Journal of Vibration Engineering & Technologies*, *5*(3), 263–268.

Leyendecker, T., Lemmer, O., Esser, S., & Ebberink, J. (1991a). The Development of the PVD Coating TiAlN as a Commercial Coating for Cutting Tools. *Surface and Coatings Technology*, *48*(2), 175–178. doi:10.1016/0257-8972(91)90142-J

Leyendecker, T., Lemmer, O., Jürgens, A., Esser, S., & Ebberink, J. (1991b). Industrial Application of Crystalline Diamond-Coated Tools. *Surface and Coatings Technology*, *48*(3), 253–260. doi:10.1016/0257-8972(91)90013-M

Liu, Z. J., Liu, Z. K., McNerny, C., Mehrotra, P., & Inspektor, A. (2005). Investigations of the Bonding Layer in Commercial CVD Coated Cemented Carbide Inserts. *Surface and Coatings Technology*, *198*(1–3), 161–164. doi:10.1016/j.surfcoat.2004.10.112

Matthews, A. (1985). Titanium Nitride PVD Coating Technology. *Surface Engineering*, *1/2*(2), 93–104. doi:10.1179ur.1985.1.2.93

McKeen, L. W. (2016). Powder Coating and Films. In L. W. McKeen (Ed.), *In Plastics Design Library, Fluorinated Coatings and Finishes Handbook* (2nd ed.). Norwich, CT: William Andrew Publishing. doi:10.1016/B978-0-323-37126-1.00011-4

Moorthy, V., & Shaw, B. A. (2012). Contact fatigue performance of helical gears with surface coatings. *Wear*, *276-277*, 130–140. doi:10.1016/j.wear.2011.12.011

Münz, W. D. (1986). Titanium Aluminium Nitride Films: A New Alternative to TiN Coatings. *Journal of Vacuum Science & Technology. A, Vacuum, Surfaces, and Films*, *4/6*(6), 2717–2725. doi:10.1116/1.573713

Oila, A., & Bull, S. J. (2005). Assessment of the factors influencing micro-pitting in rolling/sliding contacts. *Wear*, *258*(10), 1510–1524. doi:10.1016/j.wear.2004.10.012

Popov, B. N. (2015). Organic Coatings. In B. N. Popov (Ed.), Corrosion Engineering. Amsterdam, The Netherlands: Elsevier. doi:10.1016/B978-0-444-62722-3.00013-6

Porter, F. (1994). Principles of Applying Coatings by Hot Dipping. In L. L. Shreir, R. A. Jarman, & G. T. Burstein (Eds.), *Corrosion* (3rd ed.). Oxford, UK: Butterworth-Heinemann. doi:10.1016/B978-0-08-052351-4.50095-9

Schneller, T., Waser, R., Kosec, M., & Payne, D. (Eds.). (2013). *Chemical Solution Deposition of Functional Oxide Thin Films*. Vienna, Austria: Springer. doi:10.1007/978-3-211-99311-8

Schubert, A., & Nestler, A. (2009). Endbearbeitung von Aluminiummatrix-Verbundwerkstoffen durch Drehen. *Materialwissenschaft und Werkstofftechnik*, *40*(7), 580–586. doi:10.1002/mawe.200900493

Sharma, A. K., Tiwari, A. K., & Dixit, A. R. (2016). Effects of Minimum Quantity Lubrication (MQL) in Machining Processes Using Conventional and Nanofluid Based Cutting Fluids: A Comprehensive Review. *Journal of Cleaner Production*, *127*, 1–18. doi:10.1016/j.jclepro.2016.03.146

Tai, B. L., Stephenson, D. A., Furness, R. J., & Shih, A. J. (2014). Minimum Quantity Lubrication (MQL). In *Automotive Powertrain Machining*, Procedia CIRP, 14:523–528. 6th CIRP International Conference on High Performance Cutting.

Uhlmann, E., Richarz, F., Sammler, S., Heitmüller, F., & Bilz, M. (2014). Machining of Carbon Fibre Reinforced Plastics. *Procedia CIRP*, *24*, 19–24. doi:10.1016/j.procir.2014.07.135

Usuki, H., Narutaki, N., Yamane, Y., Karasuno, S., & Ito, T. (1991). A Study on Cutting Performance of Diamond Coated Tools – Tool Wear of Diamond Coated Tools in Machining of CFRP. *Bulletin of the Japan Society of Precision Engineering*, *25*(1), 35–36.

VDI-GME-Studie. (2014). Werkstoffinnovationen für nachhaltige Mobilität und Energieversorgung. Verein Deutscher Ingenieure e.V., Düsseldorf

Veprek, S., & Veprek-Heijman, M. G. J. (2008). Industrial Applications of Superhard Nanocomposite Coatings. *Surface and Coatings Technology*, *202*(21), 5063–5073. doi:10.1016/j.surfcoat.2008.05.038

Walton, D., & Goodwin, A. J. (1998). The wear of unlubricated metallic spur gears. *Wear*, *222*(2), 103–113. doi:10.1016/S0043-1648(98)00291-9

Zhang, J. (2005). *Influence of shot-peening and surface finish on the fatigue of gears*. (Doctoral Thesis, Newcastle University, UK).

Chapter 17
Property Enhancement of Carbon Fiber-Reinforced Polylactic Acid Composites Prepared by Fused-Deposition Modeling

Pravin R. Kubade
https://orcid.org/0000-0002-0580-4572
Shivaji University, India

Hrushikesh B. Kulkarni
N.B.N. Sinhgad College of Engineering, India

Vinayak C. Gavali
Shivaji University, India

ABSTRACT

Additive Manufacturing or three-dimensional printing refers to a process of building lighter, stronger three-dimensional parts, manufactured layer by layer. Additive manufacturing uses a computer and CAD software which passes the program to the printer to build the desired shape. Metals, thermoplastic polymers, and ceramics are the preferred materials used for additive manufacturing. Fused deposition modeling is one additive manufacturing technique involving the use of thermoplastic polymer for creating desired shape. Carbon fibers can be added into polymer to strengthen the composite without adding additional weight. Present work deals with the manufacturing of Carbon fiber-reinforced Polylactic Acid composites prepared using fused deposition modeling. Mechanical and thermo-mechanical properties of composites are studied as per ASTM standards and using sophisticated instruments. It is observed that there is enhancement in thermo-mechanical properties of composites due to addition reinforcement which is discussed in detail.

DOI: 10.4018/978-1-7998-0117-7.ch017

INTRODUCTION

The manufacturing industries are facing stiff competitions to sustain in the market place due to reduction in product life cycle in post era of globalization. To meet the challenging demands in the global market, the industries depend in technological advancements in the manufacturing field so that manufacturing and design lead time can be reduced. These days market demands durable product with complex profiles at reasonable manufacturing time and cost. As the complexity of the part increases, it demands more advanced manufacturing processes in order to reduce time and manufacturing cost. In the recent few years new manufacturing technologies have been developed to address these issues.

Additive manufacturing (AM) is one such process that can produce precise durable end use parts in less time with minimum cost. The manufacturing processes are categorized into two types depending upon the machining processes i.e. subtractive and additive. The subtractive manufacturing (SM) processes include, milling, grinding, cutting and turning in which material is removed from the work piece to get the final shape. In case of SM, the work piece has to pass through various machining processes which increase the wastage of material and manufacturing time. On the other hand, AM process develops product by adding material layer by layer from bottom to top in a sequential manner. Since the material is added as defined by the machining software, product development and material wastage time can be subsequently reduced as compared to SM process.

The additive manufacturing process alternatively known as Rapid Prototyping is widely appreciated for its tremendous ability in producing complex 3D geometry parts directly from computer aided design (CAD) generated models without requirements of fixture, tools and dies. The RP process can easily manufacture physical models from conceptual designs processes through computer aided design (CAD) data saved in the STL format. The rapid prototyping (RP) enables easy and quick transition from concept generation in the form of computer images to the fabrication of 3D physical models. Although the RP process can produce durable parts in less time, the availability of material type limits its widespread industrial application and daily life applications. However, the ongoing advances in the fields of manufacturing technologies and material have boosted the widespread application of RP process to produce end use parts rather than a prototype model (Yan et al., 2009).

In order to increase the industrial application of RP process, some technological advancements are needed. In this direction, overcoming the limitation of materials functionality of rapid prototyping build parts, the durability and strength of the build parts must be enhanced to face the demands of the customer. The strength of RP build parts under both dynamic loading and static loading condition must be assessed to enhance functionality. The wear behavior of the RP build parts needs to be examined to assess its durability.

Fused Deposition Modeling (FDM) is a three-dimensional printing technique pioneered in the 1990s by Stratasys. During FDM, printer starts the extrusion of raw material in the form of thin filament through heated nozzle and deposited on the base platform, where further solidifies. Layer by layer deposition continues and object is built in 3D form. *F*igure 1 shows the schematic of FDM Process.

Advantages of FDM 3D Printers Include

- Broad range of FDM printers are available in the market today
- The raw material is durable, cheap and maintains dimensional integrity
- There is a wide choice of raw material

Figure 1. Schematic of FDM Process (Ning, Cong, Qiu, Wei, & Wang, 2015)

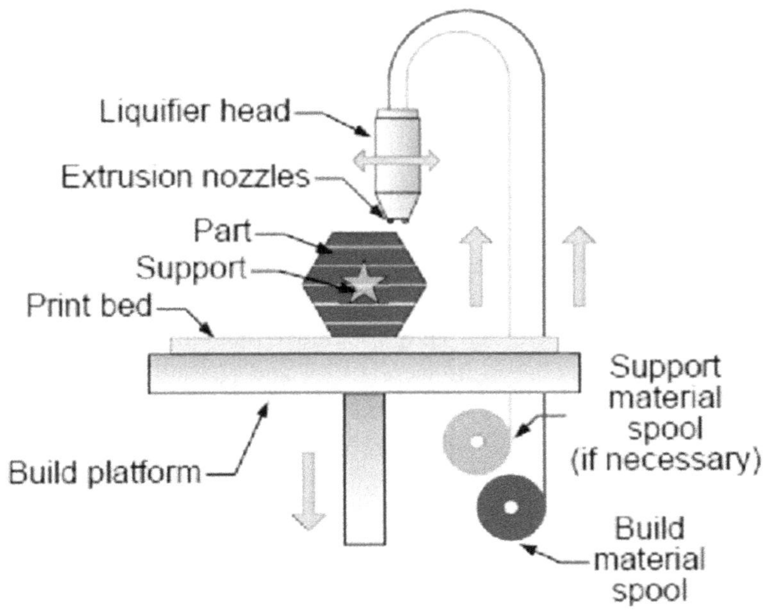

- Quick and cheap generation of models
- No worry of exposure to toxic chemicals
- FDM Printers are Affordable

Disadvantage

- For high accuracy FDM printers may be found inadequate.

FDM 3D Printers Find Application Include

- Manufacturing of prototypes
- For mould inserts and tooling patterns
- Production of precise and complex parts like jigs & fixtures
- Conceptual modeling
- Investment casting and injection molding

Sectors That Use FDM 3D Printers Include

- Automotive
- Aerospace
- Manufacturing
- Industrial
- Medical

- Architecture
- Consumer Goods

BACKGROUND

Li, Li, & Liu (2016) manufactured continuous carbon fiber reinforced polylactic acid composite by using rapid prototyping approach of 3D printing. They investigated the mechanical strength and thermodynamic properties, which were measured by using the electronic testing machine and dynamic mechanical analyzer (DMA). The design of nozzle and path control method was developed in such a way that it can satisfy the demands of continuous carbon fiber printing. They concluded that tensile strength and flexural strength of modified carbon fiber reinforced composite were 13.8% and 164% higher than original carbon fiber reinforced sample. The storage modulus of modified carbon fiber reinforced sample is higher than the PLA and original fiber reinforced sample for about 166% and 351% respectively. The scanning electron microscope (SEM) images result indicated that preferable bonding interfaces were achieved of modified carbon fiber reinforced composite.

Dickson, Barry, McDonnell, & Doweling (2017) studied the performance of continuous reinforcement of carbon, Kevlar and glass fiber into nylon, which are manufactured by using FDM. They investigated the mechanical properties of composite sample in flexural and tension and also considering fiber orientation, fiber type and volume fraction. The results were compared with that of both pure nylon specimen and known material properties. They concluded that reinforcement of carbon fiber into nylon shows higher increase in mechanical strength compared to nylon specimen increment in tensile and flexural test up to 6.3 fold and 5 fold. By comparing all composite samples, it was found that the nylon composite strength in the order of carbon fiber>Glass fiber>Kevlar fiber.

Plymill, Minneci, Greeley, & Gritton (2016) developed an improved and sustainable feedstock material for fused deposition modeling through reinforcement of polylactic acid with graphene and multi-walled carbon nanotubes. Composite with loading of 0.5, 0.2 and 0.1 wt% of reinforcement were extruded to form a filament feedstock for FDM. They investigated mechanical properties by using tensile and impact testing and fracture surface analyzed by using scanning electron microscope. By using differential scanning calorimetry thermal properties were investigated. They concluded that reinforcement of graphene in polylactic acid increases the mechanical properties by 47% in tensile strength, 17% increase in modules, and 12% increase in energy absorbed upon fracture. The 0.1% loading of multi-walled carbon nanotubes had respective increase in 41%, 16%and 9% with all reinforcement no significant change in thermal properties.

Hull et al. (2015) studies carried on carbon fiber blended with acrylonitrile butadiene styrene (ABS) thermoplastic to produce carbon fiber reinforced ABS filament in order to improve the mechanical properties of FDM printed objects. During filament extrusion three process variable show significant effect on filament diameter, expansion percentage and extrusion rate. These are process variable included carbon content, extrusion temperature and nozzle size. Their objective was to test the feasible ranges of these process variables and to investigate their effects on filament extrusion. They conclude that three process parameter in filament extrusion have shown significant effect on filament diameter, expansion percentage, and extrusion rate. They was found that the actual carbon fiber content in extruded filaments were lower than the prepared blending percentage.

Dong et al. (2014) investigated the effect of fiber content (5-30wt%) and fiber treatment on surface morphology, tensile, flexural, thermal and biodegradable properties of polylactic acid reinforced with coir fiber bio composites were evaluated via scanning electron microscopy (SEM),mechanical testing, differential scanning calorimetry, thermogravimetric analysis (TGA) and soil burial method. During his research the result is 20% treated coir fiber were to achieve optimum tensile and flexural strength of bio composites. Regardless of fiber treatment the thermal stability of bio composite was worsened with increasing fiber content. The bio composite undergo much faster degradation than PLA. The mechanical properties of alkali treated fiber bio composite better than those untreated counter parts despite being less than that of PLA in both cases. The soil burial test gave result that good biodegradability has shown in bio composite, especially to a greater extent for those with fiber treatment.

Tian et al. (2016) studied continuous carbon fiber and PLA filament were utilized as reinforcing phase and matrix respectively and simultaneously fed into the fused deposition modeling 3D printing process to forming of CFRTPCs. Investigate the interfaces and performance of printed composite by analyzing the influencing of process parameter on the temperature and pressure in the process. Forming mechanism of multiple interfaces was proposed and utilized to explain correlation between process and performance. They conclude that for impregnation of plastics into fiber bundle achieved when temperature of liquefier in the range of 200-230^{0}c. Bonding strength between layers and lines could be guaranteed with the layer thickness from 0.4mm to 0.6mm and hatch spacing of around 0.6mm. With the optimized process parameter 3D printed CFR PLA composite with fiber content of 27% can achieve maximum flexural strength of 335 MPa and flexural modulus of 30GPa.

Chumaevskii et al. (2016) investigated mechanical properties of both pure and chopped carbon fiber reinforced polyetherketone sample. Fracture surfaces have been examined using both optical and scanning electron microscopy. During their research result showed that carbon fiber reinforcement of polyether ketone matrix result in considerable improvement of the composite strength. Thus tensile ultimate strength, elasticity modulus and compression strength were increased by a factor of 2.8, 3.5 and 2.9 respectively. High mechanical characteristics achieved using carbon fiber reinforcement was provided by effective retarding the structural defect development. By using chopped fiber instead of continuous ones for reinforcement allows more isotropic mechanical characteristics.

Ning et al., (2015) studied FDM of carbon fiber reinforced thermoplastic composite, which was manufactured by adding carbon fiber pellets into ABS material and then extruded the filament. After FDM fabrication they investigated the effect reinforcement into tensile properties and Flexural properties by using varying percentage of carbon fiber. SEM micrograph was carried to find the parts fracture reasons during tensile and flexural test of CFRP composite specimen. They concluded that compared with pure plastic specimen, CFRP composite specimen with 5% wt carbon fiber content had larger flexural stress, flexural modulus, and flexural toughness with an increase of 11.82%, 16.82 and 21.86 respectively. The tensile strength and young's modulus of fabricated specimen with 5wt% or 7.5wt% carbon fiber content could increase 22.5% and 30% respectively. Porosity became severe in the specimen with 10 wt% carbon fiber content.

Ning, Cong, Hu, & Wang (2017) investigated the effect of fused deposition modeling process parameter on mechanical properties of carbon fiber reinforced plastic composite. In this experiment, carbon fiber composite parts were manufactured by FDM and tensile tests were conducted to obtain tensile properties. The effect of FDM process parameter on tensile properties of FDM fabricated carbon fiber reinforced plastic composite part was investigated. Material failure modes and reasons were observed by scanning electron microscope of part after tensile testing. They concluded that raster angle [0, 90] exhibited larger tensile strength, young's modulus and yield strength than raster angle [-45, -45]. Infill speed of 25m/s led to largest mean value of tensile properties. Nozzle temperature at 220^{0}c tensile properties first increases and then decreases. Tensile strength, young's modulus and yield strength had the largest mean values when layer thickness was 0.15mm.

Tekinalp et al. (2014) investigated short fiber reinforced acrylonitrile butadiene styrene composite as feed stock for 3D printing in terms of their processibility, mechanical performance and microstructure. Reinforcement of carbon fiber into an ABS matrix with varying weight percentage and these feed stock materials were used to fabricate composite by both FDM and compression molding processes. They concluded that tensile strength and modulus of 3D printed sample increased 115% and 700% respectively. Fiber was highly oriented in the print direction on 3D printed yielding sample on the other hand lower fiber orientation in the compressive molding process yielding sample. While no visible porosity was observed in CM samples, significant porosity was observed in FDM-printed samples. SEM micrographs show that fibers had pulled out of the matrix, indicating weak interfacial adhesion between the fibers and the matrix.

Hinchcliffe, Hess, Srubar (2016) described the effect of initial fiber prestressing on the specific tensile and flexural properties of natural fiber-reinforced polylactic acid (PLA) composite materials. They investigated the effects of fiber type (e.g., jute, flax) matrix cross sectional geometry, number of reinforcement strands, and level of initial fiber prestress on the tensile and flexural strength-to-weight and stiffness-to-weight ratios of PLA matrices. During their research, results show that utilizing 3D printing to produce more efficient structural shape can improve specific tensile and flexural properties of PLA composite and these properties are increased by post tensioning. Flax gives superior tensile properties as compared to jute. Increasing the value 116% and 62% for tensile and stiffness-to-weight respectively, and 12% and 10% for flexural strength and rigidity-to-weigh respectively compared to solid unreinforced PLA.

Tian et al. (2017) carried work on continuous carbon fiber and PLA matrix was recycled in the form of PLA impregnated carbon fiber filament from 3D printed composite component and reused as the raw material for the 3D printing process. The original printing trajectory is reversely applied, allowing for a 100% recycling of the continuous fiber without any effect on the mechanical properties. They conclude that tensile performance of recycled carbon fiber filament higher than originally printed composite. Remanufactured CFRTPCs also exhibit a 25% higher bending strength than that of original ones, which experimentally demonstrated the first non-downgrade recycling process for CFRTPCs. Material recovery rate 100% and 73% for continuous carbon fiber and PLA matrix respectively were achieved for better environmental impact. Energy consumption of 66 and 67.7MJ/kg respectively for remanufacturing and recycling processes was detected.

Zhong et al. (2001) investigates the processibility of properties of ABS and short glass fiber reinforced ABS composites for use as a feedstock filament in FDM. Study on the Acrylonitrile butadiene styrene (ABS) modified by reinforcement of short glass, fiber plasticizer and compactibilizer. Glass fibers were found to significantly improve the strength of an ABS filament at the expense of reduced flexibility and handleability. The latter two properties of glass fiber reinforced ABS filaments were improved by adding a small amount of plasticizer and compactibilizer. The resulting composite filament prepared by extrusion was found to work well with a FDM machine.

Matsuzaki et al. (2016) developed method for the 3D printing of continuous fiber reinforced thermoplastic based on FDM. In this method thermoplastic filament and continuous fiber was separately supplied to the 3D printer for printing through heated nozzle. PLA was used for matrix and carbon fiber or twisted jute yarns of natural fiber were used for reinforcement. During their research they found that carbon fiber reinforced composite shows mechanical properties superior to those of jute reinforced and unreinforced thermoplastic. The tensile modulus and strength of carbon fiber reinforced thermoplastic were 19.5GPa and 185.2MPa respectively. Flexural strength and modulus were 133 MPa and 5.93 GPa respectively for carbon fiber reinforced thermoplastic.

Melenka et al. (2016) carried work on 3D printed specimens of nylon reinforced with continuous Kevlar fibers were tested in several volume fractions. The fibers were practically oriented with the test direction. Results show increase of stiffness's and strengths with increase of fiber volume content. The experimentally determined elastic modulus was found to be 1767.2, 6920.0 and 9001.2 MPa for fiber volume fractions of 4.04, 8.08 and 10.1% respectively. The predicted elastic module was found to be 4155.7, 7380.0 and 8992.1 MPa. The model results differed from experiments by 57.5, 6.2 and 0.1% for the 4.04, 8.08 and 10.1% fiber volume fractions. The predictive model allows for the elastic properties of fiber reinforced 3D printed parts. It was also proposed a volume averaging stiffness method that is able to predict stiffness's of 3D printed composite parts of that particular case, with reasonable agreement with experimental data.

Ferreira et al. (2017) studied mechanical characterization of materials produced by 3D printing based on fused filament fabrication. The material chose for this study were polylactic acid (PLA) and a PLA reinforced with short carbon fibers in a weight fraction of 15% (PLA+CF). Only unidirectional or specially oriented specimens were used. The result of this research was in the microstructure of PLA+CF, the short carbon fibers stayed highly oriented with the material deposition direction in FFF specimens and length of the fiber, explains differences in material properties. The PLA matrix carried out the majority of the stresses at the failure load level in both PLA and PLA+CF. Tensile modulus and shear modulus were also increased by short fiber respectively 1.25 and 1.16 times unreinforced PLA.

Letcher and Waytashek (2014) investigate the printing orientation in different raster angle for polylactic acid thermoplastic material. The MakerBot Replicator 2x printer was used to print specimen to conduct tensile, flexural and fatigue testing. Studied the effect of orientation on part strength for that specimen were printed at raster orientation angles of 0°, 45° and 90°.PLA filament was also tensile tested. In his research results found that 45° raster orientation angle made the strongest specimen at an ultimate tensile strength of 64 MPa. Flexural test was conducted by using a 3-point bending fixture. For flexural testing the 0° raster orientation produced the strongest parts with an ultimate bending stress of 102 MP and in fatigue testing the 90° specimens were clearly the least resistant to fatigue loadings. The filament testing showed similar results to the printed specimen result.

Yap et al. (2016) studies strength and elasticity of the fiber reinforced polymers are dependent on the intrinsic mechanical properties of matrix and fiber, the fiber layup pattern as well as the volume percentage of the matrix and fiber. The effects of these factors on the 3D printed fiber reinforced composite materials were investigated in this paper. The fiber reinforced polymers were fabricated using multi-material inkjet printer with reinforcement of rigid strong polymer fiber into matrix of rubbery material. Two types of fiber layup configurations and fiber/matrix volume ratios were designed for this study. During his research experimental result shows that both tensile strength and elastic modulus of the fiber reinforced polymers could be largely enhanced by varying the fiber/matrix ratio and layup pattern.

Griffini et al. (2016) carried work on Glass (GFR) and carbon fiber-reinforced (CFR) dual-cure polymer composites fabricated by UV-assisted three-dimensional (UV-3D) printing are investigated. The resin material combines an acrylic-based photo curable resin with a low temperature (140^{0}C) thermally-curable resin system based on bisphenol. They conclude that UV-3D printed macrostructures, giving a clear indication of their potential use in real-life structural applications. Dynamic mechanical analysis and differential scanning calorimetry highlighted the good thermal stability and mechanical properties of the printed parts. Fiber reinforcing effects on the UV-3D printed objects were assessing by using uniaxial tensile tests. During his research study was conducted on the use of a sizing treatment on carbon fibers to improve the fiber/matrix interfacial adhesion, giving preliminary indications on the potential of this approach to improve the mechanical properties of the 3D printed carbon fiber reinforced components.

Mahajan and Cormier (2015) Study carried on 3D printed carbon-epoxy composite components in which the reinforcing carbon fibers have been preferentially aligned during the micro-extrusion process. By adding carbon fiber as a reinforcing material, properties such as, thermal conductivity, mechanical strength, and electrical conductivity can be greatly enhanced. Investigate the however these properties are significantly influenced by the degree of fiber alignment. Based on analysis of experimental results, tensile test samples were printed with fibers aligned perpendicular and parallel to the tensile axis. During his research he concludes that 44.12% increase in ultimate tensile stress and a 42.67% increase in sample modulus with carbon fiber aligned along the tensile axis. The percentage of carbon fiber translation speed and nozzle diameter was found to have the most significant effect on the degree of fiber alignment. Table 1. gives summary of literature review in present work.

Problem Identification

From the available literature it is clear that most of the researchers worked on continuous carbon fiber reinforced thermoplastic composites, however very less work is reported on fabrication of chopped carbon fiber reinforced thermoplastic using FDM method. The 3D printed object of chopped carbon fiber reinforced thermoplastic composite will have different characteristic than the continuous fiber reinforced thermoplastic composite, thus mechanical and thermo-mechanical properties of such 3D printed chopped carbon fiber reinforced thermoplastic composite for varying percentage of chopped carbon fiber needs to be analyzed. Hence there is scope to investigate the effect of chopped carbon fibers reinforced thermoplastic composite, fabricated using FDM method on mechanical and thermo-mechanical properties.

Objectives of Research Work

- Fabrication of series of chopped fiber reinforced thermoplastic composites by FDM method considering varying weight percentage of carbon fiber.
- Investigation of carbon percentage and process parameters on physical and mechanical properties of fabricated composites.
- Investigation of carbon percentage and process parameters on thermal and thermo-mechanical properties of fabricated composites.
- Optimization of carbon fiber percentage considering physical, mechanical and thermal properties of fabricated composites using multi criteria decision method.
- Comparative study of different blend systems in order to get optimal results in terms of mechanical and thermal properties of chopped carbon fiber reinforced poly lactic acid composite using fused deposition modeling method.

Table 1. Summary of literature review

Author and Ref	Work material	Method used	Process Parameters	Performance Measures	Concluding Remarks
Li et al. (2016)	PLA, CFR –PLA, modified CFR –PLA	FDM	Nozzle design, Printing path	1) Tensile strength of modifiedCFR-PLA13.8% more than CFR-PLA. 2) Flexural strength of modified CFR-PLA13.8% more than CFR-PLA.3) Storage module of modified CFR-PLA 166% more than PLA.	A modification in CFR-PLA increases mechanical properties of CFR-PLA. CFR-PLA successfully modified by methylene dichloride treatment.
Dickson et al. (2017)	Nylon, Carbon fiber, Kevlar fiber, glass fiber	FDM	Fiber Printing path,	1) Tensile and flexural strength enhanced by 6.3 fold and 5 Fold than nylon sample. 2) Carbon fiber > Glass fiber > Kevlar fiber. 3) SEM- Micrographic structure	Carbon fiber reinforcement in Nylon increases the mechanical strength than the glass fiber reinforcement. Similarly the glass fiber reinforcement increases the mechanical strength than Kevlar fiber reinforcement.
Plymill et al. (2016)	Graphene, Multiwalled carbon nanotubes, PLA	FDM		1)0.2 wt% loading of graphene showing a 47%, 17% and 12% increase in tensile strength, modulus, and energy absorbed upon fracture. 2) The 0.1 wt% loading of MWCNT showing a 41%, 16% and 9% increase in tensile strength, modulus, and energy absorbed upon fracture.	Reinforcement of Graphene and MWCNT in PLA shows enhanced tensile strength, modulus and energy absorbed upon fracture.
Hull et al. (2015)	ABS, Carbon fiber	FDM	carbon fiber content, extrusion temperature, nozzle size	Three process parameters in filament extrusion have shown significant effects on filament diameter, expansion percentage, and extrusion rate.	Addition of carbon fiber, change in extrusion temperature and variation in nozzle size affects the filament diameter, expansion percentage and extrusion rate.
Author and Ref	**Work material**	**Method used**	**Process Parameters**	**Performance Measures**	**Concluding Remarks**
Dong et al. (2014)	PLA, coir fiber	FDM		20 wt% treated coir fibers were determined to achieve optimum tensile and flexural strengths of biocomposites. The biocomposites undergo much faster degradation than PLA.	The mechanical properties of the alkali treatedfiber biocomposite are higher than the PLA. Untreated fiber biocomposite has lower mechanical properties than PLA.
Tian et al. (2016)	PLA, Carbon fiber	FDM	Temperature, pressure	Fiber content reached 27%, flexural strength of 335 MPa and modulus of 30 GPa were obtained for the printed composite specimens.	Fiber content of the printed specimens can be easily controlled by changing the process parameter.
Chumaevskii et al. (2016)	carbon fiber, polyetheret-herketone	FDM		Tensile ultimate strength, elasticity modulus and compression strength increased by a factor of 2.8, 3.5 and 2.9 respectively.	The carbon fiber reinforcement in polyetheretherketone matrix causes considerable improvement in strength of composite.

continued on following page

Table 1. Continued

Author and Ref	Work material	Method used	Process Parameters	Performance Measures	Concluding Remarks
Ning et al. (2015)	ABS, Carbon fiber	FDM		CFRP composite specimen with 5% wt carbon fiber content had larger flexural stress, flexural modulus, and flexural toughness with an increase of 11.82%, 16.82% and 21.86% respectively. The tensile strength and young's modulus of fabricated specimen with 5wt% or 7.5wt% carbon fiber content could increase 22.5% and 30% respectively.	Increase in the weight percentage of carbon fiber in CFRP composite specimen produces severe problems of porosity which decreases mechanical properties.
Ning, Cong, Hu, & Wang (2017)	ABS, Carbon fiber	FDM	Raster angle, Infill speed, Nozzle temperature, Layer thickness	Best tensile properties at following condition; 1) Raster angle [0,90] 2)Infill speed of 25 mm/s 3)Nozzle temperature 220^{0}C 4)Thickness 0.15mm	Reinforcement of carbon fiber in ABS by given process parameters significantly increases the tensile strength, young's modulus and yield strength.
Author and Ref	**Work material**	**Method used**	**Process Parameters**	**Performance Measures**	**Concluding Remarks**
Tekinalp et al. (2014)	ABS, carbon fiber	FDM	Fiber length, fiber orientation, porosity	The tensile strength and modulus of 3D-printed samples increased 115% and 700% respectively.	With increasing fiber content, voids inside the FDM-printed beads increases where as the voids between the beads decreases. FDM-printed samples have high fiber orientation in the printing direction.
Hinchcliffeet al. (2016)	PLA, Flax, jute	FDM		Flax gives superior tensile properties as compared to jute. Increasing the value 116% and 62% for tensile and stiffness-to-weight respectively, and 12% and 10% for flexural strength and rigidity-to-weight respectively compared to solid unreinforced PLA.	Utilizing 3D printing to produce more efficient structural shape can improve specific tensile and flexural properties of PLA composite; these properties are increased by post tensioning.
Tian et al. (2017)	PLA, Carbon fiber	FDM		Remanufactured CFRTPCs also exhibit a 25% higher bending strength than that of original. Material recovery rate 100% and 73% for continuous carbon fiber and PLA matrix respectively achieved.	Tensile performance of recycled carbon fiber reinforced PLA filament is higher than originally printed composite.
Zhong et al. (2001)	ABS, short glass fiber	FDM		The resulting composite filament prepared by extrusion was found to work well with a FDM machine.	Glass fibers reinforcement in ABS improves strength of filament at expense of reduced flexibility.

continued on following page

Table 1. Continued

Author and Ref	Work material	Method used	Process Parameters	Performance Measures	Concluding Remarks
Van Der Klift et al. (2016)	PLA, carbon fiber, jute fiber	FDM		The tensile modulus and strength of 3D printed CFRP are 19.5GPa and 185.2MPa respectively. Flexural strength and modulus was 133 MPa and 5.93 GPa respectively for CFRP.	Carbon fiber reinforced composite shows mechanical properties superior than jute reinforced and unreinforced thermoplastic.
Melenka et al. (2016)	Nylon, Kevlar fiber	FDM		Determined elastic modulus was found to be 1767.2, 6920.0 and 9001.2 MPa for fiber volume fractions of 4.04, 8.08 and 10.1% respectively. The predicted elastic moduli were found to be 4155.7, 7380.0 and 8992.1 MPa. The model results differed from experiments by 57.5, 6.2 and 0.1% for the 4.04, 8.08 and 10.1% fiber volume fractions.	Reinforcement of Kevlar fiber in nylon with increase in fiber volume increases the stiffness's and strengths of composite.
Author and Ref	**Work material**	**Method used**	**Process Parameters**	**Performance Measures**	**Concluding Remarks**
Ferreria et al. (2017)	PLA, carbon fiber	FDM	Raster angle	Tensile modulus and the shear modulus were also increased by the short fibers, respectively 1.25 and 1.16 times unreinforced PLA. Short carbon fibers stay highly oriented with the material deposition direction in the FFF specimens and the length of the fiber.	Reinforcement of carbon fiber into PLA matrix in unidirectional or specially orientation state increases tensile and shear modulus than unreinforced PLA.
Letcher & Waytashek (2014)	PLA	FDM	Raster angle	1) 45° raster orientation angle made the strongest specimen at an ultimate tensile strength of 64 MPa. 2) The 0° raster orientation produced the strongest parts with an ultimate bending stress of 102 MPa.	Raster angle or printing orientation shows greater effect on the mechanical properties of thermoplastics. Proper raster angle increases the mechanical properties of PLA.
Yap et al. (2016)	Tango plus(FLX930), ABS	PolyJet 3D printing	Raster angle, volume fraction	Both tensile strength and elastic modulus of the fiber reinforced polymers could be largely enhanced by varying the fiber/ matrix ratio and layup pattern.	Raster angle and volume fraction affects the tensile strength and elastic modulus of fiber reinforced thermoplastic.
Invernizzi et al. (2016)	Glass fiber, carbon fiber, acrylic-based photo curable resin	UV-assisted 3D printing		DSC and DMA the good thermal stability and mechanical properties of the printed parts. Sizing treatment on carbon fiber increases the mechanical property of composite.	Sizing treatment on carbon fibers improves the fiber/ matrix interfacial adhesion. This approach amends the mechanical properties of the 3D printed CFRP.
Mahajan & Cormier (2015)	carbon fibers and epoxy resin	nScrypt 3Dn Tabletop micro-extrusion system	translation speed, nozzle diameter, standoff distance, Air pressure	During his research he concludes that 44.12% increase in ultimate tensile stress, and a 42.67% increase in sample modulus with carbon fibers aligned along the tensile axis.	The percentage of carbon fibers, translation speed, and nozzle diameter significant affect on the degree of fiber alignment.

Table 2. Printing parameters

Nozzle temperature	**230 ^{0}c**
Printing velocity	60 mm/sec
Layer thickness	0.20 mm
Bed temperature	60 ^{0}c
Infill density	100%
Printing Orientation	45^{0}

Scope

Fused deposition modeling (FDM) method used for fabricating polylactic acid parts those are mainly used as rapid prototypes for functional testing with advantages of low cost, minimal wastage, and ease of material change. Due to the intrinsically limited mechanical properties of pure polylactic acid, there is a critical need to improve mechanical properties for FDM-fabricated pure polylactic acid parts. One of the possible methods is adding reinforced materials (such as carbon fibers) into polylactic acid materials to form carbon fiber reinforced plastic polylactic acid (CFRP) composites those could be directly used in the actual application areas, such as aerospace, automotive, and wind energy.

EXPERIMENTAL DESIGN

In this work raw materials used were virgin PLA thermoplastic pellets and carbon fiber powders. The carbon fiber powder has two different average carbon fiber lengths, 150 mm and 100 mm, with common fiber diameter of 7.2 μm. The pellets and carbon fiber powders were mixed in a blender with different weight percentage of carbon fiber like 10%, 12%, 15%, 18% and 20%. The plastic extruder was used to fabricate the carbon fiber filled filaments. During the extrusion processes, extrusion temperature, filament yield speed, and nozzle diameter were set at 220^{0}C, 2 m/min and 1.75 mm respectively. The filaments should be cut into small pieces and referred in the extruder for the second extrusion to make them with high bulk density, which led to more consistent flow rates and fusion on each layer. During such process filaments with more homogeneous distribution of carbon fibers could be obtain to improve the FDM fabrication process and parts performance. The FDM having number of input parameters i.e. nozzle temperature, layer thickness, nozzle speed and bed temperature that could be effect on the strength of the printed sample. To minimize this input parameters effect on PLA and CFRP composite samples there is need to find identical input parameters which show same effect on all sample. The input parameter for PLA and CFRP composite find out by printing material at different condition and a set of parameter selected which shows good print with same effect on sample. Selected printing input parameters are shown in Table 2.which gives better sample print for PLA and CFRP composite.

Fabrication of neat PLA and chopped carbon fiber reinforced PLA composites will be carried out on Pratham 3D printing machine available in the Production Engineering department of KIT's College of Engineering, Kolhapur. "CURA" software is available to do programs for printing different geometrical shapes. Experimental setup of FDM is as shown in Figure 2 The filament of PLA and CFRP composite used as a printing material for FDM. Three samples of PLA and CFRP composite for each test printed

Table 3. Specifications of Pratham 3D printing machine

Nozzle diameter	**0.4 mm**
Nozzle temperature up to	260 ^{0}c
Bed temperature up to	90 ^{0}c
Bed dimension	300x300 mm
Height of printing up to	230 mm
Filament diameter	1.75 mm

by using FDM. Here only varying the filament material for printing and input parameters of this FDM machine i.e. layer thickness, nozzle temperature, nozzle speed and bed temperature taken as same for all material.

Figure 2. Experimental setup-3D printing machine

Figure 3. Tensile Properties of carbon fiber reinforced PLA composites

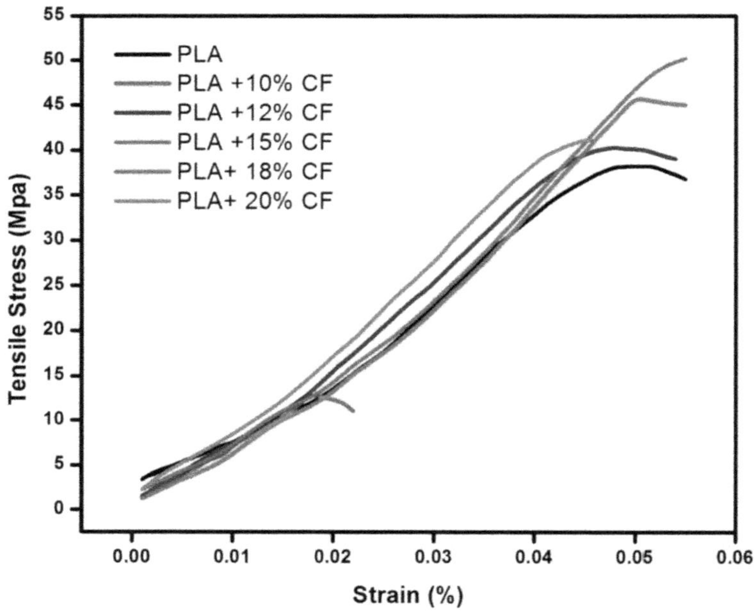

Figure 2 shows experimental setup and Table 3. Shows specification for 3D printing machine. Fabrication of neat PLA and chopped carbon fiber reinforced PLA composites will be carried out on Pratham 3D printing machine available in the Production Engineering department of KIT's College of Engineering, Kolhapur. "CURA" software is available into our college to do programs for printing different geometrical shapes

Figure 4. Effect of carbon fiber content on tensile strength of PLA composites

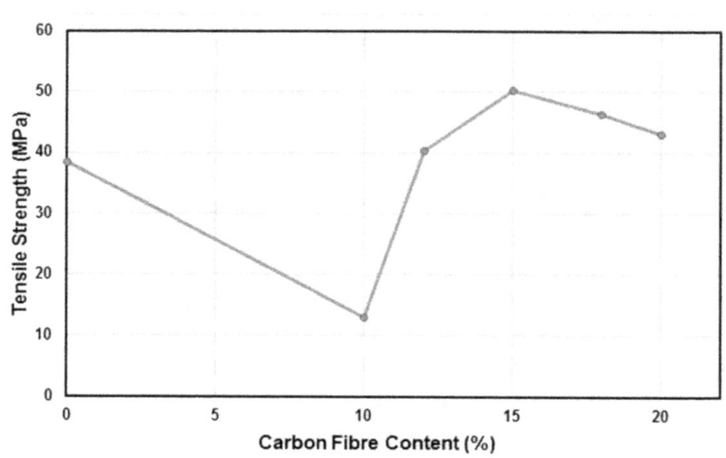

Results and Discussions

Tensile Properties

Tensile properties CFRP are studied as per ASTM D638 standard and sample are prepared accordingly. Tensile test was conducted on universal testing machine of 10 KN force transducer capacity. Three specimens at each carbon fiber content are tested to obtain proper results.

The tensile strength of neat PLA and CFRP composite was tested by fixture on the universal testing machine. A tensile property was tested by using three samples in order to ensure the repeatability of the test results. A testing speed of 10 mm/min is applied during the tensile test. Typical tensile strain-stress curves of neat PLA and CFRP composites are illustrated in Figure 3.

For the testing specimens with carbon fiber content of 15 wt% and, for example the curve was selected from the results of three specimens, depending on the maximum number of values those were the most closed to the mean value of tensile strength.

Figure 4 shows that with the increase of carbon fiber content from 0 wt% to 10 wt%, large decrement in tensile strength. With increase of carbon fiber content to 12 wt% large increments in tensile strength and it reaches to maximum with the levels of carbon fiber content increasing from 12 wt% to 15 wt%. Increment of carbon fiber above the 15 wt% tensile properties tends to decrease. Carbon fiber content reaches to 18 and 20 wt% are shows continuous decrement in tensile strength. The largest mean value (50 MPa) could be found at 15 wt% carbon fiber content, while the smallest mean value was about 13 MPa for 10 wt% CFRP composite samples. Compared with pure PLA specimen adding 15 wt% of carbon fiber up into PLA could increase tensile strength by 32%.

Effects of carbon fiber content on tensile properties of CFRP composite specimens are shown in Figure 4. The neat PLA sample contain 100% PLA thermoplastic so carbon fiber content taken as 0 wt%. The pure PLA sample has grater tensile strength and it decreases with the addition of 10% of carbon fiber.

The CFRP composite with 10 wt% carbon fiber shows lower strength than PLA and other composites. The carbon fiber content increases from 10 wt% to 12 wt% large enhancement in tensile strength, which greater than neat PLA sample. The carbon fiber content increases to 15 wt%, it shows larger value of tensile strength. Carbon fiber content up to 10 wt% decreases the tensile strength and increase in CF content up to 15 wt% improves the composite tensile strength. Further increment in CF content of 18 wt% and 20 wt%, decrease the tensile strength.

Flexural Test

Flexural or three-point bend test is conducted as per ASTM D790 standard. Three specimens at each carbon fiber content are tested to obtain proper results.

The flexural strength test is conducted on the same testing machine where tensile test conducted. In that only the fixture is changed to the three-point bending type. The diameter of the load indenter was adjusted to 5 mm and its descent velocity was 5 mm/min. Flexural test was conducted to compare flexural strength of the specimens with and without carbon fibers.

According to the ASTM D790 standard, flexural test would be terminated when the maximum strain in the outer surface of the specimen reached 5% or when break occurred prior to reaching the maximum strain. Typical flexural strain-stress curves of specimens with and without carbon fibers are illustrated in Figure 5

Figure 5. Stress-strain curve of specimens with or without of carbon fiber reinforced PLA composites

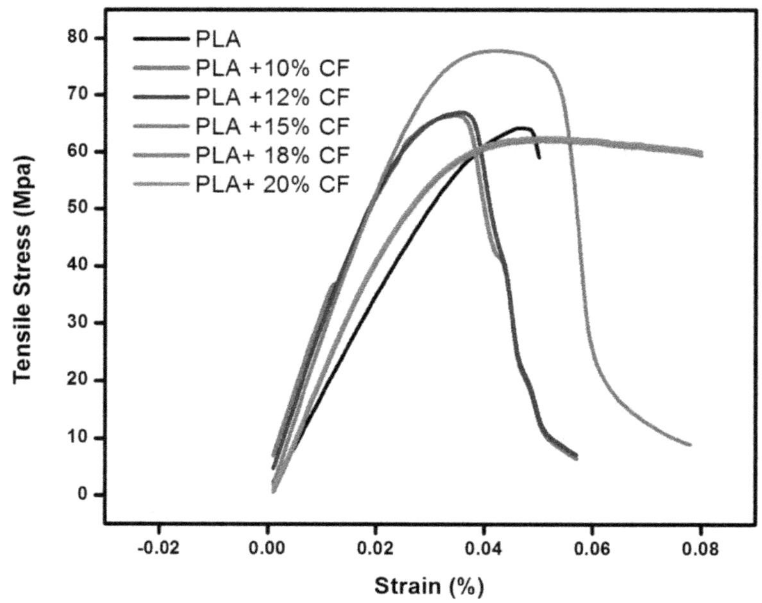

It can be seen that flexural strength of pure PLA was 64 MPa and slightly increases to 66 MPa when reinforcement of 10 wt% of chopped carbon fiber into PLA. Further increase of carbon fiber content from 10 wt% to 12 wt%, there was small increment in flexural strength. Carbon fiber content increases to 15 wt% gives maximum flexural strength 78MPa. CFRP composite specimen with 15 wt% carbon fiber content enhances the flexural strength by 22% compared with the pure PLA specimen. In CFRP composite carbon fiber content increases to 18 wt% and 20 wt% shows decrement in flexural strength by 3% compared with pure PLA sample. Effects of carbon fiber content on flexural strength of CFRP composite specimens are shown in Figure 6.

Figure 6. Flexural strength of carbon fiber reinforced PLA composites

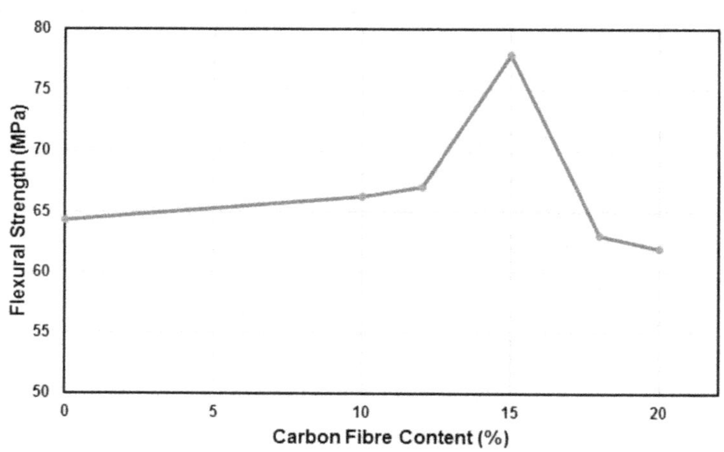

The PLA has lower value of flexural strength and increases with increase of carbon fiber content. CFRP composite with 10wt% carbon fiber increases the small amount of flexural strength of composite.

Carbon fiber content increases from 10 wt% to 12 wt% also shows enhancement in flexural strength. In CFRP composite sample 10 wt% 12 wt% carbon fiber contents could not show larger enhancement in flexural strength. In CFRP composite with 15wt% carbon significantly increase in flexural strength. The flexural strength of composite sample was enhanced by reinforcement of 15 wt% carbon fiber. Reinforcement of 18 wt% and 20 wt% carbon fiber into CFRP composite decreases the flexural strength.

Figure 7. Micro hardness properties of carbon fiber reinforced PLA composites

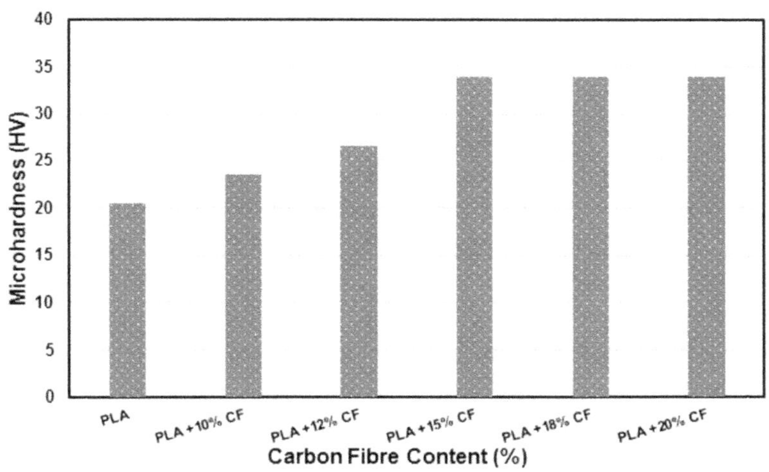

Figure 8. Impact strength of carbon fiber reinforced PLA composites

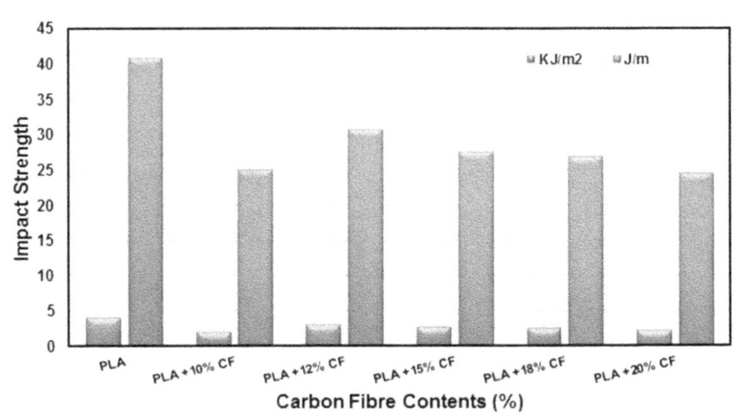

Figure 9. Storage modulus of carbon fiber reinforced PLA composites

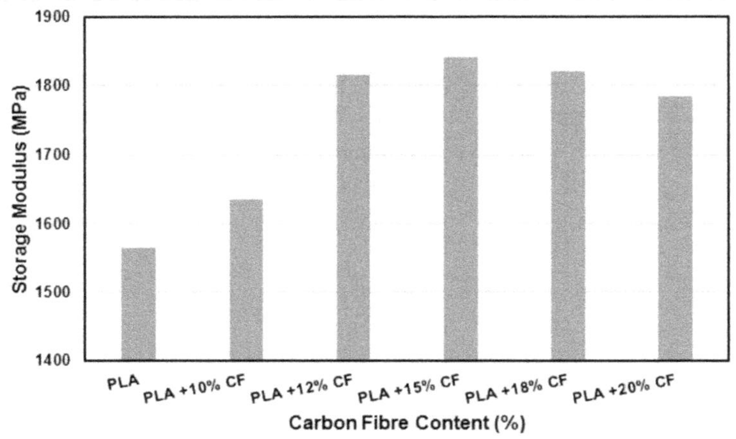

Microhardness Test

Micro hardness test is conducted as per ASTM E 384 standard. Three specimens at each carbon fiber content are tested to obtain proper results. Micro hardness test carried on neat PLA and CFRP composite of PLA to measure the effect of reinforcement of chopped carbon fiber in PLA with varying content. Micro hardness testing is a method of determining a materials hardness or resistance to penetration when test samples are very small or thin, when small regions in composite samples are to be measured.

The unit of hardness given by the test is known as the Vickers pyramid number (HV). The 0.1KN force was applied on the surface of the sample and indentation on the surface of the samples.

From the Figure 7 analyze that the hardness value of the neat PLA was lower than CFRP composite which indicate resistance to penetration lower in PLA. With the increase of wt% of carbon fiber the hardness value increases. CFRP composite with 10 wt% and 12 wt% carbon fiber shows increment in hardness value by 15% and 27% respectively as compared with PLA. Increase in carbon fiber content from 10 wt% to 12 wt%, there was small increment in hardness value. CFRP composite sample with 15

Figure 10. Glass transition temperature (^oC) of Carbon fiber reinforced PLA composites

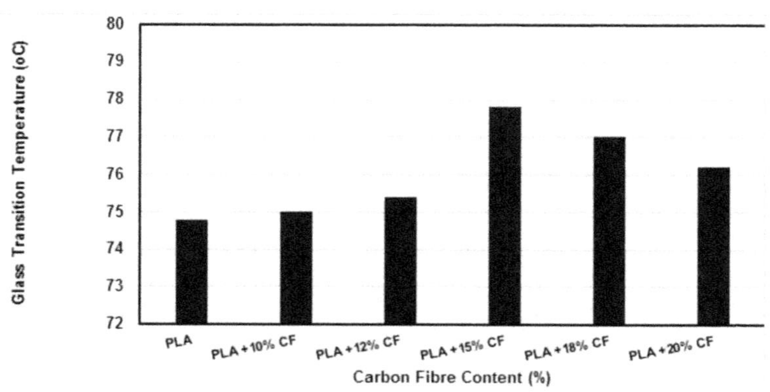

wt%, 18 wt% and 20 wt% carbon fiber content shows same enhancement in hardness value by 70% as compared with PLA.

Izod Impact Test

Izod impact test is conducted as per ASTM D 256 standard. Three specimens at each carbon fiber content are tested to obtain proper results. Impact strength of PLA and CFRP composite of PLA was measured by using Izod impact test. This test fixes one end of a notched in a cantilever position by means of a vice. Strikers on the arm of a pendulum or similar energy strike the specimen. The energy absorbed by the specimen in the breaking process is known as the breaking energy.

Figure 11. TGA of 12 wt% of carbon fiber reinforced PLA composites

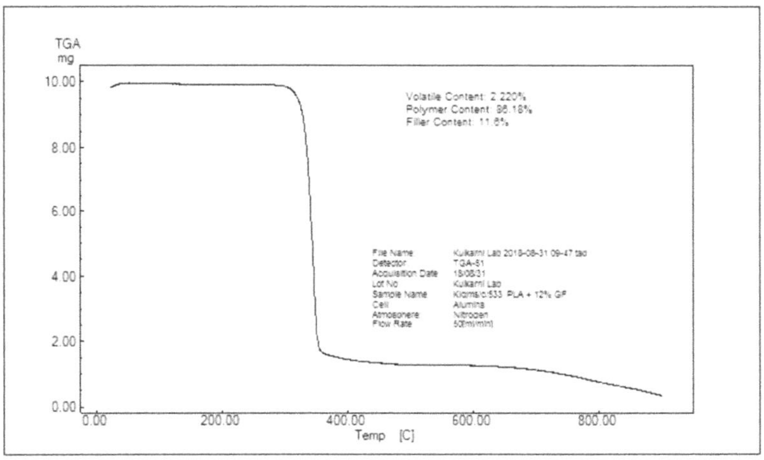

Figure 12. TGA of 15 wt% of carbon fiber reinforced PLA composites

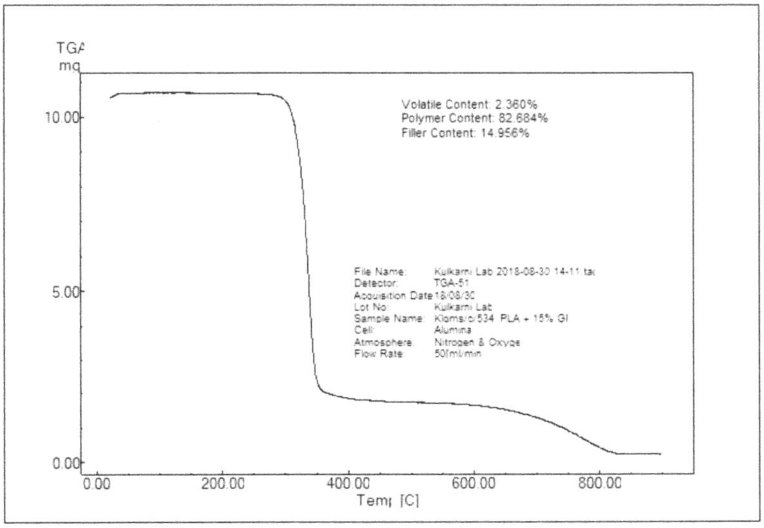

Figure 13. TGA of 20 wt% of carbon fiber reinforced PLA composites

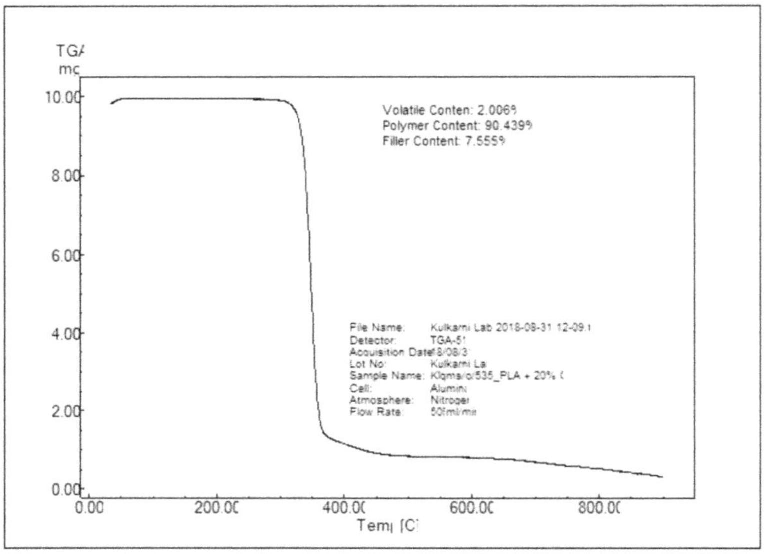

The breaking energy can be converted into an indication of a materials impact resistance using such units as joules. Impact strength of neat PLA and CFRP composite samples are shown Figure 8 The impact strength of neat PLA was 40.8 J/m which larger than 10 wt%, 12 wt%, 15 wt%, 18 wt% and 20 wt% carbon fiber reinforced PLA composites. Addition of 10 wt% and 15 wt% of carbon fiber into CFRP composite reduction in impact strength by 39% and 32% respectively as compared with PLA. Increasing in carbon fiber content from 10 wt% to 12 wt% there was small increment in impact strength up to 31 J/m. Further increment of 15 wt%, 18 wt% and 20 wt% carbon content into CFRP composites shows continuous decrease in impact strength.

Dynamic Mechanical Analysis

The statistic results of storage modulus for different printed materials are shown in Figure 9. Increase in carbon fiber content shows enhancement in storage modulus of CFRP composite sample. It is observed that the 15 wt% carbon fiber reinforced PLA samples have highest storage modulus. That means that the 15 wt% carbon fiber reinforced PLA has a better interfacial bond and strength, which lead to a more effective stress transfer between the fiber bundles. The storage modulus of 15 wt% carbon fiber reinforced samples is higher than the PLA sample by 18%. Increasing the CF content above 15 wt% in composite then storage modulus starts to decreases. The much better fiber-matrix interfacial bond is present in the 15 wt% CFRP composite sample. Effects of carbon fiber content on glass transition temperature of composites are shown in Figure 10.

The glass transition temperature (T_g) values shows the thermal stability of composites, which are improves by being reinforced with chopped carbon fiber up to 15 wt%.

Reinforcement of 10 wt% and 12 wt% carbon fiber shows small increment in glass transition temperature. It is observed that the 15 wt% carbon fiber reinforced PLA samples have highest glass transition temperature which increased by 3^0C. Furthermore reinforcement of carbon fiber content shows decrement in glass transition temperature. The much better heat conduction ability is present in the 15 wt% CFRP composite sample. From above work, it can be deduced that addition on reinforcement may improve the thermo-mechanical properties of composites[22-29].

Thermogravimetric Analysis

Figure 11, Figure 12, and Figure 13 shows the thermogravimetric analysis (TGA) curves of samples with 12 wt%, 15 wt% and 20 wt% levels of carbon fiber contents respectively. There was a small amount of volatile content available (near about weight = 2%). Overall, the TGA curves of samples with three levels of carbon fiber contents are similar, the weight loss of samples that occurred within the temperature of 50 to 300 °C were less than 2%, where as a significant amount of weight loss was observed as the temperature increased from 300 to 400 °C.

The residual weights stayed almost steadily from 400 to 600 °C without further thermal degradation. After the 600 °C there was starting small amount of thermal degradation up to 800°C.The degradation peak temperature for samples with three levels of carbon fiber contents are 365 °C (12%), 375 °C (15%), and 360 °C (20%). The thermal stability of 15 wt% carbon fiber reinforced PLA composite is higher than other two (12 wt% and 20 wt %) carbon content composites. Their corresponding percentage values after removing the residue from PLA are 11.6% (12%), 14.95% (15%), and 7.55% (20%). It was found that in 20 wt% CFRP composite, there was more carbon fiber weight loss during filament extrusion as the prepared carbon fiber content increased.

CONCLUSION

The following conclusions were drawn from this study.

- Compared with pure PLA specimen adding 15 wt% of carbon fiber up into PLA could increase tensile strength by 32%. CFRP composite specimen with 15 wt% carbon fiber content enhances the flexural strength by 22% compared with the pure PLA specimen. Specimen with 15 wt% carbon fiber content had the largest mean value of tensile strength and flexural strength, and it start to decrease when carbon fiber content increases to 20 wt %.
- Pure PLA has lower hardness value but it increases by 70% with reinforcement of 15 wt%, 18 wt% and 20 wt% of carbon fiber. Composite with 15 wt%, 18 wt% and 20 wt% of carbon fiber also shows same enhancement in hardness value. Hardness of the specimens increases with increase in the carbon fiber content into composite.
- Impact strength of pure PLA was higher than all other CFRP composites. Composite with 12 wt% of carbon fiber shows increment in impact strength by 24% as compared with 10 wt% CFRP composite. Increase in carbon fiber content up to 20 wt% in CFRP composite there was continuous reduction into impact strength
- Reinforcement of 15 wt % of carbon fiber into Polylactic acid enhances the storage modulus and glass transition temperature by 18% and 3 0C respectively as compared with pure Polylactic acid sample. Storage modulus and glass transition temperature increases with the increase of carbon fiber content up to 15 wt% and above increment of the carbon fiber content decrease storage

modulus and glass transition temperature.15 wt% CFRP composite has higher thermal stability and better fiber-matrix interfacial bond.

- Degradation temperature of 15 wt% CFRP composite was higher than other two 12 wt% and 20 wt% CFRP composites. In 12 wt% and 15 wt% CFRP composites loss of carbon fiber content were small amount. Into 20 wt% CFRP composite there was more carbon fiber weight loss during filament extrusion as the prepared carbon fiber content increased.

FUTURE SCOPE

Present research work provides ample opportunity for the investigation of mechanical and thermo-mechanical properties of carbon fiber reinforced PLA composite which fabricated using FDM process. Some scopes for future research have been given below.

- Effects of environmental variables like temperature and humidity on the part quality may be explored.
- Research on increase of build space and provision of multiple nozzles for part deposition in FDM process needs to be explored to cater to the needs of medium or large batch manufacturer.
- FDM process specific CAD modeling and analysis tools need to be developed.
- Option of depositing multiple materials in a single setting needs to be explored.
- Possibility of using different materials or modification in the present material composition may be explored.
- Furthermore, research can be extended to study the effect of process parameters on circularity of inner holes in FDM build parts.

REFERENCES

Chumaevskii, A. V., Tarasov, S. Y., Filippov, A. V., Kolubaev, E. A., Rubtsov, V. E., & Eliseev, A. A. (2016, November). Mechanical strength of additive manufactured carbon fiber reinforced polyetheretherketone. In AIP Conference Proceedings: Vol. 1783. *No. 1* (p. 020029). AIP Publishing. doi:10.1063/1.4966322

Dickson, A. N., Barry, J. N., McDonnell, K. A., & Dowling, D. P. (2017). Fabrication of continuous carbon, glass and Kevlar fibre reinforced polymer composites using additive manufacturing. *Additive Manufacturing*, *16*, 146–152. doi:10.1016/j.addma.2017.06.004

Dong, Y., Ghataura, A., Takagi, H., Haroosh, H. J., Nakagaito, A. N., & Lau, K. T. (2014). Polylactic acid (PLA) bio composites reinforced with coir fibres: Evaluation of mechanical performance and multifunctional properties. *Composites. Part A, Applied Science and Manufacturing*, *63*, 76–84. doi:10.1016/j.compositesa.2014.04.003

Ferreira, R. T. L., Amatte, I. C., Dutra, T. A., & Bürger, D. (2017). Experimental characterization and micrography of 3D printed PLA and PLA reinforced with short carbon fibers. *Composites. Part B, Engineering*, *124*, 88–100. doi:10.1016/j.compositesb.2017.05.013

Gavali, V. C., Kubade, P. R., & Hrushikesh, B. K. (2018, February). Mechanical And Thermomechanical Properties Of Carbon Fibre Reinforced Thermoplastic Composite Fabricated Using Fused Deposition Modelling (Fdm) Method: A Review. *International Journal of Mechanical and Production Engineering Research and Development*, *8*(1), 1161–1168. doi:10.24247/ijmperdfeb2018137

Hinchcliffe, S. A., Hess, K. M., & Srubar, W. V. III. (2016). Experimental and theoretical investigation of prestressed natural fiber-reinforced polylactic acid (PLA) composite materials. *Composites. Part B, Engineering*, *95*, 346–354. doi:10.1016/j.compositesb.2016.03.089

Hull, E., Grove, W., Zhang, M., Song, X., Pei, Z. J., & Cong, W. (2015, June). Effects of process variables on extrusion of carbon fiber reinforced ABS filament for additive manufacturing. In *ASME 2015 International Manufacturing Science and Engineering Conference*. American Society of Mechanical Engineers. 10.1115/MSEC2015-9396

Invernizzi, M., Natale, G., Levi, M., Turri, S., & Griffini, G. (2016). UV-assisted 3D printing of glass and carbon fiber-reinforced dual-cure polymer composites. *Materials (Basel)*, *9*(7), 583. doi:10.3390/ma9070583 PMID:28773704

Kubade, P., & Tambe, P. (2016). Influence of halloysite nanotubes (HNTs) on morphology, crystallization, mechanical and thermal behaviour of PP/ABS blends and its composites in presence and absence of dual compatibilizer. *Composite Interfaces*, *23*(5), 433–451. doi:10.1080/09276440.2016.1144392

Kubade, P., & Tambe, P. (2017). Influence of surface modification of halloysite nanotubes and its localization in PP phase on mechanical and thermal properties of PP/ABS blends. *Composite Interfaces*, *24*(5), 469–487. doi:10.1080/09276440.2016.1235442

Kubade, P. R., Tambe, P., & Kulkarni, H. B. (2017). Morphological, Thermal and Mechanical Properties of 90/10 (WT%/WT%) PP/ABS Blends and their Polymer Nanocomposites. *Advanced Composites Letters*, *26*(6). doi:10.1177/096369351702600602

Kubade, P. R., Tambe, P., & Kulkarni, H. B. (2017). Morphological, Thermal and Mechanical Properties of 90/10 (WT%/WT%) PP/ABS Blends and their Polymer Nanocomposites. *Advanced Composites Letters*, *26*(6). doi:10.1177/096369351702600602

Kulkarni, H., Tambe, P., & Joshi, G. (2017). High concentration exfoliation of graphene in ethyl alcohol using block copolymer surfactant and its influence on properties of epoxy nanocomposites. *Fullerenes, Nanotubes, and Carbon Nanostructures*, *25*(4), 241–249. doi:10.1080/1536383X.2017.1283616

Kulkarni, H. B., Mahamuni, S. S., Gaikwad, P. M., Pula, M. A., Mahamuni, S., Bansode, S. H., … Nehatrao, S. A. (2017). Enhanced Mechanical Properties Of Epoxy/Graphite Composites. *Int. J. Adv. Engg. Res. Studies, 6(1), Oct.-Dec.* 2017: 5.

Kulkarni, H. B., Tambe, P., & Joshi, G. M. (2018). Influence of covalent and non-covalent modification of graphene on the mechanical, thermal and electrical properties of epoxy/graphene nanocomposites: A review. *Composite Interfaces*, *25*(5-7), 381–414. doi:10.1080/09276440.2017.1361711

Letcher, T., & Waytashek, M. (2014, November). Material property testing of 3D-printed specimen in PLA on an entry-level 3D printer. In *ASME 2014 international mechanical engineering congress and exposition*. American Society of Mechanical Engineers. doi:10.1115/IMECE2014-39379

Li, N., Li, Y., & Liu, S. (2016). Rapid prototyping of continuous carbon fiber reinforced polylactic acid composites by 3D printing. *Journal of Materials Processing Technology*, *238*, 218–225. doi:10.1016/j.jmatprotec.2016.07.025

Mahajan, C., & Cormier, D. (2015, January). 3D printing of carbon fiber composites with preferentially aligned fibers. In *Proceedings IIE annual conference* (p. 2953). Institute of Industrial and Systems Engineers (IISE).

Melenka, G. W., Cheung, B. K., Schofield, J. S., Dawson, M. R., & Carey, J. P. (2016). Evaluation and prediction of the tensile properties of continuous fiber-reinforced 3D printed structures. *Composite Structures*, *153*, 866–875. doi:10.1016/j.compstruct.2016.07.018

Ning, F., Cong, W., Hu, Y., & Wang, H. (2017). Additive manufacturing of carbon fiber-reinforced plastic composites using fused deposition modeling: Effects of process parameters on tensile properties. *Journal of Composite Materials*, *51*(4), 451–462. doi:10.1177/0021998316646169

Ning, F., Cong, W., Qiu, J., Wei, J., & Wang, S. (2015). Additive manufacturing of carbon fiber reinforced thermoplastic composites using fused deposition modelling. *Composites. Part B, Engineering*, *80*, 369–378. doi:10.1016/j.compositesb.2015.06.013

Plymill, A., Minneci, R., Greeley, D. A., & Gritton, J. (2016). Graphene and carbon nanotube PLA composite feedstock development for fused deposition modeling.

Tekinalp, H. L., Kunc, V., Velez-Garcia, G. M., Duty, C. E., Love, L. J., Naskar, A. K., ... Ozcan, S. (2014). Highly oriented carbon fiber–polymer composites via additive manufacturing. *Composites Science and Technology*, *105*, 144–150. doi:10.1016/j.compscitech.2014.10.009

Tian, X., Liu, T., Wang, Q., Dilmurat, A., Li, D., & Ziegmann, G. (2017). Recycling and remanufacturing of 3D printed continuous carbon fiber reinforced PLA composites. *Journal of Cleaner Production*, *142*, 1609–1618. doi:10.1016/j.jclepro.2016.11.139

Tian, X., Liu, T., Yang, C., Wang, Q., & Li, D. (2016). Interface and performance of 3D printed continuous carbon fiber reinforced PLA composites. *Composites. Part A, Applied Science and Manufacturing*, *88*, 198–205. doi:10.1016/j.compositesa.2016.05.032

Van Der Klift, F., Koga, Y., Todoroki, A., Ueda, M., Hirano, Y., & Matsuzaki, R. (2016). 3D printing of continuous carbon fibre reinforced thermo-plastic (CFRTP) tensile test specimens. *Open Journal of Composite Materials*, *6*(1), 18–27. doi:10.4236/ojcm.2016.61003

Yan, Y., Li, S., Zhang, R., Lin, F., Wu, R., Lu, Q., & Wang, X. (2009). Rapid prototyping and manufacturing technology: Principle, representative technics, applications, and development trends. *Tsinghua Science and Technology*, *14*(S1), 1–12. doi:10.1016/S1007-0214(09)70059-8

Yap, Y. L., Dikshit, V., Lionar, S. P., Yang, H., Lim, J. C., Qi, X., ... Wei, J. (2016). Investigation of fiber reinforced composite using multi-material 3d printing. In *Annual International Solid Freeform Fabrication Symposium*, Austin, TX.

Zhong, W., Li, F., Zhang, Z., Song, L., & Li, Z. (2001). Short fiber reinforced composites for fused deposition modeling. *Materials Science and Engineering A*, *301*(2), 125–130. doi:10.1016/S0921-5093(00)01810-4

Chapter 18
Thermal Characterization and Improvement of Curing Stage in Resin Transfer Molding Process

Soukaina Elyoussfi
Ibn Tofail University, Morocco

Aouatif Saad
Ibn Tofail University, Morocco

Adil Echchelh
Ibn Tofail University, Morocco

Mohamed Hattabi
Hassan II University, Morocco

ABSTRACT

Resin Transfer Molding has become one of the most efficient processes to manufacture composite parts. Among the steps in composite part processing is the curing reaction. In the majority of cases, this reaction is of exothermic nature accompanied by a rise in temperature in the laminate. This leads to the appearance of a thermal gradient. This research aims to study the thermal gradient generated. The objective is to minimize the temperature excess in the composite. By means of a one-dimensional numerical study using the finite differential method, we have showed that the energy balance depends not only on the temperature and on the degree of curing but also on several other factors, namely: the volume fraction of the fibres, the temperature cycle, and the reinforcement thickness. Authors have shown in this study the effect of increasing temperature on the optimization of the curing cycle. The chapter also investigated the effect of thickness variation on temperature distribution in the composite. A comparison of the authors' results with literature achievements showed agreement.

DOI: 10.4018/978-1-7998-0117-7.ch018

INTRODUCTION

The importance of materials from industry requires the use of composite materials with good properties whose design respects the environment and contributes to sustainable development with the maintenance of a growing economy. Composite materials are known for their rigidity, their high strength and shaping of complex parts and reduction of the number of interfaces (Saad et al. (2011).), they take an increasingly important place in the industry. Already known and used in several fields of application: Aeronautics, automobile, health, sport …

Among the shaping of thermoset matrix composites, resin transfer molding (RTM), this latter is considered one of the most promising techniques available today, it is able to make a large part complex with a high mechanical performance, tolerance, dimensional, narrow and very high finish. The manufacturing of a part by the RTM process can be freely divided into four main steps (Choi et al., 1998): the manufacturing of fiber preform, the filling of the mold, the curing reaction and the release of the part. The cure process irreversibly transforms the soft fiber/resin matrix to a hard structure component. It is a critical step in that the temperature and cure histories, and their spatial variation within the layup cross section, during the process directly influence the final quality of composite products (Roger et al., 1987; Wang et al., 2018). At the initiation of a typical process, the outer layers of the laminate, which are subjected to the external heating, cure more rapidly, than the inner layers, whereas, as the cure progress, temperature of the inner layers may exceed these of the outer layers due to the exothermic cure reaction and low thermal conductivity of the composite. Uncontrolled polymerization may cause undesired and excessive thermal variation that could induce microscopic defects in the network structure of the matrix phase, and macroscopic defects such as voids, bubbles and debonded broken fibers (Halpin et al., 1983; Kenny et al., 1989). Processing of polymeric composites is based on thermoset matrixes therefore requires optimization of the cure cycle parameters as well as adequate formation of the reacting system as a function of the geometry of the parts. Mallick has studied the effect of cure cycle time, temperature, preheating and post-cooling on mechanical properties of continuous as well as chopped glass fiber reinforced polyester and vinyl ester systems. Internal heat generation due to curing reaction causes high thermal gradients across the thickness, the flexural and interlaminer shear strengths are strongly dependent on the mold cycle time (Mallick et al., 1993). Barone and Caulk studied the influence of the applied heat on the curing process of epoxy resin and proposed a thermo chemical model based on a two-dimensional heat conduction equation with internal heat generated by the exothermic chemical reaction (Barone et al., 1979).

Several attempts have been made to reduce this thermal gradient by optimizing the energy cycle. We can cite as not exhaustive, the work of Choi et al. (1998) who handled with a temperature profile that represents a constant temperature range, they therefore observed the disappearance of this gradient which is argued by the low heat transfer in this time interval, so the temperature will not exceed its value at the mold wall. Vincenza et al. (2002), who carried out a dimensional study of the energy equation, assumed in their study that the mold is completely saturated and only interested in the curing stage.

Kim et al. (1997), developed an autoclave firing cycle for thick resin matrix composites to reduce temperature overtaking. Guo et al. (2004) worked in their model on the RTM procedure to predict the temperature distribution in a thick composite, they were able to numerically calculate the temperature based on the finite element information of the energy equation in the monodimensional case by neglecting the convective term. The work of Abbassi et al. (2004) and Cheung et al. (2004) makes it possible to demonstrate the existence of a thermal gradient which is more in the center of the laminate and which always appears following the exothermic reaction of curing. Loos and Springer (1983) involved the development of a one-dimensional curing simulation of a flat plate using an implicit finite difference method. In addition, Bogetti and Gillespie (1992) have also used the finite difference method to develop a simulation analysis of two-dimensional polymerization of thick thermosetting composites. Zhang et al. (2009) studied the curing process for epoxy resin in thick pieces, using a three-dimensional finite-element model by transient heat transfer during the polymerization cycle.

Another interesting feature of the RTM process is its potential to become a high-speed process for the manufacturing of structural and semi structural components at a low cost. Therefore, the performance of produced part and reduction of cycle time are the major concerns in this process. The time it takes the resin to wet fiber tows controls both the quality of the products and the rate at which it can be produced in composite manufacturing. In a typical RTM cycle, the resin must quickly fill the mold cavity and wet all individual fibers before much reaction occurs. Computer simulation has provided a time and cost-effective tool for optimization of the process and offered important information to design the tooling and prediction of processing conditions, tracking the resin front and identifying hard-to-fill regions and regions of possible void formation. In the past decade, an extensive effort has been made to simulate the RTM process (Bruschke et al., 1990; Young et al., 1991; Chan et al., 1991; Trochu, et al., 1993; Lee et al., 1994; Bruschke et al., 1994; Kang et al., 1995; Yoo, et al., 1996; Mal et al., 1998; Lam et al., 2000; Yi et al., 1997. Of these, the methods based on finite element and control volume are the most popular to solve the filling stage because of their simplicity in handling the moving boundary problems. In these techniques, a fixed grid approach is used in which there is no need to regenerate the mesh during the flow progression. This makes the simulation rapid and effective for complicated geometries compared to moving grid approach.

The main objective of this study is on one hand gain a fundamental understanding of the unique curing process for 1 cm thick thermoset composite parts. Subsequently, a method was defined to optimize curing and a simple procedure was developed to predict the temperature profile and degree of curing during the resin polymerization step in the RTM process using the one-dimensional finite difference method for polyester/glass composite laminate. The generality of the approach is explored by changing conditions at the boundaries of the mold walls, the effects of fiber density fraction, the nature of fiber, and the effect of thickness variation on the thermal gradient. On the other hand, an attempt is made to optimize the curing cycle in resin transfer moulding process using numerical simulation by CV/FEM method. This will be done by the study and the analyse of methods enabling minimization of mold-filling time without losing the part quality.

MATHEMATICAL MODEL

The main structure of the master model is formed first by a flow model that describe the filling process of a mold and the impregnation of the reinforcement by a resin, and then by an energy balance which takes into account the accumulation of the heat in the composite, the heat generated by the chemical reaction, the heat conduction in the material and the dissipation at the composite skin. The energy balance equation is coupled with a suitable expression for the kinetic behavior of the chemical reactions accounting for diffusion control effects. The solution of the complete mathematical system gives the flow front position, the pressure in different position of the mold, the temperature and degree of reaction as function of time and position. One of the main advantages of this mathematical model is one is able to give this information, because it is possible but difficult to determine the profile of temperature from experiments and it is quite impossible to measure the profiles of state of cure within laminate.

Flow Model

In modeling the filling process of RTM, the mass balance at a point within the domain for an isothermal incompressible fluid flow inside a fiber preform is expressed as:

$$0 \leq f < 1; \tag{1}$$

It has been observed that liquid flow through porous media in the RTM process can be expressed by Darcy's law which has reasonable agreement with experimental result for low Reynolds number (Das et al., 1983). Because of having low Reynolds number in the RTM process (Re<1), Darcy's law has also become the most commonly used equation for describing the flow within the mold, and is used as the momentum equation. This equation is expressed as:

$$div\vec{V} = 0 \tag{2}$$

Equation (2) can be rewritten in tensor notation as:

$$\vec{V} = -\frac{[K]}{\mu} grad.P \tag{3}$$

where P is the pressure, u, v are the components of the velocity vector, $\begin{pmatrix} u \\ v \end{pmatrix} = -\frac{1}{\mu}\begin{bmatrix} K_{xx} & K_{xy} \\ K_{yx} & K_{yy} \end{bmatrix}\begin{pmatrix} \frac{\partial P}{\partial x} \\ \frac{\partial P}{\partial y} \end{pmatrix}$ is the viscosity of the fluid and [K] is the permeability tensor of the fiber preform. Permeability of the fiber can be isotropic or anisotropic and is a function of fiber type and porosity. In the case of an isotropic fiber preform, the permeability tensor is reduced to a single scalar. Darcy's law applies to Newtonian fluids, ignores inertia effect and gives plug flow inside the cavity. Substituting Darcy's law into continuity equation results in an equation, which can be used to solve the pressure field. This equation has been successfully used to simulate the flow of resin in a porous fiber bed, and its validity has been proven by many researchers Advani et al. (1994).

$$\mu \tag{4}$$

Heat Transfer Equation

Since the resin flows through the fibrous medium, the heat transfer takes place between the mold wall, the fiber and the resin in the non-isothermal process. The temperature distribution of the composite structure during curing can be simulated across the thickness using the one-dimensional heat transfer equation that governs it, can be written in expanded form (Halpin et al., 1983; Kenny et al., 1989; Yi et al., 1997).

$$div(\frac{[K]}{\mu} grad.P) = 0 \tag{5}$$

Where: T represents the temperature, z is the direction perpendicular to the fibers through the thickness, t is the time. q˙ is the source term for taking into account the exothermic of the reaction is given as follows Goncalvès (2005):

$$\acute{A}C_p \frac{\partial T}{\partial t} = k\frac{\partial^2 T}{\partial z^2} + \dot{q} = \dot{q} \tag{6}$$

In these equations k, ρ and $\rho_r.\left(1 - v_f\right).\Delta H_r.\frac{d\alpha}{dt}$ are thermal conductivity, density and heat capacity. The expressions of the changes of the thermal properties are as a function of the temperature (T), the degree of curing (α) and the volume fraction (C_p), also are calculated from the properties of the fiber and the resin of composites used.

In order to evaluate the role of each heat transfer equation term, Han et al. (1986) conducted a survey and adjusted the definitions of each term of overall physical properties of the material. In this heat transfer analysis, the three components namely resin, fibers and cured resin are defined as follows:

The mass fractions of the resin (v_f), the cured product (ω_r) and the fibers (ω_p) are respectively given by:

$$\omega_f \; w_r = \tag{7a}$$

$$w_{r0}.\left(1 - \alpha\right) \tag{7b}$$

$$w_p = w_{r0} \; .\alpha \tag{7c}$$

The volume fraction of the resin ($w_f = w_{f0}$), the cured product (v_r) and the fibers (v_p) are respectively given by:

$$v_f \tag{8a}$$

$$v_r = w_r \frac{\rho}{\rho_r} \tag{8b}$$

$$v_p = w_p \frac{\rho}{\rho_p} \tag{8c}$$

The mixing rule has been used in some research programs to obtain the effective properties of the composite Choi et al., 1998. The expressions of density, of specific heat capacity are given, respectively, by:

$$v_f = \frac{\rho}{\rho_f} \tag{9}$$

$$\frac{1}{\rho} = \frac{w_r}{\rho_r} + \frac{w_p}{\rho_r} + \frac{w_f}{\rho_f} \tag{10}$$

With the densities of the resin ($C_p = w_r.C_{p,r} + w_p.C_{p,p} + w_f.C_{p,f}$), the cured product (ρ_r) and the fibers (ρ_p). The same goes for the thermal capacities of the resin (ρ_f), the cured product ($C_{p,r}$)) and the fibers ($C_{p,p}$).

In addition, the thermal conductivity can be calculated by this rule and is related to the fiber volume fraction (Saad et al., 2014). Thus, the thermal conductivity of a composite is decomposed in two cases according to the structure of the fibers Choi et al. (1998), one for the fiber and the resin is formed of a parallel mode, and the other for the fiber and the resin. Is formed of a perpendicular mode, the two expressions are expressed as follows:

$$C_{p,f} \tag{11}$$

$$k_{\parallel} = V_{rp}.k_{rp} + V_f.k_f \tag{12}$$

For a simple formula of thermal conductivity, we have therefore chosen the logarithmic mean of the thermal conductivity of each component to predict the overall thermal conductivity that is written as follows Choi et al. (1998):

$$k_{\perp} = k_{rp} \cdot \frac{\left[k_f + k_{rp} + V_f \cdot \left(k_f - k_{rp}\right)\right]}{\left[k_f + k_{rp} - V_f \cdot \left(k_f - k_{rp}\right)\right]} \tag{13}$$

Chemical Reaction

As part of the modeling of the RTM process, a study of the kinetics of the chemical reaction of the thermosetting matrix is necessary, in order to establish a kinetic model allowing the prediction of the evolution of the degree of conversion and the flow of heat released by the reaction as a function of the thermal history.

The coupling of this kinetic model with the heat equation must then allow the prediction of the evolution of the conversion and temperature fields that define the properties of the reactive system during the process.

The general form of the kinetic models used in the literature describes the state of the reaction simply using the degree of conversion. Under these conditions, assuming that the chemical reaction is thermally activated as in the RTM process; it can be modeled using a differential equation (Saad (2011)).

$$lnk = V_{rp}.lnk_{rp} + V_f.lnk_f \tag{14}$$

It is generally assumed that the contributions of degree of curing and temperature are independent, hence the form:

$$\frac{d\alpha}{dt} = F(\alpha, T) \tag{15}$$

Or the $\frac{d\alpha}{dt} = \sum K_i(T) f_i(\alpha)$ called the kinetic function, each corresponds to a reaction or set of reactions and where $f_i(\pm)$ are the Arrhenius constants:

$K_i(T)$ (16)With $\left(\quad E \quad \right)$ the activation energy and E_i the frequency factor of the reaction or reaction group and R is constant of the perfect gases.

The kinetic model proposed by Kamal and Sourour (1973) has been used by several researchers (Choi et al., 1998; Vincenza et al., 2002; Cheung et al., 2004) to study curing reactions for Polyester/ Glass composite laminate . The rate equation for curing reactions is defined as (Choi et al., 1998; Cheung et al., 2004):

$$A_i \tag{17}$$

With: $\frac{d\alpha}{dt} = (K_1 + K_2\alpha^m)(1-\alpha)^n$ and $K_1 = A_1.\exp\left(-\frac{E_1}{RT}\right)$

Where T is the temperature, $K_2 = A_2.\exp\left(-\frac{E_2}{RT}\right)$ indicates The rate of the curing reaction $\frac{d\pm}{dt}$ and K_1 are reaction rate constants. K_2 and E_1 present the pre-exponential factors. R the Universal Gas Constant, m and n reaction orders.

Equations (6) and (7) were solved numerically without dimension using the explicit finite difference method. For the numerical calculation of the Polyester / Glass laminate, experimentally determined values of the kinetic parameters described in the articles found in the literature (Choi et al., 1998) have been used.

Boundary Conditions

The boundary conditions are divided into three regions: mold wall, inlet gate and flow front. At the mold walls there is no flow in the normal direction and the condition is expressed as:

E_2

$$\begin{cases} P = P_{inj}, \\ P = P_{front}, \\ \frac{\partial P}{\partial n} = 0, \end{cases}$$

Where f is the fill factor, it represents the ratio of occupied volume by the resin to its total pore volume for each control volume.

Equation (5) is solved as a function of the conditions at the boundary of the temperature of the mold wall, the lateral wall and the center of the mold (Saad et al., 2014). The possible boundary conditions could be:

Initial condition (t = 0):

$$\begin{cases} P \geq 0 \\ P = 0 \\ 0 \leq f \leq 1 \end{cases} \tag{10a}$$

The side wall:

$$T = T_0 \tag{10b}$$

In the center of the wall:

$$\mathrm{T} = \mathrm{T}_{\mathrm{w}}\left(\mathrm{t}\right) \tag{10c}$$

For the polymerization of Polyester / Glass composite laminate, the initial condition (Cheung et al., 2004): $\frac{\partial T}{\partial n} = 0$ α0 à t= ± =

NUMERICAL METHOD

The Finite Difference Method

The explicit finite difference method is used in our case to discretize both equations (5) and (17). The finite difference method is preferred to an implicit method because a substantial reduction in the calculation time is achieved (Bogetti et al., 1992; Kamal et al., 1973). The finite difference method is less sensitive to calculation errors (Kamal et al., 1973).

The discretization of the one-dimensional heat equation by the finite deference method with an explicit scheme was analyzed. Three-Point Central differentiation formulae are used (Saad et al., 2012) The second-order derivative in Equation (1) is approximated by Twardowski et al. (1995):

$$\mathrm{t}_0 \tag{13}$$

Where index i indicates the position of z.

We approach the relation (5) by a two-tiered scheme in time:

$$\frac{\partial^2 T}{\partial z^2} \cong \frac{1}{\Delta z^2}\left(T^i_{j+1} - 2T^i_j + T^i_{j-1}\right) \tag{14}$$

By substituting equations (13), (14) and (6) in equation (5), the temperatures after a time increment Δt are expressed as follows:

$$\frac{\partial \mathrm{T}}{\partial \mathrm{t}} = \frac{\mathrm{T}^{\mathrm{t}+\Delta \mathrm{t}}_{\mathrm{j}} - \mathrm{T}^{\mathrm{t}}_{\mathrm{j}}}{\Delta \mathrm{t}} + \frac{\mathrm{T}^{\mathrm{i}+1}_{\mathrm{j}} - \mathrm{T}^{\mathrm{i}}_{\mathrm{j}}}{\Delta \mathrm{t}} = \frac{\mathrm{k}}{\Delta \mathrm{z}^2}\left(\mathrm{T}^{\mathrm{i}}_{\mathrm{j}+1} - 2\mathrm{T}^{\mathrm{i}}_{\mathrm{j}} + \mathrm{T}^{\mathrm{i}}_{\mathrm{j}-1}\right) \tag{15}$$

We get:

$$\acute{\mathrm{A}}_{\mathrm{r}}.\left(1 - \mathrm{v}_{\mathrm{f}}\right).\Delta \mathrm{H}_{\mathrm{r}}\left(\frac{\mathrm{d}\pm}{\mathrm{dt}}\right)^{\mathrm{t}+\Delta \mathrm{t}}_{\mathrm{j}} + \mathrm{a}^{*} T^{i+1}_{j} = T^i_j + \Delta t / \Delta z^2\left(T^i_{j-1} - 2T^i_j + T^i_{j+1}\right) \tag{16}$$

With $\rho r.\left(1 - v_f\right).\Delta H_r.{*}\Delta t / \Delta z^2.\left(\frac{d\alpha}{dt}\right)^t_j$.

The stability condition we have used in this work is expressed as follows:

$$: a = \frac{k}{\acute{A}c_p} \tag{17}$$

The same thing for the discretization of the curing degree equation (12):

$$\Delta t \le \Delta z^2 / \left(2.a\right) = \frac{d\pm}{dt}{}_j^{t+\Delta t} \tag{18}$$

We get:

$$\frac{\pm_j^{t+\Delta t} - \pm_j^t}{\Delta t} \; \pm_j^{t+\Delta t} = \left(+\frac{d\pm}{dt}{}_j^{t+\Delta t}\right).\pm_j^t \tag{19}$$

The value of Δt is known because it has been calculated at the previous time step increment and $\pm_j^t$ is known from the expression of the empirically determined rate of curing.

The temperature and degree of curing distributions across the range are calculated at each time step as a function of the temperature history of the curing cycle. Our own model is solved by build up computer program in Matlab. The results of this survey are discussed in the next section.

THE CV/FEM METHOD

In order to solve such a moving boundary problems as the resin flow front advances using the traditional finite element method, it requires the computation domain redefinition and mesh regeneration. Mesh regeneration needs a large amount of computation time as the domain becomes complicated. Alternatively, the control volume finite element method CV/FEM, which forms and solves a set of equations for nodal control volumes as if they were finite elements, does not require mesh regeneration. Thus, the computation is more efficient.

We give here a brief description of the conventional CV/FEM method. To begin the numerical formulation, the mold cavity is first divided into a number of elements, and control volumes are constructed around the nodal points.

The control volume formulation is illustrated using the three-node triangular element configuration. As shown in Figure 1, each three-node triangular element is divided into three sub-areas by connecting the centroid to the midpoints of all three sides. A control volume is composed of six sub-areas, which have a common node at the center of the control volume.

At the flow front, a parameter f is used to represent the status of each control volume in the flow domain. If the control volume has not been occupied by the fluid, f is equal to zero. If the control volume is partially filled, f is equal to the volume fraction of the fluid occupying the control volume. f factor is set to 1 if the volume is completely filled by advancing fluid. The control volumes with f values varying between 0 and 1 are considered flow front elements. The pressure in these partially filled flow front

Figure 1. Schematic formation of control volume

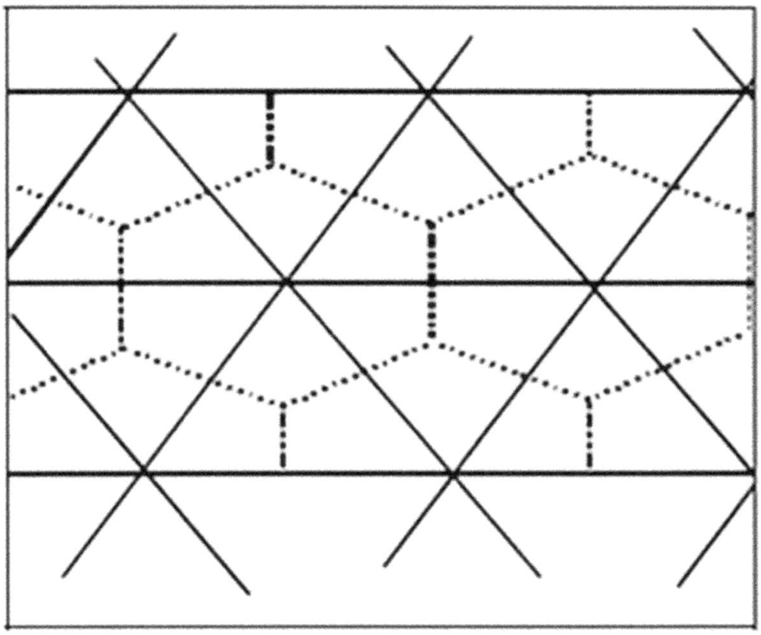

control volumes is set to the ambient pressure. With the aforementioned boundary conditions, the set of linear algebraic equations can be solved to determine the pressure field at each time step during mold filling. Based on the calculated pressure field, the velocity field can then be computed using Darcy's law.

In this approach, the finite element method is used to approximate the fluid pressure field at the nodes of the finite elements at each time step. The pressure is calculated using the Galerkin procedure, applied to the equation (4) as follows:

$$\frac{d\pm}{dt_j}^{t+\Delta t} \quad (20)$$

$$\int_{\Omega} \phi \, \nabla (K \nabla P) \, d\Omega = 0 \quad (21)$$

Where $-\int_{\Omega} \nabla \phi \left(K \, \nabla P \right) \, d\Omega + \int_{\Gamma} \phi \left(K \nabla P \right) \vec{n} \, d\Gamma = 0$, $\Omega = \bigcup_{i=1}^{k} \Omega_i$ is the domain of one element and k is the number of all elements, Ω_i is the outside normal to the domain, $\vec{n}$ is the surface of the element.

For all basis functions Γ, the pressure P has the decomposition:

$$\phi_i \quad (22)$$

The equations can then be written as follows:

$$P(x,y,t)=\sum_{j=1}^{n}P_j(t)\phi_j(x,y) \tag{23}$$

With:

$$\left[K_{ij}^{\ e}\right]\left\{P_j^e\right\}=\left\{f_j^e\right\} \qquad i,j=1,......,n \text{ and } K_{ij}^{\ e}=\int_{\Omega}K\nabla\phi_i\nabla\phi_j\,d\Omega=0 \tag{24}$$

Equation (25) is the working equation to solve the nodal pressure and can be applied to all elements within the solution domain yielding a set of linear algebraic equations by applying an appropriate assembly procedure. In matrix notation, it can be expressed as:

$$f_j^e=\int_{\Gamma}\phi\left(K\nabla P\right)\vec{n}\,d\Gamma \tag{25}$$

After the assembling of the stiffness matrix and imposing the boundary conditions, the system of equations is solved for unknown nodal pressures using an appropriate solver.

Using calculated pressure, the velocities are then calculated at the centroid of each element using Darcy's law (eqt. 2). After the velocity of the resin is obtained, the control volume technique is used to track the flow front position.

RESULTS AND DISCUSSIONS

Validation of the Numerical Method

Consider a flat rectangular mold with dimensions of 10 cm of languor and 1 cm of thickness. The mold is filled with Glass fiber preforms and saturated with Polyester resin and which is heated through the mold wall to initiate curing. For the Polyester/Glass composite laminate, we have adopted in our work the kinetic parameters obtained from the work of Choi et al. (1993) shown in Table 1, and the thermo-physical properties shown in Table 2.The composite laminate was subjected to a prescribed temperature cycle for which results were available (Choi et al., 1993). We opted in our work the temperature cycle applied to the mold wall, which they used by Choi et al. (1993) to initiate the curing to the composite laminate, which illustrates in figure (2).

Figure 3 shows the temperature distribution as a function of time at the center and at the surface of the plate for the Polyester/Glass composite laminate. According to these results, there is a large temperature overshoot at the center of the Polyester/ glass composite laminate, with an exothermic peak temperature occurring of 200 s and ending at t=600 s, resulting in a thermal gradient at the center of the composite laminate. Figure 4 shows the evolution of the degree of cure according to the temperature profiles shown in Figure 3. As illustrated from figure 4, the degree of cure ranges from 0 to 0.99 to a short period of 0 to 800 seconds. However, after 500 seconds, the reaction rate was faster in its center than in the laminate wall and complete curing was first carried out in the center. Due to the contribution of the thermal conductivity of the fiber the temperature at the center of the laminate rapidly reaches the external imposed temperature, and increases as a consequence of the imbalance between the rate of heat generation by exothermic reaction of the resin and the heat removed by the composite, because of the low thermal conductivity of the resin and fiber. The increase in the temperature at the center increases the resin reaction rate, which in turn generates more heat. The resulting 'exotherm' can cause severe damage to the part if not controlled properly. This behavior agrees with the results obtained by Choi et al. (1993), Kenny et al. (1989), (1990), and El Brouzi, Bouzon, & Vergnaud (1989).

Strategies for Reducing Thermal Gradients

Optimization Effect of the Temperature Cycle

Several researchers have proposed an approach to reduce and control thermal gradients in the RTM process for Polyester/Glass composite laminate by optimizing conditions at the edges of the mold Wall (Choi et al., 1993; Vincenza et al., 2002; Kamal et al., 1973). Choi et al. (1993) manipulated with a temperature profile that represents a constant temperature range, they observed therefore the disappearance of this gradient that is argued by the low heat transfer in this time interval, so the temperature will not exceed its value at the mold wall. Vincenza et al. (2002) defined a curing optimization methodology and developed a very simple procedure to regulate the wall temperature, which induces uniform activation of the polymerization reaction before the resin freezes. The optimization of the curing process of the

Table 1. Cure kinetics for Glass / polyester

$A\ P = F$ $k_1\ (1/s)$ $k_2\ (1/s)$ (cal/M) E_1 (cal/M) n m E_2 H (cal/M)	5,68667 Δ 8,61167 10^{12} **25570** **17930** **1,42** **0,58** **47,8**

Table 2. Material properties of Glass / polyester

Materials **Uncured Polyester** **Cured Polyester** **Class fiber**	10^8 (g/ A) **1,10** **1.20** **2.54**	cm^3 (cal/g.K) **0.45** **0.45** **0.199**	**k(cal/cm.s.K)** 4,05 C_p 1,54 $.10^{-4}$ +9,46 $.10^{-4}$ 2.07 $.10^{-7}\ T$

Figure 2. wall temperature cycle for Polyester / Glass composite laminate

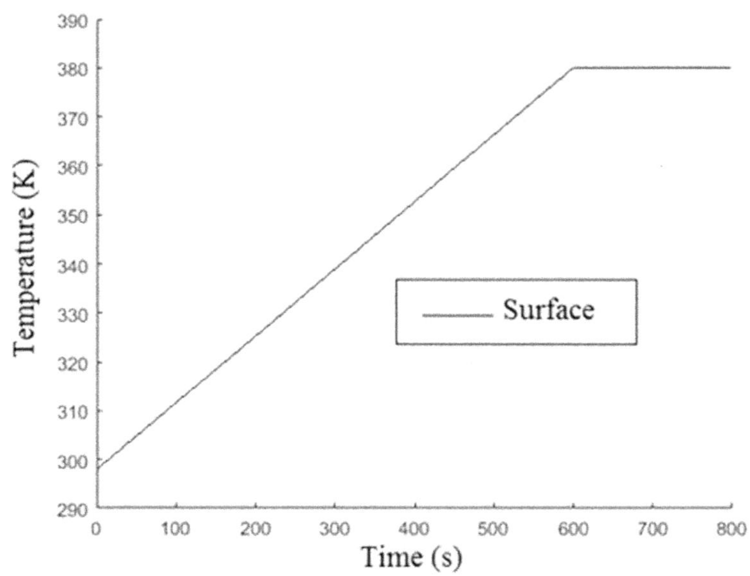

composite is based on the minimization of thermal gradients in the thickness of the composite at the moment the resin hardens. The only variables to lead the process to the scenario are the control of the curing kinetics or the temperature adaptation of the boundary conditions to influence the temperature distribution in the thickness of the composite Vincenza et al. (2002).

Figure 3. Temperature profile of the central node for the Polyester/ Glass plate

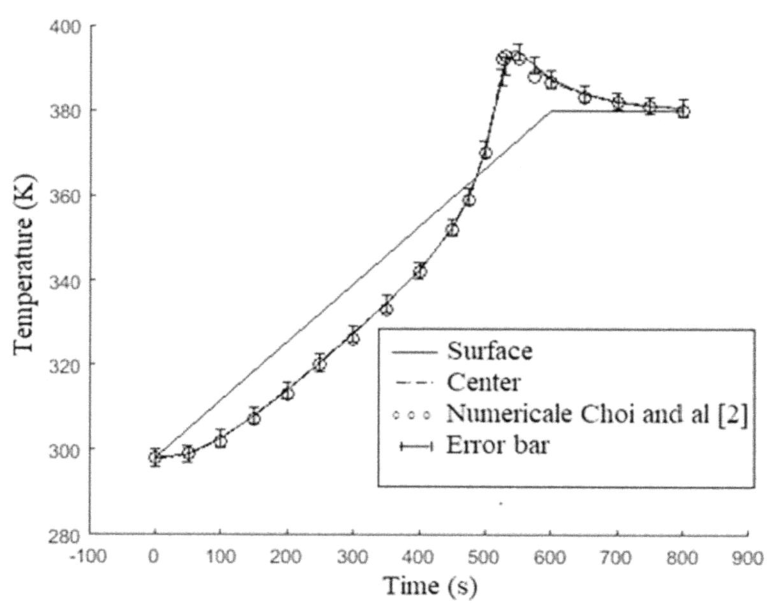

Figure 4. Degree of cure of the central node for a Polyester/Glass laminate

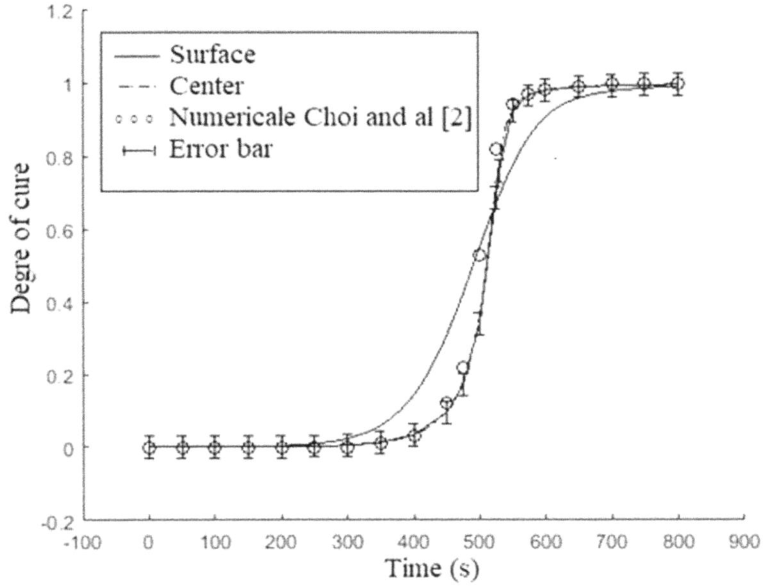

To overcome this problem of the appearance of the thermal gradient we have developed a computer code using the one-dimensional finite difference method to minimize the thermal gradient. Figure 5 shows the temperature cycle optimization effect for the Polyester/Glass composite laminate as a function of time. Figure 6 shows the degree of curing of the composite laminate as a function of time for the same laminate.

Figure 5. Temperature profile of the central node for the Polyester / Glass plate in the case of an optimized cure cycle

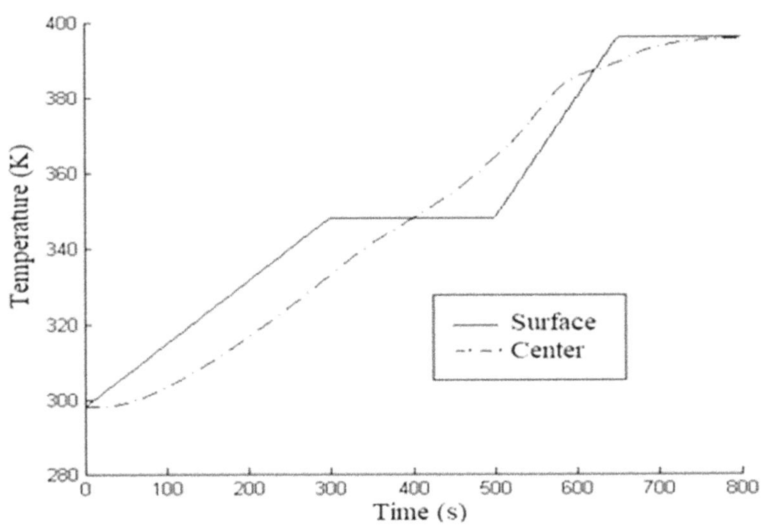

Figure 6. degree of curing of the central node for Polyester /Glass plate in the case of an optimized curing cycle

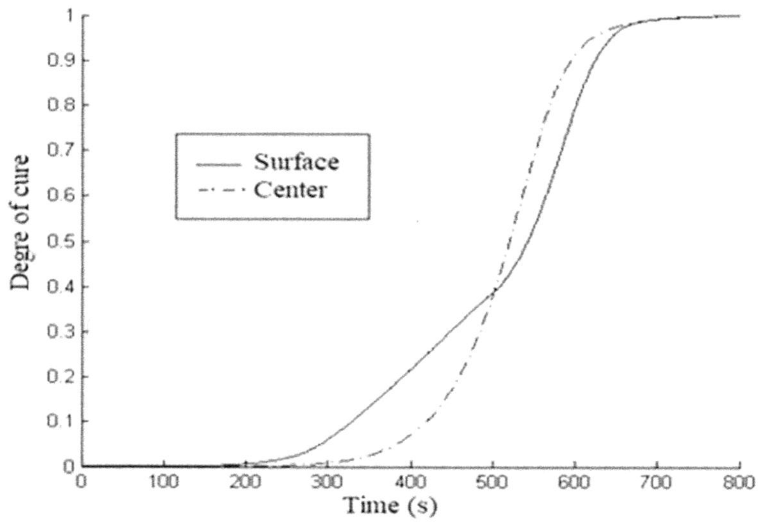

The results as presented in figures 5 and 6 show satisfactory results. No thermal gradients are observed along the room. In addition, the temperature at the center of the composite laminate did not exceed the limiting temperature, so the highest temperature was observed at the surface of the laminate.

Effect of Fiber Volume Fraction

One of the most important parameters affecting the properties of composites is the fiber density fraction. Choi et al. (1993) investigated the effect of variation in fiber content on the temperature profile and degree of curing. Figure 7 shows the temperature distribution at different Glass fiber density fractions as a function of time. From these results, we find that the ratio of fiber to resin content has a marked effect on the reduction of the thermal gradient at the center of the laminate. Note the higher the fiber density fraction, the lower the exothermic reaction peak of the temperature and vice versa (Figure 7). The same is true for the distribution of degree of curing as a function of time observed in Figure 8, the greater the volume fraction of the fiber, uniform the curing reaction becomes.

Effect of Thickness Variation

In view of the increasing importance of the use of thick composite parts, in particular the growing demand in fields such as the Navy, Aeronautics, civil and military structures that require parts of a thickness often exceeding 2 cm (Saad et al., 2011). Curing of the thick sections remains a problem affecting the temperature evolution at the center of the composite laminate, which can result in a large thermal gradient, residual stress generation, and degradation of the composite laminate. Twardowski et al. (1993) have worked in their model on the influence of thickness on temperature and curing gradients for parts of thickness> 5 cm. Kim et al. (1997) have also studied a thick thermoset resin Matrix autoclave cycle and developed to predict the temperature distribution of thick thermoset Composite laminates during

Figure 7. temperature profile of the central node for the Polyester /Glass composite laminate in the case of a curing cycle optimized at different fiber density fractions

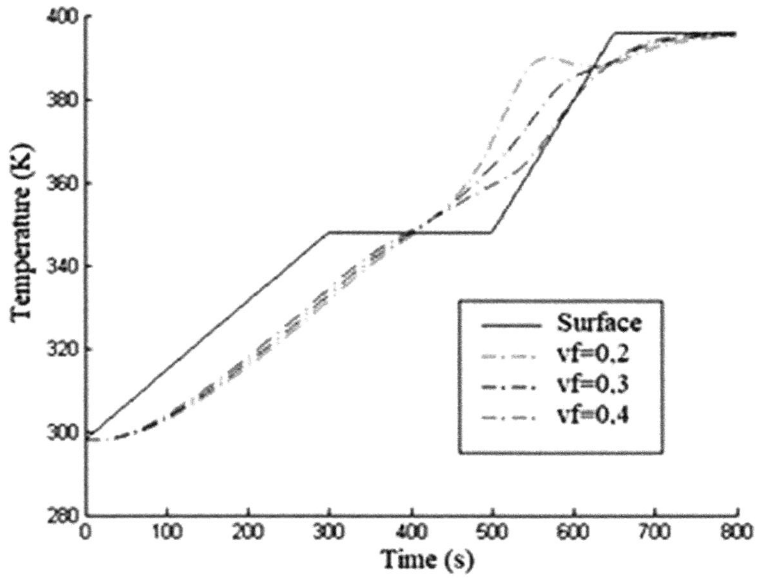

curing. In our case, we will keep the same model as before (Choi et al., 1993), but we must increase the

Figure 8. degree of curing of the central node for Polyester /Glass plate in the case of a curing cycle optimized in different cases of fiber density fraction

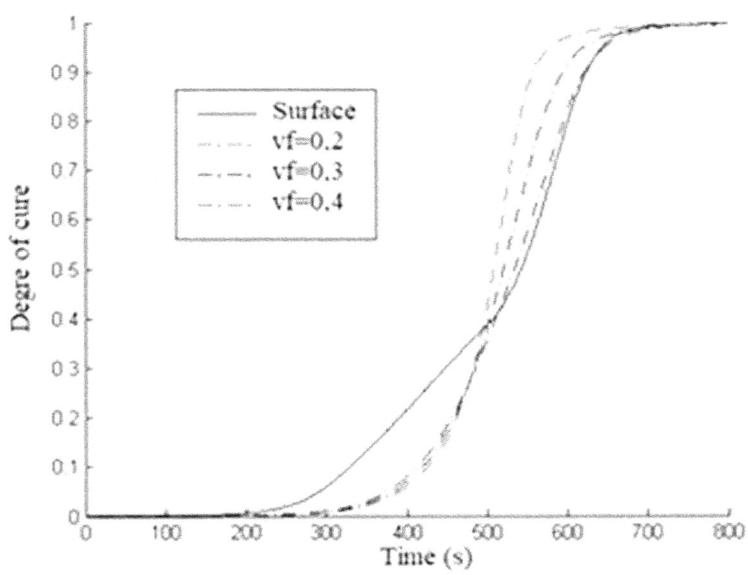

thickness of the laminate composite by about 2 cm. The temperature distribution and degree of curing will be studied numerically using MATLAB. The changes in temperature profiles and degree of cure are illustrated in Figures 9 and 10.

Figures 9 and 10 emphasize the effect of increasing the thickness in the mold, it appears that the range of exothermic peak temperature in 1 cm of thickness is between 493 and 728 seconds and for a thickness of 1.5 cm, the range of exothermic peak is between 586 and 1200 seconds. With regard to the thickness of 2 cm, the exothermic peak range of the temperature is between 697 and 1200 seconds, which means that the increase in thickness in the mold gives a wider exothermic peak range. On the other hand, the reaction time for the 2 cm and 1.5 cm temperature profiles is greater than that for the 1 cm thick temperature profile. Thus, the increase in thickness results in a complex temperature gradient that is sharper. If we compare the temperature evolution for the three curves, we note that the thickness of 2 cm, the temperature gradually increases until T =410K then decreases before being constant.

It was found in this study that the greater the thickness, the more pronounced the temperature gradients; therefore, the residual stresses are amplified. On the other hand, the increase in thickness variation increases by one-step.

Effect of the Fibers Nature

Consider the same mold used before (1 cm thick and 10 cm long), in this part we will focus on the effects of the fibre nature on the temperature profile and on the thermal gradient. We will keep the same resin, but we will change the fiber. We are replacing fiberglass with carbon fiber. The thermal properties of the reinforcements used in this study are summarized in table 3. Figure 11 shows two types of materials: Polyesters/Glass and Polyester/Carbon. From the code used on MATLAB, we obtain the results of temperature distribution and degree of curing as a function of time shown in Figure 12.

Figure 9. Temperature profiles according to the thickness of the composite

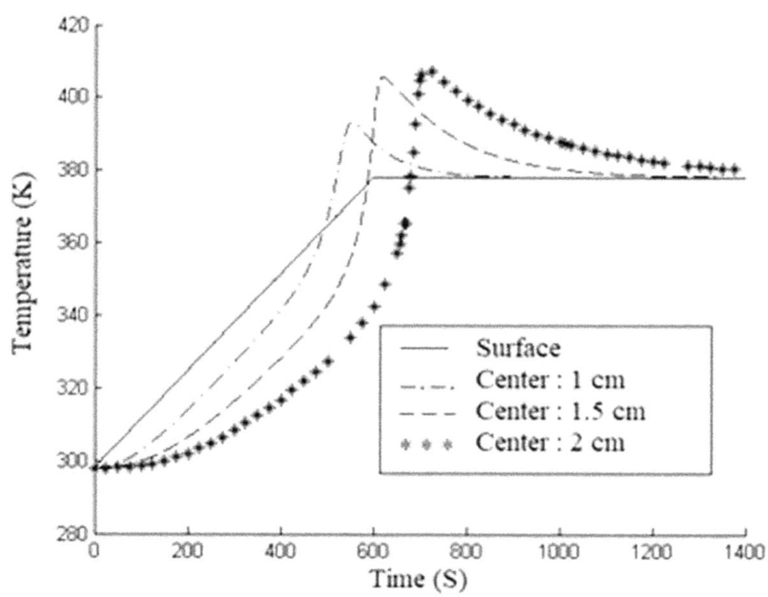

Figure 10. Profiles of degree of curing as a function of the thickness of the composite

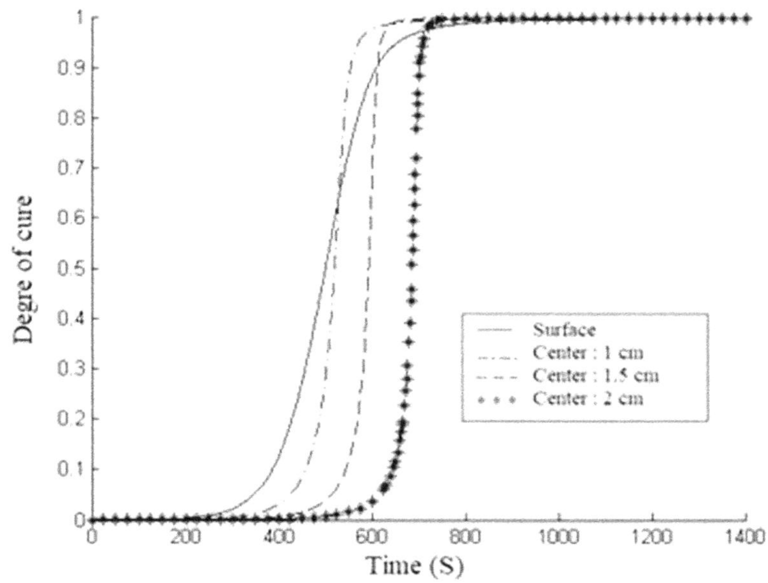

The choice of reinforcement for example Carbon fiber or Glass fiber, has a marked effect on the temperature profile during firing (Sébastien), as illustrated in Figure 11 a large temperature difference in the center of the laminate between Polyester/Glass and Polyester/Carbon composite for 1 cm thickness. Of course, we distinguish from Table 3 that carbon fiber has a higher thermal conductivity than fiber Glass. By contrast, the density and thermal capacity are higher for the fiber Glass compared to Carbon fiber.

On the other hand, we can summarize from Figures 11 and 12 that fiber Glass compatible with Polyester resin more than that of carbon fiber. Also that the physical thermo parameters of the fiber play a very important role on the temperature profile during cooking.

Optimization of the Cycle Time in RTM Process

Increasing of Temperature

In order to reduce the processing time, one of the widely used practices is to elevate the resin temperature by heating the mould. The viscosity is not a constant but follows the power-law model, for a polyester resin the temperature dependence of the resin viscosity is given by the law (Gonzalez-Romero (1985)):

Table 3. Thermal properties of glass/polyester composite and carbon/polyester composite

Materials	$.10^{-3}$ (g/ A)	cm^3 (cal/g.K)	k(cal/cm.s.K)
Class fiber	2.54	0.199	2.07 C_p
Carbon fiber	1.8	0.1701	0.0481

Figure 11. Effect of the choice of reinforcement on the temperature profile

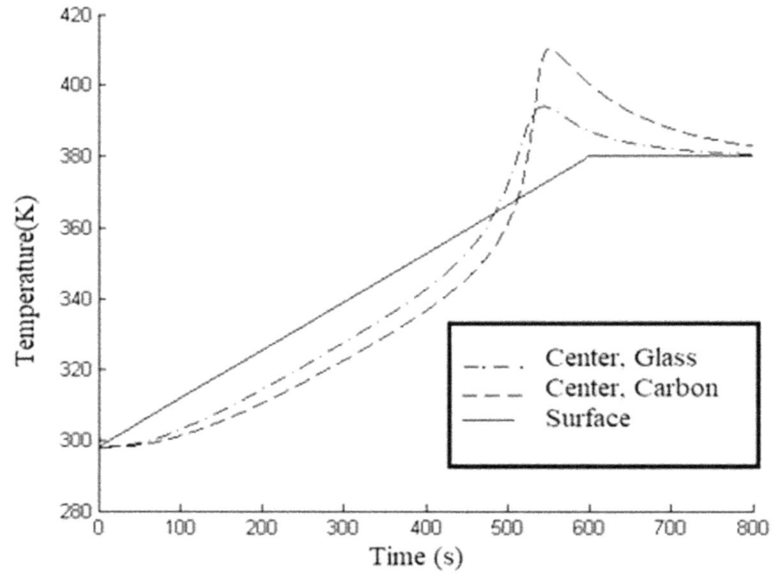

$.10^{-3}$ (26)

Where $\mu = \mu_0.\exp(\frac{\alpha_T}{T})$ and μ_0 are material properties.

Figure 12. Effect of the choice of reinforcement on the degree of curing profile

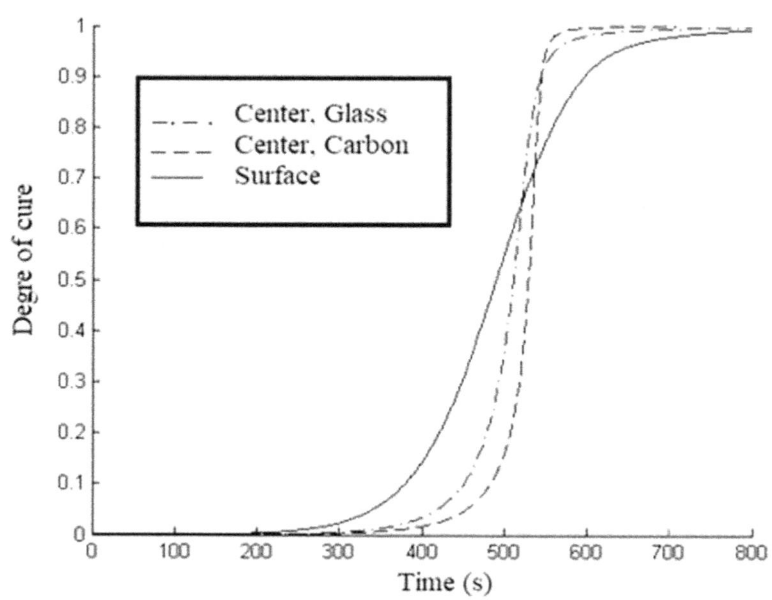

If the resin is heated the viscosity decreases and thus the flow speed is enhanced. Figure (13) show clearly that the viscosity decreases with an elevation of temperature. This contributes to a good infiltration of the resin inside fiber tows and to an increase of the resin velocity witch consequently leads to a minimization of the cycle time of the resin as shown in figure (14).

Hence, the heated mould reduces the total manufacturing time by accelerated curing as well as the reduced resin viscosity. Indeed, the flow front location change with an elevation of the temperature (Figure 15), this can be explained by the fact of the resin with higher temperature advance ahead of that with lower temperature due to its small viscosity and higher velocity.

Two important advantages of increasing temperature of the resin, in the one hand this allows one to enhance the mechanical properties of the final composite part by increasing fibre volume fraction, and in the other hand to minimize the processing cost by decreasing the time of the process.

However, if the resin temperature is raised excessively, the gelation of resin may be reached too fast and the resin flow may prematurely stop before the completion of mould filling.

Optimization of Injection Gate Positioning

RTM process is devoted to produce pieces with complex geometries. In the industry of the composite, the plates employed often consist of reinforcements with a variable number of plies and stacking sequences. A correct simulation of this process requires taking into account all these parameters.

Figure 16 and 17 present our numerical results, where we have proceeded to a filling of three mould cavity with uniform thickness and variable thickness of reinforcement (case 1: Thickness 1 <Thickness 2) and (case 2: Thickness 1 >Thickness 2).

One can deduce from the results presented in figure (8 and 9) that in the mold with variable thickness of reinforcement, the flow front is retarded with regards to the one with uniform thickness of reinforce-

Figure 13. Viscosity variation with temperature

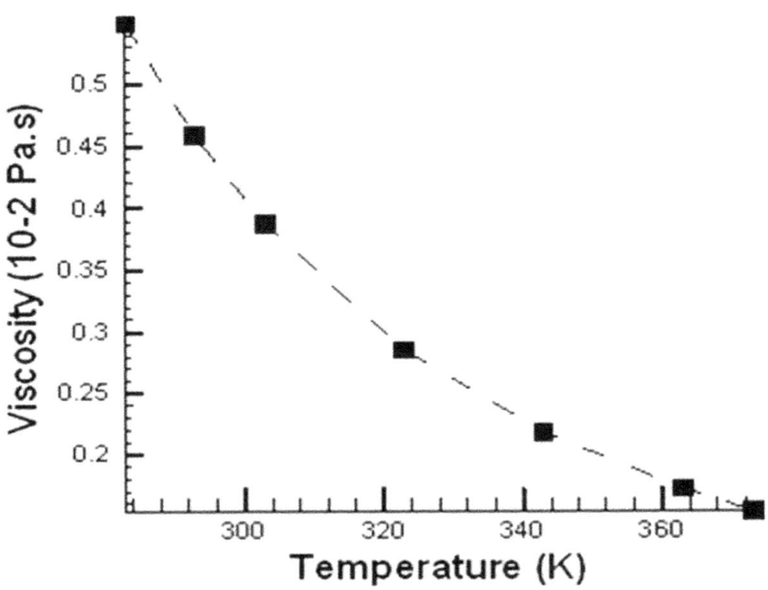

Figure 14. Time process evolution with temperature

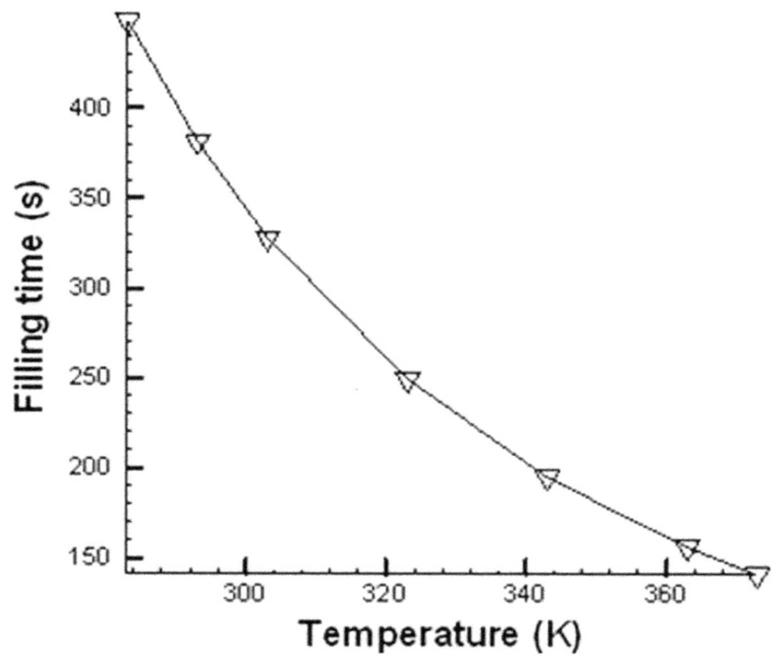

Figure 15. Flow front position with different temperature of the mould

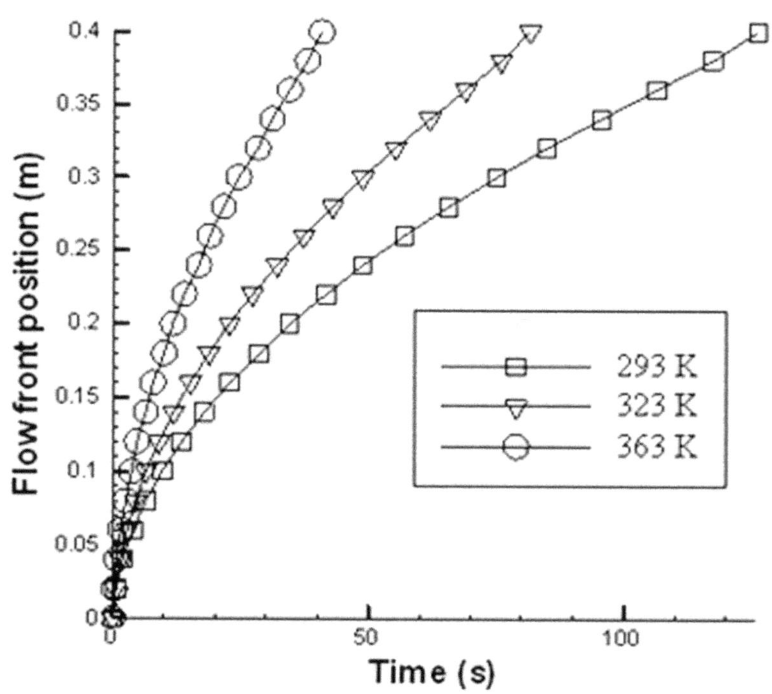

Figure 16. Comparison of flow front position in a mold with variable and uniform reinforcement thickness Thickness 1 < Thickness 2

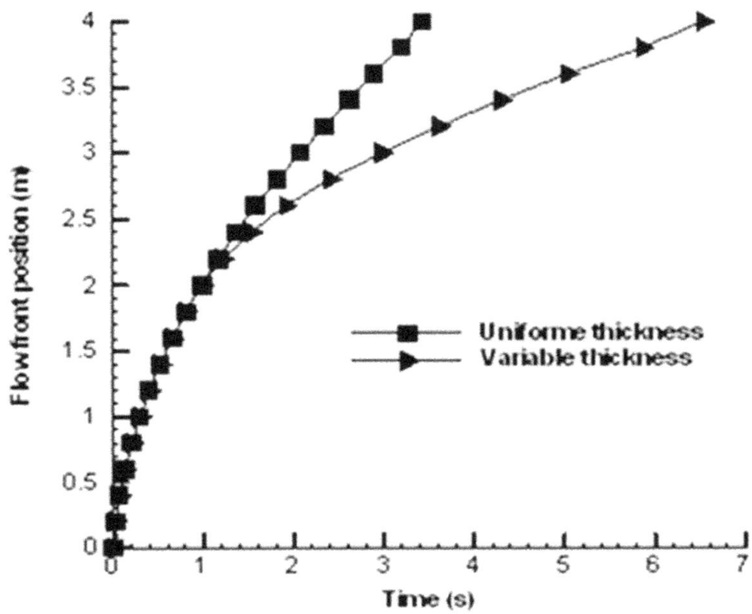

ment, indeed the flow front in the first case reaches the outlet of the mould while in the later it is still inside the mould.

This increase in the process time is principally due to the fact that when the thickness of the reinforcement is increased the volume of the total pore in the mold decreases when the mold is closed, by consequent the permeability of the reinforcement become smaller leading automatically to increase the resistance proved by the fiber to the flow advancement and thus to an increase in the time of mould filling.

This thickness variation leads finally to an increase in the time of the process "the cycle time", where it is found that this time is equal to 1613.6s (case 1: Thickness 1 <Thickness 2) and 3238.2s (case 2: Thickness 1 <Thickness 2) against 807.09s in the case of uniform thickness. This variation in the time of filling in the two cases is explained by Saad et al. (2012), who imputed this to a variation of the pressure at the interface of the two thickness variable regions.

As a consequence, to optimize the filling process of a mold with two different reinforcement thicknesses, the injection ports must be located at the surface with lower thickness (case1) in order to minimize the cycle time and to reduce processing cost of the composite part.

CONCLUSION

In this work, we have developed a program of calculation by a one-dimensional finite differences method applied to the case of Polyester / Glass composite at the macroscopic scale, this code has the ability to predict the temperature profile and the degree of curing in the mold. It was found that the temperature at the central of the laminate increases up to the external imposed temperature, because of the low thermal

Figure 17. Comparison of flow front position in a mold with variable and uniform reinforcement thickness Thickness 1 > Thickness 2.

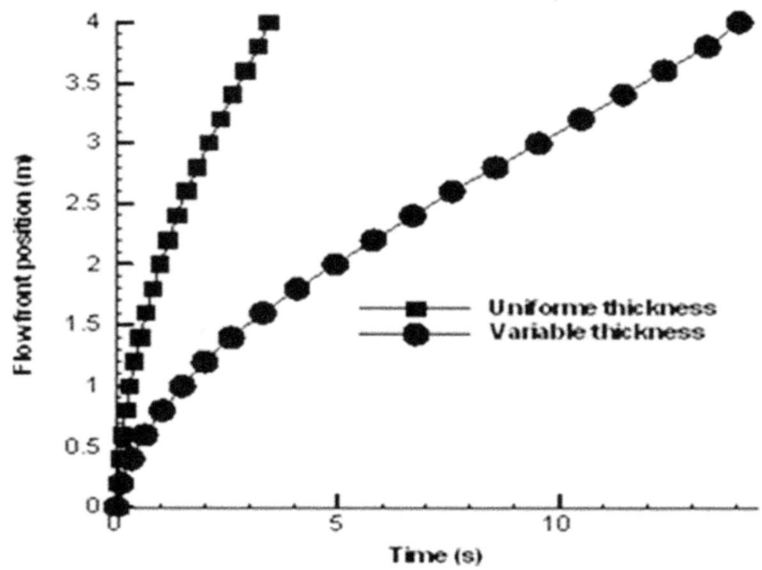

conductivity of the resin and fiber. The heat generated by the exothermic reaction of the resin is not adequately removed; the increase in the temperature at the center increases the resins rate reaction, which in turn generates more heat. The resulting 'exotherm' can cause severe damage to the part if not controlled properly. Our numerical study allows us to highlight and to reduce the thermal gradient apparition in the RTM process and therfore to improve the quality of composite parts manufactured by the RTM process.

Thanks to our numerical simulations, we have found that the thermal gradient, which is detrimental to the quality of the final product, can be reduced considerably through a series of preventive measures, namely:

- The choice of the fibers type used, we have shown that the thermo-physical properties of the materials used play an important role in the evolution of the thermal gradient in the mold.
- The effects of the reinforcement thickness variation in the mold. A reduction in thickness reduces the thermal gradients within the mold.
- Optimization of the temperature cycle in the mold contributes to significantly reduce the thermal gradient.
- The effects of fiber volume fraction. Thus, the increase in fibers fraction volume fraction present in the system led to a minimum thermal gradient in the composite material and vice versa.

With the results obtained by our calculation code, the methodology developed proved to be able to effectively solve the problem of thermal gradients.

We can also summarise from this study, that the preheating of the mould was proven to enhance the product quality and reduce the injection time by a big margin. However, excessive increase in resin temperature can lead to gelation of the resin during mold filling and stop the filling process especially for large parts.

In complex geometries with different reinforcement thickness, optimization of injection ports is an incontournable step to enhance the time process.

These results are in good agreement with the experimental and numerical results available in the literature.

REFERENCES

Abbassi, A., & Shahnazari, M. R. (2004). Numerical modeling of mold filling and curing in non-isothermal RTM process. *Applied Thermal Engineering*, 1–13.

Advani, S. G., & Bruschke, M. V. (1994). Resin Transfer Moulding Flow Phenomena in Polymeric Composites, In S. G. Advani (Ed.), Flow and Rheology in Polymer Composites Manufacturing, Elsevier Science BV, Amsterdam, The Netherlands.

Antonucci, V., Giordano, M., Ting, H. K., & Advani, S. G. (2002). A Methodology to Reduce Thermal Gradients due to the Exothermic Reactions in Composites Processing. *International Journal of Heat and Mass Transfer*, *45*(8), 1675–1684. doi:10.1016/S0017-9310(01)00266-6

Barone, M. R., & Caulk, D. A. (1979). The effect of deformation and thermoset cure on heat conduction in a chopped-fiber reinforced polyester during compression molding. *International Journal of Heat and Mass Transfer*, *22*(7), 1021–1032. doi:10.1016/0017-9310(79)90175-3

Bogetti, T. A., & Gillespie, J. W. Jr. (1992). Process-Induced Stress and Deformation in Thick-Section Thermoset Composite Laminates. *Journal of Composite Materials*, *26*(5), 627–660. doi:10.1177/002199839202600502

Bruschke, M. V., & Advani, S. G. (1990). A finite element/control volume approach to mold filling in anisotropic porous media. *Polymer Composites*, *11*(6), 398–405. doi:10.1002/pc.750110613

Bruschke, M. V., & Advani, S. G. (1994). A numerical approach to model nonisothermal viscous flow through fibrous media with free surfaces. *International Journal for Numerical Methods in Fluids*, *19*(7), 575–603. doi:10.1002/fld.1650190704

Chan, A. W., & Hwang, S. T. (1991). Modeling of the impregnation process during resin transfer molding. *Polymer Engineering and Science*, *31*(15), 1149–1156. doi:10.1002/pen.760311512

Cheung, A., Yu, Y., & Pochiraju, K. (2004). Three-dimensional finite element simulation of curing of polymer composites. *Finite Elements in Analysis and Design*, *40*(8), 895–912. doi:10.1016/S0168-874X(03)00119-7

Choi, M. A., Lee, M. H., & Lee, S. J., (1998). Numerical studies during cure in the resin transfer molding, *Korea polymer journal, 6,* 211-218.

Das, B. M. (1983). Advanced Soil Mechanics. New York, NY: McGraw-Hill.

El Brouzi, A., Bouzon, J., & Vergnaud, J. M., (1989). Plastics and Rubber Processing and Application, *12*.

Goncalvès, E. (2005). *Resolution Numerique*. Discretisation des EDP et EDO, Institut National Polytechnique de Grenoble.

Gonzalez-Romero, J. F., & Mascosko, C. W. (1985). Viscosity rise during radical crosslinking polymerization with inhibition. *Journal of Rheology (New York, N.Y.)*, *29*(3), 259–272. doi:10.1122/1.549790

Guo, Z. S., Du S., & Zhang, B. (2004). Temperature field of thick thermoset composite laminates during cure process. *Composites Science and Technology*, 1–7.

Guo, Z. S. Du S., & Zhang, B. (2004). Temperature distribution of thick thermoset composites, *Modelling and simulation in materials science and engineering, 12*, 443–452 .

Halpin, J. C., Kardos, J. L., & Dudukovic, M. P. (1983). Processing science: An approach for prepreg composite systems. *Pure and Applied Chemistry*, *55*(5), 893–906. doi:10.1351/pac198355050893

Han, C., Lee, D., & Chin, H. (1986). *Development of a Mathematical Model for the Pultrusion Process*, *26*, 393–404.

Kamal, M. R., & Sourour, S. (1973). Kinetic modelling of epoxy-amine cure process. *Polymer Engineering and Science*, *13*, 59–64. doi:10.1002/pen.760130110

Kang, M. K., Lee, W. I., Yoo, J. Y., & Cho, S. M. (1995). Simulation of mold filling process during resin transfer molding. *Journal of Materials Processing & Manufacturing Science*, *3*, 297–313.

Kenny, J. M., Apicella, A., & Nicolais, L. A. (1989). A model for the thermal and chemorheological behavior of thermosets. I: Processing of epoxy-based composites. *Polymer Engineering and Science*, *9*(15), 973–983. doi:10.1002/pen.760291502

Kenny, J. M., Maffezzoli, A., & Nicolais, L. (1990). A model for the thermal and chemorheological behavior of thermoset processing: (II) Unsaturated polyester based composites. *Composites Science and Technology*, *38*(4), 339–358. doi:10.1016/0266-3538(90)90020-6

Kim, J. S., & Lee, D. G. (1997). Development of an Autoclave Cure Cycle with Cooling and Reheating Steps for Thick Thermoset Composite Laminates. *Journal of Composite Materials*, *31*(22), 2264–2282. doi:10.1177/002199839703102203

Lam, Y. C., Joshi, S. C., & Liu, X. L. (2000). Numerical simulation of the mold filling process in resin-transfer molding. *Composites Science and Technology*, *60*(6), 845–855. doi:10.1016/S0266-3538(99)00192-X

Lee, L. J., Young, W. B., & Lin, R. J. (1994). Mold filling and cure modeling of RTM and SRIM processes. *Composite Structures*, *27*(1-2), 109–120. doi:10.1016/0263-8223(94)90072-8

Loos, A. C., & Springer, G. S. (1983). Curing of epoxy matrix composites. *Journal of Composite Materials*, *17*(2), 135–169. doi:10.1177/002199838301700204

Mal, O., Couniot, A., & Dupret, F. (1998). Nonisothermal simulation of the resin transfer molding process. *Composites. Part A, Applied Science and Manufacturing*, *29*(1-2), 189–198. doi:10.1016/S1359-835X(97)00065-1

Mallick, P. K. (1993). Fiber-Reinforced Composites: Materials, Manufacturing, and Design, 2nd Ed. New York, NY: Marcel Dekker.

Wang, D., Deng, Y., Liu, Y. M., & Liu, Y. (2018). Numerical investigation of temperature gradient-induced thermal stress for steel–concrete composite bridge deck in suspension bridges. *Journal of Central South University*, *25*(1), 185–195. doi:10.100711771-018-3728-5

Sébastien Riopel, Etude expérimentale de la cuisson d'un composite unidirectionnel verre/polyester de grande épaisseur. Mémoire présenté en vue de l'obtention du diplome de maîtrise ès sciences appliquées (génie &canique): Université de Montréal.

Roger, J., & Mones, E. T. (1987). *Journal of Applied Polymer Science*, *3*, 999–1020.

Saad, A. (2011). Contribution à l'étude numérique des procédès de fabrication des matériaux composites-cas du moulage par transfert de résine (rtm), thèse de doctorat de la faculté des sciences de Kenitra: Université Ibn Tofail, Maroc.

Saad, A., Echchelh, A., Hattabi, M., & El Ganaoui, M. (2012). Optimization of the cycle time in resin transfer molding process by numerical Simulation. *Journal of Reinforced Plastics and Composites*, *31*(20), 1388–1399. doi:10.1177/0731684412458166

Saad, A., Echchelh, A., Hattabi, M., & El Ganaoui, M. (2012). Numerical simulation of thickness variation effect on resin transfer moulding process. *Journal of Polymer & Composites*, *33*(1), 10–21. doi:10.1002/pc.21226

Saad, A., Echchelh, A., Hattabi, M., & El Ganaoui, M. (2014). The identification of effective thermal conductivity for fibrous reinforcement composite by inverse method. *Journal of Reinforced Plastics and Composites*, *33*(23), 2183–2191. doi:10.1177/0731684414556011

Trochu, F., Gauvin, R., & Gao, D. M. (1993). Numerical analysis of resin transfer molding process by the finite element method. *Advances in Polymer Technology*, *12*(4), 329–342. doi:10.1002/adv.1993.060120401

Twardowski, T. E., Lin, S. E., & Geil, P. H. (1993). Curing in Thick Composite Laminates: Experiment and Simulation. *Journal of Composite Materials*, 27.

Yi, S., Hilton, H. H., & Ahmad, M. F. (1997). A finite element approach for cure simulation of thermo-setting matrix composites. *Civil-Comp and Elsevier Science*, *64*, 383–388.

Yoo, Y. E., & Lee, W. I. (1996). Numerical simulation of the resin transfer mold filling process using the boundary element method. *Polymer Composites*, *17*(3), 368–374. doi:10.1002/pc.10623

Young, W. B., Rupel, K., Han, K., Lee, L. J., & Liou, M. (1991). Analysis of resin injection molding in molds with preplaced fiber mats. II: Numerical simulation and experiments of mold filling. *Polymer Composites*, 128–138.

Zhang, J., Xu, Y. C., & Huang, P. (2009). Effect of cure cycle on curing process and hardness for epoxy resin. *Express Polymer Letters*, *3*(9), 534–541. doi:10.3144/expresspolymlett.2009.67

Compilation of References

Abbassi, A., & Shahnazari, M. R. (2004). Numerical modeling of mold filling and curing in non-isothermal RTM process. *Applied Thermal Engineering*, 1–13.

Abouzid, H., & Chakkor, O. (2017, November). Blind source separation approach for audio signals based on support vector machine classification. In *Proceedings of the 2nd international conference on computing and wireless communication systems* (p. 39). ACM.

Abouzid, H., & Chakkor, O. (2018, April). Dimension Reduction Techniques for Signal Separation Algorithms. In *International Conference on Big Data, Cloud and Applications* (pp. 326-340). Cham, Switzerland: Springer.

Abouzid, H., Chakkor, O., Reyes, O. G., & Ventura, S. (n.d.). Signal speech reconstruction and noise removal using convolutional denoising audioencoders with neural deep learning. *Analog Integrated Circuits and Signal Processing,* 1-12.

Abreu, R., Kamm, J., & Reiß, A. S. (2017). Micropolar modelling of rotational waves in seismology. *Geophysical Journal International*, *210*(2), 1021–1046. doi:10.1093/gji/ggx211

Abrial, J. R., Butler, M., Hallerstede, S., & Voisin, L. (2006, November). An open extensible tool environment for Event-B. In *International Conference on Formal Engineering Methods* (pp. 588-605). Berlin, Germany: Springer. 10.1007/11901433_32

Abrial, J.-R. (2010). Modeling in Event-B: System and Software Engineering. New York, NY: Cambridge University Press. doi:10.1017/CBO9781139195881

Abrial, J. R., Butler, M., Hallerstede, S., Hoang, T. S., Mehta, F., & Voisin, L. (2010). Rodin: An open toolset for modelling and reasoning in Event-B. *International Journal of Software Tools for Technology Transfer*, *12*(6), 447–466. doi:10.100710009-010-0145-y

Achenbach, J. D. (1973). *Wave propagation in elastic solids*. Amsterdam, The Netherlands: North Holland.

Adamiec, M., Dziubiński, M., Siemionek, E., & Drozd, A. (2018, September). Diagnostics of ABS and ESP systems. [IOP Publishing]. *IOP Conference Series. Materials Science and Engineering*, *421*(3).

Aderohunmu, F. A., & Deng, J. D. (2009). *An enhanced stable election protocol (sep) for clustered heterogeneous wsn*. New Zealand: Department of Information Science, University of Otago.

Advanced Design System (ADS). (n.d.). Retrieved from http://www.home.agilent.com/agilent/home.jsp

Advani, S. G., & Bruschke, M. V. (1994). Resin Transfer Moulding Flow Phenomena in Polymeric Composites, In S. G. Advani (Ed.), Flow and Rheology in Polymer Composites Manufacturing, Elsevier Science BV, Amsterdam, The Netherlands.

Agarwal, M., Behera, A. K., & Meshram, M. K. (2016). Wide-angle quad-band polarisation-insensitive metamaterial absorber. *Electronics Letters*, *52*(5), 340–342. doi:10.1049/el.2015.4134

Ahn, D., Park, J. S., Kim, C. S., Kim, J. Y., Qian, X., & Itoh, T. (2001). A design of the low-pass filter using the novel microstrip defected ground structure. *IEEE Transactions on Microwave Theory and Techniques*, *49*(1), 86–93. doi:10.1109/22.899965

Ahn, H.-A., Hong, S.-K., & Kwon, O.-K. (2016). A fast switching current regulator using slewing time reduction method for high dimming ratio of LED backlight drivers. *IEEE Transactions on Circuits and Systems. II, Express Briefs*, *63*(11), 1014–1018. doi:10.1109/TCSII.2016.2548158

Alahmad, M. A., & Hess, H. L. (2008). Evaluation and Analysis of a New Solid State Rechargeable Microscale Lithium Battery. *IEEE Transactions on Industrial Electronics*, *55*(9), 3391–3401. doi:10.1109/TIE.2008.925318

Al-Areqi, N. N., & You, K. Y. (2018). Broadband and compact complex ratio measuring unit. *Microwave and Optical Technology Letters*, *60*(12), 3039–3045. doi:10.1002/mop.31445

Al-Areqi, N. N., You, K. Y., Dimon, M. N., Khamis, N. H., & Lee, C. Y. (2018). *Progress in Electromagnetics Research, C*, 82, 199–207. Miniaturization of three-section branch-line coupler using diamond-series stubs microstrip line.

Al-Karaki, J. N., Ul-Mustafa, R., & Kamal, A. E. (2004). Data aggregation in wireless sensor networks-exact and approximate algorithms. In *HPSR. 2004 Workshop on High Performance Switching and Routing, 2004.* (pp. 241–245). IEEE. 10.1109/HPSR.2004.1303478

Al-Karaki, J. N., & Kamal, A. E. (2004). Routing techniques in wireless sensor networks: A survey. *IEEE Wireless Communications*, *11*(6), 6–28. doi:10.1109/MWC.2004.1368893

Alkurt, F. O., Altintas, O., Bakir, M., Tamer, A., Karadag, F., Bagmanci, M., ... Akgol, O. (2018). Octagonal shaped metamaterial absorber based energy harvester. *Materials Science*, *24*(3), 253–259. doi:10.5755/j01.ms.24.3.18625

Almeida, P. S., Camponogara, D., Dalla Costa, M., Braga, H., & Alonso, J. M. (2015). Matching LED and driver life spans: A review of different techniques. *IEEE Industrial Electronics Magazine*, *9*(2), 36–47. doi:10.1109/MIE.2014.2352861

Almeida, P. S., Dalla Costa, M. A., Alonso, J., & Braga, H. A. C. (2013). Application of series resonant converters to reduce ripple transmission to LED arrays in offline drivers. *Electronics Letters*, *49*(6), 414–415. doi:10.1049/el.2012.4412

Almoneef, T. S., Erkmen, F., Al Dhaeebi, M. A., & Ramahi, O. M. (2018). Dual polarized metasurface for microwave energy harvesting and wireless power transfer. *12th European Conference on Antennas and Propagation (EuCAP 2018)*, 337, 1-3. 10.1049/cp.2018.0696

Almoneef, T. S., & Ramahi, O. M. (2015). Metamaterial electromagnetic energy harvester with near unity efficiency. *Applied Physics Letters*, *106*(15). doi:10.1063/1.4916232

Alonso, J. M., Calleja, A. J., Gacio, D., Cardesín, J., Lopez, E., Dalla Costa, M. A., ... Do Prado, R. N. (2013). High-power-factor light-emitting diode lamp power supply without electrolytic capacitors for high-pressure-sodium lamp retrofit applications. *IET Power Electronics*, *6*(8), 1502–1515. doi:10.1049/iet-pel.2012.0142

Alsanousi, A. A., Ashoka, A., & Jeberson, W. (2013). A Compact Microstrip Low Pass Filter Based on DGS with Shaped Microstrip Line. *IACSIT International Journal of Engineering and Technology*, *3*(6).

Altenbach, H., & Eremeyev, V. A. (Eds.). (2012). Generalized Continua-from the Theory to Engineering Applications (Vol. 541). Berlin, Germany: Springer Science & Business Media.

Amamra, S. A., Ahmed, H., & El-Sehiemy, R. A. (2017). Firefly algorithm optimized robust protection scheme for DC microgrid. *Electric Power Components and Systems*, *45*(10), 1141–1151.

Amari, S. I., Cichocki, A., & Yang, H. H. (1996). A new learning algorithm for blind signal separation. In Advances in neural information processing systems (pp. 757-763).

Ammerman, R. F., Gammon, T., Sen, P. K., & Nelson, J. P. (2010). DC-arc models and incident-energy calculations. *IEEE Transactions on Industry Applications*, *46*(5), 1810–1819.

Andrea, J., Schweitzer, P., & Tisserand, E. (2010, October). A new DC and AC arc fault electrical model. In *2010 Proceedings of the 56th IEEE Holm Conference on Electrical Contacts* (pp. 1-6). IEEE.

Anon. (2019). [online] Available at https://sites.google.com/site/mrstevensonstechclassroom/hl-topics-only/4a-robotics-ai/neural-networks-computational-intelligence.

Anon. (2019). [online] Available at https://www.quora.com/How-do-artificial-neural-networks-work

Antonucci, V., Giordano, M., Ting, H. K., & Advani, S. G. (2002). A Methodology to Reduce Thermal Gradients due to the Exothermic Reactions in Composites Processing. *International Journal of Heat and Mass Transfer*, *45*(8), 1675–1684. doi:10.1016/S0017-9310(01)00266-6

Arif, A., Wang, Z., Wang, J., Mather, B., Bashualdo, H., & Zhao, D. (2018). Load modeling—A review. *IEEE Transactions on Smart Grid*, *9*(6), 5986–5999.

Ariman, T. (1972). Wave propagation in a micropolar elastic half-space. *Acta Mechanica*, *13*(1-2), 11–20. doi:10.1007/BF01179655

Arriola, W. A., Lee, J. Y., & Kim, I. S. (2011). Wideband 3 dB branch line coupler based on λ/4 open circuited coupled lines. *IEEE Microwave and Wireless Components Letters*, *21*(9), 486–488. doi:10.1109/LMWC.2011.2138687

Askar, A. (1972). Molecular crystals and the polar theories of the continua experimental values of material coefficients for KNO3. *International Journal of Engineering Science*, *10*(3), 293–300. doi:10.1016/0020-7225(72)90044-4

Augustine, S., Quiroz, J. E., Reno, M. J., & Brahma, S. (2018). *DC Microgrid Protection: Review and Challenges (No. SAND2018-8853). Sandia National Lab.* Albuquerque, NM: SNL-NM.

Aytouna, F., Zbitou, J., Aghoutane, M., Touhami, N., Tribak, A., & Latrach, M. (2016). A Novel CPW Low Cost Lowpass Filter Integrating Periodic Structures. [IJECE]. *Iranian Journal of Electrical and Computer Engineering*, *6*(3), 1106–1111.

Bahl, I. J. (2009). *Fundamentals of RF and Microwave Transistor Amplifiers*. Hoboken, NJ: John Wiley & Sons. doi:10.1002/9780470462348

Bahl, I., & Bhartia, P. (2003). *Microwave Solid State Circuit Design*. Hoboken, NJ: John Wiley & Sons.

Bakır, M., Karaaslan, M., Unal, E., Akgol, O., & Sabah, C. (2017). Microwave metamaterial absorber for sensing applications. *Opto-Electronics Review*, *25*(4), 318–325. doi:10.1016/j.opelre.2017.10.002

Bakır, M., Karadağ, F., Tetik, E., Bağmancı, M., Altıntaş, O., & Karaaslan, M. (2017). Wideband metamaterial absorber based on CRRs with lumped elements for microwave energy harvesting. *The Journal of Microwave Power and Electromagnetic Energy*, *52*(1), 45–59.

Balanis, C. A. (2005). *Antenna Theory, Analysis and Design* (3rd ed.). New York, NY: John Wiley & Sons.

Bancroft, R. (2006). *Microstrip and Printed Antenna Design*. New Delhi, India: Prentice-Hall of India.

Bannister, D. J., Ide, J. P., & Perkins, M. (1992). Improved six-port reflectometer based on symmetrical five-port junction. *Electronics Letters*, *28*(4), 406–408. doi:10.1049/el:19920255

Barone, M. R., & Caulk, D. A. (1979). The effect of deformation and thermoset cure on heat conduction in a chopped-fiber reinforced polyester during compression molding. *International Journal of Heat and Mass Transfer*, *22*(7), 1021–1032. doi:10.1016/0017-9310(79)90175-3

Basiniuk, U. L., Levantsevich, M. A., Maksimchenko, N. N., & Mardasevich, A. I. (2013). Improvement of Triboengineering Properties and Noise Reduction of Tooth Gears by Cladding Functional Coatings on Working Surfaces of Interfaced Teeth. *Journal of Friction and Wear*, *34*(6), 575–582. doi:10.3103/S1068366613060044

Basri, R. & Chiu, W. K. (2004). Numerical analysis on the interaction of guided Lamb waves with a local elastic stiffness reduction in quasi-isotropic composite plate structures. Composite Structures, 87–99.

Batayneh, W., Jaradat, M., & Bataineh, A. (2018, November). Intelligent Adaptive Control for Anti-Lock Braking System. In *ASME 2018 International Mechanical Engineering Congress and Exposition.* American Society of Mechanical Engineers. 10.1115/IMECE2018-87659

Bayati, N., Hajizadeh, A., & Soltani, M. (2018, June). Impact of Faults and Protection Methods on DC Microgrids Operation. *In 2018 IEEE International Conference on Environment and Electrical Engineering and 2018 IEEE Industrial and Commercial Power Systems Europe* (EEEIC/I&CPS Europe) (pp. 1-6). IEEE.

Bayati, N., Dadkhah, A., Sadeghi, S. H. H., Vahidi, B., & Milani, A. E. (2017, June). Considering variations of network topology in optimal relay coordination using time-current-voltage characteristic. In *2017 IEEE International Conference on Environment and Electrical Engineering and 2017 IEEE Industrial and Commercial Power Systems Europe (EEEIC/I&CPS Europe)* (pp. 1-5). IEEE.

Bayati, N., Hajizadeh, A., & Soltani, M. (2018). Protection in DC microgrids: A comparative review. *IET Smart Grid*, *1*(3), 66–75.

Bayati, N., Hajizadeh, A., & Soltani, M. (2018, September). Accurate Modeling of DC Microgrid for Fault and Protection Studies. *In 2018 International Conference on Smart Energy Systems and Technologies (SEST)* (pp. 1-6). IEEE.

Bažant, Z. P., & Christensen, M. (1972). Analogy between micropolar continuum and grid frameworks under initial stress. *International Journal of Solids and Structures*, *8*(3), 327–346. doi:10.1016/0020-7683(72)90093-5

Bekasiewicz, A., & Koziel, S. (2015). Miniaturised dual-band branch-line coupler. *Electronics Letters*, *51*(10), 769–771. doi:10.1049/el.2015.0751

Belfort, A. J., & Cullen, A. L. (1982). First-order theory of the five-port symmetrical star junction. *Electronics Letters*, *18*(19), 841–842. doi:10.1049/el:19820572

Belouchrani, A., Abed-Meraim, K., Cardoso, J., & Moulines, E. (1997). A blind source separation technique using second-order statistics. *IEEE Transactions on Signal Processing*, *45*(2), 434–444. doi:10.1109/78.554307

Bender, V. C., Marchesan, T. B., & Alonso, J. M. (2015). Solid-state lighting: A concise review of the state of the art on LED and OLED modeling. *IEEE Industrial Electronics Magazine*, *9*(2), 6–16. doi:10.1109/MIE.2014.2360324

Bergeault, E., Huyart, B., Geneves, G., & Jallet, L. (1991). Characterization of diode detectors used in six-port reflectometers. *IEEE Transactions on Instrumentation and Measurement*, *40*(6), 1041–1043. doi:10.1109/19.119790

Bergmann, D., Fiege, L., Freitag, G., Gerlich, M., Grieser-schmitz, S., Lanze, E., & Robison, R. (2018). Vehicle with an anti-lock brake system and a method for braking a vehicle, U.S. Patent Application No. 14/345,904.

Bezek, J., Raffaj, V., & Bilik, V. (1991). Application of the lumped six-port reflectometer to water level measurement. *Proc. 9th Microwave Conf, MIKON-91*, Poland, 374-378.

Bharj, S. S., Camisa, R., Grober, S., Wozniak, F., & Pendleton, E. (1992). High efficiency Cband 1000 element rectenna array for microwave powered applications. *IEEE MTTS Int. Microwave Symp. Dig., Albuquerque, NM*, 301.

Bhat, B., & Koul, S. K. (1989). *Stripline-Like Transmission Lines for Microwave Integrated Circuits*. New York, NY: Wiley.

Bhattacharya, J. (1962). On the dispersion curve for Love wave due to irregularity in the thickness of the transversely isotropic crustal layer. *Gerlands Beitrage zur Geophysik*, *6*, 324–334.

Bilík, V. (2002). Six-port measurement technique: principles, impact, applications. Radioelektronika-2002 Conference, Slovakia. 123-155.

Bilik, V. (1992a). Calibrating the six-port reflectometer using a matched load and four unity-reflection standards. Part 1: Perfectly matched load. *Wuxiandian Gongcheng*, *1*, 10–13.

Bilik, V. (1992b). Calibrating the six-port reflectometer using a matched load and four unity-reflection standards. Part 2: Sliding termination. *Wuxiandian Gongcheng*, *1*, 2–8.

Bilik, V., & Bezek, J. (1995). Thin-film lumped six-port vector reflectometer for 10-1900MHz. *Proc.MITEKO 1995*, Pardubice, Czech Republic. 181-184.

Bilik, V., Raffaj, V., & Bezek, J. (1991). Miniature broadband lumped six-port reflectometer. *Journal of Communication*, *42*, 7–14.

Bilik, V., Raffaj, V., & Bezek, J. A. (1990). New extremely wideband lumped six-port reflectometer. *20th European Microwave Conference*. 1473-1478.

Bobzin, K. (2017). High-performance coatings for cutting tools. *CIRP Journal of Manufacturing Science and Technology*, *18*, 1–9. doi:10.1016/j.cirpj.2016.11.004

Böck, S., & Widmer, G. (2013, September). Maximum filter vibrato suppression for onset detection. In *Proc. of the 16th Int. Conf. on Digital Audio Effects (DAFx)*. Maynooth, Ireland.

Boere, L. D., & da Silva, A. L. (2018). *Sizing of anti-lock braking system for Formula SAE vehicles (No. 2018-36-0201)*. SAE Technical Paper.

Bogetti, T. A., & Gillespie, J. W. Jr. (1992). Process-Induced Stress and Deformation in Thick-Section Thermoset Composite Laminates. *Journal of Composite Materials*, *26*(5), 627–660. doi:10.1177/002199839202600502

Bonnet, G. (1983). Au-delà d'une vitesse de groupe: vitesse d'onde et vitesse de signal. Première partie: déformation de l'amplitude et influence de l'affaiblissement. Annales des Télécommunications, 1–17.

Bonnet, G. (1983). Au-delà d'une vitesse de groupe: vitesse d'onde et vitesse de signal. Première partie: l'opérateur vitesse de groupe en l'absence d'affaiblissement. Annales des Télécommunications, 1–22.

Bouchon, M., Campillo, M., & Gaffet, S. (1989). A boundary integral equation-discrete wavenumber representation method to study wave propagation in multilayered media having irregular interfaces. *Geophysics*, *54*(9), 1134–1140. doi:10.1190/1.1442748

Bourasseau, N., Moulin, E., Delabarre, C., & Bonniau, P. (2000). Radome health monitoring with Lamb waves: Experimental approach. NDT & E International, 33(6), 393–400. doi:10.1016/S0963-8695(00)00007-4 doi:10.1016/S0963-8695(00)00007-4

Boutejdar, A. (2016). Design of 5 GHz-compact reconfigurable DGS-bandpass filter using varactor-diode device and coupling matrix technique. *Microwave and Optical Technology Letters*, *58*(2), 304–309. doi:10.1002/mop.29561

Boutejdar, A., & Abd Ellatif, W. (2016). Improvement of Compactness of Low Pass Filter Using New Quasi-Yagi-DGS-Resonator and Multilayer-Technique. *Progress In Electromagnetics Research C*, *69*, 115–124. doi:10.2528/PIERC16073003

Boutejdar, A., Makkey, M., Elsherbini, A., & Omar, A. (2008). Design of compact stop band extended microstrip low pass filters by employing mutual-coupled square shaped defected ground structures. *Microwave and Optical Technology Letters*, *50*(4), 1107–1111. doi:10.1002/mop.23273

Bozanic, M., & Sinha, S. (2016). *Power Amplifier for the S-, C-, X- and ku bands*. Switzerland: Springer International. doi:10.1007/978-3-319-28376-0

Bree, G. (2009). Transmission line and lumped element quadrature couplers. *High Frequency Electronics*. 44-48.

Britton, R. D., Elcoate, C. D., Alanou, M. P., Evans, H. P., & Snidle, R. W. (2000). Effect of surface finish on gear tooth friction. *Journal of Tribology*, *122*(1), 354–360. doi:10.1115/1.555367

Browne, J. (1987). Broadband Amps Sport Coplanar Waveguide. *Microwaves & RF*, *26*(2), 131–134.

Browne, J. (1987). Coplanar MIC Amplifier Bridges 0.5 To 18.0 GHz. *Microwaves & RF*, *26*(6), 194–195.

Browne, J. (1989). Coplanar Waveguide Supports Integrated Multiplier Systems. *Microwaves & RF*, *28*(3), 137–138.

Browne, J. (1990). Coplanar Circuits Arm Limiting Amp with 100-dB Gain. *Microwaves & RF*, *29*(4), 213–220.

Browne, J. (1992). Broadband Amp Drops through Noise Floor. *Microwaves & RF*, *31*(2), 141–144.

Brown, W. C. (1965). Experimental airborne microwave supported platform. *Raytheon, Burlington, MA, USA, Tech. Rep. RADC-TR-65-188. Contract (New York, N.Y.)*, *AF30*(602), 3481.

Brown, W. C. (1984). The history of power transmission by radio waves. *IEEE Transactions on Microwave Theory and Techniques*, *MTT-32*(9), 1230–1242. doi:10.1109/TMTT.1984.1132833

Brown, W. C., & Triner, J. F. (1982). Experimental thin-film, etched-circuit rectenna. *IEEE MTT-S Int. Microwave Symp. Dig., Dallas, TX*, pp. 185-187. 10.1109/MWSYM.1982.1130655

Brunton, S. L., Rowley, C. W., Kulkarni, S. R., & Clarkson, C. (2010). Maximum Power Point Tracking for Photovoltaic Optimization Using Ripple-Based Extremum Seeking Control. *IEEE Transactions on Power Electronics*, *25*(10), 2531–2540.

Bruschke, M. V., & Advani, S. G. (1990). A finite element/control volume approach to mold filling in anisotropic porous media. *Polymer Composites*, *11*(6), 398–405. doi:10.1002/pc.750110613

Bruschke, M. V., & Advani, S. G. (1994). A numerical approach to model nonisothermal viscous flow through fibrous media with free surfaces. *International Journal for Numerical Methods in Fluids*, *19*(7), 575–603. doi:10.1002/fld.1650190704

Cairoli, P., Kondratiev, I., & Dougal, R. A. (2013). Coordinated control of the bus tie switches and power supply converters for fault protection in DC microgrids. *IEEE Transactions on Power Electronics*, *28*(4), 2037–2047.

Caldecott, R. (1973). The generalized multiprobe reflectometer and its application to automated transmission line measurements. *IEEE Transactions on Antennas and Propagation*, *21*(4), 550–554. doi:10.1109/TAP.1973.1140527

Cao, F., Chen, C., & Zhou, L. (2018, May). Research on Simulation of Anti-lock Braking System Based on MATLAB. In *2018 3rd International Conference on Automation, Mechanical Control and Computational Engineering (AMCCE 2018)*. Atlantis Press. 10.2991/amcce-18.2018.147

Carlsson, J. O., & Martin, P. M. (2010). Chemical Vapor Deposition. In P. M. Martin (Ed.), *Handbook of Deposition Technologies for Films and Coatings* (3rd ed.). Norwich, CT: William Andrew Publishing. doi:10.1016/B978-0-8155-2031-3.00007-7

Cawley, P., & Alleyne, D. (1991). A two-dimensional Fourier transform method for the measurement of propagating multimode signals. J. Acoust. Soc.1159-1168.

Chan, A. W., & Hwang, S. T. (1991). Modeling of the impregnation process during resin transfer molding. *Polymer Engineering and Science*, *31*(15), 1149–1156. doi:10.1002/pen.760311512

Chang, K., Li, M. Y., & Sauter, T. H. (1990). Low-cost microwave/millimeter-wave impedance measuring scheme using a three-probe microstrip circuit. *IEEE Transactions on Microwave Theory and Techniques*, *38*(10), 1455–1460. doi:10.1109/22.58685

Chattopadhyay, A., Gupta, S., Sharma, V. K., & Kumari, P. (2008). Propagation of SH waves in an irregular monoclinic crustal layer. *Archive of Applied Mechanics*, *78*(12), 989–999. doi:10.100700419-008-0209-6

Chaurasiya, D., Ghost, S., Bhattacharyya, S., & Srivastava, K. V. (2015). An ultrathin quad-band polarization-insensitive wide-angle metamaterial absorber. *Microwave and Optical Technology Letters*, *57*(3), 697–702. doi:10.1002/mop.28928

Chaurey, A., & Kandpal, T. C. (2010). A techno-economic comparison of rural electrification based on solar home systems and PV microgrids. *Energy Policy*, *38*(6), 3118–3129.

Chen, C. Q., Hao, M. L., Li, Z. Q., Du, Z. B., & Yang, H. (2016). A 1.8–2.8 GHz Highly Linear Broadband Power Amplifier for LTE-A Application. *Progress in Electromagnetics Research C.*, *66*, 47–54. doi:10.2528/PIERC16050503

Chen, D., Han, S., Ye, A., Ren, X., Wang, W., & Wang, T. (2019). Prediction of tire–pavement friction based on asphalt mixture surface texture level and its distributions. *Road Materials and Pavement Design*, 1–20. doi:10.1080/14680629.2018.1560354

Cheng, Y. Z., Fang, C., Zhang, Z., Wang, B., Chen, J., & Gong, R. Z. (2016). A compact and polarization-insensitive perfect metamaterial absorber for electromagnetic energy harvesting application. *2016 Progress In Electromagnetics Research Symposium, PIERS 2016 - Proceedings*, 1, 1910–1914. 10.1109/PIERS.2016.7734828

Chen, H. J., Huang, T. H., & Chang, C. S. (2006). A novel cross-shape DGS applied to design ultra-wide stop-band low-pass filters. *IEEE Microwave and Wireless Components Letters*, *16*(5), 252–254. doi:10.1109/LMWC.2006.873594

Chen, N., & Chung, H. S.-H. (2011). A driving technology for retrofit LED lamp for fluorescent lighting fixtures with electronic ballasts. *IEEE Transactions on Power Electronics*, *26*(2), 588–601. doi:10.1109/TPEL.2010.2066579

Chen, P. H., Ishida, K., Zhang, X., Okuma, Y., Ryu, Y., Takamiya, M., & Sakurai, T. (2010). 0. 18-V Input Charge Pump with Forward Body Biasing in Startup Circuit using 65nm CMOS. In *Custom Integrated Circuits Conference (CICC), IEEE*, pp. 2–5.

Chen, W. L., Wang, G. M., & Zhang, C. X. (2009). Miniaturization of wideband branch-line couplers using fractal-shaped geometry. *Microwave and Optical Technology Letters*, *51*(1), 26–29. doi:10.1002/mop.24002

Chen, Z., & Xu, B. (1987). Linearization of Diode Detector Characteristics. *IEEE MTT-S International Microwave Symposium Digest*. 265-267. 10.1109/MWSYM.1987.1132380

Cheung, A., Yu, Y., & Pochiraju, K. (2004). Three-dimensional finite element simulation of curing of polymer composites. *Finite Elements in Analysis and Design*, *40*(8), 895–912. doi:10.1016/S0168-874X(03)00119-7

Chien, H., Diaz-Jimenez, C., Rohrer, G. S., Ban, Z., Prichard, P., & Liu, Y. (2013). The Influence of Residual Thermal Stresses on the Mechanical Properties of Multilayer α-Al_2O_3/$TiCxN_{1-x}$ Coatings on WC/Co Cutting Tools. *Surface and Coatings Technology*, *215*, 119–126. doi:10.1016/j.surfcoat.2012.07.088

Chin, V. J., Salam, Z., & Ishaque, K. (2015). Cell modelling and model parameters estimation techniques for photovoltaic simulator application: A review. *Applied Energy*, *154*, 500–519.

Chiu, L., & Xue, Q. (2010). Investigation of a wideband 90° hybrid coupler with an arbitrary coupling level. *IEEE Transactions on Microwave Theory and Techniques*. *58*(4), 1022-1029.

Chiua, W. K., Roseb, L. R. F., & Nadarajaha, N. (2017). Scattering of the fundamental anti –symmetric Lamb wave by a midplane edge delamination in a fiber-composite laminate. Procedia Engineering, 188, 317–324. doi:10.1016/j.proeng.2017.04.490 doi:10.1016/j.proeng.2017.04.490

Choi, M. A., Lee, M. H., & Lee, S. J., (1998). Numerical studies during cure in the resin transfer molding, *Korea polymer journal, 6,* 211-218.

Chow, L. A. (2018). Equipment and Manufacturability Issues in Chemical Vapor Deposition Processes. In K. Seshan, & D. Schepis (Eds.), *Handbook of Thin Film Deposition* (4th ed.). Norwich, CT: William Andrew Publishing. doi:10.1016/B978-0-12-812311-9.00009-8

Chuc, D. H., & Duong, B. G. (2016). Design and Fabrication of Rectifying Antenna Circuit for Wireless Power Transmission System Operating at ISM Band. *Iranian Journal of Electrical and Computer Engineering*, *6*(4), 1522–1528.

Chumaevskii, A. V., Tarasov, S. Y., Filippov, A. V., Kolubaev, E. A., Rubtsov, V. E., & Eliseev, A. A. (2016, November). Mechanical strength of additive manufactured carbon fiber reinforced polyetheretherketone. In AIP Conference Proceedings: Vol. 1783. *No. 1* (p. 020029). AIP Publishing. doi:10.1063/1.4966322

Chun, Y. H., & Hong, J. S. (2006). Compact wide-band branch-line hybrids. *IEEE Transactions on Microwave Theory and Techniques*. *54*(2), 704–709.

Chunhui, Y. (2006). Some aspects of numerical simulation for Lamb wave propagation in composite laminates. Composite Structures, 267–275.

Ciminelli, C., D'Orazio, A., De Sario, M., Petruzzelli, V., & Prudenzano, F. (2000). A new structure for a six port reflectometer by using a five port ring. *10th Mediterranean Electrotechnical Conference, MELECON 2000*. 181-184.

Clarke, E. M., & Wing, J. M. (1996). Formal methods: State of the art and future directions. [CSUR]. *ACM Computing Surveys*, *28*(4), 626–643. doi:10.1145/242223.242257

Close-Up, T. (1988, April). *Microwaves & RF*, *27*(4), 79.

Colantonio, P., Giannini, F., & Limiti, E. (2009). *High Efficiency RF and Microwave Solid State Power Amplifiers*. UK: John Wiley & Sons. doi:10.1002/9780470746547

Collier, R. J., & El-Deeb, N. A. (1979). On the use of a microstrip three-line system as a six-port reflectometer. *IEEE Transactions on Microwave Theory and Techniques*, *27*(10), 847–853. doi:10.1109/TMTT.1979.1129747

Collin, R. E. (1992). *Foundations for Microwave Engineering* (2nd ed.). New York, NY: IEEE Press.

Connelly, R., Pattanaik, A. K., & Sarin, V. K. (2005). Development of Moderate Temperature CVD Al_2O_3 Coatings. *International Journal of Refractory Metals & Hard Materials*, *23*(4–6), 317–321. doi:10.1016/j.ijrmhm.2005.05.017

Conroy, J., & Watson, R. (2009). Aggregate modelling of wind farms containing full-converter wind turbine generators with permanent magnet synchronous machines: Transient stability studies. *IET Renewable Power Generation*, *3*(1), 39–52.

Cosserat, E., & Cosserat, F. (1909). *Theorie des Corps Deformables*. Paris, France: Hermann et Fils.

Crane Aerospace & Electronics Microwave Solutions. (2012). *Quadrature hybrids*. West Caldwell, NJ: Merrimac Industries, 1–6.

Cripps, S. C. (2007). Microwave bytes - VNA tales. *IEEE Microwave Magazine*, *8*(5), 28–44. doi:10.1109/MMM.2007.904719

Cullen, A. L., & Yeo, S. P. (1986). Six-port reflectometer theory for symmetrical five-port junction with directional coupler. *IEE Proceedings H-Microwaves, Antennas and Propagation. 133*(4), 277-287.

D'Avico, L. U. C. A. (2019). *Analysis and design of advanced anti-Lock braking systems.* (Doctoral dissertation, Italy).

Das, B. M. (1983). Advanced Soil Mechanics. New York, NY: McGraw-Hill.

Database https://www.neonscience.org/calc-ndvi-py

de Albuquerque, V. H. C., de Alexandria, A. R., Cortez, P. C., & Tavares, J. M. R. (2009). Evaluation of multilayer perceptron and self-organizing map neural network topologies applied on microstructure segmentation from metallographic images. *NDT & E International, 42*(7), 644-651.

de Oliveira, T. R., Bolzon, A. S., & Donoso-Garcia, P. F. (2014, October). Grounding and safety considerations for residential DC microgrids. *In IECON 2014-40th Annual Conference of the IEEE Industrial Electronics Society* (pp. 5526-5532). IEEE.

De Ronde, F. C. (1982). Octave-wide matched symmetrical, reciprocal, 4- and 5 ports. *IEEE MTT-S International Microwave Symposium Digest*. 521-523. 10.1109/MWSYM.1982.1130778

Debbech, S., Bon, P., & Collart-Dutilleul, S. (2019). A model-based system engineering approach to manage railway safety-related decisions. *International Journal of Transport Development and Integration*, *3*(1), 30–43. doi:10.2495/TDI-V3-N1-30-43

Deichmann, U., Meisner, C., Murray, S., & Wheeler, D. (2011). The economics of renewable energy expansion in rural Sub-Saharan Africa. *Energy Policy*, *39*(1), 215–227.

Demcenko, A., & Mazeika, L. (2010). Calculation of lamb waves dispersion curves in multi-layered planar structures. Journal of Sound and Vibration, 1435–1449.

Deng, J., Zhang, Z., Marchi, E., & Schuller, B. (2013, September). Sparse autoencoder-based feature transfer learning for speech emotion recognition. In *2013 Humaine Association Conference on Affective Computing and Intelligent Interaction* (pp. 511-516). IEEE. 10.1109/ACII.2013.90

Deng, L., & Han, M. (2007). Microwave absorbing performances of multiwalled carbon nanotube composites with negative permeability. *Applied Physics Letters*, *91*(2). doi:10.1063/1.2755875

Deshmukh, A., Shendge, P. D., & Phadke, S. B. (2018, June). Effect of Second Order Actuator on Performance of Anti-Lock Braking System with Multiple Sliding Surface. In *2018 Second International Conference on Intelligent Computing and Control Systems (ICICCS)* (pp. 313-317). IEEE. 10.1109/ICCONS.2018.8662886

Deville, Y., & Hosseini, S. (2009). Recurrent networks for separating extractable-target nonlinear mixtures. Part I: Non-blind configurations. *Signal Processing*, *89*(4), 378–393. doi:10.1016/j.sigpro.2008.09.016

Dhar, S., Patnaik, R. K., & Dash, P. K. (2017). Fault detection and location of photovoltaic based DC microgrid using differential protection strategy. *IEEE Transactions on Smart Grid*, *9*(5), 4303–4312.

Diamanti,K., Hodgkinson, J. M. & Soutis, C. (2004). Application of a Lamb Wave Technique for the Non Destructive Inspection of Composite Structures.

Dickson, A. N., Barry, J. N., McDonnell, K. A., & Dowling, D. P. (2017). Fabrication of continuous carbon, glass and Kevlar fibre reinforced polymer composites using additive manufacturing. *Additive Manufacturing*, *16*, 146–152. doi:10.1016/j.addma.2017.06.004

Dilbag, S. (2008). *Some dynamical problems in micropolar elasticity*, (Doctoral dissertation, Department of Mathematics, Panjab University, India).

Directive 2002/95/EC of the European Parliament and of the Council of 27 January 2002 on *the restriction of the use of certain hazardous substances in electrical and electronic equipment.*

Dolgov, A., Zane, R., & Popovic, Z. (2010). Power Management System for Online.

Dong, I. K., Araki, K., & Naito, Y. (1984). Properties of the symmetrical circuit and its broad-band five-port design. *IEEE Transactions on Microwave Theory and Techniques*, *32*(1), 51–57. doi:10.1109/TMTT.1984.1132611

Dong, Y., Ghataura, A., Takagi, H., Haroosh, H. J., Nakagaito, A. N., & Lau, K. T. (2014). Polylactic acid (PLA) bio composites reinforced with coir fibres: Evaluation of mechanical performance and multifunctional properties. *Composites. Part A, Applied Science and Manufacturing*, *63*, 76–84. doi:10.1016/j.compositesa.2014.04.003

Douyére, A. (2008). *Methodology for synthesis and optimization of rectenna bounded to an operating domain. Physics*. France: University of Reunion.

Dutt, S., Agrawal, S., & Vig, R. (2018). Cluster-Head Restricted Energy Efficient Protocol (CREEP) for Routing in Heterogeneous Wireless Sensor Networks. *Wireless Personal Communications*, *100*(4), 1477–1497. doi:10.100711277-018-5649-x

Eccleston, K. W., & Ong, S. H. M. (2003). Compact planar microstripline branch-line and rat-race couplers. *IEEE Transactions on Microwave Theory and Techniques*, *51*(10), 2119–2125. doi:10.1109/TMTT.2003.817442

Egmont-Petersen, M., de Ridder, D., & Handels, H. (2002). Image processing with neural networks—a review. *Pattern recognition, 35*(10), 2279-2301.

El Brouzi, A., Bouzon, J., & Vergnaud, J. M., (1989). Plastics and Rubber Processing and Application, *12.*

El Hamraoui, A., Abdelmounim, E. H., Zbitou, J., Bennis, H., & Latrach, M. (2016). A new compact CPW-fed dual-band monopole antenna for RFID applications. *In Proceedings of The 5th International Conference on Multimedia Computing and Systems (ICMCS)* (pp. 492-496). Marrakech, Morocco. IEEE.

Elbhiri, B., Saadane, R., & Aboutajdine, D. (2010). Developed Distributed Energy-Efficient Clustering (DDEEC) for heterogeneous wireless sensor networks. In *2010 5th International Symposium on I/V Communications and Mobile Network (ISVC),* (pp. 1–4). Citeseer.

El-Deeb, N. A. (1983). The calibration and performance of a microstrip six-port reflectometer. *IEEE Transactions on Microwave Theory and Techniques*, *31*(7), 509–514. doi:10.1109/TMTT.1983.1131538

Elhadji, B. N. (2014). Contrôle santé de structures sandwichs composites, caractérisation et évaluation non destructives de l’adhésion et du vieillissement – Simulations par éléments finis et expérimentation par mesure d’impédance et par ultrasons multiéléments. Université du Havre, École Doctorale S.P.M.I.I.

Elhadji, B. N., & Hugues, D. (2012). Non Destructive Testing of sandwich composites: adhesion defects evaluation; Experimental and Finite Element Method simulation comparison. Proceedings of the Acoustics Nantes Conference.

Elhami, G., & Weber, R. M. (2019). *Audio Feature Extraction with Convolutional Neural Autoencoders with Application to Voice Conversion*. No. CONF.

Elmahjouby, S., Errkik, A., Mandry, R., Tajmouati, A., & Latrach, M. (2016). A New CPW Compact Stopband Filter Using DGS Loaded. In *proceedings of International conference on computing wireless and communication systems.* (pp 33-35). Settat, Morocco: University of Hassan 1st.

Elmahjouby, S., Errkik, A., Zbitou, J., Tajmouati, A., El Abdellaoui, L., & Latrach, M. (2017). A Novel Compact CPW LowPass Filter Integrating Periodic Triangle DGS Cells. *Transactions on Machine Learning and Artificial Intelligence*, *5*(4), 457–462.

Elsayed, A. T., Mohamed, A. A., & Mohammed, O. A. (2015). DC microgrids and distribution systems: An overview. *Electric Power Systems Research*, *119*, 407–417.

Engen, G. F. (1977a). An improved circuit for implementing the six-port technique of microwave measurements. *IEEE MTT-S International Microwave Symposium Digest*. 53-55. 10.1109/MWSYM.1977.1124355

Engen, G. F. (1977b). The six-port reflectometer: Alternative network analyzer. *IEEE Transactions on Microwave Theory and Techniques*, *25*(12), 1075–1080. doi:10.1109/TMTT.1977.1129277

Engen, G. F. (1978). Calibrating the six-port reflectometer by means of sliding terminations. *IEEE Transactions on Microwave Theory and Techniques*, *26*(12), 951–957. doi:10.1109/TMTT.1978.1129527

Engen, G. F., & Hoer, C. A. (1972). Application of an arbitrary 6-port junction to power-measurement problems. *IEEE Transactions on Instrumentation and Measurement*, *21*(4), 470–474. doi:10.1109/TIM.1972.4314069

Eremeyev, V. A., Lebedev, L. P., & Altenbach, H. (2012). Foundations of micropolar mechanics. Berlin, Germany: Springer Science & Business Media.

Ericsen, T., Hingorani, N., & Khersonsky, Y. (2006). Power electronics and future marine electrical systems. *IEEE Transactions on Industry Applications*, *42*(1), 155–163.

Eringen, A. C. (1966). Linear theory of micropolar elasticity. *Journal of Mathematics and Mechanics,* 909-923.

Eringen, A. C. (1984). Plane waves in nonlocal micropolar elasticity. *International Journal of Engineering Science*, *22*(8-10), 1113–1121. doi:10.1016/0020-7225(84)90112-5

Eringen, A. C. (1999). *Microcontinuum Field Theories-I, Foundations and Solids*. New York, NY: Springer-Verlag. doi:10.1007/978-1-4612-0555-5

Eringen, A. C. (2003a). Micropolar mixture theory of porous media. *Journal of Applied Physics*, *94*(6), 4184–4190. doi:10.1063/1.1598640

Eringen, A. C. (2003b). Continuum theory of micromorphic electromagnetic thermoelastic solids. *International Journal of Engineering Science*, *41*(7), 653–665. doi:10.1016/S0020-7225(02)00274-4

Eringen, A. C., & Suhubi, E. S. (1964). Nonlinear theory of simple micro-elastic solids-I. *International Journal of Engineering Science*, *2*(2), 189–203. doi:10.1016/0020-7225(64)90004-7

Erlich, I., Kretschmann, J., Mueller-Engelhardt, S., Koch, F., & Fortmann, J. (2008, July). Modeling of wind turbines based on doubly-fed induction generators for power system stability studies. In 2008 IEEE Power and Energy Society General Meeting-Conversion and Delivery of Electrical Energy in the 21st Century (pp. 1-8). IEEE.

Ettmayer, P., Kolaska, H., & Ortner, H. M. (2014). History of Hardmetals. In V. K. Sarin (Ed.), Comprehensive Hard Materials. Amsterdam, The Netherlands: Elsevier Ltd. doi:10.1016/B978-0-08-096527-7.00001-5

European Commission. (2006). Regulation (EC) No 1907/2006 of the European Parliament and of the Council of 18 December 2006 concerning the Registration, Evaluation, Authorisation and Restriction of Chemicals (REACH), establishing a European Chemicals Agency.

Ewing, W. M., Zardetzky, W. S., & Press, F. (1957). *Elastic Waves in Layered Media*. New York, NY: McGraw-Hill Book Co. doi:10.1063/1.3060203

Faisal, S., Javaid, N., Javaid, A., Khan, M. A., Bouk, S. H., & Khan, Z. A. (2013). Z-SEP: Zonal-stable election protocol for wireless sensor networks. *ArXiv Preprint ArXiv:1303.5364*.

Fang, X. T., Zhang, X. C., & Tong, C. M. (2011). A novel miniaturized microstrip six-port junction. *Progress In Electromagnetics Research Letters*, *23*, 129–135. doi:10.2528/PIERL11032302

Farrokhabadi, M., König, S., Cañizares, C. A., Bhattacharya, K., & Leibfried, T. (2018). Battery energy storage system models for microgrid stability analysis and dynamic simulation. *IEEE Transactions on Power Systems*, *33*(2), 2301–2312.

Faruque, M. R. I., Islam, M. T., & Misran, N. (2012). Design analysis of new metamaterial for EM absorption reduction. *Progress in Electromagnetics Research*, *124*, 119–135. doi:10.2528/PIER11112301

Fedor, S., & Collier, M. (2007). On the problem of energy efficiency of multi-hop vs one-hop routing in wireless sensor networks. In *21st International Conference on Advanced Information Networking and Applications Workshops (AINAW'07)* (Vol. 2, pp. 380–385). IEEE. 10.1109/AINAW.2007.272

Feng, B., Ribeiro, A. L., & Ramos, H. G. (2017). Interaction of Lamb waves with the edges of a delamination in CFRP composites and a reference-free localization method for delamination. IEEE.

Feng, Y. U., & Jun, X. I. E. (2019). Simulation analysis of efficiency and stability for vehicle's ABS control on combined steering and braking maneuvers. In *MATEC Web of Conferences* (*Vol. 272*, p. 01024). EDP Sciences. 10.1051/matecconf/201927201024

Feng, L., Wang, Z., & Shen, W. (2019). Effects of complex graphite-like carbon coating on gear vibration noise. *Results in Physics*, *12*, 1495–1499. doi:10.1016/j.rinp.2019.01.048

Ferreira, R. T. L., Amatte, I. C., Dutra, T. A., & Bürger, D. (2017). Experimental characterization and micrography of 3D printed PLA and PLA reinforced with short carbon fibers. *Composites. Part B, Engineering*, *124*, 88–100. doi:10.1016/j.compositesb.2017.05.013

Fitzpatrick, J. (1978). Error models for systems measurement. *Microwave Journal*, *21*(5), 63–66.

Fletcher, S. D. A., Elders, I. M., Norman, P. J., Galloway, S. J., Booth, C. D., Burt, G. M., ... Hill, J. E. (2010). The impact of incorporating skin effect on the fault analysis and protection system performance of DC marine and aerospace power systems.

Frances, A., Asensi, R., García, Ó., Prieto, R., & Uceda, J. (2018). Modeling electronic power converters in smart DC microgrids—An overview. *IEEE Transactions on Smart Grid*, *9*(6), 6274–6287.

Francois, B., & Reynaert, P. (2015). Highly linear fully integrated wideband RF PA for LTE advanced in 180-nm SOI. *IEEE Transactions on Microwave Theory and Techniques*, *63*(3), 649–658. doi:10.1109/TMTT.2014.2380319

Frank, M., Breidt, D., & Cremer, R. (2007). Nanocrystalline Diamond Coatings for Machining. *Vakuum in Forschung und Praxis*, *19*(S1), 34–38. doi:10.1002/vipr.200790041

Frazer, R. C., Shaw, B. A., Palmer, D., & Fish, M. Optimizing gear geometry for minimum transmission error, mesh friction, losses and scuffing risk through computer aided engineering. In: AGMA Technical Paper 09FTM07.

Fu, C., & Zheng, K. (2018, November). Patterns for Modeling Task-Level Timing Constraints with Event-B. In *2018 IEEE 9th International Conference on Software Engineering and Service Science (ICSESS)* (pp. 260-266). IEEE. 10.1109/ICSESS.2018.8663926

Fung, A., & Eom, H. (1981). A theory of wave scattering from an inhomogeneous layer with an irregular interface. *IEEE Transactions on Antennas and Propagation*, *29*(6), 899–910. doi:10.1109/TAP.1981.1142679

Gao, F., Bozhko, S., Asher, G., Wheeler, P., & Patel, C. (2016). An improved voltage compensation approach in a droop-controlled DC power system for the more electric aircraft. *IEEE Transactions on Power Electronics*, *31*(10), 7369–7383.

Gao, L., Liu, Y., Ren, H., & Guerrero, J. (2017). A DC microgrid coordinated control strategy based on integrator current-sharing. *Energies*, *10*(8), 1116.

Gauthier, R. D. (1982). Experimental investigations on micropolar media. *Mechanics of micropolar media*, 395-463.

Gavali, V. C., Kubade, P. R., & Hrushikesh, B. K. (2018, February). Mechanical And Thermomechanical Properties Of Carbon Fibre Reinforced Thermoplastic Composite Fabricated Using Fused Deposition Modelling (Fdm) Method: A Review. *International Journal of Mechanical and Production Engineering Research and Development*, *8*(1), 1161–1168. doi:10.24247/ijmperdfeb2018137

Gayathri, P., Umesh, K., & Ganguli, R. (2010). Effect of matrix cracking and material uncertainty on composite plates. Reliability Engineering & System Safety, 95(7), 716–728. doi:10.1016/j.ress.2010.02.004doi:10.1016/j.ress.2010.02.004

George, K. (2006). DC power production, delivery and utilization. *Electric Power Research Institute White Paper*.

Gerke, B. F., Ngo, A. T., Alstone, A. L., & Fisseha, K. S. (2014). *The evolving price of household LED lamps: Recent trends and historical comparisons for the US market: Lawrence Berkeley National Lab*. Berkeley, CA: LBNL.

Getsinger, W. J. (1993). End-effects in quasi-TEM transmission lines. *IEEE Transactions on Microwave Theory and Techniques, 41*(4), 666–672.

Ge, Z., & Chen, X. (2008). An efficient approach for simulating wave propagation with the boundary element method in multilayered media with irregular interfaces. *Bulletin of the Seismological Society of America*, *98*(6), 3007–3016. doi:10.1785/0120080920

Ghannouchi, F. M., & Mohammadi, A. (2009). *The six-port technique with microwave and wireless applications*. Norwood, MA: Artech House.

Ghosh, D., & Kumar, G. (2015). A four branch microstrip coupler with improved bandwidth and isolation. In *Twenty First National Conference on Communications*, Mumbai, India.

Ghosh, S., Bhattacharyya, S., & Srivastava, K. V. (2014). Bandwidth-enhancement of a polarization insensitive metamaterial absorber. *Microwave and Optical Technology Letters*, *56*(2), 350–355. doi:10.1002/mop.28122

Ghosh, S., Nguyen, T. T., & Lim, S. (2019). Recent progress in angle-insensitive narrowband and broadband metamaterial absorbers. *EPJ Applied Metamaterials*, *6*(12), 1–15.

Gibson, W. C. (2008). The Method of Moments in Electromagnetics. *Chapman &* Hall/CRC.

Glamočić, L. (2018). The Influence of Modern Technical Systems in Vehicles on Road Safety. *JTTTP-Journal of Traffic and Transport Theory and Practice, 3*(1-2).

Goncalvès, E. (2005). *Resolution Numerique*. Discretisation des EDP et EDO, Institut National Polytechnique de Grenoble.

Gonzalez, G. (1997). Microwave Transistor Amplifiers Analysis and Design (2nd ed.). Upper Saddle River, NJ: Prentice Hall.

Gonzalez-Romero, J. F., & Mascosko, C. W. (1985). Viscosity rise during radical crosslinking polymerization with inhibition. *Journal of Rheology (New York, N.Y.)*, *29*(3), 259–272. doi:10.1122/1.549790

Graham, D. (2014). Diamonds Can Be A Tool's Best Friend. *Manufacturing Engineering*, (Supplement 4), 27.

Grainger, S., & Blunt, J. (1998). *Engineering Coatings: Design and Application* (2nd ed.). Cambridge, UK: Woodhead Publishing Ltd. doi:10.1533/9781845698577

Grais, E. M., & Plumbley, M. D. (2017, November). Single channel audio source separation using convolutional denoising autoencoders. In *2017 IEEE Global Conference on Signal and Information Processing (GlobalSIP)* (pp. 1265-1269). IEEE. 10.1109/GlobalSIP.2017.8309164

Gudimetla, P., & Kharidi, A. (2009). Simulation of delaminations in composite laminates. International Conference on Precision, Meso, Micro and Nano Engineering, 11 -12.

Gunasekaran, M., Mohamed Ismail, H., Chokkalingam, B., Mihet-Popa, L., & Padmanaban, S. (2018). Energy management strategy for rural communities' DC Micro Grid power system structure with maximum penetration of renewable energy sources. *Applied Sciences*, *8*(4), 585.

Guo, N., & Cawley, P. (1993). The interaction of Lamb waves with delaminations in composite laminates. The Journal of the Acoustical Society of America, 94(4), 2240–2246. doi:10.1121/1.407495 doi:10.1121/1.407495

Guo, X., Liu, X., Zhu, E., & Yin, J. (2017, November). Deep clustering with convolutional autoencoders. In *International Conference on Neural Information Processing* (pp. 373-382). Cham, Switzerland: Springer.

Guo, Z. S. Du S., & Zhang, B. (2004). Temperature distribution of thick thermoset composites, *Modelling and simulation in materials science and engineering, 12*, 443–452 .

Guo, Z. S., Du S., & Zhang, B. (2004). Temperature field of thick thermoset composite laminates during cure process. *Composites Science and Technology*, 1–7.

Gupta, K., Laubscher, R. F., Davim, J. P., & Jain, N. K. (2016). Recent Developments in Sustainable Manufacturing of Gears: A Review. *Journal of Cleaner Production*, *112*(4), 3320–3330. doi:10.1016/j.jclepro.2015.09.133

Gupta, S., Dhillon, S. S., Khera, P., & Marwaha, A. (2013). Dual Band U-Slotted Microstrip Patch Antenna for C Band and X Band Radar Applications. *5th International Conference on Computational Intelligence and Communication Networks (CICN)*, pp. 41-45. 10.1109/CICN.2013.18

Hajimiragha, A., Canizares, C. A., Fowler, M. W., & Elkamel, A. (2010). Optimal transition to plug-in hybrid electric vehicles in Ontario, Canada, considering the electricity-grid limitations. *IEEE Transactions on Industrial Electronics*, *57*(2), 690–701.

Halpin, J. C., Kardos, J. L., & Dudukovic, M. P. (1983). Processing science: An approach for prepreg composite systems. *Pure and Applied Chemistry*, *55*(5), 893–906. doi:10.1351/pac198355050893

Han, C., Lee, D., & Chin, H. (1986). *Development of a Mathematical Model for the Pultrusion Process*, *26*, 393–404.

Hanfei, M., Asaad, M., Mohammad Faisal, H., Roshan, J., Md Yeasin, B., & Victor, G. (2019). Vibration-Based In-Situ Detection and Quantification of Delamination in Composite Plates. Sensors (Basel), 19(7), 1734. doi:10.3390/s19071734 PubMed doi:10.339019071734 PMID:30978968

Harb, S., Kedia, M., Zhang, H., & Balog, R. S. (2013). Microinverter and string inverter grid connected photovoltaic system A comprehensive study, Photovoltaic Specialists Conference (PVSC), pp. 2885-2890, June 2013. 10.1109/PVSC.2013.6745072

Harrist, D. W. (2001). Wireless Battery Charging System Using Radio Frequency Energy Harvesting. *Thesis BS, University of Pittsburgh.*

Harsh, K. B. (2012). Detection of Damage in a Composite Structure Using Guided Waves. Los Angeles, CA: University Of California.

Hasan, A., Helaoui, M., & Ghannouchi, F. M. (2012). Dynamic linearization of diodes for high speed and peak power detection applications. *IEEE MTT-S International Microwave Symposium Digest.* 1-3. 10.1109/MWSYM.2012.6259710

Hatziargyriou, N., Asano, H., Iravani, R., & Marnay, C. (2007). Microgrids. *IEEE Power & Energy Magazine*, *5*(4), 78–94.

Haubner, R., Lessiak, M., Pitonak, R., Köpf, A., & Weissenbacher, R. (2017). Evolution of Conventional Hard Coatings for its Use on Cutting Tools. *International Journal of Refractory Metals and Hard Materials*, *62*(B), 210-218.

Hayashi, T., & Kawashima, K. (2002). Multiple reflections of Lamb waves at a delamination. Ultrasonics, 44(1-8), 193–197. doi:10.1016/S0041-624X(02)00136-1 PubMed doi:10.1016/S0041-624X(02)00136-1 PMID:12159931

Hedayati, M. K., Javaherirahim, M., Mozooni, B., Abdelaziz, R., Tavassolizadeh, A., Sai, V., ... Elbahri, M. (2011). Design of a perfect black absorber at visible frequencies Using Plasmonic Metamaterials. *Advanced Materials*, *23*(45), 5410–5414. doi:10.1002/adma.201102646 PMID:21997378

Heinzelman, W. (2000). *Application-Specific protocol architectures for wireless networks* (Doctoral dissertation, Massachusetts Institute of Technology).

Heinzelman, W. R., Chandrakasan, A., & Balakrishnan, H. (2000). Energy-efficient communication protocol for wireless microsensor networks. In *Proceedings of the 33rd annual Hawaii international conference on System sciences, 2000.* (pp. 10-pp). IEEE. 10.1109/HICSS.2000.926982

Hérault, J., Jutten, C., & Ans, B. (1985). Détection de grandeurs primitives dans un message composite par une architecture de calcul neuromimétique en apprentissage non supervisé. In 10 Colloque sur le traitement du signal et des images, FRA, 1985. GRETSI, Groupe d'Etudes du Traitement du Signal et des Images.

Hesselbarth, J., Wiedmann, F., & Huyart, B. (1997). Two new six-port reflectometers covering very large bandwidths. *IEEE Transactions on Instrumentation and Measurement*, *46*(4), 966–969. doi:10.1109/19.650809

Hinchcliffe, S. A., Hess, K. M., & Srubar, W. V. III. (2016). Experimental and theoretical investigation of prestressed natural fiber-reinforced polylactic acid (PLA) composite materials. *Composites. Part B, Engineering*, *95*, 346–354. doi:10.1016/j.compositesb.2016.03.089

Hoer, C. A. (1979). Calibrating a six-port reflectometer with four impedance standards. United States: National Bureau of Standards (NBS) Technical Note 1012.

Hoer, C. A. (1972). The six-port coupler: A new approach to measuring voltage, current, power, impedance, and phase. *IEEE Transactions on Instrumentation and Measurement*, *21*(4), 466–470. doi:10.1109/TIM.1972.4314068

Hoer, C. A., Roe, K. C., & Allred, C. M. (1976). Measuring and minimizing diode detector nonlinearity. *IEEE Transactions on Instrumentation and Measurement*, *25*(4), 324–329. doi:10.1109/TIM.1976.6312236

Holleck, H. (1988). Metastable Layer Materials – Prediction of Constitution and Structure. *Surface and Coatings Technology*, *36*, 151–159. doi:10.1016/0257-8972(88)90145-4

Hooshyar, H., & Baran, M. E. (2013). Fault analysis on distribution feeders with high penetration of PV systems. *IEEE Transactions on Power Systems*, *28*(3), 2890–2896.

Horestani, A. K., Member, S., Fumeaux, C., Member, S., Al-sarawi, S. F., & Abbott, D. (2013). Displacement sensor based on diamond-shaped tapered split ring resonator. *IEEE Sensors Journal*, *13*(4), 1153–1160. doi:10.1109/JSEN.2012.2231065

Horestani, A. K., Naqui, J., Shaterian, Z., Abbott, D., Fumeaux, C., & Martín, F. (2014). Two-dimensional alignment and displacement sensor based on movable broadside-coupled split ring resonators. *Sensors and Actuators A: Physical*, *210*, 18–24. doi:10.1016/j.sna.2014.01.030

Hossain, T. M., Jamlos, M. F., Jamlos, M. A., Soh, P. J., Ibrahim, S. Z., Isa, M. M., ... Ghazali, M. F. (2019). Compact broadband triple-ring five-port reflectometer for microwave brain imaging applications. *IEEE Access: Practical Innovations, Open Solutions*, *7*. doi:10.1109/ACCESS.2019.2891058

Hosseini, S., & Deville, Y. (2013). Recurrent networks for separating extractable-target nonlinear mixtures. Part II. Blind configurations. *Signal Processing*, *93*(4), 671–683. doi:10.1016/j.sigpro.2012.08.027

HSMS-282X series. (2001). Surface mount RF Schottky barrier diode. *Agilent Technol.,* Palo Alto, CA, Tech. Data.

http://neuralnetworksanddeeplearning.com/chap6.html. (n.d.). Retrieved from https://www.kisspng.com/png-deep-learning-machine-learning-artificial-intellig-1401501/preview.html

https://www.natural-solutions.eu/blog/la-reconnaissance-dimage-avec-les-rseaux-de-neurones-convolutifs. (n.d.). Retrieved from https://www.kisspng.com/png-deep-learning-machine-learning-artificial-intellig-1401501/preview.html

Hu, H. W. (2007). Physical Aging in Long Term Creep of Polymeric Composite Laminates. Journal of Mechanisms, 23(3), 245–252. doi:10.1017/S1727719100001283 doi:10.1017/S1727719100001283

Huang, M., Tsai, Y., & Chen, K. (2010). Sub-1 V Input Single-Inductor Dual-Output (SIDO) DC-DC Converter with Adaptive Load-Tracking Control (ALTC) for Single-Cell-Powered Systems. *IEEE Transactions on Power Electronics*, *25*(7), 1713–1724.

Huang, X., Yang, H., Yu, S., Wang, J., Li, M., & Ye, Q. (2013). Triple-band polarization-insensitive wide-angle ultra-thin planar spiral metamaterial absorber. *Journal of Applied Physics*, *113*(21). doi:10.1063/1.4809655

Hu, C. J. (1980). A novel approach to the design of multiple-probe high-power microwave automatic impedance measuring schemes. *IEEE Transactions on Microwave Theory and Techniques*, *28*(12), 1422–1428. doi:10.1109/TMTT.1980.1130260

Hu, C. J. (1983). Microwave automatic impedance measuring schemes using three fixed probes. *IEEE Transactions on Microwave Theory and Techniques*, *31*(9), 756–762. doi:10.1109/TMTT.1983.1131586

Hughes, D., & Correll, N. (2016). Distributed machine learning in materials that couple sensing, actuation, computation and communication. *arXiv preprint arXiv:1606.03508.*

Hu, K. W., & Liaw, C. M. (2016). Incorporated operation control of DC microgrid and electric vehicle. *IEEE Transactions on Industrial Electronics*, *63*(1), 202–215.

Hull, E., Grove, W., Zhang, M., Song, X., Pei, Z. J., & Cong, W. (2015, June). Effects of process variables on extrusion of carbon fiber reinforced ABS filament for additive manufacturing. In *ASME 2015 International Manufacturing Science and Engineering Conference*. American Society of Mechanical Engineers. 10.1115/MSEC2015-9396

Hunter, J. D., & Somlo, P. I. (1985). An explicit six-port calibration method using five standards (short papers). *IEEE Transactions on Microwave Theory and Techniques*, *33*(1), 69–72. doi:10.1109/TMTT.1985.1132950

Hu, Q., & Zane, R. (2011). Minimizing required energy storage in off-line LED drivers based on series-input converter modules. *IEEE Transactions on Power Electronics*, *26*(10), 2887–2895. doi:10.1109/TPEL.2010.2087772

IEEE Std 521-1984. (1984). *IEEE Standard Letter Designations for Radar-Frequency Bands*.

Ieşan, D. (2007). Thermoelasticity of bodies with microstructure and microtemperatures. *International Journal of Solids and Structures*, *44*(25-26), 8648–8662. doi:10.1016/j.ijsolstr.2007.06.027

Igel, H., Schreiber, U., Flaws, A., Schuberth, B., Velikoseltsev, A., & Cochard, A. (2005). Rotational motions induced by the M8.1 Tokachioki earthquake, Sept. 25, 2003. *Geophysical Research Letters*, *32*(8), L08309. doi:10.1029/2004GL022336

Innok, A., Uthansakul, P., & Uthansakul, M. (2012). Angular beamforming technique for MIMO beamforming system. *International Journal of Antennas and Propagation*, *2012*, 1–10. doi:10.1155/2012/638150

Inspektor, A., & Salvador, P. A. (2014). Architecture of PVD Coatings for Metal cutting Applications: A Review. *Surface and Coatings Technology*, *257*, 138–153. doi:10.1016/j.surfcoat.2014.08.068

Invernizzi, M., Natale, G., Levi, M., Turri, S., & Griffini, G. (2016). UV-assisted 3D printing of glass and carbon fiber-reinforced dual-cure polymer composites. *Materials (Basel)*, *9*(7), 583. doi:10.3390/ma9070583 PMID:28773704

Ip, K. H., & Mai, Y. W. (2004). Delamination detection in smart composite beam using Lamb waves. Smart Materials and Structures, 13(3), 544–551. doi:10.1088/0964-1726/13/3/013 doi:10.1088/0964-1726/13/3/013

Iran-Nejad, V., Lotfi-Neyestanak, A. A., & Shahzadi, A. (2012). Compact broadband quadrature hybrid coupler using planar artificial transmission line. *Electronics Letters*, *48*(25), 1602–1603. doi:10.1049/el.2012.3252

Ishaque, K., Salam, Z., & Taheri, H. (2011). Simple, fast and accurate two-diode model for photovoltaic modules. *Solar Energy Materials and Solar Cells*, *95*(2), 586–594.

Jackson, R. W. (1989). Mode conversion at discontinuities in finite-width conductor-backed coplanar waveguide. *IEEE Transactions on Microwave Theory and Techniques*, *37*(10), 1582–1589. doi:10.1109/22.41005

Jansson, A., Humphrey, E., Montecchio, N., Bittner, R., Kumar, A., & Weyde, T. (2017). Singing voice separation with deep U-net convolutional networks. In *Proceeds of 18th International Society for Music Information Retrieval Conference* (pp. 23-27).

Jarrar, A., & Balouki, Y. (2018). Formal modeling of a complex adaptive air traffic control system. *Complex Adaptive Systems Modeling*, *6*(1), 6. doi:10.118640294-018-0056-4

Jarry, P., & Beneat, J. N. (2016). *Microwave Amplifier and Active Device Design Using the Real Frequency Technique*. Hoboken, NJ: John Wiley & Sons. doi:10.1002/9781119073093

Jastram, M., & Butler, M. (2011). Rodin User's Handbook: Covers Rodin v. 2.8.

Jeremy Jordan. (2019). *Introduction to autoencoders.* Available at https://www.jeremyjordan.me/autoencoders/

Jin, Z., Sulligoi, G., Cuzner, R., Meng, L., Vasquez, J. C., & Guerrero, J. M. (2016). Next-generation shipboard dc power system: Introduction smart grid and dc microgrid technologies into maritime electrical networks. *IEEE Electrification Magazine*, *4*(2), 45–57.

Joly-Pottuz, L., Dassenoy, F., Belin, M., Vacher, B., Martin, J. M., & Fleischer, N. (2005). Ultralow-Friction and Wear Properties of IF-WS2 under Boundary Lubrication. *Tribology Letters*, *18*(4), 477–485. doi:10.100711249-005-3607-8

Ju, Z., Xu, G., Wei, Z., Li, J., Zhao, Q., & Huang, J. (2018). A single-patterned five-band terahertz metamaterial absorber based on multiple resonance mechanisms. *Modern Physics Letters B, 32*(3).

Judah, S. K. (1985). Calibration of multiport reflectometers. *IEE Proceedings H-Microwaves, Antennas and Propagation. 132*(7), 468-470.

Kakigano, H., Nomura, M., & Ise, T. (2010, June). Loss evaluation of DC distribution for residential houses compared with AC system. In The 2010 International Power Electronics Conference-ECCE ASIA- (pp. 480-486). IEEE.

Kalesar, B. M., Noshahr, J. B., Kermani, M., Bavandsavadkoohi, H., & Ahbab, F. (2018, June). Effect of Angles of Harmonic Components of Back to Back Converter of Distributed Generation Resources on Current Behavior of Distribution Networks. In *2018 IEEE International Conference on Environment and Electrical Engineering and 2018 IEEE Industrial and Commercial Power Systems Europe (EEEIC/I&CPS Europe)* (pp. 1-4). IEEE.

Kamal, M. R., & Sourour, S. (1973). Kinetic modelling of epoxy-amine cure process. *Polymer Engineering and Science, 13*, 59–64. doi:10.1002/pen.760130110

Kang, M. K., Lee, W. I., Yoo, J. Y., & Cho, S. M. (1995). Simulation of mold filling process during resin transfer molding. *Journal of Materials Processing & Manufacturing Science, 3*, 297–313.

Karaaslan, M., Bağmancı, M., Ünal, E., Akgol, O., Altıntaş, O., & Sabah, C. (2018). Broad band metamaterial absorber based on wheel resonators with lumped elements for microwave energy harvesting. *Optical and Quantum Electronics, 50*(5), 1–18. doi:10.100711082-018-1484-2

Karaaslan, M., Bağmancı, M., Ünal, E., Akgol, O., & Sabah, C. (2017). Microwave energy harvesting based on metamaterial absorbers with multi-layered square split rings for wireless communications. *Optics Communications, 392*, 31–38. doi:10.1016/j.optcom.2017.01.043

Karim, M. F., Liu, A. Q., Alphones, A., Zhang, X. J., & Yu, A. B. (2005). CPW band-stop filter using unloaded and loaded EBG structures. *IEE Proceedings, Microwaves, Antennas and Propagation, 152*(6), 434. doi:10.1049/ip-map:20050096

Kashaf, A., Javaid, N., Khan, Z. A., & Khan, I. A. (2012). TSEP: Threshold-sensitive stable election protocol for WSNs. In *2012 10th International Conference on Frontiers of Information Technology (FIT)* (pp. 164–168). IEEE.

Kaur, T., Sharma, S. K., & Singh, A. K. (2016). Influence of imperfectly bonded micropolar elastic half-space with non-homogeneous viscoelastic layer on propagation behavior of shear wave. *Waves in Random and Complex Media, 26*(4), 650–670. doi:10.1080/17455030.2016.1185191

Kaur, T., Sharma, S. K., & Singh, A. K. (2017). Shear wave propagation in vertically heterogeneous viscoelastic layer over a micropolar elastic half-space. *Mechanics of Advanced Materials and Structures, 24*(2), 149–156. doi:10.1080/15376494.2015.1124948

Kazimierczuk, M. K. (2015). *RF Power Amplifiers* (2nd ed.). Hoboken, NJ: John Wiley & Sons.

Kenny, J. M., Apicella, A., & Nicolais, L. A. (1989). A model for the thermal and chemorheological behavior of thermosets. I: Processing of epoxy-based composites. *Polymer Engineering and Science, 9*(15), 973–983. doi:10.1002/pen.760291502

Kenny, J. M., Maffezzoli, A., & Nicolais, L. (1990). A model for the thermal and chemorheological behavior of thermoset processing: (II) Unsaturated polyester based composites. *Composites Science and Technology, 38*(4), 339–358. doi:10.1016/0266-3538(90)90020-6

Khlopotin, A., Olsson, P., & Larsson, F. (2015). Transformational cloaking from seismic surface waves by micropolar metamaterials with finite couple stiffness. *Wave Motion, 58*, 53–67. doi:10.1016/j.wavemoti.2015.07.002

Khodadoost, A. A. A., Bayati, N., Reza, M., Gharehpetian, G. B., & Sadeghi, S. H. (2018). Fault Current Limiter optimal sizing considering different Microgrid operational modes using Bat and Cuckoo Search Algorithm. *Archives of Electrical Engineering.*

Khurana, A., & Tomar, S. K. (2009). Longitudinal wave response of a chiral slab interposed between micropolar solid half-spaces. *International Journal of Solids and Structures*, *46*(1), 135–150. doi:10.1016/j.ijsolstr.2008.08.018

Kim, H. C., Lee, B. J., & Park, M. J. (2010). Dual-band branch-line coupler with port extensions. *IEEE Transactions on Microwave Theory and Techniques, 58*(3), 651–655.

Kim, J. S., & Lee, D. G. (1997). Development of an Autoclave Cure Cycle with Cooling and Reheating Steps for Thick Thermoset Composite Laminates. *Journal of Composite Materials*, *31*(22), 2264–2282. doi:10.1177/002199839703102203

Kim, T. H., & Olver, A. V. (1998). Stress history in rolling-sliding contact of rough surface. *Tribology International*, *31*(12), 727–736. doi:10.1016/S0301-679X(98)00085-1

Kingma, D. P., & Welling, M. (2013). Auto-encoding variational bayes. *arXiv preprint arXiv:1312.6114.*

Kirschning, M., Jansen, R. H., & Koster, N. H. L. (1981). Accurate model for open end effect of microstrip lines. Electronics Letters, *17*(3), 123–125.

KissPNG - HD png images and illustrations. Free unlimited download. (n.d.). Retrieved from https://www.kisspng.com/png-deep-learning-machine-learning-artificial-intellig-1401501/preview.html

Knöchel, R. (1999). Broadband flat coupling two-branch and multibranch directional couplers. *IEEE MTT-S International Microwave Symposium Digest*, *3*, 1327–1330.

Knotek, O., & Leyendecker, T. (1987). On the Structure of (Ti,Al)N-PVD-Coatings. *Journal of Solid State Chemistry*, *70*(2), 318–322. doi:10.1016/0022-4596(87)90071-5

Knott, E. F., & Lunden, C. D. (1995). The two-sheet capacitive Jaumann absorber. *IEEE Transactions on Antennas and Propagation*, *43*(11), 1339–1343. doi:10.1109/8.475112

Kobayashi, T., Ishikawa, F., & Honiden, S. (2019). Consistency-preserving refactoring of refinement structures in Event-B models. *Formal Aspects of Computing*, 1–34.

Kock, W. E. (1948). Metallic Delay Lenses. *The Bell System Technical Journal*, *27*(1), 58–82. doi:10.1002/j.1538-7305.1948.tb01331.x

Koelpin, A., Vinci, G., Laemmle, B., Kissinger, D., & Weigel, R. (2010). The six-port in modern society. *IEEE Microwave Magazine*, *11*(7), 35–43. doi:10.1109/MMM.2010.938584

Koketsu, K. (2000). Finite difference traveltime calculation for head waves travelling along an irregular interface. *Geophysical Journal International*, *143*(3), 729–734. doi:10.1046/j.1365-246X.2000.00269.x

Koo, J. J., Oh, S. M., & Hwang, M. S., Park, C., Jeong, Y., Lim, J., … Ahn, D. (2007, October). A new DGS unequal power divider. In *Proceedings of European Microwave Conference.* (pp. 556-559). IEEE.

Korka, Z. (2009). *Cercetari privind reducerea vibratiilor in functionarea reductoarelor cu roti dintate cilindrice (Research on reducing the vibration of cylindrical gearboxes).* (Unpublished doctoral dissertation). "Eftimie Murgu" University of Resita, Romania

Korka, Z. (2018). Precision Coatings. In K. Gupta (Ed.), *Micro and Precision Manufacturing. Engineering Materials.* Cham, Switzerland: Springer. doi:10.1007/978-3-319-68801-5_8

Korka, Z. I., Cojocaru, V., Gillich, N., & Miclosina, C. O. (2017). Comparative Assessment of Noise Reduction in Helical Gearboxes. *Journal of Vibration Engineering & Technologies*, *5*(3), 263–268.

Korka, Z., Gillich, G. R., Mituletu, I. C., & Tufoi, M. (2015). Gearboxes noise reduction by applying a fluoropolymer coating procedure. *Environmental Engineering and Management Journal*, *14*(6), 1433–1439. doi:10.30638/eemj.2015.155

Koschny, T., Markos, P., Smith, D. R., & Soukoulis, C. M. (2003). Resonant and anti-resonant frequency dependence of the effective parameters of metamaterials. *Physical Review E*, *68*(6), 1–4. doi:10.1103/PhysRevE.68.065602

Koster, N. H. L., Koslowski, S., Bertenburg, R., Heinen, S., & Wolff, I. (1989). Investigations on Airbridges used for MMICs in CPW technique. In Proceedings of 19th EuMC, (pp. 666-671).

Krishnaswami, H. (2011). Photovoltaic Microinverter using Single-stage Isolated High-frequency link Series Resonant Topology, IEEE Energy Conversion and Congress Exposition (ECCE), pp. 495-500, Sep. 2011.

Krishna, V. V., Patel, B., & Sanyal, S. (2010). Harmonic suppressed compact wideband branch-line coupler using unequal length open-stub units. *International Journal of RF and Microwave Computer-Aided Engineering*, *21*(1), 115–119.

Kubade, P. R., Tambe, P., & Kulkarni, H. B. (2017). Morphological, Thermal and Mechanical Properties of 90/10 (WT%/WT%) PP/ABS Blends and their Polymer Nanocomposites. *Advanced Composites Letters*, *26*(6). doi:10.1177/096369351702600602

Kubade, P., & Tambe, P. (2016). Influence of halloysite nanotubes (HNTs) on morphology, crystallization, mechanical and thermal behaviour of PP/ABS blends and its composites in presence and absence of dual compatibilizer. *Composite Interfaces*, *23*(5), 433–451. doi:10.1080/09276440.2016.1144392

Kubade, P., & Tambe, P. (2017). Influence of surface modification of halloysite nanotubes and its localization in PP phase on mechanical and thermal properties of PP/ABS blends. *Composite Interfaces*, *24*(5), 469–487. doi:10.1080/09276440.2016.1235442

Kulkarni, H. B., Mahamuni, S. S., Gaikwad, P. M., Pula, M. A., Mahamuni, S., Bansode, S. H., … Nehatrao, S. A. (2017). Enhanced Mechanical Properties Of Epoxy/Graphite Composites. *Int. J. Adv. Engg. Res. Studies, 6(1), Oct.-Dec.* 2017: 5.

Kulkarni, H. B., Tambe, P., & Joshi, G. M. (2018). Influence of covalent and non-covalent modification of graphene on the mechanical, thermal and electrical properties of epoxy/graphene nanocomposites: A review. *Composite Interfaces*, *25*(5-7), 381–414. doi:10.1080/09276440.2017.1361711

Kulkarni, H., Tambe, P., & Joshi, G. (2017). High concentration exfoliation of graphene in ethyl alcohol using block copolymer surfactant and its influence on properties of epoxy nanocomposites. *Fullerenes, Nanotubes, and Carbon Nanostructures*, *25*(4), 241–249. doi:10.1080/1536383X.2017.1283616

Kumar, N., & Grebennikov, A. (2015). Distributed Power Amplifiers for RF and Microwave Communications. Norwood, MA: Artech House.

Kumar, R., & Sharma, N. (2008). Propagation of waves in micropolar viscoelastic generalized thermoelastic solids having interficial imperfections. *Theoretical and Applied Fracture Mechanics*, *50*(3), 226–234. doi:10.1016/j.tafmec.2008.07.010

Kumar, R., & Singh, B. (1997). Reflection and transmission of elastic waves at a loosely bonded interface between an elastic and micropolar elastic solid. *Indian Journal of Pure and Applied Mathematics*, *28*, 1133–115.

Kurgan, P., Filipcewicz, J., & Kitlinski, M. (2012). Development of a compact microstrip resonant cell aimed at efficient microwave component size reduction. *IET Microwaves, Antennas & Propagation, 6*(12), 1291-1298.

Kurgan, P., & Kitliński, M. (2009). Novel doubly perforated broadband microstrip branch-line couplers. *Microwave and Optical Technology Letters*, *51*(9), 2149–2152. doi:10.1002/mop.24566

Lam, Y. C., Joshi, S. C., & Liu, X. L. (2000). Numerical simulation of the mold filling process in resin-transfer molding. *Composites Science and Technology*, *60*(6), 845–855. doi:10.1016/S0266-3538(99)00192-X

Lasance, C. J., & Poppe, A. (2014). Thermal management for LED applications. New York, NY: Springer. doi:10.1007/978-1-4614-5091-7

Lasnier, F. (2017). *Photovoltaic engineering handbook*. Abingdon, UK: Routledge.

LeCun, Y., Bengio, Y., & Hinton, G. (2015). Deep learning. nature, 521(7553), 436.

Lee, C. Y., You, K. Y., Tan, T. S., Then, Y. L., Lee, Zahid, L., Lim, W. L., & Lee, C. H. (2016). Enhanced five-port rin circuit reflectometer for synthetic breast tissue dielectric determination. *Progress In Electromagnetic Research C*. *69,* 83-95.

Lee, S. K., & Lee, Y. S. (2012). Wideband branch-line couplers with single-section quarter-wave transformers for arbitrary coupling levels. *IEEE Microwave and Wireless Components Letters, 22*(1), 19-21.

Lee, C. Y., You, K. Y., Abbas, Z., Lee, K. Y., Lee, Y. S., & Cheng, E. M. (2018). S-band five-port ring reflectometer-probe system for *in vitro* breast tumor detection. *International Journal of RF and Microwave Computer-Aided Engineering*, *28*(3), 1–13. doi:10.1002/mmce.21198

Lee, K. F., & Luk, K. M. (2010). *Microstrip Patch Antennas*. London, U.K.: Imperial College Express. doi:10.1142/p669

Lee, K. F., Yang, S. L. S., & Kishk, A. (2008). Dual- and multi-band U-slot patch antennas. *IEEE Antennas Wireless Commun. Lett.*, *7*, 645–648. doi:10.1109/LAWP.2008.2010342

Lee, K. Y., Abbas, Z., Jusoh, M. A., You, K. Y., & Cheng, E. M. (2011). Determination of moisture content in oil palm fruits using a five-port reflectometer. *Sensors (Basel)*, (11), 4073–4085. PMID:22163837

Lee, L. J., Young, W. B., & Lin, R. J. (1994). Mold filling and cure modeling of RTM and SRIM processes. *Composite Structures*, *27*(1-2), 109–120. doi:10.1016/0263-8223(94)90072-8

Lee, W. H., Igel, H., & Trifunac, M. D. (2009). Recent advances in rotational seismology. *Seismological Research Letters*, *80*(3), 479–490. doi:10.1785/gssrl.80.3.479

Leger, A., & Deschamps, M. (2009). Ultrasonic wave propagation in non-homogeneous media. Springer Science & Business Media. doi:10.1007/978-3-540-89105-5 doi:10.1007/978-3-540-89105-5

Letcher, T., & Waytashek, M. (2014, November). Material property testing of 3D-printed specimen in PLA on an entry-level 3D printer. In *ASME 2014 international mechanical engineering congress and exposition*. American Society of Mechanical Engineers. doi:10.1115/IMECE2014-39379

Leterme, W., & Van Hertem, D. (2014). Reduced modular multilevel converter model to evaluate fault transients in DC grids.

Levin, T., & Thomas, V. M. (2012). Least-cost network evaluation of centralized and decentralized contributions to global electrification. *Energy Policy*, *41*, 286–302.

Levin, T., & Thomas, V. M. (2016). Can developing countries leapfrog the centralized electrification paradigm? *Energy for Sustainable Development*, *31*, 97–107.

Levy, R., & Lind, L. F. (1968). Synthesis of symmetrical branch-guide directional couplers. *IEEE Transactions on Microwave Theory and Techniques*, *16*(2), 80–89. doi:10.1109/TMTT.1968.1126612

Leyendecker, T., Lemmer, O., Esser, S., & Ebberink, J. (1991a). The Development of the PVD Coating TiAlN as a Commercial Coating for Cutting Tools. *Surface and Coatings Technology*, *48*(2), 175–178. doi:10.1016/0257-8972(91)90142-J

Leyendecker, T., Lemmer, O., Jürgens, A., Esser, S., & Ebberink, J. (1991b). Industrial Application of Crystalline Diamond-Coated Tools. *Surface and Coatings Technology*, *48*(3), 253–260. doi:10.1016/0257-8972(91)90013-M

Li, S., Tan, S.-C., Hui, S., & Tse, C. (2013). *A review and classification of LED ballasts.* Paper presented at the 2013 IEEE Energy Conversion Congress and Exposition. 10.1109/ECCE.2013.6647106

Liao, S. S., Sun, P. T., Chin, N. C., & Peng, J. T. (2005). A novel compact-size branch-line coupler. *IEEE Transactions on Microwave and Wireless Components Letters*, *15*(9), 588–590. doi:10.1109/LMWC.2005.855378

Li, C. J., Zhang, H. Q., & Wang, P. S. (2010). A novel six-port circuit based on four quadrature hybrids. *International Journal of RF and Microwave Computer-Aided Engineering*, *20*(1), 128–132.

Li, L., Wang, J., Du, H., Wang, J., Qu, S., & Xu, Z. (2015). A band enhanced metamaterial absorber based on E-shaped all-dielectric resonators. *AIP Advances*, *5*(1). doi:10.1063/1.4907050

Li, L., & Xie, S. (2004). Finite element method for linear micropolar elasticity and numerical study of some scale effects phenomena in MEMS. *International Journal of Mechanical Sciences*, *46*(11), 1571–1587. doi:10.1016/j.ijmecsci.2004.10.004

Lim, J. S., Kim, C. S., Park, J. S., Ahn, D., & Nam, S. W. (2000). Design of 10 dB 90° branch line coupler using microstrip line with defected ground structure. *Electronics Letters*, *36*(21), 1784–1785. doi:10.1049/el:20001238

Lim, J., Kim, S. C. S., & Lee, Y. T. (2002). A spiral-shaped defected ground structure for coplanar waveguide. *IEEE Microwave and Wireless Components Letters*, *12*(9), 330–332. doi:10.1109/LMWC.2002.803208

Li, N., Li, Y., & Liu, S. (2016). Rapid prototyping of continuous carbon fiber reinforced polylactic acid composites by 3D printing. *Journal of Materials Processing Technology*, *238*, 218–225. doi:10.1016/j.jmatprotec.2016.07.025

Lindsey, S., & Raghavendra, C. S. (2002). PEGASIS: Power-efficient gathering in sensor information systems. In *Aerospace conference proceedings, 2002. IEEE* (Vol. 3, p. 3). Citeseer.

Lin, F., Chu, Q. X., & Wong, S. W. (2013). Dual-band planar crossover with two-section branch-line structure. *IEEE Transactions on Microwave Theory and Techniques*, *61*(6), 2309–2316. doi:10.1109/TMTT.2013.2261084

Li, S., & Bosisio, R. G. (1982). Calibration of multiport reflectometers by means of four open/short circuits. *IEEE Transactions on Microwave Theory and Techniques*, *30*(7), 1085–1090. doi:10.1109/TMTT.1982.1131200

Li, S., Tan, S.-C., Lee, C. K., Waffenschmidt, E., Hui, S. R., & Chi, K. T. (2016). A survey, classification, and critical review of light-emitting diode drivers. *IEEE Transactions on Power Electronics*, *31*(2), 1503–1516. doi:10.1109/TPEL.2015.2417563

Liu, J. Y., & Yang, Y. H. (2018, December). Denoising Auto-encoder with Recurrent Skip Connections and Residual Regression for Music Source Separation. In *2018 17th IEEE International Conference on Machine Learning and Applications (ICMLA)* (pp. 773-778). IEEE. 10.1109/ICMLA.2018.00123

Liu, D., Smaragdis, P., & Kim, M. (2014). Experiments on deep learning for speech denoising. In *Fifteenth Annual Conference of the International Speech Communication Association.*

Liu, T., Cao, X., Gao, J., Zheng, Q., Li, W., & Yang, H. (2013). RCS Reduction of Waveguide Slot Antenna With Metamaterial Absorber. *IEEE Transactions on Antennas and Propagation*, *61*(3), 1479–1484. doi:10.1109/TAP.2012.2231922

Liu, Z. J., Liu, Z. K., McNerny, C., Mehrotra, P., & Inspektor, A. (2005). Investigations of the Bonding Layer in Commercial CVD Coated Cemented Carbide Inserts. *Surface and Coatings Technology*, *198*(1–3), 161–164. doi:10.1016/j.surfcoat.2004.10.112

Li, W., Luo, M., Monti, A., & Ponci, F. (2012, May). Wavelet based method for fault detection in medium voltage DC shipboard power systems. In *2012 IEEE International Instrumentation and Measurement Technology Conference Proceedings* (pp. 2155-2160). IEEE.

Li, Z., & Ito, K. (2006). *The immersed interface method: numerical solutions of PDEs involving interfaces and irregular domains, 33*. SIAM. doi:10.1137/1.9780898717464

Long, X., He, J., Zhou, J., Fang, L., Zhou, X., Ren, F., & Xu, T. (2015). A review on light-emitting diode based automotive headlamps. *Renewable & Sustainable Energy Reviews*, *41*, 29–41. doi:10.1016/j.rser.2014.08.028

Loos, A. C., & Springer, G. S. (1983). Curing of epoxy matrix composites. *Journal of Composite Materials*, *17*(2), 135–169. doi:10.1177/002199838301700204

Love, A. E. H. (Ed.). (2013). *A treatise on the mathematical theory of elasticity*. Cambridge, UK: Cambridge University Press.

Low Power RF Energy Harvesting Optimization. (n.d.). *IEEE Transactions on Circuits and Systems. I, Regular Papers*, *57*(7), 1802–1811. doi:10.1109/TCSI.2009.2034891

Luca, A. D. (2016). Numerical simulation of the Lamb wave propagation in impacted CFRP laminate. Procedia Engineering, 167, 109–115. doi:10.1016/j.proeng.2016.11.676 doi:10.1016/j.proeng.2016.11.676

Lu, H. C., & Chu, T. H. (1999). Microwave diversity imaging using six-port reflectometer. *IEEE Transactions on Microwave Theory and Techniques*, *47*(1), 84–87. doi:10.1109/22.740082

Madeo, A., Neff, P., Ghiba, I. D., Placidi, L., & Rosi, G. (2015). Wave propagation in relaxed micromorphic continua: Modeling metamaterials with frequency band-gaps. *Continuum Mechanics and Thermodynamics*, *27*(4-5), 551–570. doi:10.100700161-013-0329-2

Madonna, G., Ferrero, A., & Pirola, M. (1999). Design of a broadband multiprobe reflectometer. *IEEE Transactions on Instrumentation and Measurement*, *48*(2), 622–625. doi:10.1109/19.769672

Maghsoodi, A., Ohadi, A., & Sadighi, M. (2014). Calculation of Wave Dispersion Curves in Multilayered Composite-Metal Plates. Shock and Vibration. Hindawi Publishing.

Magne, P., Nahid-Mobarakeh, B., & Pierfederici, S. (2013). Active stabilization of dc microgrids without remote sensors for more electric aircraft. *IEEE Transactions on Industry Applications*, *49*(5), 2352–2360.

Ma, H., Lai, J.-S., Feng, Q., Yu, W., Zheng, C., & Zhao, Z. (2012). A novel valley-fill SEPIC-derived power supply without electrolytic capacitor for LED lighting application. *IEEE Transactions on Power Electronics*, *27*(6), 3057–3071. doi:10.1109/TPEL.2011.2174446

Mahajan, C., & Cormier, D. (2015, January). 3D printing of carbon fiber composites with preferentially aligned fibers. In *Proceedings IIE annual conference* (p. 2953). Institute of Industrial and Systems Engineers (IISE).

Mal, A. K. (1962). On the frequency equation for Love waves due to abrupt thickening of the crustal layer. *Geofisica Pura e Applicata*, *52*(1), 59–68. doi:10.1007/BF01996000

Mallick, P. K. (1993). Fiber-Reinforced Composites: Materials, Manufacturing, and Design, 2nd Ed. New York, NY: Marcel Dekker.

Mal, O., Couniot, A., & Dupret, F. (1998). Nonisothermal simulation of the resin transfer molding process. *Composites. Part A, Applied Science and Manufacturing*, *29*(1-2), 189–198. doi:10.1016/S1359-835X(97)00065-1

Mammosser, D., Boisvert, M., & Micheau, P. (2013, December). Designing regenerative braking strategies for electric vehicles with an efficiency map. In Congrès français de mécanique. AFM, Maison de la Mécanique, 39/41 rue Louis Blanc, 92400 Courbevoie, France.

Manjeshwar, A., & Agrawal, D. P. (2001). TEEN: a routing protocol for enhanced efficiency in wireless sensor networks. In null (p. 30189a). IEEE. doi:10.1109/IPDPS.2001.925197

Mao, J., Xu, W., Yang, Y., Wang, J., & Yuille, A. L. (2014). Explain images with multimodal recurrent neural networks. *arXiv preprint arXiv:1410.1090*.

Mao, X., Shen, C., & Yang, Y. B. (2016). Image restoration using very deep convolutional encoder-decoder networks with symmetric skip connections. In Advances in neural information processing systems (pp. 2802-2810).

Marian, V., Vollaire, C., Verdier, J., & Allard, B. (2011). Rectenna circuit topologies for contactless energy transfer. Eindhoven, The Netherlands.

Mark, E. O. (2001). Experimental Investigation of Lamb Waves in Transversely Isotropic Composite Plates. Cambridge, MA: Massachusetts Institute of Technology.

Martin, E., Margineda, J., & Zamarro, J. M. (1982). An automatic network analyzer using a slotted line reflectometer. *IEEE Transactions on Microwave Theory and Techniques*, *30*(5), 667–670. doi:10.1109/TMTT.1982.1131119

Matsumoto, H. (2002). Research on solar power satellites and microwave power transmission in Japan. *IEEE Microwave Magazine*, *3*(4), 36–45. doi:10.1109/MMW.2002.1145674

Matsunaga, T., Nishiyama, E., & Toyoda, I. (2015). 5.8-GHz Stacked Differential Rectenna Suitable for Large-Scale Rectenna Arrays With DC Connection. *IEEE Transactions on Antennas and Propagation*, *63*(12), 5944–5949. doi:10.1109/TAP.2015.2491319

Matthews, A. (1985). Titanium Nitride PVD Coating Technology. *Surface Engineering*, *1/2*(2), 93–104. doi:10.1179ur.1985.1.2.93

Mayer, B. (1992). *Wide-band branch line coupler*. USA: U.S. Patent, Number: 5,132,645.

Mayer, B., & Knöchel, R. (1990). Branchline-couplers with improved design flexibility and broad bandwidth. IEEE International Digest on Microwave Symposium. 391-394. IEEE.

Mayer, B. (1990). New broadband branch-line coupler. *Electronics Letters*, *26*(18), 1477–1478. doi:10.1049/el:19900948

McKeen, L. W. (2016). Powder Coating and Films. In L. W. McKeen (Ed.), *In Plastics Design Library, Fluorinated Coatings and Finishes Handbook* (2nd ed.). Norwich, CT: William Andrew Publishing. doi:10.1016/B978-0-323-37126-1.00011-4

McSpadden, J. O., Fan, L., & Chang, K. (1998). Design and experiments of a high conversion-efficiency 5.8-GHz rectenna. IEEE Trans. On Microwave Theory and Techniques, 46(12), 2053-2060.

Mehra, P. S., Doja, M. N., & Alam, B. (2017). Zonal based approach for clustering in heterogeneous WSN. *International Journal of Information Technology*, 1–9. doi:10.100741870-017-0071-2

Melenka, G. W., Cheung, B. K., Schofield, J. S., Dawson, M. R., & Carey, J. P. (2016). Evaluation and prediction of the tensile properties of continuous fiber-reinforced 3D printed structures. *Composite Structures*, *153*, 866–875. doi:10.1016/j.compstruct.2016.07.018

Merabet, B., Cirio, L., Takhedmit, H., Costa, F., Vollaire, C., Allard, B., & Picon, O. (2009a). Low-cost converter for harvesting of microwave electromagnetic energy. In 2009 IEEE Energy Conversion Congress and Exposition, pp. 2592–2599.

Merkel, A., Tournat, V., & Gusev, V. (2011). Experimental evidence of rotational elastic waves in granular phononic crystals. *Physical Review Letters 107*(22), 225, 502.

Michalcová, L., & Hron, R. (2018). IOP Conf. Series: Materials Science and Engineering 326, doi:10.1088/1757-899X/326/1/012006

Midya, G. K. (2004). On Love-type surface waves in homogeneous micropolar elastic media. *International Journal of Engineering Science*, *42*(11-12), 1275–1288. doi:10.1016/j.ijengsci.2004.03.002

Mindlin, R. D., & Tiersten, H. F. (1962). Effects of couple-stresses in linear elasticity. *Archive for Rational Mechanics and Analysis*, *11*(1), 415–448. doi:10.1007/BF00253946

Mohammed, S. S., Ramasamy, K., & Shanmuganantham, T. (2010). Wireless Power Transmission – A Next Generation Power Transmission System. *International Journal of Computers and Applications*, *1*(13), 100–103.

Mohra, A. S. (2004). Six-port reflectometer based on four 0°/180° microstrip ring couplers. *Microwave and Optical Technology Letters*, *40*(2), 167–170. doi:10.1002/mop.11318

Monadi, M., Gavriluta, C., Luna, A., Candela, J. I., & Rodriguez, P. (2016). Centralized protection strategy for medium voltage DC microgrids. *IEEE Transactions on Power Delivery*, *32*(1), 430–440.

Moon, J., & Oh, K. (2015). Effects of light-emitting diode (LED) configuration on luminance and color of an edge-lit backlight unit. *Journal of Display Technology*, *11*(9), 768–775. doi:10.1109/JDT.2015.2443374

Moorthy, V., & Shaw, B. A. (2012). Contact fatigue performance of helical gears with surface coatings. *Wear*, *276-277*, 130–140. doi:10.1016/j.wear.2011.12.011

Morawski, T., & Sypniewski, M. (1987). AM/PM switched three port reflectometer. *Proc. 2nd Intern. Symp. on Measurement of Electrical Quantities IMEKO 87*. 33-40.

Morozumi, S. (2007, April). Micro-grid demonstration projects in Japan. In 2007 Power Conversion Conference-Nagoya (pp. 635-642). IEEE.

Mukesh, K. K., Binod, K. K., & Sachin, K. (2017). Defected Ground Structure: Fundamentals, Analysis, and Applications in Modern Wireless Trends. *International Journal (Toronto, Ont.)* of Antennas and Propagation, Volume 2017, Article ID 2018527. p 22.

Münz, W. D. (1986). Titanium Aluminium Nitride Films: A New Alternative to TiN Coatings. *Journal of Vacuum Science & Technology. A, Vacuum, Surfaces, and Films*, *4/6*(6), 2717–2725. doi:10.1116/1.573713

Muraguchi, M., Yukitake, T., & Naito, Y. (1983). Optimum design of 3-dB branch-line couplers using microstrip lines. *IEEE Transactions on Microwave Theory and Techniques*, *31*(8), 674–678. doi:10.1109/TMTT.1983.1131568

Murty, G. S. (1976). Reflection, transmission and attenuation of elastic waves at a loosely-bonded interface of two half spaces. *Geophysical Journal International*, *44*(2), 389–404. doi:10.1111/j.1365-246X.1976.tb03663.x

Mustapha, S., & Ye, L. (2015). Propagation behavior of guided waves in tapered sandwich structures and debonding identification using time reversal. Wave Motion, 57, 154–170. doi:10.1016/j.wavemoti.2015.03.010 doi:10.1016/j.wavemoti.2015.03.010

Nadarajah, N., Vien, B. S. & Chiu, W. K. (2015) Computational Study of the A0 Scattered Field Due to an Edge Delamination. Journal of Mechanics Engineering and Automation. 229-236.

Nagy, P. B. (1992). Ultrasonic classification of imperfect interfaces. *Journal of Nondestructive Evaluation*, *11*(3-4), 127-139.

Naik, J., Dhar, S., & Dash, P. K. (2019). Adaptive differential relay coordination for PV DC microgrid using a new kernel based time-frequency transform. *International Journal of Electrical Power & Energy Systems*, *106*, 56–67.

Naqui, J., Coromina, J., Karami-Horestani, A., Fumeaux, C., & Martín, F. (2015). Angular displacement and velocity sensors based on coplanar waveguides (CPWs) loaded with s-shaped split ring resonators (S-SRR). *Sensors (Basel)*, *15*(5), 9628–9650. doi:10.3390150509628 PMID:25915590

Naqui, J., & Martı, F. (2013). Transmission lines loaded with bisymmetric resonators and their application to angular displacement and velocity sensors. *IEEE Transactions on Microwave Theory and Techniques*, *61*(12), 4700–4713. doi:10.1109/TMTT.2013.2285356

Naqui, J., & Martín, F. (2016, May). Application of broadside-coupled split ring resonator (BC-SRR) loaded transmission lines to the design of rotary encoders for space applications. In *2016 IEEE MTT-S International Microwave Symposium (IMS)* (pp. 1-4). IEEE. 10.1109/MWSYM.2016.7540017

Nardelli, A., Deuschle, E., de Azevedo, L. D., Pessoa, J. L. N., & Ghisi, E. (2017). Assessment of Light Emitting Diodes technology for general lighting: A critical review. *Renewable & Sustainable Energy Reviews*, *75*, 368–379. doi:10.1016/j.rser.2016.11.002

Nayfeh, A., & Chimenti, D. (1989). Free wave propagation in plate of general anisotropic media. Journal of Applied Mechanics, 56(4), 881–886. doi:10.1115/1.3176186 doi:10.1115/1.3176186

Neumeyer, B. (1990). A new analytical method for complete six-port reflectometer calibration. *IEEE Transactions on Instrumentation and Measurement*, *39*(2), 376–379. doi:10.1109/19.52518

Ng, A. (2011). Sparse autoencoder. CS294A Lecture notes, 72(2011), 1-19.

Ng, C. T., & Veidta, M. (2012). Scattering of the fundamental antisymmetric Lamb wave at delaminations in composite laminates. Acoustical Society of America, 115.

Ng, C. T., Veidt, M., Rose, L. R. F., & Wang, C. H. (2012). Analytical and finite element prediction of Lamb wave scattering at delaminations in quasi-isotropic composite laminates. Journal of Sound and Vibration, 331(22), 4870–4883. doi:10.1016/j.jsv.2012.06.002 doi:10.1016/j.jsv.2012.06.002

Nigbor, R. L. (1994). Six-degree-of-freedom ground-motion measurement. *Bulletin of the Seismological Society of America*, *84*(5), 1665–1669.

Ning, F., Cong, W., Hu, Y., & Wang, H. (2017). Additive manufacturing of carbon fiber-reinforced plastic composites using fused deposition modeling: Effects of process parameters on tensile properties. *Journal of Composite Materials*, *51*(4), 451–462. doi:10.1177/0021998316646169

Ning, F., Cong, W., Qiu, J., Wei, J., & Wang, S. (2015). Additive manufacturing of carbon fiber reinforced thermoplastic composites using fused deposition modelling. *Composites. Part B, Engineering*, *80*, 369–378. doi:10.1016/j.compositesb.2015.06.013

Nowacki, W. (1974). *Micropolar Elasticity, International Center for Mechanical Sciences, Courses and Lectures No. 151*. New York, NY: Springer-Verlag.

Ohira, T. (2013). Power efficiency and optimum load formulas on RF rectifiers featuring flow-angle equations. *IEICE Electronics Express*, *10*(11), 1–9. doi:10.1587/elex.10.20130230

Oila, A., & Bull, S. J. (2005). Assessment of the factors influencing micro-pitting in rolling/sliding contacts. *Wear*, *258*(10), 1510–1524. doi:10.1016/j.wear.2004.10.012

Ojha, J. R., & Peters, M. (2010). *Patch Antennas and Microstrip Lines*. InTech Germany.

Ondruš, J., & Kolla, E. (2018, September). The impact of ABS system on the braking characteristics of the specified motorcycle on the dry road surface. [IOP Publishing]. *IOP Conference Series. Materials Science and Engineering*, *421*(2). doi:10.1088/1757-899X/421/2/022024

Onsy, A. M., Sharaf, A. M., Ashrey, M. M., & Eldemerdash, S. M. (2018). Influence of Intelligent Active Suspension System Controller Design Techniques on Vehicle Braking Characteristics. *SAE International Journal of Vehicle Dynamics, Stability, and NVH, 3*(10-03-01-0003).

Otremba, F., Romero Navarrete, J., & Lozano Guzmán, A. (2018, September). Modelling of a Partially Loaded Road Tanker during a Braking-in-a-Turn Maneuver. Actuators 7(3), (p. 45). Multidisciplinary Digital Publishing Institute. doi:10.3390/act7030045

Palermo, A., & Marzani, A. (2018). Control of Love waves by resonant metasurfaces. *Scientific Reports*, *8*(1), 7234. doi:10.103841598-018-25503-8 PMID:29740063

Panda, R. S. (2013). Lamb wave interactions with delaminations in composite laminates using air-coupled ultrasonic visualization. 14th Asia-Pacific Conference on Non-Destructive Testing.

Pang, Y., Wang, J., Cheng, Q., Xia, S., Zhou, X. Y., Xu, Z., ... Qu, S. (2017). Thermally tunable water-substrate broadband metamaterial absorbers. *Applied Physics Letters*, *110*(10), 104103. doi:10.1063/1.4978205

Paris, D. T., & Hurd, F. K. (1969). *Basic Electromagnetic Theory*. New York, NY: McGraw-Hill Book Co.

Park, J. I., Kim, C. S., Kim, J., Park, J. S., Qian, Y., Ahn, D., & Itoh, T. (1999, November). Modeling of a photonic bandgap and its application for the low-pass filter design. In 1999 Asia Pacific Microwave Conference. Microwaves Enter the 21st Century. Conference Proceedings (Vol. 2, pp. 331-334). IEEE.

Park, J. D., Candelaria, J., Ma, L., & Dunn, K. (2013). DC ring-bus microgrid fault protection and identification of fault location. *IEEE Transactions on Power Delivery*, *28*(4), 2574–2584.

Park, M. J. (2013). Comments on 'quasi-arbitrary phase-difference hybrid coupler'. *IEEE Transactions on Microwave Theory and Techniques*, *61*(3), 1397–1398. doi:10.1109/TMTT.2013.2241779

Patel, J. A., & Solanki, P. D. (2014). Comparative Analysis of String Inverter and Micro Inverter for Solar Based Power System, *International Journal of Advanced Research in Electrical, Electronics and Instrumentation Engineering,* April.

Patterson, B. T. (2012). Dc, come home: Dc microgrids and the birth of the enernet. *IEEE Power & Energy Magazine*, *10*(6), 60–69.

Paul, D. K., Gardner, P., & Prasetyo, B. Y. (1991). Broadband branchline coupler for S band. *Electronics Letters*, *27*(15), 1318–1319. doi:10.1049/el:19910828

Pavlovsky, V. (2019). Introduction To Artificial Neural Networks. [online] Vojtech Pavlovsky. Available at https://www.vaetas.cz/posts/introduction-artificial-neural-networks/

Plymill, A., Minneci, R., Greeley, D. A., & Gritton, J. (2016). Graphene and carbon nanotube PLA composite feedstock development for fused deposition modeling.

Poole, C., & Darwazeh, I. (2016). Microwave Active Circuit Analysis and Design. Amsterdam, The Netherlands: Elsevier Ltd.

Popov, B. N. (2015). Organic Coatings. In B. N. Popov (Ed.), Corrosion Engineering. Amsterdam, The Netherlands: Elsevier. doi:10.1016/B978-0-444-62722-3.00013-6

Porter, F. (1994). Principles of Applying Coatings by Hot Dipping. In L. L. Shreir, R. A. Jarman, & G. T. Burstein (Eds.), *Corrosion* (3rd ed.). Oxford, UK: Butterworth-Heinemann. doi:10.1016/B978-0-08-052351-4.50095-9

Potapov, A. S., Tarasova, T. A., Svirbutovich, O. A., & Fedyaev, M. Y. (2018, May). Complex for experimental studies of anti-lock braking system operation. In *International Conference Aviamechanical engineering and transport (AVENT 2018).* Atlantis Press.

Potter, C. M., & Hjipieris, G. (1993). A robust six- to four-port reduction algorithm. *IEEE MTT-S International Microwave Symposium Digest*. 1263-1266. 10.1109/MWSYM.1993.277103

Potter, C., & Bullock, A. (1993). Nonlinearity correction of microwave diode detectors using a repeatable attenuation step. *Microwave Journal*, *36*, 272, 274, 277–279.

Pozar, D. M. (2012). *Microwave engineering* (4th ed.). Hoboken, NJ: John Wiley & Sons.

Pozar, D. M. (2012). *Microwave Engineering* (4th ed., pp. 147–149). USA: Wiley.

Qiao, L., & Yeo, S. P. (1995). Improved implementation of four-standard procedure for calibrating six-port reflectometers. *IEEE Transactions on Instrumentation and Measurement*, *44*(3), 632–636. doi:10.1109/19.387297

Qing, L., Zhu, Q., & Wang, M. (2006). Design of a distributed energy-efficient clustering algorithm for heterogeneous wireless sensor networks. *Computer Communications*, *29*(12), 2230–2237. doi:10.1016/j.comcom.2006.02.017

Rachakh, A., Abdellaoui, L., Zbitou, J., & Errkik, A., Tajmouati, A., & Mohamed, L. (2018). A Novel Configuration of a Microstrip Microwave Wideband Power Amplifier for Wireless Application. *Telkomnika*, *16*(1), 2014–2031. doi:10.12928/telkomnika.v16i1.7369

Radisic, V., Qian, Y., Coccioli, R., & Itoh, T. (1998). A novel 2-D photonic bandgap Structure for microstrip lines. *IEEE Trans. Microwave and Guided Wave Lett*, *8*(2), 69–71. doi:10.1109/75.658644

Raffaj, V., Bezek, J., Bilik, V., & Beles, M. (1988). Experiments with lumped 6-port reflectometer in the 100 kHz - 100 MHz range. *Proc. Elektrotechnika*, *88*, 101–103.

Raffaj, V., & Bilik, V. (1993). All-resistive six-port reflectometer with two-phase feeding. *Proc.MIOP-93*. 60-64.

Ramadas, C. (2009). Interaction of the primary anti-symmetric Lamb mode (A0) with symmetric delaminations: numerical and experimental studies. Smart Materials and Structures.

Ramahi, O. M., Almoneef, T. S., AlShareef, M., & Boybay, M. S. (2012). Metamaterial particles for electromagnetic energy harvesting. *Applied Physics Letters*, *101*(17). doi:10.1063/1.4764054

Ramakrishna, S. A. (2005). Physics of negative refractive index materials. *Reports on Progress in Physics*, *68*(2), 449–521. doi:10.1088/0034-4885/68/2/R06

Rani, R., Kakkar, D., Kakkar, P., & Raman, A. (2019). *Distance Based Enhanced Threshold Sensitive Stable Election Routing Protocol for Heterogeneous Wireless Sensor Network* (pp. 101–122). Berlin, Germany: Springer. doi:10.1007/978-3-662-57277-1_5

Raypah, M. E., Dheepan, M., Devarajan, M., & Sulaiman, F. (2016). Investigation on thermal characterization of low power SMD LED mounted on different substrate packages. *Applied Thermal Engineering*, *101*, 19–29. doi:10.1016/j.applthermaleng.2016.02.092

Raytheon Company, Wayland. (1977). Electronic and mechanical improvement of the receiving terminal of a free-space microwave power transmission system. *MA, Tech. Rep. PT-4964, NASA Rep.CR135-194.*

Reed, J. (1958). The multiple branch waveguide coupler. *I.R.E. Transactions on Microwave Theory and Techniques*, *6*(4), 398–403. doi:10.1109/TMTT.1958.1125213

Reed, J., & Wheeler, G. J. (1956). A method of analysis of symmetrical four-port networks. *I.R.E. Transactions on Microwave Theory and Techniques*, *4*(4), 246–252. doi:10.1109/TMTT.1956.1125071

Ren, Y. J., Li, M. Y., & Chang, K., (2007). 35 GHz rectifying antenna for wireless power transmission. *Electronics Letters*, *43*(11).

Riaziat, M., Majidi-Ahy, R., & Jaung-Feng, I. (1990). Propagation modes and dispersion characteristics of coplanar waveguides. *IEEE Transactions on Microwave Theory and Techniques*, *38*(3), 245–251. doi:10.1109/22.45333

Ribate, M., Mandry, R., Latrach, M., Errkik, A., & El Abdellaoui, L. (2017). GaAs FET Broadband Power Amplifier for L and S Bands Applications. In *Proceedings of the 2nd International Conference on Computing and Wireless Communication Systems* (p.42). Academic Press. 10.1145/3167486.3167530

Ribate, M., Zbitou, J., Mandry, R., Errkik, A., & Latrach, M. (2018). Broadband GaAs FET Power Amplifier for L and S Bands Applications. *International Journal of Intelligent Engineering and Systems*, *11*(5), 96–105. doi:10.22266/ijies2018.1031.09

Riblet, G. P. (1978). A directional coupler with very flat coupling. *IEEE Transactions on Microwave Theory and Techniques*, *26*(2), 70–74. doi:10.1109/TMTT.1978.1129315

Riblet, G. P., & Hansson, E. R. B. (1981). The use of a matched symmetrical five-port junction to make six-port measurements. *IEEE MTT-S International Microwave Symposium Digest*. 151-153. 10.1109/MWSYM.1981.1129852

Riblet, G. P., & Hansson, E. R. B. (1982). Aspects of the calibration of a single six-port using a load and offset reflection standards. *IEEE Transactions on Microwave Theory and Techniques*, *30*(12), 2120–2125. doi:10.1109/TMTT.1982.1131394

Richelli, A., Colalongo, L., Tonoli, S., & Kovacs-Vajna, Z. (2009). A 0.2-1.2V DC/DC Boost Converter for Power Harvesting Applications. *IEEE Transactions on Power Electronics*, *24*(5-6), 1541–1546. doi:10.1109/TPEL.2009.2013224

Roger, J., & Mones, E. T. (1987). *Journal of Applied Polymer Science*, *3*, 999–1020.

Saad, A. (2011). Contribution à l'étude numérique des procédès de fabrication des matériaux composites-cas du moulage par transfert de résine (rtm), thèse de doctorat de la faculté des sciences de Kenitra: Université Ibn Tofail, Maroc.

Saad, A., Echchelh, A., Hattabi, M., & El Ganaoui, M. (2012). Numerical simulation of thickness variation effect on resin transfer moulding process. *Journal of Polymer & Composites*, *33*(1), 10–21. doi:10.1002/pc.21226

Saad, A., Echchelh, A., Hattabi, M., & El Ganaoui, M. (2012). Optimization of the cycle time in resin transfer molding process by numerical Simulation. *Journal of Reinforced Plastics and Composites*, *31*(20), 1388–1399. doi:10.1177/0731684412458166

Saad, A., Echchelh, A., Hattabi, M., & El Ganaoui, M. (2014). The identification of effective thermal conductivity for fibrous reinforcement composite by inverse method. *Journal of Reinforced Plastics and Composites*, *33*(23), 2183–2191. doi:10.1177/0731684414556011

Sadovskii, V., & Sadovskaya, O. (2015). Modeling of elastic waves in a blocky medium based on equations of the cosserat continuum. *Wave Motion*, *52*, 138–150. doi:10.1016/j.wavemoti.2014.09.008

Sahli, M., Correia, J. P. M., Ahzi, S., & Touchal, S. (2018). Multi-physics modeling and simulation of heat and electrical yield generation in photovoltaics. *Solar Energy Materials and Solar Cells*, *180*, 358–372.

Saini, P., & Sharma, A. K. (2010a). E-DEEC-enhanced distributed energy efficient clustering scheme for heterogeneous WSN. In *2010 1st International Conference on Parallel Distributed and Grid Computing (PDGC)* (pp. 205–210). IEEE. 10.1109/PDGC.2010.5679898

Saini, P., & Sharma, A. K. (2010b). Energy efficient scheme for clustering protocol prolonging the lifetime of heterogeneous wireless sensor networks. *International Journal of Computers and Applications*, *6*(2).

Saleh, K. A., Hooshyar, A., & El-Saadany, E. F. (2017). Hybrid passive-overcurrent relay for detection of faults in low-voltage DC grids. *IEEE Transactions on Smart Grid*, *8*(3), 1129–1138.

Sannino, A., Postiglione, G., & Bollen, M. H. (2002, October). Feasibility of a DC network for commercial facilities. In *37th IAS Annual Meeting Record of the 2002 IEEE Industry Applications Conference.* (Vol. 3, pp. 1710-1717). IEEE.

Schelkunoff, S. A., & Friis, H. T. (1952). *Antennas: theory and practice* (Vol. 639). New York, NY: Wiley.

Schlüter, J., & Böck, S. (2014, May). Improved musical onset detection with convolutional neural networks. In *2014 Ieee International Conference on Acoustics, Speech and Signal Processing* (pp. 6979-6983). IEEE.

Schneider, M. V. (1969). Microstrip lines for microwave integrated circuits. *The Bell System Technical Journal*, *48*(5), 1421–1444. doi:10.1002/j.1538-7305.1969.tb04274.x

Schneller, T., Waser, R., Kosec, M., & Payne, D. (Eds.). (2013). *Chemical Solution Deposition of Functional Oxide Thin Films*. Vienna, Austria: Springer. doi:10.1007/978-3-211-99311-8

Schubert, A., & Nestler, A. (2009). Endbearbeitung von Aluminiummatrix-Verbundwerkstoffen durch Drehen. *Materialwissenschaft und Werkstofftechnik*, *40*(7), 580–586. doi:10.1002/mawe.200900493

Sébastien Riopel, Etude expérimentale de la cuisson d'un composite unidirectionnel verre/polyester de grande épaisseur. Mémoire présenté en vue de l'obtention du diplome de maîtrise ès sciences appliquées (génie &canique): Université de Montréal.

Sebechlebsky, S., Raffaj, V., Bilik, V. (1991). Synthesis of reflectors for wide-band lumped six-port reflectometers. *Elektrotechnicky casopis. 42,* 449-470.

Sechilariu, M., Wang, B. C., Locment, F., & Jouglet, A. (2014). DC microgrid power flow optimization by multi-layer supervision control. Design and experimental validation. *Energy Conversion and Management*, *82*, 1–10.

Shams, K. M. Z., & Ali, M. (2007). Wireless power transmission to a buried sensor in concrete. *IEEE Sensors Journal*, *7*(12), 1573–1577. doi:10.1109/JSEN.2007.908230

Sharma, S. K., Ghosh, S., & Srivastava, K. V. (2016). An ultra-thin triple-band polarization-insensitive metamaterial absorber for S, C and X band applications. *Applied Physics A, 122*(12).

Sharma, A. K., & Itoh, T. (1993). Special Issue on Modeling and Design of Coplanar Monolithic Microwave and Millimeter-Wave Integrated Circuits. *IEEE Transactions on Microwave Theory and Techniques*, *41*(9).

Sharma, A. K., Tiwari, A. K., & Dixit, A. R. (2016). Effects of Minimum Quantity Lubrication (MQL) in Machining Processes Using Conventional and Nanofluid Based Cutting Fluids: A Comprehensive Review. *Journal of Cleaner Production*, *127*, 1–18. doi:10.1016/j.jclepro.2016.03.146

Shashank, P. (2014). Lamb Wave Propagation and Material Characterization of Metallic and Composite Aerospace Structures for Improved Structural Health Monitoring (SHM). Ontario, Canada: Carleton University Ottawa.

Shaver, L. (2017). *Implementation of a DC Microgrid*. University of Wisconsin-Madison.

Sichirollo, F., Alonso, J. M., & Spiazzi, G. (2015). A novel double integrated buck offline power supply for solid-state lighting applications. *IEEE Transactions on Industry Applications*, *51*(2), 1268–1276. doi:10.1109/TIA.2014.2350071

Sikha Simon, K., Chakyar, S. P., Andrews, J., & Joseph, V. P. (1849). Metamaterial split ring resonator as a sensitive mechanical vibration sensor. In AIP Conference Proceedings (Vol. 20021, No. 2017).

Sim, M. S., You, K. Y., Esa, F., Dimon, M. N., & Khamis, N. H. (2018). Multiband metamaterial microwave absorbers using split ring and multiwidth slot structure. *International Journal of RF and Microwave Computer-Aided Engineering*, *28*(7), 1–13. doi:10.1002/mmce.21473

Singh, A. K., Chaki, M. S., & Chattopadhyay, A. (2018). Remarks on impact of irregularity on SH-type wave propagation in micropolar elastic composite structure. *International Journal of Mechanical Sciences*, *135*, 325–341. doi:10.1016/j.ijmecsci.2017.11.032

Singh, S. P., & Sharma, S. C. (2015). A Survey on Cluster Based Routing Protocols in Wireless Sensor Networks. *Procedia Computer Science*, *45*, 687–695. doi:10.1016/j.procs.2015.03.133

Slavin, L. M., & Wolf, B. (1970). Scattering of Love waves in a surface layer with an irregular boundary for the case of a rigid underlying half-space. *Bulletin of the Seismological Society of America*, *60*(3), 859–877.

Smaragdakis, G., Matta, I., & Bestavros, A. (2004). SEP: A stable election protocol for clustered heterogeneous wireless sensor networks. Boston, MA: Boston University Computer Science Department.

Smith, D. R., & Kroll, N. (2000). Negative refractive index in left-handed materials. *Physical Review Letters*, *85*(14), 2933–2936. doi:10.1103/PhysRevLett.85.2933 PMID:11005971

Sobczyk, K., & Kirkner, D. J. (2012). Stochastic modeling of microstructures. Berlin, Germany: Springer Science & Business Media.

Soheilifar, M. R., & Sadeghzadeh, R. A. (2014). Design, fabrication and characterization of stacked layers planar broadband metamaterial absorber at microwave frequency. *AEÜ International Journal of Electronics and Communications*, *69*(1), 126–132. doi:10.1016/j.aeue.2014.08.005

Solanki, C. S. (2012). *Solar photovoltaics fundamentals, technologies and applications* (2nd ed.).

Soleimanpour, R. & Ng, C. T. (2015). Mode conversion and scattering analysis of guided waves at delaminations in laminated composite beams. Structural Monitoring and Maintenance, 213-236.

Somlo, P. I., & Hunter, J. D. (1982). A six-port reflectometer and its complete characterization by convenient calibration procedures. *IEEE Transactions on Microwave Theory and Techniques*, *30*(2), 186–192. doi:10.1109/TMTT.1982.1131045

Sor, J., Quin, Y., & Itoh, T. (2011). Miniature Low-Loss CPW Periodic Structures for Filter Applications. *IEEE Transactions on Microwave theory and techniques, 49*(12).

Sporkmann, T. (1998). The Current State of the Art in Coplanar MMICs. *Microwave Journal*, *41*(8), 60–74.

Sporkmann, T. (1998). The Evolution of Coplanar MMICs over the past 30 Years. *Microwave Journal*, *41*(7), 96–111.

Stankaitis, P., Iliasov, A., Ait-Ameur, Y., Kobayashi, T., Ishikawa, F., & Romanovsky, A. (2019, January). A Refinement Based Method for Developing Distributed Protocols. In *2019 IEEE 19th International Symposium on High Assurance Systems Engineering (HASE)* (pp. 90-97). IEEE. 10.1109/HASE.2019.00023

Staszek, K., Gruszczynski, S., & Wincza, K. (2013). Theoretical limits and accuracy improvement of reflection-coefficient measurements in six-port reflectometers. *IEEE Transactions on Microwave Theory and Techniques*, *61*(8), 2966–2974. doi:10.1109/TMTT.2013.2269053

Staszek, K., Gruszczynski, S., & Wincza, K. (2015). Measurement accuracy enhancement in six-port reflectometers. *IEEE Microwave and Wireless Components Letters*, *25*(8), 553–555. doi:10.1109/LMWC.2015.2440784

Stegens, R. E., & Alliss, D. N. (1987). Coplanar Microwave Integrated Circuit for Integrated Subsystems. *Microwave Sys. News Comm. Tech.*, *17*(11), 84–96.

Stober, S., Cameron, D. J., & Grahn, J. A. (2014, October). Does the beat go on?: Identifying rhythms from brain waves recorded after their auditory presentation. In *Proceedings of the 9th Audio Mostly: A Conference on Interaction with Sound* (p. 23). ACM.

Strobel, A., Eickhoff, R., Ziroff, A., & Ellinger, F. (2010). Comparison of Pulse and FMCW based Radiolocation for Indoor Tracking Systems. In 2010 Future Network & Mobile Summit (pp. 1–8). IEEE.

Stumper, U. (1990). Finding initial estimates needed for the Engen method of calibrating single six-port reflectometers. *IEEE Transactions on Microwave Theory and Techniques*, *38*(7), 946–949. doi:10.1109/22.55790

Su, L., Mata-Contreras, J., Vélez, P., & Martín, F. (2016). Configurations of splitter/combiner microstrip sections loaded with stepped impedance resonators (SIRs) for sensing applications. *Sensors, 16*(12).

Su, L., Mata-Contreras, J., Vélez, P., Fernández-Prieto, A., & Martín, F. (2018). Analytical method to estimate the complex permittivity of oil samples. *Sensors, 18*(4).

Su, L., Mata-Contreras, J., Vélez, P., & Martín, F. (2017). A review of sensing strategies for microwave sensors based on metamaterial-inspired resonators: Dielectric characterization, displacement, and angular velocity measurements for health diagnosis, telecommunication, and space applications. *International Journal of Antennas and Propagation*, *2017*, 1–13. doi:10.1155/2017/5619728

Su, L., & Vélez, P. (2016). Splitter/combiner microstrip sections loaded with pairs of complementary split ring resonators differential sensing applications. *IEEE Transactions on Microwave Theory and Techniques*, *64*(12), 4362–4370. doi:10.1109/TMTT.2016.2623311

Sun, K. O., Ho, S. J., Yen, C. C., & van der Weide, D. (2005). A compact branch-line coupler using discontinuous microstrip lines. *IEEE Microwave and Wireless Components Letters, 15*(8), 519-520.

Susman, L. (1984). Calibration of a six-port reflectometer using projective geometry concepts. *Electronics Letters*, *20*(1), 9. doi:10.1049/el:19840007

Szymanski, P., & Wiatr, W. (2004). A broadband multistate interferometer for impedance measurement. *15th International Conference on Microwaves, Radar and Wireless Communications, MIKON-2004*,755-758. 10.1109/MIKON.2004.1357150

Tabone, M., Azzopardi, J. P., & Farrugia, M. (2018, December). Emulation of Vehicle Speed and Impact on Vehicle Event Data Recorder. In *2018 18th International Conference on Mechatronics-Mechatronika (ME)* (pp. 1-7). IEEE.

Tai, B. L., Stephenson, D. A., Furness, R. J., & Shih, A. J. (2014). Minimum Quantity Lubrication (MQL). In *Automotive Powertrain Machining*, Procedia CIRP, 14:523–528. 6th CIRP International Conference on High Performance Cutting.

Tak, J., Jin, Y., & Choi, J. (2016). A dual-band metamaterial microwave absorber. *Microwave and Optical Technology Letters*, *58*(9), 2052–2057. doi:10.1002/mop.29977

Tang, C. W., Chen, M. G., Lin, Y. S., & Wu, J. W. (2006). Broadband microstrip branch-line coupler with defected ground structure. *Electronics Letters*, *42*(25), 1458–1460. doi:10.1049/el:20063025

Tang, C. W., Chen, M. G., & Tsai, C. H. (2008). Miniaturization of microstrip branch-line coupler with dual transmission lines. *IEEE Microwave and Wireless Components Letters*, *18*(3), 185–187. doi:10.1109/LMWC.2008.916798

Tang, C. W., Tseng, C. T., & Hsu, K. C. (2014). Design of wide passband microstrip branch-line couplers with multiple sections. *IEEE Transactions on Components, Packaging, and Manufacturing Technology*, *4*(7), 1222–1227. doi:10.1109/TCPMT.2014.2320499

Tanuj Kumar Mandal. (2006). Wireless Transmission of Electricity – Development and Possibility. *Sixth International Symposium Nikola TESLA*, SASA, Belgrade, Serbia, Oct. 18–20.

Taybi, A., Tajmouati, A., Zbitou, J., Errkik, A., & Latrach, M. (2019). A new design of a microstrip rectenna at 5.8 GHz for wireless power transmission applications. *Iranian Journal of Electrical and Computer Engineering*, *9*(2), 1258–1266.

Taybi, A., Tajmouati, A., Zbitou, J., Errkik, A., Latrach, M., & El Abdellaoui, L. (2018). A New Configuration of a High Output Voltage 2.45 GHz Rectifier for Wireless Power Transmission Applications. *Telecommunication, Computing, Electronics and Control Journal*, *16*(5), 1939–1946.

Taybi, A., Tajmouati, A., Zbitou, J., Errkik, A., Latrach, M., & El Abdellaoui, L. (2019). A new design of high output voltage rectifier for rectenna system at 2.45 GHz. *Indonesian Journal of Electrical Engineering and Computer Science*, *13*(1), 226–234. doi:10.11591/ijeecs.v13.i1.pp226-234

Tekinalp, H. L., Kunc, V., Velez-Garcia, G. M., Duty, C. E., Love, L. J., Naskar, A. K., ... Ozcan, S. (2014). Highly oriented carbon fiber–polymer composites via additive manufacturing. *Composites Science and Technology*, *105*, 144–150. doi:10.1016/j.compscitech.2014.10.009

Termonia, Y. (1990). Fibre coating as a means to compensate for poor adhesion in fibre-reinforced materials. *Journal of Materials Science*, *25*(1), 103–106. doi:10.1007/BF00544192

Thakur, J. P., & Jun-Seok, P. (2006). A new design approach for circular polarize antenna with DGS under the unbalanced feed-lines. *36th European Microwave Conference*, (pp. 1483-1485). 10.1109/EUMC.2006.281339

Then, Y. L., You, K. Y., Dimon, M. N., Chong, J. C., & Tan, T. S. (2013). Compact microstrip S-band 90° hybrid coupler. *IEEE Symposium on Wireless Technology & Applications*, 202-206.

Tian, X., Liu, T., Wang, Q., Dilmurat, A., Li, D., & Ziegmann, G. (2017). Recycling and remanufacturing of 3D printed continuous carbon fiber reinforced PLA composites. *Journal of Cleaner Production*, *142*, 1609–1618. doi:10.1016/j.jclepro.2016.11.139

Tian, X., Liu, T., Yang, C., Wang, Q., & Li, D. (2016). Interface and performance of 3D printed continuous carbon fiber reinforced PLA composites. *Composites. Part A, Applied Science and Manufacturing*, *88*, 198–205. doi:10.1016/j.compositesa.2016.05.032

Toker, C., Saglam, M., Ozme, M., & Gunalp, N. (2001). Branch-line couplers using unequal line lengths. *IEEE Transactions on Microwave Theory and Techniques*, *49*(4), 718–721. doi:10.1109/22.915448

Toupin, R. A. (1962). Elastic materials with couple-stresses. *Archive for Rational Mechanics and Analysis*, *11*(1), 385–414. doi:10.1007/BF00253945

Tranter, C. J. (1966). *Integral transform in mathematical physics*. London, UK: Methuen and Company Limited.

Trochu, F., Gauvin, R., & Gao, D. M. (1993). Numerical analysis of resin transfer molding process by the finite element method. *Advances in Polymer Technology*, *12*(4), 329–342. doi:10.1002/adv.1993.060120401

Tsang, K. F., Lee, K. Y., Chan, S. P., & Chan, W. S. (1991). Design of broadband five-port circuits. *China., 1991 International Conference on Circuits and Systems*. 980-983. 10.1109/CICCAS.1991.184529

Turco, E. (2018). In-plane shear loading of granular membranes modeled as a Lagrangian assembly of rotating elastic particles. *Mechanics Research Communications*, *92*, 61–66. doi:10.1016/j.mechrescom.2018.07.007

Twardowski, T. E., Lin, S. E., & Geil, P. H. (1993). Curing in Thick Composite Laminates: Experiment and Simulation. *Journal of Composite Materials*, 27.

Tyagi, S., & Kumar, N. (2013). A systematic review on clustering and routing techniques based upon LEACH protocol for wireless sensor networks. *Journal of Network and Computer Applications*, *36*(2), 623–645. doi:10.1016/j.jnca.2012.12.001

Uhlmann, E., Richarz, F., Sammler, S., Heitmüller, F., & Bilz, M. (2014). Machining of Carbon Fibre Reinforced Plastics. *Procedia CIRP*, *24*, 19–24. doi:10.1016/j.procir.2014.07.135

Ülker, S., & Weikle, R. M. (2001). A millimeter-wave six-port reflectometer based on the sampled-transmission line architecture. *IEEE Microwave and Wireless Components Letters*, *11*(8), 340–342. doi:10.1109/7260.941783

Usuki, H., Narutaki, N., Yamane, Y., Karasuno, S., & Ito, T. (1991). A Study on Cutting Performance of Diamond Coated Tools – Tool Wear of Diamond Coated Tools in Machining of CFRP. *Bulletin of the Japan Society of Precision Engineering*, *25*(1), 35–36.

Van Der Klift, F., Koga, Y., Todoroki, A., Ueda, M., Hirano, Y., & Matsuzaki, R. (2016). 3D printing of continuous carbon fibre reinforced thermo-plastic (CFRTP) tensile test specimens. *Open Journal of Composite Materials*, *6*(1), 18–27. doi:10.4236/ojcm.2016.61003

Varadan, V. K., Varadan, V. V., & Lakhtakia, A. (1987). On the possibility of designing anti-reflection coatings using chiral composites. *Journal of Wave-Material Interaction*, *2*(3), 71–81.

VDI-GME-Studie. (2014). Werkstoffinnovationen für nachhaltige Mobilität und Energieversorgung. Verein Deutscher Ingenieure e.V., Düsseldorf

Veidt, M., & Ng, C. T. (2011). Influence of stacking sequence on scattering characteristics of the fundamental anti-symmetric Lamb wave at through holes in composite laminates. J. Acoust, 129.

Venkataramanan, G., & Illindala, M. (2002). Microgrids and sensitive loads. In *Conference Proceedings of 2002 IEEE Power Engineering Society Winter Meeting*. (Vol. 1, pp. 315-322). IEEE.

Venneri, F., Costanzo, S., & Di Massa, G. (2018). Multi-band fractal microwave absorbers. In *World Conference on Information Systems and Technologies* (pp. 1488–1493). Springer International Publishing.

Veprek, S., & Veprek-Heijman, M. G. J. (2008). Industrial Applications of Superhard Nanocomposite Coatings. *Surface and Coatings Technology*, *202*(21), 5063–5073. doi:10.1016/j.surfcoat.2008.05.038

Vladimir, V. B. (1994). Delaminations in composite structures: its origin, buckling, growth and stability. First International Conference on Composites Engineering (ICCE/1), New Orleans, LA. 28-31.

Voevodin, E. S., Kashura, A. S., Fomin, E. V., Pulyanova, K. V., Askhabov, A. M., & Golub, N. V. (2018, May). Reduction of accidents on public roads by increasing informativeness of drivers. In International Conference Aviamechanical engineering and transport (AVENT 2018). Atlantis Press.

Voigt, W. (1887). Theoritiscke studien uber die elastizitats verhaltnisse der krystalle. *Abhandlungen der Braunschweigischen Wissenschaftlichen Gesellschaft*, *34*, 3–51.

Vu, D. Q., Le, D. H., Dinh, H. T., Trinh, T. G., Yue, L., Le, D. T., & Vu, D. L. (2018). Broadening the absorption bandwidth of metamaterial absorber by coupling three dipole resonances. *Physica B, Condensed Matter*, *532*, 90–94. doi:10.1016/j.physb.2017.03.046

Walker, J. L. B. (1993). A Survey of European Activity on Coplanar Waveguide: Atlanta, Georgia. *IEEE MTT-S International Microwave Symposium Digest*, *2*, 693–696.

Walton, D., & Goodwin, A. J. (1998). The wear of unlubricated metallic spur gears. *Wear*, *222*(2), 103–113. doi:10.1016/S0043-1648(98)00291-9

Wang, D., Deng, Y., Liu, Y. M., & Liu, Y. (2018). Numerical investigation of temperature gradient-induced thermal stress for steel–concrete composite bridge deck in suspension bridges. *Journal of Central South University*, *25*(1), 185–195. doi:10.100711771-018-3728-5

Wang, J. F., Qu, S. B., Xu, Z., Zhang, J. Q., Yang, Y. M., Ma, H., & Gu, C. (2008). A candidate three-dimensional GHz left-handed metamaterial composed of coplanar magnetic and electric resonators. *Photonics and Nanostructures-fundamentals and Applications, 6*(3-4), 183-187.

Wang, B., Ruan, X., Yao, K., & Xu, M. (2010). A method of reducing the peak-to-average ratio of LED current for electrolytic capacitor-less AC–DC drivers. *IEEE Transactions on Power Electronics*, *25*(3), 592–601. doi:10.1109/TPEL.2009.2031319

Wang, G. D., Liu, M. H., Hu, X. W., Kong, L. H., Cheng, L. L., & Chen, Z. Q. (2014). Multi-band microwave metamaterial absorber based on coplanar Jerusalem crosses. *Chinese Physics B*, *23*(1). doi:10.1088/1674-1056/23/1/017802

Wang, J., Gao, Y., Liu, W., Sangaiah, A., Kim, H.-J., Wang, J., ... Kim, H.-J. (2019). An Improved Routing Schema with Special Clustering Using PSO Algorithm for Heterogeneous Wireless Sensor Network. *Sensors (Basel)*, *19*(3), 671. doi:10.339019030671 PMID:30736392

Wang, S. Y., Xu, P., Geyi, W., & Ma, Z. (2016). Split-loop resonator array for microwave energy harvesting. *Applied Physics Letters*, *109*(20). doi:10.1063/1.4967917

Wang, S., Ruan, X., Yao, K., Tan, S.-C., Yang, Y., & Ye, Z. (2012). A flicker-free electrolytic capacitor-less AC–DC LED driver. *IEEE Transactions on Power Electronics*, *27*(11), 4540–4548. doi:10.1109/TPEL.2011.2180026

Wei, X. X., & Li, Y. (2006, May). A high-frequency transmission line model for the low voltage power distribution network. In *2006 1ST IEEE Conference on Industrial Electronics and Applications* (pp. 1-7). IEEE.

Wen, C. P. (1969). Coplanar waveguide: A surface strip transmission line suitable for non reciprocal gyromagnetic device application. *IEEE Transactions on Microwave Theory and Techniques*, *17*(12), 1087–1090. doi:10.1109/TMTT.1969.1127105

Weng, L. H., Guo, Y. C., Shi, X. W., & Chen, X.-Q. (2008). An overview on defencted ground structure. *Progress in Electromagnetics Research B*, *7*, 173–189. doi:10.2528/PIERB08031401

Wiedmann, F., Huyart, B., Bergeault, E., & Jallet, L. (1997). New structure for a six-port reflectometer in monolithic microwave integrated-circuit technology. *IEEE Transactions on Instrumentation and Measurement*, *46*(2), 527–530. doi:10.1109/19.571902

Wiedmann, F., Huyart, B., Bergeault, E., & Jallet, L. (1999). A new robust method for six-port reflectometer calibration. *IEEE Transactions on Instrumentation and Measurement*, *48*(5), 927–931. doi:10.1109/19.799649

Wight, J. S., Chudobiak, W. J., & Makios, V. (1976). A microstrip and stripline crossover structure. *IEEE Transactions on Microwave Theory and Techniques*, *24*(5), 270. doi:10.1109/TMTT.1976.1128838

WildML. (2019). Implementing a Neural Network from Scratch in Python – An Introduction. [online]

Willis, H. F. (1948). LV. A formula for expanding an integral as a series. *The London, Edinburgh and Dublin Philosophical Magazine and Journal of Science*, *39*(293), 455–459. doi:10.1080/14786444808521694

Withayachumnankul, W., Jaruwongrungsee, K., Tuantranont, A., Fumeaux, C., & Abbott, D. (2013). Metamaterial-based microfluidic sensor for dielectric characterization. *Sensors and Actuators. A, Physical*, *189*, 233–237. doi:10.1016/j.sna.2012.10.027

Wolff, I. (2006). *Coplanar Microwave Integrated Circuit*. Hoboken, NJ: John Wiley & Sons.

Wong, Y. S., Zheng, S. Y., & Chan, W. S. (2012). Quasi-arbitrary phase-difference hybrid coupler. *IEEE Transactions on Microwave Theory and Techniques*, *60*(6), 1530–1539. doi:10.1109/TMTT.2012.2187918

Woods, D. (1979). Analysis and calibration theory of the general 6-port reflectometor employing four amplitude detectors. *Proceedings of the Institution of Electrical Engineers*, *126*(2), 221–228. doi:10.1049/piee.1979.0051

Wu, Y. L., Shen, J. Y., & Liu, Y. N. (2013). Comments on 'quasi-arbitrary phase-difference hybrid coupler. *IEEE Transactions on Microwave Theory and Techniques*. *61*(4), 1725–1727.

Xu, Y., Du, J., Dai, L. R., & Lee, C. H. (2014). An experimental study on speech enhancement based on deep neural networks. *IEEE Signal Processing Letters*, *21*(1), 65–68. doi:10.1109/LSP.2013.2291240

Yang, B., Xuan, F.-Z., Chen, S., Zhou, S., Gao, Y., & Xiao, B. (2017). Damage localization and identification in WGF/epoxy composite laminates by using Lamb waves: Experiment and simulation. Composite Structures, 165, 138–147. doi:10.1016/j.compstruct.2017.01.015 doi:10.1016/j.compstruct.2017.01.015

Yang, B., Li, W., Zhao, Y., & He, X. (2010). Design and analysis of a grid-connected photovoltaic power system. *IEEE Transactions on Power Electronics*, *25*(4), 992–1000.

Yang, J. F. C., & Lakes, R. S. (1982). Experimental study of micropolar and couple stress elasticity in compact bone in bending. *Journal of Biomechanics*, *15*(2), 91–98. doi:10.1016/0021-9290(82)90040-9 PMID:7076689

Yang, J., Fletcher, J. E., & O'Reilly, J. (2012). Short-circuit and ground fault analyses and location in VSC-based DC network cables. *IEEE Transactions on Industrial Electronics*, *59*(10), 3827–3837.

Yan, J., Zhou, M., & Ding, Z. (2016). Recent Advances in Energy-Efficient Routing Protocols for Wireless Sensor Networks: A Review. *IEEE Access: Practical Innovations, Open Solutions*, *4*, 5673–5686. doi:10.1109/ACCESS.2016.2598719

Yan, Y., Li, S., Zhang, R., Lin, F., Wu, R., Lu, Q., & Wang, X. (2009). Rapid prototyping and manufacturing technology: Principle, representative technics, applications, and development trends. *Tsinghua Science and Technology*, *14*(S1), 1–12. doi:10.1016/S1007-0214(09)70059-8

Yao, J. J. (2010). Nonstandard hybrid and crossover design with branch-line structures. *IEEE Transactions on Microwave Theory and Techniques*, *58*(12), 3801–3808.

Yao, J. J., Lee, C., & Yeo, S. P. (2011). Microstrip branch-line couplers for crossover application. *IEEE Transactions on Microwave Theory and Techniques*, *59*(1), 87–92. doi:10.1109/TMTT.2010.2090695

Yao, J. J., & Yeo, S. P. (2008). Six-port reflectometer based on modified hybrid couplers. *IEEE Transactions on Microwave Theory and Techniques*, *56*(2), 493–498. doi:10.1109/TMTT.2007.914626

Yap, Y. L., Dikshit, V., Lionar, S. P., Yang, H., Lim, J. C., Qi, X., ... Wei, J. (2016). Investigation of fiber reinforced composite using multi-material 3d printing. In *Annual International Solid Freeform Fabrication Symposium*, Austin, TX.

Yazid, Y., Ez-zazi, I., Salhaoui, M., Arioua, M., Ahmed, E. O., & González, A. (2019). Extensive Analysis of Clustered Routing Protocols For Heteregeneous Sensor Networks. In *Proceedings of the Third International Conference on Computing and Wireless Communication Systems, ICCWCS 2019, Faculty of Sciences, Ibn Tofaïl University -Kénitra- Morocco.* EAI. 10.4108/eai.24-4-2019.2284208

Yeo, S. P., & Ang, A. L. (1988). Alternative approach to the calibration of six-port reflectometers using four standards. *IEE Proceedings H-Microwaves, Antennas and Propagation.* 135, 353-355.

Yeom, K. W. (2015). Microwave Circuit Design, a Practical Approach Using ADS. Upper Saddle River, NJ: Prentice Hall.

Yeo, S. P., & Choong, F. C. (2001). Matched symmetrical five-port microstrip coupler. *IEEE Transactions on Microwave Theory and Techniques*, *49*(8), 1498–1500. doi:10.1109/22.939932

Yeo, S. P., & Lee, K. H. (1990). Improvements in design of six-port reflectometer comprising symmetrical five-port waveguide junction and directional coupler. *IEEE Transactions on Instrumentation and Measurement*, *39*(1), 184–188. doi:10.1109/19.50441

Yeo, S. P., Yeo, T. S., & Ng, K. M. (1989). First-order eigenmode analysis of symmetrical five-port microstrip ring coupler. *Microwave and Optical Technology Letters*, *2*(3), 91–94. doi:10.1002/mop.4650020305

Yi, S., Hilton, H. H., & Ahmad, M. F. (1997). A finite element approach for cure simulation of thermosetting matrix composites. *Civil-Comp and Elsevier Science*, *64*, 383–388.

Yoo, M., Kim, H. K., & Lim, S. (2016). Chemical Electromagnetic-based ethanol chemical sensor using metamaterial absorber. *Sensors and Actuators. B, Chemical*, *222*, 173–180. doi:10.1016/j.snb.2015.08.074

Yoon, H. J., & Min, B. W. (2017). Two section wideband 90° hybrid coupler using parallel-coupled three-line. *IEEE Microwave and Wireless Components Letters*, *27*(6), 548–550. doi:10.1109/LMWC.2017.2701304

Yoo, T. W., & Chang, K., (1991). 35 GHz Integrated Circuit Rectifying Antenna with 33% Efficiency. *Electronics Letters*, *27*(23), 2117.

Yoo, Y. E., & Lee, W. I. (1996). Numerical simulation of the resin transfer mold filling process using the boundary element method. *Polymer Composites*, *17*(3), 368–374. doi:10.1002/pc.10623

You, K. Y., Abbas, Z., & Khalid, K. (2010). Application of Microwave Moisture Sensor for Determination of Oil Palm Fruit Ripeness. *Measurement Science Review*, *10*(1), 7–14.

You, K. Y., Chung, J. C., Malek, M. F. A., Lee, Y. S., & Mirza, S. (2019). Miniaturized two-section branch-line coupler using open-stub slow-wave structure. *8th International Conference on Innovation in Electronics and Communication Engineering* (Accepted).

You, K. Y., Lee, C. W., & Lee, C. Y. (2018). Commercial and open source electromagnetic simulators for education, research and industrial design. *International Journal of Advances in Microwave Technology*, *2*(3), 117–121.

You, K. Y., Nadera, A. A., Chong, J. C., Lee, K. Y., Cheng, E. M., & Lee, Y. S. (2018). Analytical modeling of conventional and miniaturization three-section branch-line couplers. *Journal of Electrical Engineering & Technology*, *13*(2), 858–867.

Young, L. (1962). Synchronous branch guide directional couplers for low and high power applications. *I.R.E. Transactions on Microwave Theory and Techniques*, *10*(6), 459–475. doi:10.1109/TMTT.1962.1125554

Young, W. B., Rupel, K., Han, K., Lee, L. J., & Liou, M. (1991). Analysis of resin injection molding in molds with preplaced fiber mats. II: Numerical simulation and experiments of mold filling. *Polymer Composites*, 128–138.

Younis, O., & Fahmy, S. (2004). HEED: A hybrid, energy-efficient, distributed clustering approach for ad hoc sensor networks. *IEEE Transactions on Mobile Computing*, *3*(4), 366–379. doi:10.1109/TMC.2004.41

Yuan, H., Zhu, B. O., Feng, Y., Yuan, H., Zhu, B. O., & Feng, Y. (2015). A frequency and bandwidth tunable metamaterial absorber in x-band. *Journal of Applied Physics*, *117*(17). doi:10.1063/1.4919753

Yu, D., Liu, P., Dong, Y., Zhou, D., & Zhou, Q. (2017). A sextuple-band ultra-thin metamaterial absorber with perfect absorption. *Optics Communications*, *396*, 28–35. doi:10.1016/j.optcom.2017.03.026

Zbitou, J., Latrach, M., & Toutain, S. (2006). Hybrid Rectenna and Monolithic Integrated Zero-Bias Microwave Rectifier. *IEEE Transactions on Microwave Theory and Techniques*, *54*(1), 147–152. doi:10.1109/TMTT.2005.860509

Zhang, J. (2005). *Influence of shot-peening and surface finish on the fatigue of gears*. (Doctoral Thesis, Newcastle University, UK).

Zhang, Y., Zhao, J., Cao, J., & Mao, B. (2018). Microwave Metamaterial Absorber for Non-Destructive Sensing Applications of Grain. *Sensors, 18*(1912), 1–10.

Zhang, B. H., Zhang, J. W., Wu, Z. P., Liu, C. G., & Zhang, B. (2017). A 2.45 GHz Dielectric Resonator Rectenna for Wireless Power Transmission. *International conference of Electronics, Communication and Aerospace Technology (ICECA)*. Coimbatore, India. pp. 20-22. 10.1109/APCAP.2017.8420562

Zhang, J., Xu, Y. C., & Huang, P. (2009). Effect of cure cycle on curing process and hardness for epoxy resin. *Express Polymer Letters*, *3*(9), 534–541. doi:10.3144/expresspolymlett.2009.67

Zhang, N., Zhou, P., Cheng, D., Weng, X., Xie, J., & Deng, L. (2013). Dual-band absorption of mid-infrared metamaterial absorber based on distinct dielectric spacing layers. *Optics Letters*, *38*(7), 1125–1127. doi:10.1364/OL.38.001125 PMID:23546265

Zhao, H., Wu, Q., Guo, Q., Sun, H., & Xue, Y. (2015). Distributed model predictive control of a wind farm for optimal active power control part I: Clustering-based wind turbine model linearization. *IEEE transactions on sustainable energy, 6*(3), 831-839.

Zhao, X., Fan, K., Zhang, J., Seren, H. R., Metcalfe, G. D., Wraback, M., ... Zhang, X. (2015). Optically tunable metamaterial perfect absorber on highly flexible substrate. *Sensors and Actuators. A, Physical*, *231*, 74–80. doi:10.1016/j.sna.2015.02.040

Zhong, H. T., Yang, X. X., Tan, C., & Yu, K. (2016). Triple-band polarization-insensitive and wide-angle metamaterial array for electromagnetic energy harvesting. *Applied Physics Letters*, *109*(25), 1–5. doi:10.1063/1.4973282

Zhong, J., Huang, Y., Wen, G., Sun, H., Wang, P., & Gordon, O. (2012). Single-/dual-band metamaterial absorber based on cross-circular-loop resonator with shorted stubs. *Applied Physics. A, Materials Science & Processing*, *108*(2), 329–335. doi:10.100700339-012-6989-0

Zhong, W., Li, F., Zhang, Z., Song, L., & Li, Z. (2001). Short fiber reinforced composites for fused deposition modeling. *Materials Science and Engineering A*, *301*(2), 125–130. doi:10.1016/S0921-5093(00)01810-4

Zhou, B., Wang, H., & Sheng, W. X. (2010). A modified UWB wilkinson power divider using delta stub. *Progress In Electromagnetics Research Letters*, *19*, 49–55. doi:10.2528/PIERL10101805

Zhou, Y. J., Yang, L., Xiao, Q. X., Pan, T. Y., Ma, H. F., & Tan, C. (2016, July). Plasmonic metamaterials based subwavelength multiband antenna for wireless energy harvesting. In *2016 IEEE MTT-S International Microwave Workshop Series on Advanced Materials and Processes for RF and THz Applications (IMWS-AMP)* (pp. 1-4). IEEE. 10.1109/IMWS-AMP.2016.7588367

Zhu, B., Huang, C., Feng, Y., Zhao, J., & Jiang, T. (2010). Dual band switchable metamaterial electromagnetic absorber. *Progress In Electromagnetics Research B*, *24*, 121–129. doi:10.2528/PIERB10070802

Zhu, W., Huang, Y., Rukhlenko, I. D., Wen, G., & Premaratne, M. (2012). Configurable metamaterial absorber with pseudo wideband spectrum. *Optics Express*, *12*(19), 1563–1568. PMID:22418545

Ziani, A., & Moulai, H. (2009). Extinction properties of electric arcs in high voltage circuit breakers. *Journal of Physics. D, Applied Physics*, *42*(10), 105205.

Zong, B. F., Wang, G. M., Zhang, C. X., & Wang, Y. W. (2014). Miniaturised branch-line coupler with ultra-wide high suppression stopband. *Electronics Letters*, *50*(19), 1365–1367. doi:10.1049/el.2014.1150

Zulhilmi, I. M., Peeie, M. H., Eiman, R. I. M., Izhar, I. M., & Asyraf, S. M. (2019). Investigation on Vehicle Dynamic Behaviour During Emergency Braking at Different Speed. *International Journal of Automotive and Mechanical Engineering*, *16*(1).

About the Contributors

Catalin Pruncu applies mechanical principles and design methodologies to solve problems in material behaviour produced by metal-forming processes, particularly on hot stamping and hot forging. His research is dedicated to further develop and validate hot form quench (HFQ®) Technology for use in the global automotive industry. Catalin has great experience in Microstructural characterization, Solid Mechanics, Fracture Mechanics, Tribology, Design Mechanics, & Finite Element Modelling determined both numerically and experimentally. He spends some time in industry to create high-integrity products for naval and nuclear application.

***

Houda Abouzid was born in Tetouan, Morocco. She is a Ph.D. student at Abdelmalek Essaadi University, National School of Applied Sciences, Tetouan. She received her degree of Engineering of telecommunication systems and networks in 2013 at the same school. Since 2014, she has provided several vocational training courses on matlab programming, networks, and embedded systems. She is also a vacancy teacher at the National School of Applied Sciences (ENSA-Te). She is the president and member of creation of a telecom generation club at ENSA-Te since 2016, providing different educational courses and workshops dedicated to all Ph.D. students of Abdelmalek Essaadi University. She is interested in signal processing, Neural network, and machine learning techniques.

Nadera Najib Al-Areqi was born in 1990. She obtained her B.Sc. Computer Engineering (Honours) degree in Hodeida Universiti Yemen in 2009. She received her M.Sc. in Electrical-Electronic & Telecommunication (Honours) in the University of Technology, Malaysia (UTM), in 2015. She pursued her Ph.D. in Microwave Engineering at the Electrical Engineering department in the University of Technology, Malaysia (UTM), in 2018. Her research interest is in simulation and development of microwave passive devices for agricultural applications.

Navid Bayati was born in Arak, Iran, in 1992. He received the B.Sc. degree in Electrical Engineering from Arak University, Iran, in 2014. Also, he received the M.Sc. degree in Electrical Engineering from Amirkabir University of Technology, Tehran, in 2017. He is studying for a PhD in power system at Aalborg University, Denmark. His research interests are power system protection and DC Microgrids.

El-Hassan Boutyour is at the Research Laboratory: Laboratory of Mechanics, Energy, Electronics and Telecommunications (Science and Technic Faculty, Settat, Morocco). In 1983, he earned a doctorate in University Metz, France. In 2006, he had habilitation at Science and Technic Faculty, Settat, Morocco, and is now a professor there.

Sharmeela C. (CS) earned a B.E. in Electrical and Electronics Engineering, and an M.E. in Power Systems Engineering, from Annamalai University, India, and a Ph.D in Electrical Engineering from Anna University, India, in 1999, 2000 & 2009 respectively. She did P.G Diploma in Electrical Energy Management and Energy Audit from Annamalai University, Chidambaram, in 2010. At present, she holds the post of Assistant Professor (Sr.Gr.) in department of EEE, CEG campus, Anna University, India. She has 19 years of teaching experience and taught various subjects to undergraduate and post graduate students. She did a number of research projects and consultancy work in renewable energy, power quality, & design of PQ compensators for various industries. She published more than 30 research publications in peer-reviewed journals, and more than 50 research papers in international and national conferences in India and abroad. CS has talked of importance of power quality for more than 10 years. She guides 14 PhDs (five completed and nine ongoing), 58 Master's level, 10 Undergraduate's level, and four AMIE students. She co-ordinated and organized a short-term course on power quality for Tamilnadu Electricity Board Engineers in various years. She received best paper award in IEEE POWERCON 2004 conference held in Singapore. She is a Life-member of the Institution of Engineers (India), ISTE, and SSI.

Mriganka Shekhar Chaki is working as Senior Research Fellow at Dept. of Applied Mathematics of IIT (ISM) Dhanbad, India. He obtained his B.Sc (Honours) in Mathematics from St. Xavier's College, University of Calcutta, West Bengal, India in 2013 and received his Master's degree in Mathematics & Computing from Indian School of Mines, Dhanbad, India in 2015. His areas of research interest are Solid mechanics, Microcontinuum mechanics, Wave propagation.

Otman Chakkor was born in 1977 in Tangier, Morocco. He is a Professor (Head of Telecommunication Department) at Abdelmalek Essaadi University, in National School of Applied Sciences ENSA in Morocco, and is a member of the RSAID Laboratory. He received his Ph.D.-Thesis from the University of Granada (Spain), Department of Architecture and Technologies of Computers, in the area of Signal processing, He obtained his degree of Electric Engineering IEEA from Abdelmalek Essaadi University (Morocco) in 1999.

Ee Meng Cheng was born in 1980. He obtained his B.Sc (Honours)-Instrumentation Science in Universiti Putra Malaysia in 2004. He pursued his M.Sc. in Wave Propagation at the Institute for Mathematical Research on 2005 in Universiti Putra Malaysia and his Ph.D in Microwave at the Faculty of Science in 2007 in Universiti Putra Malaysia. Recently, he has been a Senior Lecturer in School of Mechatronic Engineering, Universiti Malaysia Perlis. He is also Chartered Engineer from Institution of Engineering and Technology, as well as Engineering Council in the UK. His main personnel research interest is in the computational electromagnetic modeling, microwave dielectric spectroscopy, and wave propagation in RF & microwave and microwave sensors development for food and agricultural applications.

Soukaina Elyoussfi was born in Oulmes, Morocco, in 1993. She has a Master's degree in energy mechanics of fluids and is a doctoral student at the Laboratory of Energetic Engineering and Materials.

Fahmiruddin Esa was born in 1984. He graduated from Universiti Putra Malaysia (UPM) with a Bachelor's degree in Physics in 2006. Two years later, he furthered his studies at Universiti Teknologi Malaysia (UTM) and obtained Master of Science (Physics) by coursework in 2010. He began his Ph.D. study a year later in Microwave at the Faculty of Science, UPM, and completed it in 2015. He is currently a lecturer at Department of Physics and Chemistry, Faculty of Applied Sciences and Technology, Universiti Tun Hussein Onn Malaysia (UTHM), Pagoh campus, Malaysia. His research interest is in experimental and simulation for the characterisation techniques of microwave-absorbing materials.

Amin Hajizadeh received the B.Sc. degree from the Ferdowsi University of Mashhad, Iran, in 2002, and the M.Sc. and Ph.D. degrees from the K. N. Toosi University of Technology Tehran, Iran, in 2005 and 2010, all in electrical engineering. In 2009, he was a Visiting Ph.D. student with the Department of Electrical Power Engineering, Norwegian University of Science and Technology, Trondheim, Norway. He was an Assistant Professor with the Shahrood University of Technology, Shahrood, Iran, from 2010 to 2014. He held a post-doctoral position with the Norwegian University of Science and Technology, Trondheim, Norway, from 2015 to 2016. Since 2016, he has been an Associate Professor with the Department of Energy Technology, Aalborg University, Denmark. His current research interests include control of distributed energy resources, design, and control of power electronic converters for microgrid and marine power systems.

Abdessamad Jarrar is a Ph.D. candidate at IIMCS Laboratory, which stands for Informatics, Imaging, and Modeling of Complex Systems, at Faculty of Sciences and Technologies Hassan 1st University, Settat, Morocco. His research interests center around modeling of complex systems using formal methods. He is in his fourth year of his PhD. His current research focuses on modeling complex adaptive systems, and he is also working on a method to solve the problem of infinite cycles in complex systems.

Zoltan Iosif Korka was born in Caransebes, Romania, in 1967. He received the Ph.D. degree in mechanical engineering from the University "Eftimie Murgu" from Resita in 2009. Since 2012 he is a Lecturer at the before mentioned university, the field of interest being vibrodiagnosis of gears and precision coatings.

El Abdellaoui Larbi was born in Tiflet, Morocco, in 1961. He received the Ph.D. degree in electronics from the University of Metz, France, in 1994. He is currently an associate Professor of Electronics in Faculty of sciences and techniques, University Hassan 1st, Settat, Morocco. He is involved in the design of hybrid, monolithic active, and passive microwave electronic circuits.

Mohamed Latrach was born in Douar Ksiba, Sless, Morocco, in 1958. He received the Ph.D. degree in electronics from the University of Limoges, France, in 1990. He is a Professor of microwave engineering with the Ecole Suprieure d'Electronique de l'Ouest (ESEO), Angers, France, where his research involves RF and microwaves. His field of interest is the design of hybrid, monolithic active, and passive microwave circuits, metamaterials, LH materials, antennas, and their applications in wireless Communications.

Rachid Mandry was born in Casablanca, Morocco, in 1967. He received the Ph.D. in nuclear physics and particles from Claude Bernard University, France, in 1993. He is an associate Professor of Physics in FST, University Hassan 1st, Settat, Morocco. He is involved in the design of hybrid, active, and passive microwave circuits.

Salah Nissabouri is a Ph.D. researcher at the FST of Settat (University Hassan 1). He graduated from the ENSAM national engineering school and obtained his engineering degree in 2012. He followed a one-year training in the CRMEF Education. He has been a professor of engineering sciences for five years. His Research team is Mechanics and energetics, Laboratory: Laboratory of Mechanics, Energy, Electronics and Telecommunications (Science and Technic Faculty, Settat, Morocco). His research axis includes Non Destructive Testing (NDT) using Ultrasound Lamb waves. Research topics include: Study of the Lamb modes interaction with a defect in homogeneous materials, and in composite structures.

Mohamed Ribate was born in Khouribga, Morocco, in 1991. He received B.S degree in Electronic Automatic and Telecommunication from Polydisciplinary Faculty of Khouribga, Hassan 1st University, Morocco, in 2013, and Master Eng. in Networking and Telecommunication engineering from Faculty of Sciences and Techniques (FST), Hassan 1st University, Settat, Morocco, in 2015. He is working towards his Ph.D. in Microwave and RF Electronics at FST of Settat, University Hassan 1st, Morocco, since November 2016. His research interest includes study, design, and implementation of Broadband Power Amplifier using Micro Strip Technology.

Aouatif Saad has since 2018 been an associate professor at the university Ibn Tofail, Kenitra, Morocco. Areas of research include, but are not limited to, fluid mechanics, porous media, optimization and characterisation of composite material, and numerical simulation of composite material processing.

Lee Yeng Seng is a Senior Lecturer at the Department of Electronic Engineering Technology, Faculty of Engineering Technology, Universiti Malaysia Perlis (UniMAP), Malaysia. He received his bachelor's degree Communication Engineering (Honours) degree and Ph.D. in communication engineering from the School of Computer and Communication Engineering at UniMAP, Malaysia, in 2012 and 2016. He is also a member of Bioelectromagnetic Research Group (BioEM), IET, BEM, and IEEE Microwave Theory and Techniques Society. His research interests focus on Dielectric Material Characterization, Microwave absorber, Food security, Electromagnetic wave, Antenna, Microwave Measurement, and Frequency Selective Surface (FSS).

Elmahjouby Sghir was born in Outat Elhaj, Morocco, in 1989. He received a Masters degree in Telecommunication and Microwave Devices from The National School of Applied Sciences of Fez, the University of Sidi Mohamed Ben Abdellah (USMBA), Fez, Morocco, in 2013. He is working toward a Ph.D degree in Electronic and Telecommunication in LMEET Laboratory, FST, Hassan First University, Morocco. His research interests include the Study and design of filter structures in coplanar technology for telecommunication applications.

Man Seng Sim was born in 1993. He received the Bachelor of Science (Applied Physics) with Honours from Universiti Tun Hussein Onn Malaysia (UTHM), Batu Pahat, Johor, Malaysia, in 2017. He is pursuing a Master of Philosophy in Electrical Engineering at the School of Electrical Engineering in Universiti Teknologi Malaysia (UTM), Skudai, Johor, Malaysia. His research interest is in electromagnetic metamaterials.

Abhishek Kumar Singh is an Associate Professor at Department of Applied Mathematics in Indian Institute of Technology (ISM) Dhanbad, India. His areas of research include Elastodynamics, Wave motion, Solid mechanics, and Mathematical Modeling.

Mohsen Nourbakhsh Soltani received the M.Sc. degree in electrical engineering from the Sharif University of Technology, Iran, in 2004, and the Ph.D. degree in electrical and electronic engineering, from Aalborg University, Denmark, in 2008. He was a Visiting Researcher with the Eindhoven University of Technology in 2007. He fulfilled a Postdoctoral and an Assistant Professor program at Aalborg University 2008–2012. In 2010, he was a Visiting Scholar with Stanford University. Since 2012, he has been an Associate Professor with the Department of Energy Technology, Aalborg University. In 2015, he completed a research leadership training program with the Harvard Business School. His research interests include modeling, control, optimization, estimation, and fault detection, and their applications to electromechanical and energy conversion systems, wind turbines, and wind farms. Some of his recent projects involve modeling, control, and estimation in large-scale wind farms and wind turbines.

Abdelali Tajmouati was born in Morocco in 1962. He received the Ph.D. degree in science engineering from the University of Perpignan, France, in 1992. He is an associate Professor of Electronics, thermal transfer and thermodynamic in Faculty of sciences and techniques, University Hassan 1st, Morocco. He is involved in the design of hybrid, monolithic active, and passive microwave electronic circuits.

Abdellah Taybi was born in Tafrout, Morocco, in 1989. He received the master's degree in electronic aviation safety from the International Academy Mohamed 6 of Civil Aviation, Morocco. He is working toward a Ph.D. degree in physics and engineering sciences at Faculty of Sciences and Technologies, Morocco. His research interests include the analysis and design of rectenna systems for wireless power transmission applications.

Muhammad Wasif Umar received his B.Sc in Electronics Engineering from Bahauddin Zakariya University, Multan, Pakistan, in 2009. In 2011, he received his M.Sc in Control and Electronics from Teesside University, Middlesbrough, UK. From 2011 to 2013, he worked with different organizations in UK. He is working towards the PhD degree at Universiti Teknologi Petronas, Malaysia. His main research interests include design and modeling of electrical drives, dc-dc power converters, high-brightness LED drivers, and non-linear control.

Nor Zaihar Yahaya was born in Lumut, Malaysia. He received his B.Sc degree in Electrical Engineering from University of Missouri-Kansas City, MO, in 1996. He then served five years in the industry in Malaysia. In 2002, he was awarded M.Sc in Microelectronics from the University of Newcastle Upon Tyne, UK. In 2011, he received Ph.D from Universiti Teknologi Petronas, Malaysia. His main teaching/

research areas are the study of Solid State Electronics, Microelectronics & Devices, and Power Electronics & Drives. He has published more than 70 papers in proceedings, journals, and book chapters with several Patents and Copyrights. He is working on projects related to power electronics/ renewable energy, and for oil & gas/Petronas.

Yassine Yazid received his Master's degree in Communication Systems and Embedded Electronics from National School of Applied Sciences, Abdelmalek Essaadi University, in 2018. He is a first-year PhD student at the Department of Information and Communication Systems, Laboratory of Information and Communication Technologies (LabTIC), National School of Applied Sciences and Abdelmalek Essaadi University (Tangier, Morocco). His research interests are in Wireless sensor Network (WSN) communication protocols, Low Power Wide Area Networks (LPWAN), and IoT applications.

Lee Kim Yee was born in 1978. He received his BSc Physics, Master and PhD degree from the Universiti Putra Malaysia (UPM) in the year 2002, 2004, and 2008, respectively. He joined Faculty of Engineering and Science, Universiti Tunku Abdul Rahman (UTAR), in 2007. He is an associate professor under Department of Electrical and Electronic Engineering. His research interests include microwave measurement technique, microwave circuit and instrumentation, control and automation, dielectric measurement, microwave sensor, and RFID.

Kok Yeow You was born in 1977. He obtained his B.Sc. Physics (Honours) degree in Universiti Kebangsaan Malaysia (UKM), in 2001. He pursued his M.Sc. in Microwave at the Faculty of Science in 2003, and his Ph.D. in Wave Propagation at the Institute for Mathematical Research in 2006 in Universiti Putra Malaysia (UPM), Serdang, Selangor, Malaysia. He has recently worked as a senior lecturer at the Communication Engineering Department, Faculty of Electrical Engineering, Universiti Teknologi Malaysia (UTM), Skudai, Johor, Malaysia. His main personal research interest is in the theory, simulation, and instrumentation of electromagnetic wave propagation at microwave frequencies focusing on the development of microwave passive devices and sensors for medical and agricultural applications.

Index

www.ingramcontent.com/pod-product-compliance
Lightning Source LLC
LaVergne TN
LVHW080405011026
849161LV00009B/2352